Advanced Mathematical Methods for Finance

Giulia Di Nunno • Bernt Øksendal
Editors

Advanced Mathematical Methods for Finance

Springer

Editors
Giulia Di Nunno
Bernt Øksendal
CMA, Department of Mathematics
University of Oslo
P.O. Box 1053, Blindern 0316
Oslo, Norway
and
Norwegian School of Economics and
Business Administration
Helleveien 30
5045 Bergen, Norway
giulian@math.uio.no
oksendal@math.uio.no

ISBN 978-3-642-43551-5 ISBN 978-3-642-18412-3 (eBook)
DOI 10.1007/978-3-642-18412-3
Springer Heidelberg Dordrecht London New York

Mathematics Subject Classification (2010): 91Gxx, 91G10, 91G20, 91G40, 91G70, 91G80, 91B16, 91B30, 91B70, 93E11, 93E20, 60E15, 60G15, 60G22, 60G40, 60G44, 60G51, 60G57, 60G60, 60H05, 60H07, 60H10, 60H15, 60H20, 60H30, 60H40, 60J65, 60K15, 62G07, 62G08, 62M07, 62P20, 41A25, 46B70, 94Axx, 35F20, 35Q35

Cover design: deblik

Printed on acid-free paper

Springer is part of Springer Science+Business Media (www.springer.com)

Preface

The title of this volume "Advanced Mathematical Methods for Finance," AMaMeF for short, originates from the European network of the European Science Foundation with the same name that started its activity in 2005. The goals of its program have been the development and the use of advanced mathematical tools for finance, from theory to practice.

This book was born in the same spirit of the program. It presents innovations in the mathematical methods in various research areas representing the broad spectrum of AMaMeF itself. It covers the mathematical foundations of financial analysis, numerical methods, and the modeling of risk. The topics selected include measures of risk, credit contagion, insider trading, information in finance, stochastic control and its applications to portfolio choices and liquidation, models of liquidity, pricing, and hedging. The models presented are based on the use of Brownian motion, Lévy processes and jump diffusions. Moreover, fractional Brownian motion and ambit processes are also introduced at various levels. The chosen blending of topics gives a large view of the up-to-date frontiers of the mathematics for finance. This volume represents the joint work of European experts in the various fields and linked to the program AMaMeF.

After five years of activity, AMaMeF has reached many of its goals, among which the creation and enhancement of the relationships among European research teams in the sixteen participating countries: Austria, Belgium, Denmark, Finland, France, Germany, Italy, The Netherlands, Norway, Poland, Romania, Slovenia, Sweden, Switzerland, Turkey, and United Kingdom.

We are grateful to all the researchers and practitioners in the financial industry for their valuable input to the program and for having participated to the proposed activities, either conferences, or workshops, or exchange research visits these may have been. We are also grateful to Carole Mabrouk for her administrative assistance.

It was an honor to be chairing this program during these years and to have worked together with an engaged team as the AMaMeF Steering Committee, whose members, in addition to ourselves, have been (in alphabetic order): Ole Barndorff-Nielsen, Tomas Björk, Vasili Brinzanescu, Mark Davis, Arnoldo Frigessi, Lane Hughston, Hayri Körezlioglu, Claudia Klüppelberg, Damien Lamberton, Marco

Papi, Benedetto Piccoli, Uwe Schmock, Christoph Schwab, Mete Soner, Peter Spreij, Lukasz Stettner, Johan Tysk, Esko Valkeila, and Michèle Vanmaele. We thank them all for the important work done together and the cooperative and friendly atmosphere.

Oslo
30th August 2010

Giulia Di Nunno
Bernt Øksendal

Contents

Chapter 1
Dynamic Risk Measures

Beatrice Acciaio and Irina Penner

Abstract This paper gives an overview of the theory of dynamic convex risk measures for random variables in discrete-time setting. We summarize robust representation results of conditional convex risk measures, and we characterize various time consistency properties of dynamic risk measures in terms of acceptance sets, penalty functions, and by supermartingale properties of risk processes and penalty functions.

Keywords Dynamic convex risk measure · Robust representation · Penalty function · Time consistency · Entropic risk measure

Mathematics Subject Classification (2010) 91B30 · 91B16

1.1 Introduction

Risk measures are quantitative tools developed to determine minimum capital reserves that are required to be maintained by financial institutions in order to ensure their financial stability. An axiomatic analysis of risk assessment in terms of capital requirements was initiated by Artzner, Delbaen, Eber, and Heath [2, 3], who introduced coherent risk measures. Föllmer and Schied [23] and Frittelli and Rosazza

Financial support from the European Science Foundation (ESF) "Advanced Mathematical Methods for Finance" (AMaMeF) under the exchange grant 2281 and hospitality of Vienna University of Technology are gratefully acknowledged by B. Acciaio.
I. Penner was supported by the DFG Research Center MATHEON "Mathematics for key technologies." Financial support from the European Science Foundation (ESF) "Advanced Mathematical Methods for Finance" (AMaMeF) under the short visit grant 2854 is gratefully acknowledged.

B. Acciaio (✉)
Department of Economy, Finance and Statistics, University of Perugia, Via A. Pascoli 20, 06123 Perugia, Italy
e-mail: beatrice.acciaio@stat.unipg.it

I. Penner
Institut für Mathematik, Humboldt-Universität zu Berlin, Unter den Linden 6, 10099 Berlin, Germany
e-mail: penner@math.hu-berlin.de

G. Di Nunno, B. Øksendal (eds.), *Advanced Mathematical Methods for Finance*,
DOI 10.1007/978-3-642-18412-3_1,

Gianin [25] replaced subadditivity and positive homogeneity by convexity in the set of axioms and established the more general concept of a convex risk measure. Since then, convex and coherent risk measures and their applications have attracted a growing interest both in mathematical finance research and among practitioners.

One of the most appealing properties of a convex risk measure is its robustness against model uncertainty. Under some regularity condition, it can be represented as a suitably modified worst expected loss over a whole class of probabilistic models. This was initially observed in [3, 23, 25] in the static setting, where financial positions are described by random variables on some probability space, and a risk measure is a real-valued functional. For a comprehensive presentation of the theory of static coherent and convex risk measures, we refer to Delbaen [15] and Föllmer and Schied [24, Chap. 4].

A natural extension of a static risk measure is given by a conditional risk measure, which takes into account the information available at the time of risk assessment. As its static counterpart, a conditional convex risk measure can be represented as the worst conditional expected loss over a class of suitably penalized probability measures; see [6, 12, 18, 26, 29, 34, 37]. In the dynamical setting described by some filtered probability space, risk assessment is updated over the time in accordance with the new information. This leads to the notion of dynamic risk measure, which is a sequence of conditional risk measures adapted to the underlying filtration.

A crucial question in the dynamical framework is how risk evaluations at different times are interrelated. Several notions of time consistency were introduced and studied in the literature. One of today's most used notions is strong time consistency, which corresponds to the dynamic programming principle; see [4, 7, 12, 13, 16–18, 22, 26, 29] and references therein. As shown in [7, 16, 22], strong time consistency can be characterized by additivity of the acceptance sets and penalty functions, and also by a supermartingale property of the risk process and the penalty function process. Similar characterizations of the weaker notions of time consistency, so-called rejection and acceptance consistency, were given in [19, 33]. Rejection consistency, also called prudence in [33], seems to be a particularly suitable property from the point of view of a regulator, since it ensures that one always stays on the safe side when updating risk assessment. The weakest notions of time consistency considered in the literature are weak acceptance and weak rejection consistency, which require that if some position is accepted (or rejected) for any scenario tomorrow, it should be already accepted (or rejected) today; see [4, 9, 35, 41, 43].

As pointed out in [21, 28], risk assessment in the multiperiod setting should also account for uncertainty about the time value of money. This requires to consider entire cash flow processes rather than total amounts at terminal dates as risky objects, and it leads to a further extension of the notion of risk measure. Risk measures for processes were studied in [1, 4, 10–13, 27, 28, 34]. The new feature in this framework is that not only the amounts but also the timing of payments matters; cf. [1, 12, 13, 28]. However, as shown in [4] in the static and in [1] in the dynamical setting, risk measures for processes can be identified with risk measures for random variables on an appropriate product space. This allows a natural translation of results obtained in the framework of risk measures for random variables to the framework of processes; see [1].

The aim of this paper it to give an overview of the current theory of dynamic convex risk measures for random variables in discrete-time setting; the corresponding results for risk measures for processes are given in [1]. The paper is organized as follows. Section 1.2 recalls the definition of a conditional convex risk measure and its interpretation as the minimal capital requirement from [18]. Section 1.3 summarizes robust representation results from [8, 18, 22]. In Sect. 1.4 we first give an overview of different time consistency properties based on [40]. Then we focus on the strong notion of time consistency in Sect. 1.4.1, and we characterize it by supermartingale properties of risk processes and penalty functions. The results of this subsection are mainly based on [22], with the difference that here we give characterizations of time consistency also in terms of absolutely continuous probability measures, similar to [8]. In addition, we relate the martingale property of a risk process with the worst-case measure, and we provide explicit forms of the Doob and Riesz decompositions of the penalty function process. Section 1.4.2 generalizes [33, Sects. 2.4 and 2.5] and characterizes rejection and acceptance consistency in terms of acceptance sets, penalty functions, and, in case of rejection consistency, by a supermartingale property of risk processes and one-step penalty functions. Section 1.4.3 recalls characterizations of weak time consistency from [9, 41, 43], and Sect. 1.4.4 characterizes the recursive construction of time consistent risk measures suggested in [12, 13]. Finally, the dynamic entropic risk measure with a nonconstant risk aversion parameter is studied in Sect. 1.5.

1.2 Setup and Notation

Let $T \in \mathbb{N} \cup \{\infty\}$ be the time horizon, $\mathbb{T} := \{0, \dots, T\}$ for $T < \infty$, and $\mathbb{T} := \mathbb{N}_0$ for $T = \infty$. We consider a discrete-time setting given by a filtered probability space $(\Omega, \mathcal{F}, (\mathcal{F}_t)_{t \in \mathbb{T}}, P)$ with $\mathcal{F}_0 = \{\emptyset, \Omega\}$, $\mathcal{F} = \mathcal{F}_T$ for $T < \infty$, and $\mathcal{F} = \sigma(\bigcup_{t \geq 0} \mathcal{F}_t)$ for $T = \infty$. For $t \in \mathbb{T}$, $L_t^\infty := L^\infty(\Omega, \mathcal{F}_t, P)$ is the space of all essentially bounded $\mathcal{F}_t$-measurable random variables, and $L^\infty := L^\infty(\Omega, \mathcal{F}_T, P)$. All equalities and inequalities between random variables and between sets are understood to hold P-almost surely, unless stated otherwise. We denote by $\mathcal{M}_1(P)$ (resp. by $\mathcal{M}^e(P)$) the set of all probability measures on $(\Omega, \mathcal{F})$ that are absolutely continuous with respect to P (resp. equivalent to P).

In this work we consider risk measures defined on the set L^∞, which is understood as the set of discounted terminal values of financial positions. In the dynamical setting, a conditional risk measure ρ_t assigns to each terminal payoff X an $\mathcal{F}_t$-measurable random variable $\rho_t(X)$ that quantifies the risk of the position X given the information $\mathcal{F}_t$. A rigorous definition of a conditional convex risk measure was given in [18, Definition 2].

Definition 1.1 A map $\rho_t : L^\infty \to L_t^\infty$ is called a *conditional convex risk measure* if it satisfies the following properties for all $X, Y \in L^\infty$:

(i) Conditional cash invariance: For all $m_t \in L_t^\infty$,

$$\rho_t(X + m_t) = \rho_t(X) - m_t;$$

(ii) Monotonicity: $X \leq Y \Rightarrow \rho_t(X) \geq \rho_t(Y)$;
(iii) Conditional convexity: for all $\lambda \in L_t^\infty$, $0 \leq \lambda \leq 1$,

$$\rho_t\big(\lambda X + (1-\lambda)Y\big) \leq \lambda\rho_t(X) + (1-\lambda)\rho_t(Y);$$

(iv) Normalization: $\rho_t(0) = 0$.

A conditional convex risk measure is called a *conditional coherent risk measure* if it has in addition the following property:

(v) Conditional positive homogeneity: for all $\lambda \in L_t^\infty$, $\lambda \geq 0$,

$$\rho_t(\lambda X) = \lambda\rho_t(X).$$

In the dynamical framework one can also analyze risk assessment for cumulated cash flow *processes* rather than just for terminal payoffs, i.e., one can consider a risk measure that accounts not only for the amounts but also for the timing of payments. Such risk measures were studied in [1, 10–13, 27, 28]. As shown in [4] in the static and in [1] in the dynamical setting, convex risk measures for processes can be identified with convex risk measures for random variables on an appropriate product space. This allows one to extend results obtained in our present setting to the framework of processes; cf. [1].

If ρ_t is a conditional convex risk measure, the function $\phi_t := -\rho_t$ defines a conditional monetary utility function in the sense of [12, 13]. The term "monetary" refers to conditional cash invariance of the utility function, the only property in Definition 1.1 that does not come from the classical utility theory. Conditional cash invariance is a natural request in view of the interpretation of ρ_t as a conditional capital requirement. In order to formalize this aspect, we first recall the notion of the *acceptance set* of a conditional convex risk measure ρ_t:

$$\mathcal{A}_t := \big\{X \in L^\infty \;\big|\; \rho_t(X) \leq 0\big\}.$$

The following properties of the acceptance set were given in [18, Proposition 3].

Proposition 1.2 *The acceptance set $\mathcal{A}_t$ of a conditional convex risk measure ρ_t is*

1. *conditionally convex, i.e., $\alpha X + (1-\alpha)Y \in \mathcal{A}_t$ for all $X, Y \in \mathcal{A}_t$ and $\mathcal{F}_t$-measurable α such that $0 \leq \alpha \leq 1$;*
2. *solid, i.e., $Y \in \mathcal{A}_t$ whenever $Y \geq X$ for some $X \in \mathcal{A}_t$;*
3. *such that $0 \in \mathcal{A}_t$ and $\operatorname{ess\,inf}\{X \in L_t^\infty \mid X \in \mathcal{A}_t\} = 0$.*

Moreover, ρ_t is uniquely determined through its acceptance set, since

$$\rho_t(X) = \operatorname{ess\,inf}\big\{Y \in L_t^\infty \;\big|\; X + Y \in \mathcal{A}_t\big\}. \tag{1.1}$$

Conversely, if some set $\mathcal{A}_t \subseteq L^\infty$ satisfies conditions (1)–(3), *then the functional $\rho_t : L^\infty \to L_t^\infty$ defined via* (1.1) *is a conditional convex risk measure.*

Proof Properties (1)–(3) of the acceptance set follow easily from properties (i)–(iv) in Definition 1.1. To prove (1.1), note that by cash invariance $\rho_t(X) + X \in \mathcal{A}_t$ for all X, and this implies "$\geq$" in (1.1). On the other hand, for all $Z \in \{Y \in L_t^\infty \mid X + Y \in \mathcal{A}_t\}$, we have

$$0 \geq \rho_t(Z + X) = \rho_t(X) - Z,$$

and thus $\rho_t(X) \leq \operatorname{ess\,inf}\{Y \in L_t^\infty \mid X + Y \in \mathcal{A}_t\}$.

For the proof of the last part of the assertion, we refer to [18, Proposition 3]. □

Due to (1.1), the value $\rho_t(X)$ can be viewed as the minimal conditional capital requirement needed to be added to the position X in order to make it acceptable at time t. Moreover, (1.1) can be used to define risk measures; cf. Example 1.8.

1.3 Robust Representation

As observed in [3, 24, 25] in the static setting, the axiomatic properties of a convex risk measure yield, under some regularity condition, a representation of the minimal capital requirement as a suitably modified worst expected loss over a whole class of probabilistic models. In the dynamical setting, such robust representations of conditional coherent risk measures were obtained in [6, 8, 18, 22, 29, 37] for random variables and in [12, 34] for stochastic processes. In this section we mainly summarize the results from [8, 18, 22].

The alternative probability measures in a robust representation of a risk measure ρ_t contribute to the risk evaluation to a different degree. To formalize this aspect, we use the notion of the minimal penalty function $\alpha_t^{\min}$, defined for each $Q \in \mathcal{M}_1(P)$ as

$$\alpha_t^{\min}(Q) = Q\text{-}\operatorname*{ess\,sup}_{X \in \mathcal{A}_t} E_Q[-X|\mathcal{F}_t]. \tag{1.2}$$

The following property of the minimal penalty function is a standard result that will be used in the proof of Theorem 1.4.

Lemma 1.3 *For $Q \in \mathcal{M}_1(P)$ and $0 \leq s \leq t$,*

$$E_Q\big[\alpha_t^{\min}(Q)\big|\mathcal{F}_s\big] = Q\text{-}\operatorname*{ess\,sup}_{Y \in \mathcal{A}_t} E_Q[-Y|\mathcal{F}_s] \quad Q\text{-a.s.}$$

and in particular

$$E_Q\big[\alpha_t^{\min}(Q)\big] = \sup_{Y \in \mathcal{A}_t} E_Q[-Y].$$

Proof First we claim that the set

$$\big\{E_Q[-X|\mathcal{F}_t] \,\big|\, X \in \mathcal{A}_t\big\}$$

is directed upward for any $Q \in \mathcal{M}_1(P)$. Indeed, for $X, Y \in \mathcal{A}_t$, we can define $Z := XI_A + YI_{A^c}$, where $A := \{E_Q[-X|\mathcal{F}_t] \geq E_Q[-Y|\mathcal{F}_t]\} \in \mathcal{F}_t$. Conditional convexity of ρ_t implies that $Z \in \mathcal{A}_t$, and by definition of Z,

$$E_Q[-Z|\mathcal{F}_t] = \max\big(E_Q[-X|\mathcal{F}_t], E_Q[-Y|\mathcal{F}_t]\big) \quad Q\text{-a.s.}$$

Hence, there exists a sequence $(X_n^Q)_{n\in\mathbb{N}}$ in $\mathcal{A}_t$ such that

$$\alpha_t^{\min}(Q) = \lim_n E_Q\big[-X_n^Q|\mathcal{F}_t\big] \quad Q\text{-a.s.}, \tag{1.3}$$

and by monotone convergence we get

$$\begin{aligned} E_Q\big[\alpha_t^{\min}(Q)\big|\mathcal{F}_s\big] &= \lim_n E_Q\big[E_Q[-X_n^Q|\mathcal{F}_t]\big|\mathcal{F}_s\big] \\ &\leq Q\text{-}\operatorname*{ess\,sup}_{Y\in\mathcal{A}_t} E_Q[-Y|\mathcal{F}_s] \quad Q\text{-a.s.} \end{aligned}$$

The converse inequality follows directly from the definition of $\alpha_t^{\min}(Q)$. □

The following theorem relates robust representations to some continuity properties of conditional convex risk measures. It combines [18, Theorem 1] with [22, Corollary 2.4]; similar results can be found in [6, 12, 29].

Theorem 1.4 *For a conditional convex risk measure ρ_t, the following are equivalent:*

1. *ρ_t has a robust representation*

$$\rho_t(X) = \operatorname*{ess\,sup}_{Q\in\mathcal{Q}_t}\big(E_Q[-X|\mathcal{F}_t] - \alpha_t(Q)\big), \quad X \in L^\infty, \tag{1.4}$$

where

$$\mathcal{Q}_t := \big\{Q \in \mathcal{M}_1(P) \,\big|\, Q = P|_{\mathcal{F}_t}\big\},$$

and α_t is a map from $\mathcal{Q}_t$ to the set of $\mathcal{F}_t$-measurable random variables with values in $\mathbb{R} \cup \{+\infty\}$ such that $\operatorname{ess\,sup}_{Q\in\mathcal{Q}_t}(-\alpha_t(Q)) = 0$.

2. *ρ_t has the robust representation in terms of the minimal penalty function, i.e.,*

$$\rho_t(X) = \operatorname*{ess\,sup}_{Q\in\mathcal{Q}_t}\big(E_Q[-X|\mathcal{F}_t] - \alpha_t^{\min}(Q)\big), \quad X \in L^\infty, \tag{1.5}$$

where $\alpha_t^{\min}$ is given in (1.2).

3. *ρ_t has the robust representation*

$$\rho_t(X) = \operatorname*{ess\,sup}_{Q\in\mathcal{Q}_t^f}\big(E_Q[-X|\mathcal{F}_t] - \alpha_t^{\min}(Q)\big) \quad P\text{-a.s.}, \quad X \in L^\infty, \tag{1.6}$$

where

$$\mathcal{Q}_t^f := \big\{Q \in \mathcal{M}_1(P) \,\big|\, Q = P|_{\mathcal{F}_t},\ E_Q\big[\alpha_t^{\min}(Q)\big] < \infty\big\}.$$

4. *ρ_t has the "Fatou-property": for any bounded sequence $(X_n)_{n\in\mathbb{N}}$ which converges P-a.s. to some X,*

$$\rho_t(X) \le \liminf_{n\to\infty} \rho_t(X_n) \quad P\text{-}a.s.$$

5. *ρ_t is continuous from above, i.e.,*

$$X_n \searrow X \quad P\text{-}a.s \quad \Longrightarrow \quad \rho_t(X_n) \nearrow \rho_t(X) \quad P\text{-}a.s$$

for any sequence $(X_n)_n \subseteq L^\infty$ and $X \in L^\infty$.

Proof (3) ⇒ (1) and (2) ⇒ (1) are obvious. (1) ⇒ (4): Dominated convergence implies that $E_Q[X_n|\mathcal{F}_t] \to E_Q[X|\mathcal{F}_t]$ for each $Q \in \mathcal{Q}_t$, and $\liminf_{n\to\infty} \rho_t(X_n) \ge \rho_t(X)$ follows by using the robust representation of ρ_t as in the unconditional setting, see, e.g., [24, Lemma 4.20].

(4) ⇒ (5): Monotonicity implies $\limsup_{n\to\infty} \rho_t(X_n) \le \rho_t(X)$, and $\liminf_{n\to\infty} \rho_t(X_n) \ge \rho_t(X)$ follows by (4).

(5) ⇒ (2): The inequality

$$\rho_t(X) \ge \operatorname*{ess\,sup}_{Q\in\mathcal{Q}_t}\big(E_Q[-X|\mathcal{F}_t] - \alpha_t^{\min}(Q)\big) \tag{1.7}$$

follows from the definition of $\alpha_t^{\min}$. In order to prove the equality, we will show that

$$E_P\big[\rho_t(X)\big] \le E_P\Big[\operatorname*{ess\,sup}_{Q\in\mathcal{Q}_t}\big(E_Q[-X|\mathcal{F}_t] - \alpha_t^{\min}(Q)\big)\Big].$$

To this end, consider the map $\rho^P : L^\infty \to \mathbb{R}$ defined by $\rho^P(X) := E_P[\rho_t(X)]$. It is easy to check that ρ^P is a convex risk measure which is continuous from above. Hence [24, Theorem 4.31] implies that ρ^P has the robust representation

$$\rho^P(X) = \sup_{Q\in\mathcal{M}_1(P)} \big(E_Q[-X] - \alpha(Q)\big), \quad X \in L^\infty,$$

where the penalty function $\alpha(Q)$ is given by

$$\alpha(Q) = \sup_{X\in L^\infty:\rho^P(X)\le 0} E_Q[-X].$$

Next we will prove that $Q \in \mathcal{Q}_t$ if $\alpha(Q) < \infty$. Indeed, let $A \in \mathcal{F}_t$ and $\lambda > 0$. Then

$$-\lambda P[A] = E_P\big[\rho_t(\lambda I_A)\big] = \rho^P(\lambda I_A) \ge E_Q[-\lambda I_A] - \alpha(Q),$$

so

$$P[A] \le Q[A] + \frac{1}{\lambda}\alpha(Q) \quad \text{for all } \lambda > 0,$$

and hence $P[A] \le Q[A]$ if $\alpha(Q) < \infty$. The same reasoning with $\lambda < 0$ implies $P[A] \ge Q[A]$, and thus $P = Q$ on $\mathcal{F}_t$ if $\alpha(Q) < \infty$. By Lemma 1.3, we have for every $Q \in \mathcal{Q}_t$,

$$E_P\big[\alpha_t^{\min}(Q)\big] = \sup_{Y \in \mathcal{A}_t} E_P[-Y].$$

Since $\rho^P(Y) \le 0$ for all $Y \in \mathcal{A}_t$, this implies

$$E_P\big[\alpha_t^{\min}(Q)\big] \le \alpha(Q)$$

for all $Q \in \mathcal{Q}_t$, by definition of the penalty function $\alpha(Q)$.

Finally we obtain

$$\begin{aligned} E_P\big[\rho_t(X)\big] = \rho^P(X) &= \sup_{Q \in \mathcal{M}_1(P), \alpha(Q) < \infty} \big(E_Q[-X] - \alpha(Q)\big) \\ &\le \sup_{Q \in \mathcal{Q}_t, E_P[\alpha_t^{\min}(Q)] < \infty} \big(E_Q[-X] - \alpha(Q)\big) \\ &\le \sup_{Q \in \mathcal{Q}_t, E_P[\alpha_t^{\min}(Q)] < \infty} E_P\big[E_Q[-X|\mathcal{F}_t] - \alpha_t^{\min}(Q)\big] \\ &\le E_P\Big[\operatorname*{ess\,sup}_{Q \in \mathcal{Q}_t, E_P[\alpha_t^{\min}(Q)] < \infty} \big(E_Q[-X|\mathcal{F}_t] - \alpha_t^{\min}(Q)\big)\Big] \\ &\le E_P\Big[\operatorname*{ess\,sup}_{Q \in \mathcal{Q}_t} \big(E_Q[-X|\mathcal{F}_t] - \alpha_t^{\min}(Q)\big)\Big], \end{aligned} \tag{1.8}$$

proving (1.5).

(5) $\Rightarrow$ (3) The inequality

$$\rho_t(X) \ge \operatorname*{ess\,sup}_{Q \in \mathcal{Q}_t^f} \big(E_Q[-X|\mathcal{F}_t] - \alpha_t^{\min}(Q)\big)$$

follows from (1.7) since $\mathcal{Q}_t^f \subseteq \mathcal{Q}_t$, and (1.8) proves the equality. □

Remark 1.5 The penalty function $\alpha_t^{\min}(Q)$ is minimal in the sense that any other function α_t in a robust representation (1.4) of ρ_t satisfies

$$\alpha_t^{\min}(Q) \le \alpha_t(Q) \quad P\text{-a.s.}$$

for all $Q \in \mathcal{Q}_t$. An alternative formula for the minimal penalty function is given by

$$\alpha_t^{\min}(Q) = \operatorname*{ess\,sup}_{X \in L^\infty} \big(E_Q[-X|\mathcal{F}_t] - \rho_t(X)\big) \quad \text{for all } Q \in \mathcal{Q}_t.$$

This follows as in the unconditional case; see, e.g., [24, Theorem 4.15, Remark 4.16].

In the *coherent* case the penalty function $\alpha_t^{\min}(Q)$ can only take values 0 or ∞ due to positive homogeneity of ρ_t. Thus representation (1.12) takes the following form.

Corollary 1.6 *A conditional coherent risk measure ρ_t is continuous from above if and only if it is representable in the form*

$$\rho_t(X) = \operatorname*{ess\,sup}_{Q \in \mathcal{Q}_t^0} E_Q[-X|\mathcal{F}_t], \quad X \in L^\infty, \tag{1.9}$$

where

$$\mathcal{Q}_t^0 := \{Q \in \mathcal{Q}_t \mid \alpha_t^{\min}(Q) = 0 \ Q\text{-a.s.}\}.$$

Remark 1.7 Another characterization of a conditional convex risk measure ρ_t that is equivalent to properties (1)–(5) of Theorem 1.4 is the following: The acceptance set $\mathcal{A}_t$ is weak*-closed, i.e., it is closed in L^∞ with respect to the topology $\sigma(L^\infty, L^1(\Omega, \mathcal{F}, P))$. This equivalence was shown in [12] in the context of risk measures for processes and in [29] for risk measures for random variables. Though in [29] a slightly different definition of a conditional risk measure is used, the reasoning given there works just the same in our case; cf. [29, Theorem 3.16].

Example 1.8 A class of examples of conditional convex risk measures can be obtained by considering a conditional robust version of a *shortfall risk* introduced in [24, Sect. 4.9]. To this end, let $l_t : \mathbb{R} \to \mathbb{R}$ be a convex and strictly increasing loss function, and let $\mathcal{R}_t$ be some convex subset of $\mathcal{Q}_t$. Then the set

$$\mathcal{A}_t := \{X \in L^\infty \mid E_Q[l_t(-X)|\mathcal{F}_t] \le l_t(0) \ \forall Q \in \mathcal{R}_t\} \tag{1.10}$$

satisfies properties (1)–(3) of Proposition 1.2 and thus induces a conditional convex risk measure. Such risk measures were introduced and studied in [41, Sect. 5], where they are called *conditional robust shortfall risk measures*.

A conditional robust shortfall risk measure is continuous from above by Remark 1.7. Indeed, if $(X_n)_{n\in\mathbb{N}}$ is a bounded sequence in $\mathcal{A}_t$ converging to some X, then $X \in \mathcal{A}_t$ due to Lebesgue convergence theorem, and thus the set $\mathcal{A}_t$ is weak*-closed by Krein–Šmulian theorem; cf., e.g., [24, Theorem A.63, Lemma A.64]. Moreover, if $P \in \mathcal{R}_t$ (or if there exists $Q^* \approx P$ such that $Q^* \in \mathcal{R}_t$), then the set of equivalent probability measures is dense in $\mathcal{R}_t$, and representation (1.10) can be written as

$$\mathcal{A}_t = \{X \in L^\infty \mid E_Q[l_t(-X)|\mathcal{F}_t] \le l_t(0) \ \forall Q \in \mathcal{R}_t^e\}, \tag{1.11}$$

where $\mathcal{R}_t^e$ denotes the set of all $Q \in \mathcal{M}^e(P)$ such that the corresponding $\mathcal{F}_t$-normalized measure $\tilde{Q}$ defined by $\frac{d\tilde{Q}}{dP} := \frac{Z_T}{Z_t}$ belongs to $\mathcal{R}_t$. Here Z_s denotes the density of Q with respect to P on $\mathcal{F}_s$, $s \in \mathbb{T}$.

Example 1.9 If one takes $\mathcal{R}_t = \{P\}$ and the exponential loss function $l_t(x) = \exp(\gamma_t x) - 1$ with $\gamma_t > 0$ in the previous example, one obtains the well-known *conditional entropic risk measure*

$$\rho_t(X) = \frac{1}{\gamma_t} \log E\big[\exp(-\gamma_t X)\big|\mathcal{F}_t\big], \quad X \in L^\infty.$$

The entropic risk measure was introduced in [24] in the static setting; in the dynamical setting it appeared in [5, 12, 13, 18, 22, 31]. We characterize the dynamic entropic risk measure in Sect. 1.5 in a slightly more general setting, where the risk aversion parameter γ_t might be random.

Example 1.10 Example 1.8 with a linear loss function $l_t(x) = x$ and

$$\mathcal{R}_t := \left\{ Q \in \mathcal{Q}_t \,\middle|\, \frac{dQ}{dP} \leq \lambda_t^{-1} \right\}$$

for some $\lambda_t \in L_t^\infty$, $0 < \lambda_t \leq 1$, yields an important example of a conditional coherent risk measure, the *conditional Average Value-at-Risk*

$$AV@R_{t,\lambda_t}(X) := \operatorname{ess\,sup}\big\{E_Q[-X|\mathcal{F}_t] \;\big|\; Q \in \mathcal{R}_t\big\}.$$

Static Average Value-at-Risk was introduced in [3] as a valid alternative to the widely used yet criticized Value-at-Risk. The conditional version of Average Value-at-Risk appeared in [4] and was also studied in [19, 42].

For the characterization of time consistency in Sect. 1.4, we will need a robust representation of a conditional convex risk measure ρ_t under any measure $Q \in \mathcal{M}_1(P)$, where possibly $Q \notin \mathcal{Q}_t$. Such representation can be obtained as in Theorem 1.4 by considering ρ_t as a risk measure under Q, as shown in the next corollary. This result is a version of [8, Proposition 1].

Corollary 1.11 *A conditional convex risk measure ρ_t is continuous from above if and only if it has the robust representations*

$$\rho_t(X) = Q\text{-}\operatorname*{ess\,sup}_{R \in \mathcal{Q}_t(Q)} \big(E_R[-X|\mathcal{F}_t] - \alpha_t^{\min}(R)\big) \tag{1.12}$$

$$= Q\text{-}\operatorname*{ess\,sup}_{R \in \mathcal{Q}_t^f(Q)} \big(E_R[-X|\mathcal{F}_t] - \alpha_t^{\min}(R)\big) \quad Q\text{-a.s.}, \quad X \in L^\infty, \tag{1.13}$$

for all $Q \in \mathcal{M}_1(P)$, where

$$\mathcal{Q}_t(Q) = \big\{R \in \mathcal{M}_1(P) \;\big|\; R = Q|_{\mathcal{F}_t}\big\}$$

and

$$\mathcal{Q}_t^f(Q) = \big\{R \in \mathcal{M}_1(P) \;\big|\; R = Q|_{\mathcal{F}_t}, \, E_R\big[\alpha_t^{\min}(R)\big] < \infty\big\}.$$

Proof To show that continuity from above implies representation (1.12), we can replace P by a probability measure $Q \in \mathcal{M}_1(P)$ and repeat all the reasoning of the proof of (5) $\Rightarrow$ (2) in Theorem 1.4. In this case we consider the static convex risk measure

$$\rho^Q(X) = E_Q\big[\rho_t(X)\big] = \sup_{R\in\mathcal{M}_1(P)} \big(E_R[-X] - \alpha(R)\big), \quad X \in L^\infty,$$

instead of ρ^P. The proof of (1.13) follows in the same way from [22, Corollary 2.4]. Conversely, continuity from above follows from Theorem 1.4 since representation (1.12) holds under P. □

Remark 1.12 One can easily see that the set $\mathcal{Q}_t$ in representations (1.4) and (1.5) can be replaced by $\mathcal{P}_t := \{Q \in \mathcal{M}_1(P) \mid Q \approx P \text{ on } \mathcal{F}_t\}$. Moreover, representation (1.4) is also equivalent to

$$\rho_t(X) = \operatorname*{ess\,sup}_{Q\in\mathcal{M}_1(P)} \big(E_Q[-X|\mathcal{F}_t] - \hat{\alpha}_t(Q)\big), \quad X \in L^\infty,$$

where the conditional expectation under $Q \in \mathcal{M}_1(P)$ is defined under P as

$$E_Q[X|\mathcal{F}_t] := \frac{E_P[Z_T X|\mathcal{F}_t]}{Z_t} I_{\{Z_t>0\}}$$

with $Z_s := \frac{dQ}{dP}|_{\mathcal{F}_s}$, $s \in \mathbb{T}$, and the extended penalty function $\hat{\alpha}_t$ is given by

$$\hat{\alpha}_t(Q) = \begin{cases} \alpha_t(Q) & \text{on } \{Z_t > 0\}, \\ +\infty & \text{otherwise.} \end{cases}$$

As observed, e.g., in [12, Remark 3.13], the minimal penalty function has the local property. In our context it means that for any $Q^1, Q^2 \in \mathcal{Q}_t(Q)$ with the corresponding density processes Z^1 and Z^2 with respect to P and for any $A \in \mathcal{F}_t$, the probability measure R defined via $\frac{dR}{dP} := I_A Z_T^1 + I_{A^c} Z_T^2$ has the penalty function value

$$\alpha_t^{\min}(R) = I_A \alpha_t^{\min}\big(Q^1\big) + I_{A^c} \alpha_t^{\min}\big(Q^2\big) \quad Q\text{-a.s.}$$

In particular, $R \in \mathcal{Q}_t^f(Q)$ if $Q^1, Q^2 \in \mathcal{Q}_t^f(Q)$. Standard arguments (cf., e.g., [18, Lemma 1]) imply then that the set

$$\big\{E_R[-X|\mathcal{F}_t] - \alpha_t^{\min}(R) \bigm| R \in \mathcal{Q}_t^f(Q)\big\}$$

is directed upward, and thus

$$E_Q\big[\rho_t(X)\big|\mathcal{F}_s\big] = Q\text{-}\operatorname*{ess\,sup}_{R\in\mathcal{Q}_t^f(Q)} \big(E_R[-X|\mathcal{F}_s] - E_R\big[\alpha_t^{\min}(R)\big|\mathcal{F}_s\big]\big) \tag{1.14}$$

for all $Q \in \mathcal{M}_1(P)$, $X \in L^\infty(\Omega, \mathcal{F}, P)$ and $0 \le s \le t$.

1.4 Time Consistency Properties

In the dynamical setting, risk assessment of a financial position is updated when new information is released. This leads to the notion of a dynamic risk measure.

Definition 1.13 A sequence $(\rho_t)_{t\in\mathbb{T}}$ is called a *dynamic convex risk measure* if ρ_t is a conditional convex risk measure for each $t \in \mathbb{T}$.

A key question in the dynamical setting is how the conditional risk assessments at different times are interrelated. This question has led to several notions of time consistency discussed in the literature. A unifying view was suggested in [40].

Definition 1.14 Assume that $(\rho_t)_{t\in\mathbb{T}}$ is a dynamic convex risk measure and let $\mathcal{Y}_t$ be a subset of L^∞ such that $0 \in \mathcal{Y}_t$ and $\mathcal{Y}_t + \mathbb{R} = \mathcal{Y}_t$ for each $t \in \mathbb{T}$. Then $(\rho_t)_{t\in\mathbb{T}}$ is called *acceptance (resp. rejection) consistent with respect to* $(\mathcal{Y}_t)_{t\in\mathbb{T}}$ if for all $t \in \mathbb{T}$ such that $t < T$ and for any $X \in L^\infty$ and $Y \in \mathcal{Y}_{t+1}$, the following condition holds:

$$\rho_{t+1}(X) \le \rho_{t+1}(Y) \quad (\text{resp.} \ge) \quad \Longrightarrow \quad \rho_t(X) \le \rho_t(Y) \quad (\text{resp.} \ge). \tag{1.15}$$

The idea is that the degree of time consistency is determined by a sequence of benchmark sets $(\mathcal{Y}_t)_{t\in\mathbb{T}}$: if a financial position at some future time is always preferable to some element of the benchmark set, then it should also be preferable today. The bigger the benchmark set, the stronger is the resulting notion of time consistency. In the following we focus on three cases.

Definition 1.15 We call a dynamic convex risk measure $(\rho_t)_{t\in\mathbb{T}}$

1. *strongly time consistent* if it is either acceptance consistent or rejection consistent with respect to $\mathcal{Y}_t = L^\infty$ for all t in the sense of Definition 1.14;
2. *middle acceptance (resp. middle rejection) consistent* if for all t, we have $\mathcal{Y}_t = L_t^\infty$ in Definition 1.14;
3. *weakly acceptance (resp. weakly rejection) consistent* if for all t, we have $\mathcal{Y}_t = \mathbb{R}$ in Definition 1.14.

Note that there is no difference between rejection consistency and acceptance consistency with respect to L^∞, since the role of X and Y is symmetric in that case. Obviously strong time consistency implies both middle rejection and middle acceptance consistency, and middle rejection (resp. middle acceptance) consistency implies weak rejection (resp. weak acceptance) consistency. In the rest of the paper we drop the terms "middle" and "strong" in order to simplify the terminology.

1.4.1 Time Consistency

Time consistency has been studied extensively in the recent work on dynamic risk measures, see [4, 8, 9, 12, 13, 16–18, 22, 29, 33, 34] and the references therein. In the next proposition we recall some equivalent characterizations of time consistency.

Proposition 1.16 *A dynamic convex risk measure* $(\rho_t)_{t\in\mathbb{T}}$ *is time consistent if and only if any of the following conditions holds*:

1. *for all* $t \in \mathbb{T}$ *such that* $t < T$ *and for all* $X, Y \in L^\infty$,

$$\rho_{t+1}(X) \le \rho_{t+1}(Y) \quad P\text{-}a.s \quad \Longrightarrow \quad \rho_t(X) \le \rho_t(Y) \quad P\text{-}a.s.; \tag{1.16}$$

2. *for all* $t \in \mathbb{T}$ *such that* $t < T$ *and for all* $X, Y \in L^\infty$,

$$\rho_{t+1}(X) = \rho_{t+1}(Y) \quad P\text{-}a.s \quad \Longrightarrow \quad \rho_t(X) = \rho_t(Y) \quad P\text{-}a.s.; \tag{1.17}$$

3. $(\rho_t)_{t\in\mathbb{T}}$ *is recursive, i.e.*,

$$\rho_t = \rho_t(-\rho_{t+s}) \quad P\text{-}a.s.$$

for all $t, s \ge 0$ *such that* $t, t+s \in \mathbb{T}$.

Proof It is obvious that time consistency implies condition (1.16) and that (1.16) implies (1.17). By cash invariance we have $\rho_{t+1}(-\rho_{t+1}(X)) = \rho_{t+1}(X)$, and hence one-step recursiveness follows from (1.17). We prove that one-step recursiveness implies recursiveness by induction on s. For $s = 1$, the claim is true for all t. Assume that the induction hypothesis holds for each t and all $k \le s$ for some $s \ge 1$. Then we obtain

$$\begin{aligned}\rho_t\big(-\rho_{t+s+1}(X)\big) &= \rho_t\big(-\rho_{t+s}\big(-\rho_{t+s+1}(X)\big)\big)\\ &= \rho_t\big(-\rho_{t+s}(X)\big)\\ &= \rho_t(X),\end{aligned}$$

where we have applied the induction hypothesis to the random variable $-\rho_{t+s+1}(X)$. Hence the claim follows. Finally, due to monotonicity, recursiveness implies time consistency. □

Remark 1.17 The recursivity property (3) of Proposition 1.16 corresponds to the dynamic programming principle, and it is crucial for many applications. In continuous time and in Brownian setting, it allows one to relate time consistent dynamic risk measures to the solutions of a certain type of backward stochastic differential equations, so-called g-*expectations*; cf. [20, 26, 32, 38]. Indeed, as shown in [38, Proposition 19], a conditional g-expectation defines a time consistent dynamic convex risk measure on $L^2(P)$ if the BSDE generator g is convex (and satisfies the usual assumptions ensuring existence of a solution). Conversely, as shown in [38, Proposition 20], if $(\rho_t)_{t\in[0,T]}$ is a strictly monotone time consistent dynamic convex risk measure in Brownian setting and if ρ_0 satisfies a certain boundedness condition, then (ρ_t) can be identified as a conditional g-expectation. This relation allows one in particular to characterize penalty functions of time consistent dynamic convex risk measures in Brownian setting; cf. [17].

If we restrict a conditional convex risk measure ρ_t to the space L^∞_{t+s} for some $s \geq 0$, the corresponding acceptance set is given by

$$\mathcal{A}_{t,t+s} := \{X \in L^\infty_{t+s} \mid \rho_t(X) \leq 0 \ P\text{-a.s.}\},$$

and the minimal penalty function by

$$\alpha^{\min}_{t,t+s}(Q) := Q\text{-}\operatorname*{ess\,sup}_{X \in \mathcal{A}_{t,t+s}} E_Q[-X|\mathcal{F}_t], \quad Q \in \mathcal{M}_1(P). \tag{1.18}$$

The following lemma recalls equivalent characterizations of recursive inequalities in terms of acceptance sets from [22, Lemma 4.6]; property (1.19) was shown in [16].

Lemma 1.18 *Let $(\rho_t)_{t\in\mathbb{T}}$ be a dynamic convex risk measure. Then the following equivalences hold for all s, t such that $t, t+s \in \mathbb{T}$ and all $X \in L^\infty$:*

$$X \in \mathcal{A}_{t,t+s} + \mathcal{A}_{t+s} \iff -\rho_{t+s}(X) \in \mathcal{A}_{t,t+s}, \tag{1.19}$$

$$\mathcal{A}_t \subseteq \mathcal{A}_{t,t+s} + \mathcal{A}_{t+s} \iff \rho_t(-\rho_{t+s}) \leq \rho_t \quad P\text{-a.s.}, \tag{1.20}$$

$$\mathcal{A}_t \supseteq \mathcal{A}_{t,t+s} + \mathcal{A}_{t+s} \iff \rho_t(-\rho_{t+s}) \geq \rho_t \quad P\text{-a.s.} \tag{1.21}$$

Proof To prove "$\Rightarrow$" in (1.19), let $X = X_{t,t+s} + X_{t+s}$ with $X_{t,t+s} \in \mathcal{A}_{t,t+s}$ and $X_{t+s} \in \mathcal{A}_{t+s}$. Then

$$\rho_{t+s}(X) = \rho_{t+s}(X_{t+s}) - X_{t,t+s} \leq -X_{t,t+s}$$

by cash invariance, and monotonicity implies

$$\rho_t(-\rho_{t+s}(X)) \leq \rho_t(X_{t,t+s}) \leq 0.$$

The converse direction follows immediately from $X = X + \rho_{t+s}(X) - \rho_{t+s}(X)$ and $X + \rho_{t+s}(X) \in \mathcal{A}_{t+s}$ for all $X \in L^\infty$.

In order to show "$\Rightarrow$" in (1.20), fix $X \in L^\infty$. Since $X + \rho_t(X) \in \mathcal{A}_t \subseteq \mathcal{A}_{t,t+s} + \mathcal{A}_{t+s}$, we obtain

$$\rho_{t+s}(X) - \rho_t(X) = \rho_{t+s}(X + \rho_t(X)) \in -\mathcal{A}_{t,t+s},$$

by (1.19) and cash invariance. Hence,

$$\rho_t(-\rho_{t+s}(X)) - \rho_t(X) = \rho_t(-(\rho_{t+s}(X) - \rho_t(X))) \leq 0 \quad P\text{-a.s.}$$

To prove "$\Leftarrow$", let $X \in \mathcal{A}_t$. Then $-\rho_{t+s}(X) \in \mathcal{A}_{t,t+s}$ by the right-hand side of (1.20), and hence $X \in \mathcal{A}_{t,t+s} + \mathcal{A}_{t+s}$ by (1.19).

Now let $X \in L^\infty$ and assume $\mathcal{A}_t \supseteq \mathcal{A}_{t,t+s} + \mathcal{A}_{t+s}$. Then

$$\begin{aligned}\rho_t(-\rho_{t+s}(X)) + X &= \rho_t(-\rho_{t+s}(X)) - \rho_{t+s}(X) + \rho_{t+s}(X) + X \\ &\in \mathcal{A}_{t,t+s} + \mathcal{A}_{t+s} \subseteq \mathcal{A}_t.\end{aligned}$$

Hence,

$$\rho_t(X) - \rho_t\big(-\rho_{t+s}(X)\big) = \rho_t\big(X + \rho_t\big(-\rho_{t+s}(X)\big)\big) \le 0$$

by cash invariance, and this proves "⇒" in (1.21). For the converse direction, let $X \in \mathcal{A}_{t,t+s} + \mathcal{A}_{t+s}$. Since $-\rho_{t+s}(X) \in \mathcal{A}_{t,t+s}$ by (1.19), we obtain

$$\rho_t(X) \le \rho_t\big(-\rho_{t+s}(X)\big) \le 0,$$

and hence, $X \in \mathcal{A}_t$. □

We also have the following relation between acceptance sets and penalty functions; cf. [33, Lemma 2.2.5].

Lemma 1.19 *Let $(\rho_t)_{t\in\mathbb{T}}$ be a dynamic convex risk measures. Then the following implications hold for all t, s such that $t, t+s \in \mathbb{T}$ and for all $Q \in \mathcal{M}_1(P)$:*

$$\mathcal{A}_t \subseteq \mathcal{A}_{t,t+s} + \mathcal{A}_{t+s} \quad \Longrightarrow \quad \alpha_t^{\min}(Q) \le \alpha_{t,t+s}^{\min}(Q) + E_Q\big[\alpha_{t+s}^{\min}(Q)\big|\mathcal{F}_t\big] \quad Q\text{-a.s.},$$

$$\mathcal{A}_t \supseteq \mathcal{A}_{t,t+s} + \mathcal{A}_{t+s} \quad \Longrightarrow \quad \alpha_t^{\min}(Q) \ge \alpha_{t,t+s}^{\min}(Q) + E_Q\big[\alpha_{t+s}^{\min}(Q)\big|\mathcal{F}_t\big] \quad Q\text{-a.s.}$$

Proof Straightforward from the definition of the minimal penalty function and Lemma 1.3. □

The following theorem gives equivalent characterizations of time consistency in terms of acceptance sets, penalty functions, and a supermartingale property of the risk process.

Theorem 1.20 *Let $(\rho_t)_{t\in\mathbb{T}}$ be a dynamic convex risk measure such that each ρ_t is continuous from above. Then the following conditions are equivalent:*

1. *$(\rho_t)_{t\in\mathbb{T}}$ is time consistent.*
2. *$\mathcal{A}_t = \mathcal{A}_{t,t+s} + \mathcal{A}_{t+s}$ for all t, s such that $t, t+s \in \mathbb{T}$.*
3. *$\alpha_t^{\min}(Q) = \alpha_{t,t+s}^{\min}(Q) + E_Q[\alpha_{t+s}^{\min}(Q)|\mathcal{F}_t]$ Q-a.s. for all t, s such that $t, t+s \in \mathbb{T}$ and all $Q \in \mathcal{M}_1(P)$.*
4. *For all $X \in L^\infty(\Omega, \mathcal{F}, P)$ and all t, s such that $t, t+s \in \mathbb{T}$ and all $Q \in \mathcal{M}_1(P)$, we have*

$$E_Q\big[\rho_{t+s}(X) + \alpha_{t+s}^{\min}(Q)\big|\mathcal{F}_t\big] \le \rho_t(X) + \alpha_t^{\min}(Q) \quad Q\text{-a.s.}$$

The equivalence of properties (1) and (2) of Theorem 1.20 was proved in [16]. Characterizations of time consistency in terms of penalty functions as in (3) of Theorem 1.20 appeared in [7, 8, 13, 22]; similar results for risk measures for processes were given in [12, 13]. In [7, 8] property (3) is called *cocycle property*. The supermartingale property as in (4) of Theorem 1.20 was obtained in [22]; cf. also [8] for continuous-time setting.

Proof The proof of (1) ⇒ (2) ⇒ (3) follows from Lemmas 1.18 and 1.19. To prove (3) ⇒ (4), fix $Q \in \mathcal{M}_1(P)$. By (1.14) we have

$$E_Q\big[\rho_{t+s}(X)\big|\mathcal{F}_t\big] = Q\text{-}\operatorname*{ess\,sup}_{R\in\mathcal{Q}^f_{t+s}(Q)}\big(E_R[-X|\mathcal{F}_t] - E_R\big[\alpha^{\min}_{t+s}(R)\big|\mathcal{F}_t\big]\big).$$

On the set $\{\alpha^{\min}_t(Q) = \infty\}$ property (4) holds trivially. On the set $\{\alpha^{\min}_t(Q) < \infty\}$ property (3) implies $E_Q[\alpha^{\min}_{t+s}(Q)|\mathcal{F}_t] < \infty$ and $\alpha^{\min}_{t,t+s}(Q) < \infty$; then for $R \in \mathcal{Q}^f_{t+s}(Q)$,

$$\alpha^{\min}_t(R) = \alpha^{\min}_{t,t+s}(Q) + E_R\big[\alpha^{\min}_{t+s}(R)\big|\mathcal{F}_t\big] < \infty \quad Q\text{-a.s.}$$

Thus,

$$E_Q\big[\rho_{t+s}(X) + \alpha^{\min}_{t+s}(Q)\big|\mathcal{F}_t\big] = Q\text{-}\operatorname*{ess\,sup}_{R\in\mathcal{Q}^f_{t+s}(Q)}\big(E_R[-X|\mathcal{F}_t] - \alpha^{\min}_t(R)\big) + \alpha^{\min}_t(Q)$$

on $\{\alpha^{\min}_t(Q) < \infty\}$. Moreover, since $\mathcal{Q}^f_{t+s}(Q) \subseteq \mathcal{Q}_t(Q)$, (1.12) implies

$$\begin{aligned} E_Q\big[\rho_{t+s}(X) + \alpha^{\min}_{t+s}(Q)\big|\mathcal{F}_t\big] &\le Q\text{-}\operatorname*{ess\,sup}_{R\in\mathcal{Q}_t(Q)}\big(E_R[-X|\mathcal{F}_t] - \alpha^{\min}_t(R)\big) + \alpha^{\min}_t(Q) \\ &= \rho_t(X) + \alpha^{\min}_t(Q) \quad Q\text{-a.s.} \end{aligned}$$

It remains to prove (4) ⇒ (1). To this end, fix $Q \in \mathcal{Q}^f_t$ and $X, Y \in L^\infty$ such that $\rho_{t+1}(X) \le \rho_{t+1}(Y)$. Note that $E_Q[\alpha_{t+s}(Q)] < \infty$ due to (4), and hence $Q \in \mathcal{Q}^f_{t+s}(Q)$. Using (4) and representation (1.13) for ρ_{t+s} under Q, we obtain

$$\begin{aligned} \rho_t(Y) + \alpha^{\min}_t(Q) &\ge E_Q\big[\rho_{t+1}(Y) + \alpha^{\min}_{t+1}(Q)\big|\mathcal{F}_t\big] \\ &\ge E_Q\big[\rho_{t+1}(X) + \alpha^{\min}_{t+1}(Q)\big|\mathcal{F}_t\big] \\ &\ge E_Q\big[E_Q[-X|\mathcal{F}_{t+1}] - \alpha^{\min}_{t+1}(Q) + \alpha^{\min}_{t+1}(Q)\big|\mathcal{F}_t\big] \\ &= E_Q[-X|\mathcal{F}_t]. \end{aligned}$$

Hence representation (1.6) yields $\rho_t(Y) \ge \rho_t(X)$, and time consistency follows from Proposition 1.16. □

Properties (3) and (4) of Theorem 1.20 imply in particular supermartingale properties of penalty function processes and risk processes. This allows one to apply martingale theory for characterization of the dynamics of these processes, as we do in Propositions 1.21 and 1.24; cf. also [8, 16, 17, 22, 33].

Proposition 1.21 *Let $(\rho_t)_{t\in\mathbb{T}}$ be a time consistent dynamic convex risk measure such that each ρ_t is continuous from above. Then the process*

$$V^Q_t(X) := \rho_t(X) + \alpha^{\min}_t(Q), \quad t \in \mathbb{T},$$

is a Q-supermartingale for all $X \in L^\infty$ *and all* $Q \in \mathcal{Q}_0$, *where*

$$\mathcal{Q}_0 := \left\{ Q \in \mathcal{M}_1(P) \,\middle|\, \alpha_0^{\min}(Q) < \infty \right\}.$$

Moreover, $(V_t^Q(X))_{t\in\mathbb{T}}$ *is a Q-martingale if* $Q \in \mathcal{Q}_0$ *is a "worst-case" measure for X at time* 0, *i.e., if the supremum in the robust representation of* $\rho_0(X)$ *is attained at Q:*

$$\rho_0(X) = E_Q[-X] - \alpha_0^{\min}(Q).$$

In this case Q is a "worst-case" measure for X at any time t, i.e.,

$$\rho_t(X) = E_Q[-X|\mathcal{F}_t] - \alpha_t^{\min}(Q) \quad Q\text{-a.s.} \quad \text{for all } t \in \mathbb{T}.$$

The converse holds if $T < \infty$ *or* $\lim_{t\to\infty} \rho_t(X) = -X$ *P-a.s. (what is called asymptotic precision in* [22]*): If* $(V_t^Q(X))_{t\in\mathbb{T}}$ *is a Q-martingale, then* $Q \in \mathcal{Q}_0$ *is a "worst-case" measure for X at any time* $t \in \mathbb{T}$.

Proof The supermartingale property of $(V_t^Q(X))_{t\in\mathbb{T}}$ under each $Q \in \mathcal{Q}_0$ follows directly from properties (3) and (4) of Theorem 1.20. To prove the remaining part of the claim, fix $Q \in \mathcal{Q}_0$ and $X \in L^\infty$. If Q is a "worst-case" measure for X at time 0, the process

$$U_t(X) := V_t^Q(X) - E_Q[-X|\mathcal{F}_t], \quad t \in \mathbb{T},$$

is a nonnegative Q-supermartingale beginning at 0. Indeed, the supermartingale property follows from that of $(V_t^Q(X))_{t\in\mathbb{T}}$, and nonnegativity follows from representation (1.13), since $Q \in \mathcal{Q}_t^f(Q)$. Thus, $U_t = 0$ Q-a.s. for all t, and this proves the "if" part of the claim. To prove the converse direction, note that if $(V_t^Q(X))_{t\in\mathbb{T}}$ is a Q-martingale and $\rho_T(X) = -X$ (resp. $\lim_{t\to\infty} \rho_t(X) = -X$ P-a.s.), the process $U(X)$ is a Q-martingale ending at 0 (resp. converging to 0 in $L^1(Q)$), and thus $U_t(X) = 0$ Q-a.s. for all $t \in \mathbb{T}$. □

Remark 1.22 The fact that a worst-case measure for X at time 0, if it exists, remains a worst-case measure for X at any time $t \in \mathbb{T}$ was also shown in [13, Theorem 3.9] for a time consistent dynamic risk measure in finite time horizon without using the supermartingale property from Proposition 1.21.

Remark 1.23 In difference to [22, Theorem 4.5], without the additional assumption that the set

$$\mathcal{Q}^* := \left\{ Q \in \mathcal{M}^e(P) \,\middle|\, \alpha_0^{\min}(Q) < \infty \right\} \tag{1.22}$$

is nonempty, the supermartingale property of $(V_t^Q(X))_{t\in\mathbb{T}}$ for all $X \in L^\infty$ and all $Q \in \mathcal{Q}^*$ is not sufficient to prove time consistency. In this case we also do not have the robust representation of ρ_t in terms of the set $\mathcal{Q}^*$.

The process $(\alpha_t^{\min}(Q))_{t\in\mathbb{T}}$ is a Q-supermartingale for all $Q \in \mathcal{Q}_0$ due to Property (3) of Theorem 1.20. The next proposition provides explicit forms of its Doob and Riesz decompositions; cf. also [33, Proposition 2.3.2].

Proposition 1.24 *Let $(\rho_t)_{t\in\mathbb{T}}$ be a time consistent dynamic convex risk measure such that each ρ_t is continuous from above. Then for each $Q \in \mathcal{Q}_0$, the process $(\alpha_t^{\min}(Q))_{t\in\mathbb{T}}$ is a nonnegative Q-supermartingale with the Riesz decomposition*

$$\alpha_t^{\min}(Q) = Z_t^Q + M_t^Q \quad Q\text{-a.s.}, \quad t \in \mathbb{T},$$

where

$$Z_t^Q := E_Q\left[\sum_{k=t}^{T-1} \alpha_{k,k+1}^{\min}(Q)\middle|\mathcal{F}_t\right] \quad Q\text{-a.s.}, \quad t \in \mathbb{T},$$

is a Q-potential, and

$$M_t^Q := \begin{cases} 0 & \text{if } T < \infty, \\ \lim_{s\to\infty} E_Q[\alpha_s(Q)|\mathcal{F}_t] & \text{if } T = \infty, \end{cases} \quad Q\text{-a.s.}, \quad t \in \mathbb{T},$$

is a nonnegative Q-martingale.

Moreover, the Doob decomposition of $(\alpha_t^{\min}(Q))_{t\in\mathbb{T}}$ is given by

$$\alpha_t^{\min}(Q) = E_Q\left[\sum_{k=0}^{T-1} \alpha_{k,k+1}^{\min}(Q)\middle|\mathcal{F}_t\right] + M_t^Q - \sum_{k=0}^{t-1} \alpha_{k,k+1}^{\min}(Q), \quad t \in \mathbb{T},$$

with the Q-martingale

$$E_Q\left[\sum_{k=0}^{T-1} \alpha_{k,k+1}^{\min}(Q)\middle|\mathcal{F}_t\right] + M_t^Q, \quad t \in \mathbb{T},$$

and the nondecreasing predictable process $(\sum_{k=0}^{t-1} \alpha_{k,k+1}^{\min}(Q))_{t\in\mathbb{T}}$.

Proof We fix $Q \in \mathcal{M}_1(P)$ and applying property (3) of Theorem 1.20 step by step, we obtain

$$\alpha_t^{\min}(Q) = E_Q\left[\sum_{k=t}^{t+s-1} \alpha_{k,k+1}^{\min}(Q)\middle|\mathcal{F}_t\right] + E_Q\left[\alpha_{t+s}^{\min}(Q)\middle|\mathcal{F}_t\right] \quad Q\text{-a.s.} \tag{1.23}$$

for all t, s such that $t, t+s \in \mathbb{T}$. If $T < \infty$, the Doob and Riesz decompositions follow immediately from (1.23), since $\alpha_T(Q) = 0$ Q-a.s. If $T = \infty$, by monotonicity there exists the limit

$$Z_t^Q = \lim_{s\to\infty} E_Q\left[\sum_{k=t}^{s} \alpha_{k,k+1}^{\min}(Q)\middle|\mathcal{F}_t\right] = E_Q\left[\sum_{k=t}^{\infty} \alpha_{k,k+1}^{\min}(Q)\middle|\mathcal{F}_t\right] \quad Q\text{-a.s.}$$

for all $t \in \mathbb{T}$, where we have used the monotone convergence theorem for the second equality. Equality (1.23) implies then that there exists

$$M_t^Q = \lim_{s\to\infty} E_Q\big[\alpha_{t+s}^{\min}(Q)\big|\mathcal{F}_t\big] \quad Q\text{-a.s.}, \quad t \in \mathbb{T},$$

and

$$\alpha_t^{\min}(Q) = Z_t^Q + M_t^Q \quad Q\text{-a.s.}$$

for all $t \in \mathbb{T}$.

The process $(Z_t^Q)_{t\in\mathbb{T}}$ is a nonnegative Q-supermartingale. Indeed,

$$E_Q\big[Z_t^Q\big] \le E_Q\left[\sum_{k=0}^{\infty} \alpha_{k,k+1}^{\min}(Q)\right] \le \alpha_0^{\min}(Q) < \infty \tag{1.24}$$

and $E_Q[Z_{t+1}^Q|\mathcal{F}_t] \le Z_t^Q$ Q-a.s. for all $t \in \mathbb{T}$ by definition. Moreover, monotone convergence implies

$$\lim_{t\to\infty} E_Q\big[Z_t^Q\big] = E_Q\left[\lim_{t\to\infty}\sum_{k=t}^{\infty} \alpha_{k,k+1}^{\min}(Q)\right] = 0 \quad Q\text{-a.s.},$$

since $\sum_{k=0}^{\infty} \alpha_{k,k+1}^{\min}(Q) < \infty$ Q-a.s. by (1.24). Hence the process $(Z_t^Q)_{t\in\mathbb{T}}$ is a Q-potential.

The process $(M_t^Q)_{t\in\mathbb{T}}$ is a nonnegative Q-martingale, since

$$E_Q\big[M_t^Q\big] \le E_Q\big[\alpha_t^{\min}(Q)\big] \le \alpha_0^{\min}(Q) < \infty$$

and

$$\begin{aligned} E_Q\big[M_{t+1}^Q - M_t^Q\big|\mathcal{F}_t\big] &= E_Q\big[\alpha_{t+1}^{\min}(Q)\big|\mathcal{F}_t\big] - \alpha_t^{\min}(Q) - E_Q\big[Z_{t+1}^Q - Z_t^Q\big|\mathcal{F}_t\big] \\ &= \alpha_{t,t+1}^{\min}(Q) - \alpha_{t,t+1}^{\min}(Q) = 0 \quad Q\text{-a.s.} \end{aligned}$$

for all $t \in \mathbb{T}$ by property (3) of Theorem 1.20 and the definition of $(Z_t^Q)_{t\in\mathbb{T}}$.

The Doob decomposition follows straightforward from the Riesz decomposition. □

Remark 1.25 It was shown in [22, Theorem 5.4] that the martingale M^Q in the Riesz decomposition of $(\alpha_t^{\min}(Q))_{t\in\mathbb{T}}$ vanishes if and only if $\lim_{t\to\infty}\rho_t(X) \ge -X$ P-a.s., i.e., the dynamic risk measure $(\rho_t)_{t\in\mathbb{T}}$ is asymptotically safe. This is not always the case; see [22, Example 5.5].

For a *coherent* risk measure, we have

$$\mathcal{Q}_t^f(Q) = \mathcal{Q}_t^0(Q) := \big\{R \in \mathcal{M}^1(P)\;\big|\; R = Q|_{\mathcal{F}_t},\; \alpha_t^{\min}(R) = 0\ Q\text{-a.s.}\big\}.$$

In order to give an equivalent characterization of property (3) of Theorem 1.20 in the coherent case, we introduce the sets

$$\mathcal{Q}^0_{t,t+s}(Q) = \{R \ll P|_{\mathcal{F}_{t+s}} \mid R = Q|_{\mathcal{F}_t},\ \alpha^{\min}_{t,t+s}(R) = 0\ Q\text{-a.s.}\}$$

for all $t, s \geq 0$ such that $t, t+s \in \mathbb{T}$. For $Q^1 \in \mathcal{Q}^0_{t,t+s}(Q)$ and $Q^2 \in \mathcal{Q}^0_{t+s}(Q)$, we denote by $Q^1 \oplus^{t+s} Q^2$ the pasting of Q^1 and Q^2 in $t+s$ via Ω, i.e., the measure $\widetilde{Q}$ defined via

$$\widetilde{Q}(A) = E_{Q^1}\big[E_{Q^2}[I_A|\mathcal{F}_{t+s}]\big], \quad A \in \mathcal{F}. \tag{1.25}$$

The relation between stability under pasting and time consistency of coherent risk measures that can be represented in terms of equivalent probability measures was studied in [4, 16, 22, 29]. In our present setting, Theorem 1.20 applied to a coherent risk measure takes the following form.

Corollary 1.26 *Let $(\rho_t)_{t\in\mathbb{T}}$ be a dynamic coherent risk measure such that each ρ_t is continuous from above. Then the following conditions are equivalent:*

1. *$(\rho_t)_{t\in\mathbb{T}}$ is time consistent.*
2. *For all $Q \in \mathcal{M}_1(P)$ and all t, s such that $t, t+s \in \mathbb{T}$,*

$$\mathcal{Q}^0_t(Q) = \{Q^1 \oplus^{t+s} Q^2 \mid Q^1 \in \mathcal{Q}^0_{t,t+s}(Q),\ Q^2 \in \mathcal{Q}^0_{t+s}(Q^1)\}.$$

3. *For all $Q \in \mathcal{M}_1(P)$ such that $\alpha^{\min}_t(Q) = 0$ Q-a.s.,*

$$E_Q\big[\rho_{t+s}(X)\big|\mathcal{F}_t\big] \leq \rho_t(X) \quad \text{and} \quad \alpha^{\min}_{t+s}(Q) = 0 \quad Q\text{-a.s.}$$

for all $X \in L^\infty(\Omega, \mathcal{F}, P)$ and for all t, s such that $t, t+s \in \mathbb{T}$.

Proof (1) $\Rightarrow$ (2): Time consistency implies property (3) of Theorem 1.20, and we will show that this implies property (2) of Corollary 1.26. Fix $Q \in \mathcal{M}_1(P)$. To prove "$\supseteq$", let $Q^1 \in \mathcal{Q}^0_t(Q)$, $Q^2 \in \mathcal{Q}^0_{t+s}(Q^1)$, and consider $\widetilde{Q}$ defined as in (1.25). Note that $\widetilde{Q} = Q^1$ on $\mathcal{F}_{t+s}$ and

$$E_{\widetilde{Q}}[X|\mathcal{F}_{t+s}] = E_{Q^2}[X|\mathcal{F}_{t+s}] \quad Q^1\text{-a.s.} \quad \text{for all } X \in L^\infty(\Omega, \mathcal{F}, P).$$

Hence, using (3) of Theorem 1.20, we obtain

$$\begin{aligned}\alpha^{\min}_t(\widetilde{Q}) &= \alpha^{\min}_{t,t+s}(\widetilde{Q}) + E_{\widetilde{Q}}\big[\alpha^{\min}_{t+s}(\widetilde{Q})\big|\mathcal{F}_t\big] \\ &= \alpha^{\min}_{t,t+s}(Q^1) + E_{Q^1}\big[\alpha^{\min}_{t+s}(Q^2)\big|\mathcal{F}_t\big] = 0 \quad Q\text{-a.s.,}\end{aligned}$$

and thus $\widetilde{Q} \in \mathcal{Q}^0_t(Q)$. Conversely, for every $\widetilde{Q} \in \mathcal{Q}^0_t(Q)$, we have $\alpha^{\min}_{t+s}(\widetilde{Q}) = \alpha^{\min}_{t,t+s}(\widetilde{Q}) = 0$ $\widetilde{Q}$-a.s. by (3) of Theorem 1.20, and $\widetilde{Q} = \widetilde{Q} \oplus \widetilde{Q}$. This proves "$\subseteq$".

(2) $\Rightarrow$ (3): Let $R \in \mathcal{M}_1(P)$ with $\alpha^{\min}_t(R) = 0$ R-a.s.. Then $R \in \mathcal{Q}^0_t(R)$, and thus $R = Q^1 \oplus^{t+s} Q^2$ for some $Q^1 \in \mathcal{Q}^0_{t,t+s}(R)$ and $Q^2 \in \mathcal{Q}^0_{t+s}(Q^1)$. This implies

$R = Q^1$ on $\mathcal{F}_{t+s}$ and

$$E_R[X|\mathcal{F}_{t+s}] = E_{Q^2}[X|\mathcal{F}_{t+s}] \quad R\text{-a.s.}$$

Hence $\alpha_{t,t+s}^{\min}(R) = \alpha_{t,t+s}^{\min}(Q^1) = 0$, R-a.s., and $\alpha_{t+s}^{\min}(R) = \alpha_{t+s}^{\min}(Q^2) = 0$ R-a.s. To prove inequality (3), note that due to (1.14),

$$\begin{aligned} E_R\big[\rho_{t+s}(X)\big|\mathcal{F}_t\big] &= R\text{-}\operatorname*{ess\,sup}_{Q\in\mathcal{Q}^0_{t+s}(R)} E_Q[-X|\mathcal{F}_t] \\ &\le R\text{-}\operatorname*{ess\,sup}_{Q\in\mathcal{Q}^0_t(R)} E_Q[-X|\mathcal{F}_t] = \rho_t(X) \quad R\text{-a.s.}, \end{aligned}$$

where we have used that the pasting of $R|_{\mathcal{F}_{t+s}}$ and Q belongs to $\mathcal{Q}^0_t(R)$.

(3) ⇒ (1): Obviously, property (3) of Corollary 1.26 implies property (4) of Theorem 1.20 and thus time consistency. □

1.4.2 Rejection and Acceptance Consistency

Rejection and acceptance consistency were introduced and studied in [19, 33, 40, 41]. These properties can be characterized via recursive inequalities as stated in the next proposition; see [40, Theorem 3.1.5] and [19, Proposition 3.5].

Proposition 1.27 *A dynamic convex risk measure $(\rho_t)_{t\in\mathbb{T}}$ is rejection (resp. acceptance) consistent if and only if for all $t \in \mathbb{T}$ such that $t < T$,*

$$\rho_t(-\rho_{t+1}) \le \rho_t \quad (\textit{resp.} \ge) \quad P\text{-a.s.} \tag{1.26}$$

Proof We argue for the case of rejection consistency; the case of acceptance consistency follows in the same manner. Assume first that $(\rho_t)_{t\in\mathbb{T}}$ satisfies (1.26) and let $X \in L^\infty$ and $Y \in L^\infty(\mathcal{F}_{t+1})$ such that $\rho_{t+1}(X) \ge \rho_{t+1}(Y)$. Using cash invariance, (1.26), and monotonicity, we obtain

$$\rho_t(X) \ge \rho_t\big(-\rho_{t+1}(X)\big) \ge \rho_t\big(-\rho_{t+1}(Y)\big) = \rho_t(Y).$$

The converse implication follows due to cash invariance by applying (1.15) to $Y = -\rho_{t+1}(X)$. □

Remark 1.28 As shown in [19, Proposition 3.9], for a dynamic *coherent* risk measure, weak acceptance consistency and acceptance consistency are equivalent. Indeed, let $(\rho_t)_{t\in\mathbb{T}}$ be a coherent dynamic risk measure that is weakly acceptance consistent. Then

$$\rho_t(X) \le \rho_t\big(X + \rho_{t+1}(X)\big) + \rho_t\big(-\rho_{t+1}(X)\big) \quad \forall X \in L^\infty$$

due to subadditivity. Since $\rho_{t+1}(X + \rho_{t+1}(X)) = 0$, weak acceptance consistency implies $\rho_t(X + \rho_{t+1}(X)) \le 0$, and thus $\rho_t(X) \le \rho_t(-\rho_{t+1}(X))$ for all t and all $X \in L^\infty$.

Example 1.29 One obtains acceptance-consistent dynamic risk measures by taking suprema over families of time consistent dynamic risk measures. Indeed, if $\mathcal{R}$ is a collection of time consistent dynamic convex risk measures, then

$$\widehat{\rho}_t(X) := \operatorname*{ess\,sup}_{\rho\in\mathcal{R}} \rho_t(X), \qquad t \in \mathbb{T},\ X \in L^\infty,$$

defines a dynamic convex risk measure. Moreover, monotonicity of $(\widehat{\rho}_t)$ and time consistency of (ρ_t) imply $\widehat{\rho}_t(X) \le \widehat{\rho}_t(-\widehat{\rho}_{t+1}(X))$ for all t, i.e., $(\widehat{\rho}_t)_{t\in\mathbb{T}}$ is acceptance consistent. This was noted in [36, Lemma 7.1].

Rejection consistency can be characterized as follows.

Proposition 1.30 *A dynamic convex risk measure $(\rho_t)_{t\in\mathbb{T}}$ is rejection consistent if and only if any of the following conditions holds*:

1. *For all $t \in \mathbb{T}$ such that $t < T$ and all $X \in L^\infty$,*

$$\rho_t(X) - \rho_{t+1}(X) \in \mathcal{A}_{t,t+1}; \tag{1.27}$$

2. *For all $t \in \mathbb{T}$ such that $t < T$ and all $X \in \mathcal{A}_t$, we have $-\rho_{t+1}(X) \in \mathcal{A}_t$.*

Proof Since

$$\rho_t\big(-\rho_{t+1}(X)\big) = \rho_t\big(\rho_t(X) - \rho_{t+1}(X)\big) + \rho_t(X)$$

by cash invariance, (1.27) implies rejection consistency, and obviously rejection consistency implies condition (2). If (2) holds, then for any $X \in L^\infty$,

$$\rho_t\big(\rho_t(X) - \rho_{t+1}(X)\big) = \rho_t\big(-\rho_{t+1}\big(X + \rho_t(X)\big)\big) \le 0,$$

due to cash invariance and the fact that $X + \rho_t(X) \in \mathcal{A}_t$. □

Property (1.27) was introduced in [33] under the name *prudence*. It means that the adjustment $\rho_{t+1}(X) - \rho_t(X)$ of the minimal capital requirement for X at time $t+1$ is acceptable at time t. In other words, one stays on the safe side at each period of time by making capital reserves according to a rejection consistent dynamic risk measure.

Similar to time consistency, rejection and acceptance consistency can be characterized in terms of acceptance sets and penalty functions.

Theorem 1.31 *Let $(\rho_t)_{t\in\mathbb{T}}$ be a dynamic convex risk measure such that each ρ_t is continuous from above. Then the following properties are equivalent*:

1. *$(\rho_t)_{t\in\mathbb{T}}$ is rejection consistent (resp. acceptance consistent).*
2. *The inclusion*

$$\mathcal{A}_t \subseteq \mathcal{A}_{t,t+1} + \mathcal{A}_{t+1} \quad \textit{resp.} \quad \mathcal{A}_t \supseteq \mathcal{A}_{t,t+1} + \mathcal{A}_{t+1}$$

holds for all $t \in \mathbb{T}$ such that $t < T$.

3. *The inequality*

$$\alpha_t^{\min}(Q) \le (resp. \ge)\, \alpha_{t,t+1}^{\min}(Q) + E_Q\big[\alpha_{t+1}^{\min}(Q)\big|\mathcal{F}_t\big] \quad Q\text{-a.s.}$$

holds for all $t \in \mathbb{T}$ *such that* $t < T$ *and all* $Q \in \mathcal{M}_1(P)$.

Proof Equivalence of (1) and (2) was proved in Proposition 1.27 and Lemma 1.18, and the proof of (2) ⇒ (3) is given in Lemma 1.19.

Let us show that property (3) implies property (1). We argue for the case of rejection consistency; the case of acceptance consistency follows in the same manner. We fix $t \in \mathbb{T}$ such that $t < T$ and consider the risk measure

$$\widetilde{\rho}_t(X) := \rho_t\big(-\rho_{t+1}(X)\big), \quad X \in L^\infty.$$

It is easily seen that $\widetilde{\rho}_t$ is a conditional convex risk measure that is continuous from above. Moreover, the dynamic risk measure $(\widetilde{\rho}_t, \rho_{t+1})$ is time consistent by definition, and thus it fulfills properties (2) and (3) of Theorem 1.20. We denote by $\widetilde{\mathcal{A}}_t$ and $\widetilde{\mathcal{A}}_{t,t+1}$ the acceptance sets of the risk measure $\widetilde{\rho}_t$, and by $\widetilde{\alpha}_t^{\min}$ its penalty function. Since

$$\widetilde{\rho}_t(X) = \rho_t\big(-\rho_{t+1}(X)\big) = \rho_t(X)$$

for all $X \in L_{t+1}^\infty$, we have $\widetilde{\mathcal{A}}_{t,t+1} = \mathcal{A}_{t,t+1}$, and thus

$$\widetilde{\mathcal{A}}_t = \mathcal{A}_{t,t+1} + \mathcal{A}_{t+1}$$

by (2) of Theorem 1.20. Lemma 1.19 and property (3) then imply

$$\widetilde{\alpha}_t^{\min}(Q) = \alpha_{t,t+1}^{\min}(Q) + E_Q\big[\alpha_{t+1}^{\min}(Q)\big|\mathcal{F}_t\big] \ge \alpha_t^{\min}(Q)$$

for all $Q \in \mathcal{Q}_t$. Thus,

$$\rho_t(X) \ge \widetilde{\rho}_t(X) = \rho_t\big(-\rho_{t+1}(X)\big)$$

for all $X \in L^\infty$, due to representation (1.6). □

Remark 1.32 Similarly to Corollary 1.26, condition (3) of Theorem 1.31 can be restated for a dynamic *coherent* risk measure $(\rho_t)_{t\in\mathbb{T}}$ as follows:

$$\mathcal{Q}_t^0(Q) \supseteq \big\{Q^1 \oplus^{t+1} Q^2 \;\big|\; Q^1 \in \mathcal{Q}_{t,t+1}^0(Q),\; Q^2 \in \mathcal{Q}_{t+1}^0\big(Q^1\big)\big\} \quad (\text{resp.} \subseteq)$$

for all $t \in \mathbb{T}$ such that $t < T$ and all $Q \in \mathcal{M}_1(P)$.

The following proposition provides an additional equivalent characterization of rejection consistency that can be viewed as an analogon of the supermartingale property (4) of Theorem 1.20.

Proposition 1.33 *Let $(\rho_t)_{t\in\mathbb{T}}$ be a dynamic convex risk measure such that each ρ_t is continuous from above. Then $(\rho_t)_{t\in\mathbb{T}}$ is rejection consistent if and only if the inequality*

$$E_Q\big[\rho_{t+1}(X)\big|\mathcal{F}_t\big] \le \rho_t(X) + \alpha^{\min}_{t,t+1}(Q) \quad Q\text{-a.s.} \tag{1.28}$$

holds for all $Q \in \mathcal{M}_1(P)$ and all $t \in \mathbb{T}$ such that $t < T$. In this case the process

$$U_t^Q(X) := \rho_t(X) - \sum_{k=0}^{t-1} \alpha^{\min}_{k,k+1}(Q), \quad t \in \mathbb{T},$$

is a Q-supermartingale for all $X \in L^\infty$ and all $Q \in \mathcal{Q}^f$, where

$$\mathcal{Q}^f := \left\{ Q \in \mathcal{M}_1(P) \,\middle|\, E_Q\left[\sum_{k=0}^{t} \alpha^{\min}_{k,k+1}(Q)\right] < \infty \ \forall t \in \mathbb{T} \right\}.$$

The proof of Proposition 1.33 is a special case of Theorem 1.35, which involves the notion of sustainability; cf. [33].

Definition 1.34 Let $(\rho_t)_{t\in\mathbb{T}}$ be a dynamic convex risk measure. We call a bounded adapted process $X = (X_t)_{t\in\mathbb{T}}$ *sustainable with respect to the risk measure* $(\rho_t)_{t\in\mathbb{T}}$ if

$$\rho_t(X_t - X_{t+1}) \le 0 \quad \text{for all } t \in \mathbb{T} \text{ such that } t < T.$$

Consider X to be a cumulative investment process. If it is sustainable, then for all $t \in \mathbb{T}$, the adjustment $X_{t+1} - X_t$ is acceptable with respect to ρ_t.

The next theorem characterizes sustainable processes in terms of a supermartingale inequality; it is a generalization of [33, Corollary 2.4.10].

Theorem 1.35 *Let $(\rho_t)_{t\in\mathbb{T}}$ be a dynamic convex risk measure such that each ρ_t is continuous from above, and let $(X_t)_{t\in\mathbb{T}}$ be a bounded adapted process. Then the following properties are equivalent:*

1. *The process $(X_t)_{t\in\mathbb{T}}$ is sustainable with respect to the risk measure $(\rho_t)_{t\in\mathbb{T}}$.*
2. *For all $Q \in \mathcal{M}_1(P)$ and all $t \in \mathbb{T}, t \ge 1$, we have*

$$E_Q[X_t|\mathcal{F}_{t-1}] \le X_{t-1} + \alpha^{\min}_{t-1,t}(Q) \quad Q\text{-a.s.}. \tag{1.29}$$

Proof The proof of (1) $\Rightarrow$ (2) follows directly from the definition of sustainability and the definition of the minimal penalty function.

To prove (2) $\Rightarrow$ (1), let $(X_t)_{t\in\mathbb{T}}$ be a bounded adapted process such that (1.29) holds. In order to prove

$$X_t - X_{t-1} =: A_t \in -\mathcal{A}_{t-1,t} \quad \text{for all } t \in \mathbb{T}, t \ge 1,$$

suppose by way of contradiction that $A_t \notin -\mathcal{A}_{t-1,t}$. Since the set $\mathcal{A}_{t-1,t}$ is convex and weak*-closed due to Remark 1.7, the Hahn–Banach separation theorem (see, e.g., [24, Theorem A.56]) ensures the existence of $Z \in L^1(\Omega, \mathcal{F}_t, P)$ such that

$$a := \sup_{X \in \mathcal{A}_{t-1,t}} E\big[Z(-X)\big] < E[ZA_t] =: b < \infty. \tag{1.30}$$

Since $\lambda I_{\{Z<0\}} \in \mathcal{A}_{t-1,t}$ for every $\lambda \geq 0$, (1.30) implies $Z \geq 0$ P-a.s., and in particular $E[Z] > 0$. Define the probability measure $Q \in \mathcal{M}_1(P)$ via $\frac{dQ}{dP} := \frac{Z}{E[Z]}$ and note that, due to Lemma 1.3 and (1.30), we have

$$E_Q\big[\alpha_{t-1,t}^{\min}(Q)\big] = \sup_{X \in \mathcal{A}_{t-1,t}} E_Q\big[(-X)\big] = \sup_{X \in \mathcal{A}_{t-1,t}} E\big[Z(-X)\big]\frac{1}{E[Z]} = \frac{a}{E[Z]} < \infty. \tag{1.31}$$

Moreover, (1.30) and (1.31) imply

$$\begin{aligned} E_Q\big[\big(X_t - X_{t-1} - \alpha_{t-1,t}^{\min}(Q)\big)\big] &= E[Z]\big(E[ZA_t] - E_Q\big[\alpha_{t-1,t}^{\min}(Q)\big]\big) \\ &= E[Z](b-a) > 0, \end{aligned}$$

which cannot be true if (1.29) holds under Q. □

Remark 1.36 In particular, property (2) of Theorem 1.35 implies that the process

$$X_t - \sum_{k=0}^{t-1} \alpha_{k,k+1}^{\min}(Q), \quad t \in \mathbb{T},$$

is a Q-supermartingale for all $Q \in \mathcal{Q}^f$ if X is sustainable with respect to (ρ_t). As shown in [33, Theorem 2.4.6, Corollary 2.4.8], this supermartingale property is equivalent to the sustainability of X under some additional assumptions.

1.4.3 Weak Time Consistency

In this section we characterize the weak notions of time consistency from Definition 1.15. Due to cash invariance, they can be restated as follows: A dynamic convex risk measure $(\rho_t)_{t\in\mathbb{T}}$ is weakly acceptance (resp. weakly rejection) consistent if and only if

$$\rho_{t+1}(X) \leq 0 \quad (\text{resp.} \geq) \quad \Longrightarrow \quad \rho_t(X) \leq 0 \quad (\text{resp.} \geq)$$

for any $X \in L^\infty$ and for all $t \in \mathbb{T}$ such that $t < T$. This means that if some position is accepted (or rejected) for any scenario tomorrow, it should be already accepted (or rejected) today. In this form, weak acceptance consistency was introduced in [4]. Both weak acceptance and weak rejection consistency appeared in [35, 40, 41, 43].

Weak acceptance consistency was characterized in terms of acceptance sets in [41, Corollary 3.6] and in terms of a supermartingale property of penalty functions in [9, Lemma 3.17]. We summarize these characterizations in our present setting in the next proposition.

Proposition 1.37 *Let $(\rho_t)_{t\in\mathbb{T}}$ be a dynamic convex risk measure such that each ρ_t is continuous from above. Then the following properties are equivalent:*

1. $(\rho_t)_{t\in\mathbb{T}}$ *is weakly acceptance consistent.*
2. $\mathcal{A}_{t+1}\subseteq\mathcal{A}_t$ *for all* $t\in\mathbb{T}$ *such that* $t<T$.
3. *The inequality*

$$E_Q\big[\alpha_{t+1}^{\min}(Q)\big|F_t\big]\le\alpha_t^{\min}(Q)\quad Q\text{-a.s.} \tag{1.32}$$

holds for all $Q\in\mathcal{M}_1(P)$ *and all* $t\in\mathbb{T}$ *such that* $t<T$. *In particular,* $(\alpha_t^{\min}(Q))_{t\in\mathbb{T}}$ *is a* Q*-supermartingale for all* $Q\in\mathcal{Q}_0$.

Proof The equivalence of (1) and (2) follows directly from the definition of weak acceptance consistency. Property (2) implies (3), since by Lemma 1.3

$$\begin{aligned}E_Q\big[\alpha_{t+1}^{\min}(Q)\big|F_t\big]&=Q\text{-}\operatorname*{ess\,sup}_{X_{t+1}\in\mathcal{A}_{t+1}}E_Q[-X_{t+1}|\mathcal{F}_t]\\&\le Q\text{-}\operatorname*{ess\,sup}_{X\in\mathcal{A}_t}E_Q[-X|\mathcal{F}_t]=\alpha_t^{\min}(Q)\quad Q\text{-a.s.}\end{aligned}$$

for all $Q\in\mathcal{M}_1(P)$.

To prove that (3) implies (2), we fix $X\in\mathcal{A}_{t+1}$ and note that

$$E_Q[-X|\mathcal{F}_{t+1}]\le\alpha_{t+1}^{\min}(Q)\quad Q\text{-a.s.}\quad\text{for all } Q\in\mathcal{M}_1(P)$$

by the definition of the minimal penalty function. Using (1.32), we obtain

$$E_Q[-X|\mathcal{F}_t]\le E_Q\big[\alpha_{t+1}^{\min}(Q)\big|F_t\big]\le\alpha_t^{\min}(Q)\quad Q\text{-a.s.}$$

for all $Q\in\mathcal{M}_1(P)$ and in particular for $Q\in\mathcal{Q}_t^f(P)$. Thus, $\rho_t(X)\le 0$ by (1.6). □

Example 1.38 Consider a dynamic risk measure $(\rho_t)_{t\in\mathbb{T}}$, where each ρ_t is a conditional robust shortfall risk measure as defined in Example 1.8.

1. If $\mathcal{R}_t=\{P\}$ and $l_t=l_0$ for all t, then it is easy to see that $(\rho_t)_{t\in\mathbb{T}}$ is both weakly acceptance and weakly rejection consistent; see, e.g., [43], [39, Example 3.6], [41, Remark 5.3]. However, $(\rho_t)_{t\in\mathbb{T}}$ is in general not time consistent, as illustrated in [39, Example 3.7].
2. Assume that $l_t=l_0$ and that we have representation (1.11) in terms of equivalent probability measures for all t. Then $(\rho_t)_{t\in\mathbb{T}}$ is weakly acceptance consistent if $\mathcal{R}_t^e\subseteq\mathcal{R}_{t+1}^e$ for all t. This was noted in [41, Corollary 5.4] and follows directly from Proposition 1.37, since $A_{t+1}\subseteq\mathcal{A}_t$ for all t in this case.

   This applies in particular to dynamic Average Value-at-Risk $(AV@R_{t,\lambda_t})_{t\in\mathbb{T}}$ from Example 1.10. Indeed, in this case, $P\in\mathcal{R}_t$ for all t, and thus representation (1.11) holds. Condition $\mathcal{R}_t^e\subseteq\mathcal{R}_{t+1}^e$ is satisfied if

$$\lambda_{t+1}\le\lambda_t\operatorname*{ess\,inf}_{Q\in\mathcal{R}_t}E\left[\frac{dQ}{dP}\,\middle|\,\mathcal{F}_{t+1}\right]\quad\forall t\in\mathbb{T}.$$

Thus, $(AV@R_{t,\lambda_t})_{t\in\mathbb{T}}$ is weakly acceptance consistent in this case, and it is even acceptance consistent due to Remark 1.28. A dynamic Average Value-at-Risk with constant parameter λ is in general neither weakly acceptance nor weakly rejection consistent, see, e.g., [4, 35].

3. Consider the case where we have representation (1.11) and $\mathcal{R}_t^e = \mathcal{R}_0^e$ for all t. Assume further that all loss functions l_t are twice continuously differentiable, and let $\gamma_t := \frac{l_t''}{l_t'}$ denote the corresponding Arrow–Pratt coefficient of risk aversion. Then $(\rho_t)_{t\in\mathbb{T}}$ is weakly acceptance consistent if $\gamma_t \leq \gamma_{t+1}$ for all $t \in \mathbb{T}$. This was shown in [41, Corollary 5.5].

1.4.4 A Recursive Construction

In this section we assume that the time horizon T is finite. Then one can define a time consistent dynamic convex risk measure $(\widetilde{\rho}_t)_{t=0,\dots,T}$ in a recursive way, starting with an arbitrary dynamic convex risk measure $(\rho_t)_{t=0,\dots,T}$, via

$$\begin{aligned}\widetilde{\rho}_T(X) &:= \rho_T(X) = -X,\\ \widetilde{\rho}_t(X) &:= \rho_t\big(-\widetilde{\rho}_{t+1}(X)\big), \quad t = 0,\dots,T-1,\ X \in L^\infty.\end{aligned} \tag{1.33}$$

The recursive construction (1.33) was introduced in [12, Sect. 4.2], and also studied in [13, 19]. It is easy to see that $(\widetilde{\rho}_t)_{t=0,\dots,T}$ is indeed a time consistent dynamic convex risk measure, and each $\widetilde{\rho}_t$ is continuous from above if each ρ_t has this property.

Remark 1.39 If the original dynamic convex risk measure $(\rho_t)_{t=0,\dots,T}$ is rejection (resp. acceptance) consistent, then the time consistent dynamic convex risk measure $(\widetilde{\rho}_t)_{t=0,\dots,T}$ defined via (1.33) lies below (resp. above) $(\rho_t)_{t=0,\dots,T}$, i.e.,

$$\widetilde{\rho}_t(X) \leq (\text{resp.} \geq)\ \rho_t(X) \quad \text{for all } t = 0,\dots,T \text{ and all } X \in L^\infty.$$

This can be easily proved by backward induction using Proposition 1.27, monotonicity, and (1.33). Moreover, as shown in [19, Theorem 3.10] in the case of rejection consistency, $(\widetilde{\rho}_t)_{t=0,\dots,T}$ is the biggest time consistent dynamic convex risk measure that lies below $(\rho_t)_{t=0,\dots,T}$.

For all $X \in L^\infty$, the process $(\widetilde{\rho}_t(X))_{t=0,\dots,T}$ has the following properties: $\widetilde{\rho}_T(X) \geq -X$, and

$$\rho_t\big(\widetilde{\rho}_t(X) - \widetilde{\rho}_{t+1}(X)\big) = -\widetilde{\rho}_t(X) + \rho_t\big(-\widetilde{\rho}_{t+1}(X)\big) = 0 \quad \forall t = 0,\dots,T-1, \tag{1.34}$$

by definition and cash invariance. In other words, the process $(\widetilde{\rho}_t(X))_{t=0,\dots,T}$ covers the final loss $-X$ and is sustainable with respect to the original risk measure $(\rho_t)_{t=0,\dots,T}$. The next proposition shows that $(\widetilde{\rho}_t(X))_{t=0,\dots,T}$ is in fact the smallest process with both these properties. This result is a generalization of [33, Proposition 2.5.2] and, in the coherent case, related to [16, Theorem 6.4].

Proposition 1.40 *Let $(\rho_t)_{t=0,\dots,T}$ be a dynamic convex risk measure such that each ρ_t is continuous from above. Then, for each $X \in L^\infty$, the risk process $(\widetilde{\rho}_t(X))_{t=0,\dots,T}$ defined via* (1.33) *is the smallest bounded adapted process $(U_t)_{t=0,\dots,T}$ such that $(U_t)_{t=0,\dots,T}$ is sustainable with respect to $(\rho_t)_{t=0,\dots,T}$ and $U_T \geq -X$.*

Proof We have already seen that $\widetilde{\rho}_T(X) \geq -X$ and $(\widetilde{\rho}_t(X))_{t=0,\dots,T}$ is sustainable with respect to $(\rho_t)_{t=0,\dots,T}$ due to (1.34). Now let $(U_t)_{t=0,\dots,T}$ be another bounded adapted process with both these properties. We will show by backward induction that

$$U_t \geq \widetilde{\rho}_t(X) \quad P\text{-a.s.} \quad \forall t = 0, \dots, T. \tag{1.35}$$

Indeed, we have

$$U_T \geq -X = \widetilde{\rho}_T(X) \quad P\text{-a.s.}$$

If (1.35) holds for $t+1$, Theorem 1.35 yields for all $Q \in \mathcal{Q}_t^f$:

$$\begin{aligned} U_t &\geq E_Q\big[U_{t+1} - \alpha_{t,t+1}^{\min}(Q)\big|\mathcal{F}_t\big] \\ &\geq E_Q\big[\widetilde{\rho}_{t+1}(X) - \alpha_{t,t+1}^{\min}(Q)\big|\mathcal{F}_t\big] \quad P\text{-a.s.} \end{aligned}$$

Thus,

$$\begin{aligned} U_t &\geq \operatorname*{ess\,sup}_{Q \in \mathcal{Q}_t^f}\big(E_Q\big[\widetilde{\rho}_{t+1}(X)\big|\mathcal{F}_t\big] - \alpha_{t,t+1}^{\min}(Q)\big) \\ &= \rho_t\big(-\widetilde{\rho}_{t+1}(X)\big) = \widetilde{\rho}_t(X) \quad P\text{-a.s.}, \end{aligned}$$

where we have used representation (1.6). This proves (1.35). □

The recursive construction (1.33) can be used to construct a time consistent dynamic Average Value-at-Risk, as shown in the next example.

Example 1.41 It is well known that dynamic Average Value-at-Risk $(AV@R_{t,\lambda_t})_{t=0,\dots,T}$ (cf. Example 1.10) is not time consistent; see, e.g., [4, 14, 35]. Moreover, since $\alpha_0^{\min}(P) = 0$ in this case, the set $\mathcal{Q}^*$ in (1.22) is not empty, and [22, Corollary 4.12] implies that there exists no time consistent dynamic convex risk measure $(\rho_t)_{t\in\mathbb{T}}$ such that each ρ_t is continuous from above and $\rho_0 = AV@R_{0,\lambda_0}$. However, for $T < \infty$, the recursive construction (1.33) can be applied to $(AV@R_{t,\lambda_t})_{t=0,\dots,T}$ in order to modify it to a time consistent dynamic coherent risk measure $(\tilde{\rho}_t)_{t=0,\dots,T}$. This modified risk measure takes the form

$$\tilde{\rho}_t(X) = \operatorname{ess\,sup}\left\{E_Q[-X|\mathcal{F}_t] \,\middle|\, Q \in \mathcal{Q}_t, \frac{Z_{s+1}^Q}{Z_s^Q} \leq \lambda_s^{-1}, s = t, \dots, T-1\right\}$$

$$= \operatorname{ess\,sup}\left\{ E\left[-X \prod_{s=t+1}^{T} L_s \middle| \mathcal{F}_t \right] \middle| L_s \in L_s^\infty, 0 \le L_s \le \lambda_s^{-1}, \right.$$

$$\left. E[L_s|\mathcal{F}_{s-1}] = 1, s = t+1, \dots, T \right\}$$

for all $t = 0, \dots, T-1$, where $Z_t^Q = \frac{dQ}{dP}|_{\mathcal{F}_t}$. This was shown, e.g., in [13, Example 3.3.1].

1.5 The Dynamic Entropic Risk Measure

In this section we study time consistency properties of the dynamic entropic risk measure

$$\rho_t(X) = \frac{1}{\gamma_t} \log E\big[\exp(-\gamma_t X)\big|\mathcal{F}_t\big], \quad t \in \mathbb{T}, \ X \in L^\infty, \tag{1.36}$$

where the risk aversion parameter γ_t is random and satisfies $\gamma_t > 0$ P-a.s. and $\gamma_t, \frac{1}{\gamma_t} \in L_t^\infty$ for all $t \in \mathbb{T}$; cf. also Example 1.9.

It is well known (see, e.g., [18, 22]) that the conditional entropic risk measure ρ_t has the robust representation (1.5) with the minimal penalty function α_t given by

$$\alpha_t(Q) = \frac{1}{\gamma_t} H_t(Q|P), \quad Q \in \mathcal{Q}_t,$$

where $H_t(Q|P)$ denotes the conditional relative entropy of Q with respect to P at time t:

$$H_t(Q|P) = E_Q\left[\log \frac{dQ}{dP} \middle| \mathcal{F}_t\right], \quad Q \in \mathcal{Q}_t.$$

The dynamic entropic risk measure with constant risk aversion parameter $\gamma_t = \gamma_0 \in \mathbb{R}$ for all t was studied in [12, 13, 18, 22]. It plays a particular role, as explained in the following remark.

Remark 1.42 Kupper and Schachermayer [30] showed that the entropic risk measure with constant risk aversion parameter $\gamma_0 \in [0, \infty]$ is the only time consistent dynamic convex risk measure $(\rho_t)_{t \in \mathbb{N}_0}$ such that ρ_0 is law invariant.

In this section we consider an *adapted* risk aversion process $(\gamma_t)_{t \in \mathbb{T}}$ that depends both on time and on the available information. As shown in the next proposition, the process $(\gamma_t)_{t \in \mathbb{T}}$ determines time consistency properties of the corresponding dynamic entropic risk measure. This result corresponds to [33, Proposition 4.1.4] and generalizes [19, Proposition 3.13].

Proposition 1.43 *Let $(\rho_t)_{t\in\mathbb{T}}$ be the dynamic entropic risk measure with risk aversion given by an adapted process $(\gamma_t)_{t\in\mathbb{T}}$ such that $\gamma_t > 0$ P-a.s. and $\gamma_t, 1/\gamma_t \in L_t^\infty$. Then the following assertions hold*:

1. *$(\rho_t)_{t\in\mathbb{T}}$ is rejection consistent if $\gamma_t \geq \gamma_{t+1}$ P-a.s. for all $t \in \mathbb{T}$, $t < T$;*
2. *$(\rho_t)_{t\in\mathbb{T}}$ is acceptance consistent if $\gamma_t \leq \gamma_{t+1}$ P-a.s. for all $t \in \mathbb{T}$, $t < T$;*
3. *$(\rho_t)_{t\in\mathbb{T}}$ is time consistent if $\gamma_t = \gamma_0 \in \mathbb{R}$ P-a.s. for all $t \in \mathbb{T}$.*

Moreover, assertions (1), (2), *and* (3) *hold with "if and only if" if $\gamma_t \in \mathbb{R}$ for all t, or if the filtration $(\mathcal{F}_t)_{t\in\mathbb{T}}$ is rich enough in the sense that for all t and for all $B \in \mathcal{F}_t$ such that $P[B] > 0$, there exists $A \subset B$ such that $A \notin \mathcal{F}_t$ and $P[A] > 0$.*

Proof Fix $t \in \mathbb{T}$ and $X \in L^\infty$. Then

$$\begin{aligned}\rho_t(-\rho_{t+1}(X)) &= \frac{1}{\gamma_t}\log\left(E\left[\exp\left(\frac{\gamma_t}{\gamma_{t+1}}\log\left(E\left[\exp(-\gamma_{t+1}X)\middle|\mathcal{F}_{t+1}\right]\right)\right)\middle|\mathcal{F}_t\right]\right)\\ &= \frac{1}{\gamma_t}\log\left(E\left[E\left[\exp(-\gamma_{t+1}X)\middle|\mathcal{F}_{t+1}\right]^{\frac{\gamma_t}{\gamma_{t+1}}}\middle|\mathcal{F}_t\right]\right).\end{aligned}$$

Thus, $\rho_t(-\rho_{t+1}) = \rho_t$ if $\gamma_t = \gamma_{t+1}$, and this proves time consistency. Rejection (resp. acceptance) consistency follows by the generalized Jensen inequality that will be proved in Lemma 1.44. We apply this inequality at time $t+1$ to the bounded random variable $Y := \exp(-\gamma_{t+1}X)$ and the $\mathcal{B}((0,\infty)) \otimes \mathcal{F}_{t+1}$-measurable function

$$u : (0,\infty) \times \Omega \to \mathbb{R}, \quad u(x,\omega) := x^{\frac{\gamma_t(\omega)}{\gamma_{t+1}(\omega)}}.$$

Note that $u(\cdot,\omega)$ is convex if $\gamma_t(\omega) \geq \gamma_{t+1}(\omega)$ and concave if $\gamma_t(\omega) \leq \gamma_{t+1}(\omega)$. Moreover, $u(X,\cdot) \in L^\infty$ for all $X \in L^\infty$, and $u(\cdot,\omega)$ is differentiable on $(0,\infty)$ with

$$\left|u'(x,\cdot)\right| = \frac{\gamma_t}{\gamma_{t+1}} x^{\frac{\gamma_t}{\gamma_{t+1}}-1} \leq a x^b \quad P\text{-a.s.}$$

for some $a, b \in \mathbb{R}$ if $\gamma_t \geq \gamma_{t+1}$, due to our assumption $\frac{\gamma_t}{\gamma_{t+1}} \in L^\infty$. On the other hand, for $\gamma_t \leq \gamma_{t+1}$, we obtain

$$\left|u'(x,\cdot)\right| = \frac{\gamma_t}{\gamma_{t+1}} x^{\frac{\gamma_t}{\gamma_{t+1}}-1} \leq a\frac{1}{x^c} \quad P\text{-a.s.}$$

for some $a, c \in \mathbb{R}$. Thus the assumptions of Lemma 1.44 are satisfied, and we obtain

$$\rho_t(-\rho_{t+1}) \leq \rho_t \quad \text{if } \gamma_t \geq \gamma_{t+1} \quad P\text{-a.s. for all } t \in \mathbb{T} \text{ such that } t < T$$

and

$$\rho_t(-\rho_{t+1}) \geq \rho_t \quad \text{if } \gamma_t \leq \gamma_{t+1} \; P\text{-a.s. for all } t \in \mathbb{T} \text{ such that } t < T.$$

The "only if" direction for constant γ_t follows by the classical Jensen inequality.

Now we assume that the sequence $(\rho_t)_{t\in\mathbb{T}}$ is rejection consistent and our assumption on the filtration $(\mathcal{F}_t)_{t\in\mathbb{T}}$ holds. We will show that the sequence $(\gamma_t)_{t\in\mathbb{T}}$ is decreasing in this case. Indeed, for $t \in \mathbb{T}$ such that $t < T$, consider $B := \{\gamma_t < \gamma_{t+1}\}$ and suppose that $P[B] > 0$. Our assumption on the filtration allows us to choose $A \subset B$ with $P[B] > P[A] > 0$ and $A \notin \mathcal{F}_{t+1}$. We define the random variable $X := -xI_A$ for some $x > 0$. Then

$$\begin{aligned}\rho_t(-\rho_{t+1}(X)) &= \frac{1}{\gamma_t}\log\left(E\left[\exp\left(\frac{\gamma_t}{\gamma_{t+1}}\log\left(E\left[\exp(\gamma_{t+1}xI_A)\middle|\mathcal{F}_{t+1}\right]\right)\right)\middle|\mathcal{F}_t\right]\right)\\ &= \frac{1}{\gamma_t}\log\left(E\left[\exp\left(\frac{\gamma_t}{\gamma_{t+1}}I_B\log\left(E\left[\exp(\gamma_{t+1}xI_A)\middle|\mathcal{F}_{t+1}\right]\right)\right)\middle|\mathcal{F}_t\right]\right),\end{aligned}$$

where we have used that $A \subset B$. Setting

$$Y := E\left[\exp(\gamma_{t+1}xI_A)\middle|\mathcal{F}_{t+1}\right] = \exp(\gamma_{t+1}x)P[A|\mathcal{F}_{t+1}] + P\left[A^c\middle|\mathcal{F}_{t+1}\right]$$

and bringing $\frac{\gamma_t}{\gamma_{t+1}}$ inside of the logarithm, we obtain

$$\rho_t(-\rho_{t+1}(X)) = \frac{1}{\gamma_t}\log\left(E\left[\exp\left(I_B\log\left(Y^{\frac{\gamma_t}{\gamma_{t+1}}I_B}\right)\right)\middle|\mathcal{F}_t\right]\right). \tag{1.37}$$

The function $x \mapsto x^{\gamma_t(\omega)/\gamma_{t+1}(\omega)}$ is strictly concave for almost each $\omega \in B$, and thus,

$$\begin{aligned}Y^{\frac{\gamma_t}{\gamma_{t+1}}} &= \left(\exp(\gamma_{t+1}x)P[A|\mathcal{F}_{t+1}] + \left(1 - P[A|\mathcal{F}_{t+1}]\right)\right)^{\frac{\gamma_t}{\gamma_{t+1}}}\\ &\geq \exp(\gamma_t x)P[A|\mathcal{F}_{t+1}] + \left(1 - P[A|\mathcal{F}_{t+1}]\right) \quad P\text{-a.s. on } B,\end{aligned} \tag{1.38}$$

with strict inequality on the set

$$C := \left\{P[A|\mathcal{F}_{t+1}] > 0\right\} \cap \left\{P[A|\mathcal{F}_{t+1}] < 1\right\} \cap B.$$

Our assumptions $P[A] > 0$, $A \subset B$, and $A \notin \mathcal{F}_{t+1}$ imply $P[C] > 0$, and using

$$\exp(\gamma_t x)P[A|\mathcal{F}_{t+1}] + \left(1 - P[A|\mathcal{F}_{t+1}]\right) = E\left[\exp(\gamma_t xI_A)\middle|\mathcal{F}_{t+1}\right], \tag{1.39}$$

from (1.37), (1.38), and (1.39) we obtain

$$\rho_t(-\rho_{t+1}(X)) \geq \frac{1}{\gamma_t}\log\left(E\left[\exp\left(I_B\log\left(E[\exp(\gamma_t xI_A)|\mathcal{F}_{t+1}]\right)\right)\middle|\mathcal{F}_t\right]\right), \tag{1.40}$$

with the strict inequality on some set of positive probability due to strict monotonicity of the exponential and logarithmic functions. For the right-hand side of (1.40), we have

$$\begin{aligned}&\frac{1}{\gamma_t}\log\left(E\left[\exp\left(I_B\log\left(E\left[\exp(\gamma_t xI_A)\middle|\mathcal{F}_{t+1}\right]\right)\right)\middle|\mathcal{F}_t\right]\right)\\ &\quad = \frac{1}{\gamma_t}\log\left(E\left[I_B E\left[\exp(\gamma_t xI_A)\middle|\mathcal{F}_{t+1}\right] + I_{B^c}\middle|\mathcal{F}_t\right]\right)\end{aligned}$$

$$= \frac{1}{\gamma_t} \log\left(E\left[\exp(\gamma_t x I_A)\middle|\mathcal{F}_t\right]\right)$$
$$= \rho_t(X),$$

where we have used $A \subset B$ and $B \in \mathcal{F}_{t+1}$. This is a contradiction to rejection consistency of $(\rho_t)_{t\in\mathbb{T}}$, and we conclude that $\gamma_{t+1} \leq \gamma_t$ for all t. The proof in the case of acceptance consistency follows in the same manner. And since a time consistent dynamic risk measure is both acceptance and rejection consistent, we obtain $\gamma_{t+1} = \gamma_t$ for all t. □

The following lemma concludes the proof of Proposition 1.43.

Lemma 1.44 *Let $(\Omega, \mathcal{F}, P)$ be a probability space, and $\mathcal{F}_t \subseteq \mathcal{F}$ a σ-field. Let $I \subseteq \mathbb{R}$ be an open interval, and*

$$u : I \times \Omega \to \mathbb{R}$$

be a $\mathcal{B}(I) \otimes \mathcal{F}_t$-measurable function such that $u(\cdot, \omega)$ is convex (resp. concave) and finite on I for P-a.e. ω. Assume further that

$$\left|u'_+(x, \cdot)\right| \leq c(x) \quad P\text{-a.s. with some } c(x) \in \mathbb{R} \quad \text{for all } x \in I,$$

where $u'_+(\cdot, \omega)$ denotes the right-hand derivative of $u(\cdot, \omega)$. Let $X : \Omega \to [a, b]$, with $[a, b] \subseteq I$, be an $\mathcal{F}$-measurable bounded random variable such that $E[|u(X, \cdot)|] < \infty$. Then

$$E\left[u(X, \cdot)\middle|\mathcal{F}_t\right] \geq u\left(E[X|\mathcal{F}_t], \cdot\right) \quad (\text{resp} \leq) \quad P\text{-a.s.}$$

Proof We will prove the assertion for the convex case; the concave one follows in the same manner. Fix $\omega \in \Omega$ such that $u(\cdot, \omega)$ is convex. Due to convexity, we obtain, for all $x_0 \in I$,

$$u(x, \omega) \geq u(x_0, \omega) + u'_+(x_0, \omega)(x - x_0) \quad \text{for all } x \in I.$$

Take $x_0 = E[X|\mathcal{F}_t](\omega)$ and $x = X(\omega)$. Then

$$u\big(X(\omega), \omega\big) \geq u\big(E[X|\mathcal{F}_t](\omega), \omega\big) + u'_+\big(E[X|\mathcal{F}_t](\omega), \omega\big)\big(X(\omega) - E[X|\mathcal{F}_t](\omega)\big) \tag{1.41}$$

for P-almost all $\omega \in \Omega$. Note further that the $\mathcal{B}(I) \otimes \mathcal{F}_t$-measurability of u implies the $\mathcal{B}(I) \otimes \mathcal{F}_t$-measurability of u_+. Thus,

$$\omega \to u\big(E[X|\mathcal{F}_t](\omega), \omega\big) \quad \text{and} \quad \omega \to u'_+\big(E[X|\mathcal{F}_t](\omega), \omega\big)$$

are $\mathcal{F}_t$-measurable random variables, and $\omega \to u(X(\omega), \omega)$ is $\mathcal{F}$-measurable. Moreover, due to our assumption on X, there are constants $a, b \in I$ such that $a \leq$

$E[X|\mathcal{F}_t] \le b$ P-a.s.. Since $u'_+(\cdot,\omega)$ is increasing by convexity, by using our assumption on the boundedness of u'_+ we obtain

$$-c(a) \le u'_+(a,\omega) \le u'_+\big(E[X|\mathcal{F}_t],\omega\big) \le u'_+(b,\omega) \le c(b),$$

i.e., $u'_+(E[X|\mathcal{F}_t],\cdot)$ is bounded. Since $E[|u(X,\cdot)|] < \infty$, we can build the conditional expectation on the both sides of (1.41), and we obtain

$$\begin{aligned} E\big[u(X,\cdot)\big|\mathcal{F}_t\big] &\ge E\big[u\big(E[X|\mathcal{F}_t],\cdot\big) + u'_+\big(E[X|\mathcal{F}_t],\cdot\big)\big(X - E[X|\mathcal{F}_t]\big)\big|\mathcal{F}_t\big] \\ &= E\big[u\big(E[X|\mathcal{F}_t],\cdot\big)\big|\mathcal{F}_t\big] \quad P\text{-a.s.}, \end{aligned}$$

where we have used the $\mathcal{F}_t$-measurability of $u(E[X|\mathcal{F}_t],\cdot)$ and of $u'_+(E[X|\mathcal{F}_t],\cdot)$ and the boundedness of $u'_+(E[X|\mathcal{F}_t],\cdot)$. This proves our claim. □

References

1. B. Acciaio, H. Föllmer, I. Penner, Risk assessment for uncertain cash flows: Model ambiguity, discounting ambiguity, and the role of bubbles (2010, submitted)
2. P. Artzner, F. Delbaen, J.-M. Eber, D. Heath, Thinking coherently. Risk **10**, 68–71 (1997)
3. P. Artzner, F. Delbaen, J.-M. Eber, D. Heath, Coherent measures of risk. Math. Finance **9**(3), 203–228 (1999)
4. P. Artzner, F. Delbaen, J.-M. Eber, D. Heath, H. Ku, Coherent multiperiod risk adjusted values and Bellman's principle. Ann. Oper. Res. **152**, 5–22 (2007)
5. P. Barrieu, N. El Karoui, Optimal derivatives design under dynamic risk measures, in *Mathematics of Finance, Contemporary Mathematics (AMS Proceedings)* (2004), pp. 13–26
6. J. Bion-Nadal, Conditional risk measure and robust representation of convex conditional risk measures. CMAP preprint 557, Ecole Polytechnique Palaiseau (2004)
7. J. Bion-Nadal, Dynamic risk measures: time consistency and risk measures from BMO martingales. Finance Stoch. **12**(2), 219–244 (2008)
8. J. Bion-Nadal, Time consistent dynamic risk processes. Stoch. Process. Appl. **119**, 633–654 (2008)
9. C. Burgert, Darstellungssätze fuer statische und dynamische Risikomaße mit Anwendungen. Universität Freiburg (2005)
10. P. Cheridito, F. Delbaen, M. Kupper, Coherent and convex monetary risk measures for bounded càdlàg processes. Stoch. Process. Appl. **112**(1), 1–22 (2004)
11. P. Cheridito, F. Delbaen, M. Kupper, Coherent and convex monetary risk measures for unbounded càdlàg processes. Finance Stoch. **9**(3), 369–387 (2005)
12. P. Cheridito, F. Delbaen, M. Kupper, Dynamic monetary risk measures for bounded discrete-time processes. Electron. J. Probab. **11**(3), 57–106 (2006)
13. P. Cheridito, M. Kupper, Composition of time-consistent dynamic monetary risk measures in discrete time. Preprint (2006)
14. P. Cheridito, M. Stadje, Time-inconsistency of VaR and time-consistent alternatives. Finance Res. Lett. **6**, 40–46 (2009)
15. F. Delbaen, Coherent risk measures. Cattedra Galileiana, Scuola Normale Superiore, Classe di Scienze, Pisa (2000)
16. F. Delbaen, The structure of m-stable sets and in particular of the set of risk neutral measures, in *In memoriam Paul-André Meyer: Séminaire de Probabilités XXXIX*. Lecture Notes in Math., vol. 1874 (Springer, Berlin, 2006), pp. 215–258
17. F. Delbaen, S. Peng, E. Rosazza Gianin, Representation of the penalty term of dynamic concave utilities. Finance Stoch. **14**(3), 449–472 (2010)

18. K. Detlefsen, G. Scandolo, Conditional dynamic convex risk measures. Finance Stoch. **9**(4), 539–561 (2005)
19. S. Drapeau, Dynamics of Optimized Certainty Equivalents and φ-Divergence. Humboldt-Universität zu Berlin (2006)
20. N. El Karoui, S. Peng, M.C. Quenez, Backward stochastic differential equations in finance. Math. Finance **7**(1), 1–71 (1997)
21. N. El Karoui, C. Ravanelli, Cash subadditive risk measures and interest rate ambiguity. Math. Finance **19**(4), 561–590 (2009)
22. H. Föllmer, I. Penner, Convex risk measures and the dynamics of their penalty functions. Stat. Decis. **24**(1), 61–96 (2006)
23. H. Föllmer, A. Schied, Convex measures of risk and trading constraints. Finance Stoch. **6**(4), 429–447 (2002)
24. H. Föllmer, A. Schied, Stochastic finance, in *An Introduction in Discrete Time*. De Gruyter Studies in Mathematics, vol. 27 (De Gruyter, Berlin, 2004)
25. M. Frittelli, E. Rosazza Gianin, Putting order in risk measures. J. Bank. Finance **26**(7), 1473–1486 (2002)
26. M. Frittelli, E. Rosazza Gianin, Dynamic convex risk measures, in *New Risk Measures in the 21st Century* (2004), pp. 227–248
27. M. Frittelli, G. Scandolo, Risk measures and capital requirements for processes. Math. Finance **16**(4), 589–612 (2006)
28. A. Jobert, L.C.G. Rogers, Valuations dynamic convex risk measures. Math. Finance **18**(1), 1–22 (2008)
29. S. Klöppel, M. Schweizer, Dynamic utility indifference valuation via convex risk measure. Working Paper N 209: National Centre of Competence in Research Financial Valuation and Risk Management (2005)
30. M. Kupper, W. Schachermayer, Representation results for law invariant time consistent functions. Math. Financ. Econ. **2**(3), 189–210 (2009)
31. M. Mania, M. Schweizer, Dynamic exponential utility indifference valuation. Ann. Appl. Probab. **15**(3), 2113–2143 (2005)
32. S. Peng, Backward SDE related g-expectation, in *Backward Stochastic Differential Equations*, Paris, 1995–1996. Pitman Res. Notes Math. Ser., vol. 364 (1995), pp. 141–159
33. I. Penner, Dynamic convex risk measures: time consistency, prudence, and sustainability. Humboldt-Universität zu Berlin (2007)
34. F. Riedel, Dynamic coherent risk measures. Stoch. Process. Appl. **112**(2), 185–200 (2004)
35. B. Roorda, J.M. Schumacher, Time consistency conditions for acceptability measures, with an application to Tail Value at Risk. Insur. Math. Econ. **40**(2), 209–230 (2007)
36. B. Roorda, J.M. Schumacher, When can a risk measure be updated consistently? (2010, submitted)
37. B. Roorda, J.M. Schumacher, J. Engwerda, Coherent acceptability measures in multiperiod models. Math. Finance **15**(4), 589–612 (2005)
38. E. Rosazza Gianin, Risk measures via g-expectations. Insur. Math. Econ. **39**(1), 19–34 (2006)
39. A. Schied, Optimal investments for risk- and ambiguity-averse preferences: a duality approach. Finance Stoch. **11**(1), 107–129 (2007)
40. S. Tutsch, Konsistente und konsequente dynamische Risikomaße und das Problem der Aktualisierung. Humboldt-Universität zu Berlin (2006)
41. S. Tutsch, Update rules for convex risk measures. Quant. Finance **8**, 833–843 (2008)
42. N. Vogelpoth, Some results on dynamic risk measures. Ludwig-Maximilians-Universität München (2006)
43. S. Weber, Distribution-invariant risk measures, information, and dynamic consistency. Math. Finance **16**(2), 419–441 (2006)

Chapter 2
Ambit Processes and Stochastic Partial Differential Equations

Ole E. Barndorff-Nielsen, Fred Espen Benth, and Almut E.D. Veraart

Abstract Ambit processes are general stochastic processes based on stochastic integrals with respect to Lévy bases. Due to their flexible structure, they have great potential for providing realistic models for various applications such as in turbulence and finance. This papers studies the connection between ambit processes and solutions to stochastic partial differential equations. We investigate this relationship from two angles: from the Walsh theory of martingale measures and from the viewpoint of the Lévy noise analysis.

Keywords Ambit processes · Levy bases · Stochastic partial differential equations · White noise analysis · Martingale measures

Mathematics Subject Classification (2010) 60H05 · 60H15 · 60H40 · 60G57 · 60G60

2.1 Introduction

In physics, partial differential equations (PDEs) give a dynamic way to describe how phenomena in nature evolve over time and space. For instance, the classical

O.E. Barndorff-Nielsen (✉)
Thiele Center, Department of Mathematical Sciences and CREATES, School of Economics and Management, Aarhus University, 8000 Aarhus C, Denmark
e-mail: oebn@imf.au.dk

F.E. Benth
Centre of Mathematics for Applications, University of Oslo, P.O. Box 1053, Blindern, 0316 Oslo, Norway
e-mail: fredb@math.uio.no

F.E. Benth
Faculty of Economics, University of Agder, Serviceboks 422, 4604 Kristiansand, Norway

A.E.D. Veraart
CREATES, School of Economics and Management, Aarhus University, 8000 Aarhus C, Denmark
e-mail: averaart@creates.au.dk

G. Di Nunno, B. Øksendal (eds.), *Advanced Mathematical Methods for Finance*,
DOI 10.1007/978-3-642-18412-3_2,

heat equation of Einstein gives a dynamic model for how heat diffuses in a medium. Stochastic partial differential equations (SPDEs) add randomness to such evolution equations, where the noise source may come from uncertainties in measurements, unexplainable effects, and turbulent phenomena. The noise is usually modelled as a random field in time and space, also called white noise or, more generally, Lévy noise. We shall be mostly concerned with parabolic PDEs in this paper.

Ambit processes have been proposed and introduced by Barndorff–Nielsen and Schmiegel and have thereafter been applied in various areas such as turbulence modelling (see e.g. [6, 13]), in medical context in the form of describing tumor growth [12], and more recently for modelling energy markets [4, 5].

The solution of a parabolic differential equation is often represented as an integral over a Green's function (the fundamental solution of the PDE) convoluted with some initial condition. Such representations look very similar to the definition of stationary ambit processes of [13]. The Green's function representation is an explicit solution as long as the Green's function is known, where the deterministic space–time dynamics of the phenomena in question is encapsulated in the form of this function. It is closely linked to density functions of stochastic diffusion processes.

Introducing noise leads to complications of interpreting in what sense we have a solution. This requires a theory for stochastic integration in time and space, such as proposed in Walsh [46]. It turns out that solutions of parabolic equations with an additive source of noise can be represented as the stochastic convolution of the Green's function and the initial value, where the integration is with respect to the random field. We present the theory of Walsh [46] and link it to ambit processes.

When having a stochastic source term, one may have solutions being singular. This is the starting point for applying white noise analysis (WNA) or, more generally, Lévy noise analysis (LNA) to analyse SPDEs. We discuss the theory of LNA and link it to ambit processes. Here we will also include discussions of SPDEs and how they are related to ambit processes.

Note that ambit processes may provide a statistical approach to model physical processes in nature far simpler than SPDEs, since they provide a way to specify directly the model based on a probabilistic understanding of the phenomena in question. They also give a framework for extending the solutions of SPDEs. In order to have a solution in the sense of Walsh, often strong integrability conditions are imposed. The ambit processes are well defined under very weak conditions of integrability, and thereby we may extend the solutions of certain equations to include far more general initial conditions, say, or more general types of noise.

The main issue of this paper is to relate the use of the building stone in ambit processes, Lévy bases, to the language of Walsh and the theory of LNA. The latter talks about processes being the derivatives of Levy processes, while Walsh talks about random measures and their derivatives.

The outline for the remaining part of the paper is as follows. In Sect. 2.2 the concepts of ambit fields and processes are outlined, and the important special case of spatial dimension 0 is treated in some detail; in that case the ambit processes are referred to as Lévy semistationary ($\mathcal{LSS}$) processes or, in the Gaussian case, as Brownian semistationary ($\mathcal{BSS}$) processes. In particular, an indication of the theory and use of multipower variations for inference on the volatility process is given. Section 2.2 concludes by a brief discussion of some applications to turbulence and energy markets. Section 2.3 connects the idea of Lévy bases to the theory of random fields due to Walsh. We show how, subject to an L^2 restriction and based on the theory of Hilbert space random fields, it is possible to define Lévy noise for Lévy bases, and the associated integration theory is discussed. Finally, some applications to SPDEs and their relation to ambit processes are considered. Section 2.4 links the theory of Lévy noise analysis for Lévy processes, as developed in Holden, Øksendal, Ubøe, and Zhang [31], to that of Lévy bases and ambit processes and discusses SPDEs in that context. The concluding Sect. 2.5 briefly brings the various strands together.

2.2 Ambit Processes

2.2.1 Background

The general background setting for the concept of ambit processes consists of a stochastic field $Y = \{Y_t(x)\}$ in space–time $\mathcal{X} \times \mathbb{R}$, a curve $\tau(\theta) = (x(\theta), t(\theta))$ in $\mathcal{X} \times \mathbb{R}$, and the values $X_\theta = Y_{t(\theta)}(x(\theta))$ of the field along the curve, the focus being on the dynamic properties of the stochastic process $X = \{X_\theta\}$. Here the space $\mathcal{X}$ is often, but not necessarily, taken as $\mathbb{R}^d$ for $d = 1, 2$, or 3. The stochastic field is supposed to be generated by innovations in space–time, and the values $Y_t(x)$ are assumed to depend only on innovations that occur prior to or at time t. More precisely, at each point (x, t) only the innovations in some subset $A_t(x)$ of $\mathcal{X} \times \mathbb{R}_t$ (where $\mathbb{R}_t = (-\infty, t])$ are influencing the value of $Y_t(x)$, and we refer to $A_t(x)$ as the *ambit set* associated to (x, t), and to Y and X as an *ambit field* and an *ambit process*, respectively; see Fig. 2.1 for an illustration.

Obviously, without further structure nothing interesting can be said about the field Y and the process X, and we shall specify such a structure in mathematical

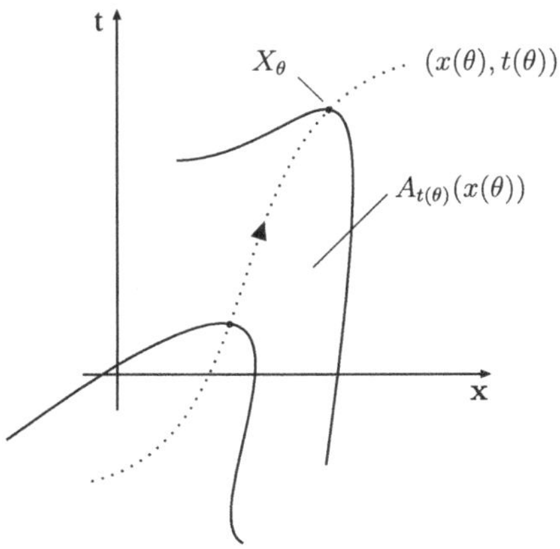

Fig. 2.1 Example of an ambit process X_θ along the curve $(x(\theta), t(\theta))$, where the ambit set is given by $A_{t(\theta)}(x(\theta))$

detail in a moment. But in verbal terms, $Y_t(x)$ will be defined in the form of a stochastic integral plus a smooth term, and the integrand in the stochastic integral will consist of a deterministic kernel times a positive random variate which is taken to embody the *volatility* or *intermittency* of the field Y. We shall mostly consider specifications under which $Y_t(x)$ is stationary in time for each fixed x.

The volatility field, denoted by σ, is given also as an ambit field, and a central issue is what can be learned about σ from observation of Y or X.

Note that, in general, ambit processes are not semimartingales. Many of the standard tools from semimartingale theory are therefore not applicable, and alternative methods are required.

The more precise mathematical specification of what is meant generally by ambit fields and processes is given in Sect. 2.2.2. In Sects. 2.2.3, 2.2.4, and 2.2.5 we focus on the null-spatial case, i.e. where $\mathcal{X}$ consists of a single point. There the concept of ambit processes specialises to that of Lévy and Brownian semistationary processes ($\mathcal{LSS}$ and $\mathcal{BSS}$ processes). Already in that setting there are many interesting questions of a nonstandard character. These have important analogues in the genuinely tempo-spatial case.

As for semimartingales, the questions of existence and properties of quadratic variations, and more generally multipower variations, are of central importance in the study of ambit fields and processes, in particular as these objects relate to the volatility/intermittency. We will review the main results in that context in Sect. 2.2.6 and refer to [8, 17], and [9] for more details.

Section 2.2.7 contains some applications of ambit processes to turbulence (Tempo-Spatial Settings in Turbulence) and energy finance (Modelling Energy Markets by Ambit Fields), respectively.

2.2.2 Ambit Fields and Processes

Generally we think of ambit fields as being of the form

$$\begin{aligned} Y_t(x) = \mu + &\int_{A_t(x)} g(\xi, s; x, t)\sigma_s(\xi)L(d\xi, ds) \\ + &\int_{D_t(x)} q(\xi, s; x, t)a_s(\xi)\, d\xi\, ds, \end{aligned} \tag{2.1}$$

where $A_t(x)$ and $D_t(x)$ are ambit sets, g and q are deterministic functions, $\sigma \geq 0$ is a stochastic field referred to as the *intermittency* or *volatility*, and L is a *Lévy basis* defined as follows (see [20, 36]): Let $\mathcal{B}(\mathbb{R}^k)$ be the Borel sets of $\mathbb{R}^k$ and denote by $\mathcal{B}_b(S)$ the bounded Borel sets of $S \in \mathcal{B}(\mathbb{R}^k)$.

Definition 2.1 A family $\{\Lambda(A) : A \in \mathcal{B}_b(S)\}$ of random vectors in $\mathbb{R}^d$ is called an $\mathbb{R}^d$-valued Lévy basis on S if the following three properties are satisfied:

1. The law of $\Lambda(A)$ is infinitely divisible for all $A \in \mathcal{B}_b(S)$.

2. If $A_1, \dots, A_n$ are disjoint subsets in $\mathcal{B}_b(S)$, then $\Lambda(A_1), \dots, \Lambda(A_n)$ are independent.
3. If $A_1, A_2, \dots$ are disjoint subsets in $\mathcal{B}_b(S)$ with $\bigcup_{i=1}^{\infty} A_i \in \mathcal{B}_b(S)$, then

$$\Lambda\left(\bigcup_{i=1}^{\infty} A_i\right) = \sum_{i=1}^{\infty} \Lambda(A_i) \quad \text{a.s.},$$

where the convergence on the right hand side is a.s.

Conditions (2) and (3) define an independently scattered random measure. Note that we use Λ when we refer to a general Lévy basis, and when we have separated out time as one dimension, we talk of Lévy bases defined on $S = \mathcal{X} \times \mathbb{R}$, and we indicate integration with respect to such bases by $L(d\xi, ds)$.

Inference on the volatility/intermittency field σ is a focal point for the study of ambit processes and fields. Often the volatility field (or the logarithmic volatility field) will itself be defined as an ambit field through

$$\sigma_t^2(x) = \int_{C_t(x)} h(\xi, s; x, t) L(d\xi, ds), \tag{2.2}$$

with h a positive function, $C_t(x)$ some ambit set, and where L is a nonnegative non-Gaussian Lévy basis.

At the present level of generality we take the integrals in (2.1) to be defined in the sense of independently scattered random measures, cf. [38], assuming that g, σ, q, and a are sufficiently regular for the integrals to exist. However, in more concrete cases it is often of interest to establish whether the definition of the integrals can be sharpened to a more dynamical version, for instance in the sense of Itô-type integrals. We return to this question later, see in particular Sects. 2.3.4 and 2.4.

Of particular interest are ambit processes that are stationary in time and nonanticipative. More specifically, they may be derived from ambit fields Y of the form

$$\begin{aligned} Y_t(x) = \mu &+ \int_{A_t(x)} g(\xi, t-s; x) \sigma_s(\xi) L(d\xi, ds) \\ &+ \int_{D_t(x)} q(\xi, t-s; x) a_s(\xi)\, d\xi\, ds. \end{aligned} \tag{2.3}$$

Here the ambit sets $A_t(x)$ and $D_t(x)$ are taken to be *homogeneous* and *nonanticipative*, i.e. $A_t(x)$ is of the form $A_t(x) = A + (x, t)$, where A only involves negative time coordinates, and similarly for $D_t(x)$. Further, we assume that $g(\xi, \tau; x) = 0$ and $q(\xi, \tau; x) = 0$ for all $\tau < 0$.

Remark Recall from [12, 36] that every Lévy basis L exhibits a Lévy–Itô decomposition. Let N denote the Poisson basis associated with the Levy basis L through such a decomposition, and let ν denote the intensity measure of N. Clearly, we have $\mathbb{E}(N(dx; d\xi, ds)) = \nu(dx; d\xi, ds)$. In the following, we are interested in *homogeneous* Lévy bases, i.e. Lévy bases which satisfy $\nu(dx; d\xi, ds) = \widetilde{\nu}(dx; d\xi)\, ds$ for a measure $\widetilde{\nu}$.

Remark Many prominent tempo-spatial models are constructed from an ordinary, partial, or fractional differential equation by adding a noise term, for instance in the form of white noise, to the equation. The solution to the equation then being often representable as an integral with respect to the noise of the Green's function of the original deterministic differential equation (see [3, 24]). Thus the solution is taking the form of an ambit process. For some examples with discussion, see Sects. 2.3.5 and 2.4.2.

Note that, in general, ambit processes involve time varying ambit sets and allow for a stochastic volatility factor. Such stochastic volatility is important in many areas in science, not only in the contexts of turbulence and finance which are in focus in this paper.

For understanding the nature of ambit processes $X_\theta = Y_{t(\theta)}(x(\theta))$, and as a step towards handling questions of inference on σ, it is useful to discuss the cores of Y and X. With the ambit field given by (2.1), the *cores* $Y_\circ$ and $X_\circ$ of Y and X are defined, respectively, by

$$Y_{\circ t}(x) = \int_{A_t(x)} g(\xi, s; x, t) L(d\xi, ds)$$

and

$$X_{\circ\theta} = \int_{A(\theta)} g\big(\xi, s; \tau(\theta)\big) L(d\xi, ds),$$

where, as above, $\tau(\theta) = (x(\theta), t(\theta))$, and where we have used $A(\theta)$ as a shorthand for $A_{t(\theta)}(x(\theta))$. In case the Lévy basis L is the Wiener basis W, we speak of a *Gaussian core*.

Remark A class of processes having some properties common with one-dimensional ambit processes is studied in [44] under the name *mixed moving averages*. More precisely, the authors study processes $X = (X_t)_{t\in\mathbb{R}}$ of the form

$$X_t = \int_{\mathcal{X}\times\mathbb{R}} f(x, t-s) \Lambda(dx, ds), \tag{2.4}$$

where $\mathcal{X}$ is a nonempty set, and Λ is a symmetric α-stable (SαS) random measure on $\mathcal{X} \times \mathbb{R}$ with Lévy measure $\nu \times$ leb, where leb is the Lebesgue measure, and ν is a σ-finite measure on $\mathcal{X}$. Note that such processes are always stationary. In the SαS non-Gaussian case, they show that this is the smallest class containing all superpositions and weak limits of ordinary SαS moving averages. Furthermore, Rosinski [39] has obtained a Wold–Karhunen-type decomposition of stationary SαS non-Gaussian processes in which mixed moving averages play a role similar to ordinary moving averages in the Gaussian case. In [40] this type of result is extended to a broad range of non-Gaussian infinitely divisible processes.

2.2.3 Null-Spatial Case: Lévy Semistationary Processes ($\mathcal{LSS}$)

When the space $\mathcal{X}$ consists of a single point (or we just consider $Y_t(x)$ of (2.1) in its dependence on t keeping x fixed) the concept of ambit processes specialises to that of *Lévy Semistationary Processes* ($\mathcal{LSS}$), introduced in [5], which are processes $Y = \{Y_t\}_{t \in \mathbb{R}}$ of the form

$$Y_t = \mu + \int_{-\infty}^{t} g(t-s)\sigma_{s-}\, dL_s + \int_{-\infty}^{t} q(t-s) a_s\, ds, \tag{2.5}$$

where μ is a constant, L is a Lévy process, g and q are nonnegative deterministic functions on $\mathbb{R}$ with $g(t) = q(t) = 0$ for $t \leq 0$, and σ and a are càdlàg processes. When σ and a are stationary, as we will require henceforth, then so is Y. Hence the name Lévy semistationary processes. It is convenient to indicate the formula for Y as

$$Y = \mu + g * \sigma \bullet L + q * a \bullet \text{leb}, \tag{2.6}$$

where leb denotes Lebesgue measure.

Generally we have taken the stochastic integrals as defined in the sense of [38]. However, in the present case of $\mathcal{LSS}$ processes, one may define the integrals in the Itô sense, relative to the filtration $\mathcal{F}^L$ generated by the increments $L_t - L_s$, $-\infty < s \leq t < \infty$. Here we adopt the latter definition, noting that the two versions agree with respect to all finite-dimensional distributions.

When $L = B$ in formula (2.5) for a standard Brownian motion B, then Y specialises to a *Brownian Semistationary Process* ($\mathcal{BSS}$), introduced in [17]. The Gaussian core of a $\mathcal{BSS}$ process is

$$Y_{\circ t} = \int_{-\infty}^{t} g(t-s)\, dB_s. \tag{2.7}$$

We consider the $\mathcal{BSS}$ processes to be the natural analogue, for stationarity-related processes, of the class $\mathcal{BSM}$ of Brownian semimartingales

$$Y_t = \int_0^t \sigma_s\, dB_s + \int_0^t a_s\, ds.$$

Already in this null-spatial case the question of drawing inference on σ^2 is highly nontrivial. The main tool is multipower variation, see [8] and [9].

2.2.4 Key Example for a $\mathcal{BSS}$ Process

An example of particular interest in the context of $\mathcal{BSS}$ processes is where

$$g(t) = t^{\nu-1} e^{-\lambda t} \quad \text{for } t \in (0, \infty), \tag{2.8}$$

for some $\lambda > 0$ and with $\nu > \frac{1}{2}$. The latter condition is needed to ensure the existence of the stochastic integral in (2.7).

Remark 2.2 For the key example (2.8), the derivative g' of g is not square integrable if $\frac{1}{2} < \nu < 1$ or $1 < \nu \leq \frac{3}{2}$; hence, in these cases, Y is not a semimartingale. For $\frac{1}{2} < \nu < 1$, we have $g(0+) = \infty$, while $g(0+) = 0$ when $1 < \nu \leq \frac{3}{2}$. These two cases are radically different in nature. Of course, for $\nu = 1$, the process $Y = \int_{-\infty}^{\cdot} g(\cdot - s)\sigma_s B(ds)$ is simply a modulated version of the Gaussian Ornstein–Uhlenbeck process, and in particular, a semimartingale. Note also that when $\nu > \frac{3}{2}$, Y is of finite variation and hence, trivially, a semimartingale. To summarise, the nonsemimartingale cases are $\nu \in (\frac{1}{2}, 1) \cup (1, \frac{3}{2}]$.

2.2.5 Generality of $\mathcal{BSS}$

As a modelling framework for continuous-time stationary processes, the specification (2.6) is quite general. In fact, the continuous-time Wold–Karhunen decomposition says that any second-order stationary stochastic process, possibly complex valued, of mean 0 and continuous in quadratic mean can be represented as

$$Z_t = \int_{-\infty}^{t} \phi(t - s)\, d\Xi_s + V_t, \tag{2.9}$$

where the deterministic function ϕ is in general a complex, deterministic square-integrable function, the process Ξ has orthogonal increments with $E\{|d\Xi_t|^2\} = \varpi\, dt$ for some constant $\varpi > 0$, and the process V is nonregular (i.e. its future values can be predicted, in the L^2 sense, by linear operations on past values without error).

Under the further condition that $\bigcap_{t\in\mathbb{R}} \overline{\mathrm{sp}}\{Z_s : s \leq t\} = \{0\}$, the function ϕ is real and uniquely determined up to a real constant of proportionality; the same is therefore true of Ξ (up to an additive constant).

In particular, if $d\Xi_s = \sigma_s\, dB_s$ with σ and B as in (2.6), then Ξ is of the above type with $\varpi = E\{\sigma_0^2\}$.

2.2.6 Multipower Variations

One of the interesting aspects in the context of $\mathcal{BSS}$ models is the question on how to estimate the stochastic volatility σ and how to make inference on it. A key tool for tackling this question is a statistic called *realised variance* and, more generally, *realised multipower variation*.

A realised multipower variation of a stochastic process X is an object of the type

$$\sum_{i=1}^{[nt]-k+1} \prod_{j=1}^{k} |\Delta_{i+j-1}^n X|^{p_j}, \tag{2.10}$$

where $\Delta_i^n X = X_{\frac{i}{n}} - X_{\frac{i-1}{n}}$ and $p_1, \ldots, p_k \geq 0$. That is, it is assumed that the process $X = (X_t)_{t\geq 0}$ is observed at times $i\delta$, where $\delta = \frac{1}{n}$ and $i = 0, 1, \ldots, [nt]$.

These concepts have been developed in the context of financial times series, see e.g. [10, 11, 18, 19, 21] for results in a framework based on Brownian semimartingales. In the presence of jumps, these quantities have been studied by [32, 33] and [45]. A detailed survey on this aspect is also given by [2]. However, in the nonsemimartingale setup the underlying theory is much more involved. We just sketch the main results here briefly and refer to [8, 17] and [9] for more details.

Consider a filtered probability space $(\Omega, \mathcal{F}, (\mathcal{F}_t)_{t\geq 0}, P)$, assuming the existence thereon of a $\mathcal{BSS}$ process Y defined as in (2.5), where $L = B$ is a standard Brownian motion. Let G denote the Gaussian core of Y as defined in (2.7), i.e.

$$G_t = Y_{\circ t} = \int_{-\infty}^{t} g(t-s)\, dB_s,$$

and let $\mathcal{G}$ be the σ-algebra generated by G. The correlation function of the increments of G is given by

$$r_n(j) = \operatorname{cov}\left(\frac{\Delta_1^n G}{\tau_n}, \frac{\Delta_{1+j}^n G}{\tau_n}\right) = \frac{\bar{R}(\frac{j+1}{n}) - 2\bar{R}(\frac{j}{n}) + \bar{R}(\frac{j-1}{n})}{2\tau_n^2}.$$

Next, we introduce a class of measures that is crucial for establishing an asymptotic theory for realised multipower variations. We define

$$\pi_\delta(A) = \frac{\int_A (g(x-\delta) - g(x))^2\, dx}{\int_0^\infty (g(x-\delta) - g(x))^2\, dx}, \qquad y \geq 0,$$

and we further set $\overline{\pi_\delta}(x) = \pi_\delta(\{y: \ y > x\})$. Note that π_δ is a probability measure on $\mathbb{R}_+$.

We are interested in the asymptotic behaviour of the *normalised multipower variations*

$$\bar{V}(Y, p_1, \ldots, p_k)_t^n = \frac{1}{n\tau_n^{p_+}} \sum_{i=1}^{[nt]-k+1} \prod_{j=1}^{k} |\Delta_{i+j-1}^n Y|^{p_j},$$

where $p_+ = \sum_{j=1}^k p_j$ and $\tau_n^2 = \bar{R}(1/n)$ with $\bar{R}(t) = E[|G_{s+t} - G_s|^2]$, $t \geq 0$.

In order to establish a weak law of large numbers, one needs the following assumption.

(LLN): There exists a sequence $r(j)$ with

$$r_n^2(j) \leq r(j), \qquad \frac{1}{n}\sum_{j=1}^{n-1} r(j) \to 0.$$

Moreover, it holds that

$$\lim_{n\to\infty} \overline{\pi_\delta}(\varepsilon) = 0$$

for any $\varepsilon > 0$.

Then the law of large numbers is given by the following proposition.

Proposition 2.3 *Assume that condition* (LLN) *holds for* $Y = g * \sigma \bullet W + q * a \bullet \text{leb}$. *Define*

$$\rho^{(n)}_{p_1,\dots,p_k} = E\left[\left|\frac{\Delta_1^n G}{\tau_n}\right|^{p_1} \cdots \left|\frac{\Delta_k^n G}{\tau_n}\right|^{p_k}\right].$$

Then we have

$$\bar{V}(Y, p_1, \dots, p_k)_t^n - \rho^{(n)}_{p_1,\dots,p_k} \int_0^t |\sigma_s|^{p_+} ds \xrightarrow{ucp} 0,$$

where the convergence is uniform on compacts in probability (*ucp*).

Furthermore, for a central limit theorem, one needs the following assumption.
(CLT): Assumption (LLN) holds, and

$$r_n(j) \to \rho(j), \quad j \geq 0,$$

where $\rho(j)$ is the correlation function of some stationary centered discrete time Gaussian process $(Q_i)_{i\geq 1}$ with $E[Q_i^2] = 1$ (as before). Moreover, for any $j, n \geq 1$, there exists a sequence $r(j)$ with

$$r_n^2(j) \leq r(j), \qquad \sum_{j=1}^{\infty} r(j) < \infty.$$

Finally, the tail mass function $\overline{\pi}^n$ is assumed to satisfy an additional mild condition.

Now, we can formulate a joint central limit theorem for a family $(\bar{V}(Y, p_1^j, \dots, p_k^j)_t^n)_{1\leq j\leq d}$ of multipower variations as follows.

Proposition 2.4 *Assume that the process* σ *is* $\mathcal{G}$*-measurable and condition* (CLT) *holds. Then we obtain the stable convergence*

$$\sqrt{n}\left(\bar{V}(Y, p_1^j, \dots, p_k^j)_t^n - \rho^{(n)}_{p_1^j,\dots,p_k^j} \int_0^t |\sigma_s|^{p_+^j} ds\right)_{1\leq j\leq d} \xrightarrow{\mathcal{G}-st} \int_0^t Z_s^{1/2} dB_s,$$

where B *is a* d*-dimensional Brownian motion that is defined on an extension of the filtered probability space* $(\Omega, \mathcal{F}, (\mathcal{F}_t)_{t\geq 0}, P)$ *and is independent of* $\mathcal{F}$, *and* Z *is a* $d \times d$*-dimensional process given by*

$$Z_s^{ij} = \beta_{ij} |\sigma_s|^{p_+^i + p_+^j}, \quad 1 \leq i, j \leq d,$$

where the $d \times d$ *matrix* β *is defined as in* [8].

Note that in order to obtain an asymptotic limit theory for a wide range of multipower variations, one is forced to consider also multipower variations of second-order differences. (For Brownian semimartingales passing to second-order differences would make no essential change in the limit theory.) Multipower variations based on second-order differences are quantities having the same form as (2.10) but using

$$\lozenge_j^n X = X_{j\delta} - 2X_{(j-1)\delta} + X_{(j-2)\delta},$$

instead of $\Delta_j^n X$. However, we shall not dwell on this aspect here but refer to [7, 9] for discussions, detailed results, and applications.

2.2.7 Applications to Turbulence and Finance

After having introduced the concept of ambit fields and ambit processes, we turn our attention to applications of such processes in turbulence and in finance.

Tempo-Spatial Settings in Turbulence

The idea of ambit processes arose out of a project aimed at establishing realistic stochastic models of the velocity fields in stationary turbulent regimes (cf. [6, 12] and also [13–17]). In turbulence the basic notion of *intermittency* refers to the fact that the energy in a turbulent field is unevenly distributed in space and time, and the paper [12] introduced stochastic models for turbulent intermittency (also referred to as *energy dissipation*) fields, in the form of ambit fields. The later paper [13] proposed a class of ambit processes for the description of the velocity field in the form

$$\begin{aligned} Y_t(x) = \mu &+ \int_{A_t(x)} g(\xi - x, t - s)\sigma_s(\xi) W(d\xi, ds) \\ &+ \int_{D_t(x)} q(\xi - x, t - s)\sigma_s^2(\xi)\, d\xi\, ds, \end{aligned} \tag{2.11}$$

for a Gaussian Lévy basis W with associated intermittency (or energy–dissipation) field

$$\sigma_t^2(x) = \int_{C_t(x)} h(\xi - x, t - s) L(d\xi, ds), \tag{2.12}$$

where L is a nonnegative Lévy basis. An alternative way of modelling σ is by defining $\log \sigma^2$ as

$$\log \sigma_t^2(x) = \int_{C_t(x)} h(\xi - x, t - s) L(d\xi, ds). \tag{2.13}$$

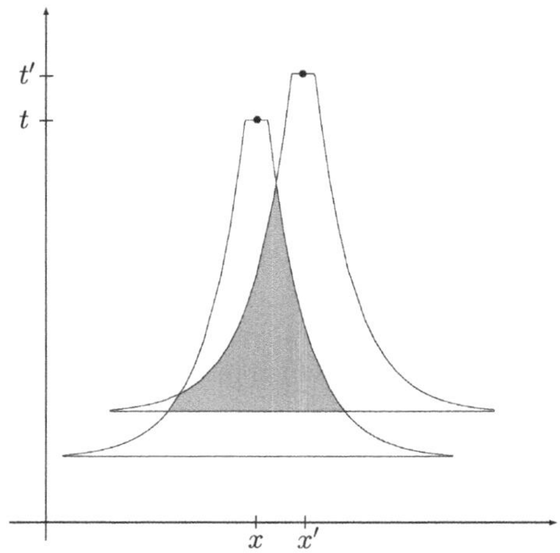

Fig. 2.2 Example of the choice of an ambit set $A_t(x)$ for turbulence modelling, see [12]

This latter specification has the advantage of allowing coupling to cascade theories in turbulence, see [43].

Clearly, the choice of the ambit sets $A_t(x)$, $D_t(x)$, $C_t(x)$ influences the behaviour of an ambit process. Therefore, it is important to investigate what shape of the ambit set reflects the empirical facts best.

In order to illustrate how such ambit sets may look, we provide a plot (see Fig. 2.2) of a particular type of ambit set, the shape of which is rooted in turbulence (see [12]).

Note that the mathematics of turbulence is inherently linked to stochastic partial differential equations (see [24]), as will be discussed in Sects. 2.3 and 2.4.

Modelling Energy Markets by Ambit Fields

Following the success in describing turbulence, it transpires that ambit fields have also great potential in financial applications. In particular, recent research, see [4, 5], has focused on using ambit fields for modelling energy markets. Due to the general structure of ambit fields, these new models are able to capture many stylised facts of energy markets in general and electricity prices in particular. Special features of those markets are e.g. strong seasonal patterns, very pronounced volatility clusters, high spikes/jumps, the existence of the so-called Samuelson effect, i.e. the fact that the volatilities of the forward price are generally smaller than the ones of the underlying spot price and converge, as time to maturity tends to zero, to the volatilities of the spot at a fast rate. Furthermore, there are strong correlations between forward contracts which are close in maturity. In the following we will describe how the structure of ambit processes can be exploited to account for these stylised facts.

Spot Price

We start with the question of how to model the electricity spot price. A natural choice of processes taken from the ambit world is the class of $\mathcal{LSS}$ processes as previously described. In [5], we propose to model the electricity spot price $S = (S_t)_{t\in\mathbb{R}}$ by

$$S_t = \Lambda(t)\exp(Y_t), \tag{2.14}$$

where $\Lambda : \mathbb{R} \to \mathbb{R}_+$ denotes a deterministic seasonal function, and

$$Y_t = \int_{-\infty}^{t} g(t-s)\omega_{s-}\, dL_s \tag{2.15}$$

for a deterministic damping function $g : \mathbb{R} \to \mathbb{R}_+$ with $g(t) = 0$ for $t < 0$ and a càdlàg, positive, stationary process $\omega = (\omega_t)_{t\in\mathbb{R}}$ which is independent of the two-sided Lévy process $L = (L_t)_{t\in\mathbb{R}}$.

There are several key features which make a model for the electricity spot price which is based on an $\mathcal{LSS}$ process both theoretically interesting and practically relevant compared to the traditional models. First and foremost, the deseasonalised, logarithmic spot price Y is modelled *directly* rather than its stochastic dynamics. By doing so, one can introduce a general damping function g, which adds much more flexibility in modelling the mean-reversion of the price process and in accounting for the well-known Samuelson effect [41].

Furthermore, we account for stochastic volatility ω since this is clearly an issue in energy markets (see e.g. Hikspoors and Jaimungal [30] and Benth [23]). A very general model for the volatility process would be that we model it itself as a Lévy–Volterra process, i.e. $\omega_t^2 = Z_t$ and $Z_t = \int_{-\infty}^{t} h(t,s)\, d\widetilde{L}_s$, where $\widetilde{L} = (\widetilde{L})_{t\in\mathbb{R}}$ is another Lévy process. The function h is assumed to satisfy the same conditions as g.

For further details on $\mathcal{LSS}$-based models for electricity spot prices, we refer to [5] and turn our attention now to models for electricity forward contracts based on ambit fields. In the context of forward modelling, we do not stick to the zero spatial case of ambit fields, but rather allow for both a temporal and a spatial component to reflect the fact that the forward price does not only depend on the current time, but also on the time to maturity.

Forward Price

In [4], for modelling the forward price of electricity, we propose to use an ambit field given by

$$f_t(x) = \int_{A_t(x)} k(\xi, t-s; x)\sigma_s(\xi) L(d\xi, ds). \tag{2.16}$$

Here, $t \geq 0$ denotes the current time, $T > 0$ denotes the time of maturity of the forward contract, and $x = T - t$ the corresponding time to maturity.

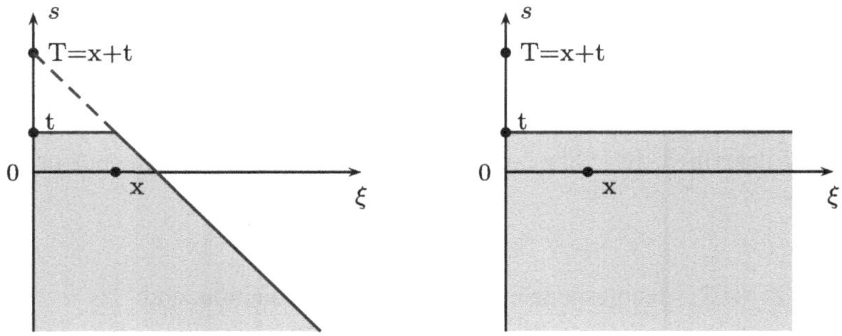

Fig. 2.3 Two relevant choices of the ambit set $A_t(x)$ in the context of modelling electricity forward prices

Clearly, in order to specify the model completely, we have to specify the ambit set $A_t(x)$, the damping or weight function k, and the stochastic volatility field $\sigma_s(\xi)$. It is important to note that in modelling terms we can vary the choice of the ambit set, the weight function h, and the volatility field σ and can still achieve the particular dependence structure we are aiming for. As such there is generally not a unique choice of the ambit set or the weight function or the volatility field to achieve a particular type of dependence structure, and the choice will be based on market intuition and considerations of mathematical/statistical tractability.

We assume that the volatility $\sigma_s(\xi) > 0$ is a stochastic field on $\mathbb{R}_+ \times \mathbb{R}$, which is stationary in the time domain, i.e. with respect to s, and which expresses the volatility on the forwards market as a whole, L is a Lévy basis (integration in the sense of [38]), and k is a damping function. For analytical tractability, we assume that σ is independent of L, and in order to ensure that $f_t(x)$ is stationary in time t, we take the ambit sets to be of the form $A_t(x) = A_0(x) + (0, t)$. Regarding the choice of ambit sets, we just illustrate, in Fig. 2.3, two possibilities of interest.

Furthermore, we suggest to model the volatility field by

$$\sigma_t^2(x) = \int_{C_t(x)} q(\xi, t - s; x)\widetilde{L}(d\xi, ds)$$

for a nonnegative Lévy basis $\widetilde{L}$, a deterministic damping function q (with $q(\xi, \tau; x) = 0$ for $\tau < 0$), and an ambit set $C_t(x) = C_0(x) + (0, t)$. In order to have that forward contracts close in maturity dates are strongly correlated with each other (as indicated by empirical studies), we could choose the Lévy kernel q such that

$$\mathrm{Cor}\big(\sigma_t^2(x), \sigma_t^2(\bar{x})\big)$$

is high for values of x and $\bar{x}$ which are close to 0 (i.e. closeness to maturity).

2.3 Lévy Bases and the Theory of Walsh

In this section we connect the notion of a Lévy basis to the theory of *white noise* random fields of Walsh [46]. Further, we show how to define the *noise* of sufficiently regular Lévy bases based on the theory of Hilbert-space random fields. We summarise the stochastic integration theory of Walsh [46] and present some applications to stochastic partial differential equations in view of ambit processes.

2.3.1 Brief Account on the Stochastic Integration Theory of Walsh

In this subsection we briefly present the approach of Walsh [46] to define stochastic integration with respect to random fields. We keep the discussion on a heuristic level, focusing on the ideas only, since we in any case will introduce the concepts of Walsh in detail below.

The purpose of Walsh [46] is to study stochastic partial differential equations rigorously. The equations are of parabolic type, meaning that the solutions are functions of time and space where their derivative in time is equal to some elliptic operator in space. The partial differential equations are perturbed by random fields, that is, stochastic processes in both time and space (or rather, derivatives of such, called the noise), and in order to make sense out of such equations, one must have available a theory for stochastic integration with respect to such processes.

The key question is how to make sense out of stochastic integrals of the form

$$\int_0^t \int_B X(s,x)M(dx,ds),$$

where B is some measurable subset of $\mathbb{R}^d$, and X is some random field in space and time. The M integrator comes from the "noise" driving the stochastic partial differential equation, and heuristically we may think of this as the space–time derivative of a random field, that is, $M(dx,ds) = \dot{M}(x,s)\,dx\,ds$. However, as is the case for classical Itô integration with respect to a Brownian motion, the time-derivative may not be well defined.

In the setting of Walsh [46], the approach is to separate the roles of time and space, and introduce a class of so-called *martingale measures* $M_t(A)$ for A being a suitable class of measurable subsets of $\mathbb{R}^d$. The martingale measures are such that for each time $t \geq 0$, M_t is a measure-valued square-integrable random variable, and for each set A, the process $t \mapsto M_t(A)$ is a martingale (with respect to a given filtration). In addition, the *covariance functional*

$$\overline{Q}_t(A,B) = \langle M(A), M(B) \rangle_t$$

plays a crucial role in the construction. Under some technical assumptions on $\overline{Q}$, Walsh [46] constructs the stochastic integral following the scheme of Itô. He shows that for elementary integrands, the stochastic integral is a martingale measure, and by limiting procedures the definition can be extended to predictable integrands X satisfying some quadratic integrability condition (yielding an extension of the Itô isometry). In fact, the stochastic integral will become a martingale measure.

As it turns out, when studying the relation between Lévy bases and the Walsh theory, so-called *orthogonal* martingale measures are the crucial objects. A martingale measure is called orthogonal if, for two disjoint sets A and B, the processes $M_t(A)$ and $M_t(B)$ are orthogonal. Orthogonal martingale measures satisfy the additional assumptions on the covariance functional, and it is moreover sufficient to study the *covariance measure*

$$Q\big([0,t]\times A\big)=\big\langle M(A)\big\rangle_t$$

instead when defining the stochastic integral. In fact, the integrands will be predictable and square integrable with respect to Q. Noteworthy is that the measure Q is closely linked to the control measure of a Lévy basis.

We now go on with a rigorous study of Lévy bases, white noise, and stochastic integration in the sense of Walsh, where many of the above concepts will be introduced and discussed in mathematical detail.

2.3.2 Lévy Bases and White Noise

In order to relate Lévy bases Λ to the white noise random fields introduced by Walsh [46], it is convenient to slightly reformulate the definition of a Lévy basis given in Definition 2.1.

We first show that a Lévy basis Λ is countably additive since its law is infinitely divisible:

Lemma 2.5 *A Lévy basis Λ is countably additive, that is, for a sequence of sets $\{A_n\}\subset\mathcal{B}_b(S)$ where $A_n\downarrow\emptyset$, it holds that*

$$\lim_{n\to\infty}P\big(\big|\Lambda(A_n)\big|\geq\varepsilon\big)=0 \tag{2.17}$$

for every $\varepsilon>0$.

Proof From the general theory of infinitely divisible laws, there exists a characteristic triplet such that the law of $\Lambda(A)$ has the triplet $(\Sigma_A,\gamma_A,\nu_A)$. One can show (see Pedersen [36, p. 3]) that $A\mapsto\gamma_A^i$, Σ_A^{ij} are signed measures for $i\neq j$ and $A\mapsto\nu_A(B)$, Σ_A^{ii} are measures for all i and $B\in\mathcal{B}(\mathbb{R}^d)$. Hence, if $A_n\downarrow\emptyset$ is a sequence of bounded Borel sets, then by standard properties of measures it holds that $(\Sigma_{A_n},\gamma_{A_n},\nu_{A_n})\to(0,0,0)$, and thus the law of $\Lambda(A_n)$ converges to δ_0. Hence, in probability and a.s. it holds that $\Lambda(A_n)$ converges to zero. The countable additivity in (2.17) follows. □

The following lemma follows from the countable additivity of Λ:

Lemma 2.6 *Condition* (3) *in Definition* 2.1 *is equivalent to the condition*: *For each pair of disjoint sets* A *and* B, *it holds a.s. that*

$$\Lambda(A \cup B) = \Lambda(A) + \Lambda(B).$$

Proof Consider $C_N = \bigcup_{i=1}^{N} A_i$ and $D_N = \bigcup_{i=N+1}^{\infty} A_i$, and use that C_N and D_N are disjoint to find that

$$\Lambda\left(\bigcup_{i=1}^{\infty} A_i\right) = \sum_{i=1}^{N} \Lambda(A_i) + \Lambda(D_N).$$

Since $D_N \downarrow \emptyset$, by the countable additivity of Λ, we can use Chebyshev's inequality to find

$$P\left(\left|\Lambda\left(\bigcup_{i=1}^{\infty} A_i\right) - \sum_{i=1}^{N} \Lambda(A_i)\right| \geq \varepsilon\right) = P\big(\left|\Lambda(D_N)\right| \geq \varepsilon\big) \leq \frac{1}{\varepsilon^2}\mathbb{E}\big[\Lambda(D_N)^2\big],$$

and the right-hand side tends to zero by countable additivity. This gives us the convergence in probability of the series $\sum_{i=1}^{N} \Lambda(A_i)$ as $N \to \infty$. But since the $\Lambda(A_i)$'s are independent random variables, we get the convergence P-a.s. by the Itô–Nisio theorem. □

Recall Condition (2) of independence for Lévy bases Λ in Definition 2.1. We note that it is equivalent to assume this condition for $n = 2$ only. To see this, let $A_1, A_2, \ldots, A_n$ be n disjoint subsets in $\mathcal{B}_b(S)$. Then, $\Lambda(A_i)$ and $\Lambda(A_j)$ are independent for any combination $i \neq j$, $i, j = 1, \ldots, n$. But then $\Lambda(A_1), \ldots, \Lambda(A_n)$ are independent.

We may give an equivalent definition of a Lévy basis Λ as follows:

Definition 2.7 A family $\{\Lambda(A) : A \in \mathcal{B}_b(S)\}$ of random vectors in $\mathbb{R}^d$ is called an $\mathbb{R}^d$-valued Lévy basis on S if the following three properties are satisfied:

1. The law of $\Lambda(A)$ is infinitely divisible for all $A \in \mathcal{B}_b(S)$.
2. If A and B are disjoint subsets in $\mathcal{B}_b(S)$, then $\Lambda(A)$ and $\Lambda(B)$ are independent.
3. If A and B are disjoint subsets in $\mathcal{B}_b(S)$, then

$$\Lambda(A \cup B) = \Lambda(A) + \Lambda(B) \quad \text{a.s.}$$

The above definition of a Lévy basis provides a natural generalisation of the object defined as *white noise* in Walsh [46]. A white noise is a random set function W on a σ-finite space $(E, \mathcal{E}, \nu)$ defined as follows:

Definition 2.8 A white noise W is a random set function on $\mathcal{E}_b$, the sets $A \in \mathcal{E}$ where $\nu(A) < \infty$, such that

1. $W(A)$ is normally distributed with zero mean and variance $\nu(A)$;

2. $W(A)$ and $W(B)$ are independent as long as A and B are disjoint;
3. $W(A \cup B) = W(A) + W(B)$ as long as A and B are disjoint.

We observe that in the case $E = \mathbb{R}^d$, this white noise concept is a very particular example of a homogeneous Lévy basis (and the definition of Lévy bases, as given in the Appendix, could easily be extended to more general spaces E). Hence, homogeneous Lévy bases provide a generalisation of white noise to *Lévy noise*.

As a note in passing, Walsh [46] concentrates on random measures which have finite variance, in the sense that for each $A \in \mathcal{B}_b(S)$, $\Lambda(A) \in L^2(P)$. Further, the following stronger countable additivity condition is introduced: Λ is said to be *countably additive* if for a sequence of sets $\{A_n\} \subset \mathcal{B}_b(S)$ where $A_n \downarrow \emptyset$, it holds that

$$\lim_{n\to\infty} \mathbb{E}\big[\Lambda(A_n)^2\big] = 0. \tag{2.18}$$

This is stronger than condition (2.17), which only holds in probability and does not require any finite variance of the random measure. However, the strong condition of Walsh [46] is suitable when defining a theory of stochastic integration which we will consider in Sect. 2.3.4.

Walsh [46] also introduces a concept of σ-finiteness of the random measures Λ. To this end, suppose that there exists an increasing sequence of sets $\{S_n\}_n \subset \mathcal{B}(S)$ such that $\bigcup_{n=1}^{\infty} S_n = S$, and for all n, it holds that $\mathcal{B}(S)|S_n \subset \mathcal{B}_b(S)$ and

$$\sup_{A\in\mathcal{B}(S)|S_n} \mathbb{E}\big[\Lambda(A)^2\big] < \infty.$$

If this is true, we say that Λ is *σ-finite*. If Λ is σ-finite, then Λ is countably additive on $\mathcal{B}(S)|S_n$ if and only if for any sequence of sets $A_n \downarrow \emptyset$ with $A_n \in \mathcal{B}(S)|S_n$, we have $\lim_{n\to\infty} \mathbb{E}[\Lambda(A_n)^2] = 0$. Walsh [46] makes this extension since, for such Λ, one may extend their domain of definition to include some new sets $A \in \mathcal{B}(S)$: If $A \in \mathcal{B}(S)$, we define

$$\Lambda(A) := \lim_{n\to\infty} \Lambda(A \cap S_n)$$

if the limit exists in $L^2(P)$ and consider $\Lambda(A)$ undefined otherwise. This leaves Λ unchanged on each $\mathcal{B}(S)|S_n$ but may change its value for sets $A \in \mathcal{B}(S)$ that are not in any $\mathcal{B}(S)|S_n$. In Walsh [46], Λ extended in this way is called a σ-finite L^2-valued random measure. Note that we can make this extension for all Lévy bases Λ trivially whenever S is bounded. For S unbounded, the σ-finiteness follows whenever Λ has mean zero. To see this, we make the following computation:

$$\begin{aligned}\mathbb{E}\big[\Lambda^2(S_n)\big] &= \mathbb{E}\big[\Lambda^2(S_n \setminus A)\big] + 2\mathbb{E}\big[\Lambda(A)\big]\mathbb{E}\big[\Lambda(S_n \setminus A)\big] + \mathbb{E}\big[\Lambda^2(A)\big] \\ &\geq \mathbb{E}\big[\Lambda^2(A)\big].\end{aligned}$$

Thus, the variance of $\Lambda(A)$ is bounded by the variance of $\Lambda(S_n)$, which is finite, and the σ-finiteness follows.

2.3.3 Lévy Bases and Random Variables in a Hilbert Space

For certain types of Lévy bases Λ, we introduce the mapping $x \mapsto \dot{\Lambda}(x)$ for $x \in S$, being the *noise* of Λ. For this purpose, it will be convenient to interpret the Lévy bases in terms of Hilbert-space-valued random variables.

To this end, let S be a bounded Borel set in $\mathbb{R}^k$ and introduce the measure space $(S, \mathcal{S}, \text{leb})$, with leb being the Lebesgue measure, and $\mathcal{S}$ the Borel sets on S. Assume that S is such that $L^2(S, \mathcal{S}, \text{leb})$ is separable and denote by $\{e_k\}_{k\in\mathbb{N}}$ a complete orthonormal system in the Hilbert space $H = L^2(S, \mathcal{S}, \text{leb})$. We suppose in addition that for all $A \in \mathcal{S}$ with $\text{leb}(A) = 0$, we have $\Lambda(A) = 0$ a.s. Finally, we assume that Λ has *nuclear covariance*,[1] that is,

$$\sum_{k=1}^{\infty} \mathbb{E}\left[\left(\int_S e_k(x)\Lambda(dx)\right)^2\right] < \infty, \tag{2.19}$$

where the integration of e_k with respect to $\Lambda(dx)$ is understood in the sense of Rajput and Rosinski as reviewed in Sect. A.3. We note that in Walsh [46], it is supposed that the integrals with respect to $\Lambda(dx)$ is in the sense of Bochner ([25] and also Chap. III in [27]), which is a stronger concept defined by convergence in variance.

The nuclear covariance condition (2.19) implies that $\Lambda(A)$ has finite variance, as the following lemma shows.

Lemma 2.9 *For every $A \in \mathcal{S}$, $\Lambda(A) \in L^2(P)$.*

Proof Let $A \in \mathcal{S}$. Since obviously $1_A(x) \in L^2(S, \text{leb})$, we have that

$$1_A(x) = \sum_{k=1}^{\infty} \int_A e_k(y)\,dy\, e_k(x),$$

and therefore

$$\Lambda(A) = \int_A \Lambda(dx) = \sum_{k=1}^{\infty} \int_A e_k(y)\,dy \int_S e_k(x)\Lambda(dx).$$

But by the Cauchy–Schwarz inequality for sums, we find

$$\begin{aligned}\mathbb{E}[\Lambda(A)^2] &\leq \sum_{k=1}^{\infty}\left(\int_A e_k(y)\,dy\right)^2 \times \sum_{k=1}^{\infty}\mathbb{E}\left[\left(\int_S e_k(x)\Lambda(dx)\right)^2\right] \\ &= |1_A|_2^2 \sum_{k=1}^{\infty}\mathbb{E}\left[\left(\int_S e_k(x)\Lambda(dx)\right)^2\right] < \infty.\end{aligned}$$

□

[1] This is in accordance with the definition of Walsh [46, p. 288].

For every $\phi \in L^2(S, \mathcal{S}, \text{leb})$, let us introduce the following functional on $L^2(S, \mathcal{S}, \text{leb})$:

$$\phi \mapsto \Lambda(\phi) := \int_S \phi(x)\Lambda(dx)\,. \tag{2.20}$$

Lemma 2.10 *The mapping* $\phi \mapsto \Lambda(\phi)$ *defined in* (2.20) *is a linear functional on* $L^2(S, \mathcal{S}, \text{leb})$.

Proof We show that the operator is bounded. We have that $\phi = \sum_{k=1}^{\infty} \phi_k e_k$ and thus

$$\int_S \phi(x)\Lambda(dx) = \sum_{k=1}^{\infty} \phi_k \int_S e_k(x)\Lambda(dx)\,.$$

The Cauchy–Schwarz inequality for sums now yields

$$\mathbb{E}\left[\left|\int_S \phi(x)\Lambda(dx)\right|^2\right] \le \sum_{k=1}^{\infty} \phi_k^2 \times \sum_{k=1}^{\infty} \mathbb{E}\left[\int_S e_k(x)\Lambda(dx)^2\right] < \infty,$$

and hence, the integral is finite a.s. Obviously, $\phi \mapsto \Lambda(\phi)$ is linear, and it therefore defines a linear functional on $L^2(S, \mathcal{S}, \text{leb})$. □

We are now ready to show that Λ has a Radon–Nikodym derivative with respect to the Lebesgue measure.

Proposition 2.11 *There exists a function* $\dot{\Lambda} \in L^2(S, \mathcal{S}, \text{leb})$ *such that*

$$\Lambda(\phi) = \int_S \dot{\Lambda}(x)\phi(x)\,dx. \tag{2.21}$$

Thus $\dot{\Lambda}$ *is the Radon–Nikodym derivative of* Λ *with respect to the Lebesgue measure on* $(S, \mathcal{S})$.

Proof Since any linear functional on a Hilbert space may be represented via the inner product with some element of the Hilbert space, we are ensured the existence of a function $\dot{\Lambda} \in L^2(S, \mathcal{S}, \text{leb})$ such that (2.21) holds. Note that since $1_A(x)$ is a function in $L^2(S, \mathcal{S}, \text{leb})$ for all $A \in \mathcal{S}$, we have

$$\Lambda(A) = \int_A \dot{\Lambda}(x)\,dx. \tag{2.22}$$

□

Moreover,

$$\Lambda(A)\Lambda(B) = \int_{A\times B} \dot{\Lambda}(x)\dot{\Lambda}(y)\,dx\,dy. \tag{2.23}$$

Note that

$$\dot{\Lambda}(x) = \sum_{k=1}^{\infty} \int_S e_k(y)\Lambda(dy)\, e_k(x).$$

Introduce

$$Q(A \times B) = \mathbb{E}\big[\Lambda(A)\Lambda(B)\big]. \tag{2.24}$$

Then we have that

$$Q(A \times B) = \int_{A\times B} \mathbb{E}\big[\dot{\Lambda}(x)\dot{\Lambda}(y)\big]\, dx\, dy.$$

We call the signed measure Q the *covariance measure* of the Lévy basis.

Define now the linear operator $\widetilde{Q}$ as

$$\widetilde{Q} f(x) = \int_S q(x, y) f(y)\, dy \tag{2.25}$$

with $q(x, y) = \mathbb{E}[\dot{\Lambda}(x)\dot{\Lambda}(y)]$. We prove that $\widetilde{Q}$ is a nonnegative, nuclear operator from $L^2(S, \mathcal{S}, \mathrm{leb})$ into itself.

Proposition 2.12 *The linear operator* $\widetilde{Q}$ *defined in* (2.25) *maps* $L^2(S, \mathcal{S}, \mathrm{leb})$ *into itself. The operator is nonnegative and nuclear.*

Proof By the Minkowski and Cauchy–Schwarz inequalities, we have

$$\begin{aligned}
\left|\int_S q(\cdot, y) f(y)\, dy\right|_2 &\le \int_S \big|q(\cdot, y) f(y)\big|_2\, dy \\
&= \int_S \left(\int_S q^2(x, y)\, dx\right)^{1/2} \big|f(y)\big|\, dy \\
&\le \left(\int_S \int_S q^2(x, y)\, dx\, dy\right)^{1/2} |f|_2 \\
&= \left(\int_S \int_S \mathbb{E}\big[\dot{\Lambda}(x)\dot{\Lambda}(y)\big]^2\, dx\, dy\right)^{1/2} |f|_2 \\
&\le \left(\int_S \int_S \mathbb{E}\big[\dot{\Lambda}^2(x)\big]\mathbb{E}\big[\dot{\Lambda}^2(y)\big]\, dx\, dy\right)^{1/2} |f|_2 \\
&= \mathbb{E}\big[|\dot{\Lambda}|_2^2\big] |f|_2 .
\end{aligned}$$

However, by Parseval's identity and the nuclear covariance condition (2.19), we have that

$$\mathbb{E}\big[|\dot{\Lambda}|_2^2\big] = \sum_{k=1}^{\infty} \mathbb{E}\left[\left(\int_S e_k(x)\Lambda(dx)\right)^2\right] < \infty,$$

and hence $\widetilde{Q}f$ is in $L^2(S, \mathcal{S}, \text{leb})$. Furthermore, we have that the operator is non-negative in the sense that $(\widetilde{Q}f, f)_2 \geq 0$ for all $f \in L^2(S, \text{leb})$. This follows since

$$(\widetilde{Q}f, f)_2 = \mathbb{E}[(f, \dot{\Lambda})_2^2] \geq 0.$$

We check whether the operator is nuclear. By using the series representation of $\dot{\Lambda}(y)$ we find

$$\begin{aligned}
\widetilde{Q}f(x) &= \int_S q(x, y) f(y)\, dy \\
&= \sum_{k=1}^{\infty} \int_S \mathbb{E}\left[\dot{\Lambda}(x) \int_S e_k(z)\, \Lambda(dz)\right] e_k(y) f(y)\, dy \\
&= \sum_{k=1}^{\infty} (e_k, f)_2 \mathbb{E}\left[\dot{\Lambda}(x) \int_S e_k(y)\, \Lambda(dy)\right].
\end{aligned}$$

This is the representation in Definition A.1 in Peszat and Zabczyk [37] of nuclear operators, where we identify $a_k(x) = e_k(x)$ and $b_k(x) = E[\dot{\Lambda}(x) \int_S e_k(y)\, \Lambda(dy)]$. Now, $\widetilde{Q}$ is nuclear if $\sum_{k=1}^{\infty} |a_k|_2 |b_k|_2 < \infty$. But this is equivalent to

$$\sum_{k=1}^{\infty} \int_S \left(\mathbb{E}\left[\dot{\Lambda}(x) \int_S e_k(y)\, \Lambda(dy)\right]\right)^2 dx < \infty,$$

since e_k is an orthonormal basis. But, by the Cauchy–Schwarz inequality, we find

$$\begin{aligned}
&\sum_{k=1}^{\infty} \int_S \left(\mathbb{E}\left[\dot{\Lambda}(x) \int_S e_k(y)\, \Lambda(dy)\right]\right)^2 dx \\
&\quad \leq \sum_{k=1}^{\infty} \int_S \mathbb{E}[\dot{\Lambda}^2(x)]\, dx\, \mathbb{E}\left[\left(\int_S e_k(y)\, \Lambda(dy)\right)^2\right] \\
&\quad = \mathbb{E}[|\dot{\Lambda}|_2^2] \sum_{k=1}^{\infty} \mathbb{E}\left[\left(\int_S e_k(y) \Lambda(dy)\right)^2\right],
\end{aligned}$$

and this is finite by the nuclear covariance condition (2.19). □

We conclude that $\widetilde{Q}$ is a covariance operator in the sense of Peszat and Zabczyk [37, p. 30], where it is defined for Gaussian random variables with values in a Hilbert space. This links the Lévy bases to the theory of square-integrable Hilbert-space-valued random variables. We note that the nuclear covariance condition (2.19) makes the Lévy basis sufficiently regular to create random fields with values in a Hilbert space, where we can define covariance operators as the crucial object to understand the covariance structure. Tracing back, we see that the covariance measure of the Lévy basis Λ can be represented by the covariance operator of

$\dot{\Lambda}$ as

$$Q(A \times B) = (\widetilde{Q}1_A, 1_B)_2 . \tag{2.26}$$

Thus, the covariance measure is representable via an integral kernel.

2.3.4 Extension of the Stochastic Integration Theory of Walsh

Let us consider a Lévy basis Λ on $[0, T] \times S \in \mathcal{B}(\mathbb{R}^{k+1})$, that is, a Lévy basis where we have separated out the first variable to denote time.

We introduce the following measure-valued process

$$M_t(A) := \Lambda\big((0, t] \times A\big) \tag{2.27}$$

for any $A \in \mathcal{B}_b(S)$. The following properties are inherited from the Lévy basis for a fixed set $A \in \mathcal{B}_b(S)$:

Proposition 2.13 *The measure-valued process $M_t(A)$ for $A \in \mathcal{B}_b(S)$ defined in* (2.27) *is an additive process,*[2] *i.e. it satisfies the following properties*:

1. *The law of $M_t(A)$ is infinitely divisible for each t.*
2. *The increments of $M_t(A)$ are independent.*
3. *The process $M_t(A)$ is stochastically continuous.*
4. *The process $M_t(A)$ is right-continuous with $M_0(A) = 0$ a.s.*

Proof The first property follows from the fact that the Lévy basis Λ is infinitely divisible. To see the second property, we observe from the additivity of Λ that

$$\Lambda\big((0, t] \times A\big) = \Lambda\big(\{(0, s] \times A\} \cup \{(s, t] \times A\}\big) = \Lambda\big((0, s] \times A\big) + \Lambda\big((s, t] \times A\big).$$

From the independence property of Λ, it holds that $\Lambda((s, t] \times A)$ is independent of $\Lambda((0, \tau] \times A)$ for all sets $(0, \tau] \times A$ where $\tau \leq s$. Hence, $M_t(A) - M_s(A)$ is independent of $M_s(A)$. We continue with proving property (3). Observe that

$$P\big(\big|M_t(A) - M_s(A)\big| > \varepsilon\big) = P\big(\big|\Lambda\big((s, t] \times A\big)\big| > \varepsilon\big),$$

and as $t \downarrow s$, we have that $(s, t] \times A \downarrow \emptyset$. Hence, from the countable additivity in probability, which holds for Lévy bases, it follows that

$$\lim_{t \downarrow s} P\big(\big|M_t(A) - M_s(A)\big| > \varepsilon\big) = 0.$$

This proves property (3). In particular, we find

$$\lim_{t \downarrow 0} P\big(\big|M_t(A)\big| > \varepsilon\big) = 0,$$

[2] More precisely, we have that $M_t(A)$ is an *additive process in law*, see Definition 1.6 in Sato [42].

and therefore $M_t(A)$ converges in probability to zero, which implies the convergence in law to δ_0. This gives that $\lim_{t\downarrow 0} M_t(A) = 0$ a.s., and we have that $M_0(A) = \lim_{t\downarrow 0} M_t(A) = 0$ a.s. Moreover, following the same argument as above, we see that for $s > t$ (using independence of Λ),

$$\Lambda\big((0,s] \times A\big) = \Lambda\big((0,t] \times A\big) + \Lambda\big((t,s] \times A\big).$$

The countable additivity of Λ yields that

$$\Lambda\big((t,s] \times A\big) \to 0$$

in probability as $s \downarrow t$ since $(t,s] \times A \downarrow \emptyset$, and therefore $\Lambda((t,s] \times A)$ converges in law to δ_0. Hence,

$$\Lambda\big((0,s] \times A\big) \to \Lambda\big((0,t] \times A\big),$$

and it follows that $M_t(A)$ is right-continuous. Hence, we have shown the last property. □

Remark To obtain a Lévy process, we would need to have stationarity of increments, i.e. the law of the increment $M_{s+t}(A) - M_s(A)$, $s, t > 0$, should be independent of s. But

$$M_{s+t}(A) - M_s(A) = \Lambda\big((s, s+t] \times A\big),$$

and the characteristic triplet for the law is thus $(\Sigma_{(s,s+t]\times A}, \gamma_{(s,s+t]\times A}, \nu_{(s,s+t]\times A})$. If there exist measures $\widetilde{\Sigma}_A$ and $\widetilde{\nu}_A$, and a signed measure $\widetilde{\gamma}_A$ such that $\Sigma_{\tau\times A} = \mathrm{leb}(\tau)\widetilde{\Sigma}_A$, $\gamma_{\tau\times A} = \mathrm{leb}(\tau)\widetilde{\gamma}_A$ and $\nu_{\tau\times A} = \mathrm{leb}(\tau)\widetilde{\nu}_A$, for a bounded Borel subset τ of the positive real line, we would have the stationarity. Such a separation property of the characteristic triplet would imply that $M_t(A)$ is a Lévy process.

We want to use $M_t(A)$ as integrators like in Walsh [46], where the Itô integration approach is used. We conveniently suppose that for each A, $M_t(A) \in L^2(\Omega, \mathcal{F}, P)$. Furthermore, we define the filtration $\mathcal{F}_t$ by $\mathcal{F}_t = \bigcap_{n=1}^{\infty} \mathcal{F}^0_{t+1/n}$, where

$$\mathcal{F}^0_t = \sigma\big\{M_s(A) : A \in \mathcal{B}_b(S), 0 < s \le t\big\} \vee \mathcal{N},$$

and where $\mathcal{N}$ denotes the P-null sets of $\mathcal{F}$. Then, $\mathcal{F}_t$ is right-continuous by construction. Finally, we suppose that the expected value of the Lévy basis Λ is equal to zero, that is, $\mathbb{E}[M_t(A)] = 0$. If this is not the case, we can always redefine the Lévy basis by subtracting its mean value in order to obtain a mean-zero process.

It turns out that $M_t(A)$ is a square-integrable martingale satisfying an orthogonality property:

Proposition 2.14 *Under the assumption of square-integrability and mean zero of $M_t(A)$, the following two properties hold:*

1. *For each A, $t \mapsto M_t(A)$ is a (square-integrable) martingale with respect to the filtration $\mathcal{F}_t$.*

2. *If A and B are two disjoint sets in $\mathcal{B}_b(S)$, then $M_t(A)$ and $M_t(B)$ are independent.*

Proof The second property holds trivially by the independence property of the Lévy basis. To see the first property, let $s \leq t$. We have by the independence property of the Lévy basis that

$$\Lambda\big((0,t] \times A\big) = \Lambda\big((0,s] \times A \cup (s,t] \times A\big) = \Lambda\big((0,s] \times A\big) + \Lambda\big((s,t] \times A\big),$$

and therefore

$$M_t(A) = M_s(A) + \Lambda\big((s,t] \times A\big).$$

Furthermore, we have that $\Lambda((s,t] \times A)$ is independent of $\mathcal{F}_s$ since any sets $[0, s_i] \times B$ will be disjoint with $(s,t] \times A$ as long as $s_i \leq s$. Therefore,

$$\mathbb{E}\big[M_t(A)\big|\mathcal{F}_s\big] = \mathbb{E}\big[M_s(A)\big|\mathcal{F}_s\big] + \mathbb{E}\big[\Lambda\big((s,t] \times A\big)\big] = M_s(A).$$

The last equality is obtained by the zero-mean assumption on the Lévy basis and the measurability of $M_s(A)$ to $\mathcal{F}_s$. □

These two properties, together with the fact that $M_0(A) = 0$ a.s., are essentially defining what is called an *orthogonal martingale measure* in Walsh [46]. Walsh [46] adds a further regularity condition on $A \mapsto M_t(A)$, which he calls the σ-finiteness to make up the definition of an orthogonal martingale measure. As we have seen earlier, the σ-finiteness follows for Lévy bases with mean zero, which is what is supposed here.

As is shown in Walsh [46] (see also [34] for a survey), for orthogonal martingale measures, we may introduce the *covariance measure* Q as

$$Q\big([0,t] \times A\big) = \big\langle M(A)\big\rangle_t \tag{2.28}$$

for $A \in \mathcal{B}_b(S)$. The covariance measure Q is positive and is used as the control measure in the Walsh sense when defining stochastic integration with respect to M. We now describe the integration procedure followed by Walsh [46], which is essentially the Itô approach to stochastic integration. To make matters slightly simpler, we suppose that S is a bounded Borel set, and we recall the notation $\mathcal{S}$ for the Borel subsets of S. Furthermore, we treat only integration up to a finite time T. Note that extensions to unbounded S and infinite time interval follow by standard arguments (see [46, p. 289]).

First, we say that a random field $f(s,x)$ is *elementary* if it has the form

$$f(s,x,\omega) = X(\omega)1_{(a,b]}(s)1_A(x), \tag{2.29}$$

where $0 \leq a < t$, X is bounded and $\mathcal{F}_a$-measurable, and $A \in \mathcal{S}$. For elementary functions, we can define stochastic integration as

$$\int_0^t \int_B f(s,x)M(dx,ds) := X\big(M_{t\wedge a}(A \cap B) - M_{t\wedge b}(A \cap B)\big) \tag{2.30}$$

for every $B \in \mathcal{S}$. In fact, the stochastic integral becomes a martingale measure as discussed earlier. The extension of stochastic integration to finite linear combinations of elementary random fields is obvious. A finite linear combinations of elementary random fields is called a *simple* random field, and the set of simple random fields is denoted $\mathcal{T}$. The *predictable* σ-algebra $\mathcal{P}$ is the σ-algebra generated by $\mathcal{T}$, and a random field is called *predictable* as long as it is $\mathcal{P}$-measurable. The norm $\|\cdot\|_M$ is defined on the predictable random fields f by

$$\|f\|_M^2 := \mathbb{E}\left[\int_{[0,T]\times S} f^2(s,x)\, Q(dx,ds)\right], \tag{2.31}$$

which determines the Hilbert space $\mathcal{P}_M := L^2(\Omega \times [0,T] \times S, \mathcal{P}, Q)$. In Walsh [46] it is proved that $\mathcal{T}$ is dense in $\mathcal{P}_M$. To define the stochastic integral of $f \in \mathcal{P}_M$, we choose an approximating sequence $\{f_n\}_n \subset \mathcal{T}$ such that $\|f - f_n\|_M \to 0$ as $n \to \infty$. It is easy to see that for each $A \in \mathcal{S}$, $\int_{[0,t]\times A} f_n(s,x)\, M(dxds)$ is a Cauchy sequence in $L^2(\Omega, \mathcal{F}, P)$, and thus there exists a limit which we define as the stochastic integral of f. It turns out that this stochastic integral is again a martingale measure, and that the "Itô isometry" holds:

$$\mathbb{E}\left[\left(\int_{[0,t]\times A} f(s,x) M(dx,ds)\right)^2\right] = \|f\|_M^2 . \tag{2.32}$$

See Walsh [46], Theorem 2.5 for the complete result and proof.[3]

The weak integration of Rajput and Rosinski [38] extends this definition of stochastic integration in the following sense. For any sequence $\{f_n\}_n \subset \mathcal{T}$ of deterministic functions converging to f in $\mathcal{P}_M$, there exists a subsequence $\{f_{n'}\}_{n'} \subset \mathcal{T}$ converging to f Q-a.e., and for this sequence, the stochastic integrals converge in probability since they converge in variance by definition. Hence, for $f \in \mathcal{P}_M$, the definition of weak integration according to Rajput and Rosinski presented in Sect. A.3 in the Appendix extends that of Walsh as long as the control measure λ of the Lévy basis Λ is absolutely continuous with respect to Q. (See Sect. A.2 in the Appendix.) However, as the following computation shows, Q and λ are equivalent: Since we have assumed that the Lévy basis Λ has zero mean, it follows from the characteristic exponent in formula (2.49) of the Appendix that

$$Q\big([0,t] \times A\big) = \int_{[0,t]\times A} \left(\sigma^2(x,s) + \int_{\mathbb{R}} z^2\, \rho(x,s,dz)\right) \lambda(dx,ds).$$

Therefore we conclude that the weak integration concept of Rajput and Rosinski is a true generalisation of that due to Walsh as long as deterministic integrands are considered. We remark in passing that the integration theory of Rajput and Rosinski

[3] Note that in Walsh [46], the argument is made for so-called *worthy* martingale measures. As argued in Walsh [46], an orthogonal martingale measure is worthy, and moreover the *control measure* used to define stochastic integrals sits in this case on the diagonal of $S \times S$. We have chosen to present that particular case.

is not restricted to square-integrable Lévy bases, as is the Walsh integration concept we have presented here.

Remark Note that we do not know if we have disintegration with the theory of Walsh. However, we know that the integral is a martingale process in time, which adds important dynamics which gives us a big advantage compared to the weaker form of integration available from Rajput and Rosinski [38].

Note also that in the definition of weak integration in the Appendix only deterministic integrands are used. The general definition of ambit processes involves stochastic integrands. This can be accommodated by further extension of the Walsh theory. Such extension is currently under development in collaboration with Andreas Basse-O'Connor, Svend Erik Graversen, and Jan Pedersen, see e.g. [22].

2.3.5 Stochastic Partial Differential Equations and Ambit Processes

In this subsection we consider a class of parabolic stochastic partial differential equations (SPDE) analysed in detail by Walsh [46]. The motivation with our presentation here is to relate the solutions of such SPDEs to ambit processes and discuss possible extensions based on these.

Letting $\dot{W}$ be a white noise in the sense of Walsh, we introduce the following nonlinear parabolic SPDE:

$$\begin{cases} \frac{\partial v}{\partial t} = \frac{\partial^2 v}{\partial x^2} - v + f(t,v)\dot{W}, & t > 0,\ 0 < x < K, \\ \frac{\partial v}{\partial x}(t,0) = \frac{\partial v}{\partial x}(t,K) = 0, & t > 0, \\ v(0,x) = v_0(x), & 0 < x < K, \end{cases} \tag{2.33}$$

where $K > 0$ is some constant, and f is a Lipschitz continuous function in x of at most linear growth. Furthermore, it is supposed that v_0 is $\mathcal{F}_0$-measurable and $\mathbb{E}[v_0^2(x)]$ is bounded. Since white noise is too rough to expect smooth solutions of the parabolic SPDE, Walsh [46] introduces a *weak solution* concept. We say that v is a *weak solution* of (2.33) if for every $\phi \in C^\infty([0,K])$ with $\phi'(0) = \phi'(K) = 0$, it holds that

$$\int_0^K \big(v(t,x) - v_0(x)\big)\phi(x)\,dx = \int_0^t \int_0^K v(s,x)\big(\phi''(x) - \phi(x)\big)\,dx\,ds + \int_0^t \int_0^K f\big(s, v(s,x)\big)\phi(x)W(dx,ds). \tag{2.34}$$

In Walsh [46], Theorem 3.2, it is proved that there exists a weak solution v to (2.33) which is bounded in variance on $[0,K] \times [0,T]$ for each $T > 0$. The proof goes by application of the Green's function and Picard iterations.

To see the connection to (2.33), note that formal differentiation of (2.34) with respect to t gives

$$\int_0^K v_t(t,x)\phi(x)\,dx = \int_0^K v(t,x)\big(\phi''(x) - \phi(x)\big)\,dx + \int_0^K f\big(t, v(t,x)\big)\phi(x)W(dx,dt).$$

An integration-by-parts applied formally to the first integral on the right-hand side and application of the initial conditions essentially leads to (2.33).

The homogeneous form of (2.33) is known as the *cable equation*, and Walsh [46] presents the Green's function of this as

$$G_t(x,y) = \frac{e^{-t}}{\sqrt{4\pi t}} \sum_{n=-\infty}^{\infty} \exp\Big(-\frac{(y-x-2nK)^2}{4t}\Big) + \exp\Big(-\frac{(y+x-2nK)^2}{4t}\Big).$$

A solution to the case $f = 1$ can be represented as

$$v(t,x) = \int_0^K G_t(x,y)v_0(y)\,dy + \int_0^t \int_0^K G_{t-s}(x,y)W(dy,ds). \tag{2.35}$$

Note that if the last integral was computed over $(-\infty, t]$ rather than over $[0,t]$, the Wold–Karhunen representation with respect to a Brownian motion could be used in principle.

The solution in (2.35) represents the solution to an SPDE which can be related to physical processes. Walsh [46] interprets the problem (2.33) to description of the nervous system, and another interpretation is diffusion of heat. These physical systems may be described directly through an ambit process rather than via an SPDE. As such, we could model the phenomena using a general Lévy basis Λ instead of the particular white noise W. Thus, a generalisation of v in (2.35) is to consider

$$v(t,x) = \int_0^K G_t(x,y)v_0(y)\,dy + \int_0^t \int_0^K G_{t-s}(x,y)L(dy,ds). \tag{2.36}$$

One may also take this further and consider "stochastic intermittency" described by a random field $\sigma(t,x)$. Thus,

$$v(t,x) = \int_0^K G_t(x,y)v_0(y)\,dy + \int_0^t \int_0^K G_{t-s}(x,y)\sigma(s,y)L(dy,ds). \tag{2.37}$$

The intermittency field σ may be defined as an ambit field, and as such, we have that $v(t,x)$ is an ambit field over the ambit set $\mathcal{A}_t(x) = [0,t] \times [0,K]$ under appropriate regularity conditions ensuring the existence of the integrals in (2.37). In fact, we have that $v(t,x)$ in (2.37) is by definition a *mild* solution of the parabolic problem

$$\begin{cases} \frac{\partial v}{\partial t} = \frac{\partial^2 v}{\partial x^2} - v + \sigma(t,x)\dot{\Lambda}, & t > 0,\ 0 < x < K, \\ \frac{\partial v}{\partial x}(t,0) = \frac{\partial v}{\partial x}(t,K) = 0, & t > 0, \\ v(x,0) = v_0(x), & 0 < x < K. \end{cases} \tag{2.38}$$

Here, $\dot{\Lambda}$ is a suggestive notation for the noise of the Lévy basis L (see Sect. 2.4 for a mathematical formulation of this). The definition of a mild solution of a parabolic stochastic partial differential equation is introduced in Da Prato and Zabczyk [26, p. 152] and is in general weaker than a weak solution. By Theorem 6.5 in Da Prato and Zabczyk [26], we have that the mild solution $v(t,x)$ in (2.37) is a weak solution under natural integrability conditions on σ and v_0.

It is important to notice that we can generalise the solution $v(t,x)$ in (2.37) to hold for very general specifications of σ; in fact, by going to the general integration concept of Rajput and Rosinski [38], we can make sense of $v(t,x)$ as an ambit field. By weakening the integration, we can still interpret v as a mild solution to the parabolic problem. A further generalisation is of course to allow for more general ambit sets $\mathcal{A}_t(x)$, leaving the specification $\mathcal{A}_t(x) = [0,t] \times [0,K]$. This will allow for a great deal of flexibility in modelling the physical phenomena in question, in particular how the dependency structure in time and space evolves.

2.4 Lévy Noise Analysis

The white noise analysis introduced by Hida in the 1980s has become a popular tool for analysing SPDEs that are singular in the sense of not admitting regular solutions. Hida proposed an analysis based on white noise, that is, the time-derivative of Brownian motion, with applications from quantum mechanics and Feynman path integrals in mind. In Hida, Kuo, Potthoff, and Streit [29] one can find a detailed account of the so-called *white noise analysis* and its applications to physics. In this paper we are concerned with SPDEs and will base our further discussion on the Lévy noise analysis for Lévy processes introduced in Holden, Øksendal, Ubøe, and Zhang [31]. In particular, we link Lévy bases and ambit processes with the Lévy noise analysis framework and finally discuss SPDEs in this context.

2.4.1 Lévy Bases and Lévy Noise

Let $\mathcal{S}(\mathbb{R}^d)$ be the Schwartz space of rapidly decreasing functions on $\mathbb{R}^d$, and define $\Omega = \mathcal{S}'(\mathbb{R}^d)$ to be its dual. Denote by $\mathcal{F}$ the Borel σ-algebra on Ω, and let ℓ be a Lévy measure on $\mathbb{R} \setminus \{0\}$ satisfying the condition of square-integrability

$$C := \int_{\mathbb{R}\setminus\{0\}} z^2 \ell(dz) < \infty . \tag{2.39}$$

By the Bochner–Minlos theorem (see Definition 5.4.1 in [31]) there exists a probability measure P on $(\Omega, \mathcal{F})$ such that

$$\begin{aligned}\int_{\Omega} e^{i\langle\omega,\phi\rangle}\,dP(\omega) \\ = \exp\bigg(-\frac{1}{2}\sigma^2|\phi|_2^2 + \int_{\mathbb{R}^d}\int_{\mathbb{R}\setminus\{0\}}\big\{e^{i\phi(y)z} - 1 - i\phi(y)z\big\}\ell(dz)\,dy\bigg),\end{aligned} \tag{2.40}$$

where $\langle\omega, \phi\rangle := \omega(\phi)$, that is, the action of $\omega \in \mathcal{S}'(\mathbb{R}^d)$ on $\phi \in \mathcal{S}(\mathbb{R}^d)$, and $|\cdot|_2$ is the norm in $L^2(\mathbb{R}^d)$. The probability space $(\Omega, \mathcal{F}, P)$ is called the d-parameter *Lévy noise probability space* by Holden et al. [31].[4] This probability space will support a d-parameter Lévy process and is the basis for defining its derivative, the *Lévy noise*.[5]

Introduce the cylindrical random variables N_ϕ by

$$N_\phi(\omega) = \langle\omega, \phi\rangle \tag{2.41}$$

for $\phi \in \mathcal{S}(\mathbb{R}^d)$. Observe, that since (2.40) gives an explicit form of the characteristic function of N_ϕ in terms of the Lévy measure, we easily find that

$$\mathbb{E}[N_\phi] = 0$$

and

$$\operatorname{Var}[N_\phi] = \big(\sigma^2 + C\big)\int_{\mathbb{R}^d}\phi^2(y)\,dy.$$

We can extend these random variables to $\phi \in L^2(\mathbb{R}^d)$ by a standard limit argument choosing a sequence $\{\phi_n\} \subset \mathcal{S}(\mathbb{R}^d)$ converging in $L^2(\mathbb{R}^d)$ to ϕ. The limit of N_{ϕ_n} exists in $L^2(P)$ and will be denoted N_ϕ. The limit is independent of the choice of approximating sequence. In particular, we can define $N_A := N_{\mathbf{1}_A}$ for bounded Borel sets $A \subset \mathbb{R}^d$. We make the following definition.

Definition 2.15 For every bounded Borel subset A of $\mathbb{R}^d$, define the random measure

$$\Lambda(A) = N_A.$$

We show that Λ defines a Lévy basis (see Proposition 2.16) and that it is homogeneous (see Proposition 2.17).

---

[4]We note that in Holden et al. [31] one constructs this probability space for Brownian motion and a pure-jump Lévy process separately. We merge this into a more general Lévy process with both jumps and continuous martingale part. Further note that the representation result (2.40) was originally introduced in [28]. See also [1] for related work.

[5]Note that Holden et al. [31] call such noise *Lévy coloured noise*.

Proposition 2.16 *The random measure Λ is a Lévy basis with mean zero and variance $(\sigma^2 + C) \cdot \mathrm{leb}(A)$, where $\mathrm{leb}(A)$ is the Lebesgue measure of A, and the associated control measure of Λ is*

$$\lambda(A) = \sigma^2 \,\mathrm{leb}(A) + \int_{\mathbb{R}} \min\bigl(1, z^2\bigr)\ell(dz)\,\mathrm{leb}(A).$$

Proof The random measure $\Lambda(A)$ has mean zero and variance equal to $M\ \mathrm{leb}(A)$, where $\mathrm{leb}(A)$ is the Lebesgue measure of the set A. We show that Λ has the additivity and independence properties.

Let A and B be two disjoint bounded Borel sets, and let $\phi_n \to 1_A$ and $\psi_n \to 1_B$ in $L^2(\mathbb{R}^d)$. Since obviously $1_{A\cup B} = 1_A + 1_B$ and $\phi_n + \xi_n$ converges to $1_A + 1_B$ in $L^2(\mathbb{R}^d)$, $\phi_n + \xi_n$ converges to $1_{A\cup B}$ in $L^2(\mathbb{R}^d)$. Hence, by the independence of the approximating sequence, we find that $N_{\phi+\xi_n}$ converges in $L^2(P)$ to $\Lambda(A\cup B)$, and since

$$N_{\phi+\xi_n}(\omega) = \langle\omega, \phi_n + \psi_n\rangle = \langle\omega, \phi_n\rangle + \langle\omega, \psi_n\rangle = N_{\phi_n}(\omega) + N_{\psi_n}(\omega),$$

it holds that

$$\Lambda(A\cup B) = \Lambda(A) + \Lambda(B).$$

This proves the additivity. To prove the independence, we have to show that for two disjoint bounded sets A and B, $\Lambda(A)$ is independent of $\Lambda(B)$, or equivalently, N_A is independent of N_B. To this end, choose two approximating sequences ϕ_n and ξ_n in $\mathcal{S}(\mathbb{R}^d)$ converging to 1_A and 1_B, respectively, in $L^2(\mathbb{R}^d)$. Use the characteristic function of N_{ϕ_n} and N_{ξ_x} to find

$$\begin{aligned}
&\ln \mathbb{E}\bigl[e^{i\theta N_{\phi_n}} e^{i\eta N_{\xi_n}}\bigr] \\
&\quad = \ln \mathbb{E}\bigl[e^{i\langle\cdot,\theta\phi_n + \eta\xi_n\rangle}\bigr] \\
&\quad = -\frac{1}{2}\sigma^2 |\theta\phi_n + \eta\xi_n|_2^2 \\
&\qquad + \int_{\mathbb{R}^d}\int_{\mathbb{R}\setminus\{0\}} \bigl\{e^{i(\theta\phi_n(y)+\eta\xi_n(y))z} - 1 - i\bigl(\theta\phi_n(y) + \eta\xi_n(y)\bigr)z\bigr\}\ell(dz)\,dy.
\end{aligned}$$

We can write the $L^2(\mathbb{R}^d)$-norm as follows:

$$\begin{aligned}
&|\theta\phi_n + \eta\xi_n|_2^2 \\
&\quad = \theta^2 \int_{\mathrm{supp}\,\phi_n \setminus \mathrm{supp}\,\xi_n} \phi_n^2(y)\,dy \\
&\qquad + \int_{\mathrm{supp}\,\phi_n \cap \mathrm{supp}\,\xi_n} \bigl(\theta\phi_n(y) + \eta\xi_n(y)\bigr)^2\,dy + \eta^2 \int_{\mathrm{supp}\,\xi_n \setminus \mathrm{supp}\,\phi_n} \xi_n^2(y)\,dy.
\end{aligned}$$

The set $\mathrm{supp}\,\phi_n \cap \mathrm{supp}\,\xi_n$ must go to a set of Lebesgue measure zero since $A \cap B = \emptyset$; otherwise the two sequences will not converge to their respective indicator functions in $L^2(\mathbb{R}^d)$. Hence, passing to the limit, we find that

$$\lim_{n\to\infty} |\theta\phi_n + \eta\xi_n|_2^2 = \theta^2 \operatorname{leb}(A) + \eta^2 \operatorname{leb}(B).$$

A similar argument shows that

$$\lim_{n\to\infty} \int_{\mathbb{R}^d} \int_{\mathbb{R}\setminus\{0\}} \left\{e^{i(\theta\phi_n(y)+\eta\xi_n(y))z} - 1 - i\big(\theta\phi_n(y) + \eta\xi_n(y)\big)z\right\}\ell(dz)\,dy$$
$$= \operatorname{leb}(A) \int_{\mathbb{R}\setminus\{0\}} \left\{e^{i\theta z} - 1 - i\theta z\right\}\ell(dz) + \operatorname{leb}(B) \int_{\mathbb{R}\setminus\{0\}} \left\{e^{i\theta z} - 1 - i\theta z\right\}\ell(dz).$$

Thus, after taking limits, we find

$$\mathbb{E}\big[e^{i\theta\Lambda(A)}e^{i\eta\Lambda(B)}\big] = \mathbb{E}\big[e^{i\theta\Lambda(A)}\big] \times \mathbb{E}\big[e^{i\eta\Lambda(B)}\big].$$

This proves the independence.

In fact, the above limit argument shows that the (log-)characteristic function of $\Lambda(A)$ is

$$\ln \mathbb{E}\big[e^{i\theta\Lambda(A)}\big] = \left(-\frac{1}{2}\theta^2\sigma^2 + \int_{\mathbb{R}\setminus\{0\}} \left\{e^{i\theta z} - 1 - i\theta z\right\}\ell(dz)\right) \operatorname{leb}(A),$$

This is the Lévy–Kintchine formula, where we can read off the control measure for the Lévy basis as being

$$\lambda(A) = \sigma^2 \operatorname{leb}(A) + \int_{\mathbb{R}} \min\big(1, z^2\big)\ell(dz) \operatorname{leb}(A)$$

(see Appendix for the definition of the control measure for a Lévy basis). □

By letting $\ell(dz) = 0$ and $\sigma = 1$, we recover the case of white noise and the setting for the white noise analysis. Note that here we consider only Lévy bases with no drift and being square integrable.

The Lévy basis has a stationarity property, as shown in the next Proposition.

Proposition 2.17 *For each $x \in \mathbb{R}^d$, $\Lambda(\cdot)$ and $\Lambda(\cdot + x)$ has the same distribution, i.e. Λ is homogeneous.*

Proof Given $\phi \in \mathcal{S}(\mathbb{R}^d)$, we prove that N_ϕ and N_{ϕ_x} have the same distribution, where $\phi_x(y) = \phi(y - x)$. It follows from the translation invariance of the Lebesgue measure that

$$\int_{\mathbb{R}^d} \int_{\mathbb{R}\setminus\{0\}} \left\{e^{i\phi(y-x)z} - 1 - i\phi(y - x)z\right\}\ell(dz)\,dy$$
$$= \int_{\mathbb{R}^d} \int_{\mathbb{R}\setminus\{0\}} \left\{e^{i\phi(y)z} - 1 - i\phi(y)z\right\}\ell(dz)\,dy.$$

Similarly we have that $|\phi|_2 = |\phi_x|_2$. Hence, the characteristic function of N_ϕ and of N_{ϕ_x} is the same. By a limit argument, it follows that N_A and N_{A+x} has the same characteristic function as well, implying that their distributions are coinciding. The proposition is proved. □

In Lévy noise analysis, one is interested in the noise process of the smoothed random variables N_ϕ. Introduce the object $\dot{N}_x$ for $x \in \mathbb{R}^d$ by

$$\dot{N}_x(\omega) = \langle \omega, \delta_x \rangle, \tag{2.42}$$

where δ_x is the Dirac δ-function. Obviously, δ_x is not an element of $L^2(\mathbb{R}^d)$ (and definitely not a Schwartz function); however, it is a tempered distribution. The notation $\langle \omega, \delta_x \rangle$ is just suggestive, since it only makes sense in an operator context as we now discuss. By conveniently introducing spaces of *smooth* random variables as certain subspaces of $L^2(P)$ one can look at their duals and in fact manage to embed $\dot{N}_x$ into one of these. Thus, if X is a smooth random variable, then $\dot{N}_x$ makes sense as a linear functional on this (we refer to [31] for details). As a simple example, we have that N_ϕ is a smooth random variable, and in this case

$$\langle\langle \dot{N}_x, N_\phi \rangle\rangle = \langle \delta_x, \phi \rangle = \phi(x).$$

From this we can do the following: Interpreting the integral in the sense of Pettis or Bochner, we can define, for $\phi \in \mathcal{S}(\mathbb{R}^d)$,

$$\int_{\mathbb{R}^d} \phi(x) \dot{N}_x \, dx \tag{2.43}$$

as an integral with values in a suitable space of linear functionals on smooth random variables. However, as it turns out, this integral will coincide with a smoothed white noise,

$$\int_{\mathbb{R}^d} \phi(x) \dot{N}_x \, dx = N_\phi.$$

But then we can interpret $\dot{N}_x$ as the noise of Λ, since we can write, by limit arguments,

$$\Lambda(A) = \int_A \dot{N}_x \, dx.$$

So, $\dot{N}$ is an extension of the previously introduced object $\dot{\Lambda}$. Note that there is no nuclear condition given here in order to introduce $\dot{N}_x$. Indeed, we have that

$$\int_{\mathbb{R}^d} \phi(x) \Lambda(dx) = \int_{\mathbb{R}^d} \phi(x) \dot{N}_x \, dx = N_\phi$$

for a function $\phi \in L^2(\mathbb{R})$, and thus,

$$\sum_{k=1}^{\infty} \mathbb{E}\left[\left(\int_{\mathbb{R}^d} e_k(x) \, \Lambda(dx)\right)^2\right] = \sum_{k=1}^{\infty} \mathbb{E}[N_{e_k}^2] = \sum_{k=1}^{\infty} |e_k|_2^2 = \sum_{k=1}^{\infty} 1 = \infty.$$

Here, $\{e_k\}$ is a complete orthonormal system in $L^2(\mathbb{R}^d)$. Hence, we have that the nuclear covariance condition does not hold. This means that we have a Lévy basis which has finite variance but is not sufficiently smooth to admit a Hilbert-space-valued Radon–Nikodym derivative $\dot{N}_x$. This links Lévy bases to the Lévy noise analysis.

2.4.2 Stochastic Partial Differential Equations and Lévy Noise Analysis

Consider the stochastic Poisson equation

$$\Delta u(x) = -\dot{N}_x, \quad x \in D,$$
$$u(x) = 0, \quad x \in \partial D,$$

where $D \subset \mathbb{R}^d$ is a bounded domain with regular boundary, and Δ is the Laplace operator in $\mathbb{R}^d$. In order to make sense out of this equation, Holden et al. [31] introduce the space of Hida distributions $(\mathcal{S})^*$, which plays much the same role for stochastic processes as the space of tempered Schwartz distributions plays for functions. The space of Hida distributions is the dual of the space of Hida test functions $(\mathcal{S})$, which is the space of smooth random variables. This space consists of square-integrable random variables for which the terms in the chaos expansion decays rapidly in variance. A precise definition of $(\mathcal{S})$ and $(\mathcal{S})^*$ is found in Holden et al. [31], but it is important to notice that $(\mathcal{S})^*$ consists of linear operators on the space $(\mathcal{S})$, and as such cannot be understood as random variables (i.e., if $X \in (\mathcal{S})^*$, $X(\omega)$ does not make sense in general for $\omega \in \Omega$). A prominent example is $\dot{N}_x \in (\mathcal{S})^*$. As is well known, the noise of a Lévy process cannot be regarded as a classical random variable.

The Poisson equation is interpreted as an SPDE in $(\mathcal{S})^*$. More precisely, we say that u is a generalised solution of the stochastic Poisson equation if $u : \overline{D} \mapsto (\mathcal{S})^*$ is twice differentiable, satisfies the boundary conditions, and the SPDE. By the differentiability of an $(\mathcal{S})^*$-valued mapping from D we mean that the limit $(u(x+h) - u(x))/h$ exists in $(\mathcal{S})^*$.

Letting $G(x, y)$ be the Green's function of Δ on D with zero boundary conditions, Løkka, Øksendal, and Proske [35] show that the unique solution is

$$u(x) = \int_D G(x, y)\dot{N}_y \, dy. \tag{2.44}$$

Note that the integral is interpreted as a Pettis integral, that is, defining an operator on the space of smooth random variables $(\mathcal{S})$. If $d \leq 3$, it is shown in Løkka et al. [35] that $u \in L^2(P)$, but in general dimensions we have to interpret the solution in a weak sense.

Since for $d \leq 3$, the solution u is square-integrable, we may write the solution as

$$u(x) = \int_D G(x, y) \Lambda(dy). \tag{2.45}$$

Therefore, u is in fact an ambit process with the ambit set being the domain D. The reason for u losing its square-integrability when going beyond dimension 3 lies in the fact that $G(x, y)$ has a singularity at $x = y$ of order $|x - y|^{2-d}$ for $d \geq 3$. By using ambit processes, we may define more general expressions

$$\widetilde{u}(x) = \int_{D_x} G(x, y) \sigma(y) \Lambda(dy) \tag{2.46}$$

for general random fields $\sigma(x)$ sufficiently regular to make the stochastic integral well defined. The set D_x denotes some ambit set which can be defined to incorporate complex spatial dependency structures. In fact, such a specification $\widetilde{u}(x)$ may go beyond what can be linked to a stochastic partial differential equation and still make sense as a random field (in particular, a real-valued random field).

Note that the theory of white noise permits the study of SPDEs driven by noise in both time and space, and provides a theory for defining the noise of Lévy processes (or, in our context, Lévy bases). Hence, one can interpret the SPDEs in a strong sense, with the price that the solutions must be understood as operators rather than random fields. This is in contrast to the theory of Walsh presented above, where the solution is formulated in terms of an integral equation moving all derivatives to test functions. Ambit processes appear as a natural object in the theory of Lévy noise as well.

2.5 Conclusions

We have considered ambit processes and their building blocks, Lévy bases, in view of two classical theories for studying stochastic partial differential equations: the Walsh theory of martingale measures and the Lévy noise analysis. Lévy bases can be naturally connected to both theories by introducing concepts of noise of Lévy bases and processes. We show that the solutions of some stochastic partial differential equations can be represented by integrals of random fields with respect to Lévy bases, naturally relating to ambit processes. In this respect, ambit processes provide a class of random fields which generalise the solutions of these physical dynamical systems and provide new and interesting models that include the additional elements of volatility fields and time-dependent ambit sets. A further key point is that the extended integration theory allows the handling of objects such as the main term in (2.3) by means of integration w.r.t. martingale measures.

Acknowledgements We would like to thank Andreas Basse-O'Connor and Jan Pedersen for helpful discussions and constructive comments. Financial support by the Center for Research in Econometric Analysis of Time Series, CREATES, funded by the Danish National Research Foundation, is gratefully acknowledged by the third author.

Appendix: Lévy Bases and Integration

This section reviews the integration theory of [38] (for a survey, see also [40]), since this concept of integration is used for defining stochastic integrals in the context of ambit fields.

A.1 Introduction

Throughout the text, let S denote a nonempty set, and let $\mathcal{A}$ denote a σ-finite δ-ring on S, i.e. $\mathcal{A}$ is a family of subsets of S such that for every pair of sets in $\mathcal{A}$, the union, the intersection, and the set difference is in $\mathcal{A}$ (hence $\mathcal{A}$ is a ring), and if $(A_n)_{n\geq 1} \subseteq \mathcal{A}$, then $\bigcap A_n \in \mathcal{A}$; also, there exists a sequence $(A_n^*)_{n\geq 1} \subseteq \mathcal{A}$ such that $\bigcup A_n^* = S$.

Note that we call a real stochastic process $\Lambda = \{\Lambda(A) : A \in \mathcal{A}\}$ on some probability space $(\Omega, \mathcal{F}, \mathbb{P})$ an *independently scattered random measure* if for every sequence of disjoint sets $(A_n)_{n\geq 1}$, the random variables $\Lambda(A_n)$, $n = 1, 2, \ldots$, are independent, and if $\bigcup_n A_n \in \mathcal{S}$, then $\Lambda(\bigcup_n A_n) = \sum_n \Lambda(A_n)$ almost surely.

A.2 Representation of the Characteristic Function of a Lévy Basis

If $\Lambda(A)$ is infinitely divisible for every $A \in \mathcal{A}$, we call it a *Lévy basis*. Its characteristic function for $A \in \mathcal{A}$ is then given by

$$\mathbb{E}\big(\exp\big(it\Lambda(A)\big)\big) \tag{2.47}$$

$$= \exp\bigg(it\nu_0(A) - \frac{1}{2}t^2\nu_1(A) + \int_{\mathbb{R}} \big(e^{itx} - 1 - it\tau(x)\big) F_A(dx)\bigg), \tag{2.48}$$

where $\nu_0 : \mathcal{S} \to \mathbb{R}$ is a signed measure, $\nu_1 : \mathcal{A} \to [0, \infty)$ is a measure, and F_A is a Lévy measure on $\mathbb{R}$ for every $A \in \mathcal{A}$, while $A \mapsto F_A(B) \in [0, \infty)$ is a measure for every $B \in \mathcal{B}(\mathbb{R})$ whenever $0 \notin \overline{B}$. Also, the centering function τ is defined by $\tau(x) = x$ if $\|x\| \leq 1$ and by $\tau(x) = x/\|x\|$ if $\|x\| > 1$.

Further, let

$$\lambda(A) = |\nu_0|(A) + \nu_1(A) + \int_{\mathbb{R}} \min\big(1, x^2\big) F_A(dx), \qquad A \in \mathcal{A}.$$

It can be shown that $\lambda : \mathcal{A} \to [0, \infty)$ is a measure on $\mathcal{A}$ such that if, for every $(A_n)_{n\geq 1} \subset \mathcal{A}$, $\lambda(A_n) \to 0$, then $\Lambda(A_n) \to 0$ in probability. Also, if, for every sequence $(A'_n)_{n\geq 1} \subset \mathcal{A}$ with $A'_n \subset A_n \in \mathcal{A}$, we have $\Lambda(A'_n) \to 0$ in probability, then $\lambda(A_n) \to 0$.

Note that the measure λ satisfies $\lambda(A_n^*) < \infty$ for $n = 1, 2, \ldots$. Hence, it can be extended to a σ-finite measure on $(S, \sigma(\mathcal{A}))$. This measure is then called the *control measure* of Λ.

It turns out that the characteristic function of an infinitely divisible random measure has also an alternative representation than the one given above.

In order to state it, we first need a preliminary result (see [38, Lemma 2.3]). Let $F_{\cdot}$ be as above. Then there exists a unique σ-finite measure F on $\sigma(\mathcal{A}) \times \mathcal{B}(\mathbb{R})$ such that $F(A \times B) = F_A(B)$ for all $A \in \mathcal{A}$, $B \in \mathcal{B}(\mathbb{R})$. Furthermore, there exists a function $\rho : S \times \mathcal{B}(\mathbb{R}) \to [0, \infty]$ such that

1. $\rho(s, \cdot)$ is a Lévy measure on $\mathcal{B}(\mathbb{R})$ for every $s \in S$,
2. $\rho(\cdot, B)$ is a Borel measurable function for every $B \in \mathcal{B}(\mathbb{R})$,
3. $\int_{S\times\mathbb{R}} h(s, x) F(ds, dx) = \int_S (\int_{\mathbb{R}} h(s, x)\rho(s, dx))\lambda(ds)$ for every $\sigma(\mathcal{A}) \times \mathcal{B}(\mathbb{R})$-measurable function $h : S \times \mathbb{R} \to [0, \infty]$. Under some restrictions regarding the behaviour at $\pm\infty$, this equality can be extended to real and complex-valued functions h.

Using the above notation, we can now rewrite the characteristic function of $\Lambda(A)$ (see [38, Proposition 2.4]):

$$\mathbb{E}\big(\exp\big(it\Lambda(A)\big)\big) = \exp\bigg(\int_A K(t, s)\lambda(ds)\bigg), \quad t \in \mathbb{R}, A \in \mathcal{A}, \tag{2.49}$$

where

$$K(t, s) = ita(s) - \frac{1}{2}t^2\sigma^2(s) + \int_{\mathbb{R}} \big(e^{itx} - 1 - it\tau(x)\big)\rho(s, dx),$$

where $a(s) = \frac{d\nu_0}{d\lambda}(s)$, $\sigma^2(s) = \frac{d\nu_1}{d\lambda}(s)$, and ρ is defined as above. Furthermore,

$$\big|a(s)\big| + \sigma^2(s) + \int_{\mathbb{R}} \min\big(1, x^2\big)\rho(s, dx) = 1, \quad \lambda\text{-a.e.}$$

A.3 Integration with Respect to a Lévy Basis

Next, we review the definition of a stochastic integral with respect to an infinitely divisible random measure Λ as defined in [38].

First, we define integration of a real simple function on S, which is given by $f = \sum_{j=1}^n x_j \mathbf{1}_{A_j}$ for disjoint $A_j \in \mathcal{A}$. Then, for every $A \in \sigma(\mathcal{A})$, the stochastic integral with respect to Λ is defined by

$$\int_A f\, d\Lambda = \sum_{j=1}^n x_j \Lambda(A \cap A_j).$$

The generalisation to general functions works as follows. We call a measurable function $f : (S, \sigma(\mathcal{A})) \to (\mathbb{R}, \mathcal{B}(\mathbb{R}))$ Λ-integrable if there exists a sequence of simple functions $(f_n)_{n\geq 1}$ such that $f_n \to f$ λ-a.e. and for every $A \in \sigma(\mathcal{A})$, the sequence

$(\int_A f_n\, d\Lambda)_{n\geq 1}$ converges in probability as $n \to \infty$. In that case, we define

$$\int_A f\, d\Lambda = \mathbb{P}\text{-}\lim_{n\to\infty} \int_A f_n\, d\Lambda.$$

The above integral is well defined in the sense that it does not depend on the approximating sequence $(f_n)_{n\geq 1}$.

A.4 Criteria for Integrability

Now we provide a characterisation of Λ-integrable functions. Necessary and sufficient conditions will depend on the characteristics given in the Lévy form of the characteristic function of Λ.

According to [38, Theorem 2.7], the integrability conditions are as follows.

Let $f : S \to \mathbb{R}$ be a $\sigma(\mathcal{A})$-measurable function. Then f is integrable w.r.t. Λ if and only if the following three conditions are satisfied:

1. $\int_S |U(f(s), s)|\lambda(ds) < \infty$,
2. $\int_S |f(s)|^2\sigma^2(s)\lambda(ds) < \infty$, and
3. $\int_S V_0(f(s), s)\lambda(ds) < \infty$, where

$$U(u, s) = ua(s) + \int_{\mathbb{R}} \big(\tau(xu) - u\tau(x)\big)\rho(s, dx),$$
$$V_0(u, s) = \int_{\mathbb{R}} \min\big(1, |xu|^2\big)\rho(s, dx).$$

Further, if f is integrable w.r.t. Λ, then the characteristic function of $\int_S f d\Lambda$ can be expressed as

$$\mathbb{E}\left(\exp\left(it\int_S f\, d\Lambda\right)\right) = \exp\left(ita_f - \frac{1}{2}t^2\sigma_f^2 + \int_{\mathbb{R}} (e^{itx} - 1 - it\tau(x))F_f(dx)\right),$$

where

$$a_f = \int_S U\big(f(s), s\big)\lambda(ds), \qquad \sigma_f = \int_S \big|f(s)\big|^2\sigma^2(s)\lambda(ds),$$

and

$$F_f(B) = F\big(\{(s,x) \in S \times \mathbb{R} : f(s)x \in B \setminus \{0\}\}\big), \quad B \in \mathcal{B}(\mathbb{R}).$$

References

1. S. Albeverio, J.-L. Wu, Euclidean random fields obtained by convolution from generalised white noise. J. Math. Phys. **36**(10), 5217–5245 (1995)
2. T. Andersen, V. Todorov, Realized volatility and multipower variation, in *Encyclopedia of Quantitative Finance*, ed. by R. Cont (Wiley, New York, 2010), pp. 1494–1500
3. V. Anh, C. Heyde, N. Leonenko, Dynamic models of long memory processes driven by Lévy noise. J. Appl. Probab. **39**, 730–747 (2002)
4. O. Barndorff-Nielsen, F. Benth, A. Veraart, Modelling electricity forward markets by ambit fields. Preprint (2010)
5. O. Barndorff-Nielsen, F. Benth, A. Veraart, Modelling energy spot prices by Lévy semistationary processes. Preprint (2010)
6. O.E. Barndorff-Nielsen, P. Blæsild, J. Schmiegel, A parsimonious and universal description of turbulent velocity increments. Eur. Phys. J. B **41**, 345–363 (2004)
7. O.E. Barndorff-Nielsen, J. Corcuera, E. Hedevang, M. Podolskij, J. Schmiegel, Multipower variations and variation ratios (2011, in preparation)
8. O.E. Barndorff-Nielsen, J. Corcuera, M. Podolskij, Multipower variation for Brownian semistationary processes. Working Paper (2009)
9. O.E. Barndorff-Nielsen, J. Corcuera, M. Podolskij, Second order difference multipower variations for BSS processes. Working Paper (2010)
10. O.E. Barndorff-Nielsen, S. Graversen, J. Jacod, M. Podolskij, N. Shephard, A central limit theorem for realised power and bipower variations of continuous semimartingales, in *From Stochastic Calculus to Mathematical Finance: Festschrift in Honour of A.N. Shiryaev*, ed. by Y.K. Kabanov, R. Liptser, J. Stoyanov (Springer, Heidelberg, 2006), pp. 33–68
11. O.E. Barndorff-Nielsen, S. Graversen, J. Jacod, N. Shephard, Limit theorems for bipower variation in financial econometrics. Econom. Theory **22**, 677–719 (2006)
12. O.E. Barndorff-Nielsen, J. Schmiegel, Lévy-based tempo-spatial modelling; with applications to turbulence. Usp. Mat. Nauk **59**, 65–91 (2004)
13. O.E. Barndorff-Nielsen, J. Schmiegel, Ambit processes; with applications to turbulence and cancer growth, in *Stochastic Analysis and Applications: The Abel Symposium 2005*, ed. by F. Benth, G. Di Nunno, T. Lindstrøm, B. Øksendal, T. Zhang (Springer, Heidelberg, 2007), pp. 93–124
14. O.E. Barndorff-Nielsen, J. Schmiegel, A stochastic differential equation framework for the timewise dynamics of turbulent velocities. Theory Probab. Appl. **52**, 372–388 (2008)
15. O.E. Barndorff-Nielsen, J. Schmiegel, Time change and universality in turbulence. Research Report 2007-8, Thiele Centre for Applied Mathematics in Natural Science, Aarhus University (2008)
16. O.E. Barndorff-Nielsen, J. Schmiegel, Time change, volatility and turbulence, in *Proceedings of the Workshop on Mathematical Control Theory and Finance, Lisbon 2007*, ed. by A. Sarychev, A. Shiryaev, M. Guerra, M. Grossinho (Springer, Berlin, 2008), pp. 29–53
17. O.E. Barndorff-Nielsen, J. Schmiegel, Brownian semistationary processes and volatility/intermittency, in *Advanced Financial Modelling*, ed. by H. Albrecher, W. Rungaldier, W. Schachermeyer. Radon Series on Computational and Applied Mathematics, vol. 8 (De Gruyter, Berlin, 2009), pp. 1–26
18. O.E. Barndorff-Nielsen, N. Shephard, Power and bipower variation with stochastic volatility and jumps. J. Financ. Econom. **2**, 1–37 (2004)
19. O.E. Barndorff-Nielsen, N. Shephard, Impact of jumps on returns and realised variances: econometric analysis of time-deformed Lévy processes. J. Econom. **131**, 217–252 (2006)
20. O.E. Barndorff-Nielsen, N. Shephard, *Financial Volatility in Continuous Time* (Cambridge University Press, Cambridge, 2011, to appear)
21. O.E. Barndorff-Nielsen, N. Shephard, M. Winkel, Limit theorems for multipower variation in the presence of jumps. Stoch. Process. Appl. **116**, 796–806 (2006)
22. A. Basse-O'Connor, S.-E. Graversen, J. Pedersen, A unified approach to stochastic integration on the real line. Preprint (2010)

23. F. Benth, The stochastic volatility model of Barndorff–Nielsen and Shephard in commodity markets. Math. Finance (2011, forthcoming)
24. B. Birnir, Turbulence of a unidirectional flow, in *Probability, Geometry and Integrable Systems*. Mathematical Sciences Research Institute Publications, vol. 55 (Cambridge University Press, Cambridge, 2008), pp. 29–52
25. S. Bochner, Integration von Funktionen, deren Werte die Elemente eines Vektorraumes sind. Fundam. Math. **20**, 262–276 (1933)
26. G. Da Prato, J. Zabczyk, *Stochastic Equations in Infinite Dimensions* (Cambridge University Press, Cambridge, 1992)
27. N. Dunford, J. Schwartz, *Linear Operators: Part 1: General Theory* (Wiley, New York, 1957)
28. I. Gelfand, N. Vilenkin, *Generalised Functions* (Academic Press, San Diego, 1964)
29. T. Hida, H.-H. Kuo, J. Potthoff, L. Streit, *White Noise—An Infinite Dimensional Calculus* (Kluwer Academic, Dordrecht, 1993)
30. S. Hikspoors, S. Jaimungal, Asymptotic pricing of commodity derivatives for stochastic volatility spot models. Appl. Math. Finance **15**(5–6), 449–467 (2008)
31. H. Holden, B. Øksendal, J. Ubøe, T. Zhang, *Stochastic Partial Differential Equations—A Modeling, White Noise Functional Approach*. Universitext (Springer, Berlin, 2009)
32. J. Jacod, Asymptotic properties of realized power variations and related functionals of semimartingales. Stoch. Process. Appl. **118**, 517–559 (2008)
33. J. Jacod, Statistics and High Frequency Data. Lecture Notes (2008)
34. D. Khoshnevisan, A primer on stochastic partial differential equations, in *A Minicourse on Stochastic Partial Differential Equations*. Lecture Notes in Mathematics, ed. by D. Khoshnevisan, F. Rassoul-Agha (Springer, Berlin, 2009), pp. 1–38
35. A. Løkka, B. Øksendal, F. Proske, Stochastic partial differential equations driven by Lévy space–time white noise. Ann. Appl. Probab. **14**, 1506–1528 (2004)
36. J. Pedersen, The Lévy–Itô decomposition of an independently scattered random measure. MaPhySto preprint MPS-RR 2003-2 (2003)
37. S. Peszat, J. Zabczyk, *Stochastic Partial Differential Equations with Lévy Noise* (Cambridge University Press, Cambridge, 2007)
38. B. Rajput, J. Rosinski, Spectral representation of infinitely divisible distributions. Probab. Theory Relat. Fields **82**, 451–487 (1989)
39. J. Rosinski, On the structure of stationary processes. Ann. Probab. **23**(3), 1163–1187 (1995)
40. J. Rosinski, Spectral representation of infinitely divisible processes and injectivity of the υ-transformation, in *5th International Conference on Lévy Processes: Theory and Applications*, Copenhagen (2007)
41. P. Samuelson, Proof that properly anticipated prices fluctuate randomly. Ind. Manage. Rev. **6**, 41–44 (1965)
42. K. Sato, *Lévy Processes and Infinitely Divisible Distributions* (Cambridge University Press, Cambridge, 1999)
43. J. Schmiegel, J. Cleve, H.C. Eggers, B.R. Pearson, M. Greiner, Stochastic energy-cascade model for $(1+1)$-dimensional fully developed turbulence. Phys. Lett. A **320**, 247–253 (2004)
44. D. Surgailis, J. Rosinski, V. Mandrekar, S. Cambanis, Stable mixed moving averages. Probab. Theory Relat. Fields **97**, 543–558 (1993)
45. A. Veraart, Inference for the jump part of quadratic variation of Itô semimartingales. Econom. Theory **26**(2), 331–368 (2010)
46. J. Walsh, An introduction to stochastic partial differential equations, in *Ecole d'Eté de Probabilités de Saint-Flour XIV—1984*, ed. by R. Carmona, H. Kesten, J. Walsh. Lecture Notes in Mathematics, vol. 1180 (Springer, Berlin, 1986)

Chapter 3
Fractional Processes as Models in Stochastic Finance

Christian Bender, Tommi Sottinen, and Esko Valkeila

Abstract We survey some new progress on the pricing models driven by fractional Brownian motion or mixed fractional Brownian motion. In particular, we give results on arbitrage opportunities, hedging, and option pricing in these models. We summarize some recent results on fractional Black & Scholes pricing model with transaction costs. We end the paper by giving some approximation results and indicating some open problems related to the paper.

Keywords Fractional Brownian motion · Arbitrage · Hedging in fractional models · Approximation of geometric fractional Brownian motion

Mathematics Subject Classification (2010) 91Gxx · 91B70 · 60G15 · 60H05

3.1 Introduction

The classical Black–Scholes pricing model is based on standard geometric Brownian motion. The log-returns of this model are independent and Gaussian. Various empirical studies on the statistical properties of log-returns show that the log-returns are not necessarily independent and also not Gaussian. One way to a more realistic modeling is to change the geometric Brownian motion to a geometric fractional Brownian motion: the dependence of the log-return increments can now be modeled

C. Bender
Department of Mathematics, Saarland University, P.O. Box 151150, 66041 Saarbrücken, Germany
e-mail: bender@math.uni-sb.de

T. Sottinen
Department of Mathematics and Statistics, University of Vaasa, P.O. Box 700, 65101 Vaasa, Finland
e-mail: tommi.sottinen@uwasa.fi

E. Valkeila (✉)
Department of Mathematics and Systems Analysis, Aalto University, P.O. Box 11100, 00076 Aalto, Finland
e-mail: esko.valkeila@tkk.fi

G. Di Nunno, B. Øksendal (eds.), *Advanced Mathematical Methods for Finance*,
DOI 10.1007/978-3-642-18412-3_3, © Springer-Verlag Berlin Heidelberg 2011

with the Hurst parameter of the fractional Brownian motion. But then the pricing model admits arbitrage possibilities with continuous trading and also with certain discrete type trading strategies.

The arbitrage possibilities with continuous trading depend on the notion of stochastic integration theory used in the definition of trading strategy. If these stochastic integrals are interpreted as Skorokhod integrals, then the arbitrage possibilities with continuous trading disappear. We will not consider this approach in what follows. For a summary of the results obtained in this area, we refer to two recent monographs on fractional Brownian motion [10] and [31]. If one uses Skorokhod integration theory, then one has several problems with the financial interpretation of these continuous trading strategies. We refer to the above two monographs for more details on these issues; see also [11] and [41] for the critical remarks concerning the Skorokhod approach from the finance point of view.

In this work we discuss the arbitrage possibilities in the fractional Black–Scholes pricing model and in the related mixed Brownian–fractional Brownian pricing model. Then we consider hedging of options in these models. The fractional Black–Scholes model admits strong arbitrage, and this implies that the initial wealth for the exact hedging strategy cannot be interpreted as a price of the option. But these replication results are interesting from the mathematical point of view. With proportional transaction costs the arbitrage possibilities disappear in the fractional Black–Scholes pricing model. Hence it is of some interest to know the hedging strategy without transaction costs. For the mixed Brownian–fractional Brownian pricing models, the arbitrage possibilities are not that obvious, and the hedging price can be sometimes interpreted as the price of the option. We shall review some recent results related to these questions.

One possibility to study the properties of the fractional Black–Scholes pricing model is to approximate it with simpler pricing models. We will present some results on the approximation at the end of this work.

3.2 Models and Notions of Arbitrage

Definition 3.1 The *fractional Brownian motion* (fBm) with *Hurst index* $H \in (0,1)$ is the centered Gaussian process $B = (B_t)_{t\in[0,T]}$ with $B_0 = 0$ and

$$\mathrm{Cov}[B_t, B_s] = \frac{1}{2}\left(t^{2H} + s^{2H} - |t-s|^{2H}\right).$$

Remark 3.2 Some well-known properties of the fBm are:

1. The fBm has stationary increments.
2. For $H = 1/2$, the fBm is the standard *Brownian motion* (Bm) W.
3. If $H \neq 1/2$, the fBm is not a semimartingale (see [14, Theorem 3.2] or Example 3.16 later).

4. If $H > 1/2$, the fBm has zero *quadratic variation* (QV) (see Definition 3.31 later). If $H < 1/2$, the QV is $+\infty$. For the Bm case $H = 1/2$, the QV is identity.
5. For $H > 1/2$, the fBm has *long-range dependence* (LRD) in the sense that

$$\rho(n) = \mathrm{Cov}[B_k - B_{k-1}, B_{k+n} - B_{k+n-1}]$$

satisfies

$$\sum_{n=1}^{\infty} |\rho(n)| = +\infty.$$

6. The paths of the fBm are a.s. Hölder continuous with index $H - \varepsilon$, where H is the Hurst index, and ε is any positive constant, but not Hölder continuous with index H. The first claim follows from the Kolmogorov–Chentsov criterion, and the second claim follows from the law of iterated logarithm of [2]:

$$\limsup_{t \downarrow 0} \frac{B_t}{t^H \sqrt{2 \ln \ln 1/t}} = 1 \quad \text{a.s.}$$

7. The fBm is *self-similar* with index H, i.e., for all $a > 0$,

$$\mathrm{Law}\big((a^H B_{at})_{t\in[0,T/a]}\big) = \mathrm{Law}\big((B_t)_{t\in[0,T]}\big).$$

Actually, the fBm is the (up to a multiplicative constant) unique centered Gaussian self-similar process with stationary increments.

In this survey we shall consider the following three discounted stock-price models in parallel:

Definition 3.3 Let $S = (S_t)_{t\in[0,T]}$ be the discounted stock price.

1. In the *Black–Scholes model* (BS model),

$$S_t = s_0 e^{\mu t + \sigma W_t - \frac{1}{2}\sigma^2 t},$$

where W is a Bm, and $s_0, \sigma > 0$, $\mu \in \mathbb{R}$.
2. In the *fractional Black–Scholes model* (fBS model),

$$S_t = s_0 e^{\mu t + \nu B_t},$$

where B is an fBm with $H \neq 1/2$, and $s_0, \nu > 0$, $\mu \in \mathbb{R}$.
3. In the *mixed fractional Black–Scholes model* (mfBS model),

$$S_t = s_0 e^{\mu t + \sigma W_t - \frac{1}{2}\sigma^2 t + \nu B_t},$$

where W is a Bm, B is an fBm with $H \neq 1/2$, W and B are independent, and $s_0, \sigma, \nu > 0$, $\mu \in \mathbb{R}$.

Remark 3.4 We shall often, for the sake of simplicity and without loss of any real generality, assume that $\mu = 0$ and $\sigma = \nu = s_0 = 1$.

Remark 3.5

1. The mfBS model is similar to the fBS model in the sense that they have essentially the same covariance structure. So, in particular, if $H > 1/2$, they both have LRD characterized by the Hurst index H.
2. The fBS model and the mfBS are different in the sense that the mfBS model has the same QV as the BS model (see Proposition 3.32) when $H > 1/2$. But the fBS model has zero QV for $H > 1/2$. So, while the fBS model and the mfBS model have the same statistical LRD property, the pricing in these models is different; in the fBS model, it might even be impossible.

We shall work, except in Sect. 3.7, in the canonical stochastic basis $(\Omega, \mathcal{F}, (\mathcal{F}_t)_{t\in[0,T]}, \mathbf{P})$. So, $\Omega = C^+_{s_0}([0,T])$ is the space of positive continuous functions over $[0,T]$ starting from s_0, and the stock price is the coordinate process $S_t(\omega) = \omega_t$. The filtration $(\mathcal{F}_t)_{t\in[0,T]}$ is generated by the stock price S and augmented to satisfy the usual conditions of completeness and right-continuity, $\mathcal{F} = \mathcal{F}_T$, and the measure $\mathbf{P}$ is defined by the models in Definition 3.3.

Definition 3.6 A *portfolio*, or trading strategy, is an adapted process $\boldsymbol{\Phi} = (\boldsymbol{\Phi}_t)_{t\in[0,T]} = (\Phi^0_t, \Phi_t)_{t\in[0,T]}$, where Φ^0_t denotes the number of bonds, and Φ_t denotes the number of shares owned by the investor at time t. The *value* of the portfolio $\boldsymbol{\Phi}$ at time t is

$$V_t(\boldsymbol{\Phi}) = \Phi^0_t + \Phi_t S_t,$$

since everything is discounted by the bond. The class of portfolios is denoted by $\mathcal{A}$.

There are some slightly different versions of the notion of free lunch, or arbitrage, that in discrete time would make little or no difference. However, in continuous time the issue of arbitrage is quite subtle as can be seen from the fundamental theorem of asset pricing by Delbaen and Schachermayer [18, Theorem 1.1]. We use the following definitions:

Definition 3.7

1. A portfolio $\boldsymbol{\Phi}$ is *arbitrage* if $V_0(\boldsymbol{\Phi}) = 0$, $V_T(\boldsymbol{\Phi}) \geq 0$ a.s., and $\mathbf{P}[V_T(\boldsymbol{\Phi}) > 0] > 0$.
2. A portfolio $\boldsymbol{\Phi}$ is *strong arbitrage* if $V_0(\boldsymbol{\Phi}) = 0$, and there exists a constant $c > 0$ such that $V_T(\boldsymbol{\Phi}) \geq c$ a.s.
3. A sequence of portfolios $(\boldsymbol{\Phi}^n)_{n\in\mathbb{N}}$ is *approximate arbitrage* if $V_0(\boldsymbol{\Phi}^n) = 0$ for all n and $V^\infty_T = \lim_{n\to\infty} V_T(\boldsymbol{\Phi}^n)$ exists in probability, $V^\infty_T \geq 0$ a.s., and $\mathbf{P}[V^\infty_T > 0] > 0$.
4. A sequence of portfolios is *strong approximate arbitrage* if it is approximate arbitrage and there exists a constant $c > 0$ such that $V^\infty_T \geq c$ a.s.
5. A sequence of portfolios $(\boldsymbol{\Phi}^n)_{n\in\mathbb{N}}$ is *free lunch with vanishing risk* if it is approximate arbitrage and

$$\lim_{n\to\infty} \operatorname*{ess\,sup}_{\omega\in\Omega} \left| V_T(\boldsymbol{\Phi}^n)(\omega) \mathbf{1}_{\{V_T(\boldsymbol{\Phi}^n)<0\}} \right| = 0.$$

3.3 Trading with (Almost) Simple Strategies

In this section we consider noncontinuous trading in continuous time. The basic classes of portfolios are:

Definition 3.8

1. A portfolio is *simple* if there exists a finite number of stopping times $0 \le \tau_0 \le \dots \le \tau_n \le T$ such that the portfolio is constant on $(\tau_k, \tau_{k+1}]$, i.e.,

$$\Phi_t = \sum_{k=0}^{n-1} \phi_{\tau_k} \mathbf{1}_{(\tau_k, \tau_{k+1}]}(t),$$

   where $\phi_{\tau_k} \in \mathcal{F}_{\tau_k}$, and an analogous expression holds for Φ^0. The class of simple portfolios is denoted by $\mathcal{A}^{\mathrm{si}}$.
2. A portfolio is *almost simple* if there exists a sequence $(\tau_k)_{k\in\mathbb{N}}$ of nondecreasing $[0, T]$-valued stopping times such that $\mathbf{P}[\exists_{k\in\mathbb{N}} \tau_k = T] = 1$ and the portfolio is constant on $(\tau_k, \tau_{k+1}]$, i.e.,

$$\Phi_t = \sum_{k=0}^{N-1} \phi_{\tau_k} \mathbf{1}_{(\tau_k, \tau_{k+1}]}(t),$$

   where $\phi_{\tau_k} \in \mathcal{F}_{\tau_k}$, and N is an a.s. $\mathbb{N}$-valued random variable, and an analogous expression holds for Φ^0. The class of almost simple portfolios is denoted by $\mathcal{A}^{\mathrm{as}}$.

Remark 3.9 Obviously $\mathcal{A}^{\mathrm{si}} \subset \mathcal{A}^{\mathrm{as}}$, and the inclusion is proper. Also, note that for every ω, the position Φ is changed only finitely many times. The difference between $\mathcal{A}^{\mathrm{si}}$ and $\mathcal{A}^{\mathrm{as}}$ is that in $\mathcal{A}^{\mathrm{si}}$ the number of readjustments is bounded in Ω, while in $\mathcal{A}^{\mathrm{as}}$ the number of readjustments is generally unbounded.

The notion of self-financing is obvious with (almost) simple strategies:

Definition 3.10 A portfolio $\mathbf{\Phi} \in \mathcal{A}^{\mathrm{as}}$ is *self-financing* if, for all k, its value satisfies

$$V_{\tau_{k+1}}(\mathbf{\Phi}) - V_{\tau_k}(\mathbf{\Phi}) = \Phi_{\tau_{k+1}}(S_{\tau_{k+1}} - S_{\tau_k}),$$

or, equivalently, the *budget constraint*

$$\Phi^0_{\tau_{k+1}} + \Phi_{\tau_{k+1}} S_{\tau_k} = \Phi^0_{\tau_k} + \Phi_{\tau_k} S_{\tau_k}$$

holds for every readjustment time τ_k of the portfolio.

Henceforth, we shall always assume that the portfolios are self-financing.

Theorem 3.11 *In the BS model there is*

1. *no arbitrage in the class* $\mathcal{A}^{\mathrm{si}}$,

2. *strong approximate arbitrage in the class* $\mathcal{A}^{\mathrm{si}}$,
3. *strong arbitrage in the class* $\mathcal{A}^{\mathrm{as}}$.

Proof The claim (i) follows from the fact that the geometric Bm remains a martingale in the subfiltration $(\mathcal{F}_{\tau_k})_{k\le n}$, and thus the claim reduces to discrete-time considerations. Claims (ii) and (iii) follow from the doubling strategy of Example 3.12 below. □

Example 3.12 Consider, without loss of generality, the risk-neutral normalized BS model

$$S_t = s_0 e^{W_t - \frac{1}{2}t}.$$

Let $t_k = T(1-2^{-k})$, $c_k = e^{\sqrt{T2^{-k}}-\frac{1}{2}T2^{-k}} - 1$, and

$$\tau = \inf\left\{t_k;\ \frac{S_{t_k} - S_{t_{k-1}}}{S_{t_{k-1}}} \ge c_k\right\} = \inf\left\{t_k;\ \frac{W_{t_k} - W_{t_{k-1}}}{\sqrt{t_k - t_{k-1}}} \ge 1\right\}.$$

Define a self-financing almost simple strategy by setting $V_0(\boldsymbol{\Phi}) = 0$ and

$$\Phi_t = \sum_{k=0}^{\infty} \phi_{t_k} \mathbf{1}_{(t_k\wedge\tau, t_{k+1}\wedge\tau]}(t),$$

where, for $k = 0, 1, \dots,$

$$\phi_{t_k} = \frac{1 - V_{t_k}(\boldsymbol{\Phi})}{S_{t_k} c_{k+1}}.$$

Now, the c_k's were chosen in such a way that $\mathbf{P}[\tau < T] = 1$. So, $\tau = t_N$ a.s. for some random $N \in \mathbb{N}$, and

$$\begin{aligned} V_\tau(\boldsymbol{\Phi}) &= V_{t_{N-1}}(\boldsymbol{\Phi}) + \phi_{t_{N-1}}(S_{t_N} - S_{t_{N-1}}) \\ &\ge V_{t_{N-1}}(\boldsymbol{\Phi}) + \frac{1 - V_{t_{N-1}}(\boldsymbol{\Phi})}{S_{t_{N-1}} c_N} S_{t_{N-1}} c_N \\ &= 1 \end{aligned}$$

a.s. So, we have strong arbitrage in the class $\mathcal{A}^{\mathrm{as}}$. Also, by setting

$$\Phi_t^n = \sum_{k=0}^{n} \phi_{t_k} \mathbf{1}_{(t_k\wedge\tau, t_{k+1}\wedge\tau]}(t),$$

we see that we have strong approximate arbitrage in the class $\mathcal{A}^{\mathrm{si}}$.

In order to exclude doubling-type arbitrage strategies like Example 3.12, one traditionally assumes that the value of the portfolio is bounded from below:

Definition 3.13 A portfolio is *nds-admissible* (*n*o *d*oubling *s*trategies) if there exists a constant $a \geq 0$ such that

$$\inf_{t\in[0,T]} V_t(\Phi) \geq -a \quad \text{a.s.}$$

The class of nds-admissible portfolios is denoted by $\mathcal{A}^{\mathrm{nds}}$.

Remark 3.14 The sell-short-and-hold strategy $\Phi = -\mathbf{1}_{[0,T]} \in \mathcal{A}^{\mathrm{si}} \setminus \mathcal{A}^{\mathrm{nds}}$.

By Delbaen and Schachermayer [18, Theorem 1.1] the BS model has no free lunch with vanishing risk, and hence no arbitrage, in the class $\mathcal{A}^{\mathrm{nds}}$. The situation for fBS model is different:

Theorem 3.15 *For $H \neq \frac{1}{2}$, in the fBS model there is*

1. *free lunch with vanishing risk in the class $\mathcal{A}^{\mathrm{si}} \cap \mathcal{A}^{\mathrm{nds}}$,*
2. *strong arbitrage in the class $\mathcal{A}^{\mathrm{as}} \cap \mathcal{A}^{\mathrm{nds}}$.*

Proof The claims follow from Cheridito [14, Theorems 3.1 and 3.2]. □

Cheridito [14] constructed his arbitrage opportunities by using the trivial QV of the fBS model (0 for $H > 1/2$ and $+\infty$ for $H < 1/2$). So, his constructions do not work in the mfBS model. Also, Cheridito's arbitrage strategies are rather implicit in the sense that the stopping times they use are not constructed explicitly.

Let us also note that probably the first one to construct arbitrage in the fractional (Bachelier) model was Rogers [36]. His arbitrage was a doubling-type strategy similar to that of Example 3.12 with the twist that he avoided investing on "bad intervals" $(t_k, t_{k+1}]$ where the stock price was likely to fall. This was possible due to the memory of the fractional Brownian motion when $H \neq 1/2$. With this avoidance he was able to keep the value of his doubling strategy from falling below any predefined negative level, thus constructing an arbitrage opportunity in the class $\mathcal{A}^{\mathrm{as}} \cap \mathcal{A}^{\mathrm{nds}}$. Let us note that Rogers [36] used a representation of the fBm starting from $-\infty$. So, he used memory from time $-\infty$, while Cheridito [14] and we use memory only from time 0.

The following very explicit Example 3.16, a variant of [9, Example 7], constructs approximate arbitrage in the fBS model for $H \neq 1/2$ and in the mfBS model for $H \in (1/2, 3/4)$, where the approximating strategies are from the class $\mathcal{A}^{\mathrm{si}}$. The construction follows an easy intuition: Due to the memory of the fBm, the stock price tends to increase (decrease) in the future if it already increased (decreased) in the past if $H > 1/2$, and the other way around if $H < 1/2$. Example 3.16 also shows that forward integrals with respect to fBm with $H \neq 1/2$ and mixed fBm with $H \in (1/2, 3/4)$ are not continuous in terms of the integrands. Thus, due to the Dellacherie–Meyer–Mokobodzky–Bichteler theorem, this proves that the fBm is not a semimartingale and that the mixed fBm is not a semimartingale when $H \in (1/2, 3/4)$.

Example 3.16

1. Consider the fBS model
$$S_t = e^{B_t},$$
where $H \neq 1/2$. Let $t_k^n = Tk/n$, $\alpha_H = 1$ if $H > 1/2$, $\alpha_H = -1$ if $H < 1/2$, and
$$\Phi_t^n = \alpha_H n^{2H-1} \sum_{k=1}^{n-1} \frac{\log S_{t_k^n} - \log S_{t_{k-1}^n}}{S_{t_k^n}} \mathbf{1}_{(t_k^n, t_{k+1}^n]}(t).$$
Then, assuming that $V_0(\boldsymbol{\Phi}^n) = 0$ and applying Taylor's theorem, we have
$$\begin{aligned} V_T(\boldsymbol{\Phi}^n) &= \alpha_H n^{2H-1} \sum_{k=1}^{n-1} (B_{t_k^n} - B_{t_{k-1}^n}) \left(\frac{S_{t_{k+1}^n}}{S_{t_k^n}} - 1 \right) \\ &= \alpha_H n^{2H-1} \sum_{k=1}^{n-1} (B_{t_k^n} - B_{t_{k-1}^n})(B_{t_{k+1}^n} - B_{t_k^n}) \\ &\quad + \alpha_H n^{2H-1} \sum_{k=1}^{n-1} (B_{t_k^n} - B_{t_{k-1}^n}) e^{\xi_k^n} (B_{t_{k+1}^n} - B_{t_k^n})^2, \end{aligned}$$
where $|\xi_k^n| \in [0, |B_{t_{k+1}^n} - B_{t_k^n}|]$. Now the first term tends to $T^{2H}|2^{2H-1} - 1|$ in probability by [29, Theorem 9.5.2], and the second one vanishes as n goes to infinity using the Hölder continuity of fBm B.
2. Consider the mfBS model
$$S_t = e^{W_t - \frac{1}{2}t + B_t},$$
where $H \in (1/2, 3/4)$. The strategy of part (i) will still be strong approximate arbitrage. Indeed, after a Taylor expansion as above, we basically have to deal with the sum of the four terms
$$\int_0^T K_t^n \, dW_t, \qquad \int_0^T L_t^n \, dW_t, \qquad \int_0^T K_t^n \, dB_t, \qquad \int_0^T L_t^n \, dB_t, \tag{3.1}$$
where
$$\begin{aligned} K_t^n &= n^{2H-1} \sum_{k=1}^{n-1} \mathbf{1}_{(t_k^n, t_{k+1}^n]}(t)(W_{t_k} - W_{t_{k-1}}), \\ L_t^n &= n^{2H-1} \sum_{k=1}^{n-1} \mathbf{1}_{(t_k^n, t_{k+1}^n]}(t)(B_{t_k} - B_{t_{k-1}}), \end{aligned}$$
and the integrals are just shorthand notation for the forward Riemann sums. Note that K^n and L^n converge to zero uniformly in probability by the Hölder continuity of (fractional) Brownian motion for $H < 3/4$. Therefore, the first two terms

in (3.1) will tend to zero in probability by the Dellacherie–Meyer–Mokobodzky–Bichteler theorem [35, Theorem II.11]. The third term will tend to zero in probability because of the independence of W and B. The fourth term will go to $T^{2H}(2^{2H-1}-1)$ in probability by part (i) of this example. We also note that $\Phi^n S$ inherits the unform convergence in probability to zero from $K^n + L^n$. Hence the amount of money invested in the stock converges to zero as n tends to infinity.

For the mfBS model, the situation is the following:

Theorem 3.17 *For the mfBS model, there is*

1. *strong approximate arbitrage in the class* $\mathcal{A}^{\mathrm{si}}$ *if* $H \in (1/2, 3/4)$,
2. *no free lunch with vanishing risk in the class* $\mathcal{A}^{\mathrm{nds}}$ *if* $H \in (3/4, 1)$.

Proof Claim (i) follows from Example 3.16(ii). Claim (ii) follows from Cheridito [13]. Indeed, in [13] it is shown that in this case the mixed fBm is actually equivalent in law to a Bm. □

Although the situation is bad arbitrage-wise for the fBS and the mfBS models in the class $\mathcal{A}^{\mathrm{si}} \cap \mathcal{A}^{\mathrm{nds}}$, Cheridito [14] showed that there is no arbitrage in the fBS model if there must be a fixed positive time between the readjustments of the portfolio (later arbitrage in this class was studied by Jarrow et al. [27]):

Definition 3.18 Let $\mathcal{T}$ be a class of finite sequences of nondecreasing stopping times $\boldsymbol{\tau} = (0 \le \tau_0 \le \cdots \le \tau_n \le T)$ satisfying some additional conditions, which can be specified as in Proposition 3.19 or Definition 3.20 below. A simple portfolio Φ is $\mathcal{T}$*-simple* if it is of the form

$$\Phi_t = \sum_{k=0}^{n-1} \phi_{\tau_k} \mathbf{1}_{(\tau_k, \tau_{k+1}]}(t),$$

where $\phi_{\tau_k} \in \mathcal{F}_{\tau_k}$, $\boldsymbol{\tau} = (\tau_k)_{k=0}^n \in \mathcal{T}$. The class of $\mathcal{T}$-simple strategies is denoted by $\mathcal{A}^{\mathcal{T}-si}$.

Proposition 3.19 *Let* $\mathcal{T}_h = \bigcup_{h>0}\{\boldsymbol{\tau};\, \tau_{k+1} - \tau_k \ge h\}$.
Then there is no arbitrage in the fBS model in the class $\mathcal{A}^{\mathcal{T}_h-\mathrm{si}}$.

Proof The claim is Cheridito's [14, Theorem 4.3]. □

3.4 Trading with Delay-Simple Strategies

While Proposition 3.19 seems promising, the class $\mathcal{A}^{\mathcal{T}_h-\mathrm{si}}$ is more restrictive than it may appear at a first sight. Indeed, e.g., the archetypical stopping time $\tau =$

$\inf\{t \geq 0; S_t - S_0 \geq 1\}$ does not belong to $\mathcal{T}_h$ if S is the geometric Bm. To remedy this problem, we propose the following more general class of stopping times and simple strategies:

Definition 3.20

1. For any stopping time τ, let $C^+_{S_\tau}([\tau, T])$ be the random space of continuous positive paths $\omega = (\omega_t)_{t \in [\tau(\omega), T]}$ with $\omega_{\tau(\omega)} = S_{\tau(\omega)}(\omega)$ fixed. A sequence of nondecreasing stopping times $\boldsymbol{\tau} = (\tau_k)_{k=0}^n$ satisfies the *delay* property if for all τ_k, there are an $\mathcal{F}_{\tau_k}$-measurable open *delay set* $U_k \subset C^+_{S_{\tau_k}}([\tau_k, T])$ and an $\mathcal{F}_{\tau_k}$-measurable a.s. positive random variable ε_k such that $\tau_{k+1} - \tau_k \geq \varepsilon_k$ in the set $U_k \cap \{\tau_{k+1} > \tau_k\}$. The set of nondecreasing sequences of stopping times satisfying the delay property is denoted by $\mathcal{T}_{\mathrm{de}}$.
2. The class of *delay-simple* strategies is $\mathcal{A}^{\mathcal{T}_{\mathrm{de}}-\mathrm{si}}$.

Theorem 3.21 *All the models BS, fBS, and mfBS are free of arbitrage in the class* $\mathcal{A}^{\mathcal{T}_{\mathrm{de}}-\mathrm{si}}$.

Before we prove Theorem 3.21, we discuss the class of delay-simple strategies.

Remark 3.22 The difference between the classes $\mathcal{T}_h$ and $\mathcal{T}_{\mathrm{de}}$ is that in $\mathcal{T}_h$ there is a fixed delay $h > 0$ between the stopping times, while in $\mathcal{T}_{\mathrm{de}}$ the delay between the stopping times depend on the path one is observing: If there is a delay on the path you are observing, then there is also a delay on all the paths that are close enough of the path that one is observing.

Obviously $\mathcal{T}_h \subset \mathcal{T}_{\mathrm{de}}$, and the inclusion is proper.

Example 3.23 The following sequences of stopping times are in $\mathcal{T}_{\mathrm{de}}$:

1.

$$\tau_{k+1} = \inf\{t > \tau_k;\, S_t - S_{\tau_k} \geq b^k_t\},$$

where b^k's are continuous function with $b^k_{\tau_k} > 0$. Indeed, take

$$U_k = \{\omega;\, S_t(\omega) < \omega^0_t \text{ for all } t \in [\tau_k, T]\},$$

where ω^0 is some path for which $\tau_{k+1}(\omega^0) > \tau_k(\omega^0)$.

2.

$$\tau_{k+1} = \inf\{t > \tau_k;\, S_t - S_{\tau_k} \leq a^k_t\},$$

where a^k's are continuous function with $a^k_{\tau_k} < 0$. Indeed, take

$$U_k = \{\omega;\, S_t(\omega) > \omega^0_t \text{ for all } t \in [\tau_k, T]\},$$

where ω^0 is some path for which $\tau_{k+1}(\omega^0) > \tau_k(\omega^0)$.

3. One can show that

$$\tau_{k+1} = \inf\{t > \tau_k;\, S_t - S_{\tau_k} \le a_t^k \text{ or } S_t - S_{\tau_k} \ge b_t^k\},$$

where a^k's and b^k's are continuous with $a^k_{\tau_k} < 0 < b^k_{\tau_k}$, is in $\mathcal{T}_{\mathrm{de}}$ (see [9, Example 6(i)]).

Example 3.24 We construct a stopping time τ in the fractional Wiener space such that $(\tau_0, \tau_1) := (0, \tau)$ is not in $\mathcal{T}_{\mathrm{de}}$: $\tau = \inf\{t > 0;\, e^{B_t + t^a} = 1\}$. By the law of iterated logarithm, $\tau > 0$ a.s. if $a < H$. However, any open set $U \subset C^+_{S_0}([0, T])$ contains sequences (ω^n) for which $\tau(\omega^n) \to 0$.

Definition 3.25 A process S satisfies the *$\mathcal{T}$-conditional up'n'down* property ($\mathcal{T}$-CUD) if, for all $\boldsymbol{\tau} \in \mathcal{T}$ and all k, either

$$\mathbf{P}[S_{\tau_{k+1}} > S_{\tau_k} | \mathcal{F}_{\tau_k}] > 0 \quad \text{and} \quad \mathbf{P}[S_{\tau_{k+1}} < S_{\tau_k} | \mathcal{F}_{\tau_k}] > 0$$

or

$$\mathbf{P}[S_{\tau_{k+1}} = S_{\tau_k} | \mathcal{F}_{\tau_k}] = 1.$$

If there are no additional restrictions for $\mathcal{T}$ (except that it contains nondecreasing finite sequences of stopping times), we say simply that S satisfies CUD.

The following lemma can be proved analogously to [27, Lemma 1].

Lemma 3.26 *There is no arbitrage in the class $\mathcal{A}^{\mathcal{T}-\mathrm{si}}$ if and only if the model satisfies $\mathcal{T}$-CUD.*

CUD is related to the support of the stock-price model S. Another support-related condition we need is:

Definition 3.27 A continuous positive process S has *conditional full support* (CFS) if, for all stopping times τ,

$$\operatorname{supp} \mathbf{P}[S \in \cdot \,| \mathcal{F}_\tau] = C^+_{S_\tau}\big([\tau, T]\big) \quad \text{a.s.}$$

Remark 3.28

1. CFS is equivalent to the *conditional small-ball property*: For every stopping time τ, all the open balls contained in $C^+_{S_\tau}([\tau, T])$ have a.s. positive regular conditional probability, i.e.,

$$\mathbf{P}\Big[\sup_{t \in [\tau, T]} \big|S_t - S_t^0\big| \le \varepsilon \Big| \mathcal{F}_\tau\Big] > 0$$

a.s. for all $S^0 \in C^+_{S_\tau}([\tau, T])$ and $\mathcal{F}_\tau$-measurable a.s. positive random variables ε. For a proof of this, see Pakkanen [34, Lemma 2.3].

2. By Pakkanen [34, Lemma 2.10] a process X has CFS with respect to its own filtration $\mathcal{F}_t^X = \sigma(X_s, 0 \le s \le t)$ if and only if it has the CFS with respect to the augmentation of $\mathcal{F}_t^X$.
3. By Guasoni et al. [25, Lemma 2.9] one can replace the stopping times with deterministic times in Definition 3.27.
4. CFS is neither necessary nor sufficient for no-arbitrage in $\mathcal{A}^{\mathrm{si}}$. On the one hand, any bounded martingale satisfies no-arbitrage in $\mathcal{A}^{\mathrm{si}}$ but violates CFS. On the other hand, $W_t + t^a$, $a < 1/2$, has arbitrage in $\mathcal{A}^{\mathrm{si}}$ by the law of the iterated logarithm but satisfies CFS. However, CFS is sufficient for absence of arbitrage with the class $\mathcal{A}^{\mathcal{T}_{\mathrm{de}}-\mathrm{si}}$. This will be shown in the next lemma.

Lemma 3.29 *Suppose that S has CFS. Then there is no arbitrage in the model S in the class $\mathcal{A}^{\mathcal{T}_{\mathrm{de}}-\mathrm{si}}$.*

Proof By Lemma 3.26 we need to show that the $\mathcal{T}_{\mathrm{de}}$-CUD is satisfied. If $\tau_{k+1} = \tau_k$, this is certainly the case. So, we can assume that $\tau_{k+1} > \tau_k$.

We show that $\mathbf{P}[S_{\tau_{k+1}} > S_{\tau_k} | \mathcal{F}_{\tau_k}] > 0$ a.s.; the proof for $\mathbf{P}[S_{\tau_{k+1}} < S_{\tau_k} | \mathcal{F}_{\tau_k}] > 0$ a.s. follows analogously.

By the CFS it is enough to show that $\{S_{\tau_{k+1}} > S_{\tau_k}\} \subset C^+_{S_{\tau_k}}([\tau_k, T])$ contains an open set. Let U_k be an ε_k-delay set for τ_k, i.e., $U \subset C^+_{S_{\tau_k}}([\tau_k, T])$ is open, and $\tau_{k+1} - \tau_k \ge \varepsilon_k$ on U_k. We first assume that U_k contains a strictly increasing path ω^0 on $[\tau_k, T]$. Denote by $B_{\omega^0}(\eta_k)$ the open η_k-ball around ω_0 in $C^+_{S_{\tau_k}}([\tau_k, T])$. Choosing η_k sufficiently small, we have $B_{\omega^0}(\eta_k) \subset U_k$ (because U_k is open) and $\omega^0_{\tau_k+\varepsilon_k} > \omega^0_{\tau_k} + \eta_k$ (because ω^0 is strictly increasing). Hence, for every $\omega \in B_{\omega^0}(\eta_k)$,

$$\begin{aligned}
\omega_{\tau_{k+1}(\omega)} - S_{\tau_k} &> \omega^0_{\tau_{k+1}(\omega)} - \eta_k - S_{\tau_k} \\
&\ge \omega^0_{\tau_k+\varepsilon_k} - S_{\tau_k} - \eta_k \\
&= \omega^0_{\tau_k+\varepsilon_k} - \omega^0_{\tau_k} - \eta_k \\
&> 0.
\end{aligned}$$

So, $B_{\omega^0}(\eta_k) \subset \{S_{\tau_{k+1}} > S_{\tau_k}\}$, and the claim follows if U_k contains a strictly increasing path. If U_k does not contain a strictly increasing path, we proceed as follows. Being an open set in $C^+_{S_{\tau_k}}([\tau_k, T])$, U_k contains paths that are strictly increasing on a small enough interval $[\tau_k, \tau_k + 2\eta_k]$. Hence, there is a strictly increasing path ω^0 and an open ball B_k around ω^0 in $C^+_{S_{\tau_k}}([\tau_k, T])$ such that any $\omega \in B_k$ coincides with some path $\bar{\omega} \in U_k$ on the segment $[\tau_k, \tau_k + \eta_k]$. Hence, $\tau_{k+1}(\omega) - \tau_k \ge (\tau_{k+1}(\bar{\omega}) - \tau_k) \wedge \eta_k \ge \varepsilon_k \wedge \eta_k =: \varepsilon^0_k$ for every $\omega \in B_k$. Therefore, B_k is an ε^0_k-delay set which contains a strictly increasing path, and so the first case applies. □

Proof of Theorem 3.21 By [22, Theorem 2.1] the Bm, the fBm, and the mixed fBm all have CFS in the space $C_0([0, T])$ (with respect to the filtration generated by the respective process), since their spectral measures have heavy enough tails. For a nice proof that fBm has CFS, see also [15]. So, the BS, the fBS, and the mfBS models all have CFS in $C_{s_0}^+([0,T])$, because with any homeomorphism η on $C_0([0,T])$, the mapping $\omega \mapsto s_0 e^{\omega+\eta}$ is a homeomorphism between $C_0([0,T])$ and $C_{s_0}^+([0,T])$. So, the claim follows from Lemma 3.29. □

3.5 Continuous Trading

While the previous sections were concerned with trading strategies which can be readjusted finitely many times only, we will now admit continuous readjustment of the portfolio. A natural generalization of the self-financing property in Definition 3.10 can be given in terms of forward integrals. Here we stick to the simplest possible definition of forward integrals due to [20] but refer to [37] for the general theory.

Definition 3.30 Let $t \le T$, and let $X = (X_s)_{s\in[0,T]}$ be a continuous process. The *forward integral* of a process $Y = (Y_s)_{s\in[0,T]}$ with respect to X (along dyadic partitions) is

$$\int_0^t Y_s\, dX_s := \lim_{n\to\infty} \sum_{\substack{i=0,\ldots,2^n-1,\\ Ti/2^n \le t}} Y_{Ti/2^n}(X_{T(i+1)/2^n} - X_{Ti/2^n})$$

if the limit exists **P**-almost surely.

If necessary, we interpret the forward integral in an improper sense at $t = T$. Itô's formula for the forward integral depends on the quadratic variation of the integrator.

Definition 3.31 The pathwise *quadratic variation* (QV) of a stochastic process (along dyadic partitions) is

$$\langle X\rangle_t := \lim_{n\to\infty} \sum_{\substack{i=0,\ldots,2^n-1,\\ Ti/2^n \le t}} (X_{T(i+1)/2^n} - X_{Ti/2^n})^2,$$

if, for all $t \le T$, the limit exists **P**-almost surely.

Proposition 3.32

1. *For the fBS model and the mfBS model with* $H < 1/2$, *the limit in Definition* 3.31 *diverges to infinity.*
2. *For the fBS model with* $H > 1/2$, *the QV is constant* 0.
3. *The QV in the BS model and in the mfBS model with* $H > 1/2$ *is given by*

$$d\langle S\rangle_t = \sigma^2 S_t^2\, dt.$$

Proof It is well known that Bm has the identity map as QV. Moreover, fBm has zero quadratic variation for $H > 1/2$ and infinite quadratic variation for $H < 1/2$, see, e.g., [10], Chap. 1.8. By independence, the QV of the mixed fBm is the sum of the QV of Bm and fBm. Finally, the stock models under consideration are $\mathcal{C}^1$-functions of these processes (up to a finite variation drift), and so a result by [20], p. 148, applies. □

The following Itô formula for the forward integrals with continuous integrator can be derived by a Taylor expansion as usual, see [20].

Lemma 3.33 *Let X be a continuous process with continuous QV. Suppose that $f \in \mathcal{C}^{1,2}([0,T] \times \mathbb{R})$. Then, for $0 \le t \le T$,*

$$f(t, X_t) = f(0, X_0) + \int_0^t \frac{\partial}{\partial t} f(u, X_u)\, du + \int_0^t \frac{\partial}{\partial x} f(u, X_u)\, dX_u + \frac{1}{2} \int_0^t \frac{\partial^2}{\partial x^2} f(u, X_u)\, d\langle X \rangle_u.$$

In particular, this formula implies that the forward integral on the right-hand side exists and has a continuous modification.

After this short digression on forward integrals, we can introduce several classes of portfolios.

Definition 3.34

1. A portfolio is *self-financing* if, for all $0 \le t \le T$,

$$V_t(\boldsymbol{\Phi}) = V_0(\boldsymbol{\Phi}) + \int_0^t \Phi_t\, dS_t.$$

   The class of self-financing portfolios (without any extra constraints) is denoted by $\mathcal{A}$.
2. A self-financing portfolio is called a *spot strategy* if $\Phi_t = \varphi(t, S_t)$ for some deterministic function φ, i.e., the number of shares held in the stock depends on time and the spot only. We apply the notation $\mathcal{A}^{\text{spot}}$ for the class of spot strategies.

The following theorem discusses arbitrage with spot strategies in the BS model. It again illustrates some subtleties of arbitrage theory in continuous time, even for models which admit an equivalent martingale measure. As in the case of almost simple strategies, arbitrage is possible, if arbitrarily large losses are allowed prior to maturity.

Theorem 3.35

1. *In the BS model there is strong arbitrage in the class $\mathcal{A}^{\text{spot}}$.*
2. *In the BS model there is no free lunch with vanishing risk in the class $\mathcal{A} \cap \mathcal{A}^{\text{nds}}$.*

Proof (i) We give a direct construction making use of Itô's formula (Lemma 3.33) and the QV of the Black–Scholes model. Without loss of generality, we assume that $\sigma = 1$ and $\mu = 0$. Let

$$\Phi_t = -\frac{\frac{\partial}{\partial x} v(t, W_t)}{S_t},$$

where $v(t, x)$ is the heat kernel

$$v(t, x) = \frac{1}{\sqrt{2\pi(T-t)}} e^{-\frac{1}{2}\frac{x^2}{T-t}}.$$

By Lemma 3.33, applied to the Bm W,

$$V_T(\Phi) = \int_0^T \Phi_t \, dS_t = -\int_0^T \frac{\partial}{\partial x} v(t, W_t) \, dW_t = v(0,0) - v(T, W_T) = \frac{1}{\sqrt{2\pi T}}$$

almost surely. So, we have constructed a strong arbitrage, and it belongs to the class $\mathcal{A}^{\text{spot}}$, because the Bm W is a deterministic function of time and the Black–Scholes stock S.

(ii) The BS model has an equivalent martingale measure. Hence the fundamental theorem of asset pricing [17] ensures that there is no free lunch with vanishing risk with nds-admissible, self-financing strategies. □

The construction of the "doubling" arbitrage in the previous theorem only relied on the quadratic variation structure of the model. In the pure fractional BS model with $H > 1/2$, the QV is constant zero. This fact, combined with Itô's formula, can be exploited to construct an nds-admissible arbitrage in class $\mathcal{A}^{\text{spot}}$. The following simple example is due to Dasgupta and Kallianpur [16] and Shiryaev [39].

Example 3.36 Choosing $\Phi_t = S_t - S_0$, we obtain by Itô's formula (Lemma 3.33) and the zero QV property of the fBS model with $H > 1/2$,

$$(S_t - S_0)^2 = 2 \int_0^t \Phi_u \, dS_u.$$

Hence, $\mathbf{\Phi}$ is nds-admissible (it is bounded from below by 0) and an arbitrage. Again, this construction of an arbitrage applies to all models with zero QV and $P(S_T \neq S_0) > 0$.

We now consider hedging in the fBS model with Hurst parameter larger than a half. Although there exists strong arbitrage in the class $\mathcal{A}^{\text{nds}} \cap \mathcal{A}^{\text{as}}$ by Theorem 3.15, one can still consider the hedging problem in the fBS model. Indeed, in spite of arbitrage, one may still be interested in hedging per se. But it must be noted that hedging cannot be used as a pricing paradigm in the presence of strong arbitrage, since for any hedge, one can find a super-hedge with smaller initial capital by combining the hedge with a strong arbitrage.

By a straightforward generalization of the previous example, we observe that a smooth European style option, i.e., with pay-off $f(S_t)$ for some $f \in \mathcal{C}^1$, can be hedged with initial endowment $f(S_0)$ and the strategy $\Phi_t = f'(S_t)$. In reality many options, like vanilla options, have convex payoff functions that do not belong to class $\mathcal{C}^1$. A generalization to this situation is possible with some extra effort as outlined next.

Definition 3.37 Let $f : \mathbb{R}_+ \to \mathbb{R}_+$ be a convex function and $H > 1/2$. If we can find a self-financing strategy $\boldsymbol{\Phi}$ and a constant c^f such that

$$f(S_T) = c^f + \int_0^T \Phi_s \, dS_s, \tag{3.2}$$

then $\boldsymbol{\Phi}$ is a *hedging strategy*, and c^f is a *hedging cost* of the option $f(S_T)$.

Remark 3.38

1. Because of the strong arbitrage possibilities in the fBS model, one cannot interpret the hedging cost c^f as a minimal super-replication price.
2. The strong arbitrage possibility of the fBS model does not imply that one can take $c^f = 0$ in (3.2): One can super-hedge with zero capital, but the hedge may not be exact. While from the purely monetary point of view this does not matter, there may be situations where one is penalized for not hedging exactly.

If f is a convex function, then f_x^+ (f_x^-) is the right (left) derivative of f. The following theorem can be regarded as a generalization of the Itô formula in Lemma 3.33 for nonsmooth convex functions in the pure fractional Brownian motion setting.

Theorem 3.39 *Suppose that S is the fBS model with $H > 1/2$ and f is a convex function. Then*

$$f(S_T) = f(S_0) + \int_0^T f_x^+(S_u) \, dS_u. \tag{3.3}$$

In particular, the European option $f(S_T)$ can be perfectly hedged with cost $f(S_0)$ and the hedging strategy given by $\Phi_t = f_x^+(S_t)$.

Proof One proves Theorem 3.39 by showing that the integral exists as a generalized Lebesgue–Stieltjes integral. This is done with the help of some fractional Besov space techniques. Finally, one proves that the integral exists as a forward integral and actually even as a Riemann–Stieltjes integral. For the rigorous proof, see [4]. Note that one can replace the right derivative f_x^+ by the left derivative f_x^-, as both derivatives differ on a countable set only. □

Example 3.40 If the convex function f corresponds to the call option, i.e., $f(x) = (x - K)^+$, then we observe that the stop-loss-start-gain portfolio replicates the call option:

$$(S_T - K)^+ = (S_0 - K)^+ + \int_0^T \mathbf{1}_{\{S_t \geq K\}} \, dS_t.$$

Note that this again gives an arbitrage strategy, if the option is at-the-money or out-of-the-money.

If $H < 1/2$, stochastic integrals for typical spot strategies with respect to the fBS model fail to exist. So it makes little sense to consider continuous trading in this situation. This unfortunate property is related to the infinite QV of the fBS model for small Hurst parameter and thus applies for the mixed model with $H < 1/2$ as well.

For the remainder of the section, we shall therefore discuss the mfBS model with $H > 1/2$. In the case $H > 3/4$, the mfBS model is equivalent in law to the BS model, see [13]. Therefore, all constructions of arbitrages with doubling strategies and all results on no-arbitrage with nds-strategies directly transfer from the BS model to the mfBS model with $H > 3/4$. Moreover, the latter model inherits the completeness of the BS model. We now discuss to what extent the mixed model with $1/2 < H \leq 3/4$ differs from the BS model. The argumentation below only makes use of the fact that the mixed model has the same QV as the BS model and has conditional full support.

Theorem 3.41 *Suppose that S is the mfBS with $H > 1/2$. Then,*

1. *There is strong arbitrage in the class $\mathcal{A}^{\text{spot}}$.*
2. *There is no nds-admissible arbitrage $\mathbf{\Phi}$ of the form*

$$\Phi_t = \varphi\left(t, \max_{0 \leq u \leq t} S_u, \min_{0 \leq u \leq t} S_u, \int_0^t S_u \, du, S_t\right)$$

with $\varphi \in \mathcal{C}^1([0, T] \times \mathbb{R}_+^4)$. A strategy of this form will be called smooth *from now on.*

Proof (i) Here the same constructive example as in Theorem 3.35 applies, because the mfBS model has the same QV as the BS model.

(ii) We fix some smooth strategy $\mathbf{\Phi}$. By a slightly more general Itô formula than the one in Lemma 3.33, one can conclude that there is a continuous functional $v : [0, T] \times C_{s_0}^+([0, T]) \to \mathbb{R}$ such that $V_t(\mathbf{\Phi}) = v(t, S)$. By the full support property, the paths of the mfBS model can be approximated by paths of the BS model and vice versa. In this way, absence of arbitrage can be transferred from the BS model to the mfBS model. The details are spelled out in [9], Theorem 4.4.

We point out that in the special case $\mathbf{\Phi} = (\Phi^0, \Phi) \in \mathcal{A}$ with $\Phi_t = \varphi(t, S_t)$ and $\Phi_t^0 = \varphi^0(t, S_t)$ for some sufficiently smooth functions (φ, φ^0), the value process $V_t(\mathbf{\Phi})$ can be linked to a PDE. This was exploited in [1] in order to prove absence of arbitrage in this special case. □

Remark 3.42

1. In Theorem 3.41, (ii), the differentiability of φ at $t = T$ can be relaxed to some extent, and absence of arbitrage still holds. The resulting class of strategies contains hedges for many relevant European, Asian, and lookback options. These hedges (as functionals on the paths) and the corresponding option prices (deduced by hedging and no-arbitrage relative to this class of portfolios) are the same as in the BS model. For the details, we refer to [9]. We note that this robustness of hedging strategies was already shown by Schoenmakers and Kloeden [38] in the case of European options.
2. The no-arbitrage result in Theorem 3.41, (ii), can be extended in several directions. Additionally to the running maximum, minimum, and average, the strategy can depend on other factors, which are supposed to be of finite variation and satisfy some continuity condition as functionals on the paths. The investor also is allowed to switch between different smooth strategies at a large class of stopping times, and still absence of arbitrage holds true for these stopping-smooth strategies. For the exact conditions on the stopping times, we refer to Sect. 6 in [9], but we note that many typical ones such as the first level crossing of the stock are included.

3.6 Trading under Transaction Costs

Recently Guasoni [23] and Guasoni et al. [25] have shown that allowing transaction costs in the fBS model, the arbitrage possibilities disappear. First they introduce, following Jouini and Kallal [28], the notion of ε-consistent price system.

Definition 3.43 Let S be a continuous process with paths in $C^+_{S_0}([0, T])$. An ε-consistent price system is a pair $(\widetilde{S}, \mathbf{Q})$ of a probability $\mathbf{Q}$ equivalent to $\mathbf{P}$ and a $\mathbf{Q}$-martingale $\widetilde{S} = (\widetilde{S}_t)_{0\le t\le T}$ such that $S_0 = \widetilde{S}_0$ and, for $0 \le t \le T$ and $\varepsilon > 0$,

$$1 - \varepsilon \le \frac{\widetilde{S}_t}{S_t} \le 1 + \varepsilon \quad \text{a.s.}$$

With proportional transaction costs, one cannot use continuous trading. Denote by $\mathcal{V}(\Phi)$ the total variation of the process Φ. In this section a trading strategy Φ is predictable finite-variation $\mathbb{R}$-valued process such that $\Phi_0 = \Phi_T = 0$. The value of Φ with ε-costs $V^\varepsilon(\Phi)$ is

$$V^\varepsilon(\Phi) = \int_0^T \Phi_s \, dS_s - \varepsilon \int_0^T S_s \, d\mathcal{V}(\Phi)_s.$$

Define $V_t^\varepsilon(\Phi)$ by

$$V_t^\varepsilon(\Phi) = V^\varepsilon(\Phi 1_{(0,t)}),$$

and so $V^\varepsilon(\Phi) = V_T^\varepsilon(\Phi)$.

Next, we define the set of admissible strategies in this context, following [25]: given $M > 0$, the strategy Φ is M-admissible if for all $t \in [0, T]$, we have that

$$V_t^\varepsilon(\Phi) \geq -M(1 + S_t) \quad \text{a.s.}$$

The set of M-admissible strategies is denoted by $\mathcal{A}_M^{\text{adm}}(\varepsilon)$. Define also

$$\mathcal{A}^{\text{adm}}(\varepsilon) = \bigcup_{M>0} \mathcal{A}_M^{\text{adm}}(\varepsilon).$$

Finally, we say that S admits arbitrage with ε-transaction costs if there is $\Phi \in \mathcal{A}^{\text{adm}}(\varepsilon)$ such that $V^\varepsilon(\Phi) \geq 0$ and $\mathbf{P}(V^\varepsilon(\Phi) > 0) > 0$.

We can now state the fundamental theorem of asset pricing with ε-transaction costs given in [25, Theorem 1.11]:

Theorem 3.44 *Let $S \in C_{s_0}^+([0, T])$. Then the following two conditions are equivalent*:

1. *For each $\varepsilon > 0$, there exists an ε-consistent price system.*
2. *For each $\varepsilon > 0$, there is no arbitrage for ε-transaction costs.*

It is shown by Guasoni et al. [24] that conditional full support implies the existence of an ε-consistent price system for every $\varepsilon > 0$. Therefore, the fBS models and the mfBS models do not admit arbitrage under transaction cost with the classes of strategies $\mathcal{A}^{\text{adm}}(\varepsilon)$ for $\varepsilon > 0$.

We will study a concrete hedging problem with proportional transaction costs.

In Theorem 3.39 it was shown that the European option $f(S_T)$ can be perfectly hedged with cost $f(S_0)$ and hedging strategy $\Phi_t = f_x^-(S_t)$. Take $T = 1$, put $t_i^n = \frac{i}{n}$, $i = 0, \dots, n$, and consider the discretized hedging strategy Φ^n,

$$\Phi_t^n = \sum_{i=1}^n f_x^-(S_{t_{i-1}^n}) 1_{(t_{i-1}^n, t_i^n]}(t). \tag{3.4}$$

Consider now discrete hedging with proportional transaction costs $k_n = k_0 n^{-\alpha}$ with $\alpha > 0$, $k_0 > 0$. The value of the strategy Φ^n at time $T = 1$ is

$$V_1(\Phi^n; k_n) = f(S_0) + \int_0^1 \Phi_t^n \, dS_t - k_n \sum_{i=1}^n S_{t_{i-1}^n} \left| f_x^-(S_{t_i^n}) - f_x^-(S_{t_{i-1}^n}) \right|. \tag{3.5}$$

Note that there is no transaction costs at time $t = 0$.

In the next theorem, μ^f is the second derivative f_{xx} of the convex function f. The derivative exists in a distributional sense, and μ^f is a Radon measure. The occupation measure Γ_{B^H} of fractional Brownian motion B^H is defined by $\Gamma_{B^H}([0, t] \times A) = \lambda\{s \in [0, t] : B_s^H \in A\}$; here λ is the Lebesgue measure, and A is a Borel set. Denote by $l^H(x, t)$ the local time of fractional Brownian motion B^H; recall that local time l^H is the density of the occupation measure with respect the Lebesgue measure.

The following theorem is proved in [3]:

Theorem 3.45 *Let $V_1(\Phi; k_n)$ be the value of the discrete hedging strategy Φ^n with proportional transaction costs $k_n = k_0 n^{-\alpha}$.*

1. *If $\alpha > 1 - H$, then, as $n \to \infty$,*
$$V_1(\Phi^n; k_n) \to f(S_1) \quad \text{in probability.}$$
2. *If $\alpha = 1 - H$, then, as $n \to \infty$,*
$$V_1(\Phi^n; k_n) \to f(S_1) - \sqrt{\frac{2}{\pi}} k_0 \int_{\mathbb{R}} \int_0^1 S_t \, dl^H(\ln(a), t) \mu^f(da). \tag{3.6}$$

Remark 3.46 Note that one can write the limit result in (3.6) as

$$f(S_1) = f(S_0) + \int_0^1 f_x^-(S_u) \, dS_u + \sqrt{\frac{2}{\pi}} k_0 \int_{\mathbb{R}} \int_0^1 S_t \, dl^H(\ln(a), t) \mu^f(da);$$

if l^W is the local time for Brownian motion, then the Itô–Tanaka formula gives

$$f(W_1) = f(0) + \int_0^1 f_x^-(W_u) \, dW_u + \frac{1}{2} \int_0^1 \int_{\mathbb{R}} dl^W(a, u) \mu^f(da).$$

Hence asymptotical transaction costs with $\alpha = 1 - H$ have a similar effect as the existence of a nontrivial quadratic variation.

3.7 Approximations

3.7.1 Binary Tree Approximations

The famous Donsker's invariance principle links random walks to the Bm. By using this principle one can approximate the BS model with Cox–Ross–Rubinstein (CRR) binomial trees. To be more precise, let for all $n \in \mathbb{N}$, $(\xi_k^n)_{k \in \mathbb{N}}$ be i.i.d. random variables with $\mathbf{P}[\xi_k^n = 1] = 1/2 = \mathbf{P}[\xi_k^n = -1]$. Set

$$W_t^n = \frac{1}{\sqrt{n}} \sum_{k=1}^{\lfloor nt \rfloor} \xi_k^n.$$

Then the Donsker's invariance principle states that the processes W^n, $n \in \mathbb{N}$, converge in the Skorokhod space $D([0, T])$ to the Bm. Let S^n to be the binomial model defined by

$$S_t^n = \prod_{s \le t} (1 + \Delta W_s^n).$$

Then the processes S^n, $n \in \mathbb{N}$, converge weakly in $D([0, T])$ to the geometric Bm $S_t = e^{W_t - t/2}$, i.e., the binomial models S^n, $n \in \mathbb{N}$, approximate the BS model.

In [40] a fractional CRR model was constructed that approximates the fBS model when $H > 1/2$, and later this approximation was extended in different directions by Nieminen [33] and Mishura and Rode [32]. We give here a brief overview of the construction in [40]:

Let $(\xi_k^n)_{k\in\mathbb{N}}$ be as before, and let $k(t,s)$ be the kernel that transforms the Bm into an fBm:

$$k(t,s) = c_H s^{\frac{1}{2}-H} \int_s^t u^{H-\frac{1}{2}} (u-s)^{H-\frac{3}{2}}\, du,$$

where

$$c_H = \left(H - \frac{1}{2}\right) \sqrt{\frac{(2H+\frac{1}{2})\Gamma(\frac{1}{2}-H)}{\Gamma(H+\frac{1}{2})\Gamma(2-2H)}},$$

and $\Gamma(z) = \int_0^\infty t^{z-1} e^{-t}\, dt$ is the Gamma function. Then

$$B_t = \int_0^t k(t,s)\, dW_s.$$

To get a piecewise constant process in $D([0,T])$, one must regularize the kernel:

$$k^n(t,s) = n \int_{s-1/n}^{s} k\left(\frac{\lfloor nt \rfloor}{n}, u\right) du.$$

Set

$$B_t^n = \int_0^t k^n(t,s)\, dW_s^n$$

and

$$S_t^n = \prod_{s \le t} \left(1 + \Delta B_s^n\right).$$

Theorem 3.47 *Let $H > 1/2$.*

1. *The random walks B^n, $n \in \mathbb{N}$, converge weakly in $D([0,T])$ to the fBm B.*
2. *The binary models S^n, $n \in \mathbb{N}$, converge weakly in $D([0,T])$ to the fBS model $S = e^B$.*
3. *The fractional CRR binary models S^n, $n \in \mathbb{N}$, are complete but exhibit arbitrage opportunities if n is sufficiently large.*

Proof (i) is the "fractional invariance principle" [40, Theorem 1], (ii) follows basically from (i), the continuous mapping theorem, and a Taylor expansion of $\log(S^n)$, see the proof of [40, Theorem 3] for details. The completeness claim of (iii) is obvious, since the market models are binary. The arbitrage claim of (iii) follows from the fact that if we have only gone up in the binary tree for long enough, the stock price will increase in the next step, no matter which branch the process takes in the tree. We refer to the proof of [40, Theorem 5] for details. □

A main motivation for considering the approximation S^n is that the continuous-time process $S_t = e^{B_t}$ solves the SDE

$$dS_t = S_t\, dB_t, \qquad S_0 = 1,$$

in the sense of forward integration. Alternatively, one can build an integral on Wick–Riemann sums [6, 10, 19, 31] and examine the SDE

$$dX_t = X_t\, d^{\diamond} B_t, \qquad X_0 = 1.$$

Here, $X_t = \exp\{B_t - t^{2H}/2\}$. Thus, the processes S and X only differ by a deterministic factor. Without going into any details here, we note that the *Wick product* can be defined by

$$e^{\Phi - E[\Phi^2]/2} \diamond e^{\Psi - E[\Psi^2]/2} = e^{(\Phi+\Psi) - E[(\Phi+\Psi)^2]/2}$$

for centered Gaussian random variables Φ and Ψ and can be extended to larger classes of random variables by bilinearity and denseness arguments, see, e.g., [6, 19]. Somewhat surprisingly, there is a very simple analogue of the Wick product for the binary random variables ξ_k^n, $k = 1, \dots, n$, see [26], which gives rise to a natural binary discretization of X_t suggested by Bender and Elliott [7].

The *discrete Wick product* can be defined as $(A, B \subset \{1, \dots, n\})$

$$\prod_{i \in A} \xi_i^n \diamond_n \prod_{i \in B} \xi_i^n := \begin{cases} \prod_{i \in A \cup B} \xi_i^n & \text{if } A \cap B = \emptyset, \\ 0 & \text{otherwise,} \end{cases}$$

and extends by bilinearity to $L^2(\mathcal{F}_n)$, where $\mathcal{F}_n$ denotes the σ-field generated by $(\xi_1^n, \dots, \xi_n^n)$. A discrete version of the Wick-fractional Black–Scholes model is then defined by

$$X_t^n = \diamond_{s \le t} \bigl(1 + \Delta B_s^n\bigr).$$

Bender and Elliott [7] argue in favor of this discretization that the discrete Wick product separates influences of the drift and volatility.

Theorem 3.48 *Let $H > 1/2$.*

1. *The binary models X^n, $n \in \mathbb{N}$, converge weakly in $D([0, T])$ to the Wick-fractional Black–Scholes model X.*
2. *The Wick-fractional CRR binary models X^n, $n \in \mathbb{N}$, are complete but exhibit arbitrage opportunities if n is sufficiently large.*

The proof of (ii) is similar to the one of Theorem 3.47, (iii), and can be found in [7]. As is pointed out there, the use of the discrete Wick products kills a part of the memory as compared to the discrete-time model S^n. It turns out, however, that the remaining part of the memory is still sufficient to construct an arbitrage. Completeness again follows from the fact that the model is binary. For the proof of (i), one cannot argue by the continuous mapping theorem, because the discrete Wick product is not a pointwise operation. Instead the relation of the Wick powers to Hermite polynomials and explicit computations of the Walsh decomposition (which can be considered a discrete analogue of the chaos decomposition to some extent) can be exploited, see [8].

3.7.2 Arbitrage-Free Approximation

The results in this section are motivated by [30], where the authors give an arbitrage-free approximation to fBS model. The prelimit models in this approximation are not complete, however.

Recall the following classical result: Let $N = (N_t)_{t\in\mathbb{R}_+}$ be a Poisson process with intensity 1, and set

$$W_t^n = \frac{1}{\sqrt{n}}(N_{nt} - nt).$$

Then W^n converges to a Bm W in the Skorokhod space $D([0,T])$, the process $dS_t^n = S_{t-}^n\, dW_t^n$, $S_0^n = S_0$, converges weakly to the BS model, and the approximation is complete and arbitrage-free.

We approximate the fractional Black–Scholes model $(S, (\mathcal{F}_t)_{t\in[0,T]}, \mathbf{P})$ with a sequence $(S^n, (\mathcal{F}^n)_{t\in[0,T]})$ of models driven by scaled renewal counting processes. The prelimit models are complete and arbitrage-free. The approximation is based on the limit theorem of Gaigalas and Kaj [21]. It goes as follows: let G be a continuous distribution function with heavy tails, i.e.,

$$1 - G(t) \sim t^{-(1+\beta)} \tag{3.7}$$

as $t \to \infty$ with $\beta \in (0,1)$.

Take η_i to be the sojourn times of a renewal counting process N. Assume that $\eta_i \sim G$ for $i \geq 2$; for the first sojourn time η_1, assume that it has the distribution $G_0(t) = \frac{1}{\mu}\int_0^t (1 - G(s))\, ds$ (here μ is the normalizing constant), so that the renewal counting process

$$N_t = \sum_{k=1}^{\infty} 1_{\{\tau_k \leq t\}}$$

is stationary, where $\tau_1 = \eta_1$ and $\tau_k := \eta_1 + \cdots + \eta_k$.

Take now independent copies $N^{(i)}$ of N, numbers $a_m \geq 0$, $a_m \to \infty$, such that

$$\frac{m}{a_m^{\beta}} \to \infty; \tag{3.8}$$

using the terminology of Gaigalas and Kaj, we can speak of *fast connection rate*.

Define the *workload* process $W(m,t)$ by

$$W(m,t) = \sum_{i=1}^{m} N_t^{(i)};$$

note that the process N^m is a counting process, since the sojourn distribution is continuous. We have that $EW(m,t) = \frac{mt}{\mu}$, since $W(m,t)$ is a stationary process.

Proposition 3.49 (Gaigalas and Kaj [21]) *Assume* (3.7) *and* (3.8). *Let*

$$Y^m(t) := \mu^{\frac{3}{2}} \sqrt{\frac{\beta(1-\beta)(2-\beta)}{2}} \frac{W(m, a_m t) - m\mu^{-1} a_m t}{m^{\frac{1}{2}} a_m^{1-\frac{\beta}{2}}}.$$

Then Y^m converges weakly [*in the Skorokhod space D*] *to a fractional Brownian motion B^H, where $H = 1 - \frac{\beta}{2}$.*

Since the process Y^m is a semimartingale, it has a semimartingale decomposition

$$Y^m = M^m + H^m; \tag{3.9}$$

here $H^m = B^m - A^m$, and B^m is the compensator of the normalized aggregated counting process W. Note that the process H^m is a continuous process with bounded variation.

Up to a constant, we have that the square bracket of the martingale part M^m of the semimartingale Y^m is

$$[M^m, M^m]_t = C \frac{W(m, a_m t)}{m a_m^{2-\beta}}.$$

But our assumptions imply that $[M^m, M^m]_t \overset{L^1(P)}{\to} 0$ as $m \to \infty$. With the Doob inequality we obtain that $\sup_{s \le t} |M_s^m| \overset{P}{\to} 0$, and fBm is the limit of a sequence of continuous processes with bounded variation.

It is not difficult to check that the solution to the linear equation

$$dS_t^m = S_{t-}^m \, dY_s^m$$

converges weakly in the Skorokhod space to a geometric fractional Brownian motion.

The driving process Y^m is a scaled counting process minus the expectation. It is well known that such models are complete and arbitrage-free. Hence we have a complete and arbitrage-free approximation to fractional Black–Scholes model. See [42] for more details.

Remark 3.50 If one computes the hedging price and the hedging strategy for the European call $(S_T^m - K)^+$ in the prelimit sequence and lets $m \to \infty$, one gets in the limit the stop-loss-start-gain hedging given in Example 3.40.

3.7.3 Microeconomic Approximation

So far there has been few economic justifications to use fractional models in option-pricing. For example, the LRD of the stock price, measured by the Hurst index H, is usually given as an econometric fact (and even that is questionable). One attempt to build a microeconomic foundation for fractional models was that of Bayraktar et al. [5]. They showed how the fBS model can arise as a large time-scale many-agent limit when there are inert agents, i.e., investors who change their portfolios infrequently, and the log-price is given by the market imbalance. We will briefly explain their framework and their main result here.

Consider n agents. Each agent k has a *trading mood* $x^k = (x_t^k)_{t\in[0,\infty)}$ that takes values in a finite state space $E \subset \mathbb{R}$ containing zero: $x_t^k > 0$ means buying, $x_t^k < 0$ means selling, and $x_t^k = 0$ means inactivity at time t. The agents are homogeneous and independent. The trading mood x^k is a semi-Markov process defined as

$$x_t^k = \sum_{m=0}^{\infty} \xi_m^k \mathbf{1}_{[\tau_m^k, \tau_{m+1}^k)}(t),$$

where the E-valued random variables ξ_m^k and the stopping times τ_m^k satisfy

$$\begin{aligned}
&\mathbf{P}\big[\xi_{m+1}^k = j, \tau_{m+1}^k - \tau_m^k \le t \mid \xi_1^k, \ldots, \xi_m^k, \tau_1^k, \ldots, \tau_m^k\big] \\
&\quad = \mathbf{P}\big[\xi_{m+1}^k = j, \tau_{m+1}^k - \tau_m^k \le t \mid \xi_m^k\big] \\
&\quad = Q\big(\xi_m^k, j, t\big).
\end{aligned}$$

So, $(\xi_m^k)_{m\in\mathbb{N}}$ is a homogeneous Markov chain on E with transition probabilities $p_{ij} = \lim_{t\to\infty} Q(i, j, t)$. It is assumed that $p_{ij} > 0$ for all $i \neq j$, so that (p_{ij}) admits a unique stationary measure $\mathbf{P}^*$. On the sojourn times $\tau_{m+1}^k - \tau_m^k$ given ξ_m^k it is assumed that:

1. The average sojourn times are finite.
2. The sojourn time at the inactive state is heavy-tailed, i.e., there exist a constant $\alpha \in (1,2)$ and a locally bounded slowly varying at infinity function L such that

$$\mathbf{P}\big[\tau_{m+1}^k - \tau_m^k \ge t \mid \xi_m^k = 0\big] \sim t^{-\alpha} L(t).$$

   (L is slowly varying at infinity if, for all $x > 0$, $L(xt)/L(t) \to 1$ as $t \to \infty$.)
3. The sojourn times at the active states $i \neq 0$ are lighter-tailed than the sojourn time at the inactive state:

$$\lim_{t\to\infty} \frac{\mathbf{P}[\tau_{m+1}^k - \tau_m^k \ge t \mid \xi_m^k = i]}{t^{-(\alpha+1)} L(t)} = 0.$$

4. The distribution of the sojourn times have continuous and bounded densities with respect to the Lebesgue measure.

An agent-independent process $\Psi = (\Psi_t)_{t\in[0,\infty)}$ describes the sizes of typical trades: Agent k accumulates the asset S at the rate $\Psi_t x_t^k$ at time t. The process Ψ is assumed to be a continuous semimartingale with Doob–Meyer decomposition $\Psi = M + A$ such that $\mathbf{E}[\langle M\rangle_T] < \infty$ and $\mathbf{E}[\mathcal{V}(A)] < \infty$, and Ψ and the x^k's are independent. As before, $\mathcal{V}(A)$ denotes the total variation of A on $[0, T]$.

The log-price X^n for the asset with n agents is assumed to be given by the *market imbalance*:

$$X_t^n = X_0 + \sum_{k=1}^{n} \int_0^t \Psi_s x_s^k \, ds.$$

The *aggregate order rate* is

$$Y_t^{\varepsilon,n} = \sum_{k=1}^{n} \Psi_t x_{t/\varepsilon}^k.$$

Let $\mu \neq 0$ be the expected trading mood under the stationary measure $\mathbf{P}^*$, and define the centered aggregate order process

$$X_t^{\varepsilon,n} = \int_0^t Y_s^{\varepsilon,n} \, ds - \mu n \int_0^t \Psi_s \, ds.$$

Then, the main result [5, Theorem 2.1] states that in the limit the centered log-prices are given by a stochastic integral with respect to an fBm:

Theorem 3.51 *There exists a constant $c > 0$ such that*

$$\lim_{\varepsilon\downarrow 0} \lim_{n\to\infty} \frac{1}{\varepsilon^{1-H}\sqrt{nL(1/\varepsilon)}} X^{\varepsilon,n} = c \int_0^{\cdot} \Psi_t \, dB_t,$$

where B is an fBm with Hurst index $H = (3-\alpha)/2 > 1/2$. The limits are weak limits in the Skorokhod space $D([0, T])$.

Remark 3.52 Assume that $\Psi \equiv 1$, i.e., the trades, and consequently the log-prices, are completely determined by the agents' intrinsic trading moods. Then

$$X^{\varepsilon,n} = \varepsilon X_{t/\varepsilon}^n - \mu n t.$$

1. The limit in Theorem 3.51 is the fBS model.
2. Bayraktar et al. [5] also considered a model where there are both active and inert investors (active investors have light-tailed sojourn times at the inactive state 0). Then they get, in the limit, the mfBS model.

3.8 Conclusions

We have given some recent results on the arbitrage and hedging in some fractional pricing models. If one wants to understand the pricing of options in the fBS model, then it is not clear to what extent the hedging capital given in (3.3) can be interpreted as the price of the option. On the other hand, these exact hedging results may have some value if one studies the hedging problem in the presence of transaction costs. The mixed Brownian–fractional Brownian pricing model has less arbitrage possibilities, but it is possible to model the dependency of the log-returns in this model family. One can also modify this model to include more "stylized" properties of log-returns, but the hedging prices will be the same as without these "stylized" features.

The mixed model seems to be a good candidate to include several of the observed "stylized" facts of log-returns in the modeling of stock prices. Hence it is reasonable to study how the properties of the standard gBS model change in the mixed model. We have shown in [9] that the hedging is the same in all models having the same structural quadratic variation as a functional of the stock price path. For example, recently Bratyk and Mishura have considered quantile hedging problems in mixed models; see [12] for more details.

3.8.1 Open Problems

We finish by giving some open problems related to the present survey.

Are fractional and mixed models free of simple arbitrage?

What kind of random variables have a Riemann–Stieltjes integral representations in the fBS model?

Can one verify statistically that option prices depend only on the quadratic variation of the underlying stock prices?

Acknowledgements T.S. and E.V. acknowledge the support from Saarland University, and E.V. is grateful to the Academy of Finland, grant 127634, for partial support. We are grateful to Peter Parczewski and an anonymous referee for useful comments.

References

1. T. Androshchuk, Yu. Mishura, Mixed Brownian–fractional Brownian model: absence of arbitrage and related topics. Stochastics **78**(5), 281–300 (2006)
2. M.A. Arcones, On the law of the iterated logarithm for Gaussian processes. J. Theor. Probab. **8**(4), 877–903 (1995)
3. E. Azmoodeh, Geometric fractional Brownian motion market model with transaction costs. Preprint (2009), 12 p.
4. E. Azmoodeh, Yu. Mishura, E. Valkeila, On hedging European options in geometric fractional Brownian motion market model. Stat. Decis. **27**(2), 129–144 (2009). doi:10.1524/stnd.2009.1021

5. E. Bayraktar, U. Horst, R. Sircar, A limit theorem for financial markets with inert investors. Math. Oper. Res. **31**(4), 789–810 (2006)
6. C. Bender, An S-transform approach to integration with respect to a fractional Brownian motion. Bernoulli **9**(6), 955–983 (2003)
7. C. Bender, R.J. Elliott, Arbitrage in a discrete version of the Wick-fractional Black–Scholes market. Math. Oper. Res. **29**(4), 935–945 (2004)
8. C. Bender, P. Parczewski, Approximating a geometric fractional Brownian motion and related processes via discrete Wick calculus. Bernoulli **16**(2), 389–417 (2010). doi:10.3150/09-BEJ223
9. C. Bender, T. Sottinen, E. Valkeila, Pricing by hedging and no-arbitrage beyond semimartingales. Finance Stoch. **12**(4), 441–468 (2008)
10. F. Biagini, Y. Hu, B. Øksendal, T. Zhang, *Stochastic Calculus for Fractional Brownian Motion and Applications*. Probability and Its Applications (New York) (Springer, London, 2008)
11. T. Björk, H. Hult, A note on Wick products and the fractional Black–Scholes model. Finance Stoch. **9**(2), 197–209 (2005)
12. M. Bratyk, Y. Mishura, The generalization of the quantile hedging problem for price process model involving finite number of Brownian and fractional Brownian motions. Theory Stoch. Process. **14**(3–4), 27–38 (2008)
13. P. Cheridito, Mixed fractional Brownian motion. Bernoulli **7**(6), 913–934 (2001)
14. P. Cheridito, Arbitrage in fractional Brownian motion models. Finance Stoch. **7**(4), 533–553 (2003)
15. A. Cherny, Brownian moving averages have conditional full support. Ann. Appl. Probab. **18**(5), 1825–1830 (2008)
16. A. Dasgupta, G. Kallianpur, Arbitrage opportunities for a class of Gladyshev processes. Appl. Math. Optim. **41**(3), 377–385 (2000)
17. F. Delbaen, W. Schachermayer, A general version of the fundamental theorem of asset pricing. Math. Ann. **300**(3), 463–520 (1994)
18. F. Delbaen, W. Schachermayer, The fundamental theorem of asset pricing for unbounded stochastic processes. Math. Ann. **312**(2), 215–250 (1998)
19. T.E. Duncan, Y. Hu, B. Pasik-Duncan, Stochastic calculus for fractional Brownian motion. I: Theory. SIAM J. Control Optim. **38**(2), 582–612 (2000)
20. H. Föllmer, *Calcul d'Itô sans probabilités*. Séminaire de Probabilités, vol. XV (Springer, Berlin, 1981), pp. 143–150
21. R. Gaigalas, I. Kaj, Convergence of scaled renewal processes and a packet arrival model. Bernoulli **9**(4), 671–703 (2003)
22. D. Gasbarra, T. Sottinen, H. van Zanten, Conditional full support of Gaussian processes with stationary increments. Preprint (2008)
23. P. Guasoni, No arbitrage under transaction costs, with fractional Brownian motion and beyond. Math. Finance **16**(3), 569–582 (2006)
24. P. Guasoni, M. Rásonyi, W. Schachermayer, Consistent price systems and face-lifting pricing under transaction costs. Ann. Appl. Probab. **18**(2), 491–520 (2008)
25. P. Guasoni, M. Rásonyi, W. Schachermayer, The fundamental theorem of asset pricing for continuous processes under small transaction costs. Ann. Finance **6**(2), 157–191 (2010)
26. H. Holden, T. Lindstrøm, B. Øksendal, J. Ubøe, Discrete Wick calculus and stochastic functional equations. Potential Anal. **1**(3), 291–306 (1992)
27. R. Jarrow, P. Protter, H. Sayit, No arbitrage without semimartingales. Ann. Appl. Probab. **19**(2), 596–616 (2009)
28. E. Jouini, H. Kallal, Martingales and arbitrage in securities markets with transaction costs. J. Econ. Theory **66**, 178–197 (1995)
29. S. Karlin, H.M. Taylor, *A First Course in Stochastic Processes*, 2nd edn. (Academic Press, San Diego, 1975)
30. C. Klüppelberg, C. Kühn, Fractional Brownian motion as a weak limit of Poisson shot noise processes—with applications to finance. Stoch. Process. Appl. **113**(2), 333–351 (2004)
31. Yu.S. Mishura, *Stochastic Calculus for Fractional Brownian Motion and Related Processes*. Lecture Notes in Mathematics, vol. 1929 (Springer, Berlin, 2008)

32. Yu.S. Mishura, S.G. Rode, Weak convergence of integral functionals of random walks that weakly converge to fractional Brownian motion (Ukrainian). Ukr. Mat. Zh. **59**(8), 1040–1046 (2007). Translation in Ukr. Math. J. **59**(8), 1155–1162 (2007)
33. A. Nieminen, Fractional Brownian motion and martingale-differences. Stat. Probab. Lett. **70**(1), 1–10 (2004)
34. M.S. Pakkanen, Stochastic integrals and conditional full support. Preprint (2008, Revised Jul. 24, 2009). Available as arXiv:0811.1847
35. P. Protter, *Stochastic Integration and Differential Equations*, 2nd edn. Stochastic Modelling and Applied Probability, vol. 21 (Springer, Berlin, 2004)
36. L.C.G. Rogers, Arbitrage with fractional Brownian motion. Math. Finance **7**(1), 95–105 (1997)
37. F. Russo, P. Vallois, P. Forward, backward and symmetric stochastic integration. Probab. Theory Relat. Fields **97**(3), 403–421 (1993)
38. J.G.M. Schoenmakers, P.E. Kloeden, Robust option replication for a Black–Scholes model extended with nondeterministic trends. J. Appl. Math. Stoch. Anal. **12**(2), 113–120 (1999)
39. A. Shiryaev, On arbitrage and replication for fractal models. Research Report 30 (1998), MaPhySto, Department of Mathematical Sciences, University of Aarhus
40. T. Sottinen, Fractional Brownian motion, random walks and binary market models. Finance Stoch. **5**(3), 343–355 (2001)
41. T. Sottinen, E. Valkeila, On arbitrage and replication in the fractional Black–Scholes pricing model. Stat. Decis. **21**(2), 93–107 (2003)
42. E. Valkeila, On the approximation of geometric fractional Brownian motion, in *Optimality and Risk—Modern Trends in Mathematical Finance the Kabanov Festschrift*, ed. by F. Delbaen, M. Rásonyi, C. Stricker (2009), pp. 251–265

Chapter 4
Credit Contagion in a Long Range Dependent Macroeconomic Factor Model

Francesca Biagini, Serena Fuschini, and Claudia Klüppelberg

Abstract We propose a new default contagion model, where default may originate from the performance of a specific firm itself but can also be directly influenced by defaults of other firms. The default intensities of our model depend on smoothly varying macroeconomic variables, driven by a long-range dependent process. In particular, we focus on the pricing of defaultable derivatives whose defaults depend on the macroeconomic process and may be affected by the contagion effect. In our approach we are able to provide explicit formulas for prices of defaultable derivatives at any time $t \in [0, T]$. Finally we calculate some examples explicitly, where the macroeconomic factor process is given by a functional of the fractional Brownian motion with Hurst index $H > \frac{1}{2}$.

Keywords Credit risk · Contagion modeling · Credit intensity · Latent process · Macroeconomic variables process · Long-range dependence · Fractional Brownian motion · Pricing defaultable derivatives

Mathematics Subject Classification (2010) Primary 60G15 · 91B70 · Secondary 91Gxx · 60G22

4.1 Introduction

The financial crisis started in 2008 has been triggered by the dramatic rise in mortgage delinquencies and foreclosures in the United States. This crisis has not only

---

F. Biagini (✉)
Department of Mathematics, Ludwig-Maximilians Universität, 80333 Munich, Germany
e-mail: biagini@math.lmu.de

S. Fuschini
Department of Mathematics, University of Bologna, 40127 Bologna, Italy
e-mail: fuschini@dm.unibo.it

C. Klüppelberg
Center for Mathematical Sciences, and Institute for Advanced Study, Technische Universität München, 85748 Garching, Germany
e-mail: cklu@ma.tum.de
url: www-m4.ma.tum.de

G. Di Nunno, B. Øksendal (eds.), *Advanced Mathematical Methods for Finance*,
DOI 10.1007/978-3-642-18412-3_4, © Springer-Verlag Berlin Heidelberg 2011

manifested the weaknesses in financial industry regulation, but also of the financial models used for pricing instruments of mortgage pools like MBSs and CDOs. In particular, the systemic risk has been disastrously underestimated. It has been industry standard to model contagion within a pool of credits by an intensity model where the intensities of surviving credits may increase at default of some credit. This approach increases the probability for default of dependent credits and so has no direct effect. In a static model, Davis and Lo [8] suggested a direct contagion model which is able to capture the immediate effect of one credit default to other credits in a pool.

We investigate a dynamic version of the direct contagion model of Davis and Lo [8], which is based on interacting intensities. Each default indicator process may be influenced by the default of other firms, which is modeled by an indicator variable representing the contagion possibility. In addition, we allow the default intensities to depend on smoothly varying macroeconomic variables (for example, supply and demand, interest rates, the gross national product, or other measures of economics activities) that are often modeled by a Markov state vector leading to affine models; see, e.g., Duffie [9] and Duffie, Filipovic, and Schachermayer [10].

It is, however, well known that many macroeconomic processes show a long-range dependence effect; see, e.g., Henry and Zaffaroni [16]. Consequently, in this paper we model the latent macroeconomic process governing the default intensities by a long-range dependent process, here exemplified by a one-dimensional process which stands, for instance, for a weighted mean of a vector of macroeconomic variables.

In this paper we focus on the pricing of defaultable derivatives depending on the macroeconomic process and affected by the contagion effect. We remark that we are not assuming that the primary assets on the market are driven by a long-range dependent process. Hence no arbitrage problem arises in the use of our model. For a discussion on this topic, we refer to Björk and Hult [2] or Øksendal [17]. In our model the long-range dependent macroeconomic process enters as a progressively measurable process into the default intensity. By usual no-arbitrage arguments the price of a contingent claim at time t is given by the conditional expectation under the pricing measure, which we suppose to be given by the market.

In this, not at all standard model we are able to provide explicit formulas for the derivative price at any time $t \in [0, T]$. We discuss suitable long-range dependent models for the macroeconomic process and calculate some examples, where the macroeconomic factor is given by a functional of the fractional Brownian motion with Hurst index $H > \frac{1}{2}$.

Our paper is organized as follows. In Sect. 4.2 we present our model and the contagion mechanism for instantaneous contagion, modeling the intensity as a function of the macroeconomic process. We explain the model in detail in Sect. 4.2—it is an intensity-based model—and we present all assumptions here. We present a specific example in Sect. 4.3 and calculate its infinitesimal generators of the default indicator process and the default number process. Afterwards, we calculate a defaultable derivatives price in Sect. 4.4, at first conditionally on the latent process. We conclude the section with a specific example, calculating the prices of a defaultable bond under contagion. Finally, in Sect. 4.5, we introduce a general long-range dependent fractional macroeconomic process as intensity process. We obtain an explicit formula, which can be evaluated numerically. In Sect. 4.5 we discuss some specific macroeconomic models and give an explicit financial example.

4.2 The Credit Model

4.2.1 The Default Model

We consider a portfolio of m firms indexed by $i \in \{1, \dots, m\}$. Its default state is described by a *default indicator process*

$$Z_t = \big(Z_t(1), \dots, Z_t(m)\big), \quad t \geq 0,$$

with values in the set $\{0, 1\}^m$. For every $i \in \{1, \dots, m\}$, the random variable $Z_t(i)$ indicates if the firm i has defaulted or not by time t, i.e., $Z_t(i) = 1$ if the firm i has defaulted by time t and $Z_t(i) = 0$ otherwise.

Aiming at an extension of the idea of Davis and Lo [7] as indicated in [8], Sect. 3, to a dynamic setting, we distinguish between default *caused by itself* and default *caused by contagion*, based on the default of some other firms. To this purpose, we introduce the *self-default indicator process*

$$Y_t = \big(Y_t(1), \dots, Y_t(m)\big), \quad t \geq 0,$$

with values in $\{0, 1\}^m$, where again $Y_t(i) = 1$ if the firm i has defaulted by time t by itself and $Y_t(i) = 0$ otherwise. We denote by τ_i the default time of the ith firm for $i \in \{1, \dots, m\}$ and by $\mathcal{I}$ the indicator function; then

$$Y_t(i) = \mathcal{I}_{\{\tau_i \leq t\}}, \quad i = 1, \dots, m.$$

Next we model contagion by using a *contagion matrix indicator process*: if firm i defaults by itself at some time t, then $C_t(i, j)$ determines whether infection of default from firm i to firm j takes place or not at time t.

We assume that, if default of firm i causes default of firm j, then this happens instantaneously resulting in $C_t(i, j) = 1$. More precisely, for any time $t \geq 0$,

$$C_t(i, j) = \begin{cases} 1 & \text{if the default of firm } i \text{ causes default of firm } j \text{ at time } t, \\ 0 & \text{otherwise.} \end{cases} \tag{4.1}$$

This results in a representation of the default indicator process of firm j,

$$\begin{aligned} Z_t(j) &= Y_t(j) + \big(1 - Y_t(j)\big)\Big(1 - \prod_{i \neq j}\big(1 - C_{t \wedge \tau_i}(i, j) Y_t(i)\big)\Big) \\ &= Y_t(j) + \big(1 - Y_t(j)\big)\Big(1 - \prod_{i \neq j}\big(1 - C_{\tau_i}(i, j) Y_t(i)\big)\Big), \quad t \geq 0. \end{aligned} \tag{4.2}$$

Since firm j is influenced by itself, we define $C_t(j,j) \equiv 1$ for all $j \in \{1,\dots,m\}$ and $t \geq 0$. Then (4.2) can also be written as

$$Z_t(j) = 1 - \prod_{i=1}^{m} \bigl(1 - C_{\tau_i}(i,j) Y_t(i)\bigr), \quad t \geq 0. \tag{4.3}$$

The defaults of the portfolio, either by itself or by infection, are caused by fluctuations in the macroeconomic environment, which we model by a state variable process $\Psi = (\Psi_t)_{t \geq 0}$ with values in $\mathbb{R}^d$ for $d \in \mathbb{N}$, representing the evolution of macroeconomic variables such as supply and demand, interest rates, the gross national product, or other measures of economics activities. In the literature Ψ is usually taken to be Markovian, so that the overall model of the system, given by $(\Psi_t, Y_t, C_t)_{t \geq 0}$, is Markovian.

It is, however, well known that many macroeconomic variables show a long-range dependence effect; see, e.g., Henry and Zaffaroni [16]. Consequently, we model the macroeconomic environment by a long-range dependent process $(\Psi_t)_{t \geq 0}$ to be specified later (see Sect. 4.5).

4.2.2 The Probability Space

The overall state of our system is described by the process $(\Psi_t, Y_t, C_t)_{t \geq 0}$ on the probability space $(\Omega, \mathcal{F}, \mathbb{P})$ endowed with the filtration

$$\mathcal{F}_t := \mathcal{F}_t^{\Psi} \vee \mathcal{F}_t^{Y} \vee \mathcal{F}_t^{C}, \quad t \geq 0,$$

where $(\mathcal{F}_t^{\Psi})_{t \geq 0}$, $(\mathcal{F}_t^{Y})_{t \geq 0}$, and $(\mathcal{F}_t^{C})_{t \geq 0}$ are the natural filtrations associated to the processes Ψ, Y, and C, respectively. Here we assume that the agent on the market knows if a firm has defaulted by itself or not and the contagion structure among the firms. Moreover, we define the filtration

$$\mathcal{G}_t := \mathcal{F}_{\infty}^{\Psi} \vee \mathcal{F}_t^{Y} \vee \mathcal{F}_t^{C}, \quad t \geq 0.$$

We assume that investors have access to $(\mathcal{F}_t)_{t \geq 0}$, whereas the larger filtration $(\mathcal{G}_t)_{t \geq 0}$, which contains information about the whole path $(\Psi_t)_{t \geq 0}$ serves mainly theoretical purposes. Finally, we assume that all filtrations satisfy the usual hypotheses of completeness and right-continuity.

From now on we work under the following assumptions.

Assumption 4.1

(1) We remain in the framework of most reduced-form credit risk models in the literature and assume that the dynamic of Ψ is not affected by the evolution of the default indicator process Z. This has the advantage that we first model the dynamic of Ψ and, in a second step, the conditional distribution of the default

indicator process Z for a given realization of the macroeconomic factor process Ψ. In particular, we require that Ψ is not affected by the evolution of the default indicator process Y and the contagion matrix C. In mathematical terms this means that for every bounded $\mathcal{F}_\infty^\Psi$-measurable random variable η,

$$\mathbb{E}[\eta|\mathcal{F}_t] = \mathbb{E}[\eta|\mathcal{F}_t^\Psi], \quad t \geq 0.$$

(2) The processes $(Y_t(i))_{t\geq 0}$ for $i \in \{1,\dots,m\}$, $(C_t(i,j))_{t\geq 0}$ for $i,j \in \{1,\dots,m\}$, $i \neq j$, are conditionally independent with respect to the filtration $(\mathcal{G}_t)_{t\geq 0}$. This means that for every $\{i_1,\dots,i_k\} \subseteq \{1,\dots,m\}$ and for every choice $(\alpha_1,\beta_1),\dots,(\alpha_l,\beta_l)$ in $\{(i,j) \in \{1,\dots,m\}^2 \mid i \neq j\}$, we have, for all $t_j \geq t$, $j = 1,\dots,k$, and $s_n \geq t$, $n = 1,\dots,l$,

$$\begin{aligned}
&\mathbb{E}\left[\prod_{j=1}^{k}\prod_{n=1}^{l} f\big(Y_{t_j}(i_j)\big) g\big(C_{s_n}(\alpha_n,\beta_n)\big)\bigg|\mathcal{G}_t\right] \\
&= \prod_{j=1}^{k} \mathbb{E}\big[f\big(Y_{t_j}(i_j)\big)\big|\mathcal{G}_t\big] \prod_{n=1}^{l} \mathbb{E}\big[g\big(C_{s_n}(\alpha_n,\beta_n)\big)\big|\mathcal{G}_t\big] \\
&= \prod_{j=1}^{k} \mathbb{E}\big[f\big(Y_{t_j}(i_j)\big)\big|\mathcal{F}_\infty^\Psi \vee \mathcal{F}_t^{Y(i_j)}\big] \prod_{n=1}^{l} \mathbb{E}\big[g\big(C_{s_n}(\alpha_n,\beta_n)\big)\big|\mathcal{F}_\infty^\Psi \vee \mathcal{F}_t^{C(\alpha_n,\beta_n)}\big]
\end{aligned}$$

for $f, g : \{0,1\} \to \mathbb{R}$, with $\mathcal{F}_t^{Y(i)} := \sigma(Y_u(i) : u \leq t)$ and $\mathcal{F}_t^{C(i,j)} := \sigma(C_u(i,j) : u \leq t)$, for every $i, j \in \{1,\dots,m\}$, $i \neq j$.

(3) For every $i \in \{1,\dots,m\}$, the self-default indicator process $(Y_t(i))_{t\geq 0}$ is a doubly stochastic indicator process with respect to the filtration $(\mathcal{F}_\infty^\Psi \vee \mathcal{F}_t^Y)_{t\geq 0}$ with stochastic intensity depending only on the path of $(\Psi_t)_{t\geq 0}$. In particular, we assume that the stochastic intensity of firm i is of the form $\lambda^i(t,\Psi_t)$ with a continuous function $\lambda^i : \mathbb{R}^2 \to \mathbb{R}_+$. This means that

$$\mathbb{E}\big[1 - Y_s(i)\big|\mathcal{G}_t\big] = \big(1 - Y_t(i)\big)e^{-\int_t^s \lambda^i(u,\Psi_u)\,du}, \quad s \geq t, \tag{4.4}$$

where the last equality holds by Corollary 5.1.5 of Bielecki and Rutkowski [5].

(4) The contagion processes $(C_t(i,j))_{t\geq 0}$ for $i \neq j$ are $\mathcal{F}_\infty^\Psi$-conditionally time-inhomogeneous Markov chains; i.e., for every function $f : \{0,1\} \to \mathbb{R}$,

$$\mathbb{E}\big[f\big(C_s(i,j)\big)\big|\mathcal{F}_\infty^\Psi \vee \mathcal{F}_t^{C(i,j)}\big] = \mathbb{E}\big[f\big(C_s(i,j)\big)\big|\mathcal{F}_\infty^\Psi \vee \sigma\big(C_t(i,j)\big)\big], \quad s \geq t.$$

For all $i, j \in \{1,\dots,m\}$, $i \neq j$, and $k, h \in \{0,1\}$, we denote the conditional transition probabilities by

$$p_{ts}^{ij}(k,h) = \mathbb{P}\big[C_s(i,j) = h\big|\mathcal{F}_\infty^\Psi \vee \sigma\big(C_t(i,j) = k\big)\big],$$

and assume that $(p_{ts}^{ij}(k,h))_{s\in\mathbb{R}^+}$ is a continuous process for all $t \in \mathbb{R}^+$, $i, j \in \{1,\dots,m\}$, and $k, h \in \{0,1\}$.

In the sequel we will use the fact that, on $\{C_t(i,j)=k\}$,

$$p^{ij}_{ts}(k,h)=\frac{\mathbb{P}[C_s(i,j)=h, C_t(i,j)=k|\mathcal{F}^{\Psi}_{\infty}]}{\mathbb{P}[C_t(i,j)=k|\mathcal{F}^{\Psi}_{\infty}]}.$$

Note that, unlike the default indicator processes, the processes $(C_t(i,j))_{t\geq 0}$ are allowed to change between 0 and 1 back and forth in time. They model the presence of a business relationship between firm i and firm j, which can be present at time 0, absent at some later time, and come in force again even later.

4.3 A Portfolio with Disjoint Contagion Classes

We want to discuss our model assumptions for the simple case of a credit portfolio with group structure. For simplicity, we assume that the matrix C is *time-independent* and *deterministic*. This means that we can divide the credit portfolio of m firms into groups, which we can identify by the following assumptions.

Assumption 4.2

(1) Reflexivity: By definition $C(i,i)=1$ for all $i\in\{1,\dots,m\}$.
(2) Symmetry: $C(i,j)=C(j,i)$ for all $i,j\in\{1,\dots,m\}$.
The influence of default is symmetric.
(3) Transitivity: $C(i,h)C(h,j)\leq C(i,j)$ for all $i,j,h\in\{1,\dots,m\}$.
If the default of firm i causes firm h to default, and firm h causes firm j to default, then also firm i causes the default of firm j.

Assumptions 4.2 define an equivalence relation on the credit portfolio, i.e., $i\sim j$ if and only if $C(i,j)=1$. The equivalence relation subdivides the portfolio into disjoint equivalence classes, which we call *contagion classes* and denote by

$$[i]:=\big\{j\in\{1,\dots,m\}\;\big|\;C(i,j)=1\big\}.$$

We assume that the portfolio consists of k contagion classes $[i_1],\dots,[i_k]$, representing, for instance, business sections or local markets.

By definition (4.2) of the default indicator process we have:

$$Z_t(i)=\begin{cases}1 & \text{if } (Y_t(i)=1)\vee(\exists j\neq i\; C(i,j)=1 \text{ s.t. } Y_t(j)=1),\\ 0 & \text{if } (Y_t(i)=0)\wedge(Y_t(j)=0\;\forall j\neq i \text{ s.t. } C(i,j)=1).\end{cases}\tag{4.5}$$

Given some $i\in\{1,\dots,m\}$, from the definition of the default indicator process in (4.2) and Assumption 4.2 we have

$$Z_t(i)=0 \iff Z_t(j)=0 \quad \forall j\in[i].$$

This means that either all firms of the same contagion class default at the same time or all of them are alive. Here we see that our modeling is different from (and more drastic than) the usual credit risk contagion modeling, where the default of some firm within a group only increases the hazard of all other group members; for examples and further references, see Schönbucher [19], Chap. 10.5.

Conditionally on the macroeconomic state variable process Ψ, the default indicator process $(Z_t)_{t\geq 0}$ is Markovian. Since in this case C is supposed to be deterministic, it is to be expected that the intensities of $(Z_t)_{t\geq 0}$ are inherited in a deterministic way by the default intensities of the self-indicator process $(Y_t)_{t\geq 0}$ as given by (4.4) of Assumption 4.1(3).

This allows us to calculate the conditional generator of the default indicator process and of the default number process.

4.3.1 Conditional Infinitesimal Generator of the Default Indicator Process

We calculate the conditional infinitesimal generator of the default indicator process $(Z_t)_{t\geq 0}$, where we use Definition 2.2 of Yin and Zhang [22].

Theorem 4.3 *The infinitesimal generator $\mathcal{A}_t$ of the $\mathcal{F}^{\Psi}_{\infty}$-conditional time-inhomogeneous Markov process $(Z_t)_{t\geq 0}$ is for any test function $f : \{0,1\}^m \to \mathbb{R}$ given by*

$$\mathcal{A}^Z_t f(z) = \prod_{j=1}^{m} \prod_{u\in[j]} \bigl(1 - |z_j - z_u|\bigr) \sum_{i=1}^{m} \bigl[f\bigl(z^{(i)}\bigr) - f(z)\bigr](1 - z_i)\lambda^i(t, \Psi_t),$$
$$z \in \{0,1\}^m, \tag{4.6}$$

where

$$z^{(i)} = \bigl(z_1 + (1 - z_1)C(i,1), \ldots, z_m + (1 - z_m)C(i,m)\bigr). \tag{4.7}$$

Proof By Proposition 11.3.1 of [5] we obtain that the infinitesimal conditional generator of Z_t is given by

$$\mathcal{A}^Z_t f(z) = \sum_{w\neq z} \bigl[f(w) - f(z)\bigr]\lambda^Z_t(z, w)$$

for any $f : \{0,1\}^m \to \mathbb{R}$, where $\lambda^Z_t(z, w)$ denotes the $\mathcal{F}^{\Psi}_{\infty}$-conditional stochastic intensity of the process Z from state z to state w, given by

$$\lambda^Z_t(z, w) := \lim_{h\to 0} \frac{p^Z_{tt+h}(z, w) - p^Z_{tt}(z, w)}{h} \tag{4.8}$$

with conditional transition probabilities

$$p^Z_{tt+s}(z, w) := \frac{\mathbb{P}(Z_{t+s} = w, Z_t = z | \mathcal{F}^{\Psi}_{\infty})}{\mathbb{P}(Z_t = z | \mathcal{F}^{\Psi}_{\infty})} =: \mathbb{P}^{\Psi}(Z_{t+s} = w | Z_t = z), \quad t, s \geq 0,$$

and

$$p^Z_{tt}(z,w)=\delta_{z,w}:=\begin{cases}1 & \text{if } z=w,\\ 0 & \text{otherwise.}\end{cases}$$

Since the different contagion classes are independent, we factorize the transition probabilities as follows:

$$p^Z_{tt+s}(z,w):=\prod_{h=1}^{k}\mathbb{P}^{\Psi}\Bigg(\bigcap_{i\in[i_h]}Z_{t+s}(i)=w_i\,\Bigg|\,\bigcap_{i\in[i_h]}Z_t(i)=z_i\Bigg). \tag{4.9}$$

Recall that in each factor in (4.9) the states $w_i, z_i \in \{0,1\}$ and that 1 is the absorbing state. Because of the deterministic contagion mechanism, at any time either the whole contagion class of firms has defaulted or has not, i.e.,

$$\exists i\in[i_h]\quad\text{s.t.}\quad\big\{Z_t(i)=0\big\}\iff\bigcap_{i\in[i_h]}\big\{Z_t(i)=0\big\}. \tag{4.10}$$

Moreover, by definition (4.5) we have that

$$\bigcap_{i\in[i_h]}\big\{Z_t(i)=0\big\}\iff\bigcap_{i\in[i_h]}\big\{Y_t(i)=0\big\}. \tag{4.11}$$

Setting $\overline{Z}_t(i_h):=\prod_{i\in[i_h]}Z_t(i)$, which also is a 0–1 random variable, we have

$$\bigcap_{i\in[i_h]}\big\{Z_t(i)=0\big\}\iff\big\{\overline{Z}_t(i_h)=0\big\},$$

and by (4.4), (4.10), and (4.11) we get

$$\mathbb{E}\big[1-\overline{Z}_{t+s}(i_h)\big|\mathcal{G}_t\big]=\big(1-\overline{Z}_t(i_h)\big)e^{-\int_t^{t+s}\sum_{i\in[i_h]}\lambda^i(u,\Psi_u)\,du}. \tag{4.12}$$

Given $z=(z_1,\dots,z_m)\in\{0,1\}^m$, we define, for $h\in\{1,\dots,k\}$,

$$z^{[i_h]}:=\big(z_1+(1-z_1)C(i_h,1),\dots,z_m+(1-z_m)C(i_h,m)\big),$$

representing the fact that only group $[i_h]$ can default, and, if it does, then all other components of z remain the same. Then by (4.9) and (4.12), taking the limit in (4.8), we obtain, for $z^{[i_h]}\neq z$,

$$\lambda^Z_t\big(z,z^{[i_h]}\big)=\prod_{j\in[i_h]}(1-z_j)\sum_{i\in[i_h]}\lambda^i(t,\Psi_t)$$

and $\lambda^Z_t(z,w)=0$ for $w\neq z^{[i_h]}$ or $w=z$.

Then, for elements z such that $z_i = z_j$ if firms i, j are in the same contagion class, the infinitesimal generator is given by

$$A_t^Z f(z) = \sum_{h=1}^{k} \left[f\left(z^{[i_h]}\right) - f(z) \right] \prod_{j \in [i_h]} (1 - z_j) \sum_{i \in [i_h]} \lambda^i(t, \Psi_t),$$

which can equivalently be represented as

$$A_t^Z f(z) = \sum_{i=1}^{m} \left[f\left(z^{(i)}\right) - f(z) \right] (1 - z_i) \lambda^i(t, \Psi_t),$$

where $z^{(i)}$ is defined as in (4.7). To guarantee that at the same time only defaults in one contagion class take place, we multiply the right-hand side by $\prod_{j=1}^{m} \prod_{u \in [j]} (1 - |z_j - z_u|)$, which means that the vector z cannot have two different components which correspond to equivalent firms. This gives the form of the generator as in (4.6). □

4.3.2 Conditional Infinitesimal Generator of the Default Number Process

We invoke the previous result to calculate the generator of the default number process for the portfolio. To this end, we split the group of all firms in l homogeneous groups $G_1, \dots, G_l$, where each group contains all the firms with the same default intensity. We recall that firms belonging to the same equivalent class $[i]$ have a default intensity given by

$$\lambda_t^{[i]} = \sum_{j \in [i]} \lambda^j(t, \Psi_t).$$

It follows that each homogeneous group G_h is given by the union of a certain number s_h of contagion classes, i.e.,

$$G_h = \left[j_1^h \right] \cup \dots \cup \left[j_{s_h}^h \right].$$

For every $h \in \{1, \dots, l\}$, we denote by n_i^h the cardinality of the class $[j_i^h]$ for $i = 1, \dots, s_h$ and by $\lambda^{G_h}(t, \Psi_t)$ the intensity of every firm belonging to the group G_h. Let

$$M_t(h) := \frac{1}{s_h} \left[\sum_{i \in [j_1^h]} \frac{Z_t(i)}{n_1^h} + \dots + \sum_{i \in [j_{s_h}^h]} \frac{Z_t(i)}{n_{s_h}^h} \right] \tag{4.13}$$

be the weighted average number of defaults in the group G_h. We now consider the process $M_t := (M_t(1), \dots, M_t(l))$. Because of the conditional independence of contagion classes, the components of this process are also conditionally independent. We calculate the conditional infinitesimal generator of $(M_t)_{t \geq 0}$.

Recall first from our calculations in the proof of Theorem 4.3 that we cannot have simultaneous defaults for two different contagion classes and that inside a contagion class all firms default at the same time. Hence the counting process $(M_t)_{t\geq 0}$ can jump from a state $u = (u_1, \ldots, u_l) = (\frac{v_1}{s_1}, \ldots, \frac{v_l}{s_l})$, where $v_k \in \{0, \ldots, s_k\}$ (for $k = 1, \ldots, l$), only to a state of the form $u + \frac{1}{s_k} e_k$, where e_k is the kth element of the canonical basis of $\mathbb{R}^l$. With an analogous proof as in Lemma 3.4 of Frey and Backhaus [11], we obtain that the transition intensity of M from u into the state $u + \frac{1}{s_k} e_k$ is given by

$$\lambda_t^M\left(u, u + \frac{1}{s_k} e_k\right) = s_k(1 - u_k)\lambda^{G_k}(s, \Psi_s).$$

Then the infinitesimal conditional generator of $(M_t)_{t\geq 0}$ has the following form.

Theorem 4.4 *Let* $M_t = (M_t(1), \ldots, M_t(l))$, $t \geq 0$, *be the default number process with components defined in* (4.13). *Under Assumptions* 4.1 *and* 4.2, *the infinitesimal generator of this* $\mathcal{F}_\infty^\Psi$*-conditional Markov process is for any test function* $f : \{0, \frac{1}{s_1}, \ldots, 1\} \times \cdots \times \{0, \frac{1}{s_l}, \ldots, 1\} \to \mathbb{R}$ *given by*

$$\mathcal{A}_t f(u) = \sum_{k=1}^{l}\left[f\left(u + \frac{1}{s_k} e_k\right) - f(u)\right] s_k(1 - u_k)\lambda^{G_k}(t, \Psi_t).$$

4.4 The Price of Credit Derivatives as a Function of Ψ

We consider the problem of pricing derivatives whose values are influenced by the contagion mechanism represented by the matrix C and the underlying macroeconomics factors Ψ as described in Sect. 4.2.1.

Assumption 4.5 (Market structure; cf. Frey and Backhaus [12], Assumption 3.1)

(1) The investor information at time t is given by the default history $\mathcal{F}_t$; i.e., the investor knows the latent process Ψ, the self-default indicator process Y, and the contagion matrix C up to time t.
(2) The default-free interest rate is deterministic, so that we can w.l.o.g. set it equal to 0. This does not prevent us to include, for instance, the LIBOR rate as one of the macroeconomic variables processes.
(3) A pricing (martingale) measure $\mathbb{P}$ exists and is known. For conditions such that this assumption holds, see, for example, Lemma 13.2 of [13]. The price in t of any $\mathcal{F}_T$-measurable claim $L \in L^1(\Omega, \mathbb{P})$ with maturity $T > 0$ is given by

$$L_t = \mathbb{E}[L|\mathcal{F}_t] \quad \text{for } 0 \leq t \leq T. \tag{4.14}$$

Here we do not assume that the pricing measure $\mathbb{P}$ is necessarily unique. In an incomplete market setting in absence of arbitrage, the price process of a claim is given by formula (4.14) for some choice of a martingale measure $\mathbb{P}$. See, for example, Theorem 5.30 of [15].

We do not investigate further the issue of market completeness in this model, since it goes beyond the interests of this paper. Given a contingent claim, we then focus on the pricing issue in this setting and compute (4.14) for a given pricing measure $\mathbb{P}$.

Without further specifying the macroeconomic process Ψ, we can formulate the following result.

Theorem 4.6 *Let $f : \mathbb{R} \times \mathbb{R}^m \to \mathbb{R}$ be a bounded measurable function. Let $\alpha = (\alpha_1, \dots, \alpha_m)$, $\beta = (\beta_1, \dots, \beta_m)$, and $z = (z_1, \dots, z_m)$ be in $\{0,1\}^m$, and $h^{(i)}, k^{(i)} \in \{0,1\}^{m-1}$ for $i = 1, \dots, m$. Set $h_{ii} = k_{ii} := 1$ for $i = 1, \dots, m$, $h_{ij} := [h^{(i)}]_j$ and $k_{ij} := [k^{(i)}]_j$ for $j \neq i$. Then, for $t \in [0, T]$,*

$$\begin{aligned}
&\mathbb{E}\big[f(\Psi_T, Z_T)\big|\mathcal{F}_t\big] \\
&\quad = \sum_{z,\alpha,\beta\in\{0,1\}^m} (-1)^{\sum_{j=1}^m \alpha_j z_j} \prod_{j=1}^m z_j^{1-\alpha_j} \prod_{i=1}^m \big(Y_t(i) a_t(i)\big)^{1-\beta_i} \big(1 - Y_t(i)\big)^{\beta_i} \\
&\qquad \times \mathbb{E}\left[f(\Psi_T, z) \prod_{i=1}^m b_{t,T}(i)^{\beta_i} \middle| \mathcal{F}_t^{\Psi} \right]
\end{aligned} \tag{4.15}$$

with

$$\begin{aligned}
a_t(i) &= \sum_{h^{(i)}\in\{0,1\}^{m-1}} \mathcal{I}_{\{\tilde{h}_i(\alpha,h)=0\}} \mathcal{I}_{\{C_{\tau_i}^{(i)}=h^{(i)}\}}, \\
b_{t,T}(i) &= \sum_{h^{(i)},k^{(i)}\in\{0,1\}^{m-1}} \mathcal{I}_{\{C_t^{(i)}=k^{(i)}\}} \bigg(\int_T^\infty \lambda^i(u, \Psi_u) e^{-\int_t^u \lambda^i(s,\Psi_s)\,ds} p_{t,u}\big(k^{(i)}, h^{(i)}\big)\, du \\
&\quad + \mathcal{I}_{\{\tilde{h}_i(\alpha,h)=0\}} \int_t^T \lambda^i(u, \Psi_u) e^{-\int_t^u \lambda^i(s,\Psi_s)\,ds} p_{t,u}\big(k^{(i)}, h^{(i)}\big)\, du \bigg),
\end{aligned}$$

where

$$\tilde{h}_i(\alpha, h) := \begin{cases} 0 & \text{if } \sum_{j=1}^m \alpha_j h_{ij} = 0, \\ 1 & \text{otherwise}, \end{cases} \tag{4.16}$$

and $p_{t,\tau_i}(k^{(i)}, h^{(i)}) := \prod_{j=1}^m p_{t\tau_i}^{ij}([k^{(i)}]_j, [h^{(i)}]_j)$ denotes the joint transition probabilities of the random vector $C_{\tau_i}^{(i)}$ from time t to time τ_i.

Our proof is based on the following lemma.

Lemma 4.7 *Assume the same notation as in Theorem 4.6. Then, for all $z \in \{0,1\}^m$ and $t \in [0, T]$,*

$$
\begin{aligned}
&\mathbb{E}[\mathcal{I}_{\{Z_T=z\}}|\mathcal{G}_t] \\
&\quad = \sum_{\alpha\in\{0,1\}^m} (-1)^{\sum_{j=1}^m \alpha_j z_j} \prod_{j=1}^m z_j^{1-\alpha_j} \prod_{i=1}^m \biggl[Y_t(i) \sum_{h^{(i)}\in\{0,1\}^{m-1}} \mathcal{I}_{\{\tilde{h}_i(\alpha,h)=0\}} \mathcal{I}_{\{C_{\tau_i}^{(i)}=h^{(i)}\}} \\
&\qquad + \bigl(1-Y_t(i)\bigr) \sum_{h^{(i)},k^{(i)}\in\{0,1\}^{m-1}} \mathcal{I}_{\{C_t^{(i)}=k^{(i)}\}} \\
&\qquad \times \biggl(\int_T^{+\infty} \lambda^i(u,\Psi_u) e^{-\int_t^u \lambda^i(s,\Psi_s)\,ds} p_{t,u}\bigl(k^{(i)},h^{(i)}\bigr)\,du \\
&\qquad + \mathcal{I}_{\{\tilde{h}_i(\alpha,h)=0\}} \int_t^T \lambda^i(u,\Psi_u) e^{-\int_t^u \lambda^i(s,\Psi_s)\,ds} p_{t,u}\bigl(k^{(i)},h^{(i)}\bigr)\,du \biggr) \biggr].
\end{aligned} \tag{4.17}
$$

Proof By (4.3) we have, for $z_j \in \{0,1\}$,

$$
\mathcal{I}_{\{Z_T(j)=z_j\}} = z_j + (-1)^{z_j} \prod_{i=1}^m \bigl(1 - C_{\tau_i}(i,j) Y_T(i)\bigr).
$$

Then, for $z \in \{0,1\}^m$,

$$
\mathcal{I}_{\{Z_T=z\}} = \prod_{j=1}^m \Biggl[z_j + (-1)^{z_j} \prod_{i=1}^m \bigl(1 - C_{\tau_i}(i,j) Y_T(i)\bigr) \Biggr]. \tag{4.18}
$$

We apply the following identity, which can be proved easily, for instance, by induction on m:

$$
\prod_{j=1}^m (A_j + B_j) = \sum_{\alpha\in\{0,1\}^m} \prod_{j=1}^m A_j^{1-\alpha_j} B_j^{\alpha_j}, \tag{4.19}
$$

where $\alpha_j \in \{0,1\}$, $j = 1,\dots,m$. Setting $0^0 := 1$, the formula holds also if there exists $j \in \{1,\dots,m\}$ such that $A_j = 0$ or $B_j = 0$. We apply this formula to

$$
A_j := z_j \quad \text{and} \quad B_j := (-1)^{z_j} \prod_{i=1}^m \bigl(1 - C_{\tau_i}(i,j) Y_T(i)\bigr).
$$

Then we obtain the following expression for the indicator function in (4.18):

$$
\begin{aligned}
\mathcal{I}_{\{Z_T=z\}} &= \sum_{\alpha\in\{0,1\}^m} \prod_{j=1}^m \Biggl(z_j^{1-\alpha_j} \Biggl[(-1)^{z_j} \prod_{i=1}^m \bigl(1 - C_{\tau_i}(i,j) Y_T(i)\bigr) \Biggr]^{\alpha_j} \Biggr) \\
&= \sum_{\alpha\in\{0,1\}^m} (-1)^{\sum_{j=1}^m \alpha_j z_j}
\end{aligned}
$$

$$\times \left(\prod_{j=1}^{m} z_j^{1-\alpha_j}\right) \prod_{j=1}^{m}\prod_{i=1}^{m}\left(1 - C_{\tau_i}(i,j)Y_T(i)\right)^{\alpha_j}. \tag{4.20}$$

Then by Assumption 4.1(2) we have that

$$\begin{aligned}\mathbb{E}[\mathcal{I}_{\{Z_T=z\}}|\mathcal{G}_t] = &\sum_{\alpha\in\{0,1\}^m} (-1)^{\sum_{j=1}^m \alpha_j z_j}\left(\prod_{j=1}^{m} z_j^{1-\alpha_j}\right) \\ &\times \prod_{i=1}^{m}\mathbb{E}\left[\prod_{j=1}^{m}\left(1 - C_{\tau_i}(i,j)Y_T(i)\right)^{\alpha_j}\middle|\mathcal{G}_t\right].\end{aligned} \tag{4.21}$$

We focus now on the calculation of the conditional expectation in (4.21). The total probability theorem, by considering all the possible contagion structures for ith row $C_{\tau_i}^{(i)}$ of the random matrix C_{τ_i} (written in its vector representation and avoiding the element $C_{\tau_i}(i,i)$), yields

$$\begin{aligned}&\mathbb{E}\left[\prod_{j=1}^{m}\left(1 - C_{\tau_i}(i,j)Y_T(i)\right)^{\alpha_j}\middle|\mathcal{G}_t\right] \\ &\quad = \sum_{h^{(i)}\in\{0,1\}^{m-1}} \mathbb{E}\left[\prod_{j=1}^{m}\left(1 - h_{ij}Y_T(i)\right)^{\alpha_j}\mathcal{I}_{\{C_{\tau_i}^{(i)}=h^{(i)}\}}\middle|\mathcal{G}_t\right] \\ &\quad = \sum_{h^{(i)}\in\{0,1\}^{m-1}} \mathbb{E}\left[\left(1 - Y_T(i)\right)^{\tilde{h}_i(\alpha,h)}\mathcal{I}_{\{C_{\tau_i}^{(i)}=h^{(i)}\}}\middle|\mathcal{G}_t\right],\end{aligned} \tag{4.22}$$

where $h_{ii} := 1$ and $h_{ij} := [h^{(i)}]_j$ for $j \neq i$, and $\tilde{h}_i(\alpha, h)$ is as in (4.16). We now calculate

$$\begin{aligned}&\mathbb{E}\left[\left(1 - Y_T(i)\right)^{\tilde{h}_i(\alpha,h)}\mathcal{I}_{\{C_{\tau_i}^{(i)}=h^{(i)}\}}\middle|\mathcal{G}_t\right] \\ &\quad = \mathbb{E}\left[(\mathcal{I}_{\{T<\tau_i\}})^{\tilde{h}_i(\alpha,h)}\mathcal{I}_{\{C_{\tau_i}^{(i)}=h^{(i)}\}}\middle|\mathcal{G}_t\right] \\ &\quad = \mathbb{E}\left[\mathbb{E}\left[(\mathcal{I}_{\{T<\tau_i\}})^{\tilde{h}_i(\alpha,h)}\mathcal{I}_{\{C_{\tau_i}^{(i)}=h^{(i)}\}}\middle|\mathcal{F}_T^{Y(i)} \vee \mathcal{G}_t\right]\middle|\mathcal{G}_t\right] \\ &\quad = \mathcal{I}_{\{\tau_i\le t\}}\mathcal{I}_{\{\tilde{h}_i(\alpha,h)=0\}}\mathcal{I}_{\{C_{\tau_i}^{(i)}=h^{(i)}\}} \\ &\qquad + \mathcal{I}_{\{\tilde{h}_i(\alpha,h)=0\}}\mathbb{E}\left[\mathcal{I}_{\{t<\tau_i\}}\sum_{k^{(i)}\in\{0,1\}^{m-1}} p_{t,\tau_i}(k^{(i)},h^{(i)})\mathcal{I}_{\{C_t^{(i)}=k^{(i)}\}}\middle|\mathcal{G}_t\right] \\ &\qquad + \mathcal{I}_{\{\tilde{h}_i(\alpha,h)\neq 0\}}\mathbb{E}\left[\mathcal{I}_{\{T<\tau_i\}}\sum_{k^{(i)}\in\{0,1\}^{m-1}} p_{t,\tau_i}(k^{(i)},h^{(i)})\mathcal{I}_{\{C_t^{(i)}=k^{(i)}\}}\middle|\mathcal{G}_t\right]\end{aligned}$$

$$
\begin{aligned}
&= Y_t(i)\mathcal{I}_{\{\tilde{h}_i(\alpha,h)=0\}}\mathcal{I}_{\{C^{(i)}_{\tau_i}=h^{(i)}\}} + \sum_{k^{(i)}\in\{0,1\}^{m-1}} \mathcal{I}_{\{C^{(i)}_t=k^{(i)}\}} \\
&\quad \times \big(\mathcal{I}_{\{\tilde{h}_i(\alpha,h)=0\}}\mathbb{E}\big[\mathcal{I}_{\{t<\tau_i\}}p_{t,\tau_i}\big(k^{(i)},h^{(i)}\big)\big|\mathcal{G}_t\big] \\
&\quad + \mathcal{I}_{\{\tilde{h}_i(\alpha,h)\neq 0\}}\mathbb{E}\big[\mathcal{I}_{\{T<\tau_i\}}p_{t,\tau_i}\big(k^{(i)},h^{(i)}\big)\big|\mathcal{G}_t\big]\big),
\end{aligned} \tag{4.23}
$$

where by using Assumption 4.1(2) and (4) we have set $p_{t,\tau_i}(k^{(i)},h^{(i)}) := \prod_{j=1}^m p^{ij}_{t\tau_i}([k^{(i)}]_j,[h^{(i)}]_j)$ to denote the joint transition probabilities of the random vector $C^{(i)}_{\tau_i}$ from time t to time τ_i under the convention that $[h^{(i)}]_i = [k^{(i)}]_i = p^{ii}_{t\tau_i}([k^{(i)}]_i,[h^{(i)}]_i) := 1$. Note that in the second term of (4.23) we have $\tau_i > t$.

Since by Assumption 4.1(4) $p_{t,\cdot}(k^{(i)},h^{(i)})$ is a bounded continuous stochastic process, we can now apply Proposition 5.1.1(ii) and Corollary 5.1.1(ii) of Bielecki and Rutkowski [5] and obtain

$$
\begin{aligned}
&\mathcal{I}_{\{\tilde{h}_i(\alpha,h)=0\}}\mathbb{E}\big[\mathcal{I}_{\{t<\tau_i\}}p_{t,\tau_i}\big(k^{(i)},h^{(i)}\big)\big|\mathcal{G}_t\big] + \mathcal{I}_{\{\tilde{h}_i(\alpha,h)\neq 0\}}\mathbb{E}\big[\mathcal{I}_{\{T<\tau_i\}}p_{t,\tau_i}\big(k^{(i)},h^{(i)}\big)\big|\mathcal{G}_t\big] \\
&\quad = \mathcal{I}_{\{\tau_i>t\}}\bigg(\mathcal{I}_{\{\tilde{h}_i(\alpha,h)=0\}}\mathbb{E}\bigg[\int_t^{+\infty}\lambda^i(u,\Psi_u)e^{-\int_t^u\lambda^i(s,\Psi_s)\,ds}p_{t,u}\big(k^{(i)},h^{(i)}\big)\,du\bigg|\mathcal{F}^{\Psi}_{\infty}\bigg] \\
&\qquad + \mathcal{I}_{\{\tilde{h}_i(\alpha,h)\neq 0\}}\mathbb{E}\bigg[\int_T^{+\infty}\lambda^i(u,\Psi_u)e^{-\int_t^u\lambda^i(s,\Psi_s)\,ds}p_{t,u}\big(k^{(i)},h^{(i)}\big)\,du\bigg|\mathcal{F}^{\Psi}_{\infty}\bigg]\bigg) \\
&\quad = \big(1-Y_t(i)\big)\bigg(\int_T^{+\infty}\lambda^i(u,\Psi_u)e^{-\int_t^u\lambda^i(s,\Psi_s)\,ds}p_{t,u}\big(k^{(i)},h^{(i)}\big)\,du \\
&\qquad + \mathcal{I}_{\{\tilde{h}_i(\alpha,h)=0\}}\int_t^T\lambda^i(u,\Psi_u)e^{-\int_t^u\lambda^i(s,\Psi_s)\,ds}p_{t,u}\big(k^{(i)},h^{(i)}\big)\,du\bigg),
\end{aligned} \tag{4.24}
$$

where in the last equality we have used the fact that all terms in the conditional expectation are $\mathcal{F}^{\Psi}_{\infty}$-measurable (Assumption 4.1(4)).

By plugging now (4.23) and (4.24) into (4.22) and then into (4.21) we conclude the proof. □

We are now ready to prove Theorem 4.6.

Proof of Theorem 4.6 Iterating the conditional expectation, we get

$$
\mathbb{E}\big[f(\Psi_T,Z_T)\big|\mathcal{F}_t\big] = \mathbb{E}\big[\mathbb{E}\big[f(\Psi_T,Z_T)\big|\mathcal{G}_t\big]\big|\mathcal{F}_t\big].
$$

In order to calculate the inner conditional expectation, we use formula (4.17) of Lemma 4.7. For the sake of simplicity, we set

$$
a_t(i) = \sum_{h^{(i)}\in\{0,1\}^{m-1}} \mathcal{I}_{\{\tilde{h}_i(\alpha,h)=0\}}\mathcal{I}_{\{C^{(i)}_{\tau_i}=h^{(i)}\}}
$$

and

$$
\begin{aligned}
b_{t,T}(i) = & \sum_{h^{(i)},k^{(i)}\in\{0,1\}^{m-1}} \mathcal{I}_{\{C_t^{(i)}=k^{(i)}\}} \\
& \times \left(\int_T^{+\infty} \lambda^i(u,\Psi_u) e^{-\int_t^u \lambda^i(s,\Psi_s)\,ds} p_{t,u}\big(k^{(i)},h^{(i)}\big)\,du \right. \\
& \left. + \mathcal{I}_{\{\tilde{h}_i(\alpha,h)=0\}} \int_t^T \lambda^i(u,\Psi_u) e^{-\int_t^u \lambda^i(s,\Psi_s)\,ds} p_{t,u}\big(k^{(i)},h^{(i)}\big)\,du \right).
\end{aligned}
$$

Then by the total probability theorem it follows that

$$
\begin{aligned}
& \mathbb{E}\big[f(\Psi_T,Z_T)\big|\mathcal{F}_t\big] \\
& = \mathbb{E}\left[\sum_{z\in\{0,1\}^m} f(\Psi_T,z)\mathbb{E}[\mathcal{I}_{\{Z_T=z\}}|\mathcal{G}_t] \middle| \mathcal{F}_t \right] \\
& = \sum_{\alpha,z\in\{0,1\}^m} (-1)^{\sum_{j=1}^m \alpha_j z_j} \prod_{j=1}^m z_j^{1-\alpha_j} \\
& \quad \times \mathbb{E}\left[f(\Psi_T,z) \prod_{i=1}^m \big(Y_t(i)a_t(i) + \big(1-Y_t(i)\big)b_{t,T}(i)\big) \middle| \mathcal{F}_t \right].
\end{aligned}
\tag{4.25}
$$

We now calculate the conditional expectation appearing in (4.25). By (4.19) we have that

$$
\begin{aligned}
& \prod_{i=1}^m \big(Y_t(i)a_t(i) + \big(1-Y_t(i)\big)b_{t,T}(i)\big) \\
& \quad = \sum_{\beta\in\{0,1\}^m} \prod_{i=1}^m \big(Y_t(i)a_t(i)\big)^{1-\beta_i} \big(\big(1-Y_t(i)\big)b_{t,T}(i)\big)^{\beta_i}.
\end{aligned}
$$

Hence,

$$
\begin{aligned}
& \mathbb{E}\left[f(\Psi_T,z) \prod_{i=1}^m \big(Y_t(i)a_t(i) + \big(1-Y_t(i)\big)b_{t,T}(i)\big) \middle| \mathcal{F}_t \right] \\
& = \mathbb{E}\left[f(\Psi_T,z) \sum_{\beta\in\{0,1\}^m} \prod_{i=1}^m \big(Y_t(i)a_t(i)\big)^{1-\beta_i} \big(\big(1-Y_t(i)\big)b_{t,T}(i)\big)^{\beta_i} \middle| \mathcal{F}_t \right] \\
& = \sum_{\beta\in\{0,1\}^m} \prod_{i=1}^m \big(Y_t(i)a_t(i)\big)^{1-\beta_i} \big(1-Y_t(i)\big)^{\beta_i} \mathbb{E}\left[f(\Psi_T,z) \prod_{i=1}^m b_{t,T}(i)^{\beta_i} \middle| \mathcal{F}_t^{\Psi} \right].
\end{aligned}
\tag{4.26}
$$

Plugging (4.26) into (4.25) concludes the proof. □

Equation (4.26) shows that the final pricing formula depends on the specification of the macroeconomic process Ψ and the dynamics of the contagion matrix C.

We first comment on the contagion matrix. Recall that it simply describes for two firms if there is a business relation at time t or not. From our formulas it is clear that we only need to know C at the time of default. There is still room for more precise modeling of the contagion matrix. For the moment, we assume that a time-independent, but possibly random, contagion matrix is given by

$$C_t(i,j) = C(i,j)Y_t(i), \quad t \geq 0, \tag{4.27}$$

where C has entries $[C]_{ij} = C_{ij}(\omega)$ given by i.i.d. random variables independent of the processes Y and Ψ. In what follows, we have then $\mathcal{F}_t = \mathcal{F}_t^\Psi \vee \mathcal{F}_t^Y \vee \sigma(C)$ for $t > 0$ and $\mathcal{F}_0 := \{\emptyset, \Omega\}$.

We study now this situation. Note that we still do not specify the macroeconomic process Ψ; this will only come in Sect. 4.5, where we price derivatives under the assumption of long-range dependence for Ψ.

Theorem 4.8 *If the contagion matrix is of the form* (4.27), *the pricing formula* (4.15) *is given for* $0 < t \leq T$ *by*

$$\begin{aligned} &\mathbb{E}\big[f(\Psi_T, Z_T)\big|\mathcal{F}_t\big] \\ &= \sum_{\alpha,z\in\{0,1\}^m} \sum_{h\in\{0,1\}^{m(m-1)}} (-1)^{\sum_{i=1}^m \alpha_i z_i} \prod_{i=1}^m z_i^{1-\alpha_i} \big(1 - Y_t(i)\big)^{\tilde{h}_i(\alpha,h)} \mathcal{I}_{\{C=h\}} \\ &\quad \times \mathbb{E}\Big[f(\Psi_T, z) e^{-\int_t^T \sum_{i=1}^m \tilde{h}_i(\alpha,h)\lambda^i(u,\Psi_u)\,du} \Big| \mathcal{F}_t^\Psi\Big] \end{aligned} \tag{4.28}$$

and for $t = 0$ *by*

$$\begin{aligned} \mathbb{E}\big[f(\Psi_T, Z_T)\big] = &\sum_{\alpha,z\in\{0,1\}^m} \sum_{h\in\{0,1\}^{m(m-1)}} (-1)^{\sum_{i=1}^m \alpha_i z_i} \prod_{i=1}^m z_i^{1-\alpha_i} \mathbb{P}(C = h) \\ &\times \mathbb{E}\Big[f(\Psi_T, z) e^{-\int_0^T \sum_{i=1}^m \tilde{h}_i(\alpha,h)\lambda^i(u,\Psi_u)\,du}\Big], \end{aligned} \tag{4.29}$$

where $\tilde{h}_i(\alpha,h)$ *is as in* (4.16) *with* $h_{ii} := 1$ *for* $i = 1, \dots, m$ *and* $h_{ij} := [h]_{ij}$ *for* $i \neq j$.

Proof First we note that in this case, $C_{\tau_i}(i,j)Y_t(i) = C(i,j)Y_t(i)$. The total probability theorem, by considering all possible contagion structures for the random matrix C (again written in its vector representation and avoiding the diagonal), yields with (4.20)

$$\begin{aligned} &\mathbb{E}[\mathcal{I}_{\{Z_T=z\}}|\mathcal{G}_t] \\ &\quad = \sum_{h\in\{0,1\}^{m(m-1)}} \mathbb{E}[\mathcal{I}_{\{Z_T=z\}}\mathcal{I}_{\{C=h\}}|\mathcal{G}_t] \end{aligned}$$

$$= \sum_{h\in\{0,1\}^{m(m-1)}} \sum_{\alpha\in\{0,1\}^m} (-1)^{\sum_{j=1}^m \alpha_j z_j} \left(\prod_{j=1}^m z_j^{1-\alpha_j}\right)$$
$$\times \mathbb{E}\left[\prod_{i=1}^m \prod_{j=1}^m \big(1 - h_{ij} Y_T(i)\big)^{\alpha_j} \mathcal{I}_{\{C=h\}} \middle| \mathcal{G}_t\right].$$

We now distinguish between $t = 0$ and $t > 0$. Since $\mathcal{I}_{\{C=h\}}$ is $\mathcal{F}_t$-measurable for every $t > 0$, by Assumption 4.1(2) and by (4.22) we obtain that, for $t > 0$,

$$\mathbb{E}[\mathcal{I}_{\{Z_T=z\}}|\mathcal{G}_t]$$
$$= \sum_{h\in\{0,1\}^{m(m-1)}} \sum_{\alpha\in\{0,1\}^m} (-1)^{\sum_{j=1}^m \alpha_j z_j} \left(\prod_{j=1}^m z_j^{1-\alpha_j}\right)$$
$$\times \mathcal{I}_{\{C=h\}} \prod_{i=1}^m \mathbb{E}\left[\prod_{j=1}^m \big(1 - h_{ij} Y_T(i)\big)^{\alpha_j} \middle| \mathcal{G}_t\right]$$
$$= \sum_{h\in\{0,1\}^{m(m-1)}} \sum_{\alpha\in\{0,1\}^m} (-1)^{\sum_{j=1}^m \alpha_j z_j} \left(\prod_{j=1}^m z_j^{1-\alpha_j}\right)$$
$$\times \mathcal{I}_{\{C=h\}} \prod_{i=1}^m \mathbb{E}\big[\big(1 - Y_T(i)\big)^{\tilde{h}_i(\alpha,h)} |\mathcal{G}_t\big], \tag{4.30}$$

where $\tilde{h}_i(\alpha, h)$ is as in (4.16) with $h_{ii} := 1$, $i = 1, \dots, m$ and $h_{ij} := [h]_{ij}$, $i \neq j$. Since by (4.4)

$$\mathbb{E}\big[\big(1 - Y_T(i)\big)^{\tilde{h}_i(\alpha,h)} |\mathcal{G}_t\big] = \big(1 - Y_t(i)\big)^{\tilde{h}_i(\alpha,h)} e^{-\int_t^T \tilde{h}_i(\alpha,h)\lambda^i(u,\Psi_u)\,du},$$

we obtain that, for $t > 0$,

$$\mathbb{E}[\mathcal{I}_{\{Z_T=z\}}|\mathcal{G}_t]$$
$$= \sum_{h\in\{0,1\}^{m(m-1)}} \sum_{\alpha\in\{0,1\}^m} (-1)^{\sum_{j=1}^m \alpha_j z_j} \left(\prod_{j=1}^m z_j^{1-\alpha_j}\right) \mathcal{I}_{\{C=h\}}$$
$$\times \prod_{i=1}^m \big(1 - Y_t(i)\big)^{\tilde{h}_i(\alpha,h)} e^{-\int_t^T \tilde{h}_i(\alpha,h)\lambda^i(u,\Psi_u)\,du}. \tag{4.31}$$

To obtain the final pricing formula, we proceed analogously as in the proof of Theorem 4.6. By (4.31) and (4.25) we have that, for $t > 0$,

$$\begin{aligned}
&\mathbb{E}\big[f(\Psi_T, Z_T)\big|\mathcal{F}_t\big]\\
&\quad=\mathbb{E}\Bigg[\sum_{z\in\{0,1\}^m} f(\Psi_T, z)\mathbb{E}[\mathcal{I}_{\{Z_T=z\}}|\mathcal{G}_t]\Bigg|\mathcal{F}_t\Bigg]\\
&\quad=\sum_{\alpha,z\in\{0,1\}^m}\ \sum_{h\in\{0,1\}^{m(m-1)}}(-1)^{\sum_{j=1}^m \alpha_j z_j}\Bigg(\prod_{j=1}^m z_j^{1-\alpha_j}\Bigg)\mathcal{I}_{\{C=h\}}\\
&\qquad\times\prod_{i=1}^m\big(1-Y_t(i)\big)^{\tilde{h}_i(\alpha,h)}\mathbb{E}\Big[f(\Psi_T, z)e^{-\int_t^T\sum_{i=1}^m \tilde{h}_i(\alpha,h)\lambda^i(u,\Psi_u)\,du}\Big|\mathcal{F}_t^{\Psi}\Big]. \qquad (4.32)
\end{aligned}$$

This proves (4.28). For $t=0$, we obtain

$$\begin{aligned}
&\mathbb{E}[\mathcal{I}_{\{Z_T=z\}}|\mathcal{G}_0]\\
&\quad=\sum_{h\in\{0,1\}^{m(m-1)}}\ \sum_{\alpha\in\{0,1\}^m}(-1)^{\sum_{j=1}^m \alpha_j z_j}\Bigg(\prod_{j=1}^m z_j^{1-\alpha_j}\Bigg)\\
&\qquad\times\mathbb{P}(C=h)e^{-\int_0^T\sum_{i=1}^m \tilde{h}_i(\alpha,h)\lambda^i(u,\Psi_u)\,du}. \qquad (4.33)
\end{aligned}$$

Substituting (4.33) into (4.32) for $t=0$, we obtain formula (4.29). □

If the contagion matrix is deterministic, i.e., for all $i, j \in \{1,\dots,m\}$ and $t\geq 0$,

$$C_t(i,j)(\omega)=C_t(i,j)\in\{0,1\}\quad \forall\omega\in\Omega,$$

then we have $\mathcal{F}_t^C=\{\emptyset,\Omega\}$ for every $t\in[0,T]$.

Corollary 4.9 *Assuming that the contagion matrix is deterministic, the pricing formula* (4.15) *simplifies to*

$$\begin{aligned}
\mathbb{E}\big[f(\Psi_T, Z_T)\big|\mathcal{F}_t\big]&=\sum_{\alpha,z\in\{0,1\}^m}(-1)^{\sum_{i=1}^m \alpha_i z_i}\prod_{i=1}^m\Big(z_i^{1-\alpha_i}\big(1-Y_t(i)\big)^{\tilde{h}_i(\alpha)}\Big)\\
&\quad\times\mathbb{E}\Big[f(\Psi_T, z)e^{-\int_t^T\sum_{i=1}^m \tilde{h}_i(\alpha)\lambda^i(u,\Psi_u)\,du}\Big|\mathcal{F}_t^{\Psi}\Big], \qquad (4.34)
\end{aligned}$$

where

$$\tilde{h}_i(\alpha):=\begin{cases}0 & \textit{if } \sum_{j=1}^m \alpha_j C_T(i,j)=0,\\ 1 & \textit{otherwise}.\end{cases} \qquad (4.35)$$

In the following simple example we investigate the effect of the contagion mechanism.

Example 4.10 We assume the group model of Sect. 4.3 in its simplest form of a portfolio consisting of two classes, taking 5 firms in one group and 10 firms in the second group. We work with a deterministic contagion matrix and consider different contagion scenarios for C,

$$C = \begin{pmatrix} C_{5\times 5} & C_{5\times 10} \\ C_{10\times 5} & C_{10\times 10} \end{pmatrix}.$$

We consider the following six scenarios, where I_d denotes the identity matrix in $\mathbb{R}^d$, $\mathbf{0}_{d\times k}$ the matrix with only entries 0, and $\mathbf{1}_{d\times k}$ the matrix with only entries 1:

$$C_1 = \mathrm{I}_{15}, \qquad C_2 = \begin{pmatrix} \mathbf{1}_{5\times 5} & \mathbf{0}_{5\times 10} \\ \mathbf{0}_{10\times 5} & \mathrm{I}_{10} \end{pmatrix}, \qquad C_3 = \begin{pmatrix} \mathrm{I}_5 & \mathbf{0}_{5\times 10} \\ \mathbf{0}_{10\times 5} & \mathbf{1}_{10\times 10} \end{pmatrix},$$

$$C_4 = \begin{pmatrix} \mathbf{1}_{5\times 5} & \mathbf{1}_{5\times 10} \\ \mathbf{0}_{10\times 5} & \mathbf{1}_{10\times 10} \end{pmatrix}, \qquad C_5 = \begin{pmatrix} \mathbf{1}_{5\times 5} & \mathbf{0}_{5\times 10} \\ \mathbf{1}_{10\times 5} & \mathrm{I}_{10} \end{pmatrix}, \qquad C_6 = \mathbf{1}_{15\times 15}.$$

Obviously, C_1 corresponds to no contagion and will serve as reference scenario. C_2 models contagion within the first group, no contagion in the second, and no contagion between firms of the two groups. C_3 models the complementary situation. Contagion matrix C_4 models contagion in the first group, but also the spill-over of default of group 1 firms into the second group. C_5 models contagion within both groups and contagion from firms in the second group to the first group. Finally, C_6 models contagion between all 15 firms.

These scenarios determine the vectors $(\tilde{h}_i(\alpha), i = 1, \dots, 15)$ for all $\alpha \in \{0, 1\}^m$. We also assume that all firms in the same group have the same intensity of default, i.e., $\lambda^i = \lambda^{[1]}$ for all $i \in \{1, \dots, 5\}$ and $\lambda^i = \lambda^{[2]}$ for all $i \in \{6, \dots, 15\}$. Furthermore, we assume that $\lambda^{[2]} = 2\lambda^{[1]}$ and that both groups are exposed to the same realization of the macroeconomic process $\lambda^{[1]}$.

Now to understand precisely what the effect of the contagion is, we take as simplest example one bond of one firm of the two groups at one time. For a defaultable bond of a firm in group $i \in \{1, 2\}$, with pricing formula (4.34) we have to calculate

$$V_0^{[i]} = E\left[\left(1 - Z_T^{[i]}\right)\exp\left\{-\int_0^T \sum_{i=1}^m \tilde{h}_i(\alpha)\lambda^{[i]}(u, \psi_u)\,du\right\}\right].$$

Note that the zeros in $\tilde{h}_i(\alpha)$ correspond to no default of all firms in group 1 and the second part of the vector to arbitrary values in the second group.

It remains to specify $\lambda^{[1]}$, and we take the CIR model with a.s. positive intensities, i.e., $\lambda^{[1]}(t, B_t) = \lambda_t^{[1]}$ is the solution to

$$d\lambda_t^{[1]} = a\left(b - \lambda_t^{[1]}\right) + \sigma\sqrt{\lambda_t^{[1]}}\,dB_t, \quad t \ge 0,$$

where $(B_t)_{t\ge 0}$ is standard Brownian motion, and we take the parameters $a = 2.0$, $b = 0.05$, $\sigma = 0.4$ and initial value $\lambda^{[1]}(0) = 0.03$. Obviously, prices should decrease for higher contagion scenarios and for bonds with higher maturity.

The results of the computations for the bond prices $V_0^{[i]}$, $i = 1, 2$, for $T = 1$ and $T = 2$ and the different scenarios are summarized in Table 4.1.

Table 4.1 Bond prices $V_0^{[i]}$ for maturities $T = 1$ and $T = 2$ and the different scenarios

	Bond of firm in group 1		Bond of firm in group 2	
	$T=1$	$T=2$	$T=1$	$T=2$
C_1	0.966936	0.923076	0.935458	0.853588
C_2	0.849446	0.681479	0.935458	0.853588
C_3	0.966936	0.923076	0.550128	0.258520
C_4	0.482438	0.195017	0.935458	0.853588
C_5	0.849446	0.681479	0.482438	0.195017
C_6	0.482438	0.195017	0.482438	0.195017

4.5 Pricing Contingent Claim Depending on the Macroeconomic Process with Credit Risk Contagion

4.5.1 Modeling the Macroeconomic Process

Now we turn to the macroeconomic process Ψ. There are many examples which consider the intensity as a function of a state vector of Markov processes; see, e.g., Schönbucher [19], Chap. 7. Gaussian processes and processes driven by Brownian motion are the most prominent ones. Here we focus on the case where $\Psi = \Psi^H$ is given by a long-range dependent process with Hurst index $H > \frac{1}{2}$. This choice is motivated by the fact that macroeconomic variables like demand and supply, interest rates, or other economic activity measures often exhibit long-range dependence. In the context of fractional processes, examples include fractional geometric Brownian motion or other processes driven by fractional Brownian motion with nonnegativity guaranteed.

We recall here the definition of fractional Brownian motion.

Definition 4.11 A fractional Brownian motion (fBm) $B^H = (B_t^H)_{t \geq 0}$ with Hurst index $H \in (0, 1)$ is a continuous centered Gaussian process with covariance function

$$\operatorname{cov}\big(B_t^H, B_s^H\big) := R^H(t,s) := \frac{1}{2}\big(t^{2H} + s^{2H} - |t-s|^{2H}\big), \quad t, s \in \mathbb{R}^+.$$

In this section we focus on the case where the macroeconomic process is given by a suitable function ψ of a stochastic integral of a deterministic function with respect to fBm with Hurst index $H > \frac{1}{2}$. For examples, see Buchmann and Klüppelberg [6], and for more details concerning fractional Brownian motion and the relevant stochastic calculus, we refer to Biagini et al. [4]. Then we will compute the pricing formula (4.15) of Theorem 4.6 under the macroeconomic variables model

$$\Psi_t^H := \psi\big(I_t^H\big), \quad I_t^H := \int_0^t g(s)\, dB_s^H, \quad t \in [0, T], \tag{4.36}$$

where ψ is an invertible continuous function, and g is a deterministic function in $L^2_H([0,T])$. We recall that $L^2_H([0,T])$ is the completion of the Schwartz space $\mathcal{S}([0,T])$ equipped with the inner product

$$\langle f,g\rangle_H := H(2H-1)\int_0^T\int_0^T f(s)g(t)|s-t|^{2H-2}\,ds\,dt < \infty, \qquad f,g\in\mathcal{S}\big([0,T]\big).$$

In particular, in (4.36) we focus on deterministic integrands $g \in H^{\mu}([0,T])$ (which is a subset of $L^2_H([0,T])$), the space of the Hölder continuous functions on $[0,T]$ of order $\mu > 1-H$, and such that $1/g(s)$ is defined for all $s\in[0,T]$.

Remark 4.12 Under the above condition on ψ and g, we get the following.

(i) The stochastic integral in formula (4.36) can be understood pathwise in the Riemann–Stieltjes sense (see Sect. 5.1 in [4]).
(ii) By Theorem 4.4.2 of [23] we have also that $B^H(t)=\int_0^t \frac{1}{g(s)}\,dI^H_s$, where this integral can again be interpreted in the Riemann–Stieltjes sense. This implies that the processes I^H and B^H generate the same filtration and that $\mathcal{F}^{\Psi^H}_t = \mathcal{F}^{B^H}_t$ (because ψ is invertible and measurable).

Although a long-range dependent macroeconomic process may be more realistic than a Markovian one, it is clear that the calculations and the resulting pricing formulas become much more complicated. In this paper we shall restrict ourselves to the case where for all $i\in\{1,\dots,m\}$, the default intensities of the self-default processes $(Y_t(i))_{t\geq 0}$ are stochastic and of the form

$$\lambda^i\big(u,\Psi^H_u\big)=\beta^i(u)I^H_u+\gamma^i(u), \quad u\geq 0, \tag{4.37}$$

where β^i and γ^i are continuous functions.

Recall that the intensities are supposed to be positive. Now, because the integral I^H has fBm as integrator, obviously it can happen with positive probability that the intensity becomes negative. By the affine transformation, however, we can at least control that this probability remains small. Of course, the same problem arises when working with affine models driven by Brownian motion as, for instance, for the Ornstein–Uhlenbeck model (see Schönbucher [19], Sect. 7.1).

In this paper we work with Gaussian macroeconomic variables but allow for different covariances in time governed by the function g. For a further discussion on possible choices of g and I^H, we also refer to [3]. Other fBm-driven models for macroeconomic factors are discussed in [6]; analogous fractional models beyond Gaussian are suggested in [14].

4.5.2 Pricing Contingent Claims with a Long-Range Dependent Ψ

In the setting outlined in Sect. 4.5.1 we focus on the pricing of contingent claims written on the long-range dependent macroeconomic index Ψ^H and affected by credit risk contagion. For the sake of simplicity, we consider the case where the contagion matrix C_t is deterministic for all $0 \leq t \leq T$. Referring to Corollary 4.9, the problem is now to calculate a term of the form

$$\begin{aligned}
&\mathbb{E}\Big[f(\Psi_T^H, z)e^{-\int_t^T \sum_{i=1}^m \tilde{h}_i(\alpha)\lambda^i(u,\Psi_u^H)\,du}\Big|\mathcal{F}_t\Big] \\
&= \mathbb{E}\Big[f^\psi(I_T^H, z)e^{-\int_t^T \sum_{i=1}^m \tilde{h}_i(\alpha)(\beta^i(u)I_u^H+\gamma^i(u))\,du}\Big|\mathcal{F}_t\Big] \\
&= e^{-\int_t^T \sum_{i=1}^m \tilde{h}_i(\alpha)\gamma^i(u)\,du}e^{\int_0^t \sum_{i=1}^m \tilde{h}_i(\alpha)\beta^i(u)I_u^H\,du} \\
&\quad \times \mathbb{E}\Big[f^\psi(I_T^H, z)e^{-\int_0^T \sum_{i=1}^m \tilde{h}_i(\alpha)\beta^i(u)I_u^H\,du}\Big|\mathcal{F}_t^{\Psi^H}\Big]
\end{aligned} \tag{4.38}$$

for fixed $z \in \{0,1\}^m$, where we have set $f^\psi := f \circ \psi$. Note that in (4.38) the last equality holds by Assumption 4.1(1).

For simplicity, we omit in the sequel the index z and write simply $f(\Psi_T^H)$ and $f^\psi(I_T^H)$ instead of $f(\Psi_T^H, z)$ and $f^\psi(I_T^H, z)$, respectively.

We now proceed as follows. For $a \in \mathbb{R}$, define the function $f_a^\psi(x) := e^{-ax}f^\psi(x)$ for $x \in \mathbb{R}$ and its Fourier transform by $\widehat{f_a^\psi}(\xi) := \int_{\mathbb{R}} e^{-i\xi x} f_a^\psi(x)\,dx$ for $\xi \in \mathbb{R}$. We assume that f and ψ are such that

$$A := \big\{a \in \mathbb{R} \mid f_a^\psi(\cdot) \in L^1(\mathbb{R}) \text{ and } \widehat{f_a^\psi}(\cdot) \in L^1(\mathbb{R})\big\} \neq \emptyset.$$

Then by Theorem 9.1 of Rudin [18] the following inversion formula holds:

$$f_a^\psi(x) = \frac{1}{2\pi}\int_{\mathbb{R}} e^{i\xi x}\widehat{f_a^\psi}(\xi)\,d\xi, \quad x \in \mathbb{R}. \tag{4.39}$$

We collect some useful results in the following lemma.

Lemma 4.13 *With the same notation and assumptions as above, we have*

$$\begin{aligned}
&\mathbb{E}\Big[f(\Psi_T^H)e^{-\int_0^T \sum_{i=1}^m \tilde{h}_i(\alpha)\beta^i(u)I_u^H\,du}\Big|\mathcal{F}_t^{\Psi^H}\Big] \\
&= \frac{1}{2\pi}\int_{\mathbb{R}} \mathbb{E}\big[e^{\int_0^T \eta(s,\xi)\,dB_s^H}\big|\mathcal{F}_t^{\Psi^H}\big]\widehat{f_a^\psi}(\xi)\,d\xi,
\end{aligned} \tag{4.40}$$

where

$$\eta(s,\xi) := g(s)\left(a + i\xi - \int_s^T \sum_{i=1}^m \tilde{h}_i(\alpha)\beta^i(u)\,du\right), \quad s \in [0,T], \tag{4.41}$$

where $\tilde{h}_i(\alpha)$ for $i \in \{1,\ldots,m\}$ are defined in (4.35).

Furthermore,

$$\mathbb{E}\big[e^{\int_0^T \eta(s,\xi)\,dB_s^H}\big|\mathcal{F}_t^{\Psi^H}\big]$$
$$= \exp\left\{\frac{1}{2}\big\|\eta(\cdot,\xi)\mathcal{I}_{(t,T)}(\cdot)\big\|_H^2 - \frac{1}{2}\big\|\psi_\eta^H(\cdot,\xi,t,T)\mathcal{I}_{(0,t)}(\cdot)\big\|_H^2\right\}$$
$$\times \exp\left\{\int_0^t \big(\eta(s,\xi) + \psi_\eta^H(s,\xi,t,T)\big)\,dB_s^H\right\}, \tag{4.42}$$

where $\|f\|_H^2 := \langle f, f\rangle_H$ *for* $f \in L_H^2([0,T])$,

$$\psi_\eta^H(s,\xi,t,T) = s^{-H+\frac{1}{2}} I_{t^-}^{-(H-\frac{1}{2})}\big(I_{T^-}^{H-\frac{1}{2}}(\cdot)^{H-\frac{1}{2}}\eta(\cdot,\xi)\mathcal{I}_{[t,T]}(\cdot)\big)(s), \tag{4.43}$$

and for $\alpha = H - \frac{1}{2} \in (0, 1/2)$, *we define*

$$\big(I_{t^-}^{\alpha}\eta\big)(s) := \frac{1}{\Gamma(\alpha)}\left(\int_s^t \eta(r)(r-s)^{\alpha-1}\,dr\right), \quad 0 \le s \le t, \tag{4.44}$$

and

$$\big(I_{T^-}^{-\alpha}\eta\big)(s) := -\frac{1}{\Gamma(1-\alpha)}\frac{d}{ds}\left(\int_s^T \eta(r)(r-s)^{-\alpha}\,dr\right), \quad 0 < s < T. \tag{4.45}$$

Proof We first prove (4.40). We introduce the notation $E_T := \exp[-\int_0^T \sum_{i=1}^m \tilde{h}_i(\alpha)\beta^i(u)I_u^H\,du]$. By Theorem 6.4 of Sottinen [20] it follows that we can exchange the order of integration and obtain

$$E_T = e^{-\int_0^T g(s)(\int_s^T \sum_{i=1}^m \tilde{h}_i(\alpha)\beta^i(u)\,du)\,dB_s^H}.$$

Then by using the definition of f_a^ψ and the Fourier inversion formula (4.39) we get

$$\mathbb{E}\big[f(\Psi_T^H)E_T\big|\mathcal{F}_t^{\Psi^H}\big]$$
$$= \mathbb{E}\left[\frac{1}{2\pi}\int_{\mathbb{R}} e^{\int_0^T [a+i\xi-\int_s^T \sum_{i=1}^m \tilde{h}_i(\alpha)\beta^i(u)\,du]g(s)\,dB_s^H}\,\widehat{f_a^\psi}(\xi)\,d\xi\,\middle|\,\mathcal{F}_t^{\Psi^H}\right]$$
$$= \frac{1}{2\pi}\mathbb{E}\left[\int_{\mathbb{R}} e^{\int_0^T \eta(s,\xi)\,dB_s^H}\,\middle|\,\mathcal{F}_t^{\Psi^H}\right]\widehat{f_a^\psi}(\xi)\,d\xi, \tag{4.46}$$

where

$$\eta(s,\xi) := g(s)\left(a + i\xi - \int_s^T \sum_{i=1}^m \tilde{h}_i(\alpha)\beta^i(u)\,du\right).$$

Since $\mathbb{E}[e^{\int_0^T \eta(s,\xi)\,dB_s^H}] < \infty$, we can exchange the order of integration in (4.46) and obtain (4.40).

Next, we prove (4.42). By Remark 4.12(ii) we have

$$\begin{aligned}
&\mathbb{E}\big[e^{\int_0^T \eta(s,\xi)\,dB_s^H} \big| \mathcal{F}_t^{\Psi^H}\big] \\
&= \mathbb{E}\big[e^{\int_0^T \eta(s,\xi)\,dB_s^H} \big| \mathcal{F}_t^{B^H}\big] \\
&= \exp\left\{\int_0^t \eta(s,\xi)\,dB_s^H\right\}\mathbb{E}\left[\exp\left\{\int_t^T \eta(s,\xi)\,dB_s^H\right\}\middle| \mathcal{F}_t^{B^H}\right],
\end{aligned}$$

which is equal to (4.42) by Proposition 3.6 of Biagini, Fink, and Klüppelberg [3]. □

Example 4.14 For the special case where $g \equiv 1$ (that gives $I_t^H = B_t^H$) and $\beta^i \equiv 0$ for all $i \in \{1, \dots, m\}$, formula (4.42) can be calculated by Theorem 3.2 of Valkeila [21] as follows:

$$\begin{aligned}
&\mathbb{E}\big[e^{(a+i\xi)B_T^H} \big| \mathcal{F}_t^{B^H}\big] \\
&\quad = \exp\left\{\frac{1}{2}(a+i\xi)^2\big(T^{2H} - \langle M^H\rangle_t\big) + (\alpha + i\xi)\left(B_t^H + \int_0^t \Phi_T(t,s)\,dB_s^H\right)\right\},
\end{aligned}$$

where

$$\langle M^H\rangle_t = \int_0^t z^H(T,s)^2\,ds$$

with

$$z^H(T,s) := \left(H - \frac{1}{2}\right)c_H s^{\frac{1}{2}-H}\int_s^T u^{H-\frac{3}{2}}(u-s)^{H-\frac{1}{2}}\,du,$$

$$c_H := \left(\frac{2H\Gamma(\frac{3}{2}-H)}{\Gamma(H+\frac{1}{2})\Gamma(2-2H)}\right)^{\frac{1}{2}},$$

and

$$\Phi_T(t,s) := \frac{1}{\pi}\sin\left(\pi\left(H-\frac{1}{2}\right)\right)s^{\frac{1}{2}-H}(t-1)^{\frac{1}{2}-H}\int_t^T \frac{u^{H-\frac{1}{2}}(u-t)^{H-\frac{1}{2}}}{u-s}\,du.$$

We are now able to provide a pricing formula for a long-range dependent macroeconomic state variable process.

Theorem 4.15 *Assume that the contagion matrix C is deterministic and that for all $i \in \{1, \dots, m\}$, the intensities of the self-default processes $Y_i = (Y_t(i))_{t\geq 0}$ are of the form*

$$\lambda^i\big(t, \Psi_t^H\big) := \beta^i(t) I_t^H + \gamma^i(t), \quad t \geq 0,$$

where β^i and γ^i are continuous functions. Consider

$$I_t^H := \int_0^t g(s)\,dB^H(s), \quad t \geq 0,$$

for $g \in H^{\mu}([0,T])$ *with* $\mu > 1 - H$ *and such that* $\frac{1}{g}$ *is well defined. Let* $f(\cdot, z)$ *and* $\psi(\cdot)$ *be deterministic continuous functions and denote, for all* $z \in \{0,1\}^m$,

$$f^{\psi}(x,z) := f\big(\psi(x), z\big), \quad x \in \mathbb{R},$$

and

$$f^{\psi}_{\alpha}(x,z) := e^{-\alpha x} f^{\psi}(x,z), \quad \alpha, x \in \mathbb{R}.$$

Assume that there exists some $a \in \mathbb{R}$ *such that* $f^{\Psi}_{a}(\cdot, z)$ *and its Fourier transform* $\hat{f}^{\Psi}_{a}(\cdot, z)$ *belong to* $L^1(\mathbb{R})$ *for all* $z \in \{0,1\}^m$.

Finally, let ψ *be invertible and set*

$$\Psi^H_t := \psi\left(\int_0^t g(s)\, dB^H(s)\right), \quad t \geq 0.$$

Then the price (4.34) *at time* $t \in [0,T]$ *is given by the following formula*:

$$\begin{aligned}
&\mathbb{E}\big[f(\Psi_T, Z_T)\big|\mathcal{F}_t\big] \\
&= \sum_{\alpha, z \in \{0,1\}^m} (-1)^{\sum_{i=1}^m \alpha_i z_i} \prod_{i=1}^m \Big(z_i^{1-\alpha_i}\big(1 - Y_t(i)\big)^{\tilde{h}_i(\alpha)}\Big) e^{-\int_t^T \sum_{i=1}^m \tilde{h}_i(\alpha)\gamma^i(u)\, du} \\
&\quad \times e^{\int_0^t \sum_{i=1}^m \tilde{h}_i(\alpha)\beta^i(u) I_u^H\, du} \frac{1}{2\pi} \int_{\mathbb{R}} \exp\bigg\{\frac{1}{2}\big\|\eta(\cdot,\xi)\mathcal{I}_{(t,T)}(\cdot)\big\|_H^2 \\
&\quad - \frac{1}{2}\big\|\psi_\eta^H(\cdot,\xi,t,T)\mathcal{I}_{(0,t)}(\cdot)\big\|_H^2\bigg\} \\
&\quad \times \exp\bigg\{\int_0^t \big(\eta(s,\xi) + \psi_\eta^H(s,\xi,t,T)\big)\, dB_s^H\bigg\} \widehat{f_a^{\psi}}(\xi, z)\, d\xi,
\end{aligned} \tag{4.47}$$

where $\tilde{h}_i(\alpha)$ *is given in* (4.35), η *in* (4.41), *and* $\psi_\eta^H(\cdot,\xi,t,T)$ *in* (4.43).

A basic structural analysis for pricing formulas with long-range dependent hazard function models will be presented in Biagini, Fink, and Klüppelberg [3] together with an extended numerical analysis study.

Example 4.16 (Inflation-linked caps and floors) To illustrate how to compute $\widehat{f_a^{\Psi}}$, we introduce some examples of inflation-linked derivatives, such as inflation-linked caps and floors, that we allow to be also exposed to contagion risk. By using the notation of Theorem 4.15, we consider a payoff of the form

$$f(\Psi_T, Z_T) := (\Psi_T - k)^+ b(Z_T),$$

where Ψ_T represents here the inflation index, and $b(\cdot)$ is a positive measurable function, that describes the contagion effects. By Theorem 4.15 we have to find some

$a \in \mathbb{R}$ such that

$$(f_{\text{call}})_a^{\psi} \in L^1(\mathbb{R}) \quad \text{and} \quad \widehat{(f_{\text{call}})_a^{\psi}} \in L^1(\mathbb{R}). \tag{4.48}$$

We show (4.48) for $g \equiv 1$ and the two special cases $\psi(x) = x$ and $\psi(x) = e^x$, corresponding to fractional Brownian motion and geometric fractional Brownian motion, respectively.

(A) Let $\psi(x) = x$.
It follows immediately that (4.48) holds for $(f_{\text{call}})_a^{\psi}$, for all $a > 0$. We compute now the Fourier transform of $(f_{\text{call}})_a^{\psi}$ for $a > 0$:

$$\widehat{(f_{\text{call}})_a^{\psi}}(u) = \int_K^{\infty} e^{-x(a+iu)}(x-K)\,dx = \frac{e^{-K(a+iu)}}{(a+iu)^2}.$$

Since

$$\left|\frac{e^{-K(a+iu)}}{(a+iu)^2}\right| = \frac{e^{-Ka}}{u^2} = O\left(\frac{1}{u^2}\right), \quad u \to \infty,$$

we have that (4.48) holds also for the Fourier transform of $(f_{\text{call}})_a^{\psi}$ for all $a > 0$.

(B) Let $\psi(x) = e^x$.
It follows from the calculations in (A) that (4.48) holds for $(f_{\text{call}})_a^{\psi}$ for all $a > 1$. We compute now the Fourier transform of $(f_{\text{call}})_a^{\psi}$ for $a > 1$:

$$\widehat{(f_{\text{call}})_a^{\psi}}(u) = \int_{\ln K}^{\infty} e^{-x(a+iu)}\left(e^x - K\right)dx = \frac{e^{-(a-1+iu)\ln K}}{(a+iu)(a-1+iu)}.$$

Since

$$\left|\frac{e^{-(a-1+iu)\ln K}}{(a+iu)(a-1+iu)}\right| = \frac{e^{-(a-1)\ln K}}{|a(a-1)-u^2+iu(2a-1)|} = O\left(\frac{1}{u^2}\right), \quad u \to \infty,$$

condition (4.48) holds also for the Fourier transform of $(f_{\text{call}})_a^{\psi}$ for all $a > 1$.

4.5.3 Comparison with Markovian Ψ

Theorem 4.17 *Assume that* (4.36) *holds for standard Brownian motion as integrator and assume also* (4.37). *Then*

$$\mathbb{E}\left[f(\Psi_T, Z_T)\middle|\mathcal{F}_t\right] = \sum_{\alpha, z \in \{0,1\}^m} (-1)^{\sum_{i=1}^m \alpha_i z_i} \prod_{i=1}^m \left(z_i^{1-\alpha_i}\left(1 - Y_t(i)\right)^{\tilde{h}_i(\alpha)}\right) e^{-\int_t^T \sum_{i=1}^m \tilde{h}_i(\alpha)\gamma^i(u)\,du}$$

$$\times e^{\int_0^t \sum_{i=1}^m \tilde{h}_i(\alpha)\beta^i(u) I_u \, du} \frac{1}{2\pi} \int_{\mathbb{R}} \exp\left\{\frac{1}{2}\int_t^T \eta^2(s,\xi)\, ds + \int_0^t \eta(s,\xi)\, dB_s\right\} \widehat{f_a^\psi}(\xi, z)\, d\xi,$$

where $\tilde{h}_i(\alpha)$ *is given in* (4.35), *and* η *in* (4.41).

Proof By Theorem 4.8, (4.38), and Lemma 4.13 we obtain that calculating the price $\mathbb{E}[f(\Psi_T, Z_T)|\mathcal{F}_t]$ boils down to compute the term

$$\mathbb{E}\big[e^{\int_0^T \eta(s,\xi)\, dB_s} \big| \mathcal{F}_t^\Psi\big]$$

with $\eta(s,\xi) = g(s)(a + i\xi - \int_s^T \sum_{i=1}^m \tilde{h}_i(\alpha)\beta^i(u)\, du)$.

Since $\exp\{\frac{1}{2}\int_0^T \eta^2(s,\xi)\, ds\} < \infty$ for every fixed $\xi \in \mathbb{R}$, i.e., the Novikov condition is satisfied, we have that

$$\mathbb{E}\big[e^{\int_0^T \eta(s,\xi)\, dB_s} \big| \mathcal{F}_t^\Psi\big] = \exp\left\{\frac{1}{2}\int_t^T \eta^2(s,\xi)\, ds + \int_0^t \eta(s,\xi)\, dB_s\right\}. \qquad \square$$

Since the integrand g in (4.36) is in $L^2([0,T]) \subset L_H^2([0,T])$ (see [1] for the proof), we now compare the prices in $t = 0$ for the standard and long-range dependent cases.

(i) For the standard Brownian motion case, the price V_0 in $t = 0$ is equal to

$$V_0 = \sum_{\alpha, z \in \{0,1\}^m} (-1)^{\sum_{i=1}^m \alpha_i z_i} \prod_{i=1}^m z_i^{1-\alpha_i} e^{-\int_0^T \sum_{i=1}^m \tilde{h}_i(\alpha)\gamma^i(u)\, du} \times \frac{1}{2\pi}\int_{\mathbb{R}} \exp\left\{\frac{1}{2}\int_0^T \eta^2(s,\xi)\, ds\right\} \widehat{f_a^\psi}(\xi, z)\, d\xi;$$

(ii) For the fractional Brownian motion case, the price V_0^H in $t = 0$ is equal to

$$V_0^H = \sum_{\alpha, z \in \{0,1\}^m} (-1)^{\sum_{i=1}^m \alpha_i z_i} \prod_{i=1}^m z_i^{1-\alpha_i} e^{-\int_0^T \sum_{i=1}^m \tilde{h}_i(\alpha)\gamma^i(u)\, du} \times \frac{1}{2\pi}\int_{\mathbb{R}} \exp\left\{\frac{1}{2}\big\|\eta(\cdot,\xi)\mathcal{I}_{(0,T)}\big\|_H^2\right\} \widehat{f_a^\psi}(\xi, z)\, d\xi.$$

The difference is indeed due to the fact that for Brownian motion, we use the Itô integral and obtain consequently an Itô term, whereas for fBm, integration is pathwise. Anyway, we see that $V_0 > V_0^H$ for every $H > \frac{1}{2}$. Of course, the long-range dependence effect takes effect for prices V_t for $t > 0$, but then numerical calculations are called for.

Acknowledgements We thank Damir Filipović for interesting discussions and remarks. We gratefully acknowledge the numerical support of Vincenzo Ferrazzano, who calculated the prices for the different scenarios of Example 4.10. Finally, we thank Holger Fink und Peter Hepperger for careful reading of the paper, whose comments improved the paper considerably.

References

1. E. Alòs, D. Nualart, Stochastic integration with respect to the fractional Brownian motion. Stoch. Stoch. Rep. **75**, 129–152 (2003)
2. T. Björk, H. Hult, A note on Wick products and the fractional Black–Scholes model. Finance Stoch. **9**, 197–209 (2005)
3. F. Biagini, H. Fink, C. Klüppelberg, A fractional credit model with long range dependent default rate (2009, submitted). Available at www-m4.ma.tum.de/Papers/
4. F. Biagini, Y. Hu, B. Øksendal, T. Zhang, *Stochastic Calculus for Fractional Brownian Motion and Applications* (Springer, London, 2008)
5. T.R. Bielecki, M. Rutkowski, *Credit Risk: Modeling, Valuation and Hedging* (Springer, Berlin, 2002)
6. B. Buchmann, C. Klüppelberg, Fractional integral equations and state space transforms. Bernoulli **12**(3), 431–456 (2006)
7. M. Davis, V. Lo, Infectious defaults. Quant. Finance **1**, 382–387 (2001)
8. M. Davis, V. Lo, Modeling default correlation in bond portfolios, in *Mastering Risk*, ed. by C. Alexander. Applications, vol. 2 (Financial Times/Prentice Hall, New York, 2001), pp. 141–151
9. D. Duffie, *Credit Risk Modeling with Affine Processes* (Scuola Normale Superiore, Pisa, 2004)
10. D. Duffie, D. Filipovic, W. Schachermayer, Affine processes and applications in finance. Ann. Appl. Probab. **13**, 984–1053 (2003)
11. R. Frey, J. Backhaus, Portfolio credit risk models with interacting default intensities: a Markovian approach. Preprint, University of Leipzig (2004)
12. R. Frey, J. Backhaus, Pricing and hedging of portfolio credit derivatives with interacting default intensities. Int. J. Theor. Appl. Finance **11**(6), 611–634 (2008)
13. D. Filipovic, *Term Structure Models: A Graduate Course* (Springer, Berlin, 2009)
14. H. Fink, C. Klüppelberg, Fractional Lévy driven Ornstein–Uhlenbeck processes and stochastic differential equations. Bernoulli **17**(1), 484–506 (2011)
15. H. Föllmer, A. Schied, *Stochastic Finance: An Introduction in Discrete Time* (De Gruyter, Berlin, 2004)
16. M. Henry, P. Zaffaroni, The long-range dependence paradigm for macroeconomics and finance, in *Long-Range Dependence*, ed. by P. Doukhan, G. Oppenheim, M. Taqqu (Birkhäuser, Boston, 2003), pp. 417–438
17. B. Øksendal, Fractional Brownian motion in finance, in *Stochastic Economic Dynamics*, ed. by B.S. Jensen, T. Palokangas (Copenhagen Business School Press, Frederiksberg, 2007), pp. 11–56
18. W. Rudin, *Real and Complex Analysis* (McGraw-Hill, New York, 1987)
19. P.J. Schönbucher, *Credit Derivatives Pricing Models* (Wiley, New York, 2003)
20. T. Sottinen, Fractional Brownian motion in finance and queueing. Ph.D. Thesis, Helsinki University (2003)
21. E. Valkeila, On some properties of geometric fractional Brownian motion. Preprint number 224. Department of Mathematics, University of Helsinki (1999)
22. G.G. Yin, Q. Zhang, *Continuous-Time Markov Chains and Applications: A Singular Perturbation Approach* (Springer, New York, 1998)
23. M. Zähle, Integration with respect to fractal functions and stochastic calculus I. Probab. Theory Relat. Fields **111**, 333–374 (1998)

Chapter 5
Modelling Information Flows in Financial Markets

Dorje C. Brody, Lane P. Hughston, and Andrea Macrina

Abstract This paper presents an overview of information-based asset pricing. In the information-based approach, an asset is defined by its cash-flow structure. The market is assumed to have access to "partial" information about future cash flows. Each cash flow is determined by a collection of independent market factors called X-factors. The market filtration is generated by a set of information processes, each of which carries information about one of the X-factors, and eventually reveals the X-factor in a way that ensures that the associated cash flows have the correct measurability properties. In the models considered each information process has two terms, one of which contains a "signal" about the associated X-factor, and the other of which represents "market noise". The existence of an established pricing kernel, adapted to the market filtration, is assumed. The price of an asset is given by the expectation of the discounted cash flows in the associated risk-neutral measure, conditional on the information provided by the market. When the market noise is modelled by a Brownian bridge, one is able to construct explicit formulae for asset prices, as well as semi-analytic expressions for the prices and greeks of options and derivatives. In particular, option price data can be used to determine the information flow-rate parameters implicit in the definitions of the information processes. One consequence of the modelling framework is a specific scheme of stochastic volatility and correlation processes. Instead of imposing a volatility and correlation model upon the dynamics of a set of assets, one is able to deduce the dynamics of the volatilities and correlations of the asset price movements from more primitive assumptions involving the associated cash flows. The paper concludes with an

D.C. Brody (✉) · L.P. Hughston
Department of Mathematics, Imperial College London, London SW7 2AZ, UK
e-mail: d.brody@imperial.ac.uk

L.P. Hughston
e-mail: l.hughston@imperial.ac.uk

A. Macrina
Department of Mathematics, King's College London, London WC2R 2LS, UK
e-mail: andrea.macrina@kcl.ac.uk

A. Macrina
Institute of Economic Research, Kyoto University, Kyoto 606-8501, Japan

G. Di Nunno, B. Øksendal (eds.), *Advanced Mathematical Methods for Finance*,
DOI 10.1007/978-3-642-18412-3_5,

examination of situations involving asymmetric information. We present a simple model for informed traders and show how this can be used as a basis for so-called statistical arbitrage. Finally, we consider the problem of price formation in a heterogeneous market with multiple agents.

Keywords Information-based asset pricing · Information filter · Price formation · Statistical arbitrage

Mathematics Subject Classification (2010) 91G40 · 93E11 · 60G40 · 94Axx

5.1 Cash Flow Structures and Market Factors

In financial markets, the revelation of information is the most important factor in the determination of the price movements of financial assets. When a new piece of information (whether true, partly true, misleading, or bogus) circulates in a financial market, the prices of related assets move in response, and they move again when the information is updated. But how do we build specific models that incorporate the impact of information on asset prices? In this article we present an overview of some of the key issues involved in modelling the flow of information in financial markets and develop in some detail some elementary models for "information" in various situations. We show how information flow processes, when appropriately modelled, can be used to determine the associated price processes of financial assets. Applications to the pricing of various types of contingent claims will also be indicated. One of the contributions of the present work is to introduce a model for dynamic correlation in the situation where we consider a portfolio of assets. Rather than imposing an artificial correlation structure on the assets under consideration, we are able to infer the correlation structure from more basic assumptions. In the final section of the paper, we make some remarks about statistical arbitrage strategies and about price formation in markets characterised by inhomogeneous information flows.

When models are constructed for the pricing and risk management of complicated financial products, the price dynamics of the simpler financial assets, upon which the more complicated products are based, are often simply "assumed" (modulo some parametric or functional freedom). One can understand from a practical angle why it can be expeditious to proceed on that basis. Nevertheless, from a fundamental view we have to consider that even the basic financial assets (shares, bonds, etc.) are characterised by a number of potentially "complex" features, and so to make sense of the behaviour of such assets we need to consider what goes into the determination of their prices. To build up models for the dynamics of asset prices, it seems logical to proceed step by step along the following lines: (1) model the cash-flows arising from the asset as random variables; (2) model the market filtration (the flow of information to the market); (3) model the pricing kernel (which takes into account discounting, risk aversion, and the absence of arbitrage); and (4) work out the resulting dynamics for the price process.

We model the unfolding of chance in a financial market with the specification of a probability space $(\Omega, \mathcal{F}, \mathbb{P})$ on which we are going to construct a filtration

$\{\mathcal{F}_t\}$ representing the flow of information to market participants. Here $\mathbb{P}$ denotes the "physical" probability measure. The markets we consider will, in general, be incomplete. That is to say, although derivatives can be priced, we do not assume that they can be hedged. Since we are going to model the filtration, we say that we are working in an information-based asset pricing framework. The general approach that we describe here is that of Brody, Hughston, and Macrina [1–3].

Consider a financial instrument that delivers to its owner a set of random cash flows $\{D_{T_k}\}_{k=1,\ldots,n}$ on the dates $\{T_k\}_{k=1,\ldots,n}$. For simplicity, we assume that these dates are fixed and finite in number. The extension to random dates and to an infinite number of dates is straightforward. Let the pricing kernel be denoted $\{\pi_t\}$. At time t the value S_t of a contract that generates the cash flows $\{D_{T_k}\}_{k=1,\ldots,n}$ is given by the following *valuation formula*:

$$S_t = \frac{1}{\pi_t} \sum_{k=1}^{n} \mathbb{1}_{\{t<T_k\}} \mathbb{E}^{\mathbb{P}}[\pi_{T_k} D_{T_k} | \mathcal{F}_t]. \tag{5.1}$$

Thus, at time t, for each cash flow that has not yet occurred, we take its discounted risk-adjusted conditional expectation, and then we form the sum of such expressions to give the total value of the asset.

Sometimes it is maintained that to regard share prices as being entirely determined by expected dividends is incorrect—that other factors come into play as well, such as the value implicit in corporate control, the value of the status of being a shareholder, and so on. In our view such "implicit" dividends, to the extent that they are relevant and can be assigned a value, have to be modelled and thus enter the valuation formula alongside the tangible cash flows. Sometimes it is argued that market sentiment is also important: indeed, it clearly is; but our view is that sentiment is implicit in the imperfect information the market is receiving concerning future cash flows; that sentiment about a future share price is, in essence, information concerning cash flows (both tangible and intangible) extending beyond the date or dates to which the sentiment refers.

In order to apply the valuation formula, we need to model the market filtration $\{\mathcal{F}_t\}$ and the pricing kernel $\{\pi_t\}$. In particular, it is logical to model the filtration first since the pricing kernel has to be adapted to the filtration. To model the filtration, we proceed as follows. Let us introduce a set of independent random variables $\{X_{T_k}\}_{k=1,\ldots,n}$, which we call *market factors* or simply "X-factors". For each k, the cash flow D_{T_k} is assumed to depend on the market factors $X_{T_1}, X_{T_2}, \ldots, X_{T_k}$. Thus, in association with each date T_k, we introduce a so-called "cash-flow function" Δ_{T_k} such that

$$D_{T_k} = \Delta_{T_k}(X_{T_1}, X_{T_2}, \ldots, X_{T_k}). \tag{5.2}$$

For each asset, we need to model the X-factors, the a priori probabilities, and the cash-flow functions. In general, the X-factor associated with a given date will be a vectorial quantity. The cash-flow diagram associated with a typical asset is illustrated schematically in Fig. 5.1.

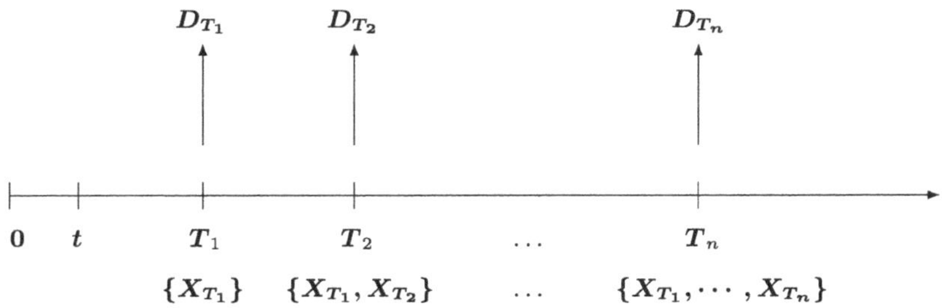

Fig. 5.1 The value S_t at time t of a security that delivers the random cash flows $D_{T_1}, D_{T_2}, \ldots$ at times $T_1, T_2, \ldots$ is determined by the valuation formula (5.1). The cash flow D_{T_1} is determined by a set of one or more independent X-factors $\{X_{T_1}\}$. Then D_{T_2} is determined by $\{X_{T_1}, X_{T_2}\}$, where X_{T_2} represents a further set of independent X-factors, and so on

5.2 X-factor Analysis

Let us look at some elementary examples of cash-flow models based on X-factors. The first example we consider is a simple credit-risky bond, with two remaining coupons to be paid and no recovery on default. Then we have the following cash-flow structure:

$$D_{T_1} = cX_{T_1}, \tag{5.3}$$

$$D_{T_2} = (c+n)X_{T_1}X_{T_2}. \tag{5.4}$$

Here c and n denote the coupon and principal, respectively, and X_{T_1} and X_{T_2} are independent digital random variables taking the values 0 or 1 with designated a priori probabilities. Evidently, if the first coupon is not paid then neither will the second. On the other hand, even if X_{T_1} takes the value unity, and the first coupon is paid, the second coupon and the principal will not be paid unless X_{T_2} also takes the value unity.

The second example is a simple model for a credit-risky coupon bond with recovery. In this case the cash-flow functions are given as follows:

$$D_{T_1} = cX_{T_1} + R_1(c+n)(1-X_{T_1}), \tag{5.5}$$

$$D_{T_2} = (c+n)X_{T_1}X_{T_2} + R_2(c+n)X_{T_1}(1-X_{T_2}). \tag{5.6}$$

Here R_1 and R_2 denote recovery rates. Thus if default occurs at the first coupon, then both the coupon and principal become immediately due, and a fixed fraction R_1 of $c+n$ is paid. But if default occurs at the second coupon date, then the recovery rate is R_2. We observe that the X-factor method allows for a rather transparent representation of the cash-flow structure of such a security and isolates the variables that underlie the various cash flows.

5.3 Information Processes

We assume that with reference to each market factor market participants will have access to information, which in general is imperfect. We model the imperfect information available to market participants concerning a typical market factor X_T with the introduction of a so-called "information process" $\{\xi_{tT}\}_{0\le t\le T}$. An information process is required to have the property

$$\xi_{TT} = f(X_T) \tag{5.7}$$

for some invertible function $f(x)$. This condition ensures that the information process "reveals" the value of the associated market factor X_T at time T. At earlier times, the value of ξ_{tT} contains "partial information" about the value of the X-factor. We shall come to some explicit examples of information processes shortly.

We are now in a position to say how we model the market filtration. In particular, we shall assume that $\{\mathcal{F}_t\}$ is generated collectively by the various market information processes $\{\xi_{tT_k}\}_{k=1,\dots,n}$. In other words, the information at time t is given by the following sigma-algebra:

$$\mathcal{F}_t = \sigma\big[\{\xi_{sT_k}\}_{0\le s\le t,\ k=1,\dots,n}\big]. \tag{5.8}$$

We thus have the following sequence of ideas: market participants are concerned with cash flows; cash flows are dependent on a set of independent market factors; market participants have partial access to the market factors; and this imperfect information generates the market filtration.

We are left with the problem of taking the conditional expectation of the risk-adjusted discounted cash flows to generate price processes; for this purpose, we have to model the pricing kernel. We assume that the pricing kernel is adapted to the market filtration. Thus from knowledge of the history of the information processes from time 0 up to time t one can work out the value of the pricing kernel at t (see, e.g., [6–9]). In a typical model the pricing kernel is given by the discounted marginal utility of consumption of a representative agent. It is reasonable to suppose that the consumption plan of the agent is adapted to the information filtration. The idea is that the filtration represents the flow of information available at each time t about the relevant market factors and that the consumption of the agent is determined by this information. In other words, the agent behaves "rationally", always acting optimally on the available information, in accordance with appropriate criteria. There may be an idiosyncratic element to any given agent's consumption plan that is not adapted to the market filtration and is essentially private. But the *representative* agent has no idiosyncratic consumption.

5.4 Brownian-Bridge Information

For the construction of explicit models, it is useful to transform to the risk-neutral measure $\mathbb{Q}$. This can be achieved by use of the pricing kernel, which we regard as specified. Thus for the present we confine the discussion to "microeconomic" issues: we take no notice of the informational notions implicit in the formulation of the pricing kernel and make the additional simplifying assumption in what follows that the default-free interest-rate system is deterministic. Then the valuation formula takes the following form:

$$S_t = \sum_{k=1}^{n} \mathbb{1}_{\{t<T_k\}} P_{tT_k} \mathbb{E}^{\mathbb{Q}}[D_{T_k}|\mathcal{F}_t]. \tag{5.9}$$

Absence of arbitrage implies that the discount bond system $\{P_{tT}\}_{0\le t\le T<\infty}$ is of the form $P_{tT} = P_{0T}/P_{0t}$, where $\{P_{0t}\}_{0\le t<\infty}$ is the initial term structure.

With these assumptions in place, we are in a position to specify a model for the information flow. For each X-factor X_T, we take the associated information process to be of the form

$$\xi_{tT} = \sigma t X_T + \beta_{tT}. \tag{5.10}$$

Here $\{\beta_{tT}\}$ is a $\mathbb{Q}$-Brownian bridge over the interval $[0, T]$, satisfying $\beta_{0T} = 0$, $\beta_{TT} = 0$, $\mathbb{E}[\beta_{tT}] = 0$, and $\mathbb{E}[\beta_{sT}\beta_{tT}] = s(T-t)/T$. The X-factor and the Brownian bridge are assumed to be $\mathbb{Q}$-independent. Thus the Brownian bridge represents "market noise", and only the "market signal" term involving the X-factor contains true market information. The parameter σ can be interpreted as the "information flow rate" for the factor X_T.

In the situation where we have a multiplicity of factors X_{T_k} $(k = 1, \dots, n)$, the information processes are taken to be of the form

$$\xi_{tT_k} = \sigma_k t X_{T_k} + \beta_{tT_k}, \tag{5.11}$$

where we assume that the X-factors and Brownian bridges are independent.

The motivation for the use of a bridge to represent noise is intuitively as follows. We assume that initially all available market information is taken into account in the determination of prices, or, equivalently, the a priori probability laws for the market factors. After the passage of time, new stories circulate, and we model this by taking into account that the variance of the Brownian bridge increases for the first half of its trajectory. Eventually, the variance falls to zero at T, when the "moment of truth" arrives. The parameter σ represents the rate at which the true value of X_T is "revealed" as time progresses. Thus, if σ is low, then X_T is effectively hidden until near the time T; on the other hand, if σ is high, then we can think of X_T as being revealed quickly. If the X-factor is "dimensionless", then σ has the units

$$\sigma \sim [\text{time}]^{-1/2}, \tag{5.12}$$

and a rough measure for the timescale τ over which information is revealed is

$$\tau = \frac{1}{\sigma^2 \operatorname{Var}[X_T]}. \tag{5.13}$$

In particular, if $\tau \ll T$, then the value of X_T will be revealed rather early, e.g., after the passage of a few multiples of τ. On the other hand, if $\tau \gg T$, then X_T will only be revealed at the last minute, as a "surprise".

We remark that the information process (5.10) has the Markov property. This feature implies simplifications in the resulting models. In particular, on account of relation (5.7) we find that the conditioning with respect to $\mathcal{F}_t$ in (5.9) can be replaced by conditioning with respect to the random variables ξ_{tT_k} $(k = 1, \dots)$. For a proof of the Markov property, see [1, 9].

5.5 Assets Paying a Single Dividend

Consider an asset that pays single dividend $D_T \geq 0$ at time T, and assume that there is only one market factor X_T, so $D_T = f(X_T)$. For the moment, let us assume further that $f(x) = x$. Thus, we have $D_T = X_T$, where the market factor X_T is a continuous nonnegative random variable with a priori $\mathbb{Q}$-density $p(x)$ for $x > 0$. It follows by use of the Markov property of $\{\xi_{tT}\}$ that the price of such an asset can be written in the form

$$\begin{aligned} S_t &= P_{tT}\mathbb{E}[D_T|\xi_{tT}] \\ &= P_{tT}\int_0^\infty x p_t(x)\,dx, \end{aligned} \tag{5.14}$$

where $p_t(x)$ is the conditional density of X_T. Making use of the Bayes formula, one can show that $p_t(x)$ is given more explicitly by

$$p_t(x) = \frac{p(x)\exp[\frac{T}{T-t}(\sigma x\xi_{tT} - \frac{1}{2}\sigma^2 x^2 t)]}{\int_0^\infty p(x)\exp[\frac{T}{T-t}(\sigma x\xi_{tT} - \frac{1}{2}\sigma^2 x^2 t)]\,dx}. \tag{5.15}$$

Thus, at each time $t < T$ the price of the asset is determined by the random value of the information ξ_{tT} available at that time and is given by

$$S_t = P_{tT}\frac{\int_0^\infty x p(x)\exp[\frac{T}{T-t}(\sigma x\xi_{tT} - \frac{1}{2}\sigma^2 x^2 t)]\,dx}{\int_0^\infty p(x)\exp[\frac{T}{T-t}(\sigma x\xi_{tT} - \frac{1}{2}\sigma^2 x^2 t)]\,dx}. \tag{5.16}$$

The dynamics of the price process can then be obtained by an application of Ito's lemma, with the following result:

$$dS_t = r_t S_t\,dt + P_{tT}\frac{\sigma T}{T-t}\,\mathrm{Var}_t[X_T]\,dW_t. \tag{5.17}$$

Here

$$\mathrm{Var}_t[X_T] = \int_0^\infty x^2 p_t(x)\,dx - \left(\int_0^\infty x p_t(x)\,dx\right)^2 \tag{5.18}$$

denotes the conditional variance of X_T, which by (5.15) is evidently given as a function of t and ξ_{tT}. The $\{\mathcal{F}_t\}$-adapted process $\{W_t\}$ driving the dynamics of the asset in (5.17) above is not given exogenously, but rather is defined in terms of the information process itself for $t < T$ by the following formula:

$$W_t = \xi_{tT} - \int_0^t \frac{1}{T-s}\big(\sigma T \mathbb{E}_s[X_T] - \xi_{sT}\big)\, ds. \tag{5.19}$$

Indeed, one can verify, by use of the Lévy criterion, that the process $\{W_t\}$, as thus defined, is an $\{\mathcal{F}_t\}$-Brownian motion. Hence we see that in an information-based approach we can *derive* the Brownian motions that drive the markets: they are not "inputs" to the model, but rather can be seen as arising as a "consequence" of the model.

5.6 Geometric Brownian Motion Model

A simple application of the X-factor technique arises in the case of geometric Brownian motion models. We consider a limited-liability company that makes a single cash distribution S_T at time T. Alternatively, think of a portfolio containing a single stock which will be sold off at time T for S_T, with the proceeds of the sale going to the investor. We assume that S_T has a log-normal distribution under $\mathbb{Q}$ and can be written in the form

$$S_T = S_0 \exp\left(rT + \nu\sqrt{T}X_T - \frac{1}{2}\nu^2 T\right), \tag{5.20}$$

where the market factor X_T is normally distributed with mean zero and variance one, and where $r > 0$ and $\nu > 0$ are constants. The information process $\{\xi_t\}$ is taken to be of the form (5.10), where in the present example the information flow rate is given by

$$\sigma = \frac{1}{\sqrt{T}}. \tag{5.21}$$

By use of the Bayes formula we find that the conditional probability density is Gaussian,

$$p_t(x) = \sqrt{\frac{T}{2\pi(T-t)}} \exp\left(-\frac{1}{2(T-t)}\big(\sqrt{T}x - \xi_{tT}\big)^2\right), \tag{5.22}$$

and has the following dynamics:

$$dp_t(x) = \frac{1}{T-t}\big(\sqrt{T}x - \xi_{tT}\big)p_t(x)\, d\xi_{tT}. \tag{5.23}$$

A short calculation then shows that the value of the asset in this example is given at time $t < T$ by

$$\begin{aligned} S_t &= e^{-r(T-t)}\mathbb{E}_t[S_T] \\ &= e^{-r(T-t)} \int_{-\infty}^{\infty} S_0 e^{rT+\nu\sqrt{T}x-\frac{1}{2}\nu^2 T}\, p_t(x)\, dx \\ &= S_0 \exp\left(rt + \nu\xi_{tT} - \frac{1}{2}\nu^2 t \right). \end{aligned} \tag{5.24}$$

The surprising fact is that $\{\xi_{tT}\}$ itself turns out to be an $\{\mathcal{F}_t\}$-Brownian motion. Hence, writing $W_t = \xi_{tT}$ for $0 \le t \le T$, we obtain the standard geometric Brownian motion model,

$$S_t = S_0 \exp\left(rt + \nu W_t - \frac{1}{2}\nu^2 t \right). \tag{5.25}$$

We see that starting with an information-based argument we are able to recover the familiar asset price dynamics given by (5.25). An important point to note is that the Brownian bridge process $\{\beta_{tT}\}$ arises naturally in this context. In fact, if we start with (5.25) then we can make use of the following well-known orthogonal decomposition:

$$W_t = \frac{t}{T} W_T + \left(W_t - \frac{t}{T} W_T \right). \tag{5.26}$$

The second term on the right, which is independent of the first term on the right, is a standard representation for a Brownian bridge process:

$$\beta_{tT} = W_t - \frac{t}{T} W_T. \tag{5.27}$$

Then by setting $X_T = W_T/\sqrt{T}$ and $\sigma = 1/\sqrt{T}$ we find that the right side of (5.26) is indeed the market information. In other words, when it is formulated in an information-based framework, the standard Black–Scholes–Merton theory can be expressed in terms of a normally distributed X-factor and an independent Brownian-bridge noise process.

5.7 Pricing Contingent Claims

The information-based price (5.16) of a single-dividend paying asset at first glance appears to be given by a rather complicated expression, suggesting perhaps that it would be impractical for use as a model for the pricing and hedging of contingent claims. However, there is a remarkable simplification involving a change of measure that allows one both to price and to hedge vanilla options. This can be seen as follows. Let us consider a European-style call option on the asset, with option maturity t and strike K. The value of the option at time 0 is given by

$$C_0 = P_{0t}\mathbb{E}^{\mathbb{Q}}\big[(S_t - K)^+\big]. \tag{5.28}$$

Let us define a process $\{\Phi_t\}$ by the expression appearing in the denominator of (5.16), so

$$\Phi_t = \int_0^\infty p(x)\exp\left[\frac{T}{T-t}\left(\sigma x\xi_{tT} - \frac{1}{2}\sigma^2 x^2 t\right)\right]dx. \tag{5.29}$$

Then it can be shown that $\{\Phi_t^{-1}\}$ is a positive $\mathbb{Q}$-martingale, which can be used to change the probability measure from $\mathbb{Q}$ to a new measure $\mathbb{B}$. Under the measure $\mathbb{B}$, which we call the "bridge measure", the information process itself is a Brownian bridge. More precisely, under $\mathbb{B}$ the process $\{\xi_{sT}\}_{0\le s\le t}$ has the law of a Brownian bridge spanning the interval $[0, T]$, restricted to $[0, t]$. That is to say, $\{\xi_{sT}\}_{0\le s\le t}$ is $\mathbb{B}$-Gaussian with mean zero and covariance $\mathrm{cov}[\xi_{aT}, \xi_{bT}] = a(T-b)/T$ for $0 \le a \le b \le t$. The initial value of the option is thus given by

$$\begin{aligned} C_0 = P_{0t}\mathbb{E}^{\mathbb{B}}\Bigg[\Bigg(P_{tT}\int_0^\infty x p(x) e^{\frac{T}{T-t}(\sigma x\xi_{tT} - \frac{1}{2}\sigma^2 x^2 t)}\,dx \\ - K\int_0^\infty p(x) e^{\frac{T}{T-t}(\sigma x\xi_{tT} - \frac{1}{2}\sigma^2 x^2 t)}\,dx\Bigg)^+\Bigg]. \end{aligned} \tag{5.30}$$

It can be shown that the asset price is a monotonically increasing function of the value of ξ_{tT}. It follows that there is a unique critical level ξ^* for the information such that the expression inside the max-function in (5.30) is positive. It follows that the option price can be written in terms of a single integration involving the normal distribution function:

$$C_0 = P_{0t}\int_0^\infty p(x)(P_{tT}x - K)N\left(\frac{\xi^* - \sigma x t}{\sqrt{t(T-t)/T}}\right)dx. \tag{5.31}$$

As another example, we consider the following. Suppose that the single cash flow D_T is a binary random variable taking the values $\{d_0, d_1\}$ with a priori probabilities $\{p_0, p_1\}$. The asset in this case can be thought of as a simple credit-risky discount bond that pays d_1 if there is no default and d_0 if there is a default. A short calculation allows one to verify that

$$C_0 = P_{0t}\big[p_1(P_{tT}d_1 - K)N\big(u^+\big) - p_0(K - P_{tT}d_0)N\big(u^-\big)\big], \tag{5.32}$$

where u^+ and u^- are defined by

$$u^\pm = \frac{\ln\big[\frac{p_1(P_{tT}d_1 - K)}{p_0(K - P_{tT}d_0)}\big] \pm \frac{1}{2}\sigma^2(d_1 - d_0)^2\tau}{\sigma\sqrt{\tau}(d_1 - d_0)} \tag{5.33}$$

with $\tau = tT/(T-1)$. It can be shown that the option delta at time 0, defined as usual by

$$\delta_0 = \frac{\partial C_0}{\partial S_0}, \tag{5.34}$$

can be calculated explicitly, with the following result:

$$\delta_0 = \frac{(P_{tT}d_1 - K)N(u^+) + (K - P_{tT}d_0)N(u^-)}{P_{tT}(d_1 - d_0)}. \tag{5.35}$$

We see, therefore, that the apparent complexity of (5.16) does not lead to any intractability when it comes to derivatives pricing and hedging.

5.8 Volatility and Correlation

In the case of an asset that pays multiple dividends, the price is determined by the conditional expectation given in (5.9). In terms of the cash-flow functions defined by (5.2), we thus obtain the following for the dynamics of the asset price:

$$dS_t = r_t S_t \, dt + \sum_{k=1}^{n} \Delta_{T_k} \, d\mathbb{1}_{\{t<T_k\}} + \sum_{k=1}^{n} \mathbb{1}_{\{t<T_k\}} P_{tT_k} \sum_{j=1}^{k} \frac{\sigma_j T_j}{T_j - t} \mathrm{Cov}_t[\Delta_{T_k}, X_{T_j}] \, dW_t^j. \tag{5.36}$$

The leading term in the drift is the short rate, as one might expect, and there is also a term representing the downward jump in the asset that occurs when a dividend is paid. The independent $\{\mathcal{F}_t\}$-adapted Brownian motions $\{W_t^j\}$ driving the price dynamics are given in terms of the corresponding information processes by

$$W_t^j = \xi_{tT_j} - \int_0^t \frac{1}{T_j - s} \left(\sigma_j T_j \mathbb{E}_s[X_{T_j}] - \xi_{sT_j}\right) ds. \tag{5.37}$$

We see that if an asset delivers one or more cash flows depending on two or more market factors, then it will exhibit "unhedgeable" stochastic volatility [2, 7]. That is to say, one would not expect to be able to hedge a position in an option by use of a position in the underlying. In general, if the asset cash flows depend on n X-factors in total, then to hedge a generic derivative based on the given asset one will need the underlying together with $n-1$ options as hedging instruments, i.e., n hedging instruments in total. One can read from (5.36) the generic form of the stochastic volatility implied for a given configuration of X-factors and cash-flow functions.

It follows likewise from (5.36) that two or more assets will exhibit dynamic correlation when they share one or more X-factors in common. As a specific example of dynamic correlation, let us consider a pair of credit-risky discount bonds. The first bond is defined by a cash flow D_{T_1} at T_1. The second is defined by a cash flow D_{T_2} at $T_2 > T_1$. The cash flow structure is taken to be

$$D_{T_1} = n_1 X_{T_1} + R_1 n_1 (1 - X_{T_1}) \tag{5.38}$$

and

$$\begin{aligned} D_{T_2} = {} & n_2 X_{T_1} X_{T_2} + R_2^a n_2 (1 - X_{T_1}) X_{T_2} \\ & + R_2^b n_2 X_{T_1} (1 - X_{T_2}) + R_2^c n_2 (1 - X_{T_1})(1 - X_{T_2}). \end{aligned} \tag{5.39}$$

Here, n_1 and n_2 denote the bond principals, and X_{T_1} and X_{T_2} are independent digital random variables. The possible recovery rates in the case of default are denoted by R_1, R_2^a, R_2^b, and R_2^c. One can have in mind the following story. Consider a factory with debt S_t^1. Across the street there is a little restaurant with debt S_t^2. If the factory goes bust ($X_{T_1} = 0$), then so will the restaurant, because this is where the workers have their lunch. On the other hand, even if the factory is successful ($X_{T_1} = 1$), the restaurant may still go bust on account of bad management ($X_{T_2} = 0$). The recovery rates on the restaurant bond depend on the details of what goes wrong: R_2^a (restaurant fails because factory fails); R_2^b (restaurant fails on account of bad management); R_2^c (factory fails, and bad restaurant management). One might expect $R_2^b > R_2^a$, since as long as the factory continues, the restaurant facilities could be sold at a good price. The worst scenario is that of R_2^c. For the dynamics of the first bond (the "factory"), for which the price is

$$S_t^1 = P_{tT_1} \mathbb{E}_t[D_{T_1}] \quad (t < T_1), \tag{5.40}$$

we have:

$$dS_t^1 = r_t S_t^1 \, dt + P_{tT_1} \frac{\sigma_1 T_1}{T_1 - t} \alpha \, \mathrm{Var}_t[X_{T_1}] \, dW_t^1, \tag{5.41}$$

where $\alpha = n_1(1 - R_1)$. For the dynamics of the second bond (the "restaurant"), for which the price is

$$S_t^2 = P_{tT_2} \mathbb{E}_t[D_{T_2}] \quad (t < T_2), \tag{5.42}$$

we have:

$$\begin{aligned} dS_t^2 = {} & r_t S_t^2 \, dt + P_{tT_2} \frac{\sigma_1 T_1}{T_1 - t} \big(\beta + \delta \mathbb{E}_t[X_{T_2}]\big) \, \mathrm{Var}_t[X_{T_1}] \, dW_t^1 \\ & + P_{tT_2} \frac{\sigma_2 T_2}{T_2 - t} \big(\gamma + \delta \mathbb{E}_t[X_{T_1}]\big) \, \mathrm{Var}_t[X_{T_2}] \, dW_t^2, \end{aligned} \tag{5.43}$$

where the constants β, γ, and δ are given by $\beta = n_2(R_2^b - R_2^c)$, $\gamma = n_2(R_2^a - R_2^c)$, and $\delta = n_2(1 - R_2^a - R_2^b + R_2^c)$. The filtration $\{\mathcal{F}_s\}$ is generated by the information processes $\{\xi_{sT_1}\}$ and $\{\xi_{sT_2}\}$ associated with X_{T_1} and X_{T_2}. The dynamics of the bond prices depend on a common Brownian driver $\{W_t^1\}$. The fact that the asset payoffs share a common X-factor thus gives rise to a dynamic correlation between the movements of the price processes $\{S_t^1\}$ and $\{S_t^2\}$. The instantaneous correlation between the price movements of the factory bond and the restaurant bond is given by the following expression:

$$\rho_t = \frac{dS_t^1\, dS_t^2}{\sqrt{(dS_t^1)^2 (dS_t^2)^2}}. \tag{5.44}$$

Hence, using the formulae for the dynamics of the two assets, we obtain

$$\rho_t = \frac{1}{\sqrt{\psi_t}} \frac{\sigma_1 T_1}{T_1 - t} \big(\beta + \delta \mathbb{E}_t[X_{T_1}]\big) \operatorname{Var}_t[X_{T_1}], \tag{5.45}$$

where

$$\begin{aligned} \psi_t = {} & \left(\frac{\sigma_1 T_1}{T_1 - t}\right)^2 \big(\beta + \delta \mathbb{E}_t[X_{T_2}]\big)^2 \big(\operatorname{Var}_t[X_{T_1}]\big)^2 \\ & + \left(\frac{\sigma_2 T_2}{T_2 - t}\right)^2 \big(\gamma + \delta \mathbb{E}_t[X_{T_1}]\big)^2 \big(\operatorname{Var}_t[X_{T_2}]\big)^2. \end{aligned} \tag{5.46}$$

We see from (5.45) that we are able to calculate explicitly the dynamics of the correlation between the movements of the two asset prices.

5.9 Amount of Information about the Future Cash Flow Contained in the Price Process

Since we are modelling the flow of information in an explicit manner, we are able to quantify how much information regarding the value of the cash flow D_T is contained in the value ξ_t at time t of the associated information process. For simplicity, in the discussion that follows we shall assume that the cash flow D_T takes the discrete values $\{d_i\}_{i=1,\dots,n}$ with a priori probabilities $\{p_i\}_{i=1,\dots,n}$. A reasonable measure for quantifying the information content is given by the mutual information $J(\xi_t, D_T)$ between the two random variables, which in the present context is given by the expression

$$J(\xi_t, D_T) = \sum_{i=1}^{n} \int_{-\infty}^{\infty} \rho(\xi, i) \ln\left(\frac{\rho(\xi, i)}{\rho(\xi)\rho(i)}\right) d\xi, \tag{5.47}$$

where

$$\rho(\xi, i) = \frac{d}{d\xi} \mathbb{Q}\big[(\xi_t < \xi) \cap (D_T = d_i)\big] \tag{5.48}$$

is the joint density of the random variables ξ_t and D_T, and $\rho(\xi)$ and $\rho(i)$ are the respective marginal probabilities. By use of the relation

$$\mathbb{Q}\big[(\xi_t < \xi) \cap (D_T = d_i)\big] = \mathbb{Q}(\xi_t < \xi | D_T = d_i)\mathbb{Q}(D_T = d_i) \tag{5.49}$$

we deduce that

$$\rho(\xi, i) = p_i \frac{1}{\sqrt{2\pi t(T-t)/T}} \exp\left(-\frac{1}{2}\frac{(x - \sigma d_i t)^2}{t(T-t)/T}\right), \tag{5.50}$$

since conditional on $D_T = d_i$ the random variable ξ_t is normally distributed with mean $\sigma t d_i$ and variance $t(T-t)/T$. From (5.50) the marginal densities

$$\rho(\xi) = \sum_{i=1}^{n} \rho(\xi, i) \quad \text{and} \quad \rho(i) = \int_{-\infty}^{\infty} \rho(\xi, i)\, d\xi \tag{5.51}$$

can be deduced. In particular, $\rho(i) = p_i$. By substituting (5.50) into (5.47), the information about the cash flow D_T contained in ξ_t can be determined.

From an information-theoretic point of view, two processes related through an invertible function, thus sharing the *same* filtration, in general possess *different* information content. On the other hand, since what is observed in the market is the price S_t, which is an invertible function of ξ_t, it is more relevant to determine the mutual information $J(S_t, D_T)$, that is, the amount of information about the future cash flow contained in the market price. It can be shown that in the present context we have $J(S_t, D_T) = J(\xi_t, D_T)$.

5.10 Information Disparity and Statistical Arbitrage

So far we have assumed that market participants have equal access to information, but one can ask what happens if some traders are more "informed" than others. Suppose that we consider a financial product that pays a single cash flow D_T at time T. We can think of this product as a kind of bond. The general market trader has access to an information process concerning D_T, but there are also "informed" traders who have access to one or more additional information processes concerning D_T. The informed trader is thus in some sense able to make a "better estimate" of the value of the asset.

To be more specific, let us suppose that while the general market trader has access to the information $\xi_t = \sigma t D_T + \beta_{tT}$, an informed trader has access in addition, say, to the information $\xi'_t = \sigma' t D_T + \beta'_{tT}$, where $\{\beta'_{tT}\}$ may or may not be correlated with the market noise $\{\beta_{tT}\}$. Thus, the information source for the informed trader is given by $\{\mathcal{G}_t\} = \sigma(\{\xi_s, \xi'_s\}_{0\le s\le t})$. The use of such extra information can, but need not, represent "insider trading" in the usual sense. That is, it may be that the informed trader merely has better access to (and better computer power for the purpose of digesting) the vast domain of publicly available information. Since we have

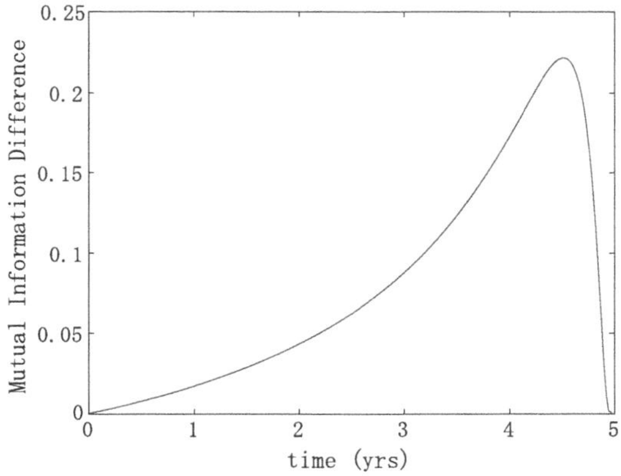

Fig. 5.2 Mutual information difference. The additional information held by an informed trader over that of the market is nonnegative. The parameters are set to be $d_1 = 0$, $d_2 = 1$, $p_1 = 0.2$, $p_2 = 0.8$, $T = 5$, $\sigma = 0.25$, $\sigma' = 0.45$, and $\rho = 0.15$ (the correlation between β_{tT} and β'_{tT})

introduced an information measure regarding impending cash flows, we can quantify the excess information held by the informed trader above that held by general market traders. This is measured by the difference of the mutual information ΔJ. In Fig. 5.2 we plot an example of ΔJ, indicating the way in which the excess information held by the informed trader changes in time.

Given that the informed trader is on average "more knowledgeable" than the general market trader, it is natural to ask how the informed trader can take advantage of the situation to seek so-called "statistical arbitrage" opportunities. We assume that the informed trader operates on a relatively small scale and that the actions of this trader do not significantly influence the market. Suppose that we consider a trading strategy such that at some designated time $t \in (0, T)$ a market trader purchases a bond if, and only if, at that time the bond price S_t is greater than the quantity $K P_{tT}$ for some specified constant K. Once a bond is purchased, it is held to maturity. The informed trader follows the same rule, but makes a better estimate of the value of the bond, and hence purchases the bond iff $\tilde{S}_t > K P_{tT}$, where

$$\tilde{S}_t = P_{tT} \mathbb{E}[D_T | \mathcal{G}_t]. \tag{5.52}$$

The significance of $\tilde{S}_t$ is that it represents the price that the informed trader knows that the market as a whole would make if the market as a whole had the same knowledge as the informed trader.

That such a strategy leads to a statistical arbitrage opportunity for the informed trader can be seen as follows [4]. We assume that the initial position of a trader is zero, i.e., purchase of a bond at t requires borrowing the amount S_t at that time and repaying the amount $P_{tT}^{-1} S_t$ at T. Thus the value of a general market trader's portfolio at T is

$$V_T = \mathbb{1}\{S_t > K P_{tT}\}\big(D_T - P_{tT}^{-1} S_t\big), \tag{5.53}$$

whereas the value of the informed trader's portfolio at T is

$$\tilde{V}_T = \mathbb{1}\{\tilde{S}_t > K P_{tT}\}\big(D_T - P_{tT}^{-1} S_{tT}\big). \tag{5.54}$$

Consider the present value $P_{0T}\mathbb{E}[\Delta V_T]$ of a security that delivers a cash flow equal to the excess profit or loss

$$\Delta V_T = \tilde{V}_T - V_T \tag{5.55}$$

generated by the strategy of the informed trader. By the tower property we have

$$\mathbb{E}[\Delta V_T] = \mathbb{E}\big[\mathbb{E}[\Delta V_T | \mathcal{G}_t]\big]. \tag{5.56}$$

However,

$$\mathbb{E}[\Delta V_T | \mathcal{G}_t] = P_{tT}^{-1}\big(\mathbb{1}\{\tilde{S}_t > K P_{tT}\} - \mathbb{1}\{S_t > K P_{tT}\}\big)(\tilde{S}_t - S_t), \tag{5.57}$$

since the random variables S_t and $\tilde{S}_t$ are $\mathcal{G}_t$-measurable. If $\tilde{S}_t > S_t$, then

$$\mathbb{1}\{\tilde{S}_t > K P_{tT}\} - \mathbb{1}\{S_t > K P_{tT}\} \geq 0; \tag{5.58}$$

whereas if $\tilde{S}_t < S_t$, then

$$\mathbb{1}\{\tilde{S}_t > K P_{tT}\} - \mathbb{1}\{S_t > K P_{tT}\} \leq 0. \tag{5.59}$$

It follows that $\mathbb{E}[\Delta V_T | \mathcal{G}_t] > 0$ with probability greater than zero, and therefore $\mathbb{E}[\Delta V_T] > 0$. We know that according to the usual no-arbitrage arguments, the present value of the payoff of the strategy of the general market trader must be zero. It follows that the informed trader can execute a transaction at zero cost that has positive value: this is what we mean by "statistical arbitrage". A simulation study of the profit arising from this trading strategy is shown in Fig. 5.3, indicating a close

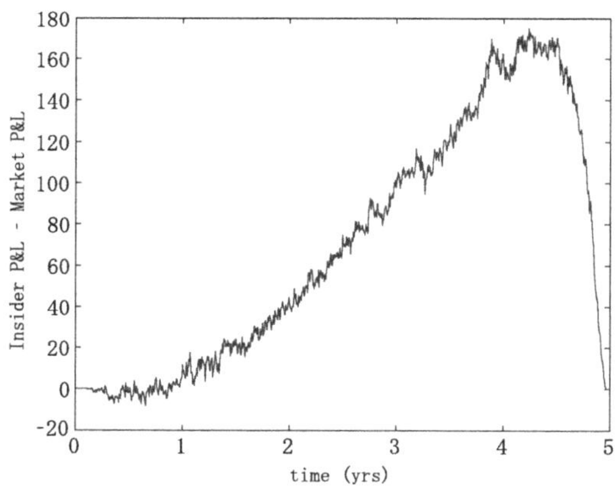

Fig. 5.3 The P&L difference for digital bonds. At each time traders purchase the bond if and only if the valuation of the bond is greater than a specified threshold. The general market trader buys if $S_t > K P_{tT}$, whereas the informed trader uses the criterion $\tilde{S}_t > K P_{tT}$. The difference in profit and loss between the informed trader and the general market trader is plotted, based on 2000 realisations, when the a priori probability of default is $p_1 = 0.2$. Other parameters are set to be $d_1 = 0$, $d_2 = 1$, $T = 5$, $\sigma = 0.25$, $\sigma' = 0.45$, $\rho = 0.15$, and $K = 0.7$

correspondence with the excess information held by the informed trader shown in Fig. 5.2.

5.11 Price Formation in Inhomogeneous Markets

The idea of "informed trading" can be extended to a market that has a number of traders operating in it, all more or less on an equal footing, but where different traders have access to different information. In other words, there is an inhomogeneous information flow in the market. This line of thinking leads naturally to the consideration of price formation in such a market, as illustrated in Fig. 5.4.

Let us consider, as an example, a market with two traders, labelled "Trader 1" and "Trader 2". As before, there is a single asset, with a single dividend D_T paid at time T. The traders have access to separate sources of information about D_T, given respectively by $\xi_t^1 = \sigma_1 t D_T + \beta_{tT}^1$ and $\xi_t^2 = \sigma_2 t D_T + \beta_{tT}^2$. Here the Brownian bridges $\{\beta_{tT}^1\}$ and $\{\beta_{tT}^2\}$ are assumed, for simplicity, to be independent. Trader 1 works out the price

$$S_t^1 = P_{tT}\mathbb{E}\big[D_T\big|\xi_t^1\big] \tag{5.60}$$

that he knows the market would have made had the market possessed the information generated by $\{\xi_t^1\}$. Likewise, Trader 2 works out the price

$$S_t^2 = P_{tT}\mathbb{E}\big[D_T\big|\xi_t^2\big] \tag{5.61}$$

that she knows the market would have made had the market possessed the information generated by $\{\xi_t^2\}$.

Traders 1 and 2 are unaware of each other's prices but can gain information by trading. The trading process works as follows. Each trader makes a spread about their price. Letting $0 < \phi^- < 1 < \phi^+$, we set

$$S_t^{1\pm} = \phi^\pm S_t^1 \tag{5.62}$$

for the buy price S_t^{1-} and sell price S_t^{1+} made by Trader 1 at time t. Thus Trader 1 is willing to buy at a price slightly below his information-based valuation S_t^1 and is willing to sell at a price that is slightly above that valuation. Likewise Trader 2 is willing to buy at a price slightly below her information-based valuation S_t^2 and is willing to sell at a price that is slightly above that valuation:

$$S_t^{2\pm} = \phi^\pm S_t^2. \tag{5.63}$$

We assume that there is an exchange that continuously monitors the prices made by the traders. The exchange effects a trade of some fixed size when the buy price of one of the traders reaches the level of the sell price of the other trader. That is to say, a trade takes place when

$$S_t^{1-} = S_t^{2+} \quad \text{or} \quad S_t^{1+} = S_t^{2-}. \tag{5.64}$$

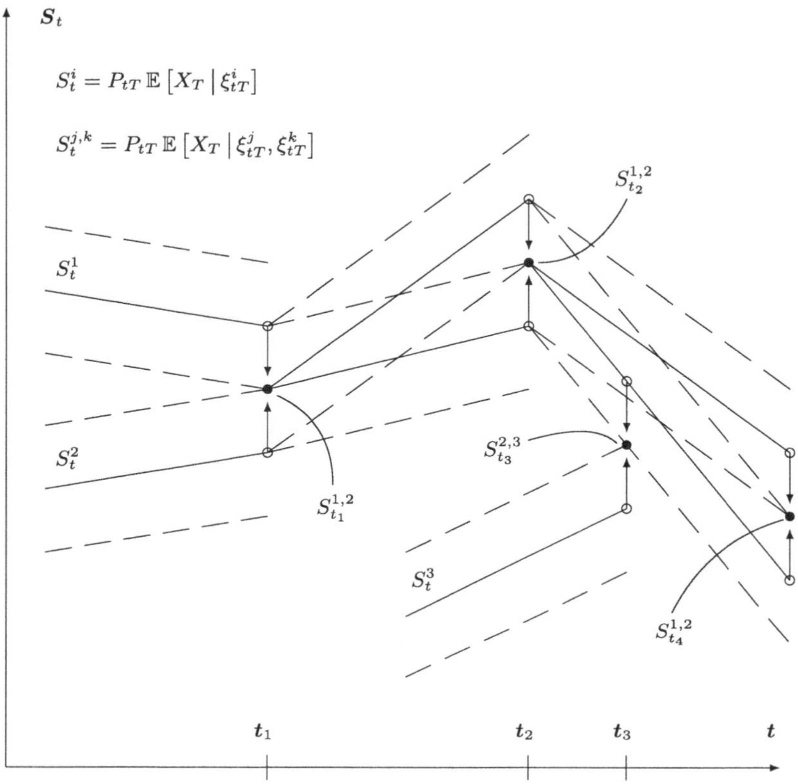

Fig. 5.4 Schematic illustration of information-based trading. An exchange executes a trade when the sell price of Trader i matches the buy price of Trader j (*dashed lines meet at filled dots*). At the execution time t_n, both traders have access to each other's valuations $S_{t_n}^i$ (*empty circles*) and thus to each other's information. As a consequence, the traders are able to update their valuations and obtain the common price $S_{t_n}^{j,k}$. After the trade has been executed and the respective asset valuations have been updated, the traders "separate" and return to their individual valuations. The respective valuations may drift in different directions. The traders will get in contact again as soon as the exchange notifies them that the respective sell and buy prices have matched again. Such an information-based trading mechanism can be extended to multiple traders, as suggested in the illustration

When a trade occurs, at that moment each trader learns the price of the other and, as a consequence, can back out the value of the corresponding information process. Therefore, when a trade occurs, the traders each briefly have access to both pieces of information and are thus in a position to make a better price, namely that given by

$$S_t^{1,2} = P_{tT}\mathbb{E}\big[D_T \big| \xi_t^1, \xi_t^2\big]. \tag{5.65}$$

We conclude that immediately after a trade the information-based prices made by each of the traders will jump to the same level and the a priori probability distribution for D_T will be updated correspondingly.

Once the trade is concluded, the link between the two traders is lost, and each trader again has access only to their own information source. Starting from the same price, the prices made by the two traders diverge as they receive different information going forward. A further trade will then occur when the buy price of one of the traders next hits the sell price of the other trader.

At the time the trade is executed, the joint information can be embodied in the value of an effective information process $\{\hat{\xi}_t\}_{0\le t\le T}$ given by

$$\hat{\xi}_t = \sqrt{(\sigma_1)^2+(\sigma_2)^2}\,t\,D_T + \frac{\sigma_1\beta^1_{tT}+\sigma_2\beta^2_{tT}}{\sqrt{(\sigma_1)^2+(\sigma_2)^2}}. \tag{5.66}$$

We note that $\{\hat{\xi}_t\}$ is indeed an information process, since it can be written in the form

$$\hat{\xi}_t = \hat{\sigma}\,t\,D_T + \hat{\beta}_{tT}, \tag{5.67}$$

where

$$\hat{\sigma} = \sqrt{(\sigma_1)^2+(\sigma_2)^2}, \quad \text{and} \quad \hat{\beta}_{tT} = \frac{\sigma_1\beta^1_{tT}+\sigma_2\beta^2_{tT}}{\sqrt{(\sigma_1)^2+(\sigma_2)^2}}. \tag{5.68}$$

One can show that $\{\hat{\beta}_{tT}\}$ is a Brownian bridge and is independent of D_T. Thus, immediately after the trade is executed, the price $S^{1,2}_t$ made by both traders is of the form

$$S^{1,2}_t = P_{tT}\frac{\int_0^\infty x p(x)\exp[\frac{T}{T-t}(\hat{\sigma}x\hat{\xi}_t - \frac{1}{2}\hat{\sigma}^2x^2t)]\,dx}{\int_0^\infty p(x)\exp[\frac{T}{T-t}(\hat{\sigma}x\hat{\xi}_t - \frac{1}{2}\hat{\sigma}^2x^2t)]\,dx}. \tag{5.69}$$

As an example, let us consider the case of a digital payout taking the values 0 and 1 with a priori probabilities p_0 and p_1. We consider the case in which the information flow rates are the same, so we set $\sigma_1=\sigma_2=\sigma$. Then for the valuations, we have

$$S^1_t = \frac{p_1\exp[\frac{T}{T-t}(\sigma\xi^1_t - \frac{1}{2}\sigma^2t)]}{p_0+p_1\exp[\frac{T}{T-t}(\sigma\xi^1_t - \frac{1}{2}\sigma^2t)]} \tag{5.70}$$

and

$$S^2_t = \frac{p_1\exp[\frac{T}{T-t}(\sigma\xi^2_t - \frac{1}{2}\sigma^2t)]}{p_0+p_1\exp[\frac{T}{T-t}(\sigma\xi^2_t - \frac{1}{2}\sigma^2t)]}. \tag{5.71}$$

For the spreads, we assume that $\phi^+=1+\delta$ and $\phi^-=1-\delta$, where δ is small. If Trader 1 is the buyer, then the condition for a trade is

$$(1-\delta)S^1_t = (1+\delta)S^2_t. \tag{5.72}$$

Given his knowledge ξ_t^1, Trader 1 can use the condition to work out the value of ξ_t^2. In particular, suppose that

$$\xi_t^2 = \xi_t^1 + \varepsilon_t, \tag{5.73}$$

where ε_t is small. Then a calculation shows that ε_t is given, to first order, by

$$\varepsilon_t = -\frac{2\delta(T-t)}{\sigma T(1 - \mathbb{E}[X_T|\xi_t^1])}. \tag{5.74}$$

The general situation, where there are a number of traders present in the market, and where the asset cash flows depend on a number of market factors, is very rich. It is evident that in the broad picture there is no universal filtration, nor a universal pricing measure. Nevertheless, by exchanging information through trading activity, market participants can maintain a "law of reasonable price range" if not a "law of one price".

Certainly, the notion that there is a universal market filtration is unrealistic. What counts is not merely "access in principle" to information, but rather "access in practice". Perhaps some broader version of market efficiency will survive, taking into account the cost of such access (cf. [5]). A subscription to the Wall Street Journal is not free, nor is a Bloomberg terminal. Access to vast information providers such as Google and Yahoo may seem free or nearly so, but from a broader perspective this is not so—someone pays, in cash or kind. What is the market price of information? And how does this depend on the "information about the information"? For the answers to these questions, we must await the development of new models.

Acknowledgements The authors are grateful to seminar participants at the Fourth General Conference on Advanced Mathematical Methods in Finance, Alesund, Norway, May 2009, the Workshop on Incomplete Information in Mathematical Finance, Chemnitz, Germany, June 2009, the Graduate School of Management, Kyoto, October 2009, and the Kyoto Institute of Economic Research, January 2010, for helpful comments. Part of this work was carried out while LPH was visiting the Aspen Center for Physics (September 2009) and Kyoto University (August, October 2009). DCB and LPH thank HSBC, Lloyds TSB, Shell International, and Yahoo Japan for research support.

References

1. D.C. Brody, L.P. Hughston, A. Macrina, Beyond hazard rates: a new framework for credit-risk modelling, in *Advances in Mathematical Finance, Festschrift Volume in Honour of Dilip Madan*, ed. by R. Elliott, M. Fu, R. Jarrow, J.-Y. Yen (Birkhäuser, Basel, 2007)
2. D.C. Brody, L.P. Hughston, A. Macrina, Information-based asset pricing. Int. J. Theor. Appl. Finance **11**, 107–142 (2008)
3. D.C. Brody, L.P. Hughston, A. Macrina, Dam rain and cumulative gain. Proc. R. Soc. Lond. A **464**, 1801–1822 (2008)
4. D.C. Brody, M.H.A. Davis, R.L. Friedman, L.P. Hughston, Informed traders. Proc. R. Soc. Lond. A **465**, 1103–1122 (2009)
5. S.J. Grossman, J.E. Stiglitz, On the impossibility of informationally efficient markets. Am. Econ. Rev. **70**, 393–408 (1980)

6. L.P. Hughston, A. Macrina, Information, inflation, and interest, in *Advances in Mathematics of Finance*. Banach Center Publications, vol. 83 (Institute of Mathematics, Polish Academy of Sciences, Warszawa, 2008)
7. A. Macrina, An information-based framework for asset pricing: X-factor theory and its applications. Ph.D. Thesis, King's College London (2006)
8. A. Macrina, P.A. Parbhoo, Securities pricing with information-sensitive discounting, in *Recent Advances in Financial Engineering 2009. Proceedings of the KIER-TMU International Workshop on Financial Engineering 2009* (World Scientific, Singapore, 2010)
9. M. Rutkowski, N. Yu, On the Brody–Hughston–Macrina approach to modelling of defaultable term structure. Int. J. Theor. Appl. Finance **10**, 557–589 (2007)

Chapter 6
An Overview of Comonotonicity and Its Applications in Finance and Insurance

Griselda Deelstra, Jan Dhaene, and Michèle Vanmaele

Abstract Over the last decade, it has been shown that the concept of comonotonicity is a helpful tool for solving several research and practical problems in the domain of finance and insurance. In this chapter, we give an extensive bibliographic overview—without claiming to be complete—of the developments of the theory of comonotonicity and its applications, with an emphasis on the achievements in the period 2004–2010. These applications range from pricing and hedging of derivatives over risk management to life insurance.

Keywords Comonotonicity · Convex order · Risk measurement · Derivatives pricing and hedging · Life insurance

Mathematics Subject Classification (2010) 60E15 · 60J65 · 91B70 · 91B30

6.1 Comonotonicity

Over the last two decades, researchers in economics, financial mathematics, and actuarial sciences have introduced results related to the concept of comonotonicity in their respective fields of interest. In this chapter, we give an overview of the relevant literature in these research fields, with the main emphasis on the development of the theory and its applications in finance and insurance over the last five years. Although it is our intention to give an extensive bibliographic overview, due to the high number of papers on applications of comonotonicity, it is impossible to present

---

G. Deelstra (✉)
Department of Mathematics, ECARES, Université Libre de Bruxelles, CP 210, 1050 Brussels, Belgium
e-mail: griselda.deelstra@ulb.ac.be

J. Dhaene
Faculty of Business and Economics, Katholieke Universiteit Leuven, 3000 Leuven, Belgium
e-mail: jan.dhaene@econ.kuleuven.be

M. Vanmaele
Department of Applied Mathematics and Computer Science, Ghent University, Krijgslaan 281 S9, 9000 Gent, Belgium
e-mail: michele.vanmaele@ugent.be

G. Di Nunno, B. Øksendal (eds.), *Advanced Mathematical Methods for Finance*,
DOI 10.1007/978-3-642-18412-3_6,

here an exhaustive overview of the recent literature. Further, we restrict this chapter to a description of how and where comonotonicity comes in and refer to the relevant papers for a detailed mathematical description. In order to make this chapter self-contained, we also provide a short overview of the basic definitions and initial main results of comonotonicity theory, hereby referring to part of the older literature on this topic.

The concept of comonotonicity is closely related to the following well-known result, which is usually attributed to both [62] and [53]: For any n-dimensional random vector $\underline{X} \equiv (X_1, X_2, \dots, X_n)$ with multivariate cumulative distribution function (cdf) $F_{\underline{X}}$ and marginal univariate cdf's $F_{X_1}, F_{X_2}, \dots, F_{X_n}$ and for any $\underline{x} \equiv (x_1, x_2, \dots, x_n) \in \mathbb{R}^n$, it holds that

$$F_{\underline{X}}(\underline{x}) \le \min\bigl(F_{X_1}(x_1), F_{X_2}(x_2), \dots, F_{X_n}(x_n)\bigr). \tag{6.1}$$

In the sequel, the notation $R_n(F_{X_1}, F_{X_2}, \dots, F_{X_n})$ will be used to denote the class of all random vectors $\underline{Y} \equiv (Y_1, Y_2, \dots, Y_n)$ with marginals F_{Y_i} equal to the respective marginals F_{X_i}. The set $R_n(F_{X_1}, F_{X_2}, \dots, F_{X_n})$ is called the *Fréchet class* related to the random vector $\underline{X}$.

The upper bound in (6.1) is reachable in the Fréchet class $R_n(F_{X_1}, F_{X_2}, \dots, F_{X_n})$ in the sense that it is the cdf of an n-dimensional random vector with marginals given by $F_{X_i}, i = 1, 2, \dots, n$. In order to prove the reachability property, consider a random variable U, uniformly distributed on the unit interval (0, 1). Then one has that

$$\bigl(F_{X_1}^{-1}(U), F_{X_2}^{-1}(U), \dots, F_{X_n}^{-1}(U)\bigr) \in R_n(F_{X_1}, F_{X_2}, \dots, F_{X_n}),$$

where the generalized inverses $F_{X_i}^{-1}$ are defined in the usual way:

$$F_{X_i}^{-1}(p) = \inf\bigl\{x \in \mathbb{R} \bigm| F_{X_i}(x) \ge p\bigr\}, \quad p \in [0, 1],$$

with $\inf\emptyset = +\infty$, by convention. Furthermore,

$$\begin{aligned} &\Pr\bigl[F_{X_1}^{-1}(U) \le x_1, F_{X_2}^{-1}(U) \le x_2, \dots, F_{X_n}^{-1}(U) \le x_n\bigr] \\ &\quad = \min\bigl(F_{X_1}(x_1), F_{X_2}(x_2), \dots, F_{X_n}(x_n)\bigr), \end{aligned}$$

which holds for any $\underline{x} \in \mathbb{R}^n$. Throughout this chapter, the notation U will uniquely be used to denote a random variable which is uniformly distributed on the unit interval (0, 1).

The random vector $(F_{X_1}^{-1}(U), F_{X_2}^{-1}(U), \dots, F_{X_n}^{-1}(U))$ is said to have the comonotonic dependence structure. More generally, a random vector $\underline{X} \equiv (X_1, \dots, X_n)$ is said to be *comonotonic* if

$$F_{\underline{X}}(\underline{x}) = \min\bigl(F_{X_1}(x_1), F_{X_2}(x_2), \dots, F_{X_n}(x_n)\bigr) \quad \text{for all } \underline{x} \in \mathbb{R}^n.$$

Other characterizations of comonotonicity can be found, e.g., in [27].

Furthermore, we will use the notation $\underline{X}^c \equiv (X_1^c, X_2^c, \dots, X_n^c)$ to indicate a comonotonic random vector belonging to the Fréchet class $R_n(F_{X_1}, F_{X_2}, \dots, F_{X_n})$.

The random vector $\underline{X}^c$ is often called a comonotonic counterpart or a comonotonic modification of $\underline{X}$. Obviously, one has that

$$\underline{X}^c \stackrel{d}{=} \left(F_{X_1}^{-1}(U), F_{X_2}^{-1}(U), \dots, F_{X_n}^{-1}(U)\right),$$

where the notation $\stackrel{d}{=}$ is used to indicate "equality in distribution". The random vector $\underline{X}^c$ is said to have the comonotonic dependence structure or copula, see, e.g., [82].

The components of the comonotonic random vector $(F_{X_1}^{-1}(U), F_{X_2}^{-1}(U), \dots, F_{X_n}^{-1}(U))$ are maximally dependent in the sense that all of them are nondecreasing functions of the same random variable. Hence, comonotonic random variables are indeed "common monotonic". From an economic point of view this means that holding a long position (or a short position) in comonotonic random variables can never lead to a hedge, as the variability of one is never tempered by counter-variability of others.

Comonotonicity corresponds with the riskiest dependence structure observed in a given Fréchet space $R_n(F_{X_1}, F_{X_2}, \dots, F_{X_n})$. A natural question which arises is whether there exists also a least risky dependence structure in $R_n(F_{X_1}, F_{X_2}, \dots, F_{X_n})$. From [62] and [53] it is known that the following bound holds in $R_n(F_{X_1}, F_{X_2}, \dots, F_{X_n})$:

$$F_{\underline{X}}(\underline{x}) \geq \max\left(\sum_{i=1}^{n} F_{X_i}(x_i) - n + 1, 0\right) \quad \text{for all } \underline{x} \in \mathbb{R}^n. \tag{6.2}$$

It is straightforward to prove that $(F_{X_1}^{-1}(U), F_{X_2}^{-1}(1-U)) \in R_2(F_{X_1}, F_{X_2})$ and that its cdf is given by $\max(F_{X_1}(x_1) + F_{X_2}(x_2) - 1, 0)$. Hence, when $n = 2$, the lower bound in (6.2) is reachable in $R_2(F_{X_1}, F_{X_2})$, and the random couple $(F_{X_1}^{-1}(U), F_{X_2}^{-1}(1-U))$ is said to have the countermonotonic dependence structure.

More generally, a bivariate random vector $\underline{X} \equiv (X_1, X_2)$ is said to be *countermonotonic* if

$$F_{\underline{X}}(\underline{x}) = \max\left(F_{X_1}(x_1) + F_{X_2}(x_2) - 1, 0\right) \quad \text{for all } \underline{x} \in \mathbb{R}^n. \tag{6.3}$$

When $n \geq 3$, the lower bound in (6.2) is not always a cdf anymore, and the concept of countermonotonicity cannot be generalized to higher dimensions without imposing additional conditions. Necessary and sufficient conditions for $\max(\sum_{i=1}^{n} F_{X_i}(x_i) - n + 1, 0)$ to be a cdf can be found, e.g., in [67].

Dhaene and Denuit [36] consider Fréchet spaces containing nonnegative mutually exclusive risks, that is, risks that cannot be strictly positive together. They show that, under some reasonable assumptions, the Fréchet lower bound is reachable in such Fréchet classes and corresponds with the mutually exclusive risks of that space.

Embrechts et al. [51] investigate the relation between comonotonicity and extremal correlations. They point out that a positive perfectly correlated random couple is comonotonic, whereas the inverse does not necessarily holds. Denuit and

Dhaene [31] investigate the relation between comonotonicity, respectively countermonotonicity, and several classical measures of association such as Pearson's correlation coefficient, Kendall's τ, Spearman's ρ, and Gini's γ.

Thanks to the works of [88, 91], and [116], comonotonicity has become an important concept in *economic theories of decision under risk and uncertainty*. Yaari developed a theory of risk dual to the classical expected utility theory of [111] by modifying the independence axiom in the latter theory. In Yaari's theory, the concept of "distorted expectations" arises as the equivalent of "expected utilities" in von Neumann and Morgenstern's theory. These distorted expectations are additive for comonotonic random variables.

6.2 Convex Bounds for Sums of Random Variables

In risk theory and finance, one is often interested in the distribution of the sum $S = X_1 + X_2 + \cdots + X_n$ of individual risks of a portfolio $\underline{X} \equiv (X_1, X_2, \ldots, X_n)$. Departing from the results of [62] and [53], stochastic order bounds have been derived for sums S of which the cdf's of the X_i are known, but the joint distribution of the random vector $(X_1, X_2, \ldots, X_n)$ is either unspecified or too cumbersome to work with. Assuming that only the marginal distributions of the random variables are given (or used), the largest sum in convex order will occur when the random variables are comonotonic.

In this section, we give a short overview of these stochastic ordering results. Early references to part of the ideas and results presented below are [80, 97], and [90]. For proofs and more details on the presented results, we refer to the overview paper of [41]. An overview of applications of these results in insurance and finance up to 2002 can be found in [40]. The current chapter is thus a complement to these 2002 overview papers [40] and [41].

6.2.1 Sums of Comonotonic Random Variables

Consider a random vector $(X_1, \ldots, X_n)$ and its comonotonic counterpart $(X_1^c, \ldots, X_n^c)$. The sum of the components of $(X_1^c, \ldots, X_n^c)$ is denoted by S^c,

$$S^c = X_1^c + \cdots + X_n^c. \tag{6.4}$$

The distribution of the comonotonic sum S^c can be determined from

$$F_{S^c}(x) = \sup\left\{ p \in [0, 1] \,\middle|\, \sum_{i=1}^{n} F_{X_i}^{-1}(p) \leq x \right\}, \quad x \in \mathbb{R}.$$

The distribution of S^c can also be specified via its quantile function $F_{S^c}^{-1}(p)$, which exhibits the following additivity property:

$$F_{S^c}^{-1}(p) = \sum_{i=1}^{n} F_{X_i}^{-1}(p), \quad p \in [0, 1].$$

Hereafter, we will always assume that all random variables X_i have finite means. The distribution of S^c can then be specified via its stop-loss transform $\mathrm{E}[(S^c - x)_+]$. Dhaene et al. [39] show that any stop-loss premium $\mathrm{E}[(S^c - x)_+]$ can be decomposed into a linear combination of stop-loss premiums $\mathrm{E}[(X_i - x_i)_+]$, $i = 1, 2, \ldots, n$, for appropriate choices of the x_i.

In order to state this decomposition formula more formally, we first introduce other types of generalized inverses of cdf's. The càdlàg inverse $F_{X_i}^{-1+}$ is defined by

$$F_{X_i}^{-1+}(p) = \sup\{x \in \mathbb{R} \mid F_{X_i}(x) \leq p\}, \quad p \in [0, 1],$$

with $\sup \emptyset = -\infty$, by convention. Following [70], for any $\alpha \in [0, 1]$, the inverse $F_{X_i}^{-1(\alpha)}$ is defined by

$$F_{X_i}^{-1(\alpha)}(p) = \alpha F_{X_i}^{-1}(p) + (1 - \alpha) F_{X_i}^{-1+}(p), \quad p \in (0, 1).$$

The decomposition formula of [39] can then be expressed as follows:

$$\mathrm{E}\big[(S^c - x)_+\big] = \sum_{i=1}^{n} \mathrm{E}\big[\big(X_i - F_{X_i}^{-1(\alpha)}\big(F_{S^c}(x)\big)\big)_+\big], \quad x \in \big(F_{S^c}^{-1+}(0), F_{S^c}^{-1}(1)\big). \tag{6.5}$$

Here, α is any element of $[0, 1]$ satisfying

$$\sum_{i=1}^{n} F_{X_i}^{-1(\alpha)}\big(F_{S^c}(x)\big) = x.$$

A special case of the decomposition formula (6.5) can be found in [66], who proves that in the Vasicek [110] model, a European option on a portfolio of pure discount bonds (in particular, an option on a coupon-bearing bond) decomposes into a portfolio of European options on the individual discount bonds in the portfolio. This holds true because in the Vasicek model, the prices of all pure discount bonds at some future time T are decreasing functions of a single random variable, namely the spot rate at that time. This implies that the price at time T of the portfolio of pure discount bonds is a comonotonic sum. Taking into account that the current price of a European option can be expressed as the discounted expected pay-off of this option, where the expectation is taken with respect to an appropriate measure, we find that for the current price of the option on the portfolio of zero-coupon bonds, a decomposition as in (6.5) holds.

6.2.2 Convex Bounds for Sums of Random Variables

Consider a random vector $\underline{X} \equiv (X_1, X_2, \ldots, X_n)$, not necessarily comonotonic, and the sum of its components

$$S = X_1 + \cdots + X_n.$$

Intuitively, one might expect that the comonotonic sum $S^c = X_1^c + \cdots + X_n^c$ of the comonotonic counterpart $\underline{X}^c$ is more variable than the original sum S. In order to state this intuitive result more formally, we need the notion of convex order.

A random variable X is said to precede a random variable Y in the *convex order* sense, notation $X \leq_{\text{cx}} Y$, if the following conditions hold:

$$\mathrm{E}\big[(x - X)_+\big] \leq \mathrm{E}\big[(x - Y)_+\big] \quad \text{for all } x$$

and

$$\mathrm{E}\big[(X - x)_+\big] \leq \mathrm{E}\big[(Y - x)_+\big] \quad \text{for all } x.$$

Other characterizations of convex order can be found, e.g., in [93] in a general context or in [34] in an actuarial context. Intuitively, the convex order relation $X \leq_{\text{cx}} Y$ states that compared to the random variable X, the random variable Y has more probability mass in its lower and upper tails. Wang and Young [114] compare the concept of ordering random variables in expected utility theory versus Yaari's dual theory of choice under risk.

One can prove that the following relation holds between the sum S and its comonotonic modification S^c:

$$X_1 + \cdots + X_n \leq_{\text{cx}} X_1^c + \cdots + X_n^c = S^c. \tag{6.6}$$

This result states that when one assumes that only the marginal distributions of the random variables are given (or used), the largest sum in convex order occurs when the random variables are comonotonic. To the best of our knowledge, this result was first mentioned in the actuarial literature in [58], who attributes it to [80]. Other early references are [97] and [90]. Tchen [97] has proven that in the class of all random vectors with given marginals the comonotonic random vectors are greater in supermodular order than any other element of this class. A simple proof for the inequality (6.6), which is based on a geometric interpretation of the support of the comonotonic distribution, is given in [71].

Since the mid-1990s, the convex order relation (6.6) has attracted a lot of attention in the actuarial literature. Authors of [6, 33, 37, 38, 56, 81] and [113] generalize (6.6) by investigating how changing the dependence structure of an insurance portfolio influences its stop-loss premiums. In any of the different situations considered in these papers, the convex order relation (6.6) corresponds with the extreme case where the comonotonic dependence structure is involved.

From the convex order relation (6.6) it follows immediately that the expectation $\mathrm{E}[(S^c-x)_+]$ can be interpreted as the solution to the following maximization problem, where we use R_n as a shorthand notation for the Fréchet space $R_n(F_{X_1},F_{X_2},\ldots,F_{X_n})$:

$$\max_{\underline{Y}\in R_n}\mathrm{E}\big[(Y_1+Y_2+\cdots+Y_n-x)_+\big]=\mathrm{E}\big[\big(X_1^c+\cdots+X_n^c-x\big)_+\big],\quad x\in\mathbb{R}.$$

This means that $\mathrm{E}[(S^c-x)_+]$ can be interpreted as an extreme-case expectation for $\mathrm{E}[(S-x)_+]$. Indeed, let us assume that the only information that is available about the distribution of the random vector $\underline{X}$ is the marginal cdf's F_{X_i} $i=1,\ldots,n$. In this case, the largest possible value for $\mathrm{E}[(S-x)_+]$ is given by $\mathrm{E}[(S^c-x)_+]$.

One can also prove that $\mathrm{E}[(S^c-x)_+]$ is the solution to the following minimization problem:

$$\min_{\sum_{i=1}^n x_i=x}\sum_{i=1}^n\mathrm{E}\big[(X_i-x_i)_+\big]=\mathrm{E}\big[\big(X_1^c+\cdots+X_n^c-x\big)_+\big],\tag{6.7}$$

where the minimum is taken over all $(x_1,x_2,\ldots,x_n)$ with $\sum_{i=1}^n x_i=x$, see [90].

To the best of our knowledge, Simon et al. [95] were the first who combined the convex order relation (6.6) and the decomposition formula (6.5) to find an upper bound for the price of an arithmetic European-type Asian option in terms of the price of an appropriate portfolio of plain vanilla European call options. Furthermore, from the optimization result (6.7) they conclude that the exercise prices of the plain vanilla options contained in their upper bound are optimal in the sense that no improvement can be obtained by considering other linear combinations of plain vanilla European options. Important to notice is that this result is model-independent. Later, Albrecher et al. [3] interpret the comonotonic upper bound of [95] as the price of a static superhedging strategy for an Asian option, where the hedging portfolio consists of plain vanilla options. Using static superhedging strategies has the advantage that it is much less sensitive to the assumption of zero transaction costs and to the hedging performance in the presence of large market movements, compared to dynamic strategies.

In order to be able to determine the upper bound in (6.6), the only information that is required about the distribution of $\underline{X}$ is its marginals. Intuitively, it is clear that it must be possible to find better convex order upper bounds for S when more information is available concerning the multivariate cdf of $\underline{X}$. Therefore, let us assume that apart from the knowledge of the marginals, there exists a random variable Λ with a given distribution function such that the conditional distributions of the random variables X_i, given $\Lambda=\lambda$, are known for all outcomes λ of Λ. Kaas et al. [70] derive the following improved convex order upper bound, denoted S^{ic}, for this particular case:

$$X_1+\cdots+X_n\le_{\mathrm{cx}}F^{-1}_{X_1|\Lambda}(U)+F^{-1}_{X_2|\Lambda}(U)+\cdots+F^{-1}_{X_n|\Lambda}(U)=S^{ic},\tag{6.8}$$

where $F^{-1}_{X_i|\Lambda}(U)$ is the notation for the random variable $f_i(U, \Lambda)$ with f_i defined by $f_i(u, \lambda) = F^{-1}_{X_i|\Lambda=\lambda}(u)$. Notice that the random vector $(F^{-1}_{X_1|\Lambda}(U), F^{-1}_{X_2|\Lambda}(U), \ldots, F^{-1}_{X_n|\Lambda}(U))$ is said to be "conditionally comonotonic".

Based on an idea that stems from mathematical physics, Kaas et al. [70] propose the following convex order lower bound for S, denoted S^ℓ, when the information available about the cdf of $\underline{X}$ is the same as the one that leads to the upper bound in (6.8):

$$S^\ell = \mathrm{E}[X_1|\Lambda] + \mathrm{E}[X_2|\Lambda] + \cdots + \mathrm{E}[X_n|\Lambda] \leq_{\mathrm{cx}} X_1 + X_2 + \cdots + X_n. \tag{6.9}$$

They remark that this lower bound has the nice property that it is a comonotonic sum, provided that all terms $\mathrm{E}[X_i|\Lambda]$ are increasing (or all are decreasing) functions of Λ. In this case, the quantiles and stop-loss premiums of $S^\ell = \sum_{i=1}^n \mathrm{E}[X_i|\Lambda]$ follow immediately from the additivity properties of comonotonic sums in Sect. 6.2.1. This property is particularly of interest in a multivariate lognormal setting. In such a setting, the lower bound turns out to be very accurate, provided that the appropriate choice is made for the conditioning random variable Λ, see, e.g., [105].

The lower bound (6.9) is applied in [40] to derive accurate approximations for European-type Asian options in a Black and Scholes setting, in case of discrete averaging of the stock price. In a lognormal setting, Rogers and Shi [89] apply a similar lower bound to derive approximations for the price of Asian options in case of continuous averaging.

6.3 Further Developments of the Theory

In this section we summarize several extensions of the theory of comonotonicity since 2004, not claiming to be exhaustive but trying to be as complete as possible, taking into account that this theory is still in development.

Inequality (6.6) implies that if a random vector with given marginal distributions is comonotonic, it has the largest sum with respect to convex order. Cheung [18] proves that the converse also holds, provided that each marginal distribution is continuous.

Defining the improved comonotonic upper bound, see relation (6.8), Kaas et al. [70] introduced implicitly the notion of *conditional comonotonicity*. This notion is later more formally considered by Jouini and Napp [68] as a generalization of the classical concept of comonotonicity. In [15], this concept is further investigated. The main result is that a random vector is comonotonic conditional to a certain sigma-field if and only if it is almost surely comonotonic locally on each atom of the conditioning sigma-field. In [17], the relationship between conditional comonotonicity and convex ordering is explored. By this notion of conditional comonotonicity it is possible to unify the classical upper bound result (inequality (6.6)) and the improved upper bound result (inequality (6.8)) in a more general framework.

The choice of the conditioning random variable Λ in (6.9) is crucial for the accuracy of the lower bound approximation S^ℓ. When S is a sum of nonindependent lognormal random variables, different alternatives for Λ have been proposed in the literature, see, e.g., [70] and [105]. These choices are "global" in the sense that Λ is chosen such that the entire distribution of the approximation $\mathrm{E}[S|\Lambda]$ is "close" to the corresponding distribution of the original sum S. In an actuarial or a financial context one is often only interested in a particular tail of the distribution of S. Therefore, Vanduffel et al. [106] propose locally optimal approximations, in the sense that the relevant tail of the distribution of $\mathrm{E}[S|\Lambda]$ is an accurate approximation for the corresponding tail of the distribution of S. Deelstra et al. [24] study sums S of the form $\sum_{i=1}^{n} w_i \alpha_i e^{\beta_i + \gamma_i Y_i}$, where the positive weights w_i sum up to one, the coefficients $\alpha_i (>0)$, β_i, γ_i are deterministic, and the Y_i's are nonindependent normally distributed random variables. In this case, Deelstra et al. [24] show that all these choices for the conditioning random variable Λ can be considered as a linear transformation of a first-order approximation of S, namely $\Lambda = \sum_{i=1}^{N} w_i \alpha_i \gamma_i Y_i \delta_i$ with δ_i taking different forms according to the different choices.

The applicability of the convex bounds (6.6), (6.8), and (6.9) to derive closed-form approximations for risk measures of a sum of nonindependent lognormal random variables with unknown dependence structure is illustrated in [40]. Valdez et al. [102] investigate to which extent the general results on convex bounds of Sect. 6.2 can be applied to sums of nonindependent *log-elliptical* random variables which incorporate sums of lognormals as a special case. First, they show that unlike the lognormal case, for general sums of logellipticals, the convex lower bound (6.9) does no longer result in closed-form approximations for the different risk measures. Second, they demonstrate how instead the weaker stop-loss order can be used to derive such closed-form approximations. In numerical illustrations they show that these newly proposed approximations are useful to measure satisfactorily the risk of discounted or compounded sums in case the stochastic log-returns are elliptically distributed.

More general, Kukush and Pupashenko [73] study comonotonic upper and lower bounds for sums under a mixture of arbitrary distributions. They also consider the case where the logarithm of the components in the sum can be represented as a mixture of normal random variables. These results may be useful to perform approximate evaluations of actuarial provisions when a regime switching model is used for the investment returns.

Yang et al. [117] investigate bivariate copula structures for modeling dependence among random variables in a distribution free way. The existence and uniqueness of a bivariate copula decomposition into a comonotonic, an independent, a countermonotonic, and an indecomposable part are proved, while the coefficients are determined from partial derivatives of the corresponding copula. Moreover, for the indecomposable part, an optimal convex approximation is provided and analyzed. The variance decomposition that they derive can be applied to find mean–variance optimal investment portfolios in finance. They also consider other applications of this decomposition in finance and insurance.

Hoedemakers et al. [61] and Ahcan et al. [2] extend the theory of convex bounds to the case of scalar products of mutually independent random vectors. This methodology allows one to obtain reliable approximations of the underlying distribution functions and very accurate estimates of quantiles and stop-loss premiums. Hua and Cheung [63] also study stochastic orders of scalar products of random vectors and derive more general conditions under which linear combinations of random variables can be ordered in the increasing convex order.

Cheung [19] introduces upper comonotonicity as a generalization of the classical notion of comonotonicity. A random vector $\underline{X} = (X_1, X_2, \ldots, X_n)$ is said to be *upper-comonotonic* if its components X_i are moving in the same direction simultaneously when their values are greater than some thresholds.

This new notion can be characterized in terms of both the joint distribution function and the underlying copula. The copula characterization allows the study of the coefficient of upper tail dependence and the distributional representation of an upper-comonotonic random vector. The additivity property of several commonly used risk measures, such as the Value-at-Risk, the Tail Value-at-Risk and the expected shortfall for sums of comonotonic risks is extended to sums of upper-comonotonic risks, provided that the level of probability is greater than a certain threshold.

For premium calculation principles or risk measures, usually only the additivity for a finite number of comonotonic risks is considered. However, a limiting status of finite additivity is the additivity for countable risks. In [115], the *countable additivity* is investigated, and new and elegant characterizations for Choquet pricing and distortion premium principles are presented. The countable exchangeability is also studied following the investigation of countable additivity for comonotonic risks.

Several multivariate extensions of comonotonicity are studied in [86]. Naive extensions do not enjoy some of the main properties of the univariate concept. In the univariate case, the definition of comonotonicity only relies on the total order structure. Hence this definition could be extended for any random vector with values in a product of totally ordered measurable spaces. Most of its properties would be valid even in this multivariate context. However, the aim of Puccetti and Scarsini [86] is to study comonotonic vectors that take values in a product of partially ordered spaces. Different definitions of *multivariate comonotonicity* are introduced, trying to extend different features of the classical definition. It is shown that no definition satisfies all the properties of the original one. Some definitions do not guarantee the existence of a comonotonic random vector for any pair of multivariate marginals. Some other definitions do not guarantee uniqueness in distribution of the comonotonic random vector with fixed marginals.

In finance, Galichon and Henry [54] and Ekeland et al. [50] propose a multivariate extension of coherent risk measures that involves a multivariate extension of the notion of comonotonicity, in the spirit of [86].

6.4 Applications of the Theory of Comonotonicity

6.4.1 Derivatives Pricing and Hedging

Several European options have a pay-off written on one or multiple underlyings combined in a weighted sum of nonindependent random variables expressing asset prices at the time of maturity or at different time points before and at maturity. Examples of this type of options with positive weights are Asian options, basket options, and Asian basket options. When the weights can be both positive and negative, one refers to these options as spread options, Asian spread options, basket spread options, and Asian basket spread options. Pricing and hedging of these products by means of comonotonicity bounds has been studied in a model-dependent and in a model-independent framework. As mentioned before, early references to this topic are [89, 95] and [40]. Hereafter, we will discuss articles published since 2004 dealing with this topic.

First we consider the *model-dependent* setting. A survey of current methods up to 2006 for pricing Asian options and computing their sensitivities to the key input parameters is provided in [7]. The methods discussed there include also the comonotonic bounds. We will focus in the present chapter only on those papers dealing with comonotonic bounds. In comparison with [7], we will also discuss more recent papers and other applications.

Schrager and Pelsser [92] use a change of numeraire technique to derive a general pricing formula for the Rate of Return Guarantees in a Regular Premium Unit Linked (UL) Insurance contract. They show that the guarantee is equivalent to a European put option on some stochastically weighted average of the stock price at maturity. They extend earlier results from [95] and [40] on pricing bounds of Asian options to UL Guarantees and stochastic interest rates in the case that the underlying sum is composed of lognormal random variables.

In [109], the pricing of European-style discrete arithmetic Asian options with fixed and floating strike is studied by deriving analytical lower and upper bounds, as explained in Sect. 6.2, and additionally combined with the ideas of [95] and of [83]. Through these bounds, a unifying framework is created for European-style discrete arithmetic Asian options that generalizes several approaches in the literature and improves existing results. Analytical and easily computable bounds are obtained under the Black and Scholes model for the asset prices. An advice of the appropriate choice of the bounds given the parameters is formulated, the effect of different conditioning variables is investigated, and their efficiency is numerically compared. Based on these approximating bounds, analytical hedging formulas are developed.

Making use of geometric arguments, Brückner [8] quantifies the maximal error in terms of truncated first moments, when a sum S is approximated by the upper bound S^c or a lower bound S^ℓ as defined in (6.6) and (6.9), respectively.

Vyncke et al. [112] construct a convex combination of the comonotonic upper bound and the lower bound for the price of a European-style arithmetic Asian option and find an approximation for this price which is such that the underlying approximate cdf has exact first and second moments.

Inspired by the ideas of [89], Chalasani and Varikooty [11] derived accurate lower and upper bounds for the price of a European-style Asian option with continuous averaging over the full lifetime of the option, using a discrete-time binary tree model. Reynaerts et al. [87] consider arithmetic Asian options with discrete sampling and generalize the method of [11] to the case of forward starting Asian options. In this case with daily time steps, that method is still very accurate, but the computation can take a very long time on a PC when the number of steps in the binomial tree is high. Reynaerts et al. [87] derive analytical lower and upper bounds based on the results presented in Sect. 6.2 and by conditioning on the value of the underlying asset at the exercise date. The comonotonic upper bound corresponds to an optimal superhedging strategy. By putting in less information than [11] the bounds lose some accuracy but are still very good, and they are easily computable, and moreover the computation on a PC is fast.

Also the price of a continuously sampled European-style Asian option with fixed exercise price can be approximated by means of the tools of Sect. 6.2. Within the Black and Scholes framework, Vanduffel et al. [107] derive analytic expressions for lower and upper bounds for such a price.

As for Asian options, determining the price of a European basket option is not a trivial task, even in the Black and Scholes model, because there is no explicit analytical expression available for the distribution of the weighted sum of prices of the assets in the basket. The upper bounds proposed in Sect. 6.2 will not always lead to good approximations since a basket of underlyings can be far from a comonotonic sum, depending on the correlations between the assets. However, by using a conditioning variable, the price of a European basket option can be decomposed in two parts, one of which can be computed exactly. For the remaining part, Deelstra et al. [22] derive a lower and some upper bounds based on the theory of comonotonicity. The lower bound obtained in this way corresponds to (6.5) with the (general) comonotonic sum (denoted by S^c in (6.5)) being replaced by the particular comonotonic sum S^ℓ introduced in (6.9). The first upper bound is based on an improved comonotonic upper bound upon the part in the pricing formula that cannot be calculated in an explicit way. The other upper bound is obtained by using the ideas of [89] and [83] upon that same part. By concentrating only upon this inexact part, much more precise approximating bounds can be obtained.

The lower bounds and some of the upper bounds discussed above are based on comonotonicity results combined with conditioning upon one variable. In a Black and Scholes setting, Vanmaele et al. [108] derive analytical expressions for comonotonic bounds of stop-loss premiums of sums of nonindependent random variables by conditioning upon two variables. They also use the idea of several conditioning variables to develop an approximation for cases for which it is cumbersome to obtain a comonotonic lower bound. The numerical application to European basket options shows that conditioning on two variables leads to very sharp results.

Combining the features of Asian and basket options, we end up with European-style discrete arithmetic Asian basket options. Deelstra et al. [23] propose pricing bounds for these options in a Black and Scholes framework. They use the general approach for deriving upper and lower bounds as in Sect. 6.2 and generalize in this way the methods of [22] and [109]. They further show how to derive an analytical closed-form expression for a lower bound in the noncomonotonic case. Finally, in numerical tests the quality of these bounds are compared to upper bounds for Asian basket options based on techniques as in [98] and [78].

When allowing also for negative weights, one can price European-style discrete arithmetic Asian basket spread options. Deelstra et al. [25] derive comonotonic lower and upper bounds for such spread options and discuss the behavior of these approximating bounds. They also develop a new hybrid moment matching method, namely a moment matching of both the positively weighted basket and the negatively weighted basket separately, combined with an improved comonotonic upper bound (6.5) with S^{ic} from (6.8) being the comonotonic sum S^c. Deelstra et al. [25] find that the improved comonotonic upper bound offers a good approximation of the price of spread options. The hybrid moment matching method based upon the improved comonotonic upper bound approach leads to a well-performing bound for Asian basket spread options. The Greeks for these two methods are explicitly derived. Moreover, the results can be extended to options denominated in foreign currency.

Deelstra et al. [26] elaborate a method for determining the optimal strike price for a put option, used to hedge a position in a financial product such as a basket of shares or a coupon-bearing bond. This strike price is optimal in the sense that it minimizes, for a given budget, a class of risk measures satisfying certain properties. Hereto they study the loss function in the worst-case scenario such that its risk is on the safe side. Formulas are derived both for one single underlying and for a weighted sum of underlyings. For the latter, two cases are considered depending on the dependence structure of the components in this weighted sum, namely the case that the components form a comonotonic vector and the case that they are not comonotonic. In the latter case comonotonic approximations based on S^c (6.4), respectively on S^ℓ (6.9), are proposed.

Now we turn to the *model-independent* bounds as the ones presented in [95] and [3] for Asian options.

In [59, 60] static-arbitrage super-replicating respectively subreplicating strategies for European-style basket options are derived. In the former article, the authors consider the set of all models which are consistent with the observed prices of vanilla options and, within this class, find the model for which the price of the basket option is largest. This price is an upper bound on the prices of the basket option which are consistent with no-arbitrage. In the absence of additional assumptions it is the lowest upper bound on the price of the basket option and is related to a comonotonic upper bound. Both the infinite market case (where prices of the plain vanilla options are available for all strikes) and the finite market case (where only a finite number of plain vanilla option prices are observed) are considered. From a pure mathematical point of view, the infinite market case results are closely related to the optimization result (6.7) presented in Sect. 6.2. In [60], subreplicating strategies are developed for European-style basket options consisting of two assets. The so-called sheeptrack portfolio has a price that can only be realized by a countermonotonic pair as defined in (6.3).

Whereas Hobson et al. [60] only concentrate on basket options, Chen et al. [12] investigate static super-replicating strategies for European-type call options written on a weighted sum of asset prices. This class of exotic options includes Asian options and basket options among others. It is assumed that there exists a market where the plain vanilla options on the different assets are traded and hence their prices can be observed in the market. Both the infinite and finite market cases are considered. It is proven that the finite market case converges to the infinite market case as the number of observed plain vanilla option prices tends to infinity. The paper shows how to construct a portfolio consisting of the plain vanilla options on the different assets whose pay-off super-replicates the pay-off of the exotic option. As a consequence, the price of the super-replicating portfolio is an upper bound for the price of the exotic option. The superhedging strategy is model-free in the sense that it is expressed in terms of the observed option prices on the individual assets, which can be, e.g., dividend paying stocks with no explicit dividend process known. As opposed to [60] who use Lagrange optimization techniques, the proofs in [12] are based on the theory of integral stochastic orders, comonotonicity, and convex bounds, see Sects. 6.1 and 6.2.

Chen et al. [13] further investigate super-replicating strategies for European-type call options written on a positively weighted sum of asset prices following the initial approach in [12]. To be more precise, three issues are proposed and investigated concerning the optimal super-replicating strategies. The first issue is the nonuniqueness of the optimal solution. The second issue is to generalize the results from a deterministic interest rate setting in the previous paper to a stochastic interest rate setting. By performing this generalization, optimal super-replicating strategies are obtained in a more general market. The third issue is about the coexistence of the comonotonicity property and the martingale property. When there is only one underlying asset, it is shown that they possibly coexist for some cases, while for some other cases there can also be a contradiction between them. As a consequence, for Asian options, the upper bound may not be reachable in an arbitrage-free market.

Distribution-free bounds in closed-form and optimal hedging strategies for spread options are derived in [76, 77]. The former article focuses on upper bounds when the spread option's joint distribution is calibrated to the information about the marginals embedded in the prices of traded options with all available strikes of a given maturity.

In the latter article, sharp distribution-free lower bounds for spread options and the corresponding optimal subreplicating portfolios are obtained. This lower bound is attained for the comonotonic distributions. Laurence and Wang [76, 77] also introduce the notion of *monotonicity gap* which can be further divided into two complementary gaps, the countermonotonicity and the comonotonicity gap. The idea is that the normalized distance of the true (quoted) market price of a spread option from the distribution free comonotonic upper bound (respectively, countermonotonic lower bound) represents a useful and new "market implied" index. This index measures how far the assets are from being countermonotonic (respectively, comonotonic) and can be used as a distribution-free complement to the so-called implied correlation that is widely used in the industry.

Finally, we draw the attention to some recent articles where comonotonicity is applied to price or hedge some other types of financial products.

Based on the positive dependence characteristic of the mortality in catastrophe areas, Shang et al. [94] develop a pricing model for catastrophe mortality bonds with comonotonicity and a jump-diffusion process. Since there is no unique risk-neutral probability in this incomplete market settings, they use the Wang transform method to price the bond.

In [10] possible bounds on CDO tranche premiums are studied. In case of a comonotonic vector of default times, a model-free lower bound on equity tranche premiums is provided, where model-free has to be understood with respect to the dependence structure between default dates. The CDO tranche premiums computations turn out to be straightforward in this comonotonic case.

Glau et al. [55] study interest rate derivatives. In particular, they consider the Lévy term structure model that extends the Heath–Jarrow–Merton model in that the instantaneous forward rate is given by a time-inhomogeneous Lévy process. Within this framework, pricing formulas based on Fourier transforms are known for the most liquid interest rate derivatives, namely caps, floors, and swaptions. Glau et al. [55] study delta-hedging and risk-minimizing hedging strategies for swaptions on the basis of zero-coupon bonds. They derive closed-form expressions for the hedging strategy in terms of the Fourier transforms by the comonotonicity property.

6.4.2 Risk Management: Risk Sharing, Optimal Investment, Capital Allocation

Risk Measures and Risk Sharing

In [40] it is shown how the convex bounds (6.6), (6.8), and (6.9) can be used to derive closed-form approximations for risk measures of a sum of nonindependent lognormal random variables. Dhaene et al. [45] further examine and summarize properties of several well-known risk measures that can be used in the framework of setting capital requirements for a risky business. Special attention is given to the class of concave distortion risk measures, also called spectral risk measures, see [1]. Note that the class of concave distortion risk measures is a subset of the more general class of coherent risk measure as introduced in [4] and [5], see also [65].

Dhaene et al. [45] investigate the relationship between distortion risk measures and theories of choice under risk. They further consider the problem of how to evaluate these risk measures for sums of nonindependent random variables and approximations for such sums, based on the concept of comonotonicity, are proposed. Another generalization of the class of concave distortion risk measures in a distribution free setting is considered in [44].

Goovaerts et al. [57] present a new axiomatic characterization of risk measures that are additive for independent random variables. The axiom of additivity for independent random variables is related to an axiom of additivity for comonotonic random variables. The risk measures characterized can be regarded as mixed exponential premiums.

The appropriateness of the subadditivity of risk measures is considered in [47]. Dhaene et al. [48] investigate the influence of the dependence between random losses on the shortfall and on the diversification benefit that arises from merging these losses. They prove that increasing the dependence between losses, expressed in terms of correlation order, has an increasing effect on the shortfall, expressed in terms of an appropriate integral stochastic order. Furthermore, increasing the dependence between losses decreases the diversification benefit. In particular, they consider merging comonotonic losses and show that even in this extreme case a nonnegative diversification benefit may arise.

Embrechts et al. [52] prove that comonotonicity gives rise to the on-average-most-adverse Value-at-Risk (VaR) scenario for a function of dependent risks when the marginal distributions are known but the dependence structure between the risks is unknown. Laeven [74] extends this result to the case where, rather than no information, partial information is available on the dependence structure between the risks. Moreover, Laeven [74] points out that the improved comonotonic or conditionally comonotonic dependence structure as introduced in (6.8) is very interesting as a worst-case scenario. Indeed, it is the most adverse dependence structure in stop-loss and supermodular order and hence in Tail-VaR-based risk management, and the on-average-most-adverse dependence structure in VaR-based risk management.

Tsanakas and Christofides [101] model an exchange economy where agents (insurers/banks) trade risks. Decision making takes place under distorted probabilities, which are used to represent either rank-dependence of preferences or ambiguity with respect to real-world probabilities. Via the construction of aggregate preferences from heterogeneous agents' utility and distortion functions, they obtain pricing formulas and risk allocations, generalizing results of [9]. In particular, in a lemma which can be viewed as a generalized version of Borch's characterization of Pareto optima, it is stated that at equilibrium the agents' risk allocations are comonotonic random variables.

Jouini et al. [69] consider the problem of optimal risk sharing of some given total risk between two economic agents characterized by law-invariant monetary utility functions or equivalently, law-invariant risk measures. In the case that both agents' utility functions are comonotone, an explicit characterization of an optimal risk sharing allocation is provided. This optimal allocation is in addition increasing in terms of the total risk.

Also in [79] the risk sharing problem is dealt with. They extend the result that a Pareto optimal risk allocation is necessarily comonotone to the case of unbounded random variables and this for certain classes of consistent risk measures. This is significant from a practical point of view where risks are often modeled as unbounded random variables.

Optimal Investment Strategies

Dhaene et al. [43] investigate multiperiod portfolio selection problems in a Black and Scholes-type market where a basket of one risk-free and m risky securities are traded continuously. They look for the optimal allocation of wealth within the class of constant-mix portfolios. First, they consider the portfolio selection problem of a decision maker who invests money at predetermined points in time in order to obtain a target capital at the end of the time period under consideration. A second problem concerns a decision maker who invests some amount of money (the initial wealth or provision) in order to be able to fulfil a series of future consumptions or payment obligations. Several optimality criteria and their interpretation within Yaari's dual theory of choice under risk are presented. For both selection problems, accurate approximations are proposed based on the concept of comonotonicity as exposed in Sect. 6.2. Similar problems are considered in the related papers [103, 104].

Cheung and Yang [14, 20] study a single-period optimal portfolio problem. It is assumed that the actual dependence structure of the asset returns is unknown or is a mixture of some common underlying source of risks. The least favorable dependence structure is first identified, then the optimal portfolio problem is analyzed as if this were the actual dependence structure. A sufficient condition to order the optimal allocations is obtained using concepts of stochastic ordering.

Capital Allocation

The Enterprise Risk Management process of a financial institution usually contains a procedure to allocate, or subdivide, the total risk capital of the company into its different business units.

In [42], an optimization argument is used to find an optimal rule for allocating the aggregate capital of a financial firm to its business units. The optimal allocation can be found using general results from the theory on comonotonicity as summarized in Sect. 6.2. Dhaene et al. [49] generalize the approach of [42] and develop a unifying framework for allocating the aggregate capital by considering more general deviation measures. Capital allocation based on the principle of comonotonicity turns out to be a special case of this general framework, as well as many other allocation rules that are described in the literature.

Taking the viewpoint of a higher authority within the financial conglomerate (typically the board of directors) by which the economic capital allocation is performed, Laeven and Goovaerts [75] propose an optimization approach to allocate economic capital, distinguishing between an allocation or raising principle, and a measure for the risk residual. The approach provides an integrated solution since it can be applied both at the aggregate (conglomerate) level and at the individual (subsidiary) level. Different degrees of information on the dependence structure between the subsidiaries are considered. When using expectations as risk measure and assuming a complete lack of information on the dependence structure between the subsidiaries, the capital allocation problem reduces to the problem considered in [42].

Dhaene et al. [46] study the CTE-based allocation rule, where the Conditional Tail Expectation (CTE) acts as risk measure to deal with the allocation problem. Comonotonicity is used to derive accurate and easy to compute closed-form approximations for the CTE-based allocation rule. Hence, the field where analytical solutions for this rule are available is extended to the case that the risks of the different units have a (log)normal distribution.

Cheung [16] studies orderings of optimal allocations of policy limits and deductibles when losses are cumulative, while Hua and Cheung [63] introduce new models to separate the effects of severities and frequencies of losses. In both of these papers the study is carried out from the viewpoint of a risk-averse policyholder and under the assumption that the dependence structure of the losses is unknown. In order to deal with this, they focus on the worst-allocation problem, the worst-dependence structure being identified as the comonotonic one. In [64] the worst allocations of policy limits and deductibles is studied from the viewpoint of an insurer. The main results of these articles are complemented and extended in [119] by applying bivariate characterizations of stochastic ordering relations.

Tsanakas et al. [99] constructs a distortion-type risk measure, which evaluates the risk of any uncertain position in the context of a portfolio that contains that position and a fixed background risk, which means that besides the specific portfolio, the holder is also exposed to a risk that he cannot (or will not) trade, control, or mitigate. The risk measure can also be used to assess the performance of individual risks within a portfolio, allowing for the portfolio's rebalancing, an area where standard capital allocation methods fail. It is shown that the properties of the risk measure depart from those of coherent distortion measures. In particular, it is shown that the presence of background risk makes risk measurement sensitive to the scale and aggregation of risk. However, the risk of an instrument X relative to a background risk Y, which is comonotone to X, is equal to the risk of X with no background risk. Further, the case of risks following elliptical distributions is examined in more detail and precise characterizations of the risk measure's aggregation properties are obtained.

Tsanakas [100] discusses the use of convex risk measures in capital allocation. He studies a flexible class of convex risk measures, namely the distortion-exponential risk measure depending on a positive real number and a concave, differentiable distortion function. For extreme cases of dependence between the risks, such as comonotonicity or countermonotonicity, see Sect. 6.1, the aggregation properties of this convex risk measure are characterized, and explicit capital allocation formulas are obtained.

6.4.3 Life Insurance and Pensions

In the classical approach to the theory of life contingencies, discounting factors and mortality tables are assumed to be deterministic. In view of the long durations of life annuity contracts, it is more realistic to take the stochastic nature of investment returns and mortality into account when investigating the risks related to annuity portfolios. Over the last two decades, a large number of papers have been published covering this stochastic approach of returns and/or mortality. In this overview, we will restrict to the subset of these papers where comonotonicity comes in.

In [72], stochastic discounting factors are introduced by considering truncated stochastic returns. Analytical results for comonotonic bounds of the present value function of a sum of discounted deterministic cash-flows are derived.

Darkiewicz et al. [21] first investigate lower and upper bounds for right tails (stop-loss premiums) of deterministic and stochastic sums of nonindependent random variables, using the concepts of Sects. 6.1 and 6.2. Then, the performance of the presented approximations is investigated numerically for individual life annuity contracts and for life annuity portfolios. The investment returns are modeled by a Brownian motion process, while the mortality is modeled by Makeham's law.

Hoedemakers et al. [61] and Ahcan et al. [2] study the distribution of a life annuity (and a portfolio of life annuities) under stochastic interest rates. They apply (6.4) and (6.9) for scalar products of mutually independent random vectors and obtain reliable approximations of the underlying distribution functions, in particular they propose very accurate estimates of quantiles and stop-loss premiums.

Zhang et al. [118] consider a homogeneous portfolio composed of n whole-life insurance policies. Since an average insurer usually has a large number of homogeneous policies, they explore the limiting properties of the convex upper bounds of the present value function of such a portfolio. These upper bounds are derived by the technique of comonotonicity under certain assumptions on the dependence structure of the residual life of the insured (i.e. independence, positive association, or negative association). The upper bounds are very informative and useful to the insurer in making conservative estimates about the risks and calculating premiums.

Denuit and Dhaene [32] and Denuit [28–30] adopt the standard Lee–Carter model for mortality projection when studying portfolios of life annuities. In these papers the discount factors are assumed to be deterministic. In the Lee–Carter model, survival probabilities depend on the future trajectory of the time index, which implies that they become random variables. In the first paper, the concept of comonotonicity is applied to obtain accurate approximations for the stochastic survival probabilities. In [29] comonotonicity-based approximations are derived for the quantiles of the conditional expected present value of the annuity payments, given the future path of the Lee–Carter time index.

Denuit, Devolder, and Goderniaux suggest in [35] securitization of longevity risk in order to offer opportunities for hedging. In particular, they propose the design of survivor bonds which could be issued directly by insurers. In order to guarantee some transparency in the product, the survivor bond is based on a public mortality index. Also here the classical Lee–Carter model for mortality forecasting is used to price a risky coupon survivor bond based on this index. The proposed pricing mechanism consists of determining the Wang risk measure of the mortality index which equals the exponential of a linear combination of correlated lognormal random variables. Taking into account comonotonic upper and lower bounds, approximate results are derived.

Spreeuw [96] applies the theory of comonotonic risks to disability annuities in a Markov model with three states (death, healthy, and disabled), where recovery from disabled to health is possible. Benefits are payable during disability, whilst premiums are only due whenever the insured is healthy. Starting from the convex upper bound (6.4) and the improved upper bound (6.8), he derives two accurate approximations for the sum of the deterministically discounted value of cash-flows involved in such a contract.

6.5 Conclusion

In this chapter, we gave an extensive—but not exhaustive—overview of the literature on the theory of comonotonicity and its applications in finance and risk theory, with an emphasis on the literature since 2004.

Taking into account the huge recent literature on this topic, we may conclude that the concept of comonotonicity indeed plays the role of a helpful tool for solving several research and practical problems in the domain of finance and insurance. It seems very reasonable to assume that the theory of comonotonicity is still in development. This observation makes us believe that in the near future more applications will follow.

In this chapter we restricted the applications to financial, actuarial, and risk management problems. Without any doubt, the concept of comonotonicity may also be a helpful tool in other domains. An example is the design of wind energy distributed power systems. The problem of defining the dependence structures in the system is tackled by modeling the statistically extreme interdependencies in the system inputs using comonotonicity theory, see, e.g., [84] and [85].

Acknowledgements Jan Dhaene acknowledges the financial support of the Onderzoeksfonds KU Leuven (GOA/07: Risk Modeling and Valuation of Insurance and Financial Cash Flows, with Applications to Pricing, Provisioning and Solvency).

References

1. C. Acerbi, Spectral measures of risk: a coherent representation of subjective risk aversion. J. Bank. Finance **26**(7), 1505–1518 (2002)
2. A. Ahcan, G. Darkiewicz, M. Goovaerts, T. Hoedemakers, Computation of convex bounds for present value functions with random payments. J. Comput. Appl. Math. **186**(1), 23–42 (2006)
3. H. Albrecher, J. Dhaene, M. Goovaerts, W. Schoutens, Static hedging of Asian options under Lévy models: the comonotonicity approach. J. Deriv. **12**(3), 63–72 (2005)
4. P. Artzner, Application of coherent capital requirements. N. Am. Actuar. J. **3**(2), 11–25 (1999)
5. P. Artzner, F. Delbaen, J.M. Eber, D. Heath, Coherent measures of risk. Math. Finance **9**, 203–228 (1999)
6. N. Bäuerle, A. Müller, Modeling and comparing dependencies in multivariate risk portfolios. ASTIN Bull. **28**, 59–76 (1998)
7. P. Boyle, A. Potapchik, Prices and sensitivities of Asian options: a survey. Insur. Math. Econ. **42**(1), 189–211 (2008)
8. K. Brückner, Quantifying the error of convex order bounds for truncated first moments. Insur. Math. Econ. **42**(1), 261–270 (2008)

9. H. Bühlmann, *Mathematical Methods in Risk Theory* (Springer, Berlin, 1970)
10. X. Burtschell, J. Gregory, J.-P. Laurent, A comparative analysis of CDO pricing models under the factor copula framework. J. Deriv. **16**(4), 9–37 (2009)
11. P. Chalasani, A. Varikooty, Accurate approximations for European-style Asian options. J. Comput. Finance **1**(4), 11–30 (1998)
12. X. Chen, G. Deelstra, J. Dhaene, M. Vanmaele, Static super-replicating strategies for a class of exotic options. Insur. Math. Econ. **42**(3), 1067–1085 (2008)
13. X. Chen, G. Deelstra, J. Dhaene, M. Vanmaele, Static super-replicating strategies for a class of exotic options: some new developments. Technical Report, KU Leuven (2009)
14. K.C. Cheung, Optimal portfolio problem with unknown dependency structure. Insur. Math. Econ. **38**(1), 167–175 (2006)
15. K.C. Cheung, Characterizations of conditional comonotonicity. J. Appl. Probab. **44**(3), 607–617 (2007)
16. K.C. Cheung, Optimal allocation of policy limits and deductibles. Insur. Math. Econ. **41**(3), 382–391 (2007)
17. K.C. Cheung, Improved convex upper bound via conditional comonotonicity. Insur. Math. Econ. **42**(2), 651–655 (2008)
18. K.C. Cheung, Characterization of comonotonicity using convex order. Insur. Math. Econ. **43**(3), 403–406 (2008)
19. K.C. Cheung, Upper comonotonicity. Insur. Math. Econ. **45**(1), 35–40 (2009)
20. K.C. Cheung, H. Yang, Ordering of optimal portfolio allocations in a model with a mixture of fundamental risks. J. Appl. Probab. **45**(1), 55–66 (2008)
21. G. Darkiewicz, G. Deelstra, J. Dhaene, T. Hoedemakers, M. Vanmaele, Bounds for right tails of deterministic and stochastic sums of random variables. J. Risk Insur. **76**(4), 847–866 (2009)
22. G. Deelstra, J. Liinev, M. Vanmaele, Pricing of arithmetic basket options by conditioning. Insur. Math. Econ. **34**(1), 55–77 (2004)
23. G. Deelstra, I. Diallo, M. Vanmaele, Bounds for Asian basket options. J. Comput. Appl. Math. **218**(2), 215–228 (2008)
24. G. Deelstra, I. Diallo, M. Vanmaele, Moment matching approximation of Asian basket option prices. J. Comput. Appl. Math. **234**(4), 1006–1016 (2010)
25. G. Deelstra, A. Petkovic, M. Vanmaele, Pricing and hedging Asian basket spread options. J. Comput. Appl. Math. **233**(11), 2814–2830 (2010)
26. G. Deelstra, M. Vanmaele, D. Vyncke, Minimizing the risk of a financial product using a put option. J. Risk Insur. **77**(4), 767–800 (2010)
27. D. Denneberg, *Non-additive Measure and Integral* (Kluwer Academic, Boston, 1994)
28. M. Denuit, Distribution of the random future life expectancies in log-bilinear mortality projection models. Lifetime Data Anal. **13**(3), 381–397 (2007)
29. M. Denuit, Comonotonic approximations to quantiles of life annuity conditional expected present value. Insur. Math. Econ. **42**(2), 831–838 (2008)
30. M. Denuit, Life annuities with stochastic survival probabilities: A review. Methodol. Comput. Appl. Probab. **11**(3, Sp. Iss. SI), 463–489 (2009)
31. M. Denuit, J. Dhaene, Simple characterizations of comonotonicity and countermonotonicity by extremal correlations. Belg. Actuar. Bull. **3**, 22–27 (2003)
32. M. Denuit, J. Dhaene, Comonotonic bounds on the survival probabilities in the Lee–Carter model for mortality projection. J. Comput. Appl. Math. **203**(1), 169–176 (2007)
33. M. Denuit, J. Dhaene, C. Ribas, Does positive dependence between individual risks increase stop-loss premiums? Insur. Math. Econ. **28**(3), 305–308 (2001)
34. M. Denuit, J. Dhaene, M. Goovaerts, R. Kaas, *Actuarial Theory for Dependent Risks: Measures, Orders and Models* (Wiley, New York, 2005)
35. M. Denuit, P. Devolder, A.-C. Goderniaux, Securitization of longevity risk: pricing survivor bonds with Wang transform in the Lee–Carter framework. J. Risk Insur. **74**(1), 87–113 (2007)
36. J. Dhaene, M. Denuit, The safest dependency structure among risks. Insur. Math. Econ. **25**(1), 11–21 (1999)

37. J. Dhaene, M. Goovaerts, Dependency of risks and stop-loss order. ASTIN Bull. **26**(2), 201–212 (1996)
38. J. Dhaene, M. Goovaerts, On the dependency of risks in the individual life model. Insur. Math. Econ. **19**(3), 243–253 (1997)
39. J. Dhaene, S.S. Wang, V.R. Young, M.J. Goovaerts, Comonotonicity and maximal stop-loss premiums. Bull. Swiss Assoc. Actuar. **2**, 99–113 (2000)
40. J. Dhaene, M. Denuit, M.J. Goovaerts, R. Kaas, D. Vyncke, The concept of comonotonicity in actuarial science and finance: application. Insur. Math. Econ. **31**(2), 133–161 (2002)
41. J. Dhaene, M. Denuit, M.J. Goovaerts, R. Kaas, D. Vyncke, The concept of comonotonicity in actuarial science and finance: theory. Insur. Math. Econ. **31**(1), 3–33 (2002)
42. J. Dhaene, M. Goovaerts, R. Kaas, Economic capital allocation derived from risk measures. N. Am. Actuar. J. **7**(2), 44–59 (2003)
43. J. Dhaene, S. Vanduffel, M.J. Goovaerts, R. Kaas, D. Vyncke, Comonotonic approximations for optimal portfolio selection problems. J. Risk Insur. **72**(2), 253–300 (2005)
44. J. Dhaene, A. Kukush, M. Pupashenko, On the structure of premium principles under pointwise comonotonicity. Theory Stoch. Process. **12**(28, N3–4), 27–45 (2006)
45. J. Dhaene, S. Vanduffel, M.J. Goovaerts, R. Kaas, Q. Tang, D. Vyncke, Risk measures and comonotonicity: a review. Stoch. Models **22**(4), 573–606 (2006)
46. J. Dhaene, L. Henrard, Z. Landsman, A. Vandendorpe, S. Vanduffel, Some results on the CTE-based capital allocation rule. Insur. Math. Econ. **42**(2), 855–863 (2008)
47. J. Dhaene, R.J.A. Laeven, S. Vanduffel, G. Darkiewicz, M.J. Goovaerts, Can a coherent risk measure be too subadditive? J. Risk Insur. **75**(2), 365–386 (2008)
48. J. Dhaene, M. Denuit, S. Vanduffel, Correlation order, merging and diversification. Insur. Math. Econ. **45**(3), 325–332 (2009)
49. J. Dhaene, A. Tsanakas, E. Valdez, S. Vanduffel, Optimal capital allocation principles. J. Risk Insur. (2011, accepted)
50. I. Ekeland, A. Galichon, M. Henry, Comonotonic measures of multivariate risks. Math. Finance. doi:10.1111/j.1467-9965.2010.00453.x
51. P. Embrechts, A. McNeil, D. Straumann, Correlation and dependency in risk management: properties and pitfalls, in *Risk Management: Value-at-Risk and Beyond*, ed. by M. Dempster, H.K. Moffat (Cambridge University Press, Cambridge, 2002), pp. 176–223
52. P. Embrechts, A. Höing, G. Puccetti, Worst VaR scenarios. Insur. Math. Econ. **37**(1), 115–134 (2005)
53. M. Fréchet, Sur les tableaux de correlation dont les marges sont donnés. Ann. Univ. Lyon Sect. A **14**, 53–77 (1951)
54. A. Galichon, M. Henry, Dual theory of choice under multivariate risks. Technical Report, SSRN eLibrary (2010)
55. K. Glau, N. Vandaele, M. Vanmaele, Hedging interest rate derivatives in a Lévy driven term structure model. Technical Report, Ghent University (2009)
56. M. Goovaerts, J. Dhaene, Supermodular ordering and stochastic annuities. Insur. Math. Econ. **24**(3), 281–290 (1999)
57. M. Goovaerts, R. Kaas, R. Laeven, Q. Tang, A comonotonic image of independence for additive risk measures. Insur. Math. Econ. **35**(3), 581–594 (2004)
58. W.R. Heilmann, On the impact of the independence of risks on stop-loss premiums. Insur. Math. Econ. **5**(3), 197–199 (1986)
59. D. Hobson, P. Laurence, T.H. Wang, Static-arbitrage optimal subreplicating strategies for basket options. Insur. Math. Econ. **37**(3), 553–572 (2005)
60. D. Hobson, P. Laurence, T.H. Wang, Static-arbitrage upper bounds for the prices of basket options. Quant. Finance **5**(4), 329–342 (2005)
61. T. Hoedemakers, G. Darkiewicz, M. Goovaerts, Approximations for life annuity contracts in a stochastic financial environment. Insur. Math. Econ. **37**(2), 239–269 (2005)
62. W. Hoeffding, Masstabinvariante Korrelationstheorie. Schr. Math. Inst. Inst. Math. Univ. Berlin **5**, 179–233 (1940)
63. L. Hua, K.C. Cheung, Stochastic orders of scalar products with applications. Insur. Math. Econ. **42**(3), 865–872 (2008)

64. L. Hua, K.C. Cheung, Worst allocations of policy limits and deductibles. Insur. Math. Econ. **43**(1), 93–98 (2008)
65. P.J. Huber, *Robust Statistics* (Wiley, New York, 1981)
66. F. Jamshidian, An exact bond option formula. J. Finance **XLIV**(1), 205–209 (1989)
67. H. Joe, *Multivariate Models and Dependence Concepts* (Chapman & Hall, London, 1997)
68. E. Jouini, C. Napp, Conditional comonotonicity. Decis. Econ. Finance **27**(2), 153–166 (2004)
69. E. Jouini, W. Schachermayer, N. Touzi, Optimal risk sharing for law invariant monetary utility functions. Math. Finance **18**(2), 269–292 (2008)
70. R. Kaas, J. Dhaene, M. Goovaerts, Upper and lower bounds for sums of random variables. Insur. Math. Econ. **27**(2), 151–168 (2000)
71. R. Kaas, J. Dhaene, D. Vyncke, M. Goovaerts, M. Denuit, A simple geometric proof that comonotonic risks have the convex-largest sum. ASTIN Bull. **32**(1), 71–80 (2002)
72. I. Koch, A. De Schepper, An application of comonotonicity and convex ordering to present values with truncated stochastic interest rates. Insur. Math. Econ. **40**(3), 386–402 (2007)
73. A. Kukush, M. Pupashenko, Bounds for a sum of random variables under a mixture of normals. Theory Stoch. Process. **13**(29, N4), 82–97 (2007)
74. R.J.A. Laeven, Worst VaR scenarios: A remark. Insur. Math. Econ. **44**(2, Sp. Iss. SI), 159–163 (2009)
75. R.J.A. Laeven, M.J. Goovaerts, An optimization approach to the dynamic allocation of economic capital. Insur. Math. Econ. **35**(2), 299–319 (2004)
76. P. Laurence, T.-H. Wang, Distribution-free upper bounds for spread options and market-implied antimonotonicity gap. Eur. J. Finance **14**(8), 717–734 (2008)
77. P. Laurence, T.-H. Wang, Sharp distribution free lower bounds for spread options and the corresponding optimal subreplicating portfolios. Insur. Math. Econ. **44**(1), 35–47 (2009)
78. R. Lord, Partially exact and bounded approximations for arithmetic Asian options. J. Comput. Finance **10**(2), 1–52 (2006)
79. M. Ludkovski, L. Rüschendorf, On comonotonicity of Pareto optimal risk sharing. Stat. Probab. Lett. **78**(10), 1181–1188 (2008)
80. I. Meilijson, A. Nadas, Convex majorization with an application to the length of critical paths. J. Appl. Probab. **16**(3), 671–676 (1979)
81. A. Müller, Stop-loss order for portfolios of dependent risks. Insur. Math. Econ. **21**(2), 219–223 (1997)
82. R.B. Nelsen, *An Introduction to Copulas*. Lecture Notes in Statistics, vol. 139 (Springer, New York, 1998)
83. J.A. Nielsen, K. Sandmann, Pricing bounds on Asian options. J. Financ. Quant. Anal. **38**(2), 449–473 (2003)
84. G. Papaefthymiou, P.H. Schavemaker, L. van der Sluis, A. Tsanakas, Design of wind energy distributed power systems: investigation of stochastic bounds using Monte Carlo simulation, in *Proceedings of the Fourth IASTED International Conference on Power and Energy Systems*, ed. by P.D. Bourkas, P. Halaris (2004), pp. 115–121
85. G. Papaefthymiou, P.H. Schavemaker, L. van der Sluis, W.L. Kling, D. Kurowicka, R.M. Cooke, Integration of stochastic generation in power systems. Int. J. Electr. Power Energy Syst. **28**(9), 655–667 (2006)
86. G. Puccetti, M. Scarsini, Multivariate comonotonicity. J. Multivar. Anal. **101**(1), 291–304 (2010)
87. H. Reynaerts, M. Vanmaele, J. Dhaene, G. Deelstra, Bounds for the price of a European-style Asian option in a binary tree model. Eur. J. Oper. Res. **168**(2), 322–332 (2006)
88. A. Roëll, Risk aversion in Quiggin and Yaari's rank-order model of choice under uncertainty. Econ. J. **97**(133a), 143–159 (1987)
89. L.C.G. Rogers, Z. Shi, The value of an Asian option. J. Appl. Probab. **32**(4), 1077–1088 (1995)
90. L. Rüschendorf, Solution of statistical optimization problem by rearrangement methods. Metrika **30**, 55–61 (1983)

91. D. Schmeidler, Integral representation without additivity. Proc. Am. Math. Soc. **97**, 255–261 (1986)
92. D.F. Schrager, A.A.J. Pelsser, Pricing rate of return guarantees in regular premium unit linked insurance. Insur. Math. Econ. **35**(2), 369–398 (2004)
93. M. Shaked, J.G. Shanthikumar, *Stochastic Orders and Their Applications* (Academic Press, San Diego, 1994)
94. Q. Shang, X. Qin, Y. Wang, Design of catastrophe mortality bonds based on the comonotonicity theory and jump-diffusion process. Int. J. Innov. Comput., Inf. Control **5**(4), 991–1000 (2009)
95. S. Simon, M.J. Goovaerts, J. Dhaene, An easy computable upper bound for the price of an arithmetic Asian option. Insur. Math. Econ. **26**(2–3), 175–184 (2000)
96. J. Spreeuw, Two approximations of the present value distribution of a disability annuity. J. Comput. Appl. Math. **186**(1), 217–231 (2006)
97. A. Tchen, Inequalities for distributions with given marginals. Ann. Probab. **8**(4), 814–827 (1980)
98. G. Thompson, Fast narrow bounds on the value of Asian options. Technical Report, University of Cambridge (1999)
99. A. Tsanakas, Risk measurement in the presence of background risk. Insur. Math. Econ. **42**(2), 520–528 (2008)
100. A. Tsanakas, To split or not to split: Capital allocation with convex risk measures. Insur. Math. Econ. **44**(2, Sp. Iss. SI), 268–277 (2009)
101. A. Tsanakas, N. Christofides, Risk exchange with distorted probabilities. ASTIN Bull. **36**(1), 219–243 (2006)
102. E.A. Valdez, J. Dhaene, M. Maj, S. Vanduffel, Bounds and approximations for sums of dependent log-elliptical random variables. Insur. Math. Econ. **44**(3), 385–397 (2009)
103. S. Vanduffel, J. Dhaene, M. Goovaerts, R. Kaas, The hurdle-race problem. Insur. Math. Econ. **33**(2), 405–413 (2003)
104. S. Vanduffel, J. Dhaene, M. Goovaerts, On the evaluation of 'saving-consumption' plans. J. Pension Econ. Finance **4**(1), 17–30 (2005)
105. S. Vanduffel, T. Hoedemakers, J. Dhaene, Comparing approximations for risk measures of sums of non-independent lognormal random variables. N. Am. Actuar. J. **9**(4), 71–82 (2005)
106. S. Vanduffel, X. Chen, J. Dhaene, M. Goovaerts, L. Henrard, R. Kaas, Optimal approximations for risk measures of sums of lognormals based on conditional expectations. J. Comput. Appl. Math. **221**(1), 202–218 (2008)
107. S. Vanduffel, Z. Shang, L. Henrard, J. Dhaene, E.A. Valdez, Analytic bounds and approximations for annuities and Asian options. Insur. Math. Econ. **42**(3), 1109–1117 (2008)
108. M. Vanmaele, G. Deelstra, J. Liinev, Approximation of stop-loss premiums involving sums of lognormals by conditioning on two variables. Insur. Math. Econ. **35**(2), 343–367 (2004)
109. M. Vanmaele, G. Deelstra, J. Liinev, J. Dhaene, M. Goovaerts, Bounds for the price of discrete arithmetic Asian options. J. Comput. Appl. Math. **185**(1), 51–90 (2006)
110. O. Vasicek, An equilibrium characterisation of the term structure. J. Financ. Econ. **5**, 177–188 (1977)
111. J. von Neumann, O. Morgenstern, *Theory of Games and Economic Behavior*, 2nd edn. (Princeton University Press, Princeton, 1947)
112. D. Vyncke, M. Goovaerts, J. Dhaene, An accurate analytical approximation for the price of a European-style arithmetic Asian option. Finance **25**, 121–139 (2004)
113. S. Wang, J. Dhaene, Comonotonicity, correlation order and premium principles. Insur. Math. Econ. **22**(3), 235–242 (1998)
114. S.S. Wang, V.R. Young, Ordering risks: expected utility theory versus Yaari's dual theory of risk. Insur. Math. Econ. **22**(2), 145–162 (1998)
115. X. Wu, X. Zhou, A new characterization of distortion premiums via countable additivity for comonotonic risks. Insur. Math. Econ. **38**(2), 324–334 (2006)

116. M.E. Yaari, The dual theory of choice under risk. Econometrica **55**(1), 95–115 (1987)
117. J. Yang, S. Cheng, L. Zhang, Bivariate copula decomposition in terms of comonotonicity, countermonotonicity and independence. Insur. Math. Econ. **39**(2), 267–284 (2006)
118. Y. Zhang, Z. Lin, C. Weng, Some limiting properties of the bounds of the present value function of a life insurance portfolio. J. Appl. Probab. **43**(4), 1155–1164 (2006)
119. W. Zhuang, Z. Chen, T. Hu, Optimal allocation of policy limits and deductibles under distortion risk measures. Insur. Math. Econ. **44**(3), 409–414 (2009)

Chapter 7
A General Maximum Principle for Anticipative Stochastic Control and Applications to Insider Trading

Giulia Di Nunno, Olivier Menoukeu Pamen, Bernt Øksendal, and Frank Proske

Abstract In this paper we suggest a general stochastic maximum principle for optimal control of anticipating stochastic differential equations driven by a Lévy-type noise. We use techniques of Malliavin calculus and forward integration. We apply our results to study a general optimal portfolio problem of an insider. In particular, we find conditions on the insider information filtration which are sufficient to give the insider an infinite wealth. We also apply the results to find the optimal consumption rate for an insider.

Keywords Anticipative stochastic control · Maximum principle · Malliavin calculus · Insider trading · Forward integrals · Skorokhod integrals

Mathematics Subject Classification (2010) 93E20 · 91G80 · 60G51 · 60H10

The research leading to these results has received funding from the European Research Council under the European Community's Seventh Framework Programme (FP7/2007-2013)/ERC grant agreement no. [228087].

G. Di Nunno (✉) · O. Menoukeu Pamen · B. Øksendal · F. Proske
CMA, Department of Mathematics, University of Oslo, P.O. Box 1053, Blindern, 316 Oslo, Norway
e-mail: giulian@math.uio.no

B. Øksendal
e-mail: oksendal@math.uio.no

F. Proske
e-mail: proske@math.uio.no

G. Di Nunno · B. Øksendal
Norwegian School of Economics and Business Administration, Helleveien 30, 5045 Bergen, Norway

O. Menoukeu Pamen
Programme in Advanced Mathematics of Finance, School of Computational and Applied Mathematics, University of the Witwatersrand, Private Bag 3, Wits, 2050, Johannesburg, South Africa
e-mail: o.m.pamen@cma.uio.no

G. Di Nunno, B. Øksendal (eds.), *Advanced Mathematical Methods for Finance*,
DOI 10.1007/978-3-642-18412-3_7,

7.1 Introduction

In the classical Black–Scholes model and in most problems of stochastic analysis applied to finance, one of the fundamental hypotheses is the homogeneity of information that market participants have. This homogeneity does not reflect reality. In fact, there exist many types of agents in the market that have different levels of information. In this paper, we are focusing on agents who have additional information (insiders) and show that it is important to understand how an optimal control is affected by particular pieces of such information.

In the following, let $\{B(t)\}_{0\leq t\leq T}$ be a Brownian motion, and $\widetilde{N}(dz,ds) = N(dz,ds) - ds\nu(dz)$ be a compensated Poisson random measure associated with a Lévy process with Lévy measure ν on the (complete) filtered probability space $(\Omega, \mathcal{F}, \mathbb{F} = \{\mathcal{F}_t\}_{0\leq t\leq T}, P)$ with fixed time horizon $T > 0$. In the sequel, we assume that the Lévy measure ν fulfills

$$\int_{\mathbb{R}_0} z^2 \nu(dz) < \infty,$$

where $\mathbb{R}_0 := \mathbb{R}\backslash\{0\}$.

Here we suppose that we are given a filtration $\mathbb{G} = \{\mathcal{G}_t\}_{0\leq t\leq T}$ with

$$\mathcal{F}_t \subseteq \mathcal{G}_t, \quad t \in [0,T], \tag{7.1}$$

representing the information available to the agent at time t. This information is used at decision making level yielding a $\mathbb{G}$-predictable strategy or control.

Suppose that the state process $X(t) = X^{(u)}(t,\omega)$; $0 \leq t \leq T, \omega \in \Omega$, characterizing the agent's wealth, is a controlled jump diffusion in $\mathbb{R}$ of the form

$$\begin{cases} d^-X(t) = b(t,X(t),u(t))\,dt + \sigma(t,X(t),u(t))\,d^-B(t) \\ \qquad\qquad + \int_{\mathbb{R}_0} \theta(t,X(t),u(t),z)\widetilde{N}(dz,d^-t); \\ X(0) = x \in \mathbb{R}. \end{cases} \tag{7.2}$$

Since $B(\cdot)$ and $\widetilde{N}(A,\cdot)$, $A \subseteq \mathbb{R}_0$ Borel, need not be semimartingales with respect to $\{\mathcal{G}_t\}_{0\leq t\leq T}$, the two last integrals in (7.2) are *anticipating* stochastic integrals that we interpret as *forward* integrals. The choice of forward integration, as an anticipative extension of the Itô integration, is motivated by the possible applications to optimal portfolio problems for insiders as in Sect. 7.6 (see, e.g., [3, 6, 7]). However, the applications are not restricted to this area and include all situations of optimization problems in anticipating environments (see, e.g., [15, 20]).

The control process

$$u : [0,T] \times \Omega \to U$$

is called an *admissible control* if (7.2) has a unique (strong) solution $X = X^{(u)}$ such that $u(\cdot)$ is predictable with respect to the filtration $\{\mathcal{G}_t\}_{0\leq t\leq T}$. We let $\mathcal{A}_{\mathbb{G}}$ denote a given family of admissible controls assumed to be $\mathbb{G}$-predictable and such that (7.2) has a strong solution.

More specifically, the problem we are dealing with is the following. Suppose that we are given a performance functional of the form

$$J(u) := E\left[\int_0^T f\big(t, X(t), u(t)\big)\,dt + g\big(X(T)\big)\right], \quad u \in \mathcal{A}_{\mathbb{G}}, \tag{7.3}$$

with

$$f : [0, T] \times \mathbb{R} \times U \times \Omega \to \mathbb{R},$$
$$g : \mathbb{R} \times \Omega \to \mathbb{R},$$

where f is an $\mathbb{F}$-adapted process for each $x \in \mathbb{R}$, $u \in U$, and g is an $\mathcal{F}_T$-measurable random variable for each $x \in \mathbb{R}$ satisfying

$$E\left[\int_0^T \left|f\big(t, X(t), u(t)\big)\right|dt + \left|g\big(X(T)\big)\right|\right] < \infty \quad \text{for all } u \in \mathcal{A}_{\mathbb{G}}.$$

The goal is to find the optimal control $u^* \in \mathcal{A}_{\mathbb{G}}$ such that

$$\Phi_{\mathbb{G}} := \sup_{u \in \mathcal{A}_{\mathbb{G}}} J(u) = J\big(u^*\big). \tag{7.4}$$

Special cases of this problem have been studied by many authors. See, e.g., [1, 3, 4, 7, 11, 12, 14, 15, 21] and the references therein.

The purpose of this paper is twofold.

First, we want to establish a general maximum principle for the optimal anticipative control problem (7.2)–(7.4), without any a priori semimartingale assumptions for the inside information filtration $\{\mathcal{G}_t\}_{0 \le t \le T}$ (see Theorems 7.13 and 7.14).

Second, we want to use these general results to investigate the following problem in insider trading: How much information does an insider need in order to generate an infinite value of $\Phi_{\mathbb{G}}$?

The following example by Pikovski and Karatzas in [14] illustrates the situation. Suppose that the financial market has two investments opportunities:

1. a risk-free asset with unit price

$$S_0(t) = 1, \quad t \in [0, T],$$

2. a risky asset with unit price

$$dS_1(t) = S_1(t)\big[\mu\,dt + \sigma\,dB(t)\big], \quad S(0) > 0,\ t \in [0, T]$$

($\mu, \sigma > 0$ constants). If the trader chooses a portfolio $\pi(t)$ representing the fraction of wealth to be invested in the risky asset at time t, the corresponding wealth process $X(t)$, $t \in [0, T]$, will have the dynamics

$$d^- X_\pi(t) = X_\pi(t)\pi(t)\big[\mu\,dt + \sigma\,d^- B(t)\big], \quad X_\pi(0) > 0.$$

If the information flow accessible to the insider trader is given by a filtration $\mathbb{G} = \{\mathcal{G}_t\}_{0 \le t \le T}$ such that $\mathcal{G}_t \supseteq \mathcal{F}_t$, this means that π is required to be $\mathbb{G}$-adapted (thus the Itô integration cannot be applied, and the forward integration is chosen to be used instead). Suppose that the insider wants to maximize the expected logarithmic utility of the terminal wealth, i.e., to find $\Phi_{\mathbb{G}}$ and π^* (if it exists) such that

$$\Phi_{\mathbb{G}} := \max_{\pi \in \mathcal{A}_{\mathbb{G}}} E\big[\ln\big(X_\pi(T)\big)\big] = E\big[\ln\big(X_{\pi^*}(T)\big)\big].$$

In [14] it is proved that if

$$\mathcal{G}_t = \mathcal{F}_t \vee \sigma\big(B(T)\big), \quad t \in [0, T],$$

then $\Phi_{\mathbb{G}} = \infty$, and π^* does not exist.

In this paper we generalize this situation in several directions:

(a) We include jumps in the risky asset model;
(b) We study more general utility functions;
(c) We study more general insider filtrations.

These points were already partially discussed in [7] from the point of view of the existence of an optimal portfolio for a given insider. The present paper, we repeat, focuses on the study of conditions on the amount of information $\mathbb{G} = \{\mathcal{G}_t\}_{0 \le t \le T}$ needed to obtain $\Phi_{\mathbb{G}} = \infty$ and the nonexistence of an optimal insider portfolio.

The main result, which represents an anticipative stochastic maximum principle, is presented in full generality (see Theorem 7.13). However it is difficult to apply because of the appearance of some terms, which all depend on the control. We then consider the special case (see Theorem 7.14) where the coefficients of the controlled process X do not depend on X; we call such processes *controlled Itô–Lévy processes*. In this case, we give a condition for the existence of an optimal control. More specific results are obtained in the cases where the insider filtration is either

(i) a *D-commutable filtration* (Sect. 7.5.1 and Theorem 7.16) or
(ii) a *smoothly anticipative filtration* (Sect. 7.5.2).

Besides the application of these results to optimal portfolio problems, we also consider applications to optimal insider consumption. In this case we show that there exists an optimal insider consumption, and in some special cases the optimal consumption can be expressed explicitly.

The paper is structured as follows: In Sect. 7.2, we briefly recall some basic concepts of Malliavin calculus and its connection to the theory of forward integration. In Sect. 7.3, we use Malliavin calculus to obtain a maximum principle for this general non-Markovian insider information stochastic control problem. Section 7.4 considers the special case of Itô–Lévy processes. In Sect. 7.5, some specific classes of insider information are considered. Finally, in Sects. 7.6 and 7.7, we apply the results from the previous sections to study optimal insider portfolio and optimal insider consumption problems, respectively.

7.2 Framework

In this section, we briefly recall some basic concepts of Malliavin calculus and its connection to the theory of forward integration. We refer to [17] or [8] for more information about Malliavin calculus. As for the theory of forward integration, the reader may consult [18, 23, 24] and [6].

7.2.1 Malliavin Calculus for Lévy Processes

In the sequel, consider a Brownian motion $\{B(t)\}_{0\le t\le T}$ on the filtered probability space

$$\big(\Omega^{(B)}, \mathcal{F}^{(B)}, \{\mathcal{F}_t^{(B)}\}_{0\le t\le T}, P^{(B)}\big),$$

where $\{\mathcal{F}_t^{(B)}\}_{0\le t\le T}$ is the $P^{(B)}$-augmented filtration generated by $\{B(t)\}_{0\le t\le T}$ with $\mathcal{F}^{(B)} = \mathcal{F}_T^{(B)}$. Further we assume that a Poisson random measure $N(dt,dz)$ associated with a Lévy process is defined on the stochastic basis

$$\big(\Omega^{(\widetilde{N})}, \mathcal{F}^{(\widetilde{N})}, \{\mathcal{F}_t^{(\widetilde{N})}\}_{0\le t\le T}, P^{(\widetilde{N})}\big).$$

We denote by $\widetilde{N}(dt,dz) = N(dt,dz) - \nu(dz)\,dt$ the compensated Poisson random measure, where ν is the Lévy measure of the Lévy process. See [2, 25] for more information about Lévy processes.

The starting point of Malliavin calculus is the following observation which goes back to Itô [13]: Square-integrable functionals of $B(t)$ and $\widetilde{N}(dt,dz)$ enjoy the chaos representation property, that is,

(i) If $F \in L^2(\mathcal{F}^{(B)}, P^{(B)})$, then

$$F = \sum_{n\ge 0} I_n^{(B)}(f_n) \tag{7.5}$$

for a unique sequence of symmetric $f_n \in L^2(\lambda^n)$, where λ is the Lebesgue measure, and

$$I_n^{(B)}(f_n) := n! \int_0^T \int_0^{t_n} \cdots \int_0^{t_2} f_n(t_1,\ldots,t_n)\,dB(t_1)\,dB(t_2)\ldots dB(t_n), \quad n\in\mathbb{N},$$

is the n-fold iterated stochastic integral with respect to $B(t)$. Here $I_n^{(B)}(f_0) := f_0$ for constants f_0.

(ii) Similarly, if $G \in L^2(\mathcal{F}^{(\widetilde{N})}, P^{(\widetilde{N})})$, then

$$G = \sum_{n\ge 0} I_n^{(\widetilde{N})}(g_n) \tag{7.6}$$

for a unique sequence of kernels g_n in $L^2((\lambda\times\nu)^n)$, which are symmetric with respect to $(t_1,z_1),\ldots,(t_n,z_n)$. Here $I_n^{(\widetilde{N})}(g_n)$ is defined as

$$I_n^{(\widetilde{N})}(g_n) := n! \int_0^T \int_{\mathbb{R}_0} \int_0^{t_n} \int_{\mathbb{R}_0} \cdots \left(\int_0^{t_2} \int_{\mathbb{R}_0} g_n(t_1,z_1,\ldots,t_n,z_n)\right) \\ \times \widetilde{N}(dt_1,dz_1)\ldots\widetilde{N}(dt_n,dz_n), \quad n\in\mathbb{N}.$$

If $F \in L^2(\mathcal{F}^{(B)}, P^{(B)})$ has chaos expansion (7.5), the *Malliavin derivative* D_t of F in the direction of the Brownian motion is defined as

$$D_t F = \sum_{n \geq 1} n I_{n-1}^{(B)}(\widetilde{f}_{n-1}), \tag{7.7}$$

where $\widetilde{f}_{n-1}(t_1, \ldots, t_{n-1}) := f_n(t_1, \ldots, t_{n-1}, t)$, provided that

$$\sum_{n \geq 0} n n! \| f_n \|^2_{L^2(\lambda^n)} < \infty. \tag{7.8}$$

Similarly, for all $G \in L^2(\mathcal{F}^{(\widetilde{N})}, P^{(\widetilde{N})})$ with chaos representation (7.6) such that

$$\sum_{n \geq 0} n n! \| g_n \|^2_{L^2((\lambda \times \nu)^n)} < \infty, \tag{7.9}$$

the Malliavin derivative $D_{t,z}$ of G in the direction of $\widetilde{N}(dt, dz)$ is introduced as

$$D_{t,z} G := \sum_{n \geq 1} n I_{n-1}^{(\widetilde{N})}(\widetilde{g}_{n-1}), \tag{7.10}$$

where $\widetilde{g}_{n-1}(t_1, z_1, \ldots, t_{n-1}, z_{n-1}) := g_n(t_1, z_1, \ldots, t_{n-1}, z_{n-1}, t, z)$.

In the following, we denote by $\mathbb{D}^B_{1,2}$ the stochastic Sobolev space of square-integrable Brownian functionals such that (7.8) is fulfilled. The symbol $\mathbb{D}^{\widetilde{N}}_{1,2}$ stands for the corresponding space with respect to $\widetilde{N}(dt, dz)$.

We recall that the *Skorokhod integral* with respect to B respectively $\widetilde{N}(\delta t, dz)$ is defined as the adjoint operator of $D_\cdot : \mathbb{D}^B_{1,2} \to L^2(\lambda \times P^{(B)})$ resp. $D_{\cdot,\cdot} : \mathbb{D}^{\widetilde{N}}_{1,2} \to L^2(\lambda \times \nu \times P^{(\widetilde{N})})$. Thus, if we denote by

$$\int_0^T (\cdot) \delta B_t \quad \text{and} \quad \int_0^T \int_{\mathbb{R}_0} (\cdot) \widetilde{N}(\delta t, dz)$$

the corresponding adjoint operators, the following duality relations are satisfied:

(i)

$$E_{P^{(B)}}\left[F \int_0^T \varphi(t) \delta B_t\right] = E_{P^{(B)}}\left[\int_0^T \varphi(t) D_t F \, dt\right] \tag{7.11}$$

for all $F \in \mathbb{D}^B_{1,2}$ and all Skorokhod-integrable $\varphi \in L^2(\lambda \times P^{(B)})$ (i.e., φ in the domain of the adjoint operator).

(ii)

$$E_{P^{(\widetilde{N})}}\left[G \int_0^T \int_{\mathbb{R}_0} \psi(t,z) \widetilde{N}(\delta t, dz)\right] = E_{P^{(\widetilde{N})}}\left[\int_0^T \int_{\mathbb{R}_0} \psi(t,z) D_{t,z} G \nu(dz) \, dt\right] \tag{7.12}$$

for all $G \in \mathbb{D}^{\widetilde{N}}_{1,2}$ and all Skorokhod-integrable $\psi \in L^2(\lambda \times \nu \times P^{(\widetilde{N})})$.

In what follows, our reference stochastic basis will be

$$\big(\Omega, \mathcal{F}, \{\mathcal{F}_t\}_{0\leq t\leq T}, P\big),$$

where $\Omega = \Omega^{(B)} \times \Omega^{(\widetilde{N})}$, $\mathcal{F} = \mathcal{F}^{(B)} \times \mathcal{F}^{(\widetilde{N})}$, $\mathcal{F}_t = \mathcal{F}_t^{(B)} \times \mathcal{F}_t^{(\widetilde{N})}$, $P = P^{(B)} \times P^{(\widetilde{N})}$.

Later on in the paper, we will employ the duality relations (7.11) and (7.12) in connection with P. We will need the following result from [9].

Theorem 7.1 (Decomposition uniqueness for Skorokhod semimartingales) *Let $\alpha(t) \in L_2(P)$ for all t and let $\beta 1_{[0,t]}$ and $\gamma 1_{[0,t]}$ be Skorokhod integrable for all t with respect to B and $\widetilde{N}$, respectively. Let $\{X(t)\}_{0\leq t\leq T}$ be a Skorokhod semimartingale of the form*

$$X_t = \zeta + \int_0^t \alpha(s)\,ds + \int_0^t \beta(s)\delta B_s + \int_0^t \int_{\mathbb{R}_0} \gamma(s,z)\widetilde{N}(dz,\delta s).$$

Then if

$$X_t = 0 \quad \textit{for all } 0 \leq t \leq T,$$

we have

$$\zeta = 0, \quad \alpha = 0, \quad \beta = 0, \quad \gamma = 0 \quad a.e.$$

7.2.2 Malliavin Calculus and Forward Integral

In this section, we briefly recall some basic concepts of Malliavin calculus and forward integrations related to this paper. We refer to [18, 23, 24] and [6] for more information about these integrals.

Forward Integral and Malliavin Calculus for $B(\cdot)$

This section constitutes a brief review of the forward integral with respect to the Brownian motion. Let $\{B(t)\}_{0\leq t\leq T}$ be a Brownian motion on a filtered probability space $(\Omega, \mathcal{F}, \{\mathcal{F}_t\}_{0\leq t\leq T}, P)$, and $T > 0$ a fixed horizon.

Definition 7.2 Let $\phi : [0,T] \times \Omega \to \mathbb{R}$ be a measurable process. The forward integral of ϕ with respect to $\{B(t)\}_{0\leq t\leq T}$ is defined by

$$\int_0^T \phi(t,\omega)\,d^- B(t) = \lim_{\epsilon\to 0} \int_0^T \phi(t,\omega)\frac{B(t+\epsilon) - B(t)}{\epsilon}\,dt \tag{7.13}$$

if the limit exists in probability, in which case ϕ is called forward integrable.

Note that if ϕ is càdlàg and forward integrable, then

$$\int_0^T \phi(t,\omega)\, d^- B(t) = \lim_{\Delta t \to 0} \sum_j \phi(t_j) \Delta B(t_j), \tag{7.14}$$

where $\Delta B(t_j) = B(t_{j+1}) - B(t_j)$, and the sum is taken over the points of a finite partition of $[0, T]$.

Definition 7.3 Let $\mathcal{M}^B$ denote the set of stochastic processes $\phi : [0, T] \times \Omega \to \mathbb{R}$ such that:

1. $\phi \in L^2(\lambda \times P)$, $\phi(t) \in \mathbb{D}^B_{1,2}$ for almost all t and satisfies

$$\mathbb{E}\left(\int_0^T |\phi(t)|^2\, dt + \int_0^T \int_0^T |D_u \phi(t)|^2\, du\, dt \right) < \infty.$$

   We will denote by $\mathbb{L}^{1,2}[0, T]$ the class of such processes.
2. $D_{t+}\phi(t) := \lim_{s \to t+} D_s \phi(t)$ exists in $L^1(\lambda \times P)$ uniformly in $t \in [0, T]$.

We let $\mathbb{M}^B_{1,2}$ be the closure of the linear span of $\mathcal{M}^B$ with respect to the norm given by

$$\|\phi\|_{\mathbb{M}^B_{1,2}} := \|\phi\|_{\mathbb{L}^{1,2}[0,T]} + \left\| D_{t+}\phi(t) \right\|_{L^1(\lambda \times P)}.$$

Then we have the relation between the forward integral and the Skorokhod integral (see [8, 15]):

Lemma 7.4 *If $\phi \in \mathbb{M}^B_{1,2}$, then it is forward integrable, and*

$$\int_0^T \phi(t)\, d^- B(t) = \int_0^T \phi(t) \delta B(t) + \int_0^T D_{t+}\phi(t)\, dt. \tag{7.15}$$

Moreover,

$$\mathbb{E}\left[\int_0^T \phi(t)\, d^- B(t) \right] = \mathbb{E}\left[\int_0^T D_{t+}\phi(t)\, dt \right]. \tag{7.16}$$

Using (7.15) and the duality formula for the Skorokhod integral (see, e.g., [8]), we deduce the following result.

Corollary 7.5 *Suppose that $\phi \in \mathbb{M}^B_{1,2}$ and $F \in \mathbb{D}^{(B)}_{1,2}$. Then*

$$\begin{aligned} \mathbb{E}\left[F \int_0^T \phi(t)\, d^- B(t) \right] &= \mathbb{E}\left[F \int_0^T \phi(t) \delta B(t) + F \int_0^T D_{t+}\phi(t)\, dt \right] \\ &= \mathbb{E}\left[\int_0^T \phi(t) D_t F\, dt + \int_0^T F D_{t+}\phi(t)\, dt \right]. \end{aligned} \tag{7.17}$$

Proposition 7.6 *Let $\mathcal{H}$ be a given fixed σ-algebra, and $\varphi : [0,T] \times \Omega \to \mathbb{R}$ be an $\mathcal{H}$-measurable process. Set $X(t) = E[B(t)|\mathcal{H}]$. Then*

$$E\left[\int_0^T \varphi(t)\, d^- B(t) \middle| \mathcal{H}\right] = E\left[\int_0^T \varphi(t)\, d^- X(t)\right]. \tag{7.18}$$

Proof Using uniform convergence on compacts in $L^1(P)$ and the definition of forward integration in the sense of Russo–Vallois (see [23]), we observe that

$$\begin{aligned}
E\left[\int_0^T \varphi(t)\, d^- B(t) \middle| \mathcal{H}\right] &= E\left[\lim_{\epsilon\to 0^+} \int_0^T \varphi(t) \frac{B(t+\epsilon) - B(t)}{\epsilon}\, dt \middle| \mathcal{H}\right] \\
&= L^1(P) - \lim_{\epsilon\to 0^+} E\left[\int_0^T \varphi(t) \frac{B(t+\epsilon) - B(t)}{\epsilon}\, dt \middle| \mathcal{H}\right] \\
&= \lim_{\epsilon\to 0^+} \int_0^T \varphi(t) E\left[\frac{B(t+\epsilon) - B(t)}{\epsilon} \middle| \mathcal{H}\right] dt \\
&= \lim_{\epsilon\to 0^+} \int_0^T \varphi(t) \frac{X(t+\epsilon) - X(t)}{\epsilon}\, dt \\
&= \int_0^T \varphi(t)\, d^- X(t), \quad \text{in the ucp sense,}
\end{aligned}$$

and the result follows. □

Definition 7.7 Let $\mathbb{H} = \{\mathcal{H}_t\}_{0\le t\le T}$ be a given filtration, and $\varphi : [0,T] \times \Omega \to \mathbb{R}$ be an $\mathbb{H}$-adapted process. The *conditional forward integral* of φ with respect to $B(\cdot)$ is defined by

$$\int_0^T \varphi(t) E\big[d^- B(t) \big| \mathcal{H}_{t^-}\big] = \lim_{\epsilon\to 0} \int_0^T \varphi(t) \frac{E[B(t+\epsilon) - B(t)|\mathcal{H}_{t^-}]}{\epsilon}\, dt \tag{7.19}$$

if the convergence holds uniformly on compacts in probability (i.e., ucp sense), where $\mathcal{H}_{t^-} = \bigvee_{s<t} \mathcal{H}_s$.

Remark 7.8 Note that Definition 7.7 is different from Proposition 7.6 except if $\mathcal{H}_t = \mathcal{H}$ for all t.

Forward Integral and Malliavin Calculus for $\widetilde{N}(\cdot,\cdot)$

In this section, we review the forward integral with respect to the Poisson random measure $\widetilde{N}$.

Definition 7.9 The forward integral

$$J(\phi) := \int_0^T \int_{\mathbb{R}_0} \phi(t,z) \widetilde{N}\big(dz, d^- t\big),$$

with respect to the Poisson random measure $\widetilde{N}$, of a càdlàg stochastic function $\phi(t,z), t \in [0,T], z \in \mathbb{R}$, with $\phi(t,z) = \phi(\omega,t,z), \omega \in \Omega$, is defined as

$$J(\phi) = \lim_{m\to\infty} \int_0^T \int_{\mathbb{R}_0} \phi(t,z) 1_{U_m}(z) \widetilde{N}(dz,dt)$$

if the limit exists in $L^2(P)$. Here $U_m, m = 1,2,\ldots$, is an increasing sequence of compact sets $U_m \subseteq \mathbb{R}\backslash\{0\}$ with $\nu(U_m) < \infty$ such that $\lim_{m\to\infty} U_m = \mathbb{R}\backslash\{0\}$. The integral on the right is for each m defined ω-wise in the usual way, as limits of integrals of simple integrands.

Definition 7.10 Let $\mathcal{M}^{\widetilde{N}}$ denote the set of stochastic functions $\phi : [0,T] \times \mathbb{R} \times \Omega \to \mathbb{R}$ in $\mathbb{D}_{1,2}^{\widetilde{N}}$ such that:

1. $\phi(t,z,\omega) = \phi_1(t,\omega)\phi_2(t,z,\omega)$, where $\phi_1(\omega,t) \in \mathbb{D}_{1,2}^{\widetilde{N}}$ is càglàd, and $\phi_2(\omega,t,z)$ is adapted such that

$$\mathbb{E}\left[\int_0^T \int_{\mathbb{R}_0} \phi_2^2(t,z)\nu(dz)\,dt\right] < \infty,$$

2. $D_{t+,z}\phi := \lim_{s\to t+} D_{s,z}\phi$ exists in $L^2(\lambda \times \nu \times P)$,
3. $\phi(t,z) + D_{t+,z}\phi(t,z)$ is Skorokhod integrable.

We let $\mathbb{M}_{1,2}^{\widetilde{N}}$ be the closure of the linear span of $\mathcal{M}^{\widetilde{N}}$ with respect to the norm given by

$$\|\phi\|_{\mathbb{M}_{1,2}^{\widetilde{N}}} := \|\phi\|_{L^2(\lambda\times\nu\times P)} + \left\|D_{t+,z}\phi(t,z)\right\|_{L^2(\lambda\times\nu\times P)}.$$

Then we have the relation between the forward integral and the Skorokhod integral (see [6, 8]):

Lemma 7.11 *If $\phi \in \mathbb{M}_{1,2}^{\widetilde{N}}$, then it is forward integrable, and*

$$\begin{aligned}\int_0^T \int_{\mathbb{R}_0} &\phi(t,z)\widetilde{N}\big(dz,d^-t\big) \\ &= \int_0^T \int_{\mathbb{R}_0} D_{t+,z}\phi(t,z)\nu(dz)\,dt + \int_0^T \int_{\mathbb{R}_0} \big(\phi(t,z) + D_{t+,z}\phi(t,z)\big)\widetilde{N}(dz,\delta t).\end{aligned} \tag{7.20}$$

Moreover,

$$\mathbb{E}\left[\int_0^T \int_{\mathbb{R}_0} \phi(t,z)\widetilde{N}\big(dz,d^-t\big)\right] = \mathbb{E}\left[\int_0^T \int_{\mathbb{R}_0} D_{t+,z}\phi(t,z)\nu(dz)\,dt\right]. \tag{7.21}$$

Then by (7.20) and duality formula for the Skorokhod integral for Poisson process (see [8]), we have

Corollary 7.12 *Suppose that* $\phi \in \mathbb{M}_{1,2}^{\widetilde{N}}$ *and* $F \in \mathbb{D}_{1,2}^{\widetilde{N}}$. *Then*

$$\begin{aligned}
&\mathbb{E}\left[F\int_0^T\int_{\mathbb{R}_0}\phi(t,z)\widetilde{N}\big(dz,d^-t\big)\right]\\
&\quad=\mathbb{E}\left[F\int_0^T\int_{\mathbb{R}_0}D_{t+,z}\phi(t,z)\nu(dz)\,dt\right]\\
&\qquad+\mathbb{E}\left[F\int_0^T\int_{\mathbb{R}_0}\big(\phi(t,z)+D_{t+,z}\phi(t,z)\big)\widetilde{N}(dz,\delta t)\right]\\
&\quad=\mathbb{E}\left[\int_0^T\int_{\mathbb{R}_0}\phi(t,z)D_{t,z}F\nu(dz)\,dt\right]\\
&\qquad+\mathbb{E}\left[\int_0^T\int_{\mathbb{R}_0}(F+D_{t,z}F)D_{t+,z}\phi(t,z)\nu(dz)\,dt\right].
\end{aligned}\tag{7.22}$$

7.3 A Stochastic Maximum Principle for Insider

We now formulate a general anticipative maximum principle for optimal control. For a presentation in the classical adapted case, see e.g. [10, 16, 21].

In view of the optimization problem (7.4), we require the following conditions 1–5 on the coefficients and on the family of admissible controls $\mathcal{A}_{\mathbb{G}}$:

1. The functions $b:[0,T]\times\mathbb{R}\times U\times\Omega\to\mathbb{R}$, $\sigma:[0,T]\times\mathbb{R}\times U\times\Omega\to\mathbb{R}$, $\theta:[0,T]\times\mathbb{R}\times U\times\mathbb{R}_0\times\Omega\to\mathbb{R}$, $f:[0,T]\times\mathbb{R}\times U\times\Omega\to\mathbb{R}$, and $g:\mathbb{R}\times\Omega\to\mathbb{R}$ are contained in C^1 with respect to the arguments $x\in\mathbb{R}$ and $u\in U$ for each $t\in\mathbb{R}$ and a.a. $\omega\in\Omega$.
2. For all $r,t\in(0,T)$, $t\le r$, and all bounded $\mathcal{G}_t$-measurable random variables $\alpha=\alpha(\omega)$, $\omega\in\Omega$, the control

$$\beta_\alpha(s):=\alpha(\omega)\chi_{[t,r]}(s),\quad 0\le s\le T,\tag{7.23}$$

   is an admissible control, i.e., belongs to $\mathcal{A}_{\mathbb{G}}$ (here $\chi_{[t,r]}$ denotes the indicator function on $[t,r]$).
3. For all $u,\beta\in\mathcal{A}_{\mathbb{G}}$ with β bounded, there exists $\delta>0$ such that

$$u+y\beta\in\mathcal{A}_{\mathbb{G}}\quad\text{for all } y\in(-\delta,\delta)\tag{7.24}$$

   and such that the family

$$\left\{\frac{\partial}{\partial x}f\big(t,X^{u+y\beta}(t),u(t)+y\beta(t)\big)\frac{d}{dy}X^{u+y\beta}(t)\right.\\
\left.+\frac{\partial}{\partial u}f\big(t,X^{u+y\beta}(t),u(t)+y\beta(t)\big)\beta(t)\right\}_{y\in(-\delta,\delta)}$$

is $\lambda \times P$-uniformly integrable and

$$\left\{g'\big(X^{u+y\beta}(T)\big)\frac{d}{dy}X^{u+y\beta}(T)\right\}_{y\in(-\delta,\delta)}$$

is P-uniformly integrable.

4. For all $u, \beta \in \mathcal{A}_{\mathbb{G}}$ with bounded β, the process

$$Y(t) = Y_\beta(t) = Y_\beta^u(t) = \frac{d}{dy}X^{(u+y\beta)}(t)\bigg|_{y=0}$$

exists and follows the stochastic differential equation

$$\begin{aligned} dY_\beta^u(t) = Y_\beta(t^-)\bigg[&\frac{\partial}{\partial x}b\big(t, X^u(t), u(t)\big)\,dt + \frac{\partial}{\partial x}\sigma\big(t, X^u(t), u(t)\big)\,d^-B(t) \\ &+ \int_{\mathbb{R}_0}\frac{\partial}{\partial x}\theta\big(t, X^u(t), u(t), z\big)\widetilde{N}\big(dz, d^-t\big)\bigg] \\ &+ \beta(t)\bigg[\frac{\partial}{\partial u}b\big(t, X^u(t), u(t)\big)\,dt + \frac{\partial}{\partial u}\sigma\big(t, X^u(t), u(t)\big)\,d^-B(t) \\ &+ \int_{\mathbb{R}_0}\frac{\partial}{\partial u}\theta\big(t, X^u(t), u(t), z\big)\widetilde{N}\big(dz, d^-t\big)\bigg], \end{aligned} \tag{7.25}$$

$$Y(0) = 0.$$

5. Suppose that for all $u \in \mathcal{A}_{\mathbb{G}}$, the processes

$$K(t) := g'\big(X(T)\big) + \int_t^T \frac{\partial}{\partial x}f\big(s, X(s), u(s)\big)\,ds, \tag{7.26}$$

$$D_t K(t) := D_t g'\big(X(T)\big) + \int_t^T D_t\frac{\partial}{\partial x}f\big(s, X(s), u(s)\big)\,ds,$$

$$D_{t,z}K(t) := D_{t,z}g'\big(X(T)\big) + \int_t^T D_{t,z}\frac{\partial}{\partial x}f\big(s, X(s), u(s)\big)\,ds,$$

$$\begin{aligned} H_0(s, x, u) := {} & K(s)\Big(b(s, x, u) + D_{s+}\sigma(s, x, u) + \int_{\mathbb{R}_0} D_{s+,z}\theta(s, x, u, z)\nu(dz)\Big) \\ & + D_s K(s)\sigma(s, x, u) \\ & + \int_{\mathbb{R}_0} D_{s,z}K(s)\big\{\theta(s, x, u, z) + D_{s+,z}\theta(s, x, u, z)\big\}\nu(dz), \end{aligned} \tag{7.27}$$

$$\begin{aligned} G(t, s) := \exp\bigg(&\int_t^s \left\{\frac{\partial b}{\partial x}\big(r, X(r), u(r)\big) - \frac{1}{2}\left(\frac{\partial \sigma}{\partial x}\right)^2\big(r, X(r), u(r)\big)\right\}dr \\ &+ \int_t^s \frac{\partial \sigma}{\partial x}\big(r, X(r), u(r)\big)\,dB^-(r) \end{aligned}$$

$$
\begin{aligned}
&+\int_t^s \int_{\mathbb{R}_0} \left\{ \ln\left(1+\frac{\partial\theta}{\partial x}(r,X(r),u(r),z)\right)\right. \\
&\left. -\frac{\partial\theta}{\partial x}(r,X(r),u(r),z)\right\} \nu(dz)\,dr \\
&+\int_t^s \int_{\mathbb{R}_0} \left\{ \ln\left(1+\frac{\partial\theta}{\partial x}(r,X(r^-),u(r^-),z)\right)\right\} \\
&\times \widetilde{N}\big(dz,d^-r\big)\Big),
\end{aligned} \tag{7.28}
$$

$$
p(t) := K(t) + \int_t^T \frac{\partial}{\partial x} H_0\big(s,X(s),u(s)\big) G(t,s)\,ds, \tag{7.29}
$$

$$
q(t) := D_t p(t), \tag{7.30}
$$

$$
r(t,z) := D_{t,z} p(t); \quad t\in[0,T], z\in\mathbb{R}_0, \tag{7.31}
$$

are well defined.

Now let us introduce the *general Hamiltonian of an insider*

$$
H : [0,T]\times\mathbb{R}\times U\times\Omega\to\mathbb{R}
$$

by

$$
\begin{aligned}
H(t,x,u,\omega) &:= p(t)\Big(b(t,x,u,\omega) + D_{t+}\sigma(t,x,u,\omega) \\
&\quad + \int_{\mathbb{R}_0} D_{t+,z}\theta(t,x,u,\omega)\nu(dz)\Big) \\
&\quad + f(t,x,u,\omega) + q(t)\sigma(t,x,u,\omega) \\
&\quad + \int_{\mathbb{R}_0} r(t,z)\big\{\theta(t,x,u,z,\omega) + D_{t+,z}\theta(t,x,u,z,\omega)\big\}\nu(dz).
\end{aligned} \tag{7.32}
$$

We can now state a general stochastic maximum principle for our control problem (7.4):

Theorem 7.13 *Retain conditions* 1–5. *Assume that* $\widehat{u}\in\mathcal{A}_{\mathbb{G}}$ *is a* critical point *of the performance functional* $J(u)$ *in* (7.4), *that is,*

$$
\frac{d}{dy} J(\widehat{u}+y\beta)\Big|_{y=0} = 0 \tag{7.33}
$$

for all bounded $\beta\in\mathcal{A}_{\mathbb{G}}$. *Then*

$$
E\left[\frac{\partial}{\partial u}\widehat{H}\big(t,\widehat{X}(t),\widehat{u}(t)\big)\Big|\mathcal{G}_t\right] + E[A] = 0 \quad a.e.\ in\ (t,\omega), \tag{7.34}
$$

where A is given by (7.88),

$$\begin{aligned}
\widehat{X}(t) &= X^{(\widehat{u})}(t),\\
\widehat{H}\big(t,\widehat{X}(t),u\big) &= p(t)\bigg(b(t,\widehat{X},u)+D_{t+}\sigma(t,\widehat{X},u)\\
&\quad+\int_{\mathbb{R}_0}D_{t+,z}\theta(t,\widehat{X},u)\nu(dz)\bigg)\\
&\quad+f(t,\widehat{X},u)+q(t)\sigma(t,\widehat{X},u)\\
&\quad+\int_{\mathbb{R}_0}r(t,z)\big\{\theta(t,\widehat{X},u,z)+D_{t+,z}\theta(t,\widehat{X},u,z)\big\}\nu(dz)
\end{aligned}\tag{7.35}$$

with

$$\begin{aligned}
\widehat{p}(t) &= \widehat{K}(t)+\int_t^T\frac{\partial}{\partial x}\widehat{H}_0\big(s,\widehat{X}(s),\widehat{u}(s)\big)\widehat{G}(t,s)\,ds,\\
\widehat{K}(t) &:= g'\big(\widehat{X}(T)\big)+\int_t^T\frac{\partial}{\partial x}f\big(s,\widehat{X}(s),\widehat{u}(s)\big)\,ds,
\end{aligned}\tag{7.36}$$

and

$$\begin{aligned}
\widehat{G}(t,s) &:= \exp\bigg(\int_t^s\bigg\{\frac{\partial b}{\partial x}\big(r,\widehat{X}(r),u(r)\big)-\frac{1}{2}\Big(\frac{\partial\sigma}{\partial x}\Big)^2\big(r,\widehat{X}(r),u(r)\big)\bigg\}dr\\
&\quad+\int_t^s\frac{\partial\sigma}{\partial x}\big(r,\widehat{X}(r),u(r)\big)\,dB^-(r)\\
&\quad+\int_t^s\int_{\mathbb{R}_0}\bigg\{\ln\Big(1+\frac{\partial\theta}{\partial x}\big(r,\widehat{X}(r),u(r),z\big)\Big)\\
&\quad-\frac{\partial\theta}{\partial x}\big(r,\widehat{X}(r),u(r),z\big)\bigg\}\nu(dz)\,dt\\
&\quad+\int_t^s\int_{\mathbb{R}_0}\bigg\{\ln\Big(1+\frac{\partial\theta}{\partial x}\big(r,\widehat{X}(r^-),u(r^-),z\big)\Big)\bigg\}\widetilde{N}\big(dz,d^-r\big)\bigg),\\
\widehat{H}(t,x,u) &= \widehat{K}(t)\bigg(b(t,x,u)+D_{t+}\sigma(t,x,u)+\int_{\mathbb{R}_0}D_{t+,z}\theta(t,x,u)\nu(dz)\bigg)\\
&\quad+D_t\widehat{K}(t)\sigma(t,x,u)+f(t,x,u)\\
&\quad+\int_{\mathbb{R}_0}D_{t,z}\widehat{K}(t)\big\{\theta(t,x,u,z)+D_{t+,z}\theta(t,x,u,z)\big\}\nu(dz).
\end{aligned}$$

Conversely, suppose that there exists $\widehat{u}\in\mathcal{A}_{\mathbb{G}}$ *such that* (7.34) *holds. Then* $\widehat{u}$ *satisfies* (7.33).

Proof See Appendix. □

7.4 Controlled Itô–Lévy Processes

The main result of the previous section (Theorem 7.13) is difficult to apply because of the appearance of the terms $Y(t)$, $D_{t+}Y(t)$, and $D_{t+,z}Y(t)$, which all depend on the control u. However, consider the special case where the coefficients do not depend on X, i.e., where

$$\begin{aligned} &b(t,x,u,\omega)=b(t,u,\omega), \qquad \sigma(t,x,u,\omega)=\sigma(t,u,\omega), \quad \text{and} \\ &\theta(t,x,u,z,\omega)=\theta(t,u,z,\omega). \end{aligned} \tag{7.37}$$

Then (7.2) gets the form

$$\begin{cases} d^-(X)(t)=b(t,u(t),\omega)\,dt+\sigma(t,u(t),\omega)\,d^-B_t \\ \qquad\qquad +\int_{\mathbb{R}_0}\theta(t,u(t),z,\omega)\widetilde{N}(dz,d^-t); \\ X(0)=x\in\mathbb{R}. \end{cases} \tag{7.38}$$

We call such processes *controlled Itô–Lévy processes*.

In this case, Theorem 7.13 simplifies to the following:

Theorem 7.14 *Let $X(t)$ be a controlled Itô–Lévy process as given in* (7.38). *Retain conditions* 1–5 *as in Theorem* 7.13.

Then the following are equivalent:

1. $\widehat{u}\in\mathcal{A}_{\mathbb{G}}$ *is a critical point of $J(u)$*,
2. $$E\Big[L(t)\alpha+M(t)D_{t+}\alpha+\int_{\mathbb{R}_0}R(t,z)D_{t+,z}\alpha\nu(dz)\Big]=0$$

   for all $\mathcal{G}_t$-measurable $\alpha\in\mathbb{D}_{1,2}$ and all $t\in[0,T]$, where

$$\begin{aligned} L(t)=K(t)\Big(&\frac{\partial b(t)}{\partial u}+D_{t+}\frac{\partial\sigma(t)}{\partial u}+\int_{\mathbb{R}_0}D_{t+,z}\frac{\partial\theta(t)}{\partial u}\nu(dz)\Big)+\frac{\partial f(t)}{\partial u} \\ &+\int_{\mathbb{R}_0}D_{t,z}K(t)\Big(\frac{\partial\theta(t)}{\partial u}+D_{t+,z}\frac{\partial\theta(t)}{\partial u}\Big)\nu(dz)+D_tK(t)\frac{\partial\sigma(t)}{\partial u}, \end{aligned} \tag{7.39}$$

$$M(t)=K(t)\frac{\partial\sigma(t)}{\partial u}, \quad \textit{and} \tag{7.40}$$

$$R(t,z)=\big\{K(t)+D_{t,z}K(t)\big\}\Big(\frac{\partial\theta(t)}{\partial u}+D_{t+,z}\frac{\partial\theta(t)}{\partial u}\Big). \tag{7.41}$$

Proof 1. It is easy to see that in this case, $p(t)=K(t), q(t)=D_tK(t), r(t,z)=D_{t,z}K(t)$, and the general Hamiltonian H given by (7.32) is reduced to H_1 given as follows:

$$\begin{aligned}
&H_1(s,x,u,\omega)\\
&:= K(s)\Big(b(s,u,\omega)+D_{s+}\sigma(s,u,\omega)+\int_{\mathbb{R}_0} D_{s+,z}\theta(s,u,\omega)\nu(dz)\Big)\\
&\quad + D_sK(s)\sigma(s,u,\omega)+f(s,x,u,\omega)\\
&\quad + \int_{\mathbb{R}_0} D_{s,z}K(s)\big\{\theta(s,u,z,\omega)+D_{s+,z}\theta(s,u,z,\omega)\big\}\nu(dz).
\end{aligned}$$

Then, performing the same calculus leads to

$$\begin{aligned}
A_1 &= A_3 = A_5 = 0,\\
A_2 &= E\bigg[\int_t^{t+h}\bigg\{K(t)\Big(\frac{\partial b(s)}{\partial u}+D_{s+}\frac{\partial\sigma(s)}{\partial u}+\int_{\mathbb{R}_0} D_{s+,z}\frac{\partial\gamma(s)}{\partial u}\nu(dz)\Big)+\frac{\partial f(s)}{\partial u}\\
&\quad + \int_{\mathbb{R}_0} D_{s,z}K(s)\Big(\frac{\partial\theta(s)}{\partial u}+D_{s+,z}\frac{\partial\gamma(s)}{\partial u}\Big)\nu(dz)+D_sK(s)\frac{\partial\sigma(s)}{\partial u}\bigg\}\alpha\,ds\bigg],\\
A_4 &= E\bigg[\int_t^{t+h} K(s)\frac{\partial\sigma(s)}{\partial u}D_{s+}\alpha\,ds\bigg],\\
A_6 &= E\bigg[\int_t^{t+h}\int_{\mathbb{R}_0}\{K(s)+D_{s,z}K(s)\}\Big(\frac{\partial\theta(s)}{\partial u}+D_{s+,z}\frac{\partial\gamma(s)}{\partial u}\Big)\\
&\quad \times\nu(dz)D_{s+,z}\alpha\,ds\bigg].
\end{aligned}$$

It follows that

$$\begin{aligned}
\frac{d}{dh}A_2\bigg|_{h=0} &= E\bigg[\bigg\{K(t)\Big(\frac{\partial b(t)}{\partial u}+D_{t+}\frac{\partial\sigma(s)}{\partial u}+\int_{\mathbb{R}_0} D_{t+,z}\frac{\partial\theta(t)}{\partial u}\nu(dz)\Big)+\frac{\partial f(t)}{\partial u}\\
&\quad + \int_{\mathbb{R}_0} D_{t,z}K(t)\Big(\frac{\partial\theta(t)}{\partial u}+D_{t+,z}\frac{\partial\gamma(t)}{\partial u}\Big)\nu(dz)+D_tK(t)\frac{\partial\sigma(t)}{\partial u}\bigg\}\alpha\bigg],\\
\frac{d}{dh}A_4\bigg|_{h=0} &= E\bigg[K(t)\frac{\partial\sigma(t)}{\partial u}D_{t+}\alpha\bigg],\\
\frac{d}{dh}A_6\bigg|_{h=0} &= E\bigg[\int_{\mathbb{R}_0}\{K(t)+D_{t,z}K(t)\}\Big(\frac{\partial\theta(t)}{\partial u}+D_{t+,z}\frac{\partial\gamma(t)}{\partial u}\Big)\nu(dz)D_{t+,z}\alpha\bigg].
\end{aligned}$$

This means that

$$\begin{aligned}
0 = E\bigg[\bigg\{&K(t)\Big(\frac{\partial b(t)}{\partial u}+D_{t+}\frac{\partial\sigma(s)}{\partial u}+\int_{\mathbb{R}_0} D_{t+,z}\frac{\partial\theta(t)}{\partial u}\nu(dz)\Big)+\frac{\partial f(t)}{\partial u}\\
&+ \int_{\mathbb{R}_0} D_{t,z}K(t)\Big(\frac{\partial\theta(t)}{\partial u}+D_{t+,z}\frac{\partial\gamma(t)}{\partial u}\Big)\nu(dz)+D_tK(t)\frac{\partial\sigma(t)}{\partial u}\bigg\}\alpha
\end{aligned}$$

$$+ K(t)\frac{\partial\sigma(t)}{\partial u}D_{t+}\alpha + \left\{\int_{\mathbb{R}_0}\{K(t) + D_{t,z}K(t)\}\right.$$
$$\left.\times\left(\frac{\partial\theta(t)}{\partial u} + D_{t+,z}\frac{\partial\gamma(t)}{\partial u}\right)\nu(dz)\right\}D_{t+,z}\alpha\Bigg],$$

and the first part of the result follows.

2. The converse part follows from the arguments used in the proof of Theorem 7.13.

By this the proof is complete. □

7.5 Applications to Some Special Cases of Filtrations

We consider the case of an insider who has an additional information compared to the standard normally informed investor.

- It can be the case of an insider who always has advanced information compared to the honest trader. This means that if $\mathbb{G} = \{\mathcal{G}_t\}_{0\le t\le T}$ and $\mathbb{F} = \{\mathcal{F}_t\}_{0\le t\le T}$ represent respectively the flows of informations of the insider and the honest investor, then we can write that $\mathcal{G}_t \supset \mathcal{F}_{t+\delta(t)}$ where $\delta(t) > 0$;
- It can also be the case of a trader who has at the initial date particular information about the future (initial enlargement of filtration). This means that if $\mathbb{G} = \{\mathcal{G}_t\}_{0\le t\le T}$ and $\mathbb{F} = \{\mathcal{F}_t\}_{0\le t\le T}$ represent respectively the flows of informations of the insider and the honest investor, then we can write that $\mathcal{G}_t = \mathcal{F}_t \vee \sigma(L)$, where L is a random variable.

7.5.1 D-commutable Filtrations

In the following, we need the notion of *D-commutativity* of a σ-algebra.

Definition 7.15 A σ-algebra $\mathcal{A} \subseteq \mathcal{F}$ is called *D-commutable* if for all $F \in \mathbb{D}_{1,2} = \mathbb{D}_{1,2}^{B} \cap \mathbb{D}_{1,2}^{\widetilde{N}}$, the conditional expectation $E[F|\mathcal{A}]$ belongs to $\mathbb{D}_{1,2}$ and

$$D_t E[F|\mathcal{A}] = E[D_t F|\mathcal{A}], \tag{7.42}$$

$$D_{t,z} E[F|\mathcal{A}] = E[D_{t,z} F|\mathcal{A}]. \tag{7.43}$$

Theorem 7.16 *Suppose that $\widehat{u} \in \mathcal{A}_{\mathbb{G}}$ is a critical point for $J(u)$. Assume that $\mathcal{G}_t$ is D-commutable for all t. Further, require that for all t, the set of smooth $\mathcal{G}_t$-measurable random variables is dense in $L^2(\mathcal{G}_t)$ and that $E[M(t)|\mathcal{G}_t]$ and $E[R(t,z)|\mathcal{G}_t]$ are Skorokhod integrable. Then for any $t_0 \in [0,T)$,*

$$0 = \int_0^T E\big[L(t)\big|\mathcal{G}_{t_0}\big]h(t)\,dt + \int_0^T E\big[M(t)\big|\mathcal{G}_{t_0}\big]h(t)\delta B_t$$
$$+ \int_0^T \int_{\mathbb{R}_0} E\big[R(t,z)\big|\mathcal{G}_{t_0}\big]h(t)\widetilde{N}(\delta t, dz) \tag{7.44}$$

for all $h \in L^2([0,T])$ *with* $\operatorname{supp} h \subseteq [t_0, T]$.

Proof Without loss of generality, we give the proof for the Brownian motion case only. The pure jump case and mixed case follow similarly. Define $\langle X, Y\rangle = E[XY]$.

Let us fix a $t_0 \in [0,T)$. Then, by assumption, it follows that for all $\mathcal{G}_{t_0}$-measurable smooth α and $h \in L^2([0,T])$ with

$$\operatorname{supp} h \subseteq [t_0, T], \quad t_0 \le t \le T,$$
$$0 = \left\langle \int_0^T E\big[L(t)\big|\mathcal{G}_{t_0}\big]h(t)\,dt, \alpha\right\rangle + \left\langle E\left[\int_0^T M(t)h(t)\delta B_t|\mathcal{G}_{t_0}\right], \alpha\right\rangle.$$

On the other hand, the duality relation (7.11) implies

$$\begin{aligned}
\left\langle E\left[\int_0^T M(t)h(t)\delta B_t\middle|\mathcal{G}_{t_0}\right], \alpha\right\rangle &= E\left[\int_0^T M(t)h(t)\delta B_t E[\alpha|\mathcal{G}_{t_0}]\right] \\
&= E\left[\int_0^T M(t)h(t)\big(D_t E[\alpha|\mathcal{G}_{t_0}]\big)\,dt\right] \\
&= E\left[\int_0^T M(t)h(t)E[D_t\alpha|\mathcal{G}_{t_0}]\,dt\right] \\
&= E\left[\int_0^T E\big[M(t)h(t)\big|\mathcal{G}_{t_0}\big]D_t\alpha\,dt\right] \\
&= \left\langle \int_0^T E\big[M(t)\big|\mathcal{G}_{t_0}\big]h(t)\delta B_t, \alpha\right\rangle
\end{aligned}$$

for all $\mathcal{G}_{t_0}$-measurable smooth α. So

$$E\left[\int_0^T M(t)h(t)\delta B_t\middle|\mathcal{G}_{t_0}\right] = \int_0^T E\big[M(t)\big|\mathcal{G}_{t_0}\big]h(t)\delta B_t.$$

Hence, by our density assumption, we obtain that

$$0 = \int_0^T E\big[L(t)\big|\mathcal{G}_{t_0}\big]h(t)\,dt + \int_0^T E\big[M(t)\big|\mathcal{G}_{t_0}\big]h(t)\delta B_t.$$

By this the proof is complete. □

To provide some concrete examples, let us confine ourselves to the following type of filtrations $\{\mathcal{G}_t\}_{0\le t\le T}$. Given an increasing family of $\{G_t\}_{t\in[0,T]}$ of Borel sets

$G_t \supset [0, t]$, define

$$\{\mathcal{G}_t\}_{0\le t\le T} \quad \text{where}$$

$$\mathcal{G}_t = \mathcal{F}_{G_t} = \sigma\left\{\int_0^T \chi_U(s)\, dB(s);\, U \subset G_t,\, U \text{ Borel}\right\} \vee \mathcal{N}, \tag{7.45}$$

where $\mathcal{N}$ is the collection of P-null sets. Then conditions (7.42) and (7.43) hold (see Proposition 3.12 in [8]). Examples of filtrations of type (7.45) are

$$\mathcal{G}_t^1 = \mathcal{F}_{t+\delta(t)},$$
$$\mathcal{G}_t^2 = \mathcal{F}_{[0,t]\cup O},$$

where O is an open set contained in $[0, T]$.

It is easily seen that filtrations of type (7.45) satisfy conditions of Theorem 7.16 as well. Hence, we have the following:

Theorem 7.17 *Suppose that* $\{\mathcal{G}_t\}_{0\le t\le T}$ *is given by* (7.45). *Then* $u = \widehat{u}$ *is a critical point for* $J(u)$ *if and only if* (7.44) *holds.*

From this, we get the following:

Theorem 7.18 *Suppose that* $\{\mathcal{G}_t\}_{0\le t\le T}$ *is of type* (7.45). *Then there exists a critical point* $u = \widehat{u}$ *for the performance functional* $J(u)$ *in* (7.3) *if and only if the following three conditions hold*:

$$\text{(i)} \quad E\big[L(t)\big|\mathcal{G}_t\big] = 0,$$
$$\text{(ii)} \quad E\big[M(t)\big|\mathcal{G}_t\big] = 0,$$
$$\text{(iii)} \quad E\big[R(t,z)\big|\mathcal{G}_t\big] = 0,$$

where L, M, *and* R *are given by* (7.39), (7.40), *and* (7.41).

Proof This follows from the uniqueness of decomposition of Skorokhod-semimartingale processes of type (7.44) (see Theorem 3.3 in [9]). □

Remark 7.19 Not all filtrations satisfy conditions (7.42) and (7.43). An important example is the following: Choose the σ-field $\mathcal{H}$ to be $\sigma(B(T))$, where $\{B(t)\}_{0\le t\le T}$ is the Wiener process (Brownian motion) starting at 0, and $T > 0$ is fixed. Then, $\mathcal{H}$ is not D-commutable. In fact, let $F = B(t_0)$ for some $t_0 < T$ and choose s such that $t_0 < s < T$. Then

$$D_s E\big[B(t_0)\big|\mathcal{H}\big] = D_s\left(\frac{t_0}{T} B(T)\right) = \frac{t_0}{T},$$

while

$$E\big[D_s B(t_0)\big|\mathcal{H}\big] = E[0|\mathcal{H}] = 0.$$

A similar argument works to prove that (7.42) and (7.43) are not satisfied for $\mathcal{G}_t = \mathcal{F}_t \vee \sigma(B_T)$ either. It follows that the technique used in the preceding section cannot be applied to the σ-algebras of the type $\mathcal{F}_t \vee \sigma(B_T)$, and hence we need a different approach to discuss such cases.

7.5.2 Smoothly Anticipative Filtrations

In this section, we consider σ-algebras which do not necessarily satisfy conditions (7.42) and (7.43). The starting point is again statement 2 of Theorem 7.14.

Definition 7.20 We say that the filtration $\{\mathcal{G}_t\}_{0\le t\le T}$ is *smoothly anticipative* if for all $t_0 \in [0,T]$, there exist a set $\mathcal{A} = \mathcal{A}_{t_0} \subseteq \mathbb{D}_{1,2} \cap L^2(\mathcal{G}_{t_0})$ and a measurable set $\mathcal{M} \subset [t_0, T]$ such that $E[L(t)|\mathcal{G}_{t_0}] \cdot \chi_{[0,T]\cap\mathcal{M}}$, $E[M(t)|\mathcal{G}_{t_0}] \cdot \chi_{[0,T]\cap\mathcal{M}}$, and $E[R(t,z)|\mathcal{G}_{t_0}] \cdot \chi_{[0,T]\cap\mathcal{M}}$, $t \in [0,T]$, $z \in \mathbb{R}_0$, are Skorokhod integrable and

(i) $D_t\alpha$ and $D_{t,z}\alpha$ are $\mathcal{G}_{t_0}$-measurable for all $\alpha \in \mathcal{A}$, $t \in \mathcal{M}$.
(ii) $D_{t+}\alpha = D_t\alpha$ and $D_{t+,z}\alpha = D_{t,z}\alpha$ for all $\alpha \in \mathcal{A}$ and a.a. $t, z, t \in \mathcal{M}$.
(iii) Span $\mathcal{A}$ is dense in $L^2(\mathcal{G}_{t_0})$.

Theorem 7.21 *Suppose that $\{\mathcal{G}_t\}_{0\le t\le T}$ is smoothly anticipative. Suppose that $\widehat{u} \in \mathcal{A}_{\mathbb{G}}$ is a critical point of $J(u)$. Then for all $h(t) = \chi_{[t_0,s)}(t)\chi_{\mathcal{M}}(t), t \in [0,T]$ (and some $s \in [0,T]$),*

$$\begin{aligned} 0 = E\bigg[\int_0^T E\big[L(t)\big|\mathcal{G}_{t_0}\big]h(t)\,dt + \int_0^T E\big[M(t)\big|\mathcal{G}_{t_0}\big]h(t)\delta B_t \\ + \int_0^T \int_{\mathbb{R}_0} E\big[R(t,z)\big|\mathcal{G}_{t_0}\big]h(t)\widetilde{N}(\delta t, dz)\bigg|\mathcal{G}_{t_0}\bigg]. \end{aligned} \tag{7.46}$$

Proof By Theorem 7.14 we know that, for every t,

$$E\bigg[L(t)\alpha + M(t)D_{t+}\alpha + \int_{\mathbb{R}_0} R(t,z)D_{t+,z}\alpha\nu(dz)\bigg] = 0.$$

Let $\alpha = E[F|\mathcal{G}_{t_0}]$ for all $F \in \mathcal{A}$. Further, choose $h \in L^2([0,T])$ with $h(t) = \chi_{[t_0,s)}(t)\chi_{\mathcal{M}}(t)$. By assumption, we see that

$$\begin{aligned} 0 = \bigg\langle\int_0^T E\big[L(t)\big|\mathcal{G}_{t_0}\big]h(t)\,dt, \alpha\bigg\rangle + \bigg\langle E\bigg[\int_0^T M(t)h(t)\delta B_t\bigg|\mathcal{G}_{t_0}\bigg], \alpha\bigg\rangle \\ + \bigg\langle E\bigg[\int_0^T \int_{\mathbb{R}_0} R(t,z)h(t)\widetilde{N}(\delta t, dz)\bigg|\mathcal{G}_{t_0}\bigg], \alpha\bigg\rangle. \end{aligned}$$

On the other hand, the duality relation (7.11) and (ii) imply that

$$\begin{aligned}
\left\langle E\left[\int_0^T M(t)h(t)\delta B_t \middle| \mathcal{G}_{t_0}\right], \alpha\right\rangle &= E\left[\int_0^T M(t)h(t)\delta B_t E[F|\mathcal{G}_{t_0}]\right] \\
&= E\left[\int_0^T M(t)h(t)\big(D_t E[F|\mathcal{G}_{t_0}]\big)\,dt\right] \\
&= E\left[\int_0^T E\big[M(t)\big|\mathcal{G}_{t_0}\big]h(t)\big(D_t E[F|\mathcal{G}_{t_0}]\big)\,dt\right] \\
&= E\left[\int_0^T E\big[M(t)\big|\mathcal{G}_{t_0}\big]h(t)\delta B_t E[F|\mathcal{G}_{t_0}]\right] \\
&= \left\langle \int_0^T E\big[M(t)\big|\mathcal{G}_{t_0}\big]h(t)\delta B_t, \alpha\right\rangle.
\end{aligned}$$

In the same way, we show that

$$\begin{aligned}
&\left\langle E\left[\int_0^T \int_{\mathbb{R}_0} R(t,z)h(t)\widetilde{N}(\delta t, dz)\middle| \mathcal{G}_{t_0}\right], \alpha\right\rangle \\
&\quad = \left\langle \int_0^T \int_{\mathbb{R}_0} E\big[R(t,z)\big|\mathcal{G}_{t_0}\big]h(t)\widetilde{N}(\delta t, dz), \alpha\right\rangle.
\end{aligned}$$

Then it follows from (iv) that

$$\begin{aligned}
0 = E\bigg[&\int_0^T E\big[L(t)\big|\mathcal{G}_{t_0}\big]h(t)\,dt + \int_0^T E\big[M(t)\big|\mathcal{G}_{t_0}\big]h(t)\delta B_t \\
&+ \int_0^T \int_{\mathbb{R}_0} E\big[R(t,z)\big|\mathcal{G}_{t_0}\big]h(t)\widetilde{N}(\delta t, dz)\bigg| \mathcal{G}_{t_0}\bigg]
\end{aligned}$$

for all $h \in L^2([0,T])$ with $\operatorname{supp} h \subseteq (t_0, T]$. □

Theorem 7.22 (Brownian motion case) *Assume that the conditions in Theorem 7.21 are in force and $\theta = 0$. In addition, we require that $E[M(t)|\mathcal{G}_{t^-}] \in \mathbb{M}_{1,2}^B$ and is forward integrable with respect to $E[d^- B(t)|\mathcal{G}_{t^-}]$. Then*

$$\begin{aligned}
0 = &\int_0^T E\big[L(t)\big|\mathcal{G}_{t^-}\big]h_0(t)\,dt + \int_0^T E\big[M(t)\big|\mathcal{G}_{t^-}\big]h_0(t)E\big[d^- B\big|\mathcal{G}_{t^-}\big] \\
&- \int_0^T D_{t^+} E\big[M(t)\big|\mathcal{G}_{t^-}\big]h_0(t)\,dt
\end{aligned} \tag{7.47}$$

for all bounded deterministic functions $h_0(t)$, $t \in [0,T]$.

Proof We apply the preceding result to $h(t) = h_0(t)\chi_{[t_i, t_{i+1}]}(t)$, where $0 = t_0 < t_1 < \cdots < t_i < t_{i+1} = T$ is a partition of $[0, T]$. From (7.46) we have

$$0 = \int_{t_i}^{t_{i+1}} E\big[L(t)\big|\mathcal{G}_{t_i}\big]h(t)\,dt + E\bigg[\int_{t_i}^{t_{i+1}} E\big[M(t)\big|\mathcal{G}_{t_i}\big]h(t)\delta B_t\bigg|\mathcal{G}_{t_i}\bigg]$$
$$+ E\bigg[\int_{t_i}^{t_{i+1}}\int_{\mathbb{R}_0} E\big[R(t,z)\big|\mathcal{G}_{t_i}\big]h(t)\widetilde{N}(\delta t, dz)\bigg|\mathcal{G}_{t_i}\bigg]. \tag{7.48}$$

By Lemma 7.4 and by assumption, we know that

$$\int_{t_i}^{t_{i+1}} E\big[M(t)\big|\mathcal{G}_{t_i}\big]h_0(t)\delta B_t = \int_{t_i}^{t_{i+1}} E\big[M(t)\big|\mathcal{G}_{t_i}\big]h_0(t)\,d^- B(t)$$
$$- \int_{t_i}^{t_{i+1}} D_{t^+} E\big[M(t)\big|\mathcal{G}_{t_i}\big]h_0(t)\,dt. \tag{7.49}$$

Substituting (7.49) into (7.48), summing over all i, and taking the limit as $\Delta t_i \to 0$, we get

$$0 = \lim_{\substack{\Delta t_i \to 0\\ n\to\infty}} \Bigg\{\sum_{i=1}^{n} \int_{t_i}^{t_{i+1}} E\big[L(t)\big|\mathcal{G}_{t_i}\big]h_0(t)\,dt$$
$$+ \sum_{i=1}^{n} \int_{t_i}^{t_{i+1}} E\big[M(t)\big|\mathcal{G}_{t_i}\big]h_0(t)\frac{E[B(t_{i+1}) - B(t_i)|\mathcal{G}_{t_i}]}{\Delta t_i}\Delta t_i$$
$$- \sum_{i=1}^{n} \int_{t_i}^{t_{i+1}} D_{t^+} E\big[M(t)\big|\mathcal{G}_{t_i}\big]h_0(t)\,dt\Bigg\}$$

in the topology of uniform convergence in probability. Hence, by Definition 7.7, we get the result. □

Important examples of filtrations satisfying the conditions of Theorem 7.21 are based on σ-algebras that are first chaos generated (see [19]). Namely, we consider σ-algebras of the form

$$\sigma\big(I_1(h_i), i \in \mathbb{N}, h_i \in L^2([0,T])\big) \vee \mathcal{N}, \tag{7.50}$$

where $\mathcal{N}$ is the collection of P-null sets. A concrete example of these σ-algebras is

$$\mathcal{G}_t^3 = \mathcal{F}_t \vee \sigma\big(B(T)\big). \tag{7.51}$$

We study the case (7.51).

Lemma 7.23 *Suppose that* $\mathcal{G}_t = \mathcal{G}_t^3 = \mathcal{F}_t \vee \sigma(B(T))$. *Then*

$$E\big[B(t)\big|\mathcal{G}_{t_0}\big] = \frac{T-t}{T-t_0}B(t_0) + \frac{t-t_0}{T-t_0}B(T) \quad \text{for all } t > t_0.$$

In particular,

$$E\big[B(t+\varepsilon)\big|\mathcal{G}_t\big] = B(t) + \frac{\varepsilon}{T-t}\big(B(T) - B(t)\big).$$

Proof We have that

$$E\big[B(t)\big|\mathcal{G}_{t_0}\big] = \int_0^{t_0} \varphi(t,s)\,dB(s) + C(t)B(T).$$

On one hand, we have

$$\begin{aligned} t = E\big[E\big[B(t)\big|\mathcal{G}_{t_0}\big]B(T)\big] &= E\bigg[\bigg(\int_0^{t_0} \varphi(t,s)\,dB(s)\bigg)B(T)\bigg] + C(t)T \\ &= \int_0^{t_0} \varphi(t,s)\,ds + C(t)T. \end{aligned} \tag{7.52}$$

On the other hand,

$$\begin{aligned} u = E\big[E\big[B(t)\big|\mathcal{G}_{t_0}\big]B(u)\big] &= E\bigg[\bigg(\int_0^{t_0} \varphi(t,s)\,dB(s)\bigg)B(u)\bigg] + C(t)u \\ &= \int_0^{u} \varphi(t,s)\,ds + C(t)u \quad \textit{for all } u < t. \end{aligned} \tag{7.53}$$

Differentiating (7.53) with respect to u, it follows that

$$\varphi(t,u) + C(t) = 1.$$

Substituting φ by its value into (7.52), we obtain $C(t) = \frac{t-t_0}{T-t_0}$ and then $\varphi(t,s) = \frac{T-t_0}{T-t_0}$. Therefore, the result follows. □

Corollary 7.24 *Suppose that* $\mathcal{G}_t = \mathcal{G}_t^3 = \mathcal{F}_t \vee \sigma(B(T))$. *Then*

$$E\big[d^- B\big|\mathcal{G}_{t^-}\big] = \frac{B(T) - B(t)}{T-t}\,dt.$$

Combining this with Theorem 7.22, we get the following:

Theorem 7.25 *Suppose that* $\mathcal{G}_t = \mathcal{G}_t^3 = \mathcal{F}_t \vee \sigma(B(T))$ *and* $\theta = 0$. *Suppose that the conditions of Theorem 7.22 hold. Then* $u = \widehat{u}$ *is a critical point for* $J(u)$ *in (7.3) if and only if*

$$E\big[L(t)\big|\mathcal{G}_{t^-}\big] + E\big[M(t)\big|\mathcal{G}_{t^-}\big]\frac{B(T) - B(t)}{T-t} - D_{t^+}E\big[M(t)\big|\mathcal{G}_{t^-}\big] = 0$$
$$\textit{for a.a. } t \in [0,T]. \tag{7.54}$$

7.6 Application to Optimal Insider Portfolio

Consider a financial market with two investments possibilities:

1. A risk-free asset, where the unit price $S_0(t)$ at time t is given by

$$dS_0(t) = r(t)S_0(t)\,dt, \qquad S_0(0) = 1. \tag{7.55}$$

2. A risky asset, where the unit price $S_1(t)$ at time t is given by the stochastic differential equation

$$\begin{aligned} dS_1(t) &= S_1\big(t^-\big)\Big[\mu(t)\,dt + \sigma_0(t)\,dB^-(t) + \int_{\mathbb{R}_0} \gamma(t,z)\widetilde{N}\big(d^-t, dz\big)\Big], \\ S_1(0) &> 0. \end{aligned} \tag{7.56}$$

Here $r(t) \geq 0$, $\mu(t)$, $\sigma_0(t)$, and $\gamma(t,z) \geq -1+\epsilon$ (for some constant $\epsilon > 0$) are given $\mathbb{G}$-predictable, forward-integrable processes, where $\mathbb{G} = \{\mathcal{G}_t\}_{0\leq t\leq T}$ is a given filtration such that

$$\mathcal{F}_t \subset \mathcal{G}_t \quad \text{for all } t \in [0,T]. \tag{7.57}$$

Suppose that a trader in this market is an insider, in the sense that she has access to the information represented by $\mathcal{G}_t$ at time t. This means that if she chooses a portfolio $u(t)$, representing the *amount* she invests in the risky asset at time t, then this portfolio is a $\mathbb{G}$-predictable stochastic process.

The corresponding wealth process $X(t) = X^{(u)}(t)$ will then satisfy the (forward) SDE

$$\begin{aligned} d^-X(t) &= \frac{X(t)-u(t)}{S_0(t)}\,dS_0(t) + \frac{u(t)}{S_1(t)}\,d^-S_1(t) \\ &= X(t)r(t)\,dt + u(t)\Big[\big(\mu(t)-r(t)\big)\,dt + \sigma_0(t)\,dB^-(t) \\ &\quad + \int_{\mathbb{R}_0} \gamma(t,z)\widetilde{N}\big(d^-t, dz\big)\Big], \qquad t \in [0,T], \end{aligned} \tag{7.58}$$

$$X(0) = x > 0. \tag{7.59}$$

By choosing $S_0(\cdot)$ as a numeraire, we can, without loss of generality, assume that

$$r(t) = 0 \tag{7.60}$$

from now on. Then (7.58) and (7.59) simplify to

$$\begin{cases} d^-X(t) = u(t)[\mu(t)\,dt + \sigma_0(t)\,dB^-(t) + \int_{\mathbb{R}_0} \gamma(t,z)\widetilde{N}(d^-t, dz)], \\ X(0) = x > 0. \end{cases} \tag{7.61}$$

This is a controlled Itô–Lévy process of the type discussed in Sect. 7.4, and we can apply the results of that section to the problem of the insider to maximize the expected utility of the terminal wealth, i.e., to find $\Phi_{\mathbb{G}}(x)$ and $u^* \in \mathcal{A}_{\mathbb{G}}$ such that

$$\Phi_{\mathbb{G}}(x) = \sup_{u \in \mathcal{A}_{\mathbb{G}}} E\big[U\big(X^{(u)}(T)\big)\big] = E\big[U\big(X^{(u^*)}(T)\big)\big], \tag{7.62}$$

where $U : \mathbb{R}_+ \to \mathbb{R}$ is a given utility function, assumed to be concave, strictly increasing, and C^1. In this case, the processes $K(t)$, $L(t)$, $M(t)$, and $R(t,z)$, given respectively by (7.26), (7.39), (7.40), and (7.41), take the form

$$K(t) = U'\big(X(T)\big), \tag{7.63}$$

$$\begin{aligned} L(t) = {} & U'\big(X(T)\big)\Big[\mu(t) + D_{t+}\sigma_0(t) + \int_{\mathbb{R}_0} D_{t+,z}\gamma(t,z)\nu(dz)\Big] \\ & + \int_{\mathbb{R}_0} D_{t,z}U'\big(X(T)\big)\big[\gamma(t,z) + D_{t+,z}\gamma(t,z)\big]\nu(dz) \\ & + D_t U'\big(X(T)\big)\sigma_0(t), \end{aligned} \tag{7.64}$$

$$M(t) = U'\big(X(T)\big)\sigma_0(t), \tag{7.65}$$

$$R(t,z) = \big\{U'\big(X(T)\big) + D_{t,z}U'\big(X(T)\big)\big\}\big\{\gamma(t,z) + D_{t+,z}\gamma(t,z)\big\}. \tag{7.66}$$

7.6.1 Case $\mathcal{G}_t = \mathcal{F}_{G_t}$, $G_t \supset [0,t]$. See (7.45)

In this case, $\mathcal{G}_t$ satisfies conditions (7.42) and (7.43). Therefore, Theorem 7.18 of Sect. 7.4 gives the following:

Theorem 7.26 *Suppose that $P\{\lambda\{t \in [0,T]; \sigma_0(t) \neq 0\} > 0\} > 0$, where λ denotes the Lebesgue measure on $\mathbb{R}$ and that $\mathcal{G}_t$ is given by* (7.45). *Then, there does not exist an optimal portfolio $u^* \in \mathcal{A}_{\mathbb{G}}$ for the insider's portfolio problem* (7.62).

Proof Suppose that an optimal portfolio exists. Then we have seen that in either case, the conclusion is that

$$E\big[L(t)\big|\mathcal{G}_t\big] = E\big[M(t)\big|\mathcal{G}_t\big] = E\big[R(t,z)\big|\mathcal{G}_t\big] = 0$$

for a.a. $t \in [0,T]$, $z \in \mathbb{R}_0$. In particular,

$$E\big[M(t)\big|\mathcal{G}_t\big] = E\big[U'\big(X(T)\big)\big|\mathcal{G}_t\big]\sigma_0(t) = 0 \quad \text{for a.a. } t \in [0,T].$$

Since $U' > 0$, this contradicts our assumption about U. Hence, an optimal portfolio cannot exist. □

Remark 7.27 In the case that $\mathcal{G}_t = \mathcal{G}_t^i, i = 1$ or $i = 3$, it is known that $B(\cdot)$ is not a semimartingale with respect to $\mathbb{G} = \{\mathcal{G}_t\}_{0 \le t \le T}$, and hence an optimal portfolio cannot exist, by Theorem 3.8 in [3] and Theorem 15 in [7]. It follows that $S_1(\cdot)$ is not a $\mathbb{G}$-semimartingale either, and hence we can even deduce that the market has an arbitrage for the insider in this case, by Theorem 7.2 in [5].

7.6.2 Case $\mathcal{G}_t = \mathcal{F}_t \vee \sigma(B(T))$. See (7.51)

In this case, $\mathcal{G}_t$ is not D-commutable (see Remark 7.19). Therefore we apply results from Sect. 7.5.2. We have seen that

$$E\big[d^- B \big| \mathcal{G}_{t^-}\big] = \frac{B(T) - B(t)}{T - t}\, dt$$

(Corollary 7.24). From this we have the following:

Theorem 7.28 *Assume that* $\mu(t) = \mu_0, \sigma_0(t) = \sigma_0$, *and* $\gamma(t, z) = 0$ *and that conditions in Theorem* 7.21 *hold. In addition, suppose that*

1. $E[M(t)|\mathcal{G}_{t^-}] \in \mathbb{M}_{1,2}^B$,
2. $\underline{\lim}_{t \uparrow T} E[|D_{t^+} E[M(t)|\mathcal{G}_{t^-}]|] < \infty$,
3. $\underline{\lim}_{t \uparrow T} E[|L(t)|] < \infty$.

Then, there does not exist a critical point of the performance functional $J(u)$ *in* (7.3).

Proof Assume that there is a critical point of the performance functional $J(u)$ in (7.3). It follows from Theorems 7.14, 7.21, and 7.22 that (7.47) holds. Replacing $K(t)$, $L(t)$, and $M(t)$ by their given expressions in (7.63), (7.64), and (7.65), (7.47) becomes

$$\begin{aligned} 0 = {} & E\big[\mu_0 U'\big(X(T)\big) + \sigma_0 D_t U'\big(X(T)\big)\big|\mathcal{G}_{t^-}\big] + E\big[U'\big(X(T)\big)\sigma_0\big|\mathcal{G}_{t^-}\big]\frac{B(T) - B(t)}{T - t} \\ & - D_{t^+} E\big[\sigma_0 U'\big(X(T)\big)\big|\mathcal{G}_{t^-}\big] \quad \text{for a.a. } t. \end{aligned} \tag{7.67}$$

Taking the limit as $t \uparrow T$, the second term in (7.67) goes to ∞. Therefore, there is no critical point for the performance functional $J(u)$ in (7.3). □

Remark 7.29 This result is a generalization of a result in [14], where the same conclusion was obtained in the special case where

$$U(x) = \ln(x).$$

7.7 Application to Optimal Insider Consumption

Suppose that we have a cash flow $X(t) = X^{(u)}(t)$ given by

$$\begin{cases} dX(t) = (\mu(t) - u(t))\,dt + \sigma(t)\,d^-B(t) + \int_{\mathbb{R}_0} \theta(t,z)\widetilde{N}(d^-t, dz), \\ X(0) = x \in \mathbb{R}. \end{cases} \tag{7.68}$$

Here $\mu(t), \sigma(t)$, and $\theta(t,z)$ are given $\mathbb{G}$-predictable processes, and $u(t) \geq 0$ is our *consumption* rate, assumed to be adapted to a given insider filtration $\mathbb{G} = \{\mathcal{G}_t\}_{0 \leq t \leq T}$, where $\mathcal{F}_t \subset \mathcal{G}_t$ for all t. Let $f(t,u,\omega); t \in [0,T], u \in \mathbb{R}, \omega \in \Omega$, be a given $\mathcal{F}_T$-measurable utility process. Assume that $u \to f(t,u,\omega)$ is strictly increasing, concave, and C^1 for a.a. (t,ω).

Let $g(x,\omega); x \in \mathbb{R}, \omega \in \Omega$, be a given $\mathcal{F}_T$-measurable random variable for each x. Assume that $x \to g(x,\omega)$ is concave for a.a. ω. Define the performance functional J by

$$J(u) = E\left[\int_0^T f\big(t, u(t), \omega\big)\,dt + g\big(X^{(u)}(T), \omega\big)\right]; \quad u \in \mathcal{A}_{\mathbb{G}}, u \geq 0. \tag{7.69}$$

Note that $u \to J(u)$ is concave, so $u = \widehat{u}$ maximizes $J(u)$ if and only if $\widehat{u}$ is a critical point of $J(u)$.

Theorem 7.30 (Optimal insider consumption I) *$\widehat{u}$ is an optimal insider consumption rate for the performance functional J in* (7.69) *if and only if*

$$E\left[\frac{\partial}{\partial u} f\big(t, \widehat{u}(t), \omega\big)\Big|\mathcal{G}_t\right] = E\big[g'\big(X^{(\widehat{u})}(T), \omega\big)\big|\mathcal{G}_t\big]. \tag{7.70}$$

Proof In this case, we have

$$\begin{aligned} K(t) &= g'\big(X^{(u)}(T)\big), \\ L(t) &= -g'\big(X^{(u)}(T)\big) + \frac{\partial}{\partial u} f\big(t, \widehat{u}(t)\big), \\ M(t) &= R(t,z) = 0. \end{aligned}$$

Therefore, Theorem 7.14 gives $\widehat{u}$ is a critical point for $J(u)$ if and only if

$$0 = E\big[L(t)\big|\mathcal{G}_t\big] = E\left[\frac{\partial}{\partial u} f\big(t, \widehat{u}(t)\big)\Big|\mathcal{G}_t\right] + E\big[-g'\big(X^{(\widehat{u})}(T)\big)\big|\mathcal{G}_t\big].$$

□

Since $X^{(\widehat{u})}(T)$ depends on $\widehat{u}$, (7.70) does not give the value of $\widehat{u}(t)$ directly. However, in some special cases, $\widehat{u}$ can be found explicitly:

Corollary 7.31 (Optimal insider consumption II) *Assume that*

$$g(x,\omega) = \lambda(\omega)x \tag{7.71}$$

for some $\mathcal{G}_T$*-measurable random variable* $\lambda > 0$.

Then the optimal consumption rate $\widehat{u}(t)$ *is given by*

$$E\left[\frac{\partial}{\partial u} f(t,u,\omega)\middle|\mathcal{G}_t\right]_{u=\widehat{u}(t)} = E[\lambda|\mathcal{G}_t]. \tag{7.72}$$

Thus, we see that an optimal consumption rate exists for *any* given insider information filtration $\{\mathcal{G}_t\}_{0\le t\le T}$. It is not necessary to be in a semimartingale setting.

Another example in the same direction is the following:

Theorem 7.32 (Complete future information) *Suppose that we have complete future information, i.e.,*

$$\mathcal{G}_t = \mathcal{F}_T \quad \textit{for all } t \in [0,T]. \tag{7.73}$$

Suppose that we have the exponential utilities, i.e.,

$$f(t,u,\omega) = -K_1(t,\omega)e^{-\alpha u}, \qquad g(x,\omega) = -K_2(\omega)e^{-\alpha x} \tag{7.74}$$

for some measurable process $K_1(t,\omega) > 0$, *some* $\mathcal{F}_T$*-measurable random variable* $K_2(\omega) > 0$, *and some constant* $\alpha > 0$.

Then the optimal consumptions rate $\widehat{u}(t)$, *if it exists, satisfies the equation*

$$\widehat{u}(t) = \frac{1}{\alpha}\ln\left(\frac{K_1(t)}{K_2}\right) + X^{(0)}(T) - \int_0^T \widehat{u}(s)\,ds, \tag{7.75}$$

where

$$X^{(0)}(T) = x + \int_0^T \mu(s)\,ds + \int_0^T \sigma(s)\,dB(s) + \int_0^T\int_{\mathbb{R}_0} \theta(s,z)\widetilde{N}(ds,dz)$$

is the terminal wealth when there is no consumption.

In particular, if $K_1(t) = K_1$ *does not depend on* t, *then* $\widehat{u}(t) = \widehat{u}$ *does not depend on* t, *and we get*

$$\widehat{u}(t) = \widehat{u} = \frac{1}{1+T}\left(\frac{1}{\alpha}\ln\left(\frac{K_1}{K_2}\right) + X^{(0)}(T)\right); \quad t \in [0,T]. \tag{7.76}$$

Proof By (7.70) we get

$$-\alpha K_1(t)e^{-\alpha\widehat{u}(t)} = -\alpha K_2 e^{-\alpha X(T)}$$

or

$$\widehat{u}(t) = \frac{1}{\alpha}\ln\left(\frac{K_1(t)}{K_2}\right) + X(T) = \widehat{u}(t) = \frac{1}{\alpha}\ln\left(\frac{K_1(t)}{K_2}\right) + X^{(0)}(T) - \int_0^T \widehat{u}(s)\,ds,$$

which proves (7.75.) If $K_1(t) = K_1$ does not depend on t, then by (7.75), $\widehat{u}(t) = u(t)$ does not depend on t either, and (7.76) follows. □

For related results (based on a different method) on optimal insider consumption, see [22].

Acknowledgement We thank José Manuel Corcuera for his valuable comments.

Appendix: Proof of Theorem 7.13

Proof 1. Since $\widehat{u} \in \mathcal{A}_{\mathbb{G}}$ is a critical point for $J(u)$, there exists a $\delta > 0$ as in (7.24) for all bounded $\beta \in \mathcal{A}_{\mathbb{G}}$. Thus,

$$\begin{aligned} 0 &= \frac{d}{dy} J(\widehat{u} + y\beta)\Big|_{y=0} \\ &= E\Big[\int_0^T \Big\{\frac{\partial}{\partial x} f\big(t, X(t), u(t)\big)\widehat{Y}(t) + \frac{\partial}{\partial u} f\big(t, X(t), u(t)\big)\beta(t)\Big\} dt \\ &\quad + g'\big(X(T)\big)\widehat{Y}(T)\Big], \end{aligned} \tag{7.77}$$

where $\widehat{Y} = Y_{\beta}^{\widehat{u}}$ is as defined in (7.25).

We study the two summands separately. By Corollaries 7.5 and 7.12 and the product rule, we get

$$\begin{aligned} &E\big[g'\big(X(T)\big)Y(T)\big] \\ &= E\Big[g'\big(X(T)\big)\Big(\int_0^T \Big\{\frac{\partial b(t)}{\partial x} Y(t) + \frac{\partial b(t)}{\partial u}\beta(t)\Big\} dt \\ &\quad + \int_0^T \Big\{\frac{\partial \sigma(t)}{\partial x} Y(t) + \frac{\partial \sigma(t)}{\partial u}\beta(t)\Big\} d^- B(t) \\ &\quad + \int_0^T \int_{\mathbb{R}_0} \Big\{\frac{\partial \theta(t)}{\partial x} Y(t) + \frac{\partial \theta(t)}{\partial u}\beta(t)\Big\}\widetilde{N}\big(dz, d^- t\big)\Big)\Big] \\ &= E\Big[\int_0^T g'\big(X(T)\big)\Big\{\frac{\partial b(t)}{\partial x} Y(t) + \frac{\partial b(t)}{\partial u}\beta(t)\Big\} dt\Big] \\ &\quad + E\Big[\int_0^T D_t g'\big(X(T)\big)\Big\{\frac{\partial \sigma(t)}{\partial x} Y(t) + \frac{\partial \sigma(t)}{\partial u}\beta(t)\Big\} dt\Big] \\ &\quad + E\Big[\int_0^T g'\big(X(T)\big) D_{t+}\Big(\frac{\partial \sigma(t)}{\partial x} Y(t) + \frac{\partial \sigma(t)}{\partial u}\beta(t)\Big) dt\Big] \\ &\quad + E\Big[\int_0^T \int_{\mathbb{R}_0} D_{t,z} g'\big(X(T)\big)\Big\{\frac{\partial \theta(t)}{\partial x} Y(t) + \frac{\partial \theta(t)}{\partial u}\beta(t)\Big\}\nu(dz)\, dt\Big] \end{aligned}$$

$$
\begin{aligned}
&+ E\bigg[\int_0^T \int_{\mathbb{R}_0} \{g'(X(T)) + D_{t,z}g'(X(T))\} \\
&\times D_{t+,z}\bigg(\frac{\partial\theta(t)}{\partial x}Y(t) + \frac{\partial\theta(t)}{\partial u}\beta(t)\bigg)\nu(dz)\,dt\bigg] \\
&= E\bigg[\int_0^T \bigg\{g'(X(T))\frac{\partial b(t)}{\partial x} + D_t g'(X(T))\frac{\partial\sigma(t)}{\partial x} \\
&+ \int_{\mathbb{R}_0} D_{t,z}g'(X(T))\frac{\partial\theta(t)}{\partial x}\nu(dz)\bigg\}Y(t)\,dt\bigg] \\
&+ E\bigg[\int_0^T \bigg\{g'(X(T))\frac{\partial b(t)}{\partial u} + D_t g'(X(T))\frac{\partial\sigma(t)}{\partial u} \\
&+ \int_{\mathbb{R}_0} D_{t,z}g'(X(T))\frac{\partial\theta(t)}{\partial u}\nu(dz)\bigg\}\beta(t)\,dt\bigg] \\
&+ E\bigg[\int_0^T g'(X(T))D_{t+}\frac{\partial\sigma(t)}{\partial x}Y(t)\,dt\bigg] \\
&+ E\bigg[\int_0^T g'(X(T))\frac{\partial\sigma(t)}{\partial x}D_{t+}Y(t)\,dt\bigg] \\
&+ E\bigg[\int_0^T g'(X(T))D_{t+}\frac{\partial\sigma(t)}{\partial u}\beta(t)\,dt\bigg] \\
&+ E\bigg[\int_0^T g'(X(T))\frac{\partial\sigma(t)}{\partial u}D_{t+}\beta(t)\,dt\bigg] \\
&+ E\bigg[\int_0^T \int_{\mathbb{R}_0} \{g'(X(T)) + D_{t,z}g'(X(T))\}D_{t+,z}\frac{\partial\theta(t)}{\partial x}Y(t)\nu(dz)\,dt\bigg] \\
&+ E\bigg[\int_0^T \int_{\mathbb{R}_0} \{g'(X(T)) + D_{t,z}g'(X(T))\}\bigg\{\frac{\partial\theta(t)}{\partial x} + D_{t+,z}\frac{\partial\theta(t)}{\partial x}\bigg\} \\
&\times D_{t+,z}Y(t)\nu(dz)\,dt\bigg] \\
&+ E\bigg[\int_0^T \int_{\mathbb{R}_0} \{g'(X(T)) + D_{t,z}g'(X(T))\}D_{t+,z}\frac{\partial\theta(t)}{\partial u}\beta(t)\nu(dz)\,dt\bigg] \\
&+ E\bigg[\int_0^T \int_{\mathbb{R}_0} \{g'(X(T)) + D_{t,z}g'(X(T))\}\bigg\{\frac{\partial\theta(t)}{\partial u} + D_{t+,z}\frac{\partial\theta(t)}{\partial u}\bigg\} \\
&\times D_{t+,z}\beta(t)\nu(dz)\,dt\bigg] \\
&= E\bigg[\int_0^T \bigg\{g'(X(T))\bigg(\frac{\partial b(t)}{\partial x} + D_{t+}\frac{\partial\sigma(t)}{\partial x} + \int_{\mathbb{R}_0} D_{t+,z}\frac{\partial\theta(t)}{\partial x}\nu(dz)\bigg)
\end{aligned}
$$

$$+ D_t g'(X(T)) \frac{\partial \sigma(t)}{\partial x}$$

$$+ \int_{\mathbb{R}_0} D_{t,z} g'(X(T)) \left(\frac{\partial \theta(t)}{\partial x} + D_{t+,z} \frac{\partial \theta(t)}{\partial x} \right) \nu(dz) \Big\} Y(t)\, dt \Big]$$

$$+ E\Big[\int_0^T \Big\{ g'(X(T)) \left(\frac{\partial b(t)}{\partial u} + D_{t+} \frac{\partial \sigma(t)}{\partial u} + \int_{\mathbb{R}_0} D_{t+,z} \frac{\partial \theta(t)}{\partial u} \nu(dz) \right)$$

$$+ D_t g'(X(T)) \frac{\partial \sigma(t)}{\partial u}$$

$$+ \int_{\mathbb{R}_0} D_{t,z} g'(X(T)) \left(\frac{\partial \theta(t)}{\partial u} + D_{t+,z} \frac{\partial \theta(t)}{\partial u} \right) \nu(dz) \Big\} \beta(t)\, dt \Big]$$

$$+ E\left[\int_0^T g'(X(T)) \frac{\partial \sigma(t)}{\partial x} D_{t+} Y(t)\, dt \right]$$

$$+ E\left[\int_0^T g'(X(T)) \frac{\partial \sigma(t)}{\partial u} D_{t+} \beta(t)\, dt \right]$$

$$+ E\Big[\int_0^T \int_{\mathbb{R}_0} \{ g'(X(T)) + D_{t,z} g'(X(T)) \} \left\{ \frac{\partial \theta(t)}{\partial x} + D_{t+,z} \frac{\partial \theta(t)}{\partial x} \right\}$$

$$\times D_{t+,z} Y(t) \nu(dz)\, dt \Big]$$

$$+ E\Big[\int_0^T \int_{\mathbb{R}_0} \{ g'(X(T)) + D_{t,z} g'(X(T)) \} \left\{ \frac{\partial \theta(t)}{\partial u} + D_{t+,z} \frac{\partial \theta(t)}{\partial u} \right\}$$

$$\times D_{t+,z} \beta(t) \nu(dz)\, dt \Big].$$

Similarly, we have using both Fubini and duality theorems,

$$E\left[\int_0^T \frac{\partial}{\partial x} f(t) Y(t)\, dt \right]$$

$$= E\Big[\int_0^T \frac{\partial}{\partial x} f(t) \Big(\int_0^t \left\{ \frac{\partial b(s)}{\partial x} Y(s) + \frac{\partial b(s)}{\partial u} \beta(s) \right\} ds$$

$$+ \int_0^t \left\{ \frac{\partial \sigma(s)}{\partial x} Y(s) + \frac{\partial \sigma(s)}{\partial u} \beta(s) \right\} d^- B(s)$$

$$+ \int_0^t \int_{\mathbb{R}_0} \left\{ \frac{\partial \theta(s)}{\partial x} Y(s) + \frac{\partial \theta(s)}{\partial u} \beta(s) \right\} \widetilde{N}(dz, d^- s) \Big) dt \Big]$$

$$= E\left[\int_0^T \left(\int_0^t \frac{\partial f(t)}{\partial x} \left\{ \frac{\partial b(s)}{\partial x} Y(s) + \frac{\partial b(s)}{\partial u} \beta(s) \right\} ds \right) dt \right]$$

$$+E\left[\int_0^T\left(\int_0^t D_s\frac{\partial f(t)}{\partial x}\left\{\frac{\partial\sigma(s)}{\partial x}Y(s)+\frac{\partial\sigma(s)}{\partial u}\beta(s)\right\}ds\right)dt\right]$$

$$+E\left[\int_0^T\left(\int_0^t \frac{\partial f(t)}{\partial x}D_{s+}\left\{\frac{\partial\sigma(s)}{\partial x}Y(s)+\frac{\partial\sigma(s)}{\partial u}\beta(s)\right\}ds\right)dt\right]$$

$$+E\left[\int_0^T\left(\int_0^t\int_{\mathbb{R}_0} D_{s,z}\frac{\partial f(t)}{\partial x}\left\{\frac{\partial\theta(s)}{\partial x}Y(s)+\frac{\partial\theta(s)}{\partial u}\beta(s)\right\}\nu(dz)\,ds\right)dt\right]$$

$$+E\left[\int_0^T\left(\int_0^t\int_{\mathbb{R}_0}\left\{\frac{\partial f(t)}{\partial x}+D_{s,z}\frac{\partial f(t)}{\partial x}\right\}\right.\right.$$

$$\left.\left.\times D_{s+,z}\left(\frac{\partial\theta(s)}{\partial x}Y(s)+\frac{\partial\theta(s)}{\partial u}\beta(s)\right)\nu(dz)\,ds\right)dt\right]$$

$$=E\left[\int_0^T\left(\int_s^T\frac{\partial f(t)}{\partial x}dt\right)\left\{\frac{\partial b(s)}{\partial x}Y(s)+\frac{\partial b(s)}{\partial u}\beta(s)\right\}ds\right]$$

$$+E\left[\int_0^T\left(\int_s^T D_s\frac{\partial f(t)}{\partial x}dt\right)\left\{\frac{\partial\sigma(s)}{\partial x}Y(s)+\frac{\partial\sigma(s)}{\partial u}\beta(s)\right\}\right]$$

$$+E\left[\int_0^T\left(\int_s^T\frac{\partial f(t)}{\partial x}dt\right)D_{s+}\left\{\frac{\partial\sigma(s)}{\partial x}Y(s)+\frac{\partial\sigma(s)}{\partial u}\beta(s)\right\}ds\right]$$

$$+E\left[\int_0^T\int_{\mathbb{R}_0}\left(\int_s^T D_{s,z}\frac{\partial f(t)}{\partial x}dt\right)\left\{\frac{\partial\theta(s)}{\partial x}Y(s)+\frac{\partial\theta(s)}{\partial u}\beta(s)\right\}\nu(dz)\,ds\right]$$

$$+E\left[\int_0^T\int_{\mathbb{R}_0}\left(\int_s^T\left\{\frac{\partial f(t)}{\partial x}+D_{s,z}\frac{\partial f(t)}{\partial x}\right\}dt\right)\right.$$

$$\left.\times D_{s+,z}\left\{\frac{\partial\theta(s)}{\partial x}Y(s)+\frac{\partial\theta(s)}{\partial u}\beta(s)\right\}\nu(dz)\,ds\right].$$

Changing the notation $s\to t$, this becomes

$$=E\left[\int_0^T\left(\int_t^T\frac{\partial f(s)}{\partial x}ds\right)\left\{\frac{\partial b(t)}{\partial x}Y(t)+\frac{\partial b(t)}{\partial u}\beta(t)\right\}dt\right]$$

$$+E\left[\int_0^T\left(\int_t^T D_t\frac{\partial f(s)}{\partial x}ds\right)\left\{\frac{\partial\sigma(t)}{\partial x}Y(t)+\frac{\partial\sigma(t)}{\partial u}\beta(t)\right\}\right]$$

$$+E\left[\int_0^T\int_{\mathbb{R}_0}\left(\int_t^T D_{t,z}\frac{\partial f(s)}{\partial x}ds\right)\left\{\frac{\partial\theta(t)}{\partial x}Y(t)+\frac{\partial\theta(t)}{\partial u}\beta(t)\right\}\nu(dz)\,dt\right]$$

$$+E\left[\int_0^T\left(\int_t^T\frac{\partial f(s)}{\partial x}ds\right)D_{t+}\left\{\frac{\partial\sigma(t)}{\partial x}Y(t)+\frac{\partial\sigma(t)}{\partial u}\beta(t)\right\}dt\right]$$

$$+E\left[\int_0^T\int_{\mathbb{R}_0}\left(\int_t^T\left\{\frac{\partial f(s)}{\partial x}+D_{t,z}\frac{\partial f(s)}{\partial x}\right\}ds\right)\right.$$

$$
\begin{aligned}
&\times \left(D_{t+,z}\left\{ \frac{\partial \theta(t)}{\partial x} Y(t) + \frac{\partial \theta(t)}{\partial u} \beta(t) \right\}\right) \nu(dz)\, dt \Bigg] \\
&= E\Bigg[\int_0^T \Bigg\{ \left(\int_t^T \frac{\partial f(s)}{\partial x} ds \right) \left(\frac{\partial b(t)}{\partial x} + D_{t+} \frac{\partial \sigma(t)}{\partial x} + \int_{\mathbb{R}_0} D_{t+,z} \frac{\partial \theta(t)}{\partial x} \nu(dz) \right) \\
&+ \left(\int_t^T D_t \frac{\partial f(s)}{\partial x} ds \right) \frac{\partial \sigma(t)}{\partial x} \\
&+ \int_{\mathbb{R}_0} \left(\int_t^T D_{t,z} \frac{\partial f(s)}{\partial x} ds \right) \left(\frac{\partial \theta(t)}{\partial x} + D_{t+,z} \frac{\partial \theta(t)}{\partial x} \right) \nu(dz) \Bigg\} Y(t)\, dt \Bigg] \\
&+ E\Bigg[\int_0^T \Bigg\{ \left(\int_t^T \frac{\partial f(s)}{\partial x} ds \right) \left(\frac{\partial b(t)}{\partial u} + D_{t+} \frac{\partial \sigma(t)}{\partial u} + \int_{\mathbb{R}_0} D_{t+,z} \frac{\partial \theta(t)}{\partial u} \nu(dz) \right) \\
&+ \left(\int_t^T D_t \frac{\partial f(s)}{\partial x} ds \right) \frac{\partial \sigma(t)}{\partial u} \\
&+ \int_{\mathbb{R}_0} \left(\int_t^T D_{t,z} \frac{\partial f(s)}{\partial x} ds \right) \left(\frac{\partial \theta(t)}{\partial u} + D_{t+,z} \frac{\partial \theta(t)}{\partial u} \right) \nu(dz) \Bigg\} \beta(t)\, dt \Bigg] \\
&+ E\Bigg[\int_0^T \left(\int_t^T \frac{\partial f(s)}{\partial x} ds \right) \frac{\partial \sigma(t)}{\partial x} D_{t+} Y(t)\, dt \Bigg] \\
&+ E\Bigg[\int_0^T \left(\int_t^T \frac{\partial f(s)}{\partial x} ds \right) \frac{\partial \sigma(t)}{\partial u} D_{t+} \beta(t)\, dt \Bigg] \\
&+ E\Bigg[\int_0^T \int_{\mathbb{R}_0} \left\{ \left(\int_t^T \frac{\partial f(s)}{\partial x} + D_{t,z} \frac{\partial f(s)}{\partial x} ds \right) \right\} \left\{ \frac{\partial \theta(t)}{\partial x} + D_{t+,z} \frac{\partial \theta(t)}{\partial x} \right\} \\
&\times D_{t+,z} Y(t) \nu(dz)\, dt \Bigg] \\
&+ E\Bigg[\int_0^T \int_{\mathbb{R}_0} \left\{ \left(\int_t^T \frac{\partial f(s)}{\partial x} + D_{t,z} \frac{\partial f(s)}{\partial x} ds \right) \right\} \left\{ \frac{\partial \theta(t)}{\partial u} + D_{t+,z} \frac{\partial \theta(t)}{\partial u} \right\} \\
&\times D_{t+,z} \beta(t) \nu(dz)\, dt \Bigg]. \qquad (7.78)
\end{aligned}
$$

Recall that

$$
K(t) := g'\big(X(T)\big) + \int_t^T \frac{\partial}{\partial x} f\big(s, X(s), u(s)\big)\, ds,
$$

and combining (7.33)–(7.78), it follows that

$$
0 = E\Bigg[\int_0^T \Bigg\{ K(t) \left(\frac{\partial b(t)}{\partial x} + D_{t+} \frac{\partial \sigma(t)}{\partial x} + \int_{\mathbb{R}_0} D_{t+,z} \frac{\partial \theta(t)}{\partial x} \nu(dz) \right)
$$

$$
\begin{aligned}
&+ D_t K(t) \frac{\partial \sigma(t)}{\partial x} \\
&+ \int_{\mathbb{R}_0} D_{t,z} K(t) \left(\frac{\partial \theta(t)}{\partial x} + D_{t+,z} \frac{\partial \theta(t)}{\partial x} \right) \nu(dz) \Big\} Y(t)\, dt \Big] \\
&+ E\Big[\int_0^T \Big\{ K(t) \left(\frac{\partial b(t)}{\partial u} + D_{t+} \frac{\partial \sigma(t)}{\partial u} + \int_{\mathbb{R}_0} D_{t+,z} \frac{\partial \theta(t)}{\partial u} \nu(dz) \right) \\
&+ D_t K(t) \frac{\partial \sigma(t)}{\partial u} \\
&+ \int_{\mathbb{R}_0} D_{t,z} K(t) \left(\frac{\partial \theta(t)}{\partial u} + D_{t+,z} \frac{\partial \theta(t)}{\partial u} \right) \nu(dz) + \frac{\partial f(t)}{\partial u} \Big\} \beta(t)\, dt \Big] \\
&+ E\Big[\int_0^T K(t) \frac{\partial \sigma(t)}{\partial x} D_{t+} Y(t)\, dt \Big] \\
&+ E\Big[\int_0^T K(t) \frac{\partial \sigma(t)}{\partial u} D_{t+} \beta(t)\, dt \Big] \\
&+ E\Big[\int_0^T \int_{\mathbb{R}_0} \{ K(t) + D_{t,z} K(t) \} \Big\{ \frac{\partial \theta(t)}{\partial x} + D_{t+,z} \frac{\partial \theta(t)}{\partial x} \Big\} \\
&\times D_{t+,z} Y(t) \nu(dz)\, dt \Big] \\
&+ E\Big[\int_0^T \int_{\mathbb{R}_0} \{ K(t) + D_{t,z} K(t) \} \Big\{ \frac{\partial \theta(t)}{\partial u} + D_{t+,z} \frac{\partial \theta(t)}{\partial u} \Big\} \\
&\times D_{t+,z} \beta(t) \nu(dz)\, dt \Big].
\end{aligned} \tag{7.79}
$$

We observe that $\mathcal{A}_{\mathbb{G}}$ contains all β_α given as $\beta_\alpha(s) := \alpha \chi_{[t,t+h]}(s)$ for some $t, h \in (0, T)$, $t + h \le T$, where $\alpha = \alpha(\omega)$ is bounded and $\mathcal{G}_t$-measurable. Then $Y^{(\beta_\alpha)}(s) = 0$ for $0 \le s \le t$, and hence (7.79) becomes

$$
A_1 + A_2 + A_3 + A_4 + A_5 + A_6 = 0, \tag{7.80}
$$

where

$$
\begin{aligned}
A_1 = E\Big[\int_t^T \Big\{ & K(t) \left(\frac{\partial b(s)}{\partial x} + D_{s+} \frac{\partial \sigma(s)}{\partial x} + \int_{\mathbb{R}_0} D_{s+,z} \frac{\partial \theta(s)}{\partial x} \nu(dz) \right) \\
&+ \int_{\mathbb{R}_0} D_{s,z} K(s) \left(\frac{\partial \theta(s)}{\partial x} + D_{s+,z} \frac{\partial \theta(s)}{\partial x} \right) \nu(dz) + D_s K(s) \frac{\partial \sigma(s)}{\partial x} \Big\} \\
&\times Y^{(\beta_\alpha)}(s)\, ds \Big],
\end{aligned}
$$

$$A_2 = E\left[\int_t^{t+h}\left\{K(t)\left(\frac{\partial b(s)}{\partial u} + D_{s+}\frac{\partial \sigma(s)}{\partial u} + \int_{\mathbb{R}_0} D_{s+,z}\frac{\partial \theta(s)}{\partial u}\nu(dz)\right) + \frac{\partial f(s)}{\partial u}\right.\right.$$
$$\left.\left.+ \int_{\mathbb{R}_0} D_{s,z}K(s)\left(\frac{\partial \theta(s)}{\partial u} + D_{s+,z}\frac{\partial \theta(s)}{\partial u}\right)\nu(dz) + D_sK(s)\frac{\partial \sigma(s)}{\partial u}\right\}\alpha\, ds\right],$$

$$A_3 = E\left[\int_t^T K(s)\frac{\partial \sigma(s)}{\partial x}D_{s+}Y^{(\beta_\alpha)}(s)\,ds\right],$$

$$A_4 = E\left[\int_t^{t+h} K(s)\frac{\partial \sigma(s)}{\partial u}D_{s+}\alpha\, ds\right],$$

$$A_5 = E\left[\int_t^T \int_{\mathbb{R}_0}\{K(s) + D_{s,z}K(s)\}\left(\frac{\partial \theta(s)}{\partial x} + D_{s+,z}\frac{\partial \theta(s)}{\partial x}\right)\right.$$
$$\left.\times\, \nu(dz)D_{s+,z}Y^{(\beta_\alpha)}(s)\,ds\right],$$

$$A_6 = E\left[\int_t^{t+h} \int_{\mathbb{R}_0}\{K(s) + D_{s,z}K(s)\}\left(\frac{\partial \theta(s)}{\partial u} + D_{s+,z}\frac{\partial \theta(s)}{\partial u}\right)\right.$$
$$\left.\times\, \nu(dz)D_{s+,z}\alpha\, ds\right].$$

Note that by the definition of Y with $Y(s) = Y^{(\beta_\alpha)}(s)$ and $s \geq t+h$, the process $Y(s)$ follows the dynamics

$$dY(s) = Y\left(s^-\right)\left[\frac{\partial b}{\partial x}(s)\,ds + \frac{\partial \sigma}{\partial x}(s)\,d^-B(s)\right.$$
$$\left.+ \int_{\mathbb{R}_0}\frac{\partial \theta}{\partial x}(s,z)\widetilde{N}\left(dz, d^-s\right)\right] \tag{7.81}$$

for $s \geq t+h$ with initial condition $Y(t+h)$ at time $t+h$. By Itô's formula for forward integral, this equation can be solved explicitly, and we get

$$Y(s) = Y(t+h)G(t+h,s), \quad s \geq t+h, \tag{7.82}$$

where, in general, for $s \geq t$,

$$G(t,s) := \exp\left(\int_t^s\left\{\frac{\partial b}{\partial x}(r, X(r), u(r), \omega) - \frac{1}{2}\left(\frac{\partial \sigma}{\partial x}\right)^2(r, X(r), u(r), \omega)\right\}dr\right.$$
$$+ \int_t^s \frac{\partial \sigma}{\partial x}(r, X(r), u(r), \omega)\,dB^-(r)$$
$$+ \int_t^s \int_{\mathbb{R}_0}\left\{\ln\left(1 + \frac{\partial \theta}{\partial x}(r, X(r), u(r), \omega)\right)\right.$$

$$- \frac{\partial \theta}{\partial x}\big(r, X(r), u(r), \omega\big)\bigg\}\nu(dz)\,dt$$

$$+ \int_t^s \int_{\mathbb{R}_0} \bigg\{ \ln\bigg(1 + \frac{\partial \theta}{\partial x}\big(r, X\big(r^-\big), u\big(r^-\big), \omega\big)\bigg)\bigg\} \widetilde{N}\big(dz, d^- r\big)\bigg).$$

Note that $G(t, s)$ does not depend on h, but $Y(s)$ does. Defining H_0 as in (7.27), it follows that

$$A_1 = E\bigg[\int_t^T \frac{\partial H_0}{\partial x}(s) Y(s)\,ds\bigg].$$

Differentiating with respect to h at $h = 0$, we get

$$\begin{aligned}
\frac{d}{dh} A_1\bigg|_{h=0} &= \frac{d}{dh} E\bigg[\int_t^{t+h} \frac{\partial H_0}{\partial x}(s) Y(s)\,ds\bigg]_{h=0} \\
&\quad + \frac{d}{dh} E\bigg[\int_{t+h}^T \frac{\partial H_0}{\partial x}(s) Y(s)\,ds\bigg]_{h=0}.
\end{aligned}$$

Since $Y(t) = 0$, we see that

$$\frac{d}{dh} E\bigg[\int_t^{t+h} \frac{\partial H_0}{\partial x}(s) Y(s)\,ds\bigg]_{h=0} = 0.$$

Therefore, by (7.82),

$$\begin{aligned}
\frac{d}{dh} A_1\bigg|_{h=0} &= \frac{d}{dh} E\bigg[\int_{t+h}^T \frac{\partial H_0}{\partial x}(s) Y(t+h) G(t+h, s)\,ds\bigg]_{h=0} \\
&= \int_t^T \frac{d}{dh} E\bigg[\frac{\partial H_0}{\partial x}(s) Y(t+h) G(t+h, s)\bigg]_{h=0} ds \\
&= \int_t^T \frac{d}{dh} E\bigg[\frac{\partial H_0}{\partial x}(s) G(t, s) Y(t+h)\bigg]_{h=0} ds,
\end{aligned}$$

where $Y(t+h)$ is given by

$$\begin{aligned}
Y(t+h) = \int_t^{t+h} & Y\big(r^-\big)\bigg[\frac{\partial b}{\partial x}(r)\,dr + \frac{\partial \sigma}{\partial x}(r)\,d^- B(r) \\
& + \int_{\mathbb{R}_0} \frac{\partial \theta}{\partial x}(r, z) \widetilde{N}\big(dz, d^- r\big)\bigg] \\
& + \alpha \int_t^{t+h} \bigg[\frac{\partial b}{\partial u}(r)\,dr + \frac{\partial \sigma}{\partial u}(r)\,d^- B(r) + \int_{\mathbb{R}_0} \frac{\partial \theta}{\partial u}(r, z) \widetilde{N}\big(dz, d^- r\big)\bigg].
\end{aligned}$$

Therefore, by the two preceding equalities,

$$\frac{d}{dh} A_1\bigg|_{h=0} = A_{1,1} + A_{1,2},$$

where

$$A_{1,1} = \int_t^T \frac{d}{dh} E\left[\frac{\partial H_0}{\partial x}(s) G(t,s) \alpha \int_t^{t+h} \left\{ \frac{\partial b}{\partial u}(r)\, dr + \frac{\partial \sigma}{\partial u}(r)\, d^- B(r) \right.\right.$$
$$\left.\left. + \int_{\mathbb{R}_0} \frac{\partial \theta}{\partial u}(r,z) \widetilde{N}\big(dz, d^- r\big) \right\}\right]_{h=0} ds,$$

and

$$A_{1,2} = \int_t^T \frac{d}{dh} E\left[\frac{\partial H_0}{\partial x}(s) G(t,s) \int_t^{t+h} Y\big(r^-\big) \left\{ \frac{\partial b}{\partial x}(r)\, dr + \frac{\partial \sigma}{\partial x}(r)\, d^- B(r) \right.\right.$$
$$\left.\left. + \int_{\mathbb{R}_0} \frac{\partial \theta}{\partial x}(r,z) \widetilde{N}\big(dz, d^- r\big) \right\}\right]_{h=0} ds.$$

Applying again the duality formula, we have

$$A_{1,1} = \int_t^T \frac{d}{dh} E\left[\alpha \int_t^{t+h} \left\{ \frac{\partial b}{\partial u}(r) F(t,s) + \frac{\partial \sigma}{\partial u}(r) D_r F(t,s) \right.\right.$$
$$+ F(t,s) D_{r^+} \frac{\partial \sigma}{\partial u}(r)$$
$$+ \int_{\mathbb{R}_0} \left\{ \left(\frac{\partial \theta}{\partial u}(r,z) + D_{r^+,z} \frac{\partial \theta}{\partial u}(r,z) \right) D_{r,z} F(t,s) \right.$$
$$\left.\left.\left. + D_{r^+,z} \frac{\partial \theta}{\partial u}(r,z) F(t,s) \right\} \nu(dz) \right\} dr \right]_{h=0} ds$$
$$= \int_t^T E\left[\alpha \left\{ \left(\frac{\partial b}{\partial u}(t) + D_{t^+} \frac{\partial \sigma}{\partial u}(t) + \int_{\mathbb{R}_0} D_{t^+,z} \frac{\partial \theta}{\partial u}(t,z) \nu(dz) \right) F(t,s) \right.\right.$$
$$+ \frac{\partial \sigma}{\partial u}(t) D_t F(t,s) + \int_{\mathbb{R}_0} \left(\frac{\partial \theta}{\partial u}(t,z) + D_{t^+,z} \frac{\partial \theta}{\partial u}(t,z) \right)$$
$$\left.\left. \times D_{t,z} F(t,s) \nu(dz) \right\}\right] ds,$$

where we have put

$$F(t,s) = \frac{\partial H_0}{\partial x}(s) G(t,s).$$

Since $Y(t) = 0$, we see that

$$A_{1,2} = 0.$$

We conclude that

$$\left.\frac{d}{dh} A_1 \right|_{h=0} = A_{1,1}. \tag{7.83}$$

Moreover, we see that

$$\begin{aligned}\frac{d}{dh}A_2\bigg|_{h=0} &= E\bigg[\bigg\{K(t)\bigg(\frac{\partial b(t)}{\partial u} + D_{t+}\frac{\partial \sigma(t)}{\partial u} + \int_{\mathbb{R}_0} D_{t+,z}\frac{\partial \theta(t,z)}{\partial u}\nu(dz)\bigg)\\ &\quad + \frac{\partial f(t)}{\partial u} + D_t K(t)\frac{\partial \sigma(t,z)}{\partial u}\\ &\quad + \int_{\mathbb{R}_0} D_{t,z}K(t)\bigg(\frac{\partial \theta(t,z)}{\partial u} + D_{t+,z}\frac{\partial \theta(t,z)}{\partial u}\bigg)\nu(dz)\bigg\}\alpha\bigg], \qquad (7.84)\\ \frac{d}{dh}A_4\bigg|_{h=0} &= E\bigg[K(t)\frac{\partial \sigma(t)}{\partial u}D_{t+}\alpha\bigg], \qquad (7.85)\\ \frac{d}{dh}A_6\bigg|_{h=0} &= E\bigg[\int_{\mathbb{R}_0}\{K(t) + D_{t,z}K(t)\}\bigg(\frac{\partial \theta(t,z)}{\partial u} + D_{t+,z}\frac{\partial \theta(t,z)}{\partial u}\bigg)\\ &\quad \times \nu(dz)D_{t+,z}\alpha\bigg]. \qquad (7.86)\end{aligned}$$

On the other hand, by differentiating A_3 with respect to h at $h = 0$, we get

$$\begin{aligned}\frac{d}{dh}A_3\bigg|_{h=0} &= \frac{d}{dh}E\bigg[\int_t^{t+h} K(s)\frac{\partial \sigma(s)}{\partial x}D_{s+}Y(s)\,ds\bigg]_{h=0}\\ &\quad + \frac{d}{dh}E\bigg[\int_{t+h}^T K(s)\frac{\partial \sigma(s)}{\partial x}D_{s+}Y(s)\,ds\bigg]_{h=0}.\end{aligned}$$

Since $Y(t) = 0$, we see that

$$\begin{aligned}\frac{d}{dh}A_3\bigg|_{h=0} &= \frac{d}{dh}E\bigg[\int_{t+h}^T K(s)\frac{\partial \sigma(s)}{\partial x}D_{s+}\big(Y(t+h)G(t+h,s)\big)\,ds\bigg]_{h=0}\\ &= \int_t^T \frac{d}{dh}E\bigg[K(s)\frac{\partial \sigma(s)}{\partial x}D_{s+}\big(Y(t+h)G(t+h,s)\big)\bigg]_{h=0}ds\\ &= \int_t^T \frac{d}{dh}E\bigg[K(s)\frac{\partial \sigma(s)}{\partial x}\big(D_{s+}G(t+h,s)\cdot Y(t+h)\\ &\quad + D_{s+}Y(t+h)\cdot G(t+h,s)\big)\bigg]_{h=0}ds\\ &= \int_t^T \frac{d}{dh}E\bigg[K(s)\frac{\partial \sigma(s)}{\partial x}\big(Y(t+h)D_{s+}G(t,s)\\ &\quad + D_{s+}Y(t+h)G(t,s)\big)\bigg]_{h=0}ds.\end{aligned}$$

Using the definition of $\widehat{p}$ and $\widehat{H}$ given respectively by (7.36) and (7.35) in the theorem, it follows by (7.80) that

$$E\left[\frac{\partial}{\partial u}\widehat{H}\big(t,\widehat{X}(t),\widehat{u}(t)\big)\bigg|\mathcal{G}_t\right]+E[A]=0 \quad \text{a.e. in } (t,\omega), \tag{7.87}$$

where

$$A=\frac{d}{dh}A_3\bigg|_{h=0}+\frac{d}{dh}A_4\bigg|_{h=0}+\frac{d}{dh}A_5\bigg|_{h=0}+\frac{d}{dh}A_6\bigg|_{h=0}. \tag{7.88}$$

2. Conversely, suppose that there exists $\widehat{u}\in\mathcal{A}_{\mathbb{G}}$ such that (7.34) holds. Then by reversing the previous arguments, we obtain that (7.80) holds for all $\beta_\alpha(s):=\alpha\chi_{[t,t+h]}(s)\in\mathcal{A}_{\mathbb{G}}$, where

$$\begin{aligned}
A_1=E\bigg[\int_t^T\bigg\{&K(t)\bigg(\frac{\partial b(s)}{\partial x}+D_{s+}\frac{\partial\sigma(s)}{\partial x}+\int_{\mathbb{R}_0}D_{s+,z}\frac{\partial\theta(s)}{\partial x}\nu(dz)\bigg)\\
&+\int_{\mathbb{R}_0}D_{s,z}K(s)\bigg(\frac{\partial\theta(s)}{\partial x}+D_{s+,z}\frac{\partial\theta(s)}{\partial x}\bigg)\nu(dz)\\
&+D_sK(s)\frac{\partial\sigma(s)}{\partial x}\bigg\}Y^{(\beta_\alpha)}(s)\,ds\bigg],\\
A_2=E\bigg[\int_t^{t+h}\bigg\{&K(t)\bigg(\frac{\partial b(s)}{\partial u}+D_{s+}\frac{\partial\sigma(s)}{\partial u}+\int_{\mathbb{R}_0}D_{s+,z}\frac{\partial\theta(s)}{\partial u}\nu(dz)\bigg)+\frac{\partial f(s)}{\partial u}\\
&+\int_{\mathbb{R}_0}D_{s,z}K(s)\bigg(\frac{\partial\theta(s)}{\partial u}+D_{s+,z}\frac{\partial\theta(s)}{\partial u}\bigg)\nu(dz)+D_sK(s)\frac{\partial\sigma(s)}{\partial u}\bigg\}\alpha\,ds\bigg],\\
A_3=E\bigg[\int_t^T&K(s)\frac{\partial\sigma(s)}{\partial x}D_{s+}Y^{(\beta_\alpha)}(s)\,ds\bigg],\\
A_4=E\bigg[\int_t^{t+h}&K(s)\frac{\partial\sigma(s)}{\partial u}D_{s+}\alpha\,ds\bigg],\\
A_5=E\bigg[\int_t^T&\int_{\mathbb{R}_0}\{K(s)+D_{s,z}K(s)\}\bigg(\frac{\partial\theta(s)}{\partial x}+D_{s+,z}\frac{\partial\theta(s)}{\partial x}\bigg)\\
&\times\nu(dz)D_{s+,z}Y^{(\beta_\alpha)}(s)\,ds\bigg],\\
A_6=E\bigg[\int_t^{t+h}&\int_{\mathbb{R}_0}\{K(s)+D_{s,z}K(s)\}\bigg(\frac{\partial\theta(s)}{\partial u}+D_{s+,z}\frac{\partial\theta(s)}{\partial u}\bigg)\\
&\times\nu(dz)D_{s+,z}\alpha\,ds\bigg]
\end{aligned}$$

for some $t,h\in(0,T)$, $t+h\le T$, where $\alpha=\alpha(\omega)$ is bounded and $\mathcal{G}_t$-measurable. Hence, these equalities hold for all linear combinations of β_α. Since all bounded

$\beta \in \mathcal{A}_{\mathbb{G}}$ can be approximated pointwise boundedly in (t, ω) by such linear combinations, it follows that (7.80) holds for all bounded $\beta \in \mathcal{A}_{\mathbb{G}}$. Hence, by reversing the remaining part of the previous proof, we conclude that

$$\frac{d}{dy} J_1(\widehat{u} + y\beta)\bigg|_{y=0} = 0 \quad \text{for all } \beta,$$

and then $\widehat{u}$ satisfies (7.33). □

References

1. J. Amendinger, P. Imkeller, M. Schweizer, Additional logarithmic utility of an insider. Stoch. Process. Appl. **75**, 263–286 (1998)
2. J. Bertoin, *Lévy Processes*. Cambridge Tracts in Mathematics, vol. 121 (Cambridge University Press, Cambridge, 1996)
3. F. Biagini, B. Øksendal, A general stochastic calculus approach to insider trading. Appl. Math. Optim. **52**, 167–181 (2005)
4. J.M. Corcuera, P. Imkeller, A. Kohatsu-Higa, D. Nualart, Additional utility of insiders with imperfect dynamical information. Finance Stoch. **8**, 437–450 (2004)
5. F. Delbaen, W. Schachermayer, A general version of the fundamental theorem of asset pricing. Math. Ann. **300**, 463–520 (1994)
6. G. Di Nunno, T. Meyer-Brandis, B. Øksendal, F. Proske, Malliavin calculus and anticipative Itô formulae for Lévy processes. Infin. Dimens. Anal. Quantum Probab. **8**, 235–258 (2005)
7. G. Di Nunno, T. Meyer-Brandis, B. Øksendal, F. Proske, Optimal portfolio for an insider in a market driven by Lévy processes. Quant. Finance **6**(1), 83–94 (2006)
8. G. Di Nunno, B. Øksendal, F. Proske, *Malliavin Calculus for Lévy Processes with Applications to Finance* (Springer, Berlin, 2008)
9. G. Di Nunno, O. Menoukeu Pamen, B. Øksendal, F. Proske, *Uniqueness of Decompositions of Skorokhod-Semimartingales*. Eprint Series in Pure Mathematics, vol. 10 (University of Oslo, Oslo, 2009)
10. N. Framstad, B. Øksendal, A. Sulem, Stochastic maximum principle for optimal control of jump diffusions and applications to finance. J. Optim. Theory Appl. **121**(1), 77–98 (2004). Errata: J. Optim. Theory Appl. **124**(2), 511–512 (2005)
11. A. Grorud, M. Pontier, Asymmetrical information and incomplete markets. Int. J. Theor. Appl. Finance **4**(2), 285–302 (2001)
12. P. Imkeller, Malliavin's calculus in insider models: additional utility and free lunches. Math. Finance **13**(1), 153–169 (2003). Conference on Applications of Malliavin Calculus in Finance (Rocquencourt, 2001)
13. Y. Itô, Generalized Poisson functionals. Probab. Theory Relat. Fields **77**, 1–28 (1988)
14. I. Karatzas, I. Pikovsky, Anticipating portfolio optimization. Adv. Appl. Probab. **28**, 1095–1122 (1996)
15. A. Kohatsu-Higa, A. Sulem, Utility maximization in an insider influenced market. Math. Finance **16**(1), 153–179 (2006)
16. T. Meyer-Brandis, B. Øksendal, X. Zhou, *A Malliavin Calculus Approach to a General Maximum Principle for Stochastic Control*. Eprint Series in Pure Mathematics, vol. 10 (University of Oslo, Oslo, 2008)
17. D. Nualart, *The Malliavin Calculus and Related Topics* (Springer, Berlin, 1995)
18. D. Nualart, E. Pardoux, Stochastic calculus with anticipating integrands. Probab. Theory Relat. Fields **78**, 535–581 (1988)
19. D. Nualart, A.S. Ustunel, E. Zakai, Some relations among classes of σ-fields on Wiener space. Probab. Theory Relat. Fields **85**, 119 (1990)

20. B. Øksendal, A. Sulem, Partial observation control in an anticipating environment. Russ. Math. Surv. **59**, 355–375 (2004)
21. B. Øksendal, A. Sulem, *Applied Stochastic Control of Jump Diffusions*, 2nd edn. (Springer, Berlin, 2007)
22. B. Øksendal, T. Zhang, *Backward Stochastic Differential Equations with Respect to General Filtrations and Applications to Insider Finance*. Eprint Series in Pure Mathematics, vol. 19 (University of Oslo, Oslo, 2009)
23. F. Russo, P. Vallois, Forward, backward and symmetric stochastic integration. Probab. Theory Relat. Fields **97**, 403–421 (1993)
24. F. Russo, P. Vallois, Stochastic calculus with respect to continuous finite variation processes. Stoch. Stoch. Rep. **70**, 140 (2000)
25. K.-I. Sato, *Lévy Processes and Infinitely Divisible Distributions*. Cambridge Studies in Advanced Mathematics, vol. 68 (Cambridge University Press, Cambridge, 1999)

Chapter 8
Analyticity of the Wiener–Hopf Factors and Valuation of Exotic Options in Lévy Models

Ernst Eberlein, Kathrin Glau, and Antonis Papapantoleon

Abstract This paper considers the valuation of exotic path-dependent options in Lévy models, in particular options on the supremum and the infimum of the asset price process. Using the Wiener–Hopf factorization, we derive expressions for the analytically extended characteristic function of the supremum and the infimum of a Lévy process. Combined with general results on Fourier methods for option pricing, we provide formulas for the valuation of one-touch options, lookback options, and equity default swaps in Lévy models.

Keywords Lévy processes · Wiener–Hopf factorization · Exotic options

Mathematics Subject Classification (2010) 91Gxx · 60G51

8.1 Introduction

The ever-increasing sophistication of derivative products offered by financial institutions, together with the failure of traditional Gaussian models to describe the

---

K. Glau would like to thank the DFG for financial support through project EB66/11-1, and the Austrian Science Fund (FWF) for an invitation under grant P18022. A. Papapantoleon gratefully acknowledges the financial support from the Austrian Science Fund (FWF grant Y328, START Prize).

E. Eberlein (✉) · K. Glau
Department of Mathematical Stochastics, University of Freiburg, Eckerstr. 1, 79104 Freiburg, Germany
e-mail: eberlein@stochastik.uni-freiburg.de

K. Glau
e-mail: glau@stochastik.uni-freiburg.de

A. Papapantoleon
Institute of Mathematics, TU Berlin, Straße des 17. Juni 136, 10623 Berlin, Germany
e-mail: papapan@math.tu-berlin.de

A. Papapantoleon
Quantitative Products Laboratory, Deutsche Bank AG, Alexanderstr. 5, 10178 Berlin, Germany

G. Di Nunno, B. Øksendal (eds.), *Advanced Mathematical Methods for Finance*,
DOI 10.1007/978-3-642-18412-3_8,

dynamics in the markets, has led to a quest for more realistic and flexible models. In fact, one of the lessons from the current financial crisis is the following: the Gaussian copula model is inappropriate to describe the interdependence between the tails of asset returns because, among other pitfalls, the tail dependence coefficient is always zero. Hence, this model cannot capture systemic risk.

In the search for appropriate alternatives, *Lévy processes* are playing a leading role, either as models for financial assets themselves or as building blocks for models, e.g., in Lévy-driven stochastic volatility models or in affine models. The field of Lévy processes has become popular in modern mathematical finance, and the interest from academics and practitioners has led to inspiring and challenging questions.

Lévy processes are attractive for applications in mathematical finance because they can describe some of the observed phenomena in the markets in a rather adequate way. This is due to the fact that their sample paths may have jumps and the generated distributions can be heavy-tailed and skewed. Another important improvement concerns the famous smile effect. See [17] and [19] for an extensive empirical justification of the non-Gaussianity of asset returns and the appropriateness of (generalized hyperbolic) Lévy processes. For an overview of the application of Lévy processes in finance, the interested reader is referred to the textbooks of Cont and Tankov [14], Schoutens [46], and the collection edited by Kyprianou et al. [38]. There are, of course, several textbooks dealing with the theory of Lévy processes; we mention [3, 9, 44], and [33], while the collection [8] contains an overview of the application of Lévy processes in different areas of research, such as quantum field theory and turbulence.

The application of Lévy processes in financial modeling, in particular for the pricing and hedging of derivatives, has led to new challenges of both *analytical* and *numerical* nature. In Lévy models simple closed-form valuation formulas are typically not available even for plain vanilla European options, let alone for exotic path-dependent options. The numerical methods which have been developed in the classical Gaussian framework lead to completely new challenges in the context of Lévy-driven models. These numerical methods can be classified roughly in three areas: probabilistic numerical methods (Monte Carlo methods), deterministic numerical methods (PIDE methods), and Fourier transform methods; for an excellent survey of these methods, their applicability and limitations, we refer to [25].

This paper focuses on the application of Fourier transform methods for the valuation of exotic path-dependent options, in particular options depending on the supremum and the infimum of Lévy processes. The bulk of the literature on this latter topic focuses on the numerical aspects. Our focus is on the analytical aspects. More specifically, we show first that the Wiener–Hopf factorization of a Lévy process possesses an analytic extension, and then we prove that the Wiener–Hopf factorization (viewed as a Laplace transform in time) can be inverted. These results allow us to derive expressions for the extended characteristic function of the supremum and the infimum of a Lévy process. This latter result, combined with general results on option pricing by Fourier methods (cf. [21]), allows us to derive pricing formulas for lookback options, one-touch options, and equity default swaps in Lévy models.

Let us briefly comment on some papers where the Wiener–Hopf factorization is used to price exotic options in Lévy models. Boyarchenko and Levendorskiĭ [11] derive valuation formulas for barrier and one-touch options for driving Lévy processes that belong to the class of so-called "regular Lévy processes of exponential type" (RLPE); see also the book [12]. The results of these authors are based on the theory of pseudodifferential operators. The numerics of this approach is pushed further in [29, 30]. Avram et al. [4, 6], Kyprianou and Pistorius [35], Alili and Kyprianou [2], and Levendorskiĭ et al. [40] consider the valuation of American and Russian options, either on a finite or an infinite time horizon. Jeannin and Pistorius [28] develop methods for the computation of prices and Greeks for various Lévy models. Central in their argumentation is the approximation of different Lévy models by the class of "generalized hyper-exponential Lévy models" that have a tractable Wiener–Hopf factorization. The same approach is also applied in [5] for the pricing of equity default swaps in Lévy models.

The major open challenge in this field is the development of analytical expressions for the Wiener–Hopf factors for general Lévy processes. In a remarkable recent development, Hubalek and Kyprianou [26] generate a family of spectrally negative Lévy processes with tractable Wiener–Hopf factors, using results from potential theory for subordinators. These results were later extended in [36] and applied to problems in actuarial mathematics in [39]. Moreover, in two very recent papers, Kuznetsov [31, 32] introduces special families of Lévy processes such that the Wiener–Hopf factors can be computed as infinite products over the roots of certain transcendental equations. These families include processes with behavior similar to the CGMY process, while the author shows that the numerical computation of the infinite products can be performed quite efficiently.

This paper is structured as follows: In Sect. 8.2, we briefly review Lévy processes and prove the analyticity of the characteristic function of the supremum. In Sect. 8.3, we review the Wiener–Hopf factorization, prove its analytic extension and invert it in time. In Sect. 8.4, we present some examples of popular Lévy models and comment on the continuity of their laws. Finally, in Sect. 8.5, we derive valuation formulas for lookback and one-touch options and for equity default swaps.

Important Remark This paper is intimately tied to, and intended to be read together with, the companion paper [21], which will be abbreviated EGP in the sequel. In particular, we will make heavy use of the notation and results from that paper.

8.2 Lévy Processes

We start by fixing the notation that will be used throughout the paper and providing some estimates on the exponential moments of a Lévy process. Then, we prove the analytic extension of the characteristic function of the supremum and the infimum of a Lévy process, sampled either at a fixed time or at an independent, exponentially distributed time.

8.2.1 Notation

Let $\mathcal{B} = (\Omega, \mathcal{F}, \mathbf{F}, P)$ be a complete stochastic basis in the sense of [27, I.1.3], where $\mathcal{F} = \mathcal{F}_T$, $0 < T \leq \infty$, and $\mathbf{F} = (\mathcal{F}_t)_{0 \leq t \leq T}$. Let $L = (L_t)_{0 \leq t \leq T}$ be a *Lévy process* on this stochastic basis, i.e., L is a semimartingale with *independent* and *stationary increments* (PIIS), and $L_0 = 0$ a.s. We denote the *triplet of predictable characteristics* of L by (B, C, ν) and the *triplet of local characteristics* by (b, c, λ); using [27, II.4.20], the two triplets are related via

$$B_t(\omega) = bt, \qquad C_t(\omega) = ct, \qquad \nu(\omega; dt, dx) = \lambda(dx)\, dt.$$

We assume that the following condition is in force.

Assumption $(\mathbb{EM})$ There exists a constant $M > 1$ such that

$$\int_{\{|x|>1\}} e^{ux} \lambda(dx) < \infty \quad \forall u \in [-M, M].$$

The triplet of predictable characteristics of a PIIS determines the law of the random variables; more specifically, for a Lévy process, we know from the Lévy–Khintchine formula that

$$E\big[e^{iuL_t}\big] = \exp\big(t \cdot \kappa(iu)\big) \tag{8.1}$$

for all $t \in [0, T]$ and all $u \in \mathbb{R}$, where the cumulant generating function is

$$\kappa(u) = ub + \frac{u^2}{2} c + \int_{\mathbb{R}} \big(e^{ux} - 1 - ux\big) \lambda(dx). \tag{8.2}$$

Assumption $(\mathbb{EM})$ entails that the Lévy process L is a *special* and *exponentially special* semimartingale; hence the use of a truncation function can be and has been omitted. Applying Theorem 25.3 in [44], we get that

$$E\big[e^{uL_t}\big] < \infty \quad \forall u \in [-M, M],\ \forall t \in [0, T].$$

Recall that for any stochastic process X, we denote by $\overline{X}$ the supremum and by $\underline{X}$ the infimum process of X, respectively.

In the sequel, we will provide the proofs of the results for the supremum process. The proofs for the infimum process can be derived analogously or using the duality between the supremum and the infimum process; see the following remark.

Remark 8.1 Let L be a Lévy process with local characteristics (b, c, λ). The *dual* of the Lévy process L, defined by $L' := -L$, has the triplet of local characteristics (b', c', λ'), where $b' = -b$, $c' = c$, and $1_A(x) * \lambda' = 1_A(-x) * \lambda$, $A \in \mathcal{B}(\mathbb{R} \setminus \{0\})$. Moreover, we have that

$$\underline{L}_t = \inf_{0 \leq s \leq t} L_s = - \sup_{0 \leq s \leq t} (-L_s) = -\overline{L'}_t.$$

8.2.2 Analytic Extension, Fixed-Time Case

In this section, we establish the existence of an analytic extension of the characteristic function of the *supremum* and the *infimum* of a Lévy process, and derive explicit bounds for the exponential moments of the supremum and infimum process.

The next lemma endows us with a link between the existence of exponential moments of a measure ϱ and the analytic extension of the characteristic function $\widehat{\varrho}$.

Lemma 8.2 *Let ϱ be a measure on the space $(\mathbb{R}, \mathcal{B}(\mathbb{R}))$. If $\int e^{ux}\varrho(dx) < \infty$ for all $u \in [-a, b]$ with $a, b \geq 0$, then the characteristic function $\widehat{\varrho}$ has an extension that is continuous on $(-\infty, \infty) \times i[-b, a]$ and is analytic in the interior of the strip $(-\infty, \infty) \times i(-b, a)$. Moreover, $\widehat{\varrho}(u) = \int e^{iux}\varrho(dx)$ for all $u \in \mathbb{C}$ with $\Im(u) \in [-b, a]$.*

Proof The function $u \mapsto e^{iux}$ clearly extends to an entire function, and the extension

$$\widehat{\varrho}(u) := \int e^{iux}\varrho(dx) \quad \big(u \in \mathbb{C} \text{ with } \Im(u) \in [-b, a]\big)$$

is well defined since

$$\left|e^{iux}\right| = e^{-\Im(u)x} \leq e^{-ax}1_{\{x\leq 0\}} + e^{bx}1_{\{x>0\}} =: h(x)$$

for $u \in \mathbb{C}$ with $\Im(u) \in [-b, a]$, and we have that $h \in L^1(\varrho)$ by assumption. Moreover, Lebesgue's dominated convergence theorem yields that this extension is continuous.

We will prove the analyticity of $\widehat{\varrho}$ in $(-\infty, \infty) \times i(-b, a)$ using the theorem of Morera (see, for example, Theorem 10.17 in [43]). Let γ be a triangle in the open set $(-\infty, \infty) \times i(-b, a)$; the theorems of Fubini and Cauchy immediately yield

$$\int_{\partial\gamma} \widehat{\varrho}(u)\,du = \int_{\partial\gamma}\int e^{iux}\varrho(dx)\,du = \int\int_{\partial\gamma} e^{iux}\,du\varrho(dx) = 0,$$

as $u \mapsto e^{iux}$ is analytic for every fixed $x \in \mathbb{R}$. Then, the analyticity of $\widehat{\varrho}$ follows from Morera's theorem. For a justification of the application of Fubini's theorem, it is enough to note that

$$\int\int_{\partial\gamma} \left|e^{iux}\right| du\varrho(dx) \leq \int\int_{\partial\gamma} h(x)\,du\varrho(dx) = \ell(\gamma)\int h(x)\varrho(dx) < \infty,$$

where $\ell(\gamma)$ denotes the length of the curve $\partial\gamma$. □

Lemma 8.3 *Let Y be a Lévy process and a special semimartingale with $E[Y_t] = 0$ for some and hence, for every $t > 0$. Then*

$$E\big[e^{Y_t^*}\big] \leq 8E\big[e^{|Y_t|}\big],$$

where $Y_t^* = \sup_{0\leq s\leq t}|Y_s|$.

Proof Using that $\frac{(Y_t^*)^n}{n!}$ is positive for every $n \geq 0$ and the monotone convergence theorem, we get

$$E\big[e^{Y_t^*}\big] = E\sum_{n=0}^{\infty}\frac{(Y_t^*)^n}{n!} = \sum_{n=0}^{\infty}E\frac{(Y_t^*)^n}{n!}.$$

Now, Remark 25.19 in [44] yields

$$E\big(Y_t^*\big)^n \leq 8E|Y_t|^n \quad \text{for every } n \geq 1,$$

while for $n = 0$, the inequality holds trivially. Hence, we get

$$\sum_{n=0}^{\infty}E\frac{(Y_t^*)^n}{n!} \leq 8\sum_{n=0}^{\infty}E\frac{|Y_t|^n}{n!} = 8E\sum_{n=0}^{\infty}\frac{|Y_t|^n}{n!} = 8E\big[e^{|Y_t|}\big].$$ □

Next, notice that under assumption ($\mathbb{EM}$), we have that

$$\int_{\mathbb{R}}\big|e^{Mx} - 1 - Mx\big|\lambda(dx) < \infty \quad \text{and} \quad \int_{\mathbb{R}}\big|e^{-Mx} - 1 + Mx\big|\lambda(dx) < \infty.$$

Let us introduce the following notation:

$$\overline{\alpha}(M) := M|b| + \frac{1}{2}cM^2 + \int_{\mathbb{R}}\big|e^{Mx} - 1 - Mx\big|\lambda(dx) \tag{8.3}$$

and

$$\underline{\alpha}(M) := M|b| + \frac{1}{2}cM^2 + \int_{\mathbb{R}}\big|e^{-Mx} - 1 + Mx\big|\lambda(dx). \tag{8.4}$$

Lemma 8.4 *Let $L = (L_t)_{0\leq t\leq T}$ be a Lévy process that satisfies assumption* ($\mathbb{EM}$). *Then we have the following estimates*:

$$E\big[e^{u\overline{L}_t}\big] \leq E\big[e^{M\overline{L}_t}\big] \leq 8\mathcal{C}(t, M) < \infty \quad (u \leq M)$$

and

$$E\big[e^{-u\underline{L}_t}\big] \leq E\big[e^{-M\underline{L}_t}\big] \leq 8\mathcal{C}(t, M) < \infty \quad (u \leq M),$$

where $\mathcal{C}(t, M) := e^{t\overline{\alpha}(M)} + e^{t\underline{\alpha}(M)}$.

Proof For $u \leq M$, we have

$$e^{u\overline{L}_t} \leq e^{M\overline{L}_t},$$

since $\overline{L}_t = \sup_{0\leq s\leq t} L_s$ is nonnegative. Further, notice that

$$\overline{L}_t = \sup_{0\leq s\leq t}\big[bs + \sqrt{c}W_s + L_s^d\big] \leq \sup_{0\leq s\leq t}\big[\sqrt{c}W_s + L_s^d\big] + \sup_{0\leq s\leq t}[bs],$$

where $L_t = bt + \sqrt{c}W_t + L_t^d$ denotes the canonical decomposition of L, with Brownian motion W and a purely discontinuous martingale $L^d = x * (\mu - \nu)$. Let us further denote

$$Y_s := \sqrt{c}W_s + L_s^d.$$

The process Y is not only a martingale but also a Lévy process and a special semimartingale with local characteristics $(0, c, \lambda)$. We have

$$\overline{L}_t \leq \sup_{0 \leq s \leq t} Y_s + |b|t \leq Y_t^* + |b|t,$$

and hence we get that

$$E\big[e^{M\overline{L}_t}\big] \leq E\big[e^{M(Y_t^* + |b|t)}\big] = e^{M|b|t} E\big[e^{MY_t^*}\big] \leq 8e^{M|b|t} E\big[e^{M|Y_t|}\big], \tag{8.5}$$

using Lemma 8.3 for the special semimartingale $Z := MY$, which is a Lévy process satisfying $E[Z_t] = 0$ for every $0 \leq t \leq T$.

Now it is sufficient to notice that

$$E\big[e^{M|Y_t|}\big] \leq E\big[e^{MY_t}\big] + E\big[e^{-MY_t}\big], \tag{8.6}$$

where Theorem 25.17 in [44] yields

$$\begin{aligned} E\big[e^{MY_t}\big] &= \exp\bigg(t\frac{cM^2}{2} + t\int_{\mathbb{R}} \big(e^{Mx} - 1 - Mx\big)\lambda(dx)\bigg) \\ &\leq e^{(\overline{\alpha}(M) - M|b|)t}; \end{aligned} \tag{8.7}$$

similarly,

$$E\big[e^{-MY_t}\big] \leq e^{(\underline{\alpha}(M) - M|b|)t}. \tag{8.8}$$

Summarizing, we can conclude from (8.5)–(8.8) that

$$\begin{aligned} E\big[e^{M\overline{L}_t}\big] &\leq 8e^{M|b|t}\big(e^{(\overline{\alpha}(M) - M|b|)t} + e^{(\underline{\alpha}(M) - M|b|)t}\big) \\ &= 8\big(e^{\overline{\alpha}(M)t} + e^{\underline{\alpha}(M)t}\big) \end{aligned}$$

and

$$E\big[e^{-M\underline{L}_t}\big] \leq 8\big(e^{\overline{\alpha}(M)t} + e^{\underline{\alpha}(M)t}\big).$$ □

A corollary of these results is the existence of an analytic continuation for the characteristic function $\varphi_{\overline{L}_t}$ of the supremum, resp. $\varphi_{\underline{L}_t}$ of the infimum, of a Lévy process.

Corollary 8.5 *Let L be a Lévy process that satisfies assumption* ($\mathbb{EM}$). *Then, the characteristic function $\varphi_{\overline{L}_t}$ of $\overline{L}_t$, resp. $\varphi_{\underline{L}_t}$ of $\underline{L}_t$, possesses a continuous extension*

$$\varphi_{\overline{L}_t}(z) = \int_{\mathbb{R}} e^{izx} P_{\overline{L}_t}(dx), \quad resp. \quad \varphi_{\underline{L}_t}(z) = \int_{\mathbb{R}} e^{izx} P_{\underline{L}_t}(dx),$$

to the half-plane $z \in \{z \in \mathbb{C} : -M \le \Im z\}$, resp. $z \in \{z \in \mathbb{C} : \Im z \le M\}$, that is analytic in the interior of the half-plane $\{z \in \mathbb{C} : -M < \Im z\}$, resp. $\{z \in \mathbb{C} : \Im z < M\}$.

Proof This is a direct consequence of Lemmas 8.2 and 8.4. □

Remark 8.6 One could derive the statement of Corollary 8.5 using the submultiplicativity of the exponential function and Theorem 25.18 in [44], see Lemma 5 in [37]. However, we will need the estimates of Lemma 8.4 in the following sections.

8.2.3 Analytic Extension, Exponential Time Case

The next step is to establish a relationship between the (analytic extension of the) characteristic function of the supremum, resp. infimum, at a *fixed* time and at an *independent* and *exponentially distributed* time. Independent exponential times play a fundamental role in the fluctuation theory of Lévy processes, since they enjoy a property similar to infinity: the time left after an exponential time is again exponentially distributed.

Let θ denote an exponentially distributed random variable with parameter $q > 0$, independent of the Lévy process L. We denote by $\overline{L}_\theta$, resp. $\underline{L}_\theta$, the supremum, resp. infimum, process of L sampled at θ, that is,

$$\overline{L}_\theta = \sup_{0 \le u \le \theta} L_u \quad \text{and} \quad \underline{L}_\theta = \inf_{0 \le u \le \theta} L_u.$$

Lemma 8.7 *Let $L = (L_t)_{0 \le t \le T}$ be a Lévy process that satisfies assumption* ($\mathbb{EM}$), *and let $\theta \sim \operatorname{Exp}(q)$ be independent of the process L.*

If $q > \overline{\alpha}(M) \vee \underline{\alpha}(M)$, then the characteristic function $\varphi_{\overline{L}_\theta}$ of $\overline{L}_\theta$ possesses a continuous extension

$$\varphi_{\overline{L}_\theta}(z) = \int_{\mathbb{R}} e^{izx} P_{\overline{L}_\theta}(dx) = q \int_0^\infty e^{-qt} E\big[e^{iz\overline{L}_t}\big] dt \tag{8.9}$$

to the half-plane $z \in \{z \in \mathbb{C} : -M \le \Im z\}$ that is analytic in the interior of the half-plane $\{z \in \mathbb{C} : -M < \Im z\}$.

If $q > \overline{\alpha}(M) \vee \underline{\alpha}(M)$, then the characteristic function $\varphi_{\underline{L}_\theta}$ of $\underline{L}_\theta$ possesses a continuous extension

$$\varphi_{\underline{L}_\theta}(z) = \int_{\mathbb{R}} e^{izx} P_{\underline{L}_\theta}(dx) = q \int_0^\infty e^{-qt} E\big[e^{iz\underline{L}_t}\big] dt \tag{8.10}$$

to the half-plane $z \in \{z \in \mathbb{C} : \Im z \le M\}$ *that is analytic in the interior of the half-plane* $\{z \in \mathbb{C} : \Im z < M\}$.

Proof We have that

$$E\big[e^{u\overline{L}_\theta}\big] = \int_0^\infty \int_0^\infty e^{ux} q e^{-qt} P_{\overline{L}_t}(dx)\,dt = \int_0^\infty E\big[e^{u\overline{L}_t}\big] q e^{-qt}\,dt,$$

and, for $q > \overline{\alpha}(M) \vee \underline{\alpha}(M)$, by Lemma 8.4 we get

$$\int_0^\infty E\big[e^{M\overline{L}_t}\big] q e^{-qt}\,dt \le 8\bigg(q\int_0^\infty e^{-t[q-\overline{\alpha}(M)]}\,dt + q\int_0^\infty e^{-t[q-\underline{\alpha}(M)]}\,dt\bigg) < \infty;$$

hence, for $u \le M$, we have

$$E\big[e^{u\overline{L}_\theta}\big] \le E\big[e^{M\overline{L}_\theta}\big] < \infty \quad \big(q > \overline{\alpha}(M) \vee \underline{\alpha}(M)\big). \tag{8.11}$$

Inequality (8.11), together with Lemma 8.2, implies that the characteristic function $\varphi_{\overline{L}_\theta}$ has a continuous extension to the half-plane $\{z \in \mathbb{C} : -M \le \Im z\}$ that is analytic in $\{z \in \mathbb{C} : -M < \Im z\}$ and is given by

$$\varphi_{\overline{L}_\theta}(z) = E\big[e^{iz\overline{L}_\theta}\big]$$

for every $z \in \mathbb{C}$ with $\Im z \ge -M$. Furthermore, Fubini's theorem yields

$$E\big[e^{iz\overline{L}_\theta}\big] = \int_0^\infty \int_0^\infty e^{izx} q e^{-qt} P_{\overline{L}_t}(dx)\,dt = q\int_0^\infty e^{-qt} E\big[e^{iz\overline{L}_t}\big]\,dt.$$

The application of Fubini's theorem is justified since, for $\Im z \ge -M$ and $q > \overline{\alpha}(M) \vee \underline{\alpha}(M)$, we have

$$E\big[\big|e^{iz\overline{L}_\theta}\big|\big] = E\big[e^{-\Im(z)\overline{L}_\theta}\big] \le E\big[e^{M\overline{L}_\theta}\big] < \infty$$

by inequality (8.11). Similarly, we prove the assertion for the infimum. □

8.3 The Wiener–Hopf Factorization

We first provide a statement and brief description of the Wiener–Hopf factorization of a Lévy process and then show that the Wiener–Hopf factorization holds for the analytically extended characteristic functions. Next, we invert the Wiener–Hopf factorization and derive an expression for the (analytically extended) characteristic function of the supremum, resp. infimum, of a Lévy process in terms of the Wiener–Hopf factors.

8.3.1 Analyticity

Fluctuation identities for Lévy processes originate from analogous results for random walks, first derived using combinatorial methods, see, e.g., [48] or [22]. Bingham [10] used this discrete-time skeleton to prove results for Lévy processes; the same approach is followed in the book of Sato [44]. Greenwood and Pitman [23, 24] proved these results for random walks and Lévy processes using excursion theory; see also the books of Bertoin [9] and Kyprianou [33].

The *Wiener–Hopf factorization*[1] serves as a common reference to a multitude of statements in the fluctuation theory for Lévy processes, regarding the distributional decomposition of the excursions of a Lévy process sampled at an independent and exponentially distributed time. The following statement relates the characteristic function of the supremum, the infimum, and the Lévy process itself. Let L be a Lévy process and θ an independent, exponentially distributed time with parameter q; then we have that

$$E\big[e^{izL_\theta}\big] = E\big[e^{iz\overline{L}_\theta}\big]E\big[e^{iz\underline{L}_\theta}\big]$$

or equivalently,

$$\frac{q}{q-\kappa(iz)} = \varphi_q^+(z)\varphi_q^-(z), \qquad z \in \mathbb{R};$$

here κ denotes the cumulant generating function of L_1, cf. (8.2), and φ_q^+, φ_q^- denote the so-called Wiener–Hopf factors.

In the sequel, we will make use of the Wiener–Hopf factorization as stated in the beautiful book of Kyprianou [33] and prove the analytic extension of the Wiener–Hopf factors to the open half-plane $\{z \in \mathbb{C} : \Im z > -M\}$.

Recall the definitions of (8.3) and (8.4), and let us denote

$$\alpha^*(M) := \max\big\{\overline{\alpha}(M), \underline{\alpha}(M)\big\}.$$

Theorem 8.8 (Wiener–Hopf factorization) *Let L be a Lévy process that satisfies assumption* ($\mathbb{EM}$) *(and is not a compound Poisson process). The Laplace transform of $\overline{L}_\theta$, resp. $\underline{L}_\theta$, at an independent and exponentially distributed time θ, $\theta \sim \mathrm{Exp}(q)$, with $q > \alpha^*(M)$, can be identified from the Wiener–Hopf factorization of L via*

$$E\big[e^{-\beta\overline{L}_\theta}\big] = \int_0^\infty qE\big[e^{-\beta\overline{L}_t}\big]e^{-qt}\,dt = \frac{\overline{\kappa}(q,0)}{\overline{\kappa}(q,\beta)} \tag{8.12}$$

and

$$E\big[e^{\beta\underline{L}_\theta}\big] = \int_0^\infty qE\big[e^{\beta\underline{L}_t}\big]e^{-qt}\,dt = \frac{\underline{\kappa}(q,0)}{\underline{\kappa}(q,\beta)} \tag{8.13}$$

---

[1]The historical reasons leading to the adoption of the terminology "Wiener–Hopf" are outlined in Sect. 6.6 in [33].

for $\beta \in \{\beta \in \mathbb{C} : \Re(\beta) > -M\}$. The Laplace exponent of the ascending, resp. descending, ladder process $\overline{\kappa}(\alpha, \beta)$, resp. $\underline{\kappa}(\alpha, \beta)$, for $\alpha \geq \alpha^*(M)$ and $\overline{k}, \underline{k} > 0$, has an analytic extension to $\beta \in \{\beta \in \mathbb{C} : \Re(\beta) > -M\}$ and is given by

$$\overline{\kappa}(\alpha, \beta) = \overline{k} \exp\left(\int_0^\infty \int_{(0,\infty)} \left(e^{-t} - e^{-\alpha t - \beta x}\right) \frac{1}{t} P_{L_t}(dx)\, dt\right) \tag{8.14}$$

and

$$\underline{\kappa}(\alpha, \beta) = \underline{k} \exp\left(\int_0^\infty \int_{(-\infty,0)} \left(e^{-t} - e^{-\alpha t + \beta x}\right) \frac{1}{t} P_{L_t}(dx)\, dt\right). \tag{8.15}$$

Remark 8.9 Note that the Wiener–Hopf factors φ_q^+ and φ_q^- are related to the Laplace exponents of the ascending and descending ladder process $\overline{\kappa}$ and $\underline{\kappa}$ via

$$\varphi_q^+(i\beta) = \frac{\overline{\kappa}(q, 0)}{\overline{\kappa}(q, \beta)} \quad \text{and} \quad \varphi_q^-(-i\beta) = \frac{\underline{\kappa}(q, 0)}{\underline{\kappa}(q, \beta)}. \tag{8.16}$$

We will prepare the proof of this theorem with an intermediate lemma. Let us denote the positive part by $a_+ := \max\{a, 0\}$.

Lemma 8.10 *Let L be a Lévy process that satisfies assumption ($\mathbb{EM}$). For $q > \kappa(M)_+$, the maps*

$$z \mapsto \int_0^\infty \int_{(0,\infty)} \left(1 - e^{izx}\right) P_{L_t}(dx) \frac{e^{-qt}}{t}\, dt \tag{8.17}$$

and

$$z \mapsto \int_0^\infty \int_{(0,\infty)} \left(e^{-t} - e^{-qt + izx}\right) P_{L_t}(dx) \frac{1}{t}\, dt \tag{8.18}$$

are well defined and analytic in the open half plane $\{z \in \mathbb{C} : \Im(z) > -M\}$.

Proof We will show that for every compact subset $K \subset \{z \in \mathbb{C} : \Im(z) > -M\}$, there is a constant $C = C(K) > 0$ such that

$$\int_0^\infty \int_{(0,\infty)} \left|e^{izx} - 1\right| P_{L_t}(dx) \frac{e^{-qt}}{t}\, dt < C(K) \tag{8.19}$$

for every $z \in K$. Then, applying Lebesgue's dominated convergence theorem yields the continuity of the function

$$z \mapsto \int_0^\infty \int_{(0,\infty)} \left(e^{izx} - 1\right) P_{L_t}(dx) \frac{e^{-qt}}{t}\, dt$$

inside the half-plane $\{z \in \mathbb{C} : \Im(z) > -M\}$. Moreover, let γ be an arbitrary triangle inside $\{z \in \mathbb{C} : \Im(z) > -M\}$; the theorems of Fubini and Cauchy yield

$$\int_{\partial\gamma} \int_0^\infty \int_{(0,\infty)} (e^{izx} - 1) P_{L_t}(dx) \frac{e^{-qt}}{t} \, dt \, dz$$
$$= \int_0^\infty \int_{(0,\infty)} \int_{\partial\gamma} (e^{izx} - 1) \, dz P_{L_t}(dx) \frac{e^{-qt}}{t} \, dt = 0. \tag{8.20}$$

Hence, applying Morera's theorem yields the analyticity of (8.17) in the open half-plane $\{z \in \mathbb{C} : \Im(z) > -M\}$.

The assertion for the second map immediately follows from the identity

$$(e^{-t} - e^{-qt+izx})t^{-1} = (1 - e^{izx})e^{-qt}t^{-1} + (e^{-t} - e^{-qt})t^{-1}$$

and the integrability of the second part, since

$$\int_\epsilon^\infty |e^{-t} - e^{-qt}| t^{-1} \, dt < \infty$$

and

$$\int_0^\epsilon |e^{-t} - e^{-qt}| t^{-1} \, dt = \int_0^\epsilon |e^{t(q-1)} - 1| e^{-qt} t^{-1} \, dt \le C|q-1| \int_0^\epsilon e^{-qt} \, dt < \infty,$$

with $C > 1$, for $\epsilon > 0$ small enough.

To show estimate (8.19), we choose a constant $k = k(K) > 0$, depending only on the compact set K, such that $|z| < k$ for every $z \in K$, and we write

$$\int_{(0,\infty)} |e^{izx} - 1| P_{L_t}(dx)$$
$$= \int_{(0,1/k]} |e^{izx} - 1| P_{L_t}(dx) + \int_{(1/k,\infty)} |e^{izx} - 1| P_{L_t}(dx)$$
$$\le \int_{(0,1/k]} |zx| P_{L_t}(dx) + \int_{(1/k,\infty)} |e^{izx}| P_{L_t}(dx) + \int_{(1/k,\infty)} P_{L_t}(dx). \tag{8.21}$$

Using inequality (30.13) of Lemma 30.3 in [44], we can deduce

$$\int_{(0,1/k]} |zx| P_{L_t}(dx) \le k \int_{(0,1/k]} |x| P_{L_t}(dx)$$
$$\le kE\big[|L_t| 1_{\{|L_t| \le 1/k\}}\big] \le C_1(K) t^{1/2}$$

with a constant $C_1(K)$ that depends only on the compact set K. Similarly, using inequality (30.10) in [44], we can estimate the last term of (8.21)

$$\int_{(1/k,\infty)} P_{L_t}(dx) = P\big(\{L_t > 1/k\}\big) \le P\big(\{|L_t| > 1/k\}\big) \le C_2(K) t$$

with a constant $C_2(K)$ that depends only on the compact set K. In order to estimate the second term of inequality (8.21), let us note that we may choose $\epsilon > 0$ small enough such that for every $z \in K$, we have $-\Im(z) < M' < M$ with $M' := M(1-\epsilon)$, and we get

$$\int_{(1/k,\infty)} \left|e^{izx}\right| P_{L_t}(dx) \le E\left[e^{M'L_t} 1_{\{|L_t|>1/k\}}\right].$$

Applying Hölder's inequality with $p := \frac{1}{1-\epsilon}$ and $q := \frac{1}{\epsilon}$, together with Lemma 30.3 in [44], yields

$$\begin{aligned} E\left[e^{M'L_t} 1_{\{|L_t|>1/k\}}\right] &\le \left(E\left[e^{pM'L_t}\right]\right)^{1/p} \left(P\left(\{|L_t| > 1/k\}\right)\right)^{1/q} \\ &\le C_3(K) t^{\epsilon} e^{(1-\epsilon)\kappa(M)t}. \end{aligned}$$

Altogether, we have

$$\int_{(0,\infty)} \left|e^{izx} - 1\right| P_{L_t}(dx) \le C_1(K) t^{1/2} + C_2(K) t + C_3(K) t^{\epsilon} e^{(1-\epsilon)\kappa(M)t}$$

with positive constants $C_1(K), C_2(K)$, and $C_3(K)$ that only depend on the compact set K. As $q > (1-\epsilon)(\kappa(M))_+$, we can conclude (8.19), which completes the proof. □

Proof of Theorem 8.8 For $\beta \in \mathbb{C}$ with $\Re\beta \ge 0$, the assertion follows directly from Theorem 6.16(ii) and (iii) in [33].

From Lemma 8.7 we know that for $q > \alpha^*(M)$, the function

$$\beta \mapsto \varphi_{\overline{L}_\theta}(i\beta) = E\left[e^{-\beta \overline{L}_\theta}\right]$$

has an analytic extension to the half-plane

$$\left\{\beta \in \mathbb{C} : \Re(\beta) > -M\right\},$$

whereas Lemma 8.10 yields that if $q > \alpha^*(M)$, the mapping

$$\beta \mapsto \frac{\overline{\kappa}(q,0)}{\overline{\kappa}(q,\beta)}$$

has an analytic extension to the half-plane

$$\left\{\beta \in \mathbb{C} : \Re(\beta) > -M\right\},$$

while identity (8.14) still holds for this extension. The identity theorem for holomorphic functions yields that (8.12) holds for every $\{\beta \in \mathbb{C} : \Re(\beta) > -M\}$ if $q > \alpha^*(M)$. The proof for (8.13) and (8.15) follows along the same lines. □

Remark 8.11 Note that, by analogous arguments, we can prove that the Laplace exponent of the ascending, resp. descending, ladder process $\overline{\kappa}(\alpha,\beta)$, resp. $\underline{\kappa}(\alpha,\beta)$, has an analytic extension to $\alpha \in \{\alpha \in \mathbb{C} : \Re(\alpha) > \alpha^*(M)\}$, which is given by (8.14), resp. (8.15).

8.3.2 Inversion

The next step is to invert the Laplace transform in the Wiener–Hopf factorization in order to recover the characteristic function of $\overline{L}_t$, at a *fixed* time t. Although the Wiener–Hopf factorization and the characteristic function of $\overline{L}_\theta$ are discussed in several textbooks, let us mention that the extended characteristic function of $\overline{L}_t$ at a fixed time has not been studied in the literature before.

The main result is Theorem 8.13, which will make use of the following auxiliary lemma.

Lemma 8.12 *The maps* $t \mapsto E[e^{-\beta \overline{L}_t}]$ *and* $t \mapsto E[e^{\beta \underline{L}_t}]$ *are continuous for all* $\beta \in \mathbb{C}$ *with* $\Re\beta \in [-M, \infty)$.

Proof Since the Lévy process L is right continuous and stochastically continuous, and $\overline{L}$ is an increasing process, we get that $\overline{L}_s \nearrow \overline{L}_t$ a.s. as $s \to t$.

As $\overline{L}_s \geq 0$, we have

$$\left|e^{-\beta \overline{L}_s}\right| = e^{-\Re(\beta)\overline{L}_s} \leq e^{M\overline{L}_s} \leq e^{M\overline{L}_t},$$

and we may apply the dominated convergence theorem to get

$$E\left[e^{-\beta \overline{L}_s}\right] \to E\left[e^{-\beta \overline{L}_t}\right] \quad \text{as } s \to t$$

for every $\beta \in \mathbb{C}$ with $\Re(\beta) \geq -M$. Analogously, taking into account that $|e^{\beta \underline{L}_s}| \leq e^{-M\underline{L}_s}$ for $\Re\beta \geq -M$, the dominated convergence theorem yields the continuity of the second map. □

Theorem 8.13 *Let L be a Lévy process that satisfies assumption* ($\mathbb{EM}$) (*and is not a compound Poisson process*). *The Laplace transforms of* $\overline{L}_t$ *and* $\underline{L}_t$ *at a fixed time* $t \in [0, T]$ *are given by*

$$E\left[e^{-\beta \overline{L}_t}\right] = \lim_{A\to\infty} \frac{1}{2\pi} \int_{-A}^{A} \frac{e^{t(Y+iv)}}{Y+iv} \frac{\overline{\kappa}(Y+iv, 0)}{\overline{\kappa}(Y+iv, \beta)} dv \tag{8.22}$$

and

$$E\left[e^{\beta \underline{L}_t}\right] = \lim_{A\to\infty} \frac{1}{2\pi} \int_{-A}^{A} \frac{e^{t(\widetilde{Y}+iv)}}{\widetilde{Y}+iv} \frac{\underline{\kappa}(\widetilde{Y}+iv, 0)}{\underline{\kappa}(\widetilde{Y}+iv, -\beta)} dv \tag{8.23}$$

for $\beta \in \mathbb{C}$ *with* $\Re\beta \in (-M, \infty)$ *and* $Y, \widetilde{Y} > \alpha^*(M)$.

Proof Theorem 8.8, together with (8.12), immediately yields

$$\int_0^\infty e^{-qt} E\left[e^{-\beta \overline{L}_t}\right] dt = \frac{1}{q} \frac{\overline{\kappa}(q, 0)}{\overline{\kappa}(q, \beta)} \tag{8.24}$$

for $\beta \in \mathbb{C}$ with $\Re(\beta) > -M$ and $q > \alpha^*(M)$.

In order to deduce that we can invert this Laplace transform, we want to verify the assumptions of Satz 4.4.3 in [15] for the real and imaginary parts of $t \mapsto E[e^{-\beta \overline{L}_t}]$. From the proof of Lemma 8.7 we get that

$$\int_0^\infty e^{-qt} \left| E\left[e^{-\beta \overline{L}_t}\right]\right| dt \le \int_0^\infty e^{-qt} E\left[e^{-\Re(\beta)\overline{L}_t}\right] dt < \infty;$$

this yields the required integrability, i.e., the absolute convergence of

$$\int_0^\infty e^{-qt} \left|\Im\left(E\left[e^{\beta \overline{L}_t}\right]\right)\right| dt \quad \text{and} \quad \int_0^\infty e^{-qt} \left|\Re\left(E\left[e^{\beta \overline{L}_t}\right]\right)\right| dt$$

for $q > \alpha^*(M)$. Further, the real and imaginary parts of $t \mapsto E[e^{-\beta \overline{L}_t}]$ are of bounded variation for $\beta \in \mathbb{C}$ with $\Re\beta \in (-M, \infty)$.

Let us verify this assertion for the imaginary part, for $-M < \Re(\beta) \le 0$ and $\Im(\beta) \le 0$. We have that

$$\Im\left(E\left[e^{-\beta \overline{L}_t}\right]\right) = i E\left[\sin\left(-\Im(\beta)\overline{L}_t\right) e^{-\Re(\beta)\overline{L}_t}\right].$$

We can decompose $\sin(x) = f(x) - g(x)$, where f and g are increasing functions with $f(0) = g(0) = 0$, and $|f(x)| \le x$ and $|g(x)| \le x$. It follows that

$$\sin\left(-\Im(\beta)\overline{L}_t\right)e^{-\Re(\beta)\overline{L}_t} = f\left(-\Im(\beta)\overline{L}_t\right)e^{-\Re(\beta)\overline{L}_t} - g\left(-\Im(\beta)\overline{L}_t\right)e^{-\Re(\beta)\overline{L}_t},$$

where both terms are increasing in time and are integrable, since

$$\begin{aligned} E\left[\left|h\left(-\Im(\beta)\overline{L}_t\right)e^{-\Re(\beta)\overline{L}_t}\right|\right] &\le \left|\Im(\beta)\right| E\left[\left|\overline{L}_t\right| e^{-\Re(\beta)\overline{L}_t}\right] \\ &\le \text{const} \cdot E\left[e^{M\overline{L}_t}\right] < \infty \end{aligned}$$

for $h = g$ and $h = f$. The assertion for the other parts follows similarly.

Now, using the continuity of the map $t \mapsto E[e^{-\beta \overline{L}_t}]$ (Lemma 8.12), we may apply Satz 4.4.3 in [15], to invert this Laplace transform, that is, to conclude that

$$\begin{aligned} E\left[e^{-\beta \overline{L}_t}\right] &= (\text{p.v.})\frac{1}{2\pi i}\int_{Y-i\infty}^{Y+i\infty} \frac{e^{tz}}{z}\frac{\overline{\kappa}(z,0)}{\overline{\kappa}(z,\beta)}\,dz \\ &= \lim_{A\to\infty}\frac{1}{2\pi}\int_{-A}^{A} \frac{e^{t(Y+iv)}}{Y+iv}\frac{\overline{\kappa}(Y+iv,0)}{\overline{\kappa}(Y+iv,\beta)}\,dv \end{aligned} \tag{8.25}$$

for all $\beta \in \mathbb{C}$ with $\Re\beta \in (-M, \infty)$ and for every $Y > \alpha^*(M)$. The proof for the infimum follows along the same lines. □

8.4 Lévy Processes: Examples and Properties

We first state some conditions for the continuity of the law of a Lévy process and the continuity of the law of the supremum of a Lévy process. Then, we describe the most popular Lévy models for financial applications and comment on their path and moment properties which are relevant for the application of Fourier transform valuation formulas.

8.4.1 Continuity Properties

The valuation theorem for discontinuous payoff functions (Theorem 2.7 in EGP) and the analysis of the properties of discontinuous payoff functions (Examples 5.2, 5.3, and 5.4 in EGP) show that if the measure of the underlying random variable does not have atoms, then the valuation formula is valid as a pointwise limit. Thus, we present sufficient conditions for the continuity of the law of a Lévy process and its supremum, and discuss these conditions for certain popular examples.

Statement 8.14 *Let L be a Lévy process with triplet (b, c, λ). Then, Theorem* 27.4 *in* [44] *yields that the law P_{L_t}, $t \in [0, T]$, is atomless iff L is a process of infinite variation or infinite activity or, in other words, if one of the following conditions holds*:

(a) $c \neq 0$ *or* $\int_{\{|x|\leq 1\}} |x|\lambda(dx) = \infty$;
(b) $c = 0$, $\lambda(\mathbb{R}) = \infty$, *and* $\int_{\{|x|\leq 1\}} |x|\lambda(dx) < \infty$.

Statement 8.15 *Let L be a Lévy process and assume that*

(a) *L has infinite variation, or*
(b) *L has infinite activity and is* regular upward. *Regular upward means that* $P(\tau_0 = 0) = 1$, *where* $\tau_0 := \inf\{t > 0 : L_t(\omega) > 0\}$.

Then, Lemma 49.3 *in* [44] *yields that $\overline{L}_t$ has a continuous distribution for every $t \in [0, T]$. The statement for the infimum of a Lévy process is analogous.*

8.4.2 Examples

Next, we describe the most popular Lévy processes for applications in mathematical finance, namely the generalized hyperbolic (GH) process, the CGMY process, and the Meixner process. We present their characteristic functions—which are essential for the application of Fourier transform methods for option pricing—and the corresponding domain of definition. We also discuss their path properties which are relevant for option pricing. For an interesting survey on the path properties of Lévy processes, we refer to [34].

Example 8.16 (GH model) Let $H = (H_t)_{0\leq t\leq T}$ be a generalized hyperbolic process with $\mathcal{L}(H_1) = \mathrm{GH}(\lambda, \alpha, \beta, \delta, \mu)$, see [16, p. 321] or [19]. The characteristic function of H_1 is

$$\varphi_{H_1}(u) = e^{iu\mu}\left(\frac{\alpha^2-\beta^2}{\alpha^2-(\beta+iu)^2}\right)^{\frac{\lambda}{2}} \frac{K_\lambda(\delta\sqrt{\alpha^2-(\beta+iu)^2})}{K_\lambda(\delta\sqrt{\alpha^2-\beta^2})}, \tag{8.26}$$

where K_λ denotes the Bessel function of the third kind with index λ (see [1]); the moment generating function exists for $u \in (-\alpha-\beta, \alpha-\beta)$. The sample paths of a generalized hyperbolic Lévy process have infinite variation. Thus, by Statements 8.14 and 8.15, we can deduce that the laws of both a GH Lévy process and its supremum do not have atoms.

The class of generalized hyperbolic distributions is not closed under convolution, and hence the distribution of H_t is no longer a generalized hyperbolic one. Nevertheless, the characteristic function of $\mathcal{L}(H_t)$ is given explicitly by

$$\varphi_{H_t}(u) = \left(\varphi_{H_1}(u)\right)^t.$$

A class closed under certain convolutions is the class of normal inverse Gaussian distributions, where $\lambda = -\frac{1}{2}$; see [7]. In that case, $\mathcal{L}(H_t) = \mathrm{NIG}(\alpha, \beta, \delta t, \mu t)$, and the characteristic function resumes the form

$$\varphi_{H_t}(u) = e^{iu\mu t}\frac{\exp(\delta t\sqrt{\alpha^2-\beta^2})}{\exp(\delta t\sqrt{\alpha^2-(\beta+iu)^2})}. \tag{8.27}$$

Another interesting subclass is given by the hyperbolic distributions which arise for $\lambda = 1$; the hyperbolic model has been introduced to finance by Eberlein and Keller [17].

Example 8.17 (CGMY model) Let $H = (H_t)_{0\le t\le T}$ be a CGMY Lévy process, see [13]; another name for this process is (generalized) tempered stable process (see, e.g., [14]). The Lévy measure of this process has the form

$$\lambda^{\mathrm{CGMY}}(dx) = C\frac{e^{-Mx}}{x^{1+Y}}1_{\{x>0\}}\,dx + C\frac{e^{Gx}}{|x|^{1+Y}}1_{\{x<0\}}\,dx,$$

where the parameter space is $C, G, M > 0$ and $Y \in (-\infty, 2)$. Moreover, the characteristic function of H_t, $t \in [0, T]$, is

$$\varphi_{H_t}(u) = \exp\bigl(tC\Gamma(-Y)\bigl[(M-iu)^Y + (G+iu)^Y - M^Y - G^Y\bigr]\bigr) \tag{8.28}$$

for $Y \neq 0$, and the moment generating function exists for $u \in [-G, M]$.

The sample paths of the CGMY process have unbounded variation if $Y \in [1, 2)$, bounded variation if $Y \in (0, 1)$, and are of compound Poisson type if $Y < 0$. Moreover, the CGMY process is regular upward if $Y > 0$; see [34]. Hence, by Statements 8.14 and 8.15, the laws of a CGMY Lévy process and its supremum do not have atoms if $Y \in (0, 2)$.

The CGMY process contains the Variance Gamma process (see [41]) as a subclass for $Y = 0$. The characteristic function of H_t, $t \in [0, T]$, is

$$\varphi_{H_t}(u) = \exp\left(tC\left[-\log\left(1 - \frac{iu}{M}\right) - \log\left(1 + \frac{iu}{G}\right)\right]\right), \tag{8.29}$$

and the moment generating function exists for $u \in [-G, M]$. The paths of the Variance Gamma process have bounded variation, infinite activity, and are regular upward. Thus, the laws of a VG Lévy process and its supremum do not have atoms.

Example 8.18 (Meixner model) Let $H = (H_t)_{0 \le t \le T}$ be a Meixner process with $\mathcal{L}(H_1) = \text{Meixner}(\alpha, \beta, \delta)$, $\alpha > 0$, $-\pi < \beta < \pi$, $\delta > 0$, see [47] and [45]. The characteristic function of H_t, $t \in [0, T]$, is

$$\varphi_{H_t}(u) = \left(\frac{\cos\frac{\beta}{2}}{\cosh\frac{\alpha u - i\beta}{2}}\right)^{2\delta t}, \tag{8.30}$$

and the moment generating function exists for $u \in (\frac{\beta - \pi}{\alpha}, \frac{\beta + \pi}{\alpha})$. The paths of a Meixner process have infinite variation. Hence, the laws of a Meixner Lévy process and its supremum do not have atoms.

8.5 Applications in Finance

In this section, we derive valuation formulas for lookback options, one-touch options, and equity default swaps in models driven by Lévy processes. We combine the results on the Wiener–Hopf factorization and the characteristic function of the supremum of a Lévy process from this paper, with the results on Fourier transform valuation formulas derived in EGP. Note that the results presented in the sequel are valid for all the examples discussed in Sect. 8.4.

We model the price process of a financial asset $S = (S_t)_{0 \le t \le T}$ as an exponential Lévy process, i.e., a stochastic process with representation

$$S_t = S_0 e^{L_t}, \quad 0 \le t \le T \tag{8.31}$$

(shortly: $S = S_0 e^L$). Every Lévy process L, subject to Assumption ($\mathbb{EM}$), has the canonical decomposition

$$L_t = bt + \sqrt{c} W_t + \int_0^t \int_{\mathbb{R}} x(\mu - \nu)(ds, dx), \tag{8.32}$$

where $W = (W_t)_{0 \le t \le T}$ denotes a P-standard Brownian motion, and μ denotes the random measure associated with the jumps of L; see [27, Chap. II].

Let $\mathcal{M}(P)$ denote the class of martingales on the stochastic basis $\mathscr{B}$. The martingale condition for an asset S is

$$S = S_0 e^L \in \mathcal{M}(P) \quad \Longleftrightarrow \quad b + \frac{c}{2} + \int_{\mathbb{R}} (e^x - 1 - x)\lambda(dx) = 0; \tag{8.33}$$

see [20] for the details. That is, throughout the rest of this paper, we will assume that P is a *martingale measure* for S.

8.5.1 Lookback Options

The results of Sect. 8.3 on the characteristic function of the supremum of a Lévy process allow us to price lookback options in models driven by Lévy processes using Fourier methods. Excluded are only compound Poisson processes. Assuming that the asset price evolves as an exponential Lévy process, a fixed strike lookback call option with payoff

$$(\overline{S}_T - K)^+ = \left(S_0 e^{\overline{L}_T} - K\right)^+ \tag{8.34}$$

can be viewed as a call option where the driving process is the *supremum* of the underlying Lévy processes L. Therefore, the price of a lookback call option is provided by the following result.

Theorem 8.19 *Let L be a Lévy process that satisfies Assumption* ($\mathbb{EM}$). *The price of a fixed strike lookback call option with payoff* (8.34) *is given by*

$$\mathbb{C}_T(\overline{S}; K) = \frac{1}{2\pi} \int_{\mathbb{R}} S_0^{R-iu} \varphi_{\overline{L}_T}(-u - iR) \frac{K^{1+iu-R}}{(iu - R)(1 + iu - R)} du, \tag{8.35}$$

where

$$\varphi_{\overline{L}_T}(-u - iR) = \lim_{A \to \infty} \frac{1}{2\pi} \int_{-A}^{A} \frac{e^{T(Y+iv)}}{Y + iv} \frac{\overline{\kappa}(Y + iv, 0)}{\overline{\kappa}(Y + iv, iu - R)} dv \tag{8.36}$$

for $R \in (1, M)$ and $Y > \alpha^*(M)$.

Proof We aim at applying Theorem 2.2 in EGP; hence we must check if conditions (C1)–(C3) (of EGP) are satisfied. Assumption ($\mathbb{EM}$), coupled with Corollary 8.5, yields that $M_{\overline{L}_T}(R)$ exists for $R \in (-\infty, M)$, and hence condition (C2) is satisfied. Now, the Fourier transform of the payoff function $f(x) = (e^x - K)^+$ is

$$\widehat{f}(u + iR) = \frac{K^{1+iu-R}}{(iu - R)(1 + iu - R)},$$

and conditions (C1) and (C3) are satisfied for $R \in (1, \infty)$; cf. Example 5.1 in EGP. Further, the extended characteristic function $\varphi_{\overline{L}_T}$ of $\overline{L}_T$ is provided by Theorem 8.13 and equals (8.36) for $R \in (-\infty, M)$ and $Y > \alpha^*(M)$. Finally, Theorem 2.2 in EGP delivers the asserted valuation formula (8.35). □

Remark 8.20 Completely analogous formulas can be derived for the fixed strike lookback put option with payoff $(K - \underline{S}_T)^+$ using the results for the infimum of a Lévy process. Moreover, floating strike lookback options can be treated by the same formulas making use of the duality relationships proved in [18] and [20].

8.5.2 One-Touch Options

Analogously, we can derive valuation formulas for one-touch options in assets driven by Lévy processes using Fourier transform methods; here, the exceptions are compound Poisson processes and nonregular upward, finite variation, Lévy processes. Assuming that the asset price evolves as an exponential Lévy process, a one-touch call option with payoff

$$1_{\{\overline{S}_T > B\}} = 1_{\{\overline{L}_T > \log(\frac{B}{S_0})\}} \tag{8.37}$$

can be valued as a digital call option where the driving process is the supremum of the underlying Lévy process.

Theorem 8.21 *Let L be a Lévy process with infinite variation, or a regular upward process with infinite activity, that satisfies Assumption* ($\mathbb{EM}$). *The price of a one-touch option with payoff* (8.37) *is given by*

$$\begin{aligned}\mathbb{DC}_T(\overline{S}; B) &= \lim_{A\to\infty} \frac{1}{2\pi} \int_{-A}^{A} S_0^{R+iu} \varphi_{\overline{L}_T}(u - iR) \frac{B^{-R-iu}}{R+iu}\, du \\ &= P\big(\overline{L}_T > \log(B/S_0)\big)\end{aligned} \tag{8.38}$$

for $R \in (0, M)$ and $Y > \alpha^*(M)$, where $\varphi_{\overline{L}_T}$ is given by (8.36).

Proof We will apply Theorem 2.7 in EGP; hence we must check conditions (D1)–(D2). As in the proof of Theorem 8.19, Assumption ($\mathbb{EM}$) shows that condition (D2) is satisfied for $R \in (-\infty, M)$, while Theorem 8.13 provides the characteristic function of $\overline{L}_T$, given by (8.36). Example 5.2 in EGP yields that the Fourier transform of the payoff function $f(x) = 1_{\{x > \log B\}}$ equals

$$\widehat{f}(iR - u) = \frac{B^{-R-iu}}{R+iu} \tag{8.39}$$

and condition (D1) is satisfied for $R \in (0, \infty)$. In addition, if the measure $P_{\overline{L}_T}$ is atomless, then the valuation function is continuous and has bounded variation. Now, by Statement 8.15, we know that the measure $P_{\overline{L}_T}$ is atomless exactly when L has infinite variation or has infinite activity and is regular upward. Therefore, Theorem 2.7 in EGP applies, and results in the valuation formula (8.38) for the one-touch call option. □

Remark 8.22 Completely analogous valuation formulas can be derived for the digital put option with payoff $1_{\{\underline{S}_T < B\}}$.

Remark 8.23 Summarizing the results of this paper and of EGP, when dealing with *continuous* payoff functions, the valuation formulas can be applied to *all* Lévy processes. When dealing with *discontinuous* payoff functions, then the valuation formulas apply to most Lévy processes *apart* from *compound Poisson* type processes without diffusion component and finite variation Lévy processes which are not *regular upward*. This is true for both non-path-dependent and *path-dependent exotic* options.

Remark 8.24 Arguing analogously to Theorems 8.19 and 8.21, we can derive the price of options with a "general" payoff function $f(\overline{L}_T)$. For example, one could consider payoffs of the form $[(\overline{S}_T - K)^+]^2$ or $\overline{S}_T 1_{\{\overline{S}_T > B\}}$; see [42, Table 3.1] and Example 5.3 in EGP for the corresponding Fourier transforms.

8.5.3 Equity Default Swaps

Equity default swaps were recently introduced in financial markets and offer a link between equity and credit risk. The structure of an equity default swap imitates that of a credit default swap: the protection buyer pays a fixed premium in exchange for an insurance payment in case of "default." In this case "default," also called the "equity event," is defined as the first time the asset price process drops below a fixed barrier, typically 30% or 50% of the initial value S_0.

Let us denote by τ_B the first passage time below the barrier level B, i.e.,

$$\tau_B = \inf\{t \geq 0;\ S_t \leq B\}.$$

The protection buyer pays a fixed premium denoted by $\mathcal{K}$ at the dates $T_1, T_2, \ldots, T_N = T$, provided that default has not occurred, i.e., $T_i < \tau_B$. In case of default, the protection seller makes the insurance payment $\mathcal{C}$, which is typically 50% of the initial value. The premium $\mathcal{K}$ is fixed such that the value of the equity default swap at inception is zero; hence we get

$$\mathcal{K} = \frac{\mathcal{C} E[e^{-r\tau_B} 1_{\{\tau_B \leq T\}}]}{\sum_{i=1}^{N} E[e^{-rT_i} 1_{\{\tau_B > T_i\}}]}, \tag{8.40}$$

where r denotes the risk-free interest rate.

Now, using that $1_{\{\tau_B \leq t\}} = 1_{\{\underline{S}_t \leq B\}}$, which immediately translates into

$$P(\tau_B \leq t) = E[1_{\{\tau_B \leq t\}}] = E[1_{\{\underline{S}_t \leq B\}}], \tag{8.41}$$

and that

$$E\left[e^{-r\tau_B} 1_{\{\tau_B \leq T\}}\right] = \int_0^T e^{-rt} P_{\tau_B}(dt),$$

the quantities in (8.40) can be calculated using the valuation formulas for one-touch options.

References

1. M. Abramowitz, I.A. Stegun (eds.), *Handbook of Mathematical Functions*, 5th edn. (Dover, New York, 1968)
2. L. Alili, A.E. Kyprianou, Some remarks on first passage of Lévy process, the American put and pasting principles. Ann. Appl. Probab. **15**, 2062–2080 (2005)
3. D. Applebaum, *Lévy Processes and Stochastic Calculus* (Cambridge University Press, Cambridge, 2004)
4. S. Asmussen, F. Avram, M.R. Pistorius, Russian and American put options under exponential phase-type Lévy models. Stoch. Process. Appl. **109**, 79–111 (2004)
5. S. Asmussen, D. Madan, M. Pistorius, Pricing equity default swaps under an approximation to the CGMY Lévy model. J. Comput. Finance **11**, 79–93 (2007)
6. F. Avram, A. Kyprianou, M.R. Pistorius, Exit problems for spectrally negative Lévy processes and applications to (Canadized) Russian options. Ann. Appl. Probab. **14**, 215–238 (2004)
7. O.E. Barndorff-Nielsen, Processes of normal inverse Gaussian type. Finance Stoch. **2**, 41–68 (1998)
8. O.E. Barndorff-Nielsen, T. Mikosch, S. Resnick (eds.), *Lévy Processes: Theory and Applications* (Birkhäuser, Basel, 2001)
9. J. Bertoin, *Lévy Processes* (Cambridge University Press, Cambridge, 1996)
10. N.H. Bingham, Fluctuation theory in continuous time. Adv. Appl. Probab. **7**, 705–766 (1975)
11. S.I. Boyarchenko, S.Z. Levendorskiĭ, Barrier options and touch-and-out options under regular Lévy processes of exponential type. Ann. Appl. Probab. **12**, 1261–1298 (2002)
12. S.I. Boyarchenko, S.Z. Levendorskiĭ, *Non-Gaussian Merton–Black–Scholes Theory* (World Scientific, Singapore, 2002)
13. P. Carr, H. Geman, D.B. Madan, M. Yor, The fine structure of asset returns: an empirical investigation. J. Bus. **75**, 305–332 (2002)
14. R. Cont, P. Tankov, *Financial Modelling with Jump Processes* (Chapman and Hall/CRC Press, Boca Raton, 2004)
15. G. Doetsch, *Handbuch der Laplace-transformation* (Birkhäuser, Basel, 1950)
16. E. Eberlein, Application of generalized hyperbolic Lévy motions to finance, in *Lévy Processes: Theory and Applications*, ed. by O.E. Barndorff-Nielsen, T. Mikosch, S.I. Resnick (Birkhäuser, Basel, 2001), pp. 319–336
17. E. Eberlein, U. Keller, Hyperbolic distributions in finance. Bernoulli **1**, 281–299 (1995)
18. E. Eberlein, A. Papapantoleon, Equivalence of floating and fixed strike Asian and lookback options. Stoch. Process. Appl. **115**, 31–40 (2005)
19. E. Eberlein, K. Prause, The generalized hyperbolic model: financial derivatives and risk measures, in *Mathematical Finance—Bachelier Congress 2000*, ed. by H. Geman, D. Madan, S. Pliska, T. Vorst (Springer, Berlin, 2002), pp. 245–267
20. E. Eberlein, A. Papapantoleon, A.N. Shiryaev, On the duality principle in option pricing: semimartingale setting. Finance Stoch. **12**, 265–292 (2008)
21. E. Eberlein, K. Glau, A. Papapantoleon, Analysis of Fourier transform valuation formulas and applications. Appl. Math. Finance **17**, 211–240 (2010)
22. W. Feller, *An Introduction to Probability Theory and Its Applications*, vol. II, 2nd edn. (Wiley, New York, 1971)
23. P. Greenwood, J. Pitman, Fluctuation identities for Lévy processes and splitting at the maximum. Adv. Appl. Probab. **12**, 893–902 (1980)
24. P. Greenwood, J. Pitman, Fluctuation identities for random walk by path decomposition at the maximum. Adv. Appl. Probab. **12**, 291–293 (1980)
25. N. Hilber, N. Reich, C. Schwab, C. Winter, Numerical methods for Lévy processes. Finance Stoch. **13**, 471–500 (2009)
26. F. Hubalek, A.E. Kyprianou, Old and new examples of scale functions for spectrally negative Lévy processes, in *Seminar on Stochastic Analysis, Random Fields and Applications VI*, ed. by R.C. Dalang, M. Dozzi, F. Russo. Progress in Probability, vol. 63 (Birkhäuser, Basel, 2010), pp. 119–145

27. J. Jacod, A.N. Shiryaev, *Limit Theorems for Stochastic Processes*, 2nd edn. (Springer, Berlin, 2003)
28. M. Jeannin, M. Pistorius, A transform approach to compute prices and greeks of barrier options driven by a class of Lévy processes. Quant. Finance **10**, 629–644 (2010)
29. O. Kudryavtsev, S. Levendorskiĭ, Pricing of first touch digitals under normal inverse Gaussian processes. Int. J. Theor. Appl. Finance **9**, 915–949 (2006)
30. O. Kudryavtsev, S. Levendorskiĭ, Fast and accurate pricing of barrier options under Lévy processes. Finance Stoch. **13**, 531–562 (2009)
31. A. Kuznetsov, Wiener–Hopf factorization and distribution of extrema for a family of Lévy processes. Ann. Appl. Probab. **20**, 1801–1830 (2010)
32. A. Kuznetsov, Wiener–Hopf factorization for a family of Lévy processes related to theta functions. J. Appl. Probab. **47**, 1023–1033 (2010)
33. A.E. Kyprianou, *Introductory Lectures on Fluctuations of Lévy Processes with Applications* (Springer, Berlin, 2006)
34. A.E. Kyprianou, R. Loeffen, Lévy processes in finance distinguished by their coarse and fine path properties, in *Exotic Option Pricing and Advanced Lévy Models*, ed. by A. Kyprianou, W. Schoutens, P. Wilmott (Wiley, New York, 2005), pp. 1–28
35. A.E. Kyprianou, M.R. Pistorius, Perpetual options and canadization through fluctuation theory. Ann. Appl. Probab. **13**, 1077–1098 (2003)
36. A.E. Kyprianou, V. Rivero, Special, conjugate and complete scale functions for spectrally negative Lévy processes. Electron. J. Probab. **13**, 1672–1701 (2008)
37. A.E. Kyprianou, B.A. Surya, On the Novikov–Shiryaev optimal stopping problem in continuous time. Electron. Commun. Probab. **10**, 146–154 (2005)
38. A.E. Kyprianou, W. Schoutens, P. Wilmott (eds.), *Exotic Option Pricing and Advanced Lévy Models* (Wiley, New York, 2005)
39. A.E. Kyprianou, V. Rivero, R. Song, Convexity and smoothness of scale functions and de Finetti's control problem. J. Theor. Probab. **23**, 547–564 (2010)
40. S. Levendorskiĭ, O. Kudryavtsev, V. Zherder, The relative efficiency of numerical methods for pricing American options under Lévy processes. J. Comput. Finance **9**, 69–98 (2005)
41. D.B. Madan, E. Seneta, The variance gamma (VG) model for share market returns. J. Bus. **63**, 511–524 (1990)
42. S. Raible, Lévy processes in finance: theory, numerics, and empirical facts. Ph.D. Thesis, Univ. Freiburg (2000)
43. W. Rudin, *Real and Complex Analysis*, 3rd edn. (McGraw-Hill, New York, 1987)
44. K. Sato, *Lévy Processes and Infinitely Divisible Distributions* (Cambridge University Press, Cambridge, 1999)
45. W. Schoutens, The Meixner process: theory and applications in finance, in *Mini-proceedings of the 2nd MaPhySto Conference on Lévy Processes*, ed. by O.E. Barndorff-Nielsen (2002), pp. 237–241
46. W. Schoutens, *Lévy Processes in Finance: Pricing Financial Derivatives* (Wiley, New York, 2003)
47. W. Schoutens, J.L. Teugels, Lévy processes, polynomials and martingales. Commun. Stat. Stoch. Models **14**, 335–349 (1998)
48. F. Spitzer, *Principles of Random Walk* (Van Nostrand, Princeton, 1964)

Chapter 9
Optimal Liquidation of a Pairs Trade

Erik Ekström, Carl Lindberg, and Johan Tysk

Abstract Pairs trading is a common strategy used by hedge funds. When the spread between two highly correlated assets is observed to deviate from historical observations, a long position is taken in the underpriced asset, and a short position in the overpriced one. If the spread narrows, both positions are closed, thus generating a profit. We study when to optimally liquidate a pairs trading strategy when the difference between the two assets is modeled by an Ornstein–Uhlenbeck process. We also provide a sensitivity analysis in the model parameters.

Keywords Pairs trading · Optimal stopping theory · Ornstein–Uhlenbeck process

Mathematics Subject Classification (2010) 91G10 · 60G40

9.1 Introduction

Consider a pair of assets having price processes with a difference fluctuating about a given level. A typical example is stocks of two companies in the same area of business. If the spread between the two price processes at some point widens, then one of the assets is underpriced relative to the other one. An investor wanting to benefit from this relative mispricing may invest in a pairs trade, i.e., the investor buys the (relatively) underpriced asset and takes a short position in the (relatively) overpriced one. When the spread narrows again, the position is liquidated, and a profit is made. Note that the holder of a pairs trade is not exposed to market risk but instead tries to benefit from relative price movements, thus making pairs trade a common hedge fund strategy.

E. Ekström (✉) · J. Tysk
Department of Mathematics, Uppsala University, Box 480, 75106 Uppsala, Sweden
e-mail: ekstrom@math.uu.se

J. Tysk
e-mail: johan.tysk@math.uu.se

C. Lindberg
Mathematical Sciences, Chalmers University of Technology, 41296 Göteborg, Sweden
e-mail: carl.lindberg@chalmers.se

G. Di Nunno, B. Øksendal (eds.), *Advanced Mathematical Methods for Finance*,
DOI 10.1007/978-3-642-18412-3_9, © Springer-Verlag Berlin Heidelberg 2011

The literature on trading strategies used by hedge funds seems to be somewhat limited compared to its practical significance. However, there are a number of recent books that treat the applied aspects of pairs trading, see [1, 5], and [6]; for a historical evaluation of pairs trading, see also [3]. The authors of [2] model pair spreads as mean-reverting Gaussian Markov chains observed in Gaussian noise. Our approach is the continuous time analogue of this since we use mean-reverting Ornstein–Uhlenbeck processes to model the spread. We thus model the difference X between the two assets as

$$dX_t = -\mu X_t \, dt + \sigma \, dW_t,$$

where μ and σ are positive constants, and W is a standard Brownian motion.

Note that there is a large model risk associated to the pairs trading strategy. Indeed, if it turns out that the difference between the assets is no longer mean-reverting, then the investor faces a considerable risk. What is typically done in practice is that the investor decides (in advance) on a stop-loss level $B < 0$, and if the value of the pair trade falls below B, then one liquidates the position and accepts the loss. The stop-loss level B can be seen as a (crude) model adjustment: if this level is reached, then the model is abandoned, and the position is closed. A natural continuation of our work would be to introduce a continuous recalibration of the model parameters to decrease the model risk.

In Sect. 9.2, we formulate and solve explicitly the optimal stopping problem of when to liquidate a pair trade in the presence of a stop-loss barrier. In Sect. 9.3, we study the dependence of the optimal liquidation level on the different model parameters, thus providing a better understanding of the consequences of possible misspecifications of the model. More precisely, we show that increasing the quotient $\alpha = 2\mu/\sigma^2$, the optimal liquidation level increases, and that the optimal liquidation level is between $-B/2$ and $-B$ for any choice of parameters μ and σ. In Sect. 9.4, we consider the optimal liquidation of a pairs trade in the presence of a discount factor. When including such a discount factor, the dependence on the model parameters becomes more delicate, and a numerical study is conducted. Finally, we also consider the optimal liquidation problem in the absence of a stop-loss barrier.

9.2 Solving the Optimal Stopping Problem

If we assume that any fraction of an asset can be traded, then there is no loss of generality to assume that the difference between the two assets fluctuates about the level 0. As explained in the introduction, we model the difference X between the two assets as a mean-reverting Ornstein–Uhlenbeck process, i.e.,

$$dX_t = -\mu X_t \, dt + \sigma \, dW_t. \tag{9.1}$$

Here μ and σ are positive constants, and W is a standard Brownian motion. For a given liquidation level $B < 0$, define the value V of the option spread by

$$V(x) = \sup_{\tau \leq \tau_B} E_x X_\tau, \tag{9.2}$$

where the supremum is taken over all stopping times that are smaller than

$$\tau_B = \inf\{t : X_t \le B\},$$

the first hitting time of the liquidation level B. The stop-loss level B is imposed to have a bound on the possible losses. Of course, if the model (9.1) is known to be true, then the spread would vanish eventually since X has a mean-reverting drift. However, in practice a stop-loss level has to be imposed to account for the risk that the model is incorrect. The stop-loss level B thus makes the risk involved in pairs trading less sensitive to a possible misspecification of the model.

If the process X is negative, then the drift is positive, and so one should not liquidate the position. If X is positive, then the negative drift works against the owner of the pair. For large values of X, this drift is substantial and should outweigh the possible benefits of the random fluctuations. This indicates that there exists a boundary $x = b$ above which liquidation is optimal and below which the pair should be kept.

General optimal stopping theory then suggests that the pair (V, b) solves

$$\begin{cases} \frac{\sigma^2}{2} V_{xx} - \mu x V_x = 0 & \text{if } x \in (B, b), \\ V(x) = x & \text{if } x \in [b, \infty), \\ V'(b) = 1, \\ V(B) = B. \end{cases} \tag{9.3}$$

The general solution to the ordinary differential equation $\frac{\sigma^2}{2} V_{xx} - \mu x V_x = 0$ is

$$V(x) = CF(x) + D.$$

Here C and D are constants,

$$F(x) = \int_0^x f(y)\, dy,$$

$$f(y) = e^{\alpha y^2/2},$$

and $\alpha = 2\mu/\sigma^2$ is the reciprocal of the variance of the stationary distribution of X. Inserting the general solution into the free-boundary problem, the equation

$$\frac{F(b) - F(B)}{b - B} = f(b) \tag{9.4}$$

for the exercise boundary b is derived.

Lemma 9.1 *Equation* (9.4) *admits a unique solution b larger than B. Moreover, $b \in (0, -B)$.*

Proof Define

$$g(x) := F(x) - F(B) - (x - B) f(x) \tag{9.5}$$

and note that $b > B$ is a solution of (9.4) if and only if $g(b) = 0$. We have $g(B) = 0$, $g'(x) = -\alpha(x - B)xf(x) \ge 0$ if $x \in [B, 0]$, and $g'(x) < 0$ if $x > 0$. Moreover,

$$g(-B) = -2F(B) + 2Bf(B) < 0$$

since F is convex. Consequently, g has a unique zero $x = b$ larger than B, and $b \in (0, -B)$. □

Now, given the unique solution b of (9.4), let

$$\hat{V}(x) = \begin{cases} \frac{F(x)}{f(b)} + B - \frac{F(B)}{f(b)}, & B \le x < b, \\ x, & x \ge b. \end{cases} \tag{9.6}$$

It is easy to check that $(\hat{V}, b)$ is the unique solution to the free-boundary problem (9.3). Moreover, it follows from the proof of Lemma 9.1 above that $\hat{V}(x) \ge x$ for all $x \ge B$.

Theorem 9.2 *The value function V coincides with the function $\hat{V}$ given in* (9.6). *Moreover, $\tau^* = \tau_B \wedge \tau_b$ is an optimal stopping time in* (9.2).

Proof Consider the process $Y_t = \hat{V}(X_{t\wedge\tau_B})$. By (a generalized version of) Itô's lemma,

$$\begin{aligned} Y_t &= \hat{V}(x) + \int_0^{t\wedge\tau_B} \left(\frac{\sigma^2}{2}\hat{V}_{xx}(X_s) - \mu X_s \hat{V}_x(X_s)\right) I(X_s \neq b)\, ds \\ &\quad + \int_0^{t\wedge\tau_B} \sigma \hat{V}_x(X_s) I(X_s \neq b)\, dW \\ &= \hat{V}(x) - \mu \int_0^{t\wedge\tau_B} X_s I(X_s > b)\, ds + \int_0^{t\wedge\tau_B} \sigma \hat{V}_x(X_s)\, dW. \end{aligned}$$

The Itô integral is a martingale since the integrand is bounded. Therefore, since b is positive, the process Y is a supermartingale. If τ is a stopping time, then the optional sampling theorem, see Problem 3.16 and Theorem 3.22 in [4], gives that

$$EX_{\tau\wedge\tau_B} \le E\hat{V}(X_{\tau\wedge\tau_B}) = EY_\tau \le EY_0 = \hat{V}(x). \tag{9.7}$$

Since τ is arbitrary, this yields

$$V(x) \le \hat{V}(x).$$

To derive the reverse inequality, note that $Y_{t\wedge\tau_b}$ is a bounded martingale and that $Y_{\tau_b} = X_{\tau_b\wedge\tau_B}$. It follows that the inequalities in (9.7) reduce to equalities if $\tau = \tau^*$, which finishes the proof. □

9.3 Dependence on Parameters

It is easy to see that the value V and the optimal threshold b are both increasing as functions of the absolute value $|B|$ of the stop-loss level. Indeed, this follows since a large $|B|$ increases the set of stopping times smaller than τ_B. The dependence on the parameters μ and σ is more delicate and given by the theorem below.

Theorem 9.3 *The optimal stopping boundary* b *is increasing as a function of* $\alpha = 2\mu/\sigma^2$, *and it satisfies* $\lim_{\alpha\downarrow 0} b(\alpha) = -B/2$ *and* $\lim_{\alpha\uparrow\infty} b(\alpha) = -B$.

Proof Define

$$g(x,\alpha) := \int_B^x e^{\alpha y^2/2}\,dy - (x-B)e^{\alpha x^2/2}$$

(compare to (9.5)). Recall that for a fixed $\alpha > 0$, the function $x \mapsto g(x,\alpha)$ satisfies $g(0,\alpha) > 0$, $g(-B,\alpha) < 0$, and $\frac{\partial g}{\partial x}(x,\alpha) < 0$ for $x > 0$. Moreover, the optimal stopping boundary $b \in (0,-B)$ is the unique positive value such that $g(b,\alpha) = 0$. Expanding into Taylor series, we have

$$\begin{aligned}\int_B^b e^{\alpha y^2/2}\,dy &= \int_B^b \sum_{k=0}^{\infty} \frac{(\alpha/2)^k}{k!} y^{2k}\,dy \\ &= \sum_{k=0}^{\infty} \frac{(\alpha/2)^k}{k!(2k+1)}\left(x^{2k+1} - B^{2k+1}\right) \\ &= \sum_{k=0}^{\infty} \frac{(\alpha/2)^k}{k!(2k+1)} x^{2k}(b-B)a_k,\end{aligned}$$

where

$$a_k = 1 + (B/b) + \cdots + (B/b)^{2k}.$$

Using

$$e^{\alpha b^2/2} = \sum_{k=0}^{\infty} \frac{(\alpha/2)^k b^{2k}}{k!},$$

we find that

$$g(b,\alpha) = (b-B)\sum_{k=1}^{\infty} \frac{(\alpha/2)^k b^{2k}}{k!}\left(\frac{a_k}{2k+1} - 1\right).$$

Note that the function $c(k) := a_k - (2k+1)$ is convex as a function of $k \geq 0$, $c(0) = 0$, and $c(k) \to \infty$ as $k \to \infty$. Consequently, there exists $k_0 \geq 0$ such that $a_k \leq 2k+1$ for $k \leq k_0$ and $a_k > 2k+1$ for $k > k_0$. Now, if $\alpha' > \alpha$, then

$$\begin{aligned}0 = g(b,\alpha) &= (b-B)(\alpha/2)^{k_0} \sum_{k=1}^{\infty} \frac{(\alpha/2)^{k-k_0} b^{2k}}{k!}\left(\frac{a_k}{2k+1} - 1\right) \\ &< (b-B)(\alpha/2)^{k_0} \sum_{k=1}^{\infty} \frac{(\alpha'/2)^{k-k_0} b^{2k}}{k!}\left(\frac{a_k}{2k+1} - 1\right) \leq g(b,\alpha').\end{aligned}$$

It follows that $g(b,\alpha') > 0$, so the unique zero $x = b'$ of $g(x,\alpha')$ satisfies $b' \geq b$, which proves the claimed monotonicity of the optimal stopping boundary b as a function of α.

Finally we consider the limits in the statement of the theorem, starting with α tending to infinity. Recall from above that $b(\alpha) < -B$. On the other hand, for a fixed $\varepsilon > 0$, we have

$$g(-B-\varepsilon,\alpha) > 0$$

for α large enough since the integral term in the definition of g is bounded below by $\frac{\epsilon}{2}e^{\alpha(-B-\epsilon/2)^2/2}$, whereas the remaining term is $O(e^{\alpha(-B-\epsilon)^2/2})$. Hence, $b(\alpha) > -B-\epsilon$, and the desired conclusion follows. Next we consider the limit as α tends to zero. The argument here is based on the approximation of e^x by $1+x$ for small x. Thus, we replace the exponential functions in the definition of g by $1+\alpha y^2/2$ and $1+\alpha x^2/2$, and define

$$\begin{aligned} h(x,\alpha) &:= \int_B^x \left(1+\alpha y^2/2\right) dy - (x-B)\left(1+\alpha x^2/2\right) \\ &= \frac{\alpha}{6}\left(3Bx^2 - 2x^3 - B^3\right). \end{aligned}$$

One finds that $h(-B/2,\alpha) = 0$ for all α. The derivative $\frac{\partial h}{\partial x}(x,\alpha) = \alpha(Bx - x^2)$ is negative and is bounded above and below by positive multiples of α in a neighborhood of $-B/2$. Since the error in the approximation of the exponential function with the linear function is of order α^2, the result follows. □

Remark 9.4 In the absence of a stop-loss level (and a discount factor), i.e., if $B = -\infty$ in the above setup, it follows from Theorem 9.3 that the optimal liquidation level $b = \infty$. Thus, the problem degenerates, and it is never optimal to liquidate the pairs trade.

9.4 Including a Discount Factor

It may be of interest to include a discounting factor in the analysis above, thus instead considering the optimal stopping problem

$$V(x) = \sup_{\tau \le \tau_B} E_x e^{-r\tau} X_\tau, \tag{9.8}$$

where $r > 0$ is a constant. This optimal stopping problem can, in principle, be studied using similar techniques as in the problem with no discounting. However, it turns out that the solution is slightly less explicit, and the parameter dependencies are more involved.

Again, it is natural to expect that the optimal stopping time takes the form of the first hitting time of a level b. The same arguments as in Sect. 9.2 suggest that the pair (V, b) solves

$$\begin{cases} \frac{\sigma^2}{2} V_{xx} - \mu x V_x - rV = 0 & \text{if } x \in (B,b), \\ V(x) = x & \text{if } x \in [b,\infty), \\ V'(b) = 1, \\ V(B) = B. \end{cases} \tag{9.9}$$

The general solution to the ordinary differential equation $\frac{\sigma^2}{2}V_{xx} - \mu x V_x - rV = 0$ is

$$V(x) = CF(x) + DG(x).$$

Here C and D are constants,

$$F(x) = \int_0^\infty u^{\beta-1} e^{\sqrt{\alpha} x u - u^2/2}\, du,$$
$$G(x) = F(-x),$$

and $\alpha = 2\mu/\sigma^2$ and $\beta = r/\mu$. Inserting the general solution into the free-boundary problem, it is easily seen that

$$C = \frac{BG(b) - bG(B)}{G(b)F(B) - G(B)F(b)} \tag{9.10}$$

and

$$D = \frac{bF(B) - BF(b)}{G(b)F(B) - G(B)F(b)}, \tag{9.11}$$

where b satisfies

$$\begin{aligned}&\big(BG(b) - bG(B)\big)F'(b) + \big(bF(B) - BF(b)\big)G'(b)\\ &\quad = G(b)F(B) - G(B)F(b).\end{aligned} \tag{9.12}$$

Arguing as in the proof of Theorem 9.2, it is straightforward to check that the value function derived above coincides with the value of the optimal stopping problem (9.8).

Theorem 9.5 *Let b be the unique solution of* (9.12) *in* $(0, -B)$, *and define C and D as in* (9.10) *and* (9.11), *respectively. The value function of the optimal stopping problem* (9.8) *is given by*

$$V(x) = \begin{cases} CF(x) + DG(x), & x \in (B, b), \\ x, & x \geq b. \end{cases}$$

Moreover, $\tau^* = \tau_B \wedge \tau_b$ is an optimal stopping time in (9.8).

In the absence of a stop-loss level, i.e., if $B = -\infty$, then

$$V(x) = \begin{cases} \frac{b}{F(b)} F(x), & x < b, \\ x, & x \geq b, \end{cases}$$

where b is the unique positive solution of $F(b) = bF'(b)$.

Proof The proof of the optimality follows along the same lines as in the proof of Theorem 9.2, and we omit the details. To prove the uniqueness of solutions to (9.12), define

$$\begin{aligned}f(x) = &\big(BG(x) - xG(B)\big)F'(x) + \big(xF(B) - BF(x)\big)G'(x)\\ &+ G(B)F(x) - G(x)F(B).\end{aligned}$$

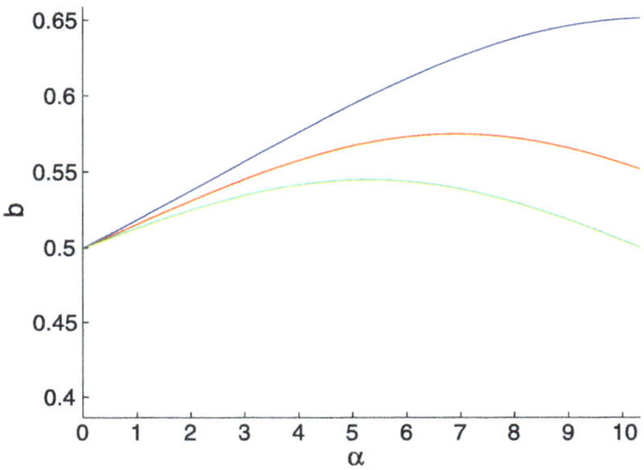

Fig. 9.1 The graph shows the optimal threshold b as a function of the parameter α for three different values of β. The values of β are 0.01 (*top*), 0.05 (*middle*), and 0.09 (*bottom*). In all three examples we used $B=-1$

Then $f(b)=0$ if and only if b solves (9.12). First note that

$$f(0)=F(0)\big(2BF'(0)+F(-B)-F(B)\big).$$

Since

$$\begin{aligned}F(-B)-F(B)&=\int_0^\infty u^{\beta-1}\big(e^{-\sqrt{\alpha}Bu}-e^{\sqrt{\alpha}Bu}\big)e^{-u^2/2}\,du\\&>-\int_0^\infty u^{\beta}2\sqrt{\alpha}Be^{-u^2/2}\,du=-2BF'(0),\end{aligned}$$

we find that $f(0)>0$. Similarly,

$$f(-B)=\big(F(B)+F(-B)\big)\big(BF'(B)+BF'(-B)+F(-B)-F(B)\big).$$

It is easy to check that $g(x):=-xF'(-x)-xF'(x)+F(x)-F(-x)$ satisfies $g(0)=0$ and $g'(x)<0$ for $x>0$. Consequently, $f(-B)<0$, so there exists a zero of f in the interval $(0,-B)$. Moreover,

$$f'(x)=\big(BF(-x)-xF(-B)\big)F''(x)+\big(xF(B)-BF(x)\big)F''(-x).$$

Since $F''(x)>F''(-x)$ and $xF(B)-BF(x)>BF(-x)-xF(-B)$ for $x\in(0,-B)$, we have $f'(x)<0$ in that interval. Thus, the function f has a unique zero in $(0,-B)$, so there exists a unique solution b to (9.12). □

Remark 9.6 As indicated above, the parameter dependencies are more involved in the presence of a discount factor, compare to Fig. 9.1. If the stop-loss level satisfies $B=-\infty$, however, then it is straightforward to check that the optimal liquidation level b is decreasing in the parameter α and in the parameter β.

References

1. D. Ehrman, *The Handbook of Pairs Trading* (Wiley, New York, 2006)

2. R. Elliott, J. van der Hoek, W. Malcolm, Pairs trading. Quant. Finance **5**, 271–276 (2005)
3. E. Gatev, W. Goetzmann, G. Rouwenhorst, Pairs trading: performance of a relative-value arbitrage. Rev. Financ. Stud. **19**, 797–827 (2006)
4. I. Karatzas, S. Shreve, *Brownian Motion and Stochastic Calculus*, 2nd edn. (Springer, New York, 2000)
5. G. Vidyamurthy, *Pairs Trading—Quantitative Methods and Analysis* (Wiley, New York, 2004)
6. M. Whistler, *Trading Pairs: Capturing Profits and Hedging Risk with Statistical Arbitrage Strategies* (Wiley, New York, 2004)

Chapter 10
A PDE-Based Approach for Pricing Mortgage-Backed Securities

Marco Papi and Maya Briani

Abstract In this paper we derive a new equilibrium model for pricing Mortgage-Backed Securities. We prove that the price can be represented as the solution of a degenerate parabolic semilinear equation, and we state existence, uniqueness, and regularity results in the framework of viscosity solutions. These results allow a complete justification of the model. We also obtain a convergence result of a numerical scheme to the solution of the valuation equation.

Keywords Degenerate parabolic equations · Viscosity solutions · Derivative pricing · Mortgages · Numerical methods

Mathematics Subject Classification (2010) 35K65 · 65M06 · 91G20 · 91G80

10.1 Introduction

In this paper, we present a new equilibrium model for pricing Mortgage-Backed Securities. In particular, we shall give a complete derivation of this model, and we study a numerical approximation. The Mortgage-Backed security (*MBS*) market plays a special role in the US economy. Originators of mortgages (S&L, saving and commercial banks) can spread risk across the economy by packaging these mortgages into investment pools through a variety of agencies. Purchasers of *MBS*s are given the opportunity to invest in interest-rate contingent claims which offer different payoff structures from US Treasury bonds.

Mortgage holders have the option to prepay the existing mortgage and refinance the property with a new mortgage. Therefore *MBS* investors are implicitly writing an American call option on a corresponding fixed-rate bond. Moreover, prepayments can also take place for reasons not related to the interest rate option. Mortgage investors are exposed to significant interest rate risk when loans are prepaid and to

M. Papi (✉)
Engineering School, Università CBM, Rome, Italy
e-mail: m.papi@unicampus.it

M. Briani
Istituto Motori, CNR, Napoli, Italy
e-mail: m.briani@im.cnr.it

G. Di Nunno, B. Øksendal (eds.), *Advanced Mathematical Methods for Finance*,
DOI 10.1007/978-3-642-18412-3_10, © Springer-Verlag Berlin Heidelberg 2011

credit risk when loans are terminated to default. Prepayments will halt the stream of cash flows that investors expect to receive. In many cases this will result in a lower than expected return on their investment. For example, if interest rates decline, there will typically be a subsequent increase in prepayments which forces investors to reinvest the unexpected additional cash-flows at the new lower interest rate level. On the other hand, if interest rates increase, there will be a decrease in prepayment activity which forces investors to wait for a longer period before they can reinvest the cash-flows at the new higher interest rate level. These complications lead to a nonlinear relation between *MBS* prices, interest rates, and coupon-specific prepayment variables. In this paper, using a risk-neutral arguments, we shall derive in a rigorous way a pricing equation based on a general equilibrium model proposed and successfully tested on data by Gabaix et al. [13].

Using the results of [25] and [26], we also consider a numerical approximation of the pricing equation, proving a convergence result and showing the qualitative behavior of the solution in a particular application of the model.

The most simple structure of *MBS*s are the pass-through securities. Investors in this kind of securities receive all payments (principal plus interest) made by mortgage holders in a particular pool (less some servicing fee). Other classes of derivative products are the stripped mortgage-backed securities (*SMBS*) which entail the ownership of either the principal or interest cash-flows arising from specific mortgages or mortgage pass-through securities. Rights to the principal are labeled *PO*s (*principal only*), and rights to the interest are labeled *IO*s (*interest only*).

Modeling and pricing *MBS*s involve three layers of complexity: (i) modeling the dynamic behavior of the term structure of interest rates, (ii) modeling the prepayment behavior of mortgage holders, and (iii) from the point of view of no-arbitrage valuation, modeling the risk premia embedded in these financial claims. Although the first two points of the valuation of an *MBS* are far from our present interest, we observe that both the rational and empirical approaches to prepayment structures and *MBS* valuation depend crucially on the correct parameterization of prepayment behavior and on the correct forecast of mortgage rates. The second one is a very important feature of a mortgage model since it explains the borrower's decision making. The characteristics of approaches used to model prepayment incentive may be used to classify existing models in the literature into two groups. The first type is related to American options. This kind of models measure the mortgagor's incentive to prepay as the difference between his liability and the outstanding principal, while the liability is defined as the present value of cash-flows that the borrower will pay off to repay his loan. This approach has been followed in [12], assuming an optimal prepayment behavior, in the sense that the borrowers terminate their mortgages if it is financially optimal; then transaction costs have been incorporated as a part of the model in [10], and Johnston and Van Drunen [17] extended the model considering different refinancing costs for the borrowers. Stanton [28] observed that borrowers fail to prepay when it is optimal to do so. Moreover, he addressed the problem of the mortgage pool heterogeneity, considering also the problem of possible unobservable heterogeneity between mortgage pools characterized by the same coupon rate, issue date, and other observable characteristics [29].

In the second class of models [6] the prepayment policy follows a comparison between the prevailing mortgage and contract rates. This comparison can be measured using the difference or a ratio of the two rates, and usually the 10 years Treasury yield is used as a proxy for the mortgage rate, see [21] and [27].

In this research paper our attention is mainly devoted to the third topic. In order to formulate the model in a continuous-time setting, which is a common feature in financial applications, we shall assume that all relevant economic factors affecting prepayments behavior may be represented by a state process $X_t = (X_t^1, \ldots, X_t^N)$ for some $N \geq 1$, following a stochastic differential equation. The state vector X_t may include factors like interest rate level and the value of houses, as in [7] and possible prepayment specific variables as transaction costs of refinancing. Using arbitrage-free arguments [20, 22], the *MBS* price at time t can be written as

$$P_t = U(X_t, t),$$

where U is a deterministic function. The challenging task here is to give a complete justification to the particular choice of the market price of risk used to derive the functional form of U. This model specification follows the work of Gabaix et al. [13]. In Sect. 10.3, we characterize U as the unique solution of a nonlinear parabolic partial differential equation, in a viscosity sense [4].

In *MBS* analysis, it seems natural to assume the market price of risk to depend directly on the value of the liability and its sensitivity to prepayments due to changes in the explanatory economic factors X_t, see formula (10.38) below. Actually, this kind of dependence has been also used in the option-based approach for modeling the prepayment intensity [15]. The distinction between the liability to the mortgagor and the asset value to the investor is that, at time of prepayment, a security holder receives the outstanding principal but the mortgagor pays it plus a transaction cost, proportional to the outstanding principal. Therefore, since a change of the borrower's liability produces a change in the value of the prepayment option and possibly in the prepayment behavior, the proportionality with the asset value, yields a natural dependence of the market price of risk on the *MBS* price (U) and on its variation ($\nabla_x U$), implying a semilinear structure for the pricing equation, see Theorem 10.13 below. The existence, uniqueness, and regularity properties of the solution to the pricing equation play an essential role. From one side, the existence of a solution yields the proof that there exists a *risk-neutral* market measure for evaluating *MBS*s, and, on the other hand, it is a fundamental result to construct stable numerical procedures, see Sect. 10.5 below.

In recent years there has been an interest in developing viscosity solution theory [4, 11] for differential equations, and the relevance of these equations can be motivated by their many applications in mathematical finance. Actually it is well known that one can reduce the computation of the price of a financial claim to the solution of a partial differential equation [20]. Unfortunately, closed-form prices for financial derivatives are available only in few special situations, and, in many cases, the pricing equation is nonlinear and degenerate. In this general setting, singularities and nonuniqueness phenomena may occur, and they require careful consideration. In this context, viscosity solutions represent a natural framework to study these problems, also from a numerical point of view. In fact, the key result in the proof of convergence of numerical schemes is the strong comparison result, which is also the main tool used to establish the existence and uniqueness of viscosity solutions, see [2].

The paper is organized as follows. In Sect. 10.2, we derive our model in the classical framework of financial modeling [20]. Sections 10.3 and 10.4 are devoted to the characterization and the properties of the solution to the pricing equation. In Sect. 10.5 we shall study a numerical scheme for an application of the model.

10.2 *MBS*s Modeling

This section describes in a more detailed fashion Mortgage-Backed securities. *MBS*s or Mortgage Pass-Throughs are claims on a portfolio of mortgages. Usually a federal agency, mortgage banker, bank, or investment company buys up mortgages of a certain type, then sells claims on the cash-flows from the portfolio (i.e., *MBS*s). In the primary market, the investors buy *MBS*s issued by agencies or private-label investment companies either directly or through dealers. Many of the investors are institutional investors. Cash-flows from *MBS*s are the cash-flows from the portfolio of mortgages (referred to as the *Collateral*). Cash-flows include: interest on principal, scheduled principal, prepaid principal.

Every pool of mortgages is characterized by the weighted average maturity (*WAM*), the weighted average coupon rate (*WAC*), which is the rate on portfolio of mortgages (*collateral*) applied to determine scheduled principal, and the Pass-Through Rate (*PTR*), which denotes the interest on principal; *PTR* is lower than *WAC*—the difference going to *MBS* issuer.

The price of *MBS*s are quoted as a percentage of the underlying mortgage balance. Let a_t be the mortgage balance at time t, and let V_t the price quote observed in the market at time t, then the market value MBS_t at time t is given by

$$MBS_t = V_t a_t. \tag{10.1}$$

The market value is the clean price—it does not take into account accrued interest AI. For *MBS*s, accrued interest is based on the time period from the settlement date. To be precise, if τ denotes the *WAC*, for monthly payments, it holds

$$AI_t = \frac{\tau}{12}\frac{t - 30[t/30]}{30}a_t, \tag{10.2}$$

$[\cdot]$ being the integer part. Therefore the full market value would be

$$MBS_t + AI_t. \tag{10.3}$$

The market price per share is the full market value divided by the number of shares in which the issued *MBS* is divided. As in the most part of the *MBS* literature [28], our model addresses the problem of the valuation of the clean price (10.1) in a continuous-time framework. The clean price (10.1) is the product of two stochastic components, the market quote V_t and the balance amount a_t, which is mainly affected by prepayments in the pool. Due to this fact, also the market quote is affected by the prepayment behavior. To simplify the presentation, we reduce the computation of the market value of an *MBS* by considering the problem facing an individual mortgage holder in a pool. Since different borrowers will in general have different characteristics (frictions, prepayment behavior, location, employment status, etc.), one may deal with this heterogeneity assuming that these characteristics are randomly distributed across borrowers in a pool, as in Stanton [28].

10.2.1 MBS *Cash-Flows*

In this section, through classical arguments of financial modeling [20], we shall present our *MBS* model in a continuous-time framework, and we derive the pricing equation for the clean price (10.1). We consider the usual information structure described by a standard d-dimensional Brownian motion,

$$B = \big(\Omega, \mathcal{F}, \{\mathcal{F}_t\}_{t\in[0,T]}, \{B_t\}_{t\in[0,T]}, \mathbb{P}\big), \qquad T > 0, \tag{10.4}$$

where $(\Omega, \mathcal{F}, \mathbb{P})$ is a complete probability space, and $\{\mathcal{F}_t\}_{t\in[0,T]}$ is a given filtration of sub-σ-algebras of $\mathcal{F}$. More precisely, for every $t \in [0, T]$, the σ-algebra $\mathcal{F}_t$ represents the information available up to time t. The filtration is an intrinsic feature of the market: this means that all traders have the same information available at any time. Here the Brownian motion represents the source of randomness of prepayments due the economic factors affecting *MBS* prices. The measure $\mathbb{P}$ is not necessarily the physical measure of the market; for instance, one can take $\mathbb{P}$ as a risk-adjusted measure after the risks, related to the economic factors, have been taken into account.

Working with a continuous-time model, we shall use the notation c_t to denote an $\mathcal{F}_t$-adapted process representing the cumulative cash-flow associated with an *MBS*.

Definition 10.1 Let us define

- $MB(t)$, the remaining principal at time t if there are not prepayments. We have

$$MB(t) = MB(0)\frac{e^{\tau' T} - e^{\tau' t}}{e^{\tau' T} - 1}, \qquad t \in [0, T], \tag{10.5}$$

  where τ' is the fixed rate paid by the mortgagor (while the interest rate received by the investor is $\tau < \tau'$).
- $S_t \geq 0$, the $\mathcal{F}_t$-adapted pure cumulative prepayment process, so that the remaining principal at time t is

$$a_t = MB(t)\exp(-S_t), \qquad t \in [0, T]. \tag{10.6}$$

- The cash-flows of the securities are:
  - For an IO, $dc_t = \tau a_t\, dt$, the interest payment on the outstanding balance.
  - For a PO, $dc_t = -da_t$, how much of the principal has been paid down in dt units of time.
  - For the pass-through, $dc_t = \tau a_t\, dt - da_t$; the cash-flow is sum of both IO, POs.

Remark 10.2 The scheduled payments in a fixed-rate mortgage imply that each mortgagor pays a fixed amount C with given frequency till the mortgage extinction. This amount is the sum of interest and principal payments. In particular, the interest part decreases in time, and consequently the principal repayment increases. At the end of each period, the remaining principal is decreased by an amount corresponding to the principal repayment. Once the frequency, the fixed interest rate, the initial pool balance, and the maturity date are fixed, the constant payment C is calculated in order to follow the previous rule and so that the final remaining balance is zero. Assuming a continuous frequency for the payments, we easily obtain the following relation for the remaining balance at time t:

$$-dMB(t) + \tau' MB(t)\, dt = C\, dt, \qquad MB(0) = 0, \tag{10.7}$$

and imposing $MB(T) = 0$, we find $C = MB(0)\tau' \frac{e^{\tau' T}}{e^{\tau' T} - 1}$. By integration of (10.7) we obtain (10.5).

Remark 10.3 According to the analysis conducted in Sect. 10.2, we may assume that τ' and τ coincide with the *WAC* and the *PRT*, respectively.

The adapted process S_t includes the pure prepayments occurring in the pool of mortgages. In current practice, practitioners often simplify their evaluations by considering a deterministic function of interest rates and time, calibrated at date t to best fit past prepayments observed. Following this approach, in Sect. 10.2.3, we shall assume that the process S_t is a deterministic function of an $\mathcal{F}_t$-adapted process X_t of economic factors affecting prepayments, namely

$$S_t = s_0(X_t, t), \quad t \in [0, T]. \tag{10.8}$$

To maintain the generality of the presentation, we do not specify any particular form for s_0 in the sequel; however we note that to be consistent with the model specification of Definition 10.1, an unbiased model for s_0 should consider a nonnegative process with nondecreasing paths.

Although general models of default have been studied, the mortgage models in literature do not implement current results of credit risk theory. However this represents a fundamental topic in order to explain termination behavior in a pool of mortgages. From a mathematical point of view, we may include default of borrowers in the prepayment function, but we need to remove the continuity of s_0, since payments due to default occur at discrete times, see, for instance, [28]. Given the introductory level of this research and our desire to present our *MBS* model in the clearest fashion possible, we do not study this topic here, and we shall consider pools of fixed-rate residential mortgages insured against default.

10.2.2 The MBS Market

In this section we present the mathematical setting of arbitrage pricing with cash-flows (or with dividends). We shall use this setting to define the *MBS* market in our model. Since *MBS*s entitle the investors to receive cash-flows during $[0, T]$, it is natural to consider the extension of the basic approach of arbitrage pricing focusing on the concept of *gain process* (G_t below). For an exhaustive description of this method, we refer the reader to Chap. 6 in [8] and to [20].

Let $\delta : [0, T] \to (0, \infty)$ be a deterministic and integrable discount rate; then the economy is made up of a representative agent (a trader in the *MBS* market) with a risk aversion $\rho > 0$. This agent can invest in the riskless asset V_t^{Riskless}, driven by the following equation:

$$dV_t^{\text{Riskless}} = \delta(t) V_t^{\text{Riskless}} dt, \qquad V_0^{\text{Riskless}} = A_0 > 0, \tag{10.9}$$

or in a finite set of *MBS*s. The vector of asset prices is

$$\left(V_t^{\text{Riskless}}, V_t^{MBS}\right) \equiv \left(V_t^{\text{Riskless}}, V_t^{MBS,1}, \dots, V_t^{MBS,k}\right) \tag{10.10}$$

for $t \in [0, T]$. The model for the *MBS* market is a financial market such that the gain process, associated with the ith asset, has the following form:

$$G_t^{MBS,i} = V_t^{MBS,i} + \int_0^t dc_s^i = V_t^{MBS,i} + \tau_i \int_0^t a_s^i \, dt - a_t^i + MB_i(0) \tag{10.11}$$

for every $i = 1, \dots, k$, where c_t^i is the ith cash-flow process, given by Definition 10.1 and with a fixed coupon rate τ_i. In the sequel we shall make the following standing assumptions:

(A1) The price V_t^{MBS} is a nonnegative Itô process w.r.t. B.
(A2) The prepayment process S_t^i of the ith *MBS* is a nonnegative Itô process w.r.t. B, for every $i = 1, \dots, k$.

In order to derive our pricing equation, we recall some basic concepts of financial modeling. Consider k traded assets whose prices are represented by a vector-valued Itô process V_t w.r.t. B, paying cash-flows (or dividends) at a given rate dC_t, that is C_t is a vector-valued Itô process describing the amount of cash-flows (or dividends) paid up to time t for each asset. Let G_t be the *gain process* associated with these securities, i.e., $G_t = V_t + C_t$. We recall that an $\mathcal{F}_t$-adapted process $\gamma_t \in \mathbb{R}^d$ satisfies Novikov's condition if $\mathbb{E}[\exp(\frac{1}{2}\int_0^T |\gamma_t|^2 dt)]$ is finite.

The following theorem, due in its first statement to Harrison and Kreps (1979), is a characterization of the absence of arbitrage opportunities in this market. We refer to [22] for a very readable proof in discrete time, to [20] for the continuous-time version, and in particular to [8] for the case with dividends.

Theorem 10.4 [8]

(**A**) *If the market described by* $(V_t^{\mathrm{Riskless}}, V_t, C_t)$ *is arbitrage-free, then there exists a measurable process* γ_t*, adapted to the filtration* $\{\mathcal{F}_t\}_t$*, taking values in* $\mathbb{R}^d$*, called the* market price of risk, *such that for almost every* $(\omega, t) \in \Omega \times [0, T]$*, the following holds*:

$$\sigma_t^G \cdot \gamma_t = \mu_t^G - \delta(t) V_t, \tag{10.12}$$

where μ_t^G *is the drift, and* σ_t^G *is the diffusion of the gain process.*

(**B**) *Conversely, if there exists a process* γ_t *which satisfies* (10.12), *Novikov's condition, and such that*

$$\xi_T^\gamma \equiv e^{-\int_0^T \gamma_s^\top dB_s - \frac{1}{2}\int_0^T |\gamma_s|^2 ds} \tag{10.13}$$

has finite variance, then there is no arbitrage in the market.

Remark 10.5 The statement (**B**) in Theorem 10.4 and the Girsanov's change-of-measure theorem [19] imply that the Radon–Nikodym derivative

$$\frac{d\mathbb{Q}}{d\mathbb{P}} = \xi_T^\gamma \tag{10.14}$$

defines a probability measure $\mathbb{Q} \sim \mathbb{P}$ equivalent to $\mathbb{P}$ on $(\Omega, \mathcal{F}_T)$ such that

$$\widehat{B}_t = B_t + \int_0^t \gamma_s \, ds, \quad t \in [0, T], \tag{10.15}$$

is an $\mathcal{F}_t$-adapted Brownian motion under $\mathbb{Q}$, and ξ_t^γ, $t \in [0, T]$, is a $\mathbb{P}$-martingale. Furthermore, the *discounted gain process*

$$G_t^Y \equiv Y_t V_t + \int_0^t Y_t \, dC_t, \quad t \in [0, T], \tag{10.16}$$

where $Y_t = [V_t^{\mathrm{Riskless}}/A_0]^{-1} = \exp(-\int_0^t \delta(s)\, ds)$, is a $\mathbb{Q}$-martingale. This is a straightforward application of (10.12) and Itô's formula. The measure $\mathbb{Q}$ is usually named an *equivalent martingale* (or *risk-neutral*) measure.

Therefore we shall consider the following assumption:

(**A3**) In the *MBS* market there exists a market price of risk $\gamma_t^{MBS} = (\gamma_t^{MBS,1}, \ldots, \gamma_t^{MBS,d})$ satisfying conditions of part (**B**) in Theorem 10.4.

Under Assumption (**A3**), the *MBS* market defined by (10.9)–(10.11) is arbitrage-free, and there exists a risk-neutral measure associated with γ_t^{MBS}. In the rest of the paper, we shall denote this measure by $\mathbb{Q}$.

Remark 10.6 Equation (10.12) defines a linear system of k equations with d unknown variables and in general is not solvable, or it can have multiple solutions. The application of (10.12) to the classical Black–Scholes (1973) framework, without dividends and with the stock price X (the state process) driven by a one-dimensional Brownian motion, and with a constant discount rate, gives a unique constant market price of risk. In fact, $G_t^1 = X_t$, $G_t^2 = P(X_t, t)$ is the price of the financial derivative, $C_t^1 = C_t^2 = 0$, $\mu_t^{G,1} = \mu X_t$, and $\sigma_t^{G,1} = \sigma X_t$, where the constants μ and $\sigma > 0$ denote the instantaneous rate of return and the volatility of the stock, respectively. Equation (10.12) leads to two scalar equations, the first one implying

$$\gamma_t = \frac{\mu_t^{G,1} - \delta X_t}{\sigma_t^{G,1}} = \frac{\mu - \delta}{\sigma}, \tag{10.17}$$

whereas the second equation yields a PDE whose solution is the price function P. The reason is that the stock price X_t reflects the m.p.r. In contrast, in term structure pricing models [5, 9] there are sources of randomness defined by economic factors (the interest rate, the inflation rate, etc.) that are not the prices of traded assets. Thus the market is clearly incomplete, and the corresponding m.p.r. and the related equivalent martingale measure $\mathbb{Q}$ (10.14) are not necessarily unique. This represents the natural setting of any *MBS* market model.

We now prove a representation theorem for the *MBS* price process and for the dynamics of the gain processes. To this end, we recall some notation and definitions of Malliavin calculus used in this paper. For a complete introduction to Malliavin calculus, we refer the reader to [24].

Let $L^p(\Omega \times [0, T])$, $p \geq 1$, be the space of $\mathcal{F}_t$-adapted measurable processes v such that $|v|^p$ is integrable w.r.t. the product measure $\mathbb{P} \otimes dt$ on $\Omega \times [0, T]$. Given an $\mathcal{F}$-measurable random variable $F : \Omega \to \mathbb{R}$, $D_t F$ denotes the Malliavin derivative at time $t \in [0, T]$ of F. The space $\mathbb{D}^{1,p}$, $p \geq 1$, is the usual space of Malliavin-differentiable random variables F in $[0, T]$ such that $|F|^p$ and $(\int_0^T |D_t F|^2 dt)^{p/2}$ have finite $\mathbb{P}$-expectations. Let $\mathbb{L}^{1,p}$ be the class of processes $v \in L^p(\Omega \times [0, T])$ such that $v_t \in \mathbb{D}^{1,p}$ for almost all t and there exists a measurable version of the two-parameter process $D_s v_t$ such that $\|Dv\|_{L^2([0,T]^2)}^p = (\int_0^T \int_0^T |D_s v_t|^2 ds\, dt)^{p/2}$ has a finite $\mathbb{P}$-expectation. In $\mathbb{L}^{1,p}$ we define the norm

$$|||v|||_{1,p} = \left(\|v\|_{L^p(\Omega\times[0,T])}^p + \mathbb{E}\left[\|Dv\|_{L^2([0,T]^2)}^p\right]\right)^{1/p}. \tag{10.18}$$

Moreover, we recall the Malliavin derivative of an Itô process w.r.t. the Brownian motion B. Consider the stochastic process

$$H_t = x + \int_0^t u_s\, ds + \int_0^t w_s\, dB_s, \quad t \in [0, T], \tag{10.19}$$

where $x \in \mathbb{R}$, $u \in \mathbb{L}^{1,1}$, and $w \in \mathbb{L}^{1,2}$. For $0 < r \leq t \leq T$, we have

$$D_r H_t = w_r + \int_r^t D_r u_s\, ds + \int_r^t D_r w_s\, dB_s, \tag{10.20}$$

and $D_r H_t = 0$ for $\mathbb{P}$-a.e. $r > t$.

Theorem 10.7 *Assume* (**A1**)–(**A3**). *Let* S^i, $\gamma^{MBS,j} \in \mathbb{L}^{1,4}$ *for* $i = 1, \dots, k$ *and* $j = 1, \dots, d$. *Then the following equations hold*:

$$V_t^{MBS,i} = a_t^i + \mathbb{E}^{\mathbb{Q}}\left[\int_t^T \left(\tau_i - \delta(s)\right)e^{-\int_t^s \delta(u)\,du} a_s^i\, ds \,\middle|\, \mathcal{F}_t\right], \tag{10.21}$$

$$dG_t^{MBS,i} = \delta(t) V_t^{MBS,i}\, dt + D_t G_t^{MBS,i}\, d\widehat{B}_t, \tag{10.22}$$

for every $i = 1, \dots, k$, *where* $D_t G_t^{MBS,i}$ *is given by the following formulas*:

1. *For the pass-through*,

$$-\mathbb{E}^{\mathbb{Q}}\left[\int_t^T \left(\tau_i - \delta(s)\right)e^{-\int_t^s \delta(u)\,du} a_s^i \left(D_t S_s^i + \int_t^s D_t \gamma_u^{MBS}\, d\widehat{B}_u\right) ds \,\middle|\, \mathcal{F}_t\right]. \tag{10.23}$$

2. *For an IO*,

$$-\mathbb{E}^{\mathbb{Q}}\left[\int_t^T \tau_i e^{-\int_t^s \delta(u)\,du} a_s^i \left(D_t S_s^i + \int_t^s D_t \gamma_u^{MBS}\, d\widehat{B}_u\right) ds \,\middle|\, \mathcal{F}_t\right]. \tag{10.24}$$

3. *For a PO*,

$$\mathbb{E}^{\mathbb{Q}}\left[\int_t^T \delta(s) e^{-\int_t^s \delta(u)\,du} a_s^i \left(D_t S_s^i + \int_t^s D_t \gamma_u^{MBS}\, d\widehat{B}_u\right) ds \,\middle|\, \mathcal{F}_t\right]. \tag{10.25}$$

Remark 10.8 In (10.22), the diffusion coefficient of $G_t^{MBS,i}$ is expressed by the Malliavin derivative of the process at time t. This is a general property that can be obtained by applying (10.20) with $t = r$. Actually, we see that the diffusion w_t coincides with $D_t H_t$.

Remark 10.9 The interpretation of the sign in (10.23) is the following: $D_t G_t^{MBS,i}$ can be understood as the impact of a shock to the prepayment process on the ith gain process. For the sake of simplicity, suppose that $d = 1$ and that S increases with B. If there has been a positive prepayment shock at time t ($dB_t > 0$), in the premium environments ($\tau - \delta(t) > 0$), this affects negatively the value of the principal, because the total impact is a higher prepayment: $D_t S_s > 0$. Therefore, we expect that $D_t G_t^{MBS} < 0$ and also

$$\mathbb{E}^{\mathbb{Q}}\left[\int_t^T \left(\tau_i - \delta(s)\right)e^{-\int_t^s \delta(u)\,du} a_s^i\, D_t S_s^i\, ds \,\middle|\, \mathcal{F}_t\right] > 0. \tag{10.26}$$

There is also a second effect. After a positive prepayment shock, the amount of high-coupon securities will decrease leading the m.p.r. γ_s^{MBS} to increase. Therefore we expect that $D_t \gamma_s^{MBS} > 0$. Hence the second indirect effect is due to current prepayments on future value of the price risk γ_s^{MBS}, $s \geq t$. This dampens the first effect. Similar explanations hold for (10.24) and (10.25).

To prove Theorem 10.7, we recall the generalized Clark–Ocone formula. This is one of the most useful results from Malliavin calculus, which allows the process in the martingale representation theorem w.r.t. $\widehat{B}$ to be identified explicitly. See [18] for a detailed proof of this result.

Theorem 10.10 *Let F be $\mathcal{F}_T$-measurable, $F, \gamma^j \in \mathbb{D}^{1,2}$ for $j = 1, \dots, d$, $\mathbb{Q}$ and $\widehat{B}$ be defined in* (10.14) *and* (10.15). *Assume that*

(i) $\mathbb{E}^{\mathbb{Q}}[|F|] < +\infty$,
(ii) $\mathbb{E}^{\mathbb{Q}}[\int_0^T |D_t F|^2 \, dt] < +\infty$,
(iii) $\mathbb{E}^{\mathbb{Q}}[|F| \int_0^T (\int_0^T D_t \gamma_s \, dB_s + \int_0^T D_t \gamma_s \gamma_s \, ds)^2 \, dt] < +\infty$.

Then $F = \mathbb{E}^{\mathbb{Q}}[F] + \int_0^T \mathbb{E}^{\mathbb{Q}}[(D_t F - F \int_t^T D_t \gamma_s \, d\widehat{B}_s) | \mathcal{F}_t] \, d\widehat{B}_t$.

Proof of Theorem 10.7 We prove the result in the case of pass-through securities. By Remark 10.5 and Assumptions (**A1**)–(**A3**), the process

$$G_t^{MBS,Y} = Y_t V_t^{MBS} + \int_0^t Y_s \, dc_s, \quad t \in [0, T], \tag{10.27}$$

with $c_t = (c_t^1, \dots, c_t^k)$ is a martingale under the risk-neutral measure $\mathbb{Q}$. This property implies the relation

$$V_t^{MBS,i} Y_t + \int_0^t Y_s \, dc_s^i = \mathbb{E}^{\mathbb{Q}}\left[V_T^{MBS,i} Y_T + \int_0^T Y_s \, dc_s^i \middle| \mathcal{F}_t \right], \quad t \le T, \ i = 1, \dots, k.$$

Since $V_T^{MBS,i} = a_T^i = 0$, integrating by parts, we get

$$\begin{aligned} V_t^{MBS,i} Y_t &= \mathbb{E}^{\mathbb{Q}}\left[\int_t^T Y_s \, dc_s^i \middle| \mathcal{F}_t \right] = \mathbb{E}^{\mathbb{Q}}\left[\int_t^T Y_s \tau_i a_s^i \, ds - \int_t^T Y_s \, da_s^i \middle| \mathcal{F}_t \right] \\ &= \mathbb{E}^{\mathbb{Q}}\left[\int_t^T Y_s \tau_i a_s^i \, ds + Y_t a_t^i - \int_t^T \delta(s) Y_s a_s^i \, ds \middle| \mathcal{F}_t \right] \\ &= Y_t a_t^i + \mathbb{E}^{\mathbb{Q}}\left[\int_t^T \big(\tau_i - \delta(s)\big) Y_s a_s^i \, ds \middle| \mathcal{F}_t \right]. \end{aligned} \tag{10.28}$$

Since $Y_s / Y_t = \exp(-\int_t^s \delta(u) \, du)$, dividing by Y_t, we obtain (10.21).

Let $\sigma_{G,i}^{MBS}$ be diffusion coefficient of the ith gain process, under B. Equation (10.12) implies that the drift of this process is $(\delta(t) V_t^{MBS,i} + \sigma_{G,i}^{MBS}(t) \cdot \gamma_t^{MBS})$. Hence,

$$\begin{aligned} dG_t^{MBS,i} &= \big(\delta(t) V_t^{MBS,i} + \sigma_{G,i}^{MBS}(t) \cdot \gamma_t^{MBS}\big) \, dt + \sigma_{G,i}^{MBS}(t) \, dB_t \\ &= \delta(t) V_t^{MBS,i} \, dt + \sigma_{G,i}^{MBS}(t) \, d\widehat{B}_t. \end{aligned} \tag{10.29}$$

Thus, (10.22) follows by Remark 10.8. We compute $D_t G_t^{MBS,i}$. Since $S^i \ge 0$ and $S^i \in \mathbb{L}^{1,4}$, also the remaining principal a_s^i in Definition 10.1 belongs to the space

$\mathbb{L}^{1,4}$. In fact, a^i is bounded from above by $MB(0)$ and $D_t a_s^i = -a_s^i D_t S_s^i$. Hence, it is easy to see that $|||a^i|||_{1,4} \leq MB(0)(T + |||S^i|||_{1,4}^4)^{1/4}$. By (10.20), (10.21) and the definition of $G_t^{MBS,i}$ (10.11), we get

$$D_t G_t^{MBS,i} = -D_t a_t^i + D_t V_t^{MBS,i} = D_t \mathbb{E}^{\mathbb{Q}}\left[\int_t^T \big(\tau_i - \delta(s)\big)e^{-\int_t^s \delta(u)\,du} a_s^i \, ds \middle| \mathcal{F}_t\right]$$
$$= \int_t^T \big(\tau_i - \delta(s)\big)e^{-\int_t^s \delta(u)\,du} D_t \mathbb{E}^{\mathbb{Q}}\big[a_s^i|\mathcal{F}_t\big]\, ds. \tag{10.30}$$

We apply the generalized Clark–Ocone formula to a_s^i. We have to verify conditions (i)–(iii) in Theorem 10.10. Condition (i)) follows from the boundedness of a^i. Let ξ_T^γ as in (10.13), with $\gamma = \gamma^{MBS}$, and denote $c_\xi = \mathbb{E}[(\xi_T^\gamma)^2]^{1/2} < +\infty$. Since $a_s^i \in \mathbb{D}^{1,4}$ for almost all s, the Cauchy–Schwarz inequality implies

$$\mathbb{E}^{\mathbb{Q}}\left[\int_0^T |D_t a_s|^2 \, dt\right] \leq c_\xi \mathbb{E}\left[\left(\int_0^T |D_t a_s|^2 \, dt\right)^2\right]^{1/2} < +\infty \tag{10.31}$$

for almost all s. The Itô isometry [19] (under the measure $\mathbb{Q}$) gives

$$\mathbb{E}^{\mathbb{Q}}\left[a_s^i \int_0^T \left(\int_0^T D_t \gamma_s^{MBS} \, dB_s + \int_0^T D_t \gamma_s^{MBS} \gamma_s^{MBS} \, ds\right)^2 dt\right]$$
$$\leq MB(0) \int_0^T \mathbb{E}^{\mathbb{Q}}\left[\left(\int_0^T D_t \gamma_s^{MBS} \, d\widehat{B}_s\right)^2\right] dt$$
$$= MB(0) \int_0^T \mathbb{E}^{\mathbb{Q}}\left[\int_0^T |D_t \gamma_s^{MBS}|^2 \, ds\right] dt$$
$$\leq MB(0) c_\xi \sum_{j=1}^d \mathbb{E}\left[\left(\int_0^T \int_0^T |D_t \gamma_s^{MBS,j}|^2 \, dt\, ds\right)^2\right]^{1/2}$$
$$= MB(0) c_\xi \sum_{j=1}^d \mathbb{E}\big[\|D\gamma^{MBS,j}\|_{L^2([0,T]^2)}^4\big]^{1/2} < +\infty, \tag{10.32}$$

which proves (iii). By Theorem 10.10, we can write

$$a_s^i = \mathbb{E}^{\mathbb{Q}}[a_s^i] + \int_0^T \mathbb{E}^{\mathbb{Q}}\left[\left(D_v a_s^i - a_s^i \int_v^T D_v \gamma_u^{MBS} \, d\widehat{B}_u\right)\middle| \mathcal{F}_v\right] d\widehat{B}_v, \tag{10.33}$$

which yields

$$\mathbb{E}^{\mathbb{Q}}[a_s^i|\mathcal{F}_t] = \mathbb{E}^{\mathbb{Q}}[a_s^i]$$
$$+ \int_0^t \mathbb{E}^{\mathbb{Q}}\left[\left(D_v a_s^i - a_s^i \int_v^T D_v \gamma_u^{MBS} \, d\widehat{B}_u\right)\middle| \mathcal{F}_v\right] d\widehat{B}_v. \tag{10.34}$$

Using (10.19) (with $r = t$), we obtain

$$D_t \mathbb{E}^{\mathbb{Q}}\big[a_s^i \big| \mathcal{F}_t\big] = \mathbb{E}^{\mathbb{Q}}\bigg[\bigg(D_t a_s^i - a_s^i \int_t^T D_t \gamma_u^{MBS}\, d\widehat{B}_u\bigg)\bigg|\mathcal{F}_t\bigg]$$
$$= -\mathbb{E}^{\mathbb{Q}}\bigg[a_s^i\bigg(D_t S_s^i + \int_t^T D_t \gamma_u^{MBS}\, d\widehat{B}_u\bigg)\bigg|\mathcal{F}_t\bigg]. \qquad (10.35)$$

Introducing (10.35) into (10.30) and exchanging the integral in time with the expectation, we get the expression (10.23) for $D_t G_t^{MBS,i}$. □

10.2.3 Pricing MBSs: A PDE Approach

In this section we derive a pricing equation for an *MBS*. In Remark 10.6 we observed that if the price of a financial instrument depends on nontradable economic factors, the m.p.r. γ_t is not uniquely defined. Like in interest rate models, the *MBS* model solves for all *MBS* prices relative to each other. The only way to tie down the prices is by invoking the exogenous parameter, the m.p.r. Usually the m.p.r. is parameterized in light of some economic reason or statistical argument, see, for instance, [5].

In our *MBS* model, the key result is the notion of equilibrium in the *MBS* market proposed by Gabaix et al. [13]. The authors assume the existence of an m.p.r. of equilibrium with a specific expression (see (10.11) below), and they give an empirical support for it.

Unfortunately, the authors are silent though about the existence of this m.p.r. and the Malliavin differentiability (see the hypotheses of Theorem 10.7). As directly reported in their work, the reason for this is that the mathematical toolbox required to rigorously prove their results are still largely to be developed. The main difficulty here is related to a nonlinear structure of the problem induced by the equilibrium they propose.

Our approach, based on the use of the theory of viscosity solutions, allows us to answer to the problem left opened in [13]. Actually, in the case of a single *MBS*, the derivation of the pricing equation as well as the existence of the equilibrium and its Malliavin differentiability will be completely justified with the results presented in Sects. 10.3 and 10.4.

The case of multiple assets requires more sophisticated tools in order to prove the existence of an equilibrium. This corresponds to proving the existence of a unique regular solution for a system of coupled and possibly strongly degenerate parabolic equations. The study of this more complicated problem is the object of a work in preparation. We focus our attention on the problem of pricing *MBS* pass-through certificates (Col[1]), however our method easily applies to *IO* and *PO* derivatives.

[1]This notation stands for the name *Collateral*, which is frequently used for pass-through certificates.

We shall adopt the notation tr for the trace of a matrix and $\langle\, , \rangle$ for the standard Euclidean product in $\mathbb{R}^N$, ∂_t (or $\frac{\partial}{\partial t}$) and ∂_i for the partial derivatives w.r.t. the time t and direction x_i in $\mathbb{R}^N$. Besides, ∇f stands for the gradient of f w.r.t. x, and $\nabla^2 f$ stands for the Hessian matrix of f.

We consider a market of k *MBS*s with the same maturity T. We assume that the market prices are affected by some economic factors whose values, observed at time $t = 0$, are denoted by $x \in \mathbb{R}^N$. Thus, the price vector at time $t \le T$ of these securities is $(V_t^{\mathrm{Col},1}(x), \dots, V_t^{\mathrm{Col},k}(x))$. In addition, we consider the following assumptions which allow us to work in a Markovian setting:

(S1) The risk-free rate δ is continuous, and there exists a collection of stochastic processes $\{X_t^x : t \in [0, T]\}_x$, $x \in \mathbb{R}^N$, which represent all the economic factors affecting *MBS* prices and satisfy

$$dX_t^x = \mu(X_t^x, T - t)\,dt + \sigma(X_t^x, T - t)\,dB_t \tag{10.36}$$

in $[0, T]$, where $X_0^x = x$, and the coefficients $\mu : \mathbb{R}^N \times [0, T] \to \mathbb{R}^N$ and $\sigma : \mathbb{R}^N \times [0, T] \to \mathbb{R}^{N\times d}$ are continuous in $\mathbb{R}^N \times [0, T]$ and x-Lipschitz continuous, uniformly in time, see [19].

(S2) For every $i = 1, \dots, k$ and any initial state $x \in \mathbb{R}^N$, the prepayment function is $S_t^i(x) = s_{0,i}(X_t^x, t)$, where $s_{0,i} \in C^{2,1}(\mathbb{R}^N \times [0, T])$ and $s_{0,i} \ge 0$.

Let $h_{0,i} = MB_i e^{-s_{0,i}}$, where the function MB_i is defined in (10.5) for $i = 1, \dots, k$.

(S3) There exist functions $u_i^{\mathrm{Col}} \in C^{2,1}(\mathbb{R}^N \times [0, T])$, $i = 1, \dots, k$, such that $u_i^{\mathrm{Col}} \ge -h_{0,i}$ in $\mathbb{R}^N \times [0, T]$ and

$$V_t^{\mathrm{Col},i}(x) = u_i^{\mathrm{Col}}(X_t^x, t) + h_{0,i}(X_t^x, t), \quad x \in \mathbb{R}^N,\ t \in [0, T]. \tag{10.37}$$

We define an equilibrium as in [13].

Definition 10.11 Let $x \in \mathbb{R}^N$ be the observed state of the economy. An equilibrium for the *MBS* market is a d-dimensional process $\gamma(x)$, adapted to the filtration of B, such that, defining $\mathbb{Q}$ as the measure (10.14) associated to $\gamma(x)$ (i.e., that makes $\widehat{B}_t = B_t + \int_0^t \gamma_s(x)\,ds$ a Brownian motion), for all dates $t \in [0, T]$, relation (10.21) holds, and

$$\gamma_t(x) = \rho \frac{\sum_{i=1}^k D_t G_t^{\mathrm{Col},i}(x)}{\sum_{i=1}^k V_t^{\mathrm{Col},i}(x) + V_t^{\mathrm{Riskless}}} \tag{10.38}$$

for every $t \in [0, T]$.

Remark 10.12 In the work of Gabaix et al. [13] there is no comprehensive theoretical motivation for defining the m.p.r. as in (10.38). However, we can give the intuition behind this formula. The investor behavior can be explained by a Merton-type problem [23], where he maximizes an objective function on his wealth (portfolio returns). For a large class of utility functions and under d-dimensional source of randomness, this leads to the following risk premium:

$$\mu_t^p - \delta(t) = \rho\langle\sigma_t^p, \sigma_t^M\rangle \tag{10.39}$$

for an asset p with expected return μ_t^p (under the measure $\mathbb{P}$) and sensitivity $\sigma_t^p \in \mathbb{R}^d$ to market price fluctuations. Here ρ is a risk aversion parameter, and $\sigma_t^M \in \mathbb{R}^d$ is the total sensitivity of market returns. If p entitles the holder to receive cash-flows, then $\sigma_t^p p_t$ represents the diffusion coefficient of the related gain process. σ_t^M is expressed by a weighted average of the sensitivities of the market as a whole, the weights being the prices of the assets. If the market is made up of k *MBS*s and the riskless asset (10.9), then it has the total sensitivity

$$\sigma_t^M = \frac{\sum_{i=1}^k D_t G_t^{\mathrm{Col},i}(x)}{\sum_{i=1}^k V_t^{\mathrm{Col},i}(x) + V_t^{\mathrm{Riskless}}}. \tag{10.40}$$

Let $p_t = V_t^{\mathrm{Col},i}$ for $i = 1, \dots, k$. Multiplying (10.39) by p_t and comparing this relation with the ith equation of the linear system (10.12) at time t, we deduce that (10.38) is an admissible m.p.r.

Theorem 10.13 *Assume* **(S1)**–**(S3)**. *Suppose that for any initial state of the economy $x \in \mathbb{R}^N$, there exists an equilibrium $\gamma(x)$ for the* MBS *market. Then $u^{\mathrm{Col}} = (u_1^{\mathrm{Col}}, \dots, u_k^{\mathrm{Col}})$ is a solution of the following system of equations*:

$$\begin{cases} \rho\left\langle\sigma_0^\top \nabla u_i, \frac{\sum_{j=1}^k \sigma_0^\top \nabla u_j}{V_s^{\mathrm{Riskless}} + \sum_{j=1}^k [h_{0,j} + u_j]}\right\rangle \\ \quad = -\delta(s)(h_{0,i} + u_i) + \tau_i h_{0,i} \\ \qquad + \langle\nabla u_i, \mu_0\rangle + \frac{\partial u_i}{\partial s} + \frac{1}{2}\operatorname{tr}(\sigma_0 \sigma_0^\top \nabla^2 u_i), \quad (x, s) \in \mathbb{R}^N \times (0, T), \\ u_i(x, T) = 0, \quad x \in \mathbb{R}^N, \end{cases}$$

for every $i = 1, \dots, k$, where $\sigma_0(x, s) = \sigma(x, T - s)$ and $\mu_0(x, s) = \mu(x, T - s)$ for $(x, s) \in \mathbb{R}^N \times [0, T]$. Furthermore, for all $i = 1, \dots, k$, $x \in \mathbb{R}^N$, and $t \in [0, T]$, the following holds:

$$u_i^{\mathrm{Col}}(X_t^x, t) = \mathbb{E}^{\mathbb{Q}}\left[\int_t^T \left(\tau_i - \delta(s)\right) e^{-\int_t^s \delta(u)\,du} h_{0,i}(X_s^x, s)\, ds \,\middle|\, \mathcal{F}_t\right] \tag{10.41}$$

$\mathbb{P}$-a.s.

Proof Let $x \in \mathbb{R}^N$ and $X_s = X_s^x$ for any $s \in [0, T]$. To simplify the presentation, in the sequel we omit the dependence on x. By Definition 10.11 and using (10.21), we get

$$\langle\sigma_{G,i}^{MBS}(s), \gamma_s(x)\rangle = \mu_{G,i}^{MBS}(s) - \delta(s) V_s^{MBS,i} \tag{10.42}$$

for every $s \in [0,T]$, $i = 1,\dots,k$. Applying Itô's lemma to (10.11) and using (**S2**)–(**S3**), we find

$$\begin{aligned} dG_s^{MBS,i} &= d\big(u_i^{\mathrm{Col}}(X_s,s) + h_{0,i}(X_s,s)\big) + \tau_i h_{0,i}(X_s,s)\,ds - dh_{0,i}(X_s,s) \\ &= \bigg[\big\langle \nabla u_i^{\mathrm{Col}}(X_s,s), \mu_0(X_s,s)\big\rangle + \frac{\partial u_i^{\mathrm{Col}}}{\partial s}(X_s,s) + \frac{1}{2}\operatorname{tr}\big(\sigma_0\sigma_0^\top \nabla^2 u_i^{\mathrm{Col}}(X_s,s)\big) \\ &\quad + \tau_i h_{0,i}(X_s,s)\bigg]ds + \nabla^\top u_i^{\mathrm{Col}}(X_s,s)\sigma_0\,dB_s. \end{aligned} \tag{10.43}$$

Therefore, $\sigma_{G,i}^{MBS}(s) = \nabla^\top u_i^{\mathrm{Col}}(X_s,s)\sigma_0(X_s,s)$, and

$$\mu_{G,i}^{MBS}(s) = \big\langle \nabla u_i^{\mathrm{Col}}, \mu_0\big\rangle + \frac{\partial u_i^{\mathrm{Col}}}{\partial s} + \frac{1}{2}\operatorname{tr}\big(\sigma_0\sigma_0^\top \nabla^2 u_i^{\mathrm{Col}}\big) + \tau_i h_{0,i}. \tag{10.44}$$

By (10.38) and (10.37) we have

$$\gamma_s(x) = \rho \frac{\sum_{j=1}^k \sigma_0^\top(X_s,s)\nabla u_j^{\mathrm{Col}}(X_s,s)}{\sum_{j=1}^k [u_j^{\mathrm{Col}}(X_s,s) + h_{0,j}(X_s,s)] + V_s^{\mathrm{Riskless}}}. \tag{10.45}$$

Introducing the previous relations into (10.42), we obtain

$$\begin{aligned} &\rho\Big\langle \sigma_0^\top(X_s,s)\nabla u_i^{\mathrm{Col}}(X_s,s), \frac{\sum_{j=1}^k \sigma_0^\top(X_s,s)\nabla u_j^{\mathrm{Col}}(X_s,s)}{\sum_{j=1}^k [u_j^{\mathrm{Col}}(X_s,s) + h_{0,j}(X_s,s)] + V_s^{\mathrm{Riskless}}}\Big\rangle \\ &\quad = \big\langle \nabla u_i^{\mathrm{Col}}(X_s,s), \mu_0(X_s,s)\big\rangle + \frac{1}{2}\operatorname{tr}\big(\sigma_0\sigma_0^\top \nabla^2 u_i^{\mathrm{Col}}(X_s,s)\big) + \tau_i h_{0,i}(X_s,s) \\ &\qquad + \frac{\partial u_i^{\mathrm{Col}}}{\partial s}(X_s,s) - \delta(s)\big(u_i^{\mathrm{Col}}(X_s,s) + h_{0,i}(X_s,s)\big) \end{aligned} \tag{10.46}$$

for any $s \in [0,T]$. Given the arbitrary choice of the initial state x, the evaluation of (10.46) at $s = 0$ proves the first statement of the theorem. Using (10.21) with $a_s^i = h_{0,i}(X_s,s)$ and (**S3**), the stochastic representation (10.41) follows by applying Itô's formula to $u_i^{\mathrm{Col}}(X_s,s)$. □

With the same arguments and using part (**B**) of Theorem 10.4, it is easy to prove the following converse result of Theorem 10.13.

Theorem 10.14 *Assume* (**S1**)–(**S2**), *where σ is bounded. Let $u = (u_1,\dots,u_k)$ be a smooth solution to system* (10.41) *such that ∇u_i is bounded and $u_i + h_{0,i} \ge 0$ for every $i = 1,\dots,k$. For any $x \in \mathbb{R}^N$, $i = 1\dots,k$, $t \in [0,T]$, define*

$$V_t^i = u_i\big(X_t^x,t\big) + h_{0,i}\big(X_t^x,t\big), \qquad G_t^i = V_t^i + \int_0^t dc_s^i. \tag{10.47}$$

Then $(V_t^{\text{Riskless}}, G_t^1, \dots, G_t^k)$ *defines an arbitrage-free market which admits and equilibrium in the sense of Definition* 10.11.

Based on Theorems 10.13 and 10.14, the results discussed in the next sections will be devoted to the case $k = 1$. Let us make the change of variable $s = T - t$ and define:

$$\begin{aligned} \xi(t) &= A_0 e^{\int_0^{T-t} \delta(s)\,ds}, \qquad r(t) = \delta(T-t), \\ h(x,t) &= h_{0,1}(x, T-t), \end{aligned} \tag{10.48}$$

$$U(x,t) = u_1^{\text{Col}}(x, T-t). \tag{10.49}$$

Then system (10.41) reduces to the following partial differential equation:

$$\begin{aligned} &\frac{\partial U}{\partial t} - \frac{1}{2}\operatorname{tr}\big(\sigma(x,t)\sigma^\top(x,t)\nabla^2 U\big) - \big\langle \mu(x,t), \nabla U\big\rangle + \rho \frac{|\sigma^\top(x,t)\nabla U|^2}{U + h(x,t) + \xi(t)} \\ &\quad - \tau h(x,t) + r(t)\big(U + h(x,t)\big) = 0 \quad \text{in } \mathbb{R}^N \times (0,T) \end{aligned} \tag{10.50}$$

with $U(x,0) = 0$ everywhere in $\mathbb{R}^N$.

Some properties of the solution U are required in order to be consistent with the financial problem. More precisely, we require that (1) $U + h$ is nonnegative (that is, $V^{\text{Col}} \geq 0$) and (2) U satisfies the stochastic representation formula

$$\begin{aligned} &U\big(X_t^x, T-t\big) \\ &\quad = \mathbb{E}^{\mathbb{Q}}\bigg[\int_t^T \big(\tau - r(T-s)\big) e^{-\int_t^s r(T-\kappa)\,d\kappa} h\big(X_s^x, T-s\big)\,ds \bigg| \mathcal{F}_t\bigg]. \end{aligned} \tag{10.51}$$

Remark 10.15

(a) The previous setting applies also to existing pools of mortgages. In this case, $MB(0)$ is the pool balance at the observation time.
(b) Since $a_t = h(X_t^x, T-t)$, by (10.1) we deduce that the *MBS* market quote is

$$\frac{a_t + U(X_t^x, T-t)}{a_t}.$$

(c) As in the theory of interest rate modeling, to calibrate our model, it suffices to assign the dynamics (10.36) under the risk-adjusted measure for the economic factors; then using market prices to match the solution of (10.50), one can estimate model parameters μ, σ, and ρ.

In the following sections we provide sufficient conditions for the existence of a unique solution to (10.50) and for the existence of an equilibrium, as in Definition 10.11.

10.3 Viscosity Solutions

Through previous sections, the interplay between *MBS* pricing and degenerate parabolic equations have been put in light. This section contains a review on viscosity solutions, which should be of use in the following sections.

Since (10.50) may be degenerate when $N > d$ and has a quadratic growth in the gradient, it seems natural to study problem (10.50) in the framework of viscosity solutions. Various existence and comparison/uniqueness results for viscosity solutions to degenerate equations of second order can be found in [14]. We are interested in solving a Cauchy problem in the domain $\mathbb{R}^N \times (0, T)$, with initial datum $u_0 \in C(\mathbb{R}^N; (a, b))$:

$$\begin{cases} \partial_t u + F(x, t, u, \nabla u, \nabla^2 u) = 0, & (x, t) \in \mathbb{R}^N \times (0, T), \\ u(x, 0) = u_0(x), & x \in \mathbb{R}^N, \end{cases} \tag{10.52}$$

where $F \in C(\mathbb{R}^N \times [0, T] \times (a, b) \times \mathbb{R}^N \times \mathcal{S}^N)$, $\mathcal{S}^N$ being the set of symmetric $N \times N$ real matrices. For a complete presentation of our study, we recall the definition of viscosity solutions. The notion of derivatives in the viscosity sense involves smooth test functions touching from below (respectively, from above) the graph of u at the point of interest.

Definition 10.16 Let $u : \mathbb{R}^N \times [0, T] \to (a, b)$ be locally bounded. The *parabolic super 2-jet* of u at the point $(x, t) \in \mathbb{R}^N \times [0, T)$ is the following subset of $\mathbb{R} \times \mathbb{R}^N \times \mathcal{S}^N$:

$$\mathcal{P}^{2,+}u(x, t) = \big\{\big(\partial_t\varphi(x, t), \nabla\varphi(x, t), \nabla^2\varphi(x, t)\big) : \varphi \in C^{2,1}\big(\mathbb{R}^N \times [0, T)\big), u - \varphi \\ \text{has a global strict maximum at } (x, t)\big\}.$$

Similarly, we define the *parabolic lower 2-jet* as $\mathcal{P}^{2,-}u = -\mathcal{P}^{2,+}(-u)$.

Definition 10.17 A locally bounded function $u : \mathbb{R}^N \times [0, T] \to (a, b)$ that is upper semicontinuous (resp. lower semicontinuous) is a *viscosity subsolution* (resp. lower *viscosity supersolution*) to (10.52) if for any $(x, t) \in \mathbb{R}^N \times (0, T)$ and for every $(b, q, A) \in \mathcal{P}^{2,+}u(x, t)$ (resp. $(b, q, A) \in \mathcal{P}^{2,-}u(x, t)$), the inequality

$$b + F\big(x, t, u(x, t), q, A\big) \leq 0 \quad (\text{resp.} \geq 0) \tag{10.53}$$

holds, and $u(x, 0) \leq u_0(x)$ (resp. $u(x, 0) \geq u_0(x)$) for any $x \in \mathbb{R}^N$.

Definition 10.18 A locally bounded function $u : \mathbb{R}^N \times [0, T) \to (a, b)$ is a *viscosity solution* to (10.52) if its upper semicontinuous envelope is a viscosity subsolution and its lower semicontinuous envelope is a viscosity supersolution.

The existence and uniqueness of a solution to (10.52) is based on a well-known technique which combines the Perron's method and a comparison principle between sub- and super-solutions of the problem. We refer to [4] for definitions of the upper and lower semicontinuous envelopes.

Unfortunately, viscosity solutions have very few regularity, being merely continuous, and regularity results have to be proved. We also notice that the possible strong degeneracy of (10.50) does not allow, in general, any smoothing effect. Therefore regularity estimates have to be global. The following theorem, whose proof is given in [25], provides an existence and uniqueness result to (10.50). Consider the following assumption:

(P) $h \in C^{2,1}(\mathbb{R}^N \times [0,T])$, $\partial_t h(\cdot,t)$, $\mathrm{tr}(\sigma\sigma^\top(t)\nabla^2 h(\cdot,t))$, $\nabla h(\cdot,t)$ are bounded and x-Lipschitz continuous, uniformly in time.

Theorem 10.19 (Theorem 4.2 and Proposition 4.6 in [25]) *Let σ be independent of x. Under assumptions* (**S1**) *and* (**P**) *and assuming that $\tau \geq \delta$ in $[0,T]$, problem* (10.50) *admits a unique bounded viscosity solution U such that $U + h \geq 0$.*

Remark 10.20 We observe that the inequality $\tau \geq \delta(\cdot)$ is a matter of course; otherwise the investment in the *MBS* market is less profitable than a bank account. Hence, we shall make this assumption in the rest of the paper.

10.4 Financial Motivations for the Regularity of *U*

We recall that in the case of one *MBS* pass-through, as in Definition 10.11, the m.p.r. can be expressed in the following form:

$$\gamma_t(x) = \rho \frac{\sigma^\top(X_t^x, T-t)\nabla U(X_t^x, T-t)}{U(X_t^x, T-t) + h(X_t^x, T-t) + \xi(T-t)} \tag{10.54}$$

for every $t \in (0,T]$, $\mathbb{P}$-a.s., implying the dependence of the risk-neutral measure $\mathbb{Q}$ on the solution U and its gradient. This generates a convoluted stochastic representation for U, precisely given by (10.51). The typical technique used in the framework of viscosity solutions to prove a representation like (10.51) is based on the use of the dynamic programming principle, see [11]. In that context, the existence of the value function is guaranteed by the existence of the expected value. Here, we cannot say that U exists because the expected value in (10.51) depends on U itself.

Actually, (10.50) does not arise from a control problem, so the solution U is not the value function of an optimization problem. Since viscosity solutions are merely continuous, the existence of a viscosity solution to (10.50) is not sufficient to prove that (10.51) holds. One possibility is to use the regularity of the process X_t to show that the functional given by the right-hand side of (10.51) is a viscosity solution of (10.50). This approach would require to express the dynamics of the process X_t in terms of the new measure $\mathbb{Q}$, that is,

$$dX_t = \big(\mu(X_t, T-t) - \sigma(X_t, T-t)\gamma_t(x)\big)\,dt + \sigma(X_t, T-t)\,d\widehat{B}_t,$$

where $\widehat{B}_t$ is given by (10.15). However, the usual regularity assumptions require that the coefficients of the process are Lipschitz continuous in the spatial variable. By (10.54), we argue that more regularity on ∇U is needed. Therefore, we have to prove that U belongs to a class of regularity not covered by the classical theory of viscosity solutions. Unfortunately, due to the quadratic growth term in (10.50), we can obtain this property only assuming higher regularity on the coefficients of the equation. Our approach is based on the use of an extension of Itô's formula for less regular functions. Some recent results about this topic involve the quadratic covariation to replace the second-order term. Due to this fact, they are not relevant for our purposes, while we follow the approach of Haussmann [16]. This result preserves the structure of the classical Itô's formula, and it applies to a weaker class of regularity. When the process X_t admits a density, this kind of regularity defines the correct class where the solution should be found in order to obtain a complete justification of (10.51).

In the sequel, we use the standard notation for Sobolev spaces denoting by $\mathbb{W}^{k,\infty}(A)$ the space of functions which are bounded in $A \subset \mathbb{R}^N$, together with their weak derivatives up to order k. We also use the notation $\mathbb{W}^{2,1,\infty}(\mathbb{R}^N \times (0,T))$ for the space of bounded functions u with weak derivatives $\partial_t u$, $\partial_i u$, $\partial^2_{ij} u$, $i, j = 1, \ldots, N$, bounded in $\mathbb{R}^N \times (0,T)$. The following is a regularity result for the *MBS* equation.

Theorem 10.21 *Let σ be independent of x and assume* (**S1**). *Let $h(\cdot,t) \in \mathbb{W}^{4,\infty}(\mathbb{R}^N)$, $\partial_t h(\cdot,t)$, $\mu(\cdot,t) \in \mathbb{W}^{2,\infty}(\mathbb{R}^N)$, uniformly in time. Then problem* (10.50) *admits a unique solution $U \in \mathbb{W}^{2,1,\infty}(\mathbb{R}^N \times (0,T))$.*

Proof Since the assumptions of Theorem 10.19 are verified, there exists a unique viscosity solution U such that $M_0 > U + h + \xi > m_0 > 0$ in $\mathbb{R}^N \times [0,T]$ for some constants m_0 and M_0. Let $v = U + h + \xi$; then v is a viscosity solution of the equation

$$\begin{aligned}\partial_t v &- \frac{1}{2}\operatorname{tr}\big(\sigma(t)\sigma^\top(t)\nabla^2 h\big) - \langle \mu, \nabla v\rangle + \lambda(v)\big|\sigma^\top(t)\nabla v\big|^2 + \eta(v)\big\langle\sigma^\top \nabla v, w\big\rangle \\ &+ f(x,t,v) = 0 \qquad (10.55)\end{aligned}$$

in $\mathbb{R}^N \times (0,T)$, where $\lambda(v) = \rho/v$, $\eta(v) = -2\rho/v$, $w(x,t) = \sigma^\top(t)\nabla h(x,t)$, and

$$\begin{aligned}f(x,t,v) = -\partial_t h &+ \frac{1}{2}\operatorname{tr}\big(\sigma(t)\sigma^\top(t)\nabla^2 h\big) + \big\langle\mu(x,t), \nabla h\big\rangle + \frac{\rho}{v}\big|\sigma^\top(t)\nabla h\big|^2 \\ &- \tau h(x,t) + r(t)v - \xi'(t) - r(t)\xi(t). \qquad (10.56)\end{aligned}$$

Since $v(\cdot,0) = \xi(0)$ and $f(\cdot,t,\cdot) \in \mathbb{W}^{2,\infty}(\mathbb{R}^N \times (m_0, M_0))$ uniformly in time, then $v \in \mathbb{W}^{2,1,\infty}(\mathbb{R}^N \times (0,T))$ by Theorem 3.4, p. 233 in [26]. Hence the same holds for $U = v - h - \xi$. □

10.4.1 Stochastic Representation

Let us back to the *MBS* model. In this section, we shall prove representation (10.51) using a fundamental property of $\mathbb{W}^{2,1,\infty}$ functions, see [3].

To rigorously derive (10.51), we need to apply Itô's formula to $U(X_t, T-t)$. In [16] it is proved that Itô's formula holds for functions in $\mathbb{W}^{2,1,\infty}(\mathbb{R}^N \times (0,T))$, provided that it is interpreted appropriately, using the generalized Hessian. Here, we shall use the regularity of the solution U and that result to obtain the equality in (10.51).

Theorem 10.22 *Assume* **(S1)**–**(S2)**, *where σ is bounded. Let $x_0 \in \mathbb{R}^N$ be the starting point of X_t. If $U \in \mathbb{W}^{2,1,\infty}(\mathbb{R}^N \times (0,T))$ is the solution of* (10.50) *satisfying $U + h \geq 0$ everywhere in $\mathbb{R}^N \times [0,T]$, then the following hold*:

(i) *For all $t \in [0,T]$ and $p \geq 2$, $\gamma_t(x_0) \in \mathbb{L}^{1,p}$.*

If for every $t \in (0,T]$, X_t admits a density $p(\cdot,t)$ w.r.t. the Lebesgue measure in $\mathbb{R}^N$ such that p is Borel measurable in both the variables x and t, then

(ii) *the representation formula* (10.51) *holds at x_0, $\mathbb{P}$-a.s.*
(iii) *$\gamma_t(x_0)$ is an equilibrium, in the sense of Definition* 10.11, *for the market $(V_t^{\mathrm{Riskless}}, G_t^{\mathrm{Col}})$, where $G_t^{\mathrm{Col}} = U(X_t, T-t) + h(X_t, T-t) + \int_0^t dc_s$, in $[0,T]$.*

Remark 10.23

(a) We observe that under the hypotheses of Theorem 10.21, conclusions of Theorem 10.22 hold, the m.p.r. (10.54) satisfies Novikov's condition, and the assumptions of Theorem 10.7 are verified.
(b) Criteria ensuring the existence of a density can be found in [24], where the absolute continuity and the smoothness of the density under Hörmander's condition are deeply analyzed.

The assertion of Haussmann in [16] interprets Itô's rule through some processes substituting the usual derivatives of the function at (X_t, t). They coincide with the usual derivatives if the underlying process belongs to some set of full Lebesgue measure. To state our formula, we have to neglect the term which corresponds to an integration over the paths of X_t which fall into a set of null Lebesgue measure. The existence of a density for the process X_t can be explained by this purpose. However, as is showed in Example 10.26, this assumption allows us to consider the case of possibly strongly degenerate diffusions ($N > d$). We need of a technical result. In the sequel, the notation 1_A will be used for the indicator function of a set A.

Lemma 10.24 *Let $\mathbb{Q}$ be a probability measure equivalent to $\mathbb{P}$, over Ω, such that $\frac{d\mathbb{Q}}{d\mathbb{P}}$ has finite variance. Let $W \subset \mathbb{R}^N \times [0,T]$ be a set of zero Lebesgue measure.*

Define

$$Z_{t,T} = \int_t^T m_s 1_W(X_s, s)\,ds + n_s 1_W(X_s, s)\,dB_s, \quad t \in [0, T], \tag{10.57}$$

where m_s and n_s are $\{\mathcal{F}_s\}_{0 \le s \le T}$-adapted bounded processes. If X_t satisfies the assumptions of Theorem 10.22, *then*

$$\mathbb{E}^{\mathbb{Q}}\big[|Z_{t,T}|\big|\mathcal{F}_t\big] = 0 \quad \mathbb{P}\text{-}a.s. \tag{10.58}$$

Proof of Theorem 10.22 (i) Let $p \ge 2$. We observe that γ can be written as $\gamma_t = \varphi(X_t, t)$, where

$$\varphi(x,t) = \rho \frac{\sigma^\top(x, T-t)\nabla U(x, T-t)}{U(x, T-t) + h(x, T-t) + \xi(T-t)}. \tag{10.59}$$

The properties of U imply the Lipschitz continuity of $\varphi(\cdot, t)$ for every t, with a Lipschitz constant less than some $K_\varphi > 0$, uniformly in t.

By the chain rule we deduce that $\gamma_t \in \mathbb{D}^{1,p}$ for any $t \in [0, T]$ and also

$$\sup_{0 \le s \le T} \mathbb{E}\Big[\sup_{s \le t \le T} |D_s \gamma_t|^p\Big] \le K_\varphi N \sqrt{d} \sup_{0 \le r \le T} \mathbb{E}\Big[\sup_{r \le t \le T} \|D_r X_t\|^p\Big], \tag{10.60}$$

the right-hand side being finite, since X_t is the solution of (10.36), see [24]. This implies that $\gamma \in \mathbb{L}^{1,p}$ for any $p \ge 2$.

(ii) Let $\mathcal{L}^{N+1}$ denote the Lebesgue measure in $\mathbb{R}^{N+1}$. Set $\Sigma(x,t) = \sigma(x, T-t)$, $\mu^\circ(x,t) = \mu(x, T-t)$, $h^\circ = h(x, T-t)$, $\xi^\circ(t) = \xi(T-t)$, and $U^\circ(x,t) = U(x, T-t)$. Then U° is a viscosity solution of the equation

$$\begin{aligned} \partial_t U^\circ + \frac{1}{2}\mathrm{tr}\big(\Sigma\Sigma^\top \nabla^2 U^\circ\big) + \langle \mu^\circ, \nabla U^\circ\rangle - \rho \frac{|\Sigma^\top \nabla U^\circ|^2}{U^\circ + h^\circ + \xi^\circ} \\ = \delta\big(U^\circ + h^\circ\big) - \tau h^\circ \end{aligned} \tag{10.61}$$

in $\mathbb{R}^N \times (0, T)$, with $U^\circ(\cdot, T) \equiv 0$. Since $U^\circ \in \mathbb{W}^{2,1,\infty}(\mathbb{R}^N \times (0,T))$, applying the results of [3] for a.e. $(x,t) \in \mathbb{R}^N \times (0,T)$, we get $(\partial_t U^\circ(x,t), \nabla U^\circ(x,t), \nabla^2 U^\circ(x,t)) \in \mathcal{P}^{2,\pm} U^\circ(x,t)$, where $\partial_t U^\circ, \nabla U^\circ, \nabla^2 U^\circ$, represent the weak derivatives of U°. Hence (10.61) holds for $\mathcal{L}^{N+1}$-a.e. $(x,t) \in \mathbb{R}^N \times (0,T)$. We now apply Theorem 3.1, p. 733 in [16], to the function $(x,s) \mapsto e^{-\int_t^s \delta(\kappa)\,d\kappa} U^\circ(x,s)$. In particular, there exist adapted processes ℓ, β, and α such that

$$\begin{aligned} 0 &= e^{-\int_t^T \delta(\kappa)\,d\kappa} U^\circ(X_T, T) \\ &= U^\circ(X_t, t) + \int_t^T \Big[\ell_s + \langle \mu^\circ(X_s, s), \beta_s\rangle \\ &\quad + \frac{1}{2}\mathrm{tr}\big(\Sigma\Sigma^\top(X_s, s)\alpha_s\big)\Big] ds + \int_t^T \beta_s^\top \Sigma(X_s, s)\,dB_s \end{aligned} \tag{10.62}$$

for every $t \in (0,T)$, $\mathbb{P}$-a.s. Furthermore, there exists a set $A \subset \mathbb{R}^N \times (0,T)$ of full Lebesgue measure such that the usual derivatives of U° exist in A and

$$\ell_s = e^{-\int_t^s \delta(\kappa)\,d\kappa}\big[\partial_s U^\circ(X_s,s) - \delta(s)U^\circ(X_s,s)\big], \tag{10.63}$$

$$\beta_s = e^{-\int_t^s \delta(\kappa)\,d\kappa}\nabla U^\circ(X_s,s), \qquad \alpha_s = e^{-\int_t^s \delta(\kappa)\,d\kappa}\nabla^2 U^\circ(X_s,s), \tag{10.64}$$

whenever $(X_s,s) \in A$. Without loss of generality, we can assume that (10.61) holds in A. The application of (10.61), (10.62), and (10.63)–(10.64) yields

$$\begin{aligned} 0 &= U^\circ(X_t,t) + \int_t^T \Big[\ell_s + \langle \mu^\circ(X_s,s), \beta_s\rangle + \frac{1}{2}\operatorname{tr}\big(\Sigma\Sigma^\top(X_s,s)\alpha_s \\ &\quad - \beta_s^\top \Sigma(X_s,s)\gamma_s(x_0)\big)\Big]1_A(X_s,s)\,ds + \int_t^T 1_A(X_s,s)\beta_s^\top\Sigma(X_s,s)\,d\widehat{B}_s + Z_{t,T} \\ &= U^\circ(X_t,t) + \int_t^T e^{-\int_t^s \delta(\kappa)\,d\kappa}\big[\delta(s)\big(U^\circ(X_s,s) + h^\circ(X_s,s)\big) - \tau h^\circ(X_s,s) \\ &\quad - \delta(s)U^\circ(X_s,s)\big]\,ds + \int_t^T \beta_s^\top\Sigma(X_s,s)1_A(X_s,s)\,d\widehat{B}_s + Z_{t,T}. \end{aligned} \tag{10.65}$$

By the boundedness of ℓ, β, and α, we recognize that the remaining term $Z_{t,T}$ has the same structure of (10.57). Since the risk-neutral measure $\mathbb{Q}$ is equivalent to $\mathbb{P}$ and $\frac{d\mathbb{Q}}{d\mathbb{P}}$ has finite variance, we can apply Lemma 10.24 to obtain that $Z_{t,T}$ has a null conditional expectation w.r.t. $\mathbb{Q}$. Taking the conditional expected value in both the left- and right-hand sides of (10.65), we get

$$U^\circ(X_t,t) = \mathbb{E}^{\mathbb{Q}}\bigg[\int_t^T e^{-\int_t^s \delta(\kappa)\,d\kappa}\big(\tau - \delta(s)\big)h^\circ(X_s,s)1_A(X_s,s)\,ds\,\bigg|\,\mathcal{F}_t\bigg]$$

$\mathbb{P}$-a.s. Since X admits a density in $\mathbb{R}^N$ and A has full Lebesgue measure, the assertion is proved.

(iii) This easily follows by using (10.50) and (i), (ii). □

Remark 10.25 For every probability measure $\mathbb{Q}$, equivalent to $\mathbb{P}$ on Ω, and for any $\mathcal{F}_T$-measurable and integrable random variable Y, the following holds [19]:

$$\mathbb{E}^{\mathbb{Q}}[Y|\mathcal{F}_t] = \frac{\mathbb{E}[Y\frac{d\mathbb{Q}}{d\mathbb{P}}|\mathcal{F}_t]}{\mathbb{E}[\frac{d\mathbb{Q}}{d\mathbb{P}}|\mathcal{F}_t]}. \tag{10.66}$$

Proof of Lemma 10.24 Let $\vartheta > 0$ be such that $|m_s| + |n_s| \le \vartheta$ for any $s \in [0,T]$, $\mathbb{P}$-a.s. Define $v_t = \mathbb{E}[(\frac{d\mathbb{Q}}{d\mathbb{P}})^2|\mathcal{F}_t]$ and $y_t = \vartheta\sqrt{v_t(T-t)}$. The Itô isometry and Jensen's inequality yield

$$\mathbb{E}\bigg[\frac{d\mathbb{Q}}{d\mathbb{P}}\bigg|\int_t^T n_s 1_W(X_s,s)\,dB_s\bigg|\,\bigg|\,\mathcal{F}_t\bigg]^2 \le v_t\vartheta^2\mathbb{E}\bigg[\int_t^T 1_W(X_s,s)\,ds\,\bigg|\,\mathcal{F}_t\bigg] \tag{10.67}$$

and

$$\mathbb{E}\left[\frac{d\mathbb{Q}}{d\mathbb{P}}\left|\int_t^T m_s 1_W(X_s, s)\,ds\right|\middle|\mathcal{F}_t\right] \le y_t \mathbb{E}\left[\int_t^T 1_W(X_s, s)\,ds\middle|\mathcal{F}_t\right]^{\frac{1}{2}} \tag{10.68}$$

for every $t \in [0, T]$, $\mathbb{P}$-a.s. By Remark 10.25, it suffices to prove that the random variable on the right-hand side of (10.67)–(10.68) is zero. Since this variable is nonnegative, it suffices to prove that its expectation is zero. By Fubini's theorem, this expected value can be expressed, using the density of X_s, in the following way:

$$\int_t^T \int_W p(x, s)\,dx\,ds. \tag{10.69}$$

Since W is a set of zero Lebesgue measure, the assertion follows. □

The following is a general model for the factors affecting prepayments, which is strongly degenerate and satisfies the assumptions of Theorem 10.22.

Example 10.26 Path dependency refers to the dependence of the prepayment function S on the trajectory followed by one or more of the underlying economic factors that affect prepayments, that is, $s_0 = s_0(y)$, where $s_0(y) = y$ for $y \ge 0$, and s_0 is extended to a smooth function for $y < 0$; the prepayment rate y_t follows $y_t = \int_0^t \eta(x_s, y_s)\,ds$ for some nonnegative function η; here x_t is a d-dimensional process which represents the economic factors. Since y_t is nonnegative, the values s_0 on $(-\infty, 0)$ do not affect the model, while they guarantee the smoothness of the function h required in Theorems 10.19 and 10.21. Let x_t be driven by $dx_t = b(x_t)\,dt + c(x_t)\,dB_t$. The underlying process is $X_t = (x_t, y_t) \in \mathbb{R}^N$, with $N = d + 1$, starting at $X_0 = (x_0, 0)$, and

$$\mu(X) = \begin{pmatrix} b(y) \\ \eta(x, y) \end{pmatrix} \in \mathbb{R}^N, \qquad \sigma(X) = \begin{pmatrix} c(x) \\ 0 \cdots 0 \end{pmatrix} \in \mathbb{R}^{N \times d} \tag{10.70}$$

for every $X = (x, y)$. It is clear that the quadratic form $\sigma\sigma^\top$ is strongly degenerate everywhere in $\mathbb{R}^N$. However sufficient conditions can be given in order to guarantee the existence of a density for X_t. Actually, let b, η, c be smooth functions, assume that $c(x_0)$ is invertible, and let $\langle \nabla_x \eta(x_0, 0), c^\lambda(x_0)\rangle \ne 0$ for some $\lambda = 1, \dots, d$, $c^\lambda(x_0)$ being the λth column of $c(x_0)$. Let $\alpha_0 = b - \frac{1}{2}\sum_{i=1}^d \partial_y c^i c^i$, and let $[\cdot, \cdot]$ denote the Lie bracket between vector fields [24]. Under the previous conditions, it is easy to see that

$$\begin{pmatrix} c^1(x_0) \\ 0 \end{pmatrix}, \begin{pmatrix} c^2(x_0) \\ 0 \end{pmatrix}, \dots, \begin{pmatrix} c^d(x_0) \\ 0 \end{pmatrix}, \begin{pmatrix} [\alpha_0, c^\lambda](x_0) \\ -\langle \nabla_x \eta(x_0, 0), c^\lambda(x_0)\rangle \end{pmatrix}$$

is a basis of $\mathbb{R}^N$. We observe that these vectors belong to the Lie algebra given by the classical Hörmander condition, see, in particular, [24]. Therefore X admits a smooth density w.r.t. the Lebesgue measure in $\mathbb{R}^N$.

10.5 The Numerical Solution

In this section, we present sufficient conditions for the numerical approximation of (10.50), proving a convergence result. We approximate the equation in all space by reducing the problem to a bounded domain and imposing the Neumann condition at the boundary. Let B_R be an open ball in $\mathbb{R}^N$ with a fixed center and radius $R > 0$. Then we focus on the problem

$$F\big(x, t, U, \partial_t U, \nabla U, \nabla^2 U\big) = 0 \quad \text{in } \overline{B_R} \times [0, T), \tag{10.71}$$

where, for every $(x, t, U, b, q, X) \in B_R \times (0, T) \times \mathbb{R} \times \mathbb{R}^N \times \mathcal{S}^N$ with $U + h(x, t) + \xi(t) > 0$, we have

$$\begin{aligned} F(x, t, U, b, q, X) = b &- \frac{1}{2}\operatorname{tr}\big(\sigma(t)\sigma^T(t)X\big) - \big\langle \mu(x, t), q\big\rangle + H(x, t, U, q) \\ &- \tau h(x, t) + r(t)\big(U + h(x, t)\big), \end{aligned} \tag{10.72}$$

while $F(\cdot) = U$ at $t = 0$ and $F(\cdot) = \langle q, n(x)\rangle$ if $(x, t) \in \partial B_R \times [0, T)$. Here n denotes the outward unit normal on ∂B_R, and $H(x, t, U, q) = \rho|\sigma^T(t)q|^2/(\xi(t) + h(x, t) + U)$. For any bounded continuous function k in $\mathbb{R}^N \times [0, T]$ and any $L > 0$, define the space

$$\mathcal{W}(k, L) = \big\{v \in C\big(\overline{B_R} \times [0, T]\big) : k \le v \le L, \text{ in } \overline{B_R} \times [0, T]\big\}. \tag{10.73}$$

The assumptions of Theorem 10.19 are sufficient to prove the comparison principle and the existence of a unique viscosity solution U_R to (10.71)–(10.72) in $\mathcal{W}(-h, L)$ for some $L \ge \sup_{x,t} h$. Actually this is just a translation of the general result proved in [25] with some additional arguments for the boundary condition as in [4]. Moreover we refer to [1] for the study of the rate of convergence of U_R to the solution U of the problem (10.50) as $R \to +\infty$.

We define a numerical grid in $\mathbb{R}^N \times [0, T]$ using the following notation: $\Delta x = (\Delta x_1, \dots, \Delta x_N)$ is the spatial grid size, Δt is the time grid size, $(x_j, t_n) = (j\Delta x, n\Delta t)$, $j = 0, \dots, K$, $n = 0, \dots, M$, are the grid points, v_j^n is the value of the function v, defined on the grid or defined for continuously varying (x, t), at the grid point (x_j, t_n), and $\tilde{v}$ is the vector of v values, $(v_j^n)_{j,n}$. A numerical scheme approximating (10.71)–(10.72) can be written as

$$S\big(\Delta x, \Delta t, x, t, v(x), v\big) = 0, \tag{10.74}$$

where $S : (0, +\infty)^{N+1} \times \overline{B_R} \times [0, T] \times \mathbb{R} \times \mathcal{W}(k, L) \to \mathbb{R}$. Let us recall that, in recent years, a great deal has been done for the numerical approximation of viscosity solutions. For second-order problems, let us refer to the fundamental paper by Barles and Souganidis [2], who first showed convergence results for a large class of numerical schemes. In this section, we shall extend this convergence result to the solution $U_R \in \mathcal{W}(-h, L)$. We assume the scheme (10.74) to satisfy the following properties:

(C1) *Consistency*. For any smooth function $\phi \in \mathcal{W}(-h, L)$, for any $(x, t) \in \overline{B_R} \times [0, T)$, and for some function $\epsilon = \epsilon(\Delta x, \Delta t) > 0$, one has

$$\lim_{\substack{\Delta x, \Delta t \to 0 \\ (y,s)\to(x,t)}} \frac{S(\Delta x, \Delta t, y, s, \phi(y), \phi)}{\epsilon(\Delta x, \Delta t)} = F\big(x, t, \phi, \partial_t \phi, \nabla\phi, \nabla^2\phi\big). \quad (10.75)$$

(C2) *Monotonicity*. For any $\Delta x, \Delta t > 0$, $(x, t) \in \overline{B_R} \times [0, T)$, and $v, w \in \mathcal{W}(-h, L)$ such that $v \geq w$ and $v(x) = w(x)$, one has

$$S\big(\Delta x, \Delta t, x, t, v(x), v\big) \leq S\big(\Delta x, \Delta t, x, t, w(x), w\big). \quad (10.76)$$

(C3) *Stability* For every $\Delta = (\Delta x, \Delta t)$, $\Delta x, \Delta t > 0$, the scheme has a solution $v^\Delta \in \mathcal{W}(-h, L)$.

Remark 10.27 The stability condition (**C3**) is equivalent to the standard stability assumption [2], i.e., if $v^\Delta \in \mathcal{W}(-h, L)$, then $-L \leq v^\Delta \leq L$ for any Δ.

Proposition 10.28 *Let assumptions* (**C1**)–(**C3**) *hold. Then, as* $\Delta = (\Delta x, \Delta t) \to 0$, *the solution* v^Δ *of the scheme* (10.74) *converges locally uniformly* (*l.u.*) *to the unique viscosity* U_R *solution of problem* (10.71)–(10.72).

Sketch of proof Let v^Δ be the solution of (10.74). Condition (**C3**) implies that the functions $\underline{v}, \overline{v}$, defined by $\underline{v}(x, t) = \liminf_{\Delta \to 0, (y,s)\to(x,t)} v^\Delta(y, s)$, $\overline{v}(x, t) = \limsup_{\Delta \to 0, (y,s)\to(x,t)} v^\Delta(y, s)$, belong to $\mathcal{W}(-h, L)$. The monotonicity and consistency assumptions on S imply that $\overline{v}$ and $\underline{v}$ are respectively sub- and supersolutions of the limiting equation (see the proof of Theorem 2.1 in [2]). By the strong comparison principle (see Theorem 4.2 in [25]), $U_R = \underline{v} = \overline{v}$ is the unique viscosity solution of (10.71)–(10.72). □

10.5.1 The Numerical Approximation

In this section, we apply previous results to approximate a two-dimensional model based on path dependency (see Example 10.26). In this case there is an absence of diffusion in the additional dimension, which causes a more difficult analysis to solve the problem, since the diffusion term contributes to the stability of the numerical scheme used to solve the pricing equation. Let $d = 1$ and $x_t \in \mathbb{R}$ be the underlying factor with drift $\mu_1(x, t)$ and volatility $\sigma(t)$. Assume the prepayment process to follow $dy_t = \mu_2(x_t, y_t, t)\,dt$. Let $\mu_2 \geq 0$ and $\sigma_2 \geq \sigma(t) \geq \sigma_1 > 0$ for any $t \in [0, T]$, with the usual regularity assumptions. For the sake of simplicity, let the discount rate δ be constant. We consider the problem on the bounded domain $Q_T = [x_{\min}, x_{\max}] \times [y_{\min}, y_{\max}] \times [0, T]$. The Hamiltonian F in (10.72) reduces to

$$F\big(x,y,t,U,\nabla U,\partial^2_{xx}U\big) = \partial_t U - \frac{1}{2}\sigma^2(t)\partial^2_{xx}U - \big\langle\big(\mu_1(x,t),\mu_2(x,y,t)\big),\nabla U\big\rangle + H(y,t,U,\partial_x U) + \delta U + (\delta-\tau)h(y,t), \tag{10.77}$$

where $H(y,t,U,q) = \sigma^2(t)\rho\frac{|q|^2}{\xi(t)+h(y,t)+U}$ and $\xi(t) = A_0 e^{\delta(T-t)}$.

Applying finite differences, we obtain $v^0_{ij} = 0$ for any $i, j = 0, \dots, K$ and

$$\begin{aligned} S\big(\Delta,n,i,j,v^{n+1}_{ij},\tilde{v}\big) &= v^{n+1}_{ij} - \tilde{S}\big(\Delta,n,i,j,\tilde{v}^n\big) = 0, \\ \tilde{S}\big(\Delta,n,i,j,\tilde{v}^n\big) &= v^n_{ij} + \frac{1}{2}\big(\sigma^2\big)^n\frac{\Delta t}{\Delta x^2}\big(v^n_{i+1,j} - 2v^n_{ij} + v^n_{i-1,j}\big) \\ &\quad + (\mu_1)^n_{ij}\frac{v^n_{i+1,j} - v^n_{i-1,j}}{2(\Delta x/\Delta t)} + (\mu_2)^n_{ij}\frac{v^n_{i,j+1} - v^n_{ij}}{(\Delta y/\Delta t)} \\ &\quad - \Delta t\delta v^n_{ij} - \Delta t(\delta-\tau)h^n_j \\ &\quad - \Delta t H\left(y_j, t_n, v^n_{ij}, \frac{v^n_{i+1,j} - v^n_{i-1,j}}{2\Delta x}\right) \end{aligned} \tag{10.78}$$

for any $i, j = 1, \dots, K-1$, $n = 0, \dots, M$, where $\tilde{v}^n = (v^n_{i-1,j}, v^n_{ij}, v^n_{i+1,j}, v^n_{i,j+1})$. We have additional conditions on the boundary points:

$$v^n_{0,j} = v^n_{1,j}, \quad v^n_{K,j} = v^n_{K-1,j}, \quad v^n_{i,K} = v^n_{i,K-1} \quad \text{for } i,j = 1,\dots,K-1, \tag{10.79}$$

that are first-order approximations of Neumann boundary conditions on the square $[x_{\min}, x_{\max}] \times [y_{\min}, y_{\max}]$. Let us assume:

(T1) $h \geq 0$ is a smooth function, nonincreasing in y and nondecreasing in t, such that $h(y,t) = MB(T-t)e^{-y}$ for $y \geq 0$ and $\max((\tau-\delta),\delta)\sup_{y,t} h \leq \delta L$.

(T2) $A_0 > 2\rho L$, and the grid steps Δx, Δy, Δt satisfy

$$\Delta x \sup_{Q_T}|\mu_1| \leq \sigma_1^2\left(1 - \frac{2\rho L}{A_0}\right), \qquad \sigma_2^2\frac{\Delta t}{\Delta x^2} + \sup_{Q_T}\mu_2\frac{\Delta t}{\Delta y} + \Delta t\delta \leq 1. \tag{10.80}$$

Remark 10.29 The upper bound L can be chosen so that $\overline{U} \equiv L$ is a supersolution of (10.71)–(10.72); then the comparison principle [25] implies $U \leq L$. It is easy to see that this choice is independent of A_0, which can be chosen as in (**T2**).

Theorem 10.30 *Under assumptions* (**T1**)–(**T2**), *the numerical scheme* (10.78)–(10.79) *has a solution* $v^\Delta \in \mathcal{W}(-h, L)$, *and, as* $\Delta \to 0$, v^Δ *converges l.u. to the viscosity solution of problem* (10.71)–(10.72), *F being defined in* (10.77).

Proof By Proposition 10.28, the result is proved if the scheme (10.78)–(10.79) satisfies **(C1)**–**(C3)**. The consistency **(C1)** is easily proved by the Taylor expansion. We then prove the monotonicity (**C2**). The scheme is monotone if, for all $i, j = 1, \dots, K-1, n = 0, \dots, M$,

$$1 - \sigma^2(\cdot)\frac{\Delta t}{\Delta x^2} - \mu_2(\cdot)\frac{\Delta t}{\Delta y} - \Delta t\frac{\partial H}{\partial U}(\cdot) - \Delta t\delta \geq 0, \qquad \mu_2(\cdot)\frac{\Delta t}{\Delta y} \geq 0, \quad (10.81)$$

$$\frac{1}{2}\sigma^2(\cdot)\frac{\Delta t}{\Delta x^2} \pm \mu_1(\cdot)\frac{\Delta t}{2\Delta x} \mp \frac{\Delta t}{2\Delta x}\frac{\partial H}{\partial q}(\cdot) \geq 0. \quad (10.82)$$

From the definition of H one has $\partial H/\partial U \leq 0$; hence, using $\sigma \leq \sigma_2$, the first inequality in (10.81) is verified if $\Delta = (\Delta t, \Delta x, \Delta y)$ satisfies the second relation in (10.80). Since $\mu_2 \geq 0$, the second inequality in (10.81) holds. Inequalities (10.81)–(10.82) are equivalent to

$$\Delta x \left| \frac{\partial H}{\partial q}\left(y_j, t_n, v_{ij}^n, \frac{v_{i+1,j}^n - v_{i-1,j}^n}{2\Delta x}\right) - (\mu_1)_{ij}^n \right| \leq \sigma^2(\cdot) \quad \forall i, j, n. \quad (10.83)$$

Since $((v)_{ij}^n)_{i,j,n}$ are the values at the points $(i\Delta x, j\Delta y, n\Delta t)$ of a function $v \in \mathcal{W}(-h, L)$, by (**T1**) we get

$$\left|\frac{\partial H}{\partial q}\right| \leq 2\sigma^2(t_n)\rho \left| \frac{1}{2\Delta x}\frac{v_{i+1,j}^n - v_{i-1,j}^n}{v_{ij}^n + h_j^n + \xi^n} \right| \leq \rho \frac{\sigma^2(t_n)(L + h_j^n)}{\Delta x \inf_t \xi(t)}$$
$$\leq 2\rho \frac{\sigma^2(t_n)L}{\Delta x A_0}. \quad (10.84)$$

Then, (10.83) reduces to $\Delta x \sup_{Q_T} |\mu_1| \leq \sigma^2(t_n)(1 - (2\rho L/A_0))$ that is implied by $\sigma(t_n) \geq \sigma_1$ and the first inequality in (10.80). The stability condition (**C3**) is a consequence of the monotonicity. Clearly, the scheme has an explicit solution v^Δ; then we prove that $v \equiv v^\Delta$ belongs to the space $\mathcal{W}(-h, L)$. Since $h \geq 0$, $-h_j^0 \leq v_{ij}^0 \equiv 0 \leq L$ for all i, j. Fixing n and supposing that $-h_j^n \leq v_{ij}^n \leq L$, for any i, j, the monotonicity implies

$$\tilde{S}(\Delta, n, i, j, \tilde{h}^n) \leq v_{ij}^{n+1} = \tilde{S}(\Delta, n, i, j, \tilde{v}^n) \leq \tilde{S}(\Delta, n, i, j, L\mathbf{1}), \quad (10.85)$$

$\mathbf{1}$ being the vector $(1, 1, 1, 1)$. Since $\tilde{S}(\Delta, n, i, j, L\mathbf{1}) = L - \delta\Delta t L - \Delta t(\delta - \tau)h_j^n$, by (**T1**) it holds $\tilde{S}(\Delta, n, i, j, L\mathbf{1}) \leq L$. On the other hand, since $\mu_2 \geq 0$, by the monotonicity assumptions on h in (**T1**) we get

$$\tilde{S}(\Delta, n, i, j, \tilde{h}^n) = -h_j^n + \mu_2\frac{\Delta t}{\Delta y}(-h_{j+1}^n + h_j^n) + \delta\Delta t h_j^n - \Delta t(\delta - \tau)h_j^n$$
$$\geq h_j^n(\tau\Delta t - 1) \geq -h_j^{n+1}. \quad (10.86)$$

We have proved by induction that, for all i, j, n, $-h_j^{n+1} \leq v_{ij}^{n+1} \leq L$; hence, $v \in \mathcal{W}(-h, L)$, and the scheme is stable. By Proposition 10.28, we conclude that v^Δ converges l.u. to the solution U_R as $\Delta \to 0$. □

10.5.2 The Numerical Tests

We consider a numerical experiment for the two-dimensional problem defined by the Hamiltonian (10.77), using the numerical scheme (10.78)–(10.79). Here x_t represents the prevailing level of mortgage rates and follows a Vasicek (1977) dynamics, i.e., $\mu_1(x) = a(\overline{x} - x)$ for some constants $a, \overline{x}$ and with a constant volatility parameter $\sigma > 0$. The prepayment process y_t is determined by $\mu_2 = \eta(r_0 - x)$, where $r_0 > 0$, and $\eta(\cdot) \geq 0$ is a nondecreasing function. We observe that y_t defines an incentive to prepay, since it measures the amount by which the interest rate x_t is below some given level r_0.

Precisely, r_0 represents a critical level which separates high refinancing levels in the pool of mortgages ($x_t \ll r_0$) from a low prepayment activity, essentially due to exogenous reasons, including the need to sell one's home in order to move to another location. Two simple models for η are the following:

$$\eta(x) = \lambda_1 + \lambda_2 \max(x, 0), \qquad \eta(x) = \lambda_1 + \lambda_2\, \eta_2(x), \tag{10.87}$$

$$\eta_2(x) = \begin{cases} \exp(-\lambda_3/x) & \text{if } x > 0, \\ 0 & \text{if } x \geq 0, \end{cases} \tag{10.88}$$

for constant parameters $\lambda_1 \geq 0$, $\lambda_2, \lambda_3 > 0$. The second expression for η is a smooth function, which satisfies $\eta'(x) > 0$ for any $x > 0$. Therefore, if the initial level of rates x_0 is below the critical rate r_0, using the arguments of Example 10.26, the process $X_t = (x_t, y_t)$ admits a density. If U is smooth enough, we are in the situation covered by Theorem 10.22, and there exists an equilibrium as in Definition 10.11.

Both from a financial and theoretical point of view, we are interested in the behavior of the solution for positive values of x_t and in a neighborhood of the critical level r_0. Hence the numerical domain is chosen according to this purpose. We also recall that the functional form for the outstanding balance at time $T - t$ is

$$h(y, t) = MB(0) \frac{e^{\tau' T} - e^{\tau'(T-t)}}{e^{\tau' T} - 1} e^{-y} \tag{10.89}$$

for any $y \geq 0$, with $\tau' > \tau$. Our tests use the first expression of η in (10.87), choosing $MB(0) = 10$ and the remaining coefficients as in Table 10.1. In Fig. 10.1, the clean price of an *MBS* pass-through (i.e., $U + h$) is plotted. The surface shows the dependence on the mortgage rate x and the prepayment rate level in the case of $T = 15$, the other figures showing the behavior of the price varying some of the coefficients. In each plot, the fixed parameters are chosen according to Table 10.1. Figures 10.2, 10.3, and 10.4 show the *MBS* price as a function of the mortgage rate on the line $y = 0$, varying the maturity, the critical level r_0, and the volatility parameter σ. In Fig. 10.5, we give the market price of prepayment risk (10.54), as a

Table 10.1 Model parameters for numerical tests

$\sigma = 0.2$	$\delta = 0.02$	$\tau = 0.05$	$\lambda_1 = 0$	$\rho = 0.5$
$a = 0.29368$	$\overline{x} = 0.07935$	$\tau' = 0.06$	$\lambda_2 = 0.05$	$r_0 = 0.04$

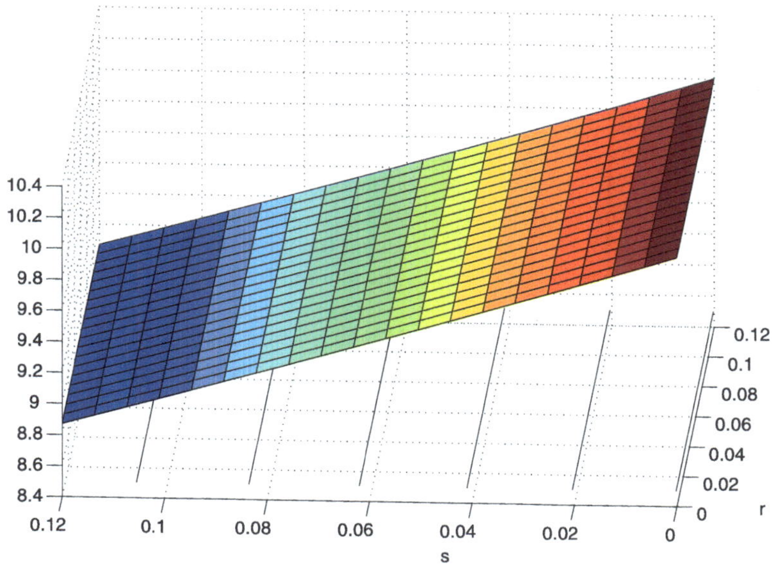

Fig. 10.1 The *MBS* price as a function of the mortgage rate r and the prepayment level S

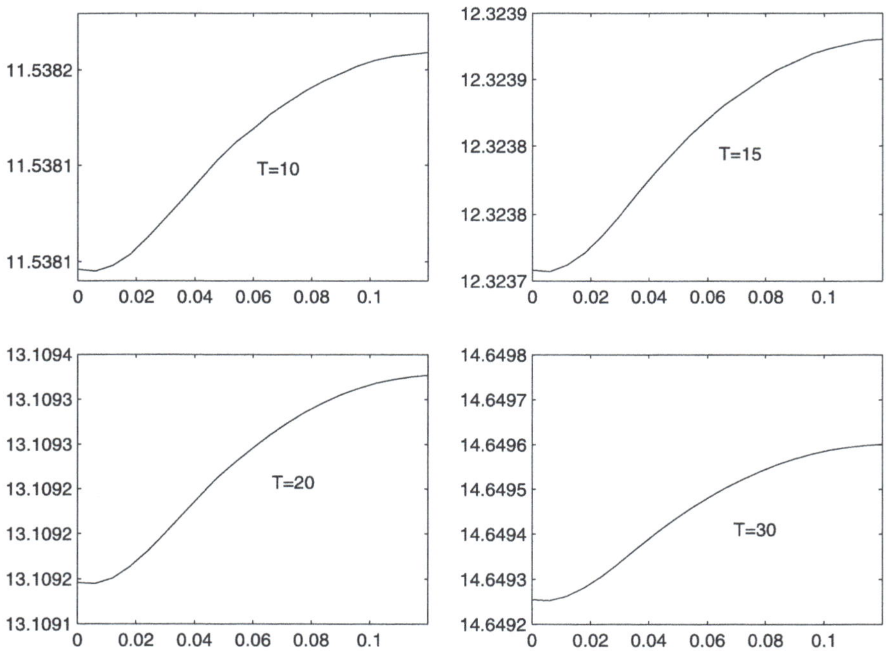

Fig. 10.2 The *MBS* price at $y = 0$, varying T

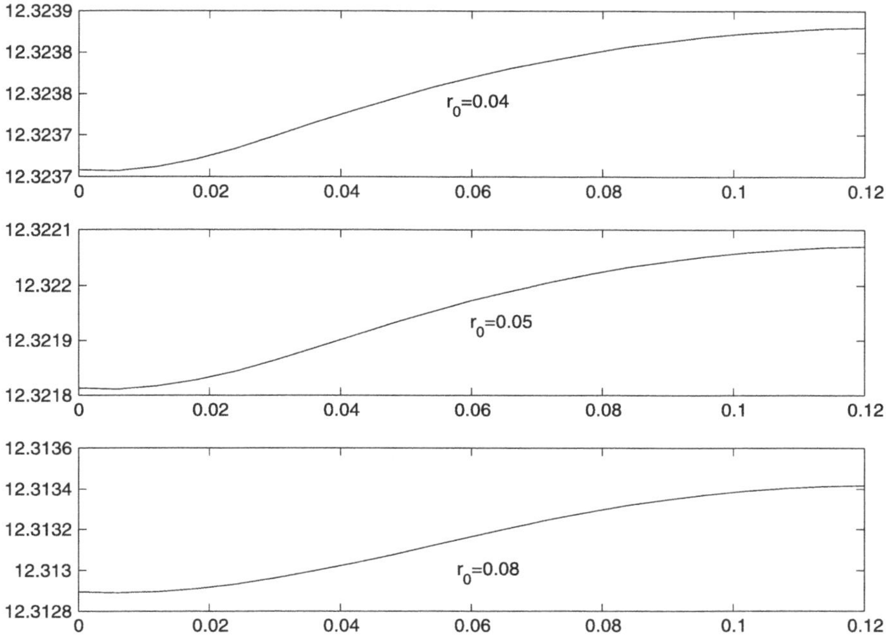

Fig. 10.3 The *MBS* price at $y = 0$, varying r_0

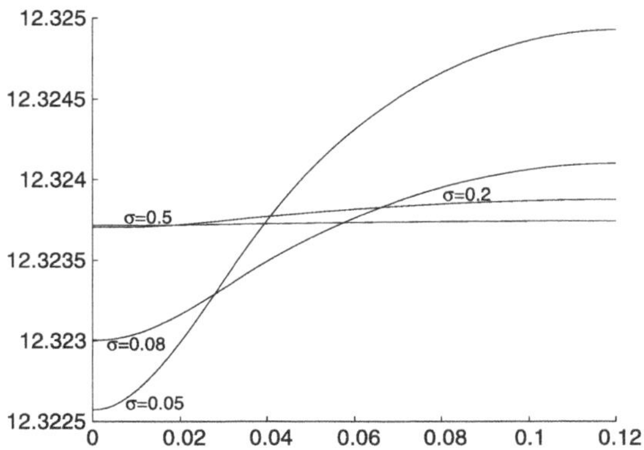

Fig. 10.4 The *MBS* price at $y = 0$, varying σ

function of x, at $y = 0$. In the large interval $[0.1\%, 20\%]$, γ is positive, and hence the price V^{Col} increases with the mortgage rate. We recall that V^{Col} is the sum of the *IO* and *PO* derivative prices, written on the same pool. Since $\partial_x V^{IO} \geq 0$ and $\partial_x V^{PO} \leq 0$, we deduce that, in our numerical example, the interest rate part of cash-flows compensates the reduction in the principal repayments. We also observe that the m.p.r. tends to be lower for extreme values of the mortgage rate, while it increases as the rate approaches the critical level r_0, which is close to the *WAC* τ. The reason is that in these cases, the behavior of mortgage holders is clear to the in-

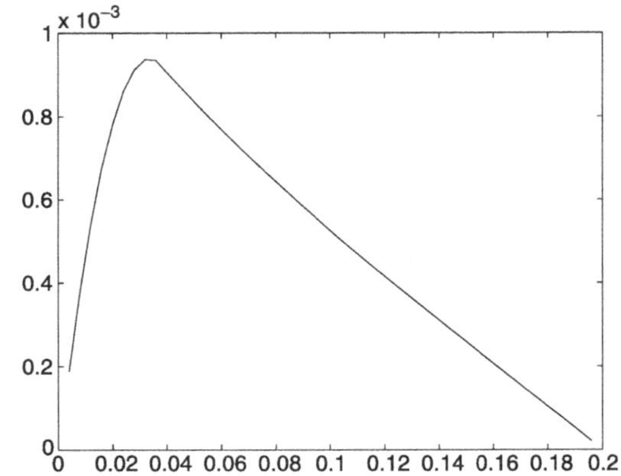

Fig. 10.5 The market price of risk γ

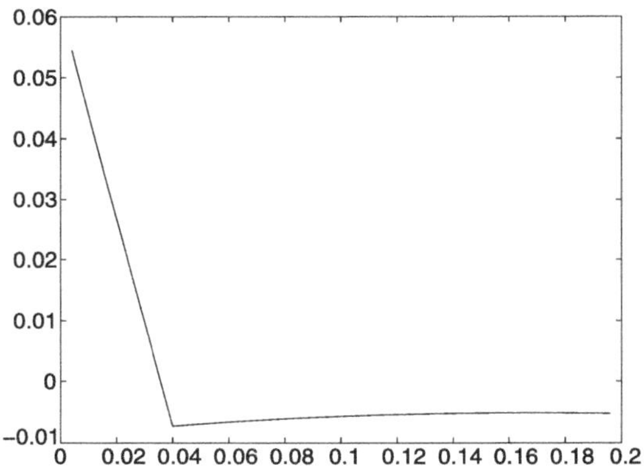

Fig. 10.6 The variation of γ with x

vestor: if $x \approx 0$, prepayments increase, and the perception of a loss in interest rates paid by the mortgagors is counterbalanced by the *PO* part. The converse happens when $x > 15\%$. Moreover, the left queue of γ is higher than the right queue. In fact, for $x \approx 0$, the previous reasoning shows that the *PO* part does not compensate the loss in the *IO*, as instead the latest makes when x is large. When x moves in a neighborhood of the critical rate r_0, the behavior in the pool is less evident, and the investor demands a higher risk premium. Figure 10.6 shows the variation of γ with x. The lack of regularity showed at $x = r_0$ depends on the regularity of η, which is merely Lipschitz continuous at r_0, and on the strong degeneracy of (10.50).

We also calculate the order γ of the numerical error under the discrete l^1-norm and l^∞-norm,

$$\gamma_{1,\infty} = \log_2\left(\frac{e^{\Delta}_{1,\infty}}{e^{\Delta/2}_{1,\infty}}\right). \tag{10.90}$$

Table 10.2 Numerical errors and order of accuracy of the scheme (10.78) with parameter chosen as in Sect. 10.5.2. $N_x \times N_y$ stands for the mesh size

$N_x \times N_y$	Error l^1	Order l^1	Error l^∞	Order l^∞
$T = 1$				
10×10	3.941407e-004		3.946165e-004	
20×20	9.178447e-005	2.1024	9.211188e-005	2.0990
40×40	1.940157e-005	2.2421	1.960494e-005	2.2322
80×80	3.065327e-006	2.6621	3.251080e-006	2.5922
$T = 5$				
10×10	2.133851e-004		2.178253e-004	
20×20	4.213563e-005	2.3403	4.274546e-005	2.3493
40×40	2.429758e-005	0.7942	2.546204e-005	0.7474
80×80	6.978259e-006	1.7999	7.527149e-006	1.7582
$T = 10$				
10×10	1.847034e-004		1.828404e-004	
20×20	8.316403e-005	1.1512	8.865463e-005	1.0443
40×40	2.751098e-005	1.5960	2.998566e-005	1.5639
80×80	1.032300e-005	1.4141	1.386634e-005	1.1127
$T = 20$				
10×10	2.396667e-004		2.718541e-004	
20×20	1.058457e-004	1.1791	1.190492e-004	1.1913
40×40	5.062764e-005	1.0640	5.567603e-005	1.0964
80×80	2.509679e-005	1.0124	4.019449e-005	0.4701
$T = 30$				
10×10	5.648831e-004		6.229993e-004	
20×20	1.631822e-004	1.7915	1.851851e-004	1.7503
40×40	8.954462e-005	0.8658	1.016767e-004	0.8650
80×80	4.749093e-005	0.9150	7.493029e-005	0.4404

The relative l^1 and l^∞ errors are calculated respectively as follows:

$$e_1^\Delta = \frac{\sum_{ij} |v_{ij}^\Delta(T) - v_{2i,2j}^{\Delta/2}(T)|}{\sum_{ij} |v_{2i,2j}^{\Delta/2}(T)|}, \qquad e_\infty^\Delta = \frac{\max_{ij} |v_{ij}^\Delta(T) - v_{2i,2j}^{\Delta/2}(T)|}{\max_{ij} |v_{2i,2j}^{\Delta/2}(T)|}. \qquad (10.91)$$

We fix A_0 and L and the discretization sizes in order to satisfy condition (**T2**). In Table 10.2, we list the l^1-norm and l^∞-norm errors and orders of accuracy for the numerical solution, increasing values of the maturity T. It comes out that the scheme is of second-order accuracy for a short time to maturity and, as we expected, first order accurate as T grows.

References

1. G. Barles, Ch. Daher, M. Romano, Convergence of numerical schemes for degenerate parabolic equations arising in finance theory. Math. Models Methods Appl. Sci. **5**, 125–143 (1995)
2. G. Barles, P.E. Souganidis, Convergence of approximation schemes for fully nonlinear equations. Asymptot. Anal. **4**, 271–283 (1991)
3. M.G. Crandall, K. Fok, M. Kocan, A. Swiech, Remarks on nonlinear uniformly parabolic equations. Indiana Univ. Math. J. **47**, 1293–1326 (1998)
4. M.G. Crandall, H. Ishii, P.L. Lions, User's guide to viscosity solutions of second order partial differential equations. Bull. Am. Math. Soc. **27**, 1–67 (1992)
5. Q. Dai, K. Singleton, Specification analysis of affine term structure models. J. Finance **55**, 1943–1978 (2000)
6. Y. Deng, J.M. Quigley, R.V. Order, Mortgage terminations, heterogeneity and the exercise of mortgage options. Econometrica **68**, 275–307 (2000)
7. C. Downing, R. Stanton, N. Wallace, An empirical test of a two-factor mortgage prepayment and valuation model: How much do house prices matter? Real Estate Econ. **33**, 681–710 (2005)
8. D. Duffie, *Dynamic Asset Pricing Theory*, 3rd edn. (Princeton University Press, New Jersey, 2001)
9. D. Duffie, R. Kan, A yield-factor model of interest rates. Math. Finance **6**, 379–406 (1996)
10. K.B. Dunn, C.S. Spatt, The effect of refinancing costs and market imperfections on the optimal call strategy and the pricing of debt contracts. Real Estate Econ. **33**, 4 (2005)
11. W.H. Fleming, H.M. Soner, *Controlled Markov Processes and Viscosity Solutions*, 2nd edn. (Springer, New York, 2005)
12. J.B. Kau, D.C. Keenan, W.J. Muller III, J.F. Epperson, A generalized valuation model for fixed-rate residential mortgages. J. Money Credit Bank. **24**, 279–299 (1992)
13. X. Gabaix, A. Krishnamurthy, O. Vigneron, Limits of arbitrage: theory and evidence from the Mortgage-Backed securities market (with technical appendix). J. Finance **62**, 2 (2007)
14. Y. Giga, S. Goto, H. Ishii, M.H. Sato, Comparison principle and convexity preserving properties for singular degenerate parabolic equations on unbounded domains. Indiana Univ. Math. J. **40**, 443–470 (1991)
15. Y. Goncharov, An intensity-based approach for valuation of mortgage contracts subject to prepayment risk. Dissertation, University of Illinois at Chicago (2003)
16. U.G. Haussmann, Generalized solutions of the Hamilton–Jacobi equation of stochastic control. SIAM J. Control Optim. **32**, 728–743 (1994)
17. E. Johnston, L. Van Drunen, Pricing mortgage pools with heterogeneous mortgagors: Empirical evidence. Working Paper, University of Utah (1988)
18. I. Karatzas, D. Ocone, A Generalized Clark representation formula, with application to optimal portfolios. Stoch. Stoch. Rep. **34**, 187–220 (1991)
19. I. Karatzas, S. Shreve, *Brownian Motion and Stochastic Calculus*, 2nd edn. (Springer, New York, 1988)
20. I. Karatzas, S.E. Shreve, *Methods of Mathematical Finance* (Springer, New York, 1998)
21. F. Kariya, S.R. Pliska, F. Ushiyama, A 3-factor valuation model for Mortgage-Backed securities. Working Paper, Kyoto University (2002)
22. D. Lamberton, B. Lapeyre, *Introduction au Calcul Stochastique Appliqué à la Finance*, 2nd edn. (Ellipses Édition Marketing, Paris, 1997)
23. R.C. Merton, Optimum consumption and portfolio rules in a continuous-time model. J. Econ. Theory **3**, 373–413 (1971)
24. D. Nualart, *The Malliavin Calculus and Related Topics* (Springer, New York, 1995)
25. M. Papi, A generalized Osgood condition for viscosity solutions to fully nonlinear parabolic degenerate equations. Adv. Differ. Equ. **7**, 1125–1151 (2002)
26. M. Papi, Regularity results for a class of semilinear parabolic degenerate equations and applications. Commun. Math. Sci. **1**, 229–244 (2003)

27. E.S. Schwartz, W.N. Torous, Prepayment, default and the valuation of Mortgage Pass-Trough securities. J. Bus. **65**, 221–239 (1992)
28. R. Stanton, Rational prepayment and the valuation of Mortgage-Backed securities. Rev. Financ. Stud. **8**, 677–708 (1995)
29. R. Stanton, Unobservable heterogeneity and rational learning: pool-specific versus generic Mortgage-Backed security prices. J. Real Estate Finance Econ. **12**, 243–263 (1996)

Chapter 11
Nonparametric Methods for Volatility Density Estimation

Bert van Es, Peter Spreij, and Harry van Zanten

Abstract Stochastic volatility modeling of financial processes has become increasingly popular. The proposed models usually contain a stationary volatility process. We will motivate and review several nonparametric methods for estimation of the density of the volatility process. Both models based on discretely sampled continuous-time processes and discrete-time models will be discussed.

The key insight for the analysis is a transformation of the volatility density estimation problem to a deconvolution model for which standard methods exist. Three types of nonparametric density estimators are reviewed: the Fourier-type deconvolution kernel density estimator, a wavelet deconvolution density estimator, and a penalized projection estimator. The performance of these estimators will be compared.

Keywords Stochastic volatility models · Deconvolution · Density estimation · Kernel estimator · Wavelets · Minimum contrast estimation · Mixing

Mathematics Subject Classification (2010) 62G07 · 62G08 · 62M07 · 62P20 · 91G70

11.1 Introduction

We discuss a number of nonparametric methods that come into play when one wants to estimate the density of the volatility process, given observations of the price process of some asset. The models that we treat are mainly formulated in continuous

B. van Es · P. Spreij (✉)
Korteweg-de Vries Institute for Mathematics, Universiteit van Amsterdam, P.O. Box 94248, 1090GE Amsterdam, The Netherlands
e-mail: spreij@uva.nl

B. van Es
e-mail: a.j.vanes@uva.nl

H. van Zanten
Department of Mathematics, Eindhoven University of Technology, P.O. Box 513, 5600MB Eindhoven, The Netherlands
e-mail: j.h.v.zanten@tue.nl

G. Di Nunno, B. Øksendal (eds.), *Advanced Mathematical Methods for Finance*,
DOI 10.1007/978-3-642-18412-3_11, © Springer-Verlag Berlin Heidelberg 2011

time, although we pay some separate attention to discrete-time models. The observations of the continuous-time models will always be in discrete time however and may occur at low frequency (fixed lag between observation instants) or high frequency (vanishing time lag). In this review, for simplicity, we focus on the univariate marginal distribution of the volatility process, although similar results can be obtained for multivariate marginal distributions.

Although the underlying models differ in the sense that they are formulated either in continuous or in discrete time, in all cases the observations are given by a discrete-time process. Moreover, as we shall see, the observation scheme can always (approximately) be cast as of "signal plus noise" type

$$Y_i = X_i + \varepsilon_i,$$

where X_i is to be interpreted as the "signal." If for fixed i, the random variables X_i and ε_i are independent, the distribution of the Y_i is a convolution of the distributions of X_i and ε_i. The density of the "signal" X_i is the object of interest, while the density of the "noise" ε_i is supposed to be known to the observer. The statistical problem is to recover the density of the signal by *deconvolution*. Classically, for such models, it was often also assumed that the processes (X_i) and (ε_i) are i.i.d. Under these conditions, Fan [12] gave lower bounds for the estimation of the unknown density f at a fixed point x_0 and showed that kernel-type estimators achieve the optimal rate. An alternative estimation method was proposed in the paper Pensky and Vidakovic [23], using wavelet methods instead of kernel estimators and where global L^2-errors were considered instead of pointwise errors.

However, for the stochastic volatility models that we consider, the i.i.d. assumption on the X_i is violated. Instead, the X_i may be modeled as stationary random variables that are allowed to exhibit some form of weak dependence, controlled by appropriate mixing properties, strongly mixing or β-mixing. These mixing conditions are justified by the fact that they are satisfied for many popular GARCH-type and stochastic volatility models (see, e.g., Carrasco and Chen [6]), as well as for continuous-time models where σ^2 solves a stochastic differential equation, see, e.g., Genon-Catalot et al. [17]. The estimators that we discuss are based on kernel methods, wavelets, and penalized contrast estimation, also referred to as penalized projection estimation. We will review the performance of these deconvolution estimators under weaker than i.i.d. assumptions and show that this essentially depends on the smoothness and mixing conditions of the underlying process and the frequency of the observations. For a survey of other nonparametric statistical problems for financial data, we refer to Franke et al. [14]

The paper is organized as follows. In Sect. 11.2 we introduce the continuous time model. In Sect. 11.3 we consider a kernel-type estimator of the invariant volatility density and apply it to a set of real data. Section 11.4 is devoted to a wavelet density estimator, and in Sect. 11.5 a minimum contrast estimator is discussed. Some related results for discrete-time models are reviewed in Sect. 11.6, and Sect. 11.7 contains some concluding remarks.

11.2 The Continuous-Time Model

Let S denote the log price process of some stock in a financial market. It is often assumed that S can be modeled as the solution of a stochastic differential equation or, more generally, as an Itô diffusion process. So we assume that we can write

$$dS_t = b_t\,dt + \sigma_t\,dW_t, \qquad S_0 = 0, \tag{11.1}$$

or, in the integral form,

$$S_t = \int_0^t b_s\,ds + \int_0^t \sigma_s\,dW_s, \tag{11.2}$$

where W is a standard Brownian motion, and the processes b and σ are assumed to satisfy certain regularity conditions (see Karatzas and Shreve [22]) to have the integrals in (11.2) well defined. In a financial context, the process σ is called the volatility process. One often takes the process σ independent of the Brownian motion W.

Adopting this common assumption throughout the paper, unless explicitly stated otherwise, we also assume that σ is a strictly stationary positive process satisfying a mixing condition, for example, an ergodic diffusion on $(0,\infty)$. The standing assumption in all what follows is that the one-dimensional marginal distribution of σ admits an invariant density w.r.t. Lebesgue measure on $(0,\infty)$. This is typically the case in virtually all stochastic volatility models that are proposed in the literature, where the evolution of σ is modeled by a stochastic differential equation, mostly in terms of σ^2 or $\log\sigma^2$ (see, e.g., Wiggins [31], Heston [20]). Often σ_t^2 is a function of a process X_t satisfying a stochastic differential equation of the type

$$dX_t = b(X_t)\,dt + a(X_t)\,dB_t \tag{11.3}$$

with Brownian motion B_t. Under regularity conditions, the invariant density of X is up to a multiplicative constant equal to

$$x \mapsto \frac{1}{a^2(x)}\exp\left(2\int_{x_0}^{x}\frac{b(y)}{a^2(y)}\,dy\right), \tag{11.4}$$

where x_0 is an arbitrary element of the state space, see, e.g., Gihman and Skorokhod [19] or Skorokhod [25]. From formula (11.4) one sees that the invariant distribution of the volatility process (take X, for instance, equal to σ^2 or $\log\sigma^2$) may take on many different forms, as is the case for the various models that have been proposed in the literature. In absence of parametric assumptions on the coefficients a and b, we will investigate nonparametric procedures to estimate the corresponding densities, even refraining from an underlying model like (11.3), partly aimed at recovering possible "stylized facts" exhibited by the observations.

For instance, one could think of *volatility clustering*. This may be cast by saying that for different time instants t_1, t_2 that are close, the corresponding values of $\sigma_{t_1}, \sigma_{t_2}$ are close again. This can partly be explained by the assumed continuity of the process σ, but it might also result from specific areas around the diagonal where the multivariate density of $(\sigma_{t_1}, \sigma_{t_2})$ assumes high values if t_1 and t_2 are relatively close. It is therefore conceivable that the density of $(\sigma_{t_1}, \sigma_{t_2})$ has high concentrations around points (ℓ, ℓ) and (h, h), with $\ell < h$, a kind of bimodality of the joint distribution, with the interpretation that clustering occurs around a low value ℓ or around a high value h. This in turn may be reflected by bimodality of the univariate marginal distribution of σ_t.

A situation in which this naturally occurs is the following. Consider a regime switching volatility process. Assume that for $i = 0, 1$, we have two stationary processes X^i having stationary densities f^i. We assume these two processes to be independent and also independent of a two-state stationary homogeneous Markov chain U with states 0, 1. The stationary distribution of U is given by $\pi_i := P(U_t = i)$. The process ξ is defined by

$$\xi_t = U_t X_t^1 + (1 - U_t) X_t^0.$$

Then ξ is stationary too, and it has the stationary density f given by

$$f(x) = \pi_1 f^1(x) + \pi_0 f^0(x).$$

Suppose that the volatility process is defined by $\sigma_t^2 = \exp(\xi_t)$ and that the X^i are both Ornstein–Uhlenbeck processes given by

$$dX_t^i = -b_i \big(X_t^i - \mu_i\big)\, dt + a_i\, dW_t^i$$

with independent Brownian motions W^1 and W^2, $\mu_1 \neq \mu_2$, and $b_1, b_2 > 0$. Suppose that the X^i start in their stationary $N(\mu_i, \frac{a_i^2}{2b_i})$ distributions. Then the stationary density f is a bimodal mixture of normal densities with μ_1 and μ_2 as the locations of the local maxima. Nonparametric procedures are able to detect such a property and are consequently by all means sensible tools to get some first insights into the shape of the invariant density.

A first object of study is the marginal univariate distribution of the stationary volatility process σ. We will also consider the invariant density of the integrated squared volatility process over an interval of length Δ. By stationarity of σ this is the density of $\int_0^\Delta \sigma_t^2\, dt$. We will consider density estimators and assess their quality by giving results on their mean squared or mean integrated squared error. For kernel estimators, we rely on Van Es et al. [10], where this problem has been studied for the marginal *univariate* density of σ. In Van Es and Spreij [9] one can find results for multivariate density estimators. Results on wavelet estimators will be taken from Van Zanten and Zareba [32]. Penalized contrast estimators have been treated in Comte and Genon-Catalot [7].

The observations of log-asset price S process are assumed to take place at the time instants $0, \Delta, 2\Delta, \ldots, n\Delta$. In case one deals with low-frequency observations, Δ is fixed. For high-frequency observations, the time gap satisfies $\Delta = \Delta_n \to 0$ as $n \to \infty$. To obtain consistency for the estimators that we will study in the latter case, we will make the additional assumption $n\Delta_n \to \infty$.

To explain the origin of the estimators that we consider in this paper, we often work with the simplified model, which is obtained from (11.1) by taking $b_t = 0$. We then suppose to have discrete-time data $S_0, S_\Delta, S_{2\Delta}, \ldots$ from a continuous-time stochastic volatility model of the form

$$dS_t = \sigma_t \, dW_t.$$

Under this additional assumption, we will see that we (approximately) deal with stationary observations Y_i that can be represented as $Y_i = X_i + \varepsilon_i$, where for each i, the random variables X_i and ε_i are independent.

11.3 Kernel Deconvolution

In this section we consider kernel deconvolution density estimators. We construct them, give expressions for bias and variance, and give an application to real data.

11.3.1 Construction of the Estimator

To motivate the construction of the estimator, we first consider (11.1) without the drift term, so we assume to have the simplified model

$$dS_t = \sigma_t \, dW_t, \qquad S_0 = 0. \tag{11.5}$$

It is assumed that we observe the process S at the discrete time instants 0, Δ, $2\Delta, \ldots, n\Delta$, satisfying $\Delta \to 0$, $n\Delta \to \infty$. For $i = 1, 2, \ldots$, we work, as in Genon-Catalot et al. [15, 16], with the normalized increments

$$X_i^\Delta = \frac{1}{\sqrt{\Delta}}(S_{i\Delta} - S_{(i-1)\Delta}).$$

For small Δ, we have the rough approximation

$$\begin{aligned} X_i^\Delta &= \frac{1}{\sqrt{\Delta}} \int_{(i-1)\Delta}^{i\Delta} \sigma_t \, dW_t \\ &\approx \sigma_{(i-1)\Delta} \frac{1}{\sqrt{\Delta}}(W_{i\Delta} - W_{(i-1)\Delta}) \\ &= \sigma_{(i-1)\Delta} Z_i^\Delta, \end{aligned} \tag{11.6}$$

where for $i = 1, 2, \ldots$, we define

$$Z_i^\Delta = \frac{1}{\sqrt{\Delta}}(W_{i\Delta} - W_{(i-1)\Delta}).$$

By the independence and stationarity of Brownian increments, the sequence $Z_1^\Delta, Z_2^\Delta, \ldots$ is an i.i.d. sequence of standard normal random variables. Moreover, the sequence is independent of the process σ by assumption.

Writing $Y_i = \log(X_i^\Delta)^2$, $\xi_i = \log\sigma^2_{(i-1)\Delta}$, $\varepsilon_i = \log(Z_i^\Delta)^2$, and taking the logarithm of the square of X_i^Δ, we get

$$Y_i \approx \xi_i + \varepsilon_i,$$

where the terms in the sum are independent. Assuming that the approximation is sufficiently accurate, we can use this approximate convolution structure to estimate the unknown density f of $\log\sigma^2_{i\Delta}$ from the transformed observed $Y_i = \log(X_i^\Delta)^2$. The characteristic functions involved are denoted by ϕ_Y, ϕ_ξ, and ϕ_k, where k is the density of the "noise" $\log(Z_i^\Delta)^2$. One obviously has $\phi_Y = \phi_\xi\phi_k$, and one easily sees that the density k is given by

$$k(x) = \frac{1}{\sqrt{2\pi}} e^{\frac{1}{2}x} e^{-\frac{1}{2}e^x}$$

and its characteristic function by

$$\phi_k(t) = \frac{1}{\sqrt{\pi}} 2^{it} \Gamma\left(\frac{1}{2} + it\right).$$

The idea of getting a deconvolution estimator of f is simple. Using a kernel function w, a bandwidth h, and the Y_i, the density g of the Y_i is estimated by

$$g_{nh}(y) = \frac{1}{nh} \sum_j w\left(\frac{y - Y_j}{h}\right).$$

Denoting by $\phi_{g,nh}$ the characteristic function of g_{nh}, one estimates ϕ_Y by $\phi_{g,nh}$ and ϕ_ξ by $\phi_{g,nh}/\phi_k$. Following a well-known approach in statistical deconvolution theory (see, e.g., Sect. 6.2.4 of Wand and Jones [30]), Fourier inversion then yields the density estimator of f. By elementary calculations from this procedure one obtains

$$f_{nh}(x) = \frac{1}{nh} \sum_{j=1}^{n} v_h\left(\frac{x - \log(X_j^\Delta)^2}{h}\right), \tag{11.7}$$

where v_h is the kernel function, depending on the bandwidth h,

$$v_h(x) = \frac{1}{2\pi} \int_{-\infty}^{\infty} \frac{\phi_w(s)}{\phi_k(s/h)} e^{-isx}\, ds. \tag{11.8}$$

One easily verifies that the estimator f_{nh} is real valued.

To justify the approximation in (11.6), we quantify a stochastic continuity property of σ^2. In addition to this, we make the mixing condition explicit. We impose the following:

Condition 11.1 The process σ^2 satisfies the following conditions.

1. It is L^1-Hölder continuous of order one half: $\mathbb{E}|\sigma_t^2 - \sigma_0^2| = O(t^{1/2})$ for $t \to 0$.
2. It is strongly mixing with coefficient $\alpha(t)$ satisfying, for some $0 < q < 1$,

$$\int_0^\infty \alpha(t)^q \, dt < \infty. \tag{11.9}$$

The kernel function w is assumed to satisfy the following conditions (an example of such a kernel is given in (11.12) below, see also Wand [29]) that include in particular the behavior of ϕ_w at the boundary of its domain.

Condition 11.2 Let w be a real symmetric function with real-valued symmetric characteristic function ϕ_w with support $[-1, 1]$. Assume further that

1. $\int_{-\infty}^\infty |w(u)|\, du < \infty$, $\int_{-\infty}^\infty w(u)\, du = 1$, $\int_{-\infty}^\infty u^2 |w(u)|\, du < \infty$,
2. $\phi_w(1-t) = At^\rho + o(t^\rho)$ as $t \downarrow 0$ for some $\rho > 0$ and $A \in \mathbb{R}$.

The first part of Condition 11.1 is motivated by the situation where $X = \sigma^2$ solves an SDE like (11.1). It is easily verified that for such processes, it holds that $\mathbb{E}|\sigma_t^2 - \sigma_0^2| = O(t^{1/2})$, provided that $b \in L_1(\mu)$ and $a \in L_2(\mu)$, where μ is the invariant probability measure. Indeed, we have $\mathbb{E}|\sigma_t^2 - \sigma_0^2| \le \mathbb{E} \int_0^t |b(\sigma_s^2)|\, ds + (\mathbb{E} \int_0^t a^2(\sigma_s^2)\, ds)^{1/2} = t\|b\|_{L_1(\mu)} + \sqrt{t}\|a\|_{L_2(\mu)}$.

The main result we present for this estimator concerns its mean squared error at a fixed point x. Although the motivation of the estimator was based on the simplified model (11.5), the result below applies to the original model (11.1). For its proof and additional technical details, see Van Es et al. [10].

Theorem 11.3 *Assume that $\mathbb{E}b_t^2$ is bounded. Let the process σ satisfy Condition* 11.1, *and let the kernel function w satisfy Condition* 11.2. *Moreover, let the density f of $\log \sigma_t^2$ be twice continuously differentiable with a bounded second derivative. Also assume that the density of σ_t^2 is bounded in a neighborhood of zero. Suppose that $\Delta = n^{-\delta}$ for given $0 < \delta < 1$ and choose $h = \gamma\pi/\log n$, where $\gamma > 4/\delta$. Then the bias of the estimator* (11.7) *satisfies*

$$\mathbb{E} f_{nh}(x) - f(x) = \frac{1}{2} h^2 f''(x) \int u^2 w(u)\, du + o(h^2), \tag{11.10}$$

whereas, the variance of the estimator satisfies the order bounds

$$\mathrm{Var}\, f_{nh}(x) = O\Big(\frac{1}{n} h^{2\rho} e^{\pi/h}\Big) + O\Big(\frac{1}{nh^{1+q}\Delta}\Big). \tag{11.11}$$

Remark 11.4 The choices $\Delta = n^{-\delta}$ with $0 < \delta < 1$ and $h = \gamma\pi/\log n$ with $\gamma > 4/\delta$ render a variance that is of order $n^{-1+1/\gamma}(1/\log n)^{2\rho}$ for the first term of (11.11) and $n^{-1+\delta}(\log n)^{1+q}$ for the second term. Since by assumption $\gamma > 4/\delta$ we have $1/\gamma < \delta/4 < \delta$, the second term dominates the first term. The order of the variance

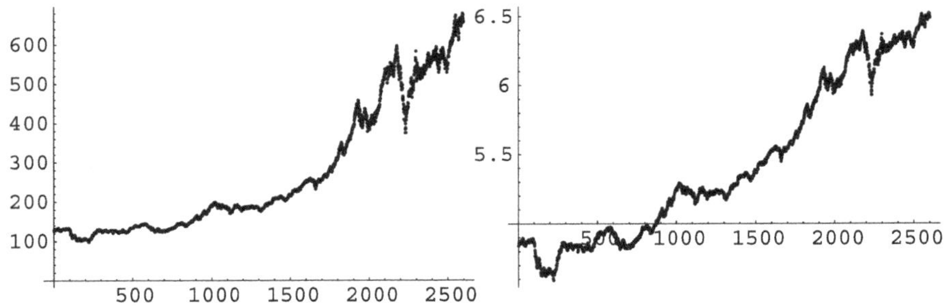

Fig. 11.1 AEX. *Left*: daily closing values. *Right*: log of the daily closing values

is thus $n^{-1+\delta}(\log n)^{1+q}$. Of course, the order of the bias is logarithmic, and hence the bias dominates the variance, and the mean squared error of $f_{nh}(x)$ is of order $(\log n)^{-4}$.

Remark 11.5 It can then be shown that for the characteristic function ϕ_k, one has the behavior

$$|\phi_k(s)| = \sqrt{2}e^{-\frac{1}{2}\pi|s|}\left(1 + O\left(\frac{1}{|s|}\right)\right), \quad |s| \to \infty.$$

This means that k is supersmooth in the terminology of Fan [12], which explains the slow logarithmic rate at which the bias vanishes. Sharper results on the variance can be obtained when σ^2 is strongly mixing, see Van Es et al. [11] for further details. The orders of the bias and of the MSE remain unchanged though.

11.3.2 An Application to the Amsterdam AEX Index

In this section we present an example using real data of the Amsterdam AEX stock exchange. We have estimated the volatility density from 2600 daily closing values of the Amsterdam stock exchange index AEX from 12/03/1990 until 14/03/2000. These data are represented in Fig. 11.1. We have centered the daily log returns, i.e., we have subtracted the mean (which equaled 0.000636), see Fig. 11.2. The deconvolution estimator is given as the left-hand picture in Fig. 11.3. Observe that the estimator strongly indicates that the underlying density is unimodal. Based on computations of the mean and variance of the estimate, with $h = 0.7$, we have also fitted a normal density by hand and compared it to the kernel deconvolution estimator. The result is given as the right-hand picture in Fig. 11.3. The resemblance is remarkable.

The kernel used to compute the estimates is a kernel from Wand [29], with $\rho = 3$ and $A = 8$,

$$w(x) = \frac{48x(x^2 - 15)\cos x - 144(2x^2 - 5)\sin x}{\pi x^7}. \tag{11.12}$$

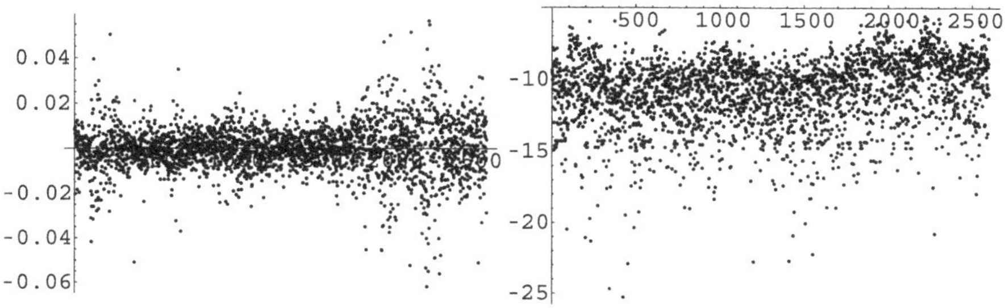

Fig. 11.2 AEX. *Left*: the values of X_t, i.e., the centered daily log returns. *Right*: $\log(X_t^2)$

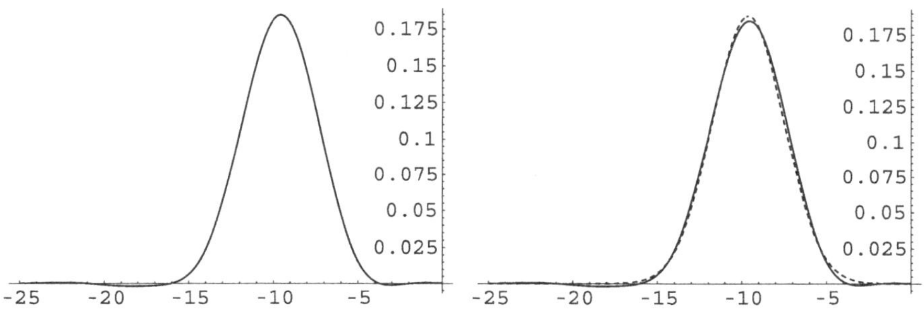

Fig. 11.3 AEX. *Left*: The estimate of the density of $\log(\sigma_t^2)$ with $h = 0.7$. *Right*: The normal fit to the $\log(\sigma_t^2)$. The *dashed line* is the normal density, and the *solid line* the kernel estimate

It has the characteristic function

$$\phi_w(t) = \left(1 - t^2\right)^3, \quad |t| \leq 1. \tag{11.13}$$

The bandwidths are chosen by hand. The estimates have been computed by fast Fourier transforms using the Mathematica 4.2 package.

This is actually the same example as in our paper Van Es et al. [11] on volatility density estimation for discrete-time models. The estimator (11.7) presented here is, as a function of the sampled data, exactly the same as the one for the discrete-time models. The difference lies in the choice of underlying model. In the present paper the model is a discretely sampled continuous-time process, while in Van Es et al. [11] it is a discrete-time process. For the latter type of models, the discretization step in the beginning of this section is not necessary since these models satisfy an exact convolution structure.

11.4 Wavelet Deconvolution

As an alternative to kernel methods, in this section we consider estimators based on wavelets. Starting point is again the simplified model (11.5). Contrary to the previous section, we are now interested in estimating the accumulated squared volatility over an interval of length Δ. We assume having observations of S at times $i\Delta$ to our disposal, but now with Δ fixed (low-frequency observations). Let, as before, $X_i^\Delta = \Delta^{-1/2}(S_{i\Delta} - S_{(i-1)\Delta})$, and let $\bar{\sigma}_i^2 = \Delta^{-1}\int_{(i-1)\Delta}^{i\Delta} \sigma_t^2\, dt$. Denote by $\mathcal{F}_\sigma$ the σ-algebra generated by the process σ. By the assumed independence of the processes σ and W, we have, for the characteristic function of X_i^Δ given $\mathcal{F}_\sigma$,

$$\mathbb{E}\big[\exp\big(isX_i^\Delta\big)\big|\mathcal{F}_\sigma\big] = \exp\Big(-\frac{1}{2}\bar{\sigma}_i^2 s^2\Big).$$

Consider also the model $\tilde{X}_i^\Delta = \bar{\sigma}_i Z_i$ with $\bar{\sigma}_i$ and Z_i independent for each i and Z_i a standard Gaussian random variable. Then

$$\mathbb{E}\big[\exp\big(is\tilde{X}_i^\Delta\big)\big|\mathcal{F}_{\sigma_i}\big] = \exp\Big(-\frac{1}{2}\bar{\sigma}_i^2 s^2\Big).$$

It follows that X_i^Δ and $\tilde{X}_i^\Delta$ are identically distributed. From this observation we conclude that the transformed increments $\log(\Delta^{-1}(S_{i\Delta} - S_{(i-1)\Delta})^2)$ are then distributed as $Y_i = \xi_i + \varepsilon_i$, where

$$\xi_i = \log \bar{\sigma}_i^2, \qquad \varepsilon_i = \log Z_i^2,$$

and Z_i is an i.i.d. sequence of standard Gaussian random variables, independent of σ. The sequence ξ_i is stationary, and we assume that its marginal density g exists, i.e., g is the density of $\log(\Delta^{-1}\int_0^\Delta \sigma_u^2\, du)$. The density of ε_i is again denoted by k. Of course, estimating g is equivalent to estimating the density of the aggregated squared volatility $\int_0^\Delta \sigma_u^2\, du$.

In the present section the main focus is on the quality of the estimator in terms of the mean integrated squared error, as opposed to establishing results for the (pointwise) mean squared error as in Sect. 11.3. At the end of this section we compare the results presented here to those of Sect. 11.3.

First we recall the construction of the wavelet estimator proposed in Pensky and Vidakovic [23]. For the necessary background on wavelet theory, see, for instance, Blatter [1], Jawerth and Sweldens [21], and the references therein. For the construction of deconvolution estimators, we need to use band-limited wavelets. As in Pensky and Vidakovic [23], we use a Meyer-type wavelet (see also Walter [27], Walter and Zayed [28]). We consider an orthogonal scaling function and wavelet φ and ψ, respectively, associated with an orthogonal multiresolution analysis of $L^2(\mathbb{R})$. We denote in this section the Fourier transform of a function f by $\tilde{f}$, i.e.,

$$\tilde{f}(\omega) = \int_{\mathbb{R}} e^{-i\omega x} f(x)\, dx,$$

and suppose that for a symmetric probability measure μ with support contained in $[-\pi/3, \pi/3]$, it holds that

$$\tilde{\varphi}(\omega) = \big(\mu(\omega - \pi, \omega + \pi]\big)^{1/2}, \qquad \tilde{\psi}(\omega) = e^{-i\omega/2}\big(\mu\big(|\omega|/2 - \pi, |\omega| - \pi\big]\big)^{1/2}.$$

Observe that the assumptions imply that φ and ψ are indeed band-limited. For the supports of their Fourier transforms, we have $\operatorname{supp}\tilde{\varphi} \subset [-4\pi/3, 4\pi/3]$ and $\operatorname{supp}\tilde{\psi} \subset [-8\pi/3, -2\pi/3] \cup [2\pi/3, 8\pi/3]$. By choosing μ smooth enough we ensure that $\tilde{\varphi}$ and $\tilde{\psi}$ are at least twice continuously differentiable.

For any integer m, the unknown density g can now be written as

$$g(x) = \sum_{l \in \mathbb{Z}} a_{m,l}\varphi_{m,l}(x) + \sum_{l \in \mathbb{Z}} \sum_{j=m}^{\infty} b_{j,l}\psi_{j,l}(x), \tag{11.14}$$

where $\varphi_{m,l}(x) = 2^{m/2}\varphi(2^m x - l)$, $\psi_{j,l}(x) = 2^{j/2}\psi(2^j x - l)$, and the coefficients are given by

$$a_{m,l} = \int_{\mathbb{R}} \varphi_{m,l}(x)g(x)\,dx, \qquad b_{j,l} = \int_{\mathbb{R}} \psi_{j,l}(x)g(x)\,dx.$$

The idea behind the linear wavelet estimator is simple. We first approximate g by the orthogonal projection given by the first term on the right-hand side of (11.14). For m large enough, the second term will be small and can be controlled by using the approximation properties of the specific family of wavelets that is being used. The projection of g is estimated by replacing the coefficients $a_{m,l}$ by consistent estimators and truncating the sum. Using the fact that the density p of an observation Y_i is the convolution of g and k, it is easily verified that

$$a_{m,l} = \int_{\mathbb{R}} 2^{m/2}U_m\big(2^m x - l\big)p(x)\,dx = 2^{m/2}\mathbb{E}U_m\big(2^m Y_i - l\big),$$

where U_m is the function with Fourier transform

$$\tilde{U}_m(\omega) = \frac{\tilde{\varphi}(\omega)}{\tilde{k}(-2^m\omega)}. \tag{11.15}$$

We estimate the coefficient $a_{m,l}$ by its empirical counterpart

$$\hat{a}_{m,l,n} = \frac{1}{n}\sum_{i=1}^{n} 2^{m/2}U_m\big(2^m Y_i - l\big).$$

Under the mixing assumptions that we will impose on the sequence Y, it will be stationary and ergodic. Hence, by the ergodic theorem, $\hat{a}_{m,l,n}$ is a consistent estimator for $a_{m,l}$. The wavelet estimator is now defined by

$$\hat{g}_n(x) = \sum_{|l| \le L_n} \hat{a}_{m_n,l,n}\varphi_{m_n,l}(x), \tag{11.16}$$

where the detail level m_n and the truncation point L_n will be chosen appropriately later.

The main results in the present section are upper bounds for the mean integrated squared error of the wavelet estimator $\hat{g}_n$, which is defined as usual by

$$\text{MISE}(\hat{g}_n) = \mathbb{E}\int_{\mathbb{R}} \big(\hat{g}_n(x) - g(x)\big)^2\,dx.$$

We will specify how to choose the detail level m_n and the truncation point L_n in (11.16) optimally in different cases, depending on the smoothness of g and k. The smoothness properties of g are described in terms of g belonging to certain Sobolev balls and by imposing a weak condition on its decay rate. The Sobolev space H^α is defined for $\alpha > 0$ by

$$H^\alpha = \left\{g : \|g\|_\alpha = \left(\int_{\mathbb{R}} |\tilde{g}(\omega)|^2 (\omega^2+1)^\alpha\,d\omega\right)^{1/2} < \infty\right\}. \tag{11.17}$$

Roughly speaking, $g \in H^\alpha$ means that the first α derivatives of g belong to $L^2(\mathbb{R})$. The Sobolev ball of radius A is defined by

$$\mathcal{S}_\alpha(A) = \big\{g \in H^\alpha : \|g\|_\alpha \le A\big\}.$$

The additional assumption on the decay rate is reflected by g belonging to

$$\mathcal{S}^*_\alpha(A, A') = \mathcal{S}_\alpha(A) \cap \left\{g : \sup_x \big|xg(x)\big| \le A'\right\}.$$

We now have the following result, see Van Zanten and Zareba [32], for the wavelet density estimator $\hat{g}_n$ of g defined by (11.16).

Theorem 11.6 *Suppose that the volatility process σ^2 is strongly mixing with mixing coefficients satisfying*

$$\sum_{k\ge 0} \alpha^p_{k\Delta} < \infty \tag{11.18}$$

for some $p \in (0, 1)$. Then with the choices

$$2^{m_n} = \frac{\log n}{1+(4\pi^2/3)}, \qquad L_n = (\log n)^r, \quad r \ge 1 + 2\alpha$$

the mean square error of the wavelet estimator satisfies

$$\sup_{g\in\mathcal{S}^*_\alpha(A,A')} \text{MISE}(\hat{g}_n) = O\big((\log n)^{-2\alpha}\big)$$

for $\alpha, A, A' > 0$. If (11.18) is satisfied for all *$p \in (0, 1)$, the same bound is true if the choice for L_n is replaced by $L_n = n$.*

Let us point out the relation with the results of Sect. 11.3 and with those in Van Es et al. [11], see also Sect. 11.6.1. In that paper kernel-type deconvolution estimators for discrete-time stochastic volatility models were considered. When applied to the present model, the results say that under the same mixing condition and assuming that g has two bounded and continuous derivatives, the (pointwise) mean squared error of the kernel estimator is of order $(\log n)^{-4}$. The analogue of g having two bounded derivatives in our setting is that $g \in \mathcal{S}_2^*(A, A')$ for some $A, A' > 0$. Indeed, the theorem yields the same bound $(\log n)^{-4}$ for the MISE in this case. The same bound is valid for the MSE when estimating the marginal density for continuous-time models, see Theorem 11.3 and its consequences in Remark 11.4. Theorem 11.6 is more general, because the smoothness level is not fixed at $\alpha = 2$, but allows for different smoothness levels of order $\alpha \neq 2$ as well. Moreover, the wavelet estimator is adaptive in the sense that it does not depend on the unknown smoothness level if the condition on the mixing coefficients holds for all $p \in (0, 1)$.

11.5 Penalized Projection Estimators

The results of the preceding sections assume that the true (integrated) volatility density has a finite degree of regularity, either in Hölder or in Sobolev sense. Under this assumption, the nonparametric estimators have logarithmic convergence rates, cf. Remark 11.4 and Theorem 11.6. Although admittedly slow, the minimax results of Fan [12] show that these rates are in fact optimal in this setting. In the paper Pensky and Vidakovic [23] it was shown however that if in a deconvolution setting the density of the unobserved variables has the same degree of smoothness as the noise density, the rates can be significantly improved, cf. also the lower bounds obtained in Butucea [4] and Butucea and Tsybakov [5]. This observation forms the starting point of the paper Comte and Genon-Catalot [7], in which a nonparametric volatility density estimator is developed that achieves better rates than logarithmic if the true density is supersmooth. In the latter paper it is assumed that there are observations $S_\Delta, S_{2\Delta}, \ldots, S_{n\Delta}$ of a process S satisfying the simple equation (11.5) with a strictly positive process $V = \sigma^2$, independent of the Brownian motion W. It is assumed that we deal with high-frequency observations, $\Delta \to 0$, and $n\Delta \to \infty$. We impose the following condition on V.

Condition 11.7 The process V is a time-homogenous, continuous Markov process, strictly stationary and ergodic. It is either β-mixing with coefficient $\beta(t)$ satisfying

$$\int_0^\infty \beta(t)\,dt < \infty$$

or is ρ-mixing. Moreover, it satisfies the Lipschitz condition

$$\mathbb{E}\Big(\log\Big(\frac{1}{\Delta}\int_0^\Delta V_t\,dt\Big) - \log V_0\Big)^2 \leq C\Delta$$

for some $C > 0$.

In addition to this, a technical assumption is necessary on the density f of $\log V_0$ we are interested in and on the density g_Δ of $\log(\frac{1}{\Delta}\int_0^\Delta V_t\,dt)$, which is assumed to exist. Contrary to the notation of the previous section, we write g_Δ instead of g, since now Δ is not fixed.

Condition 11.8 The invariant density f is bounded and has a second moment, and $g_\Delta \in L^2(\mathbb{R})$.

As a first step in the construction of the final estimator, a preliminary estimator $\hat{f}_L$ is constructed for $L \in \mathbb{N}$ fixed. Note that Condition 11.8 implies that $f \in L^2(\mathbb{R})$, and hence we can consider its orthogonal projection f_L on the subspace S_L of $L^2(\mathbb{R})$, defined as the space of functions whose Fourier transform is supported on the compact interval $[-\pi L, \pi L]$. An orthonormal basis for the latter space is formed by the Shannon basis functions $\psi_{L,j}(x) = \sqrt{L}\psi(Lx - j)$, $j \in \mathbb{Z}$, with the sinc kernel $\psi(x) = \sin(\pi x)/(\pi x)$. For integers $K_n \to \infty$ to be specified below, the space S_L is approximated by the finite-dimensional spaces $S_L^n = \operatorname{span}\{\psi_{L,j} : |j| \le K_n\}$. The function f_L is estimated by $\hat{f}_L = \operatorname{argmin}_{h\in S_L^n} \gamma_n(h)$, where the contrast function γ_n is defined for $h \in L^2(\mathbb{R}) \cap L^1(\mathbb{R})$ by

$$\gamma_n(h) = \|h\|_2^2 - \frac{2}{n}\sum_{i=1}^n u_h\bigl(\log\bigl(X_i^\Delta\bigr)^2\bigr), \qquad u_h(x) = \frac{1}{2\pi}\int_{-\infty}^{\infty} e^{ixs}\frac{\tilde{h}(-s)}{\phi_k(s)}\,ds.$$

Here, as before, ϕ_k is the characteristic function of $\log\varepsilon^2$, with ε standard normal, and $\tilde{h}$ is the Fourier transform of h. It is easily seen that

$$\hat{f}_L = \sum_{|j|\le K_n} \hat{a}_{L,j}\psi_{L,j}, \quad \hat{a}_{L,j} = \frac{1}{n}\sum_{j=1}^n u_{\psi_{L,j}}\bigl(\log\bigl(X_i^\Delta\bigr)^2\bigr).$$

Straightforward computations show that, with $\langle\cdot,\cdot\rangle$ the $L^2(\mathbb{R})$ inner product, $\mathbb{E}u_h(\log(X_i^\Delta)^2) = \langle h, g_\Delta\rangle$, and hence $\mathbb{E}\gamma_n(h) = \|h - g_\Delta\|_2^2 - \|g_\Delta\|_2^2$. So in fact, $\hat{f}_L$ is an estimator of the element of S_L^n which is closest to g_Δ. Since S_L^n approximates S_L for large n and g_Δ is close to f for small Δ, the latter element should be close to f_L.

Under Conditions 11.7 and 11.8, a bound for the mean integrated squared error, or quadratic risk $\mathrm{MISE}(\hat{f}_L) = \mathbb{E}\|\hat{f}_L - f\|_2^2$, can be derived, depending on the approximation error $\|f - f_L\|_2$, the bandwidth L, and the truncation point K_n, see Comte and Genon-Catalot [7], Theorem 1. The result implies that if f belongs to the Sobolev space H^α as defined in (11.17), then the choices $K_n = n$ and $L = L_n \sim \log n$ yield a MISE of order $(\log n)^{-2\alpha}$, provided that $\Delta = \Delta_n = n^{-\delta}$ for some $\delta \in (0,1)$. Not surprisingly, this is completely analogous to the result obtained in Theorem 11.6 for the wavelet-based estimator in the fixed Δ setting. In particular the procedure is adaptive, in that the estimator does not depend on the unknown regularity parameter α.

To obtain faster than logarithmic rates and adaptation in the case that f is supersmooth, a data-driven choice of the bandwidth L is proposed. Define

$$\hat{L} = \underset{L\in\{1,\dots,\log n\}}{\operatorname{argmin}} \left(\gamma_n(\hat{f}_L) + \operatorname{pen}_n(L)\right),$$

where the penalty term is given by

$$\operatorname{pen}_n(L) = \kappa \frac{(1+L)\Phi_k(L)}{n}$$

for a calibration constant $\kappa > 0$ and

$$\Phi_k(L) = \int_{-\pi L}^{\pi L} \frac{1}{|\phi_k(s)|^2}\, ds.$$

For the quadratic risk of the estimator $\hat{f}_{\hat{L}}$, the following result holds (Comte and Genon-Catalot [7]).

Theorem 11.9 *Under Conditions* 11.8 *and* 11.7, *we have*

$$\begin{aligned}\operatorname{MISE}(\hat{f}_{\hat{L}}) \le C_1 \inf_{L\in\{1,\dots,\log n\}} \left(\|f - f_L\|_2^2 + \frac{(1+L)\Phi_k(L)}{n} \right) \\ + C_2 \frac{\log^2 n}{K_n} + C_3 \frac{\log n}{n\Delta} + C_4 \Delta \log^3 n\end{aligned}$$

for constants $C_1, C_2, C_3, C_4 > 0$.

It can be seen that this bound is worse than the corresponding bound for the estimator $\hat{f}_L$ by a factor of order L. This is at worst a logarithmic factor which, as usual in this kind of setting, has to be paid for achieving adaptation. The examples in Sect. 6 of Comte and Genon-Catalot [7] show that indeed, the estimator $\hat{f}_{\hat{L}}$ can achieve algebraic convergence rates in case the true density f is supersmooth.

11.6 Estimation for Discrete-Time Models

Although the main focus of the present paper is on estimation procedures for continuous-time models, in the present section we also highlight some analogous results for discrete-time models. These deal with both density and regression function estimation.

11.6.1 Discrete-Time Models

The discrete time analogue of (11.5) is

$$X_t = \sigma_t Z_t, \quad t = 1, 2, \dots. \tag{11.19}$$

Here we denote by X the detrended or demeaned log-return process. Stochastic volatility models are often described in this form. The sequence Z is typically an i.i.d. noise (e.g., Gaussian), and at each time t the random variables σ_t and Z_t are independent. See the survey papers by Ghysels et al. [18] or Shephard [24]. Also in this section we assume that the process σ is strictly stationary and that the marginal distribution of σ has a density with respect to the Lebesgue measure on $(0, \infty)$. We present some results for a nonparametric estimator of the density of $\log \sigma_t^2$ and results for a nonparametric estimator of a nonlinear regression function, in case σ^2 is given by a nonlinear autoregression. The standing assumption in all what follows is that for each t, the random variables σ_t and Z_t are independent, the noise sequence is standard Gaussian, and σ is a strictly stationary, positive process satisfying a certain mixing condition.

In principle one can distinguish two classes of models. The way in which the bivariate process (σ, Z), in particular its dependence structure, is further modeled offers different possibilities. In the first class of models one assumes that the process σ is predictable with respect to the filtration $\mathcal{F}_t$ generated by the process Z and obtains that σ_t is independent of Z_t for each fixed time t. We furthermore have that (assuming that the unconditional variances are finite) σ_t^2 is equal to the conditional variance of X_t given $\mathcal{F}_{t-1}$. This class of models has become quite popular in the econometrics literature. It is well known that this class also contains the (parametric) family of GARCH-models, introduced by Bollerslev [2].

In the second class of models one assumes that the whole process σ is independent of the noise process Z, and one commonly refers to the resulting model as a stochastic volatility model. In this case, the natural underlying filtration $\mathcal{F} = \{\mathcal{F}_t\}_{t \geq 0}$ is generated by the two processes Z and σ in the following way. For each t, the σ-algebra $\mathcal{F}_t$ is generated by Z_s, $s \leq t$, and σ_s, $s \leq t+1$. This choice of the filtration enforces σ to be predictable. As in the first model, the process X becomes a martingale difference sequence, and we have again (assuming that the unconditional variances are finite) that σ_t^2 is the conditional variance of X_t given $\mathcal{F}_{t-1}$. An example of such a model is given in De Vries [26], where σ is generated as an AR(1) process with α-stable noise ($\alpha \in (0, 1)$).

As in the previous sections, we refrain from parametric modeling and review some completely nonparametric approaches. We will mainly focus on results for the second class, as it is the discrete-time analogue of the stochastic volatility models of the previous sections. At the heart of all what follows is again the convolution structure that is obtained from (11.19) by squaring and taking logarithms,

$$\log X_t^2 = \log \sigma_t^2 + \log Z_t^2.$$

11.6.2 Density Estimation

The main result of this section gives a bias expansion and a variance bound of a kernel-type density estimator of the density f of $\log \sigma_t^2$. The estimator is, analogously to (11.7),

$$f_{nh}(x) = \frac{1}{nh}\sum_{j=1}^{n} v_h\left(\frac{x-\log(X_j)^2}{h}\right), \tag{11.20}$$

where v_h is the kernel function of (11.8).

The next theorem is derived from Van Es et al. [11], where a multivariate density estimator is considered. It establishes the expansion of the bias and an order bound on the variance of our estimator under a strong mixing condition. Under broad conditions, this mixing condition is satisfied if the process σ Markov, since then convergence of the mixing coefficients to zero takes place at an *exponential rate*, see Theorems 4.2 and 4.3 of Bradley [3] for precise statements. A similar behavior occurs for ARMA processes with absolutely continuous distributions of the noise terms (Bradley [3], Example 6.1).

Theorem 11.10 *Assume that the process σ is strongly mixing with coefficient α_k satisfying*

$$\sum_{j=1}^{\infty}\alpha_j^{\beta} < \infty$$

for some $\beta \in (0,1)$. Let the kernel function w satisfy Condition 11.2, *and let the density f of* $\log\sigma_t^2$ *be bounded and twice continuously differentiable with bounded second-order partial derivatives. Assume furthermore that σ and Z are independent processes. Then we have, for the estimator of f defined as in* (11.20) *and $h \to 0$,*

$$\mathbb{E} f_{nh}(x) = f(x) + \frac{1}{2}h^2 f''(x)\int u^2\, w(u)\,du + o\big(h^2\big) \tag{11.21}$$

and

$$\operatorname{Var} f_{nh}(x) = O\left(\frac{1}{n}h^{2\rho}e^{\pi/h}\right). \tag{11.22}$$

Remark 11.11 Comparing the above results to the ones in Theorem 11.3, we observe that in the continuous-time case, the variance has an additional $O(\frac{1}{nh^{1+q}\Delta})$ term.

11.6.3 Regression Function Estimation

In this section we assume the basic model (11.19), but in addition we assume that the process σ satisfies a nonlinear autoregression, and we consider nonparametric estimation of the regression function as proposed in Franke et al. [13]. In that paper a discrete-time model was proposed as a discretization of the continuous-time model given by (11.1). In fact, Franke et al. include a mean parameter μ, but since they assume it to be known, without loss of generality we can still assume (11.19). Assume that the volatility process is strictly positive and consider $\log \sigma_t^2$. It is assumed that its evolution is governed by

$$\log \sigma_{t+1}^2 = m\big(\log \sigma_t^2\big) + \eta_t, \tag{11.23}$$

where the η_t are i.i.d. Gaussian random variables with zero mean. The regression function m is assumed to satisfy the stability condition

$$\limsup_{|x|\to\infty} \left|\frac{m(x)}{x}\right| < 1. \tag{11.24}$$

Under this condition, the process σ is exponentially ergodic and strongly mixing, see Doukhan [8], and these properties carry over to the process X as well. Moreover, the process $\log \sigma_t^2$ admits an invariant density f.

Denoting $Y_t = \log X_t^2$, we have

$$Y_t = \log \sigma_t^2 + \log Z_t^2.$$

It is common to assume that the processes Z and η are independent, the second class of models described in Sect. 11.6.1, but dependence between η_t and Z_t for fixed t can be allowed for (first model class) without changing in what follows, see Franke et al. [13].

The purpose of the present section is to estimate the function m in (11.23). To that end, we use the estimator f_{nh} as defined in (11.20). Since this estimator resembles an ordinary kernel density estimator, the important difference being that the kernel function v_h now depends on the bandwidth h, the idea is to mimic the classical Nadaraya–Watson regression estimator similarly, in order to obtain an estimator of $m(x)$. Doing so, one obtains the estimator

$$m_{nh}(x) = \frac{\frac{1}{nh}\sum_{j=1}^{n} v_h(\frac{x-Y_j}{h})Y_{j+1}}{f_{nh}(x)}. \tag{11.25}$$

It follows that

$$m_{nh}(x) - m(x) = \frac{p_{nh}(x)}{f_{nh}(x)},$$

where

$$p_{nh}(x) = \frac{1}{nh}\sum_{j=1}^{n} v_h\Big(\frac{x-Y_j}{h}\Big)\big(Y_{j+1} - m(x)\big).$$

In Franke et al. [13] bias expansions for $p_{nh}(x)$ and f_{nh} are given that fully correspond to those in Theorem 11.10. They are again of order h^2, under similar assumptions. It is also shown that the variances of p_{nh} and f_{nh} tend to zero. The main result concerning the asymptotic behavior then follows from combining the asymptotics for p_{nh} and f_{nh}.

Theorem 11.12 *Assume that m satisfies the stability condition* (11.24), *that m and f are twice differentiable, and the first of Condition* 11.2 *on the kernel w. The estimator* $m_{nh}(x)$ *satisfies* $(\log n)^2(m_{nh}(x) - m(x)) = O_p(1)$ *if* $h = \gamma / \log n$ *with* $\gamma > \pi$.

Following the proofs in Franke et al. [13], one can conclude that, e.g., the variance of p_{nh} is of order $O(\frac{\exp(\pi/h)}{nh^4})$, which tends to zero for $h = \gamma / \log n$ with $\gamma > \pi$. For the variance of f_{nh}, a similar bound holds. Comparing these order bounds to the ones in Theorem 11.10, we see that the latter ones are sharper. This is partly due to the fact that Franke et al. [13], do not impose conditions on the boundary behavior of the function ϕ_w (the second of Condition 11.2), whereas their other assumptions are the same as in Theorem 11.10.

11.7 Concluding Remarks

In recent years, many different parametric stochastic volatility models have been proposed in the literature. To investigate which of these models are best supported by observed asset price data, nonparametric methods can be useful. In this paper we reviewed a number of such methods that have recently been proposed. The overview shows that ideas from deconvolution theory can be instrumental in dealing with this statistical problem and that both for high- and for low-frequency data, methods are now available for nonparametric estimation of the (integrated) volatility density at optimal convergence rates.

On a critical note, the methods available so far all assume that the volatility process is independent of the Brownian motion driving the asset price dynamics. This is a limitation, since in several interesting models nonzero correlations are assumed between the Brownian motions driving the volatility dynamics and the asset price dynamics.

References

1. C. Blatter, *Wavelets, a Primer* (A.K. Peters, Natick, 1998)
2. T. Bollerslev, Generalized autoregressive conditional heteroscedasticity. J. Econom. **31**, 307–321 (1986)
3. R.C. Bradley, Basic properties of strong mixing conditions, in *Dependence in Probability and Statistics*, ed. by E. Eberlein, M.S. Taqqu (Birkhäuser, Boston, 1986)
4. C. Butucea, Deconvolution of supersmooth densities with smooth noise. Can. J. Stat. **32**, 181–192 (2004)
5. C. Butucea, A.B. Tsybakov, Sharp optimality in density deconvolution with dominating bias. I. Theory Probab. Appl. **52**(1), 24–39 (2008)
6. M. Carrasco, X. Chen, Mixing and moment properties of various GARCH and stochastic volatility models. Econom. Theory **18**, 17–39 (2002)
7. F. Comte, V. Genon-Catalot, Penalized projection estimator for volatility density. Scand. J. Stat. **33**(4), 875–893 (2006)

8. P. Doukhan, *Mixing: Properties and Examples*. Lecture Notes in Statistics, vol. 85 (Springer, New York, 1994)
9. A.J. van Es, P.J.C. Spreij, Multivariate nonparametric volatility density estimation (2009). Available at Mathematics ArXiv, http://arxiv.org/abs/0910.4337
10. A.J. van Es, P.J.C. Spreij, J.H. van Zanten, Nonparametric volatility density estimation. Bernoulli **9**(3), 451–465 (2003)
11. A.J. van Es, P.J.C. Spreij, J.H. van Zanten, Nonparametric volatility density estimation for discrete time models. J. Nonparametr. Stat. **17**, 237–251 (2005)
12. J. Fan, On the optimal rates of convergence of nonparametric deconvolution problems. Ann. Stat. **19**, 1257–1272 (1991)
13. J. Franke, W. Härdle, J.P. Kreiss, Nonparametric estimation in a stochastic volatility model, in *Recent Advances and Trends in Nonparametric Statistics*, ed. by M.G. Akritas, D.N. Politis (Elsevier, Amsterdam, 2003)
14. J. Franke, J.P. Kreiss, E. Mammen, Nonparametric modelling in financial time series, in *Handbook of Financial Time Series*, ed. by T. Mikosch, J.P. Kreiss, R.A. Davis, T.G. Andersen (Springer, New York, 2009)
15. V. Genon-Catalot, T. Jeantheau, C. Larédo, Limit theorems for discretely observed stochastic volatility models. Bernoulli **4**, 283–303 (1998)
16. V. Genon-Catalot, T. Jeantheau, C. Larédo, Parameter estimation for discretely observed stochastic volatility models. Bernoulli **5**, 855–872 (1999)
17. V. Genon-Catalot, T. Jeantheau, C. Larédo, Stochastic volatility models as hidden Markov models and statistical applications. Bernoulli **6**(6), 1051–1079 (2000)
18. E. Ghysels, A. Harvey, E. Renault, Stochastic volatility, in *Statistical Methods in Finance*, ed. by G.S. Maddala, C.R. Rao. Handbook of Statistics, vol. 14 (North-Holland, Amsterdam, 1996), pp. 119–191
19. I.I. Gihman, A.V. Skorohod, *Stochastic Differential Equations* (Springer, New York–Heidelberg, 1972)
20. S.L. Heston, A closed-form solution for options with stochastic volatility with applications to Bond and Currency options. Rev. Financ. Stud. **6**(2), 327–343 (1993)
21. B. Jawerth, W. Sweldens, An overview of wavelet based multiresolution analyses. SIAM Rev. **36**, 377–412 (1994)
22. I. Karatzas, S.E. Shreve, *Brownian Motion and Stochastic Calculus* (Springer, New York, 1991)
23. M. Pensky, B. Vidakovic, Adaptive wavelet estimator for nonparametric density deconvolution. Ann. Stat. **27**, 2033–2053 (1999)
24. N. Shephard, Statistical aspects of ARCH and stochastic volatility, in *Time Series Models in Econometrics, Finance and Other Fields*, ed. by D.R. Cox, D.V. Hinkley, O.E. Barndorff-Nielsen (Chapman & Hall, London, 1996), pp. 1–67
25. A.V. Skorokhod, *Asymptotic Methods in the Theory of Stochastic Differential Equations*. Translations of Mathematical Monographs, vol. 78 (American Mathematical Society, Providence, 1989)
26. C.G. de Vries, On the relation between GARCH and stable processes. J. Econom. **48**, 313–324 (1991)
27. G.G. Walter, *Wavelets and Other Orthogonal Systems with Applications* (CRC Press, Boca Raton, 1994)
28. G.G. Walter, A.I. Zayed, Characterization of analytic functions in terms of their wavelet coefficients. Complex Var. **29**, 265–276 (1996)
29. M.P. Wand, Finite sample performance of deconvolving kernel density estimators. Stat. Probab. Lett. **37**, 131–139 (1998)
30. M.P. Wand, M.C. Jones, *Kernel Smoothing* (Chapman and Hall, London, 1995)
31. J.B. Wiggins, Option valuation under stochastic volatility. J. Financ. Econ. **19**, 351–372 (1987)
32. J.H. van Zanten, P. Zareba, A note on wavelet density deconvolution for weakly dependent data. Stat. Inference Stoch. Process. **11**, 207–219 (2008)

Chapter 12
Fractional Smoothness and Applications in Finance

Stefan Geiss and Emmanuel Gobet

Abstract This overview article concerns the notion of fractional smoothness of random variables of the form $g(X_T)$, where $X = (X_t)_{t\in[0,T]}$ is a certain diffusion process. We review the connection to the real interpolation theory, give examples and applications of this concept. The applications in stochastic finance mainly concern the analysis of discrete-time hedging errors. We close the review by indicating some further developments.

Keywords Fractional smoothness · Discrete time hedging · Interpolation

Mathematics Subject Classification (2010) 41A25 · 46B70 · 60H05 · 60H07

12.1 Introduction

From the practitioners one learns that hedging an option the payoff of which is discontinuous is more difficult compared to the case of smooth payoffs: this feature appears, for instance, for digital options or barrier options (we refer the reader to [32] among others). On the one hand, for such options, the number of assets (i.e., the delta) to incorporate in the hedging portfolio is unbounded, and it may become larger and larger as one gets close to the singularity (i.e., the maturity and the strike for digital options, or the trigger level for barrier options). On the other hand, the numerical estimation of this delta becomes less and less accurate, leading to global stability issues. These heuristic observations are the starting point for deeper mathematical investigations about the concept of irregular payoffs, in order to formalize it and to quantify the payoff irregularity (with the notion of *fractional smoothness*).

S. Geiss (✉)
Department of Mathematics, University of Innsbruck, Technikerstraße 13/7, 6020 Innsbruck, Austria
e-mail: stefan.geiss@uibk.ac.at

E. Gobet
Laboratoire Jean Kuntzmann, Université de Grenoble and CNRS, BP 53, 38041 Grenoble Cedex 9, France
e-mail: emmanuel.gobet@imag.fr

G. Di Nunno, B. Øksendal (eds.), *Advanced Mathematical Methods for Finance*, DOI 10.1007/978-3-642-18412-3_12, © Springer-Verlag Berlin Heidelberg 2011

In the current contribution, we aim to give an overview of this concept and some applications in stochastic finance. Actually, the applications go beyond the financial framework, and more generally, they concern the theory of stochastic differential equations and their approximations.

The Discrete-Time Hedging Error as an Important Application Since most of the results presented here are applied to the aforementioned example of hedging possibly irregular options, we start with a brief presentation of this problem, in order to emphasize the issues to handle and to raise some natural questions. Take, for instance, an European-style option exercised at maturity $T > 0$, with a payoff of the form $h(S_T)$, where $S_t := [S_t^1, \dots, S_t^d]$ denotes the price of a d-dimensional underlying asset at time $0 \le t \le T$. Sometimes, we will use the notation $X_t^i = \log(S_t^i)$ for the log-asset, and $g(x) = h(e^{x_1}, \dots, e^{x_d})$ for the payoff in the logarithmic variables. In what follows, we assume a Markovian dynamics without jumps for the asset (solution to an SDE defined below), and we suppose that the interest rate is equal to 0 (to simplify the presentation) and that the market is complete (for details about this standard framework, see [24]). Thus, under some regularity assumptions, the payoff $h(S_T)$ can be replicated perfectly by a continuous-time strategy, where $\delta_t^S = \nabla_x H(t, S_t)$ defines the vector of number of assets to hold at time t. Here, H is the fair price of the option, that is,

$$H(t,x) = \mathbb{E}_{\mathbb{Q}}\big(h(S_T)\big|S_t = x\big),$$

where $\mathbb{Q}$ is the (unique) risk-neutral measure. In practice, only discrete-time hedging is possible at some times $\tau = (t_i)_{i=0}^n$ with $0 = t_0 < t_1 < \cdots < t_{n-1} < t_n = T$. Thus, at time $t \in [0, T]$ the option seller is left with the tracking error

$$\begin{aligned} C_t\big(h(S_T), \tau\big) &= H(t, S_t) - H(0, S_0) - \sum_i \nabla_x H(t_i, S_{t_i}) \cdot (S_{t_{i+1}\wedge t} - S_{t_i \wedge t}) \\ &= \int_0^t \big(\nabla_x H(s, S_s) - \nabla_x H\big(\phi(s), S_{\phi(s)}\big)\big) \cdot dS_s \qquad (12.1) \end{aligned}$$

with $\phi(s) = t_i$ when $t_i < s \le t_{i+1}$. We expect the tracking error (12.1) to converge to 0 as the number n of rebalancing dates goes to infinity. With the above formulation (12.1), the tracking error is naturally associated to the problem of approximation of a stochastic integral using piecewise constant integrated processes. But the delta process $(\nabla_x H(s, S_s))_{s\in[0,T)}$ may exhibit very different behaviors from payoff to payoff: if the payoff is smooth enough, then the delta may be *bounded* as time goes to maturity, while an irregular payoff usually yields an *exploding* delta as $s \to T$. This gives rise to the first question.

(Q1) Is there an intrinsic way to relate the growth rate (as $s \to T$) of the derivatives of H to the irregularity of the payoff h?
The answer will be *yes* via the notion of *fractional smoothness* introduced below, see Theorems 12.3 and 12.4.

The estimation of stochastic integrals is usually performed with L_2-norms, but in our financial setting, both measures $\mathbb{P}$ and $\mathbb{Q}$ can be considered. For practitioners, errors under the historical probability $\mathbb{P}$ are presumably more relevant, while the mathematical treatment under the risk-neutral measure $\mathbb{Q}$ is simpler in our context (because the tracking error process (12.1) is a $\mathbb{Q}$-local martingale).

(Q2) Is the definition of fractional smoothness affected by the choice of a specific measure? Do the L_2-convergence rates depend on the choice of the probability measures $\mathbb{P}$ or $\mathbb{Q}$?
In the context we consider the answer concerning the fractional smoothness is usually *no* in the sense of the comments after Theorem 12.4. Concerning the approximation rates the same is checked for examples so far (see the remarks after Theorem 12.20).

Beyond the approach to measure tracking errors in L_2, we could alternatively identify the weak limit of the renormalized tracking error.

(Q3) Do the weak convergence rates coincide with those in the L_2 sense?
The answer is *not necessarily*, as there are counterexamples in which the convergence in L_2 and in distribution hold at different rates, see Sect. 12.5.

Finally, through an efficient choice of rebalancing dates τ, one can expect to reduce tracking errors and improve the risk management of options.

(Q4) Which time nets $\tau = (t_i)_{i=0}^n$ lead to optimal convergence rates? And how to relate them to the *fractional smoothness* of the payoff?
As answer we get is that according to the index of fractional smoothness of the payoff, one can define explicitly rebalancing times achieving the optimal convergence rates, see Sect. 12.5.

These preliminary questions serve as references for the reader when reading the next sections.

Organization of the Paper First, we define the probabilistic framework and the assumptions used throughout this work. Then in Sect. 12.2, we define the *fractional smoothness* and provide basic properties: we choose a presentation that is quite illuminating regarding the previous preliminary questions. In Sect. 12.3, we take another view on *fractional smoothness* using the interpolation theory. In Sect. 12.4, we consider examples of terminal conditions and identify their fractional smoothness. Then, in Sect. 12.5, we go back to the analysis of discrete-time hedging errors and state the main results. We close by further developments and applications of the *fractional smoothness* in Sect. 12.6.

Assumptions Let us define the probabilistic setting used in the following. We fix a d-dimensional Brownian motion $W = (W_t)_{t\in[0,T]}$ defined on a complete probability space $(\Omega, \mathcal{F}_T, \mathbb{P})$, and we let $(\mathcal{F}_t)_{t\in[0,T]}$ be the augmentation of the natural filtration of W. The log-asset X is the solution of the d-dimensional forward diffusion

$$X_t = x_0 + \int_0^t b(s, X_s)\,ds + \int_0^t \sigma(s, X_s)\,dW_s.$$

To state the results, we mainly consider two types of assumptions:

(SDE) $d \geq 1$, $b, \sigma \in C_b^\infty([0,T] \times \mathbb{R}^d)$, and $\sigma\sigma^* \geq \delta I_{\mathbb{R}^d}$ for some $\delta > 0$.
(GBM) $d = 1$, and $X_t = \ln(S_t) = W_t - (t/2)$.

The smoothness conditions in (SDE) are too strong and are chosen to simplify the presentation. Whenever useful to simplify even more, we may consider the very simple case of the geometric Brownian motion (GBM) (here, the asset is a martingale, meaning that $\mathbb{P} = \mathbb{Q}$). The reader is referred to the corresponding original papers for the possible weaker conditions.

In the following, $|\cdot|$ stands for the Euclidean norm, and $A \sim_c B$ for $A/c \leq B \leq cA$ if $c \geq 1$ and $A, B \geq 0$. Expectations and conditional expectations under $\mathbb{P}$ are simply denoted by $\mathbb{E}(\cdot)$ and $\mathbb{E}(\cdot|\mathcal{F}_t)$, while under $\mathbb{Q}$, we indicate explicitly the dependency w.r.t. the probability measure by writing $\mathbb{E}_{\mathbb{Q}}(\cdot)$ and $\mathbb{E}_{\mathbb{Q}}(\cdot|\mathcal{F}_t)$.

12.2 Definition of Fractional Smoothness and Basic Properties

Fractional smoothness on the Wiener space can be defined in various ways, see [19, 35]. Our approach is motivated by the questions discussed in Sect. 12.1. Since we consider only random variables of the form $Z = g(X_T) = h(S_T)$ (a function of the process at maturity T), the time T plays a specific role in our definition. It would be necessary to modify our definition for more general dependencies like $Z = g(X_{t_1}, \ldots, X_{t_n})$, see [9].

Definition 12.1 Assume that $Z \in L_2(\mathbb{P})$.

(i) For $0 < \theta \leq 1$, we let $Z \in \widetilde{\mathbb{B}}^\theta_{2,\infty}$, provided that, for all $0 \leq t < T$,

$$\left\| Z - \mathbb{E}(Z|\mathcal{F}_t) \right\|_{L_2(\mathbb{P})} \leq c(T-t)^{\frac{\theta}{2}}.$$

(ii) For $0 < \theta < 1$, we let $Z \in \widetilde{\mathbb{B}}^\theta_{2,2}$, provided that

$$\int_0^T (T-t)^{-1-\theta} \left\| Z - \mathbb{E}(Z|\mathcal{F}_t) \right\|^2_{L_2(\mathbb{P})} dt < \infty.$$

The above spaces $\widetilde{\mathbb{B}}^\theta_{2,q}$ will always be obtained by the conditional expectation and the L_2-norm under the measure $\mathbb{P}$. Therefore we omit the dependency on $\mathbb{P}$ in the notation.

The following properties follow straight from the definition:

Proposition 12.2 *For $0 < \theta < \eta < 1$ and $p, q \in \{2, \infty\}$, we have that*

$$\widetilde{\mathbb{B}}^1_{2,\infty} \subseteq \widetilde{\mathbb{B}}^\eta_{2,p} \subseteq \widetilde{\mathbb{B}}^\theta_{2,q} \quad \textit{and} \quad \widetilde{\mathbb{B}}^\theta_{2,2} \subseteq \widetilde{\mathbb{B}}^\theta_{2,\infty}.$$

Given a bounded[1] measurable $g : \mathbb{R}^d \to \mathbb{R}$ and $(t, x) \in [0, T) \times \mathbb{R}^d$, we let

$$u(t,x) := \mathbb{E}\big(g(X_T)\big|X_t = x\big),$$
$$D^2u(t,x) := \left(\frac{\partial^2 u}{\partial x_i \partial x_j}(t,x)\right)_{i,j=1}^{d}.$$

The following equivalences are useful to exploit properties of $\widetilde{\mathbb{B}}^{\theta}_{2,2}$ and $\widetilde{\mathbb{B}}^{\theta}_{2,\infty}$.

Theorem 12.3 [15, Proposition 1.1] *Under condition* (SDE), *for* $0 < \theta < 1$ *and a bounded* g, *the following assertions are equivalent*:

(i) $g(X_T) \in \widetilde{\mathbb{B}}^{\theta}_{2,2}$;
(ii) $\int_0^T (T-t)^{-\theta}\mathbb{E}|\nabla_x u(t, X_t)|^2\, dt < \infty$;
(iii) $\int_0^T (T-t)^{1-\theta}\mathbb{E}|D^2 u(t, X_t)|^2\, dt < \infty$.

Theorem 12.4 [15, Lemma 1.2] *Under condition* (SDE), *for* $0 < \theta \le 1$ *and a bounded* g, *the following assertions are equivalent*:

(i) $g(X_T) \in \widetilde{\mathbb{B}}^{\theta}_{2,\infty}$;
(ii) $\sup_{t\in[0,T)}(T-t)^{1-\theta}\mathbb{E}|\nabla_x u(t, X_t)|^2 < \infty$;
(iii) *For* $0 < \theta < 1$, *we have that* $\sup_{t\in[0,T)}(T-t)^{2-\theta}\mathbb{E}|D^2 u(t, X_t)|^2 < \infty$.

Theorems 12.3 and 12.4 generalize results obtained in [7] and [12]. We see that the fractional smoothness index θ measures exactly the growth rate of the derivatives of the associated PDE solved by u (see question (Q1) in the introduction).

The two above theorems are also valid if u is computed using the risk-neutral measure $\mathbb{Q}$ (i.e., $u_{\mathbb{Q}}(t,x) = \mathbb{E}_{\mathbb{Q}}(g(X_T)|X_t = x)$), while the other L_2-norms are computed under $\mathbb{P}$. For instance, for $0 < \theta < 1$, the equivalence of (i) and (ii) of Theorem 12.3 becomes $g(X_T) \in \widetilde{\mathbb{B}}^{\theta,\mathbb{P}}_{2,2}$ if and only if $\int_0^T (T-t)^{-\theta} \times \mathbb{E}_{\mathbb{P}}|\nabla_x u_{\mathbb{Q}}(t, X_t)|^2\, dt < \infty$, where we have indicated explicitly if the L_2-norms or conditional expectations are computed under $\mathbb{P}$ or $\mathbb{Q}$. This property can be established following [15] and the proof of [14, Lemma 7]. This accommodates well the fact that the price functions are usually computed under the risk-neutral measure, while hedging is made under the historical probability (see question (Q2) in the introduction).

Simplified proof of Theorem 12.4 We sketch the proof in the simple case where $X = W$ is a linear Brownian motion, $d = 1$, and $\theta \in (0, 1)$. First, $(u(t, W_t) = \mathbb{E}(g(W_T)|\mathcal{F}_t))_{t\le T}$ is a martingale in $L_2(\mathbb{P})$. In addition, for any fixed $0 < \delta \le T$, the processes $(\nabla_x u(t, W_t))_{t\le T-\delta}$ and $(D^2 u(t, W_t))_{t\le T-\delta}$ are $L_2(\mathbb{P})$-martingales. This property is obtained by checking that $\nabla_x u$ and $D^2 u$ both solve the parabolic heat

[1] Here again, the boundedness assumptions on g can be weakened, and we refer to the original papers.

equation and that certain integrability assumptions are satisfied. Then by Itô's formula, one obtains, for $0 \le s \le t < T$, that

$$g(W_T) - u(t, W_t) = \int_t^T \nabla_x u(s, W_s)\, dW_s, \tag{12.2}$$

$$\nabla_x u(t, W_t) - \nabla_x u(s, W_s) = \int_s^t D^2 u(r, W_r)\, dW_r. \tag{12.3}$$

From the Itô isometry one deduces from (12.2) that $\mathbb{E}|g(W_T) - u(t, W_t)|^2 = \int_t^T \mathbb{E}|\nabla_x u(s, W_s)|^2\, ds$, and it follows that (ii) $\Rightarrow$ (i). Similarly, from (12.3) one obtains

$$\mathbb{E}\big|\nabla_x u(t, W_t)\big|^2 \le 2\mathbb{E}\big|\nabla_x u(0, W_0)\big|^2 + 2\int_0^t \mathbb{E}\big|D^2 u(r, W_r)\big|^2\, dr,$$

which proves (iii) $\Rightarrow$ (ii). Finally, we show (i) $\Rightarrow$ (iii). Standard computations give that

$$\begin{aligned}
\big(D^2 u\big)(t, W_t) &= D_z^2 \int_{\mathbb{R}} g(x) \frac{e^{-\frac{(x-z)^2}{2(T-t)}}}{\sqrt{2\pi(T-t)}}\, dx \bigg|_{z=W_t} \\
&= \int_{\mathbb{R}} g(x) \frac{(x-z)^2 - (T-t)}{(T-t)^2} \frac{e^{-\frac{(x-z)^2}{2(T-t)}}}{\sqrt{2\pi(T-t)}}\, dx \bigg|_{z=W_t} \\
&= \mathbb{E}\left(g(W_T) \frac{(W_T - W_t)^2 - (T-t)}{(T-t)^2} \bigg| \mathcal{F}_t \right) \\
&= \mathbb{E}\left(\big[g(W_T) - \mathbb{E}\big(g(W_T)\big|\mathcal{F}_t\big)\big] \frac{(W_T - W_t)^2 - (T-t)}{(T-t)^2} \bigg| \mathcal{F}_t \right),
\end{aligned}$$

which implies that

$$\big\|D^2 u(t, W_t)\big\|_{L_2(\mathbb{P})} \le \frac{\|W_1^2 - 1\|_{L_2(\mathbb{P})}}{T-t} \big\|g(W_T) - \mathbb{E}\big(g(W_T)\big|\mathcal{F}_t\big)\big\|_{L_2(\mathbb{P})},$$

so that we are done. □

12.3 Connection to Real Interpolation

Let us connect Definition 12.1 to the classical notion of fractional smoothness, which also explains the notation we have used. In particular, this connection will make clear the difference between $\widetilde{\mathbb{B}}^{\theta}_{2,\infty}$ and $\widetilde{\mathbb{B}}^{\theta}_{2,2}$.

Definition 12.5 [3, 4] Assume a couple of Banach spaces (E_0, E_1) such that E_1 is continuously embedded into E_0. Given $x \in E_0$ and $0 < \lambda < \infty$, the K-functional is given by

$$K(x,\lambda;E_0,E_1):=\inf\bigl\{\|x_0\|_{E_0}+\lambda\|x_1\|_{E_1}:x=x_0+x_1\bigr\}.$$

Moreover, given $0<\theta<1$ and $1\le q\le\infty$, we define the real interpolation norm

$$\|x\|_{\theta,q}:=\bigl\|\lambda^{-\theta}K(x,\lambda;E_0,E_1)\bigr\|_{L_q((0,\infty),\frac{d\lambda}{\lambda})}$$

and the space $(E_0,E_1)_{\theta,q}:=\{x\in E_0:\|x\|_{\theta,q}<\infty\}$.

With our setting (E_1 is continuously embedded into E_0), we obtain the following lexicographical ordering of the interpolation spaces:

$$E_1\subseteq(E_0,E_1)_{\theta,p}\subseteq(E_0,E_1)_{\theta,q}\subseteq(E_0,E_1)_{\eta,r}\subseteq E_0$$

for all $0<\eta<\theta<1$, $1\le p\le q\le\infty$, and all $1\le r\le\infty$.

We apply this concept to the analysis on the Wiener space, which needs to introduce some standard notation (see [29, Sects. 1.1 and 1.2]). Let H be a separable real Hilbert space with the scalar product denoted by $\langle\cdot,\cdot\rangle_H$, and $(\mathcal{M},\Sigma,\mu)$ be a complete probability space. We assume an isonormal family $g=\{g_h:h\in H\}$ of centered Gaussian random variables, i.e.,

$$\mathbb{E}_\mu(g_h\,g_k)=\langle h,k\rangle_H\quad\text{for all } h,k\in H,$$

and that Σ is the completed σ-field generated by the random variables $\{g_h:h\in H\}$.

For each $n\ge1$, we denote by $\mathcal{H}_n$ the closed linear subspace of $L_2(\mu)$ generated by the random variables $\{H_n(g_h):h\in H,\|h\|_H=1\}$, where

$$H_n(x)=\frac{(-1)^n}{\sqrt{n!}}e^{\frac{x^2}{2}}\frac{d^n}{dx^n}\bigl(e^{-\frac{x^2}{2}}\bigr),\tag{12.4}$$

i.e., the nth Hermite polynomial. $\mathcal{H}_0$ is the set of constants, $\mathcal{H}_n$ is the so-called Wiener chaos of order n, and we define by $J_n:L_2(\mu)\to L_2(\mu)$ the orthogonal projection onto $\mathcal{H}_n$. The following orthogonal decomposition is known as the Wiener chaos decomposition:

$$L_2(\mu)=\bigoplus_{n=0}^\infty\mathcal{H}_n.$$

Now, we are in a position to define the Malliavin Sobolev space and Malliavin Besov space.

Definition 12.6 The Malliavin Sobolev space $\mathbb{D}_{1,2}(\mu)\subseteq L_2(\mu)$ is given by

$$\mathbb{D}_{1,2}(\mu):=\left\{Z\in L_2(\mu):\|Z\|_{\mathbb{D}_{1,2}(\mu)}:=\left(\sum_{n=0}^\infty(n+1)\|J_nZ\|^2_{L_2(\mu)}\right)^{\frac12}<\infty\right\}.$$

Moreover, given $0 < \theta < 1$ and $1 \le q \le \infty$, we define the Malliavin Besov space

$$\mathbb{B}^{\theta}_{2,q}(\mu) := \big(L_2(\mu), \mathbb{D}_{1,2}(\mu)\big)_{\theta,q}$$

of fractional smoothness θ with fine parameter q.

We use this construction in the case that $H = \ell_2^d$ and $\mathcal{M} = \mathbb{R}^d$, Σ is the completion of the Borel σ-algebra on $\mathbb{R}^d$, and $\mu = \gamma_d$ is the d-dimensional standard Gaussian measure. The family of Gaussian random variables is given by

$$g_h(x) := \langle x, h \rangle \quad \text{for } x \in \mathcal{M} = \mathbb{R}^d \text{ and } h \in H = \ell_2^d.$$

To make the connection between the definitions of $\widetilde{\mathbb{B}}^{\theta}_{2,q}$ and $\mathbb{B}^{\theta}_{2,q}(\gamma_d)$ for $q \in \{2, \infty\}$, we let, as before, $(W_t)_{t \in [0,1]}$ be the standard d-dimensional Brownian motion on $(\Omega, \mathcal{F}, \mathbb{P}, (\mathcal{F}_t)_{t \in [0,1]})$. Then we have the following:

Theorem 12.7 [12, Corollary 2.3] *For $0 < \theta < 1$, $1 \le q \le \infty$, and $g \in L_2(\gamma_d)$, one has*

$$\|g\|_{\mathbb{B}^{\theta}_{2,q}(\gamma_d)} \sim_c \|g\|_{L_2(\gamma_d)} + \big\|(1-t)^{-\frac{\theta}{2}} \|M_1 - M_t\|_{L_2(\mathbb{P})}\big\|_{L_q([0,1), \frac{dt}{1-t})},$$

where $M_t := \mathbb{E}(g(W_1)|\mathcal{F}_t)$, and $c \ge 1$ depends on (θ, q) only.

Applying this theorem to $q = \infty$ gives that

$$\|g\|_{\mathbb{B}^{\theta}_{2,\infty}(\gamma_d)} \sim_c \|g\|_{L_2(\gamma_d)} + \sup_{0 \le t \le 1} (1-t)^{-\frac{\theta}{2}} \|M_1 - M_t\|_{L_2(\mathbb{P})},$$

whereas $q = 2$ gives that

$$\|g\|_{\mathbb{B}^{\theta}_{2,2}(\gamma_d)} \sim_c \|g\|_{L_2(\gamma_d)} + \left(\int_0^1 (1-t)^{-1-\theta} \|M_1 - M_t\|^2_{L_2(\mathbb{P})} \, dt\right)^{\frac{1}{2}},$$

which brings us back to Definition 12.1.

Multidimensional Black–Scholes–Samuelson Model This is a lognormal model the dynamics of which on the price and the log-price can be written as

$$dS_t^i = S_t^i \left(\sum_{j=1}^d \sigma_{ij} \, dW_t^j + \mu_i \, dt\right), \quad 1 \le i \le d,$$

$$X_t^i = \log(s_0^i) + \sum_{j=1}^d \sigma_{ij} W_t^j + \left(\mu_i - \frac{1}{2}\sigma_i^2\right) t,$$

where $\sigma_i := \sqrt{\sum_j \sigma_{ij}^2}$. Assume that $(\sigma_{ij})_{i,j=1}^d$ is invertible. To the payoff function $S \mapsto h(S)$, we associate

$$g(x_1, \ldots, x_d) := h\Big(\big(s_0^i e^{\sum_{j=1}^d \sigma_{ij} x_j + \mu_i - \frac{\sigma_i^2}{2}}\big)_{i=1}^d\Big).$$

From this we see that

$$g \in \mathbb{B}_{2,q}^{\theta}(\gamma_d) \quad \text{if and only if} \quad h(S_1) \in \widetilde{\mathbb{B}}_{2,q}^{\theta}$$

for $q \in \{2, \infty\}$ and $g \in L_2(\gamma_d)$.

Remark 12.8 In the case $\theta = 1$ we get that

$$g \in \mathbb{D}_{1,2}(\gamma_d) \quad \text{if and only if} \quad h(S_1) \in \widetilde{\mathbb{B}}_{2,\infty}^{1}$$

for all $g \in L_2(\gamma_d)$. This can be checked by using arguments from the proof of [12, Corollary 2.3].

12.4 Examples

In this section, we provide examples of random variables $Z = g(X_T)$ for which we determine the fractional smoothness.

Example 12.9 (Lipschitz function) The case where the fractional smoothness is obvious is the Lipschitz case. Assume a Lipschitz function $g : \mathbb{R}^d \to \mathbb{R}$ with constant $L \geq 0$, i.e., $|g(x) - g(y)| \leq L|x - y|$, and assume (SDE). Then one has that

$$\begin{aligned}
\mathbb{E}\big|g(X_T) - \mathbb{E}\big(g(X_T)\big|\mathcal{F}_t\big)\big|^2 &\leq \mathbb{E}\big|g(X_T) - g(X_t)\big|^2 \\
&\leq L^2 \mathbb{E}|X_T - X_t|^2 \\
&\leq L^2 c^2 (T - t),
\end{aligned}$$

using standard estimates on the increments of X. Hence, $g(X_T) \in \widetilde{\mathbb{B}}_{2,\infty}^{1}$. This example includes call and put payoffs, i.e., $g(x) = (x - K)^+$ or $g(x) = (K - x)^+$.

Exactly the same argument as above yields for θ-Hölder functions g with $\theta \in (0, 1)$ that $g(X_T) \in \widetilde{\mathbb{B}}_{2,\infty}^{\theta}$. But the situation here is not as clear as one expects as shown by the following:

Example 12.10 Assume the setting (GBM) and that

$$h_\theta(x) := (x - K)_+^\theta$$

for some $K > 0$ and $0 < \theta < 1/2$. Then it is shown in [17, Lemma 2] (under more general assumptions) that $\mathbb{E}|D^2 u(t, X_t)|^2 \leq c(T - t)^{-3/2+\theta}$, so that Theorem 12.4 gives that

$$h_\theta(S_T) \in \widetilde{\mathbb{B}}_{2,\infty}^{\theta+\frac{1}{2}}.$$

For $1/2 < \theta < 1$, one gets $h_\theta(S_T) \in \widetilde{\mathbb{B}}_{2,\infty}^{1}$.

Example 12.11 (Binary option) Generally, indicator functions yield to a fractional smoothness of order $\frac{1}{2}$. In the case $X = W$, $d = 1$, and $g(x) = \mathbf{1}_{[L,\infty)}(x)$ with $L \in \mathbb{R}$, one has

$$u(t,x) = \mathbb{P}(x + W_T - W_t \geq L) = \mathcal{N}\left(\frac{x - L}{\sqrt{T - t}}\right),$$

$$\nabla_x u(t,x) = \frac{1}{\sqrt{2\pi(T - t)}} \exp\left(-\frac{(x - L)^2}{2(T - t)}\right),$$

so that

$$\mathbb{E}\left|\nabla_x u(t, W_t)\right|^2 \sim_c \frac{1}{\sqrt{T - t}}$$

and $g(W_T) \in \widetilde{\mathbb{B}}_{2,\infty}^{\frac{1}{2}}$ because of Theorem 12.4. This can be extended to the (SDE) case as follows: Our assumption guarantees that X has a transition density Γ such that

$$\Gamma(s, x; t, y) \leq \sqrt{\frac{\kappa}{2\pi(t - s)}} e^{-\frac{1}{2}\frac{(x-y)^2}{\kappa(t-s)}} = \kappa \gamma_{\kappa(t-s)}(x - y)$$

for some $\kappa > 0$ and all $0 \leq s < t \leq T$, where γ_t is the Gaussian density with zero expectation and variance t (see [6]). Then we can compute that

$$\begin{aligned}
&\mathbb{E}\left|\mathbf{1}_{[L,\infty)}(X_T) - \mathbb{E}\left(\mathbf{1}_{[L,\infty)}(X_T)\middle|\mathcal{F}_t\right)\right|^2 \\
&\quad \leq \mathbb{E}\left|\mathbf{1}_{[L,\infty)}(X_T) - \mathbf{1}_{[L,\infty)}(X_t)\right| \\
&\quad = \mathbb{P}(X_T < L \leq X_t) + \mathbb{P}(X_t < L \leq X_T) \\
&\quad \leq \kappa^2\left[\mathbb{P}(W_{\kappa T} < L - x_0 \leq W_{\kappa t}) + \mathbb{P}(W_{\kappa t} < L - x_0 \leq W_{\kappa T})\right] \\
&\quad \leq c\sqrt{T - t},
\end{aligned}$$

where $X_0 = x_0$, so that $\mathbf{1}_{[L,\infty)}(X_T) \in \widetilde{\mathbb{B}}_{2,\infty}^{\frac{1}{2}}$. The application in financial mathematics is done via $S_t = e^{X_t}$, which gives, for a positive strike $K > 0$,

$$\mathbf{1}_{\{S_T \geq K\}} = \mathbf{1}_{\{X_T \geq \log K\}} \in \widetilde{\mathbb{B}}_{2,\infty}^{\frac{1}{2}}.$$

In our context the fractional smoothness of jump functions (under different assumptions) was considered in [7, 11, 17]. In certain multidimensional settings one can deduce for $g(x) = \mathbf{1}_{\{x_1 \geq K_1, \ldots, x_d \geq K_d\}}$ (or variants of it) the same fractional smoothness from the one-dimensional case. Finally, the indicator function $g(x) = \mathbf{1}_D(x)$ of a $\mathcal{C}^2$-domain D also leads to $g(X_T) \in \widetilde{\mathbb{B}}_{2,\infty}^{\frac{1}{2}}$ (see [16, Proposition 1.2]).

Example 12.12 (An extreme case) By the choice of the previous examples, we emphasize that random variables $g(X_T) = h(S_T)$, usually used in financial applications, belong to a space $\widetilde{\mathbb{B}}_{2,\infty}^{\theta}$ for some $\theta \in (0, 1]$. However, it is not true that $\bigcup_{\theta \in (0,1]} \widetilde{\mathbb{B}}_{2,\infty}^{\theta} = L_2(\mathbb{P})$. The following result gives a way to construct $g(W_1)$ belonging to $L_2(\mathbb{P})$ (here W is the linear Brownian motion) but $g(W_1) \notin \widetilde{\mathbb{B}}_{2,\infty}^{\theta}$ for all $\theta \in (0, 1]$:

Proposition 12.13 [12] *Let* $0 < \theta < 1$ *and* $g = \sum_{k=0}^{\infty} \alpha_k H_k \in L_2(\gamma_1)$, *where* $(H_k)_{k \geq 0}$ *is the orthogonal basis of Hermite polynomials defined in* (12.4). *Then*

$$g(W_1) \in \widetilde{\mathbb{B}}_{2,\infty}^{\theta} \quad \textit{if and only if} \quad \sup_{0 \leq t < 1} (1-t)^{1-\theta} \sum_{k=1}^{\infty} k t^{k-1} \alpha_k^2 < \infty.$$

Approximation properties as described in Sect. 12.5.2 for g with $g(W_1) \in L_2(\mathbb{P}) \setminus \bigcup_{0 < \theta \leq 1} \widetilde{\mathbb{B}}_{2,\infty}^{\theta}$ were studied in [21] and [31].

12.5 Applications

In this section we discuss some applications in stochastic finance which lead us to the fractional smoothness as introduced above. As mentioned at the beginning, a central role is played by the tracking error that arises when discrete-time hedging is used, instead of a continuous-time strategy. For the sake of convenience, we briefly recall the notation:

- the option payoff at maturity T is $Z = h(S_T)$;
- the fair price function is $H(t, x) = \mathbb{E}_{\mathbb{Q}}(h(S_T) | S_t = x)$;
- the n rebalancing dates are defined by a deterministic time net $\tau = (t_i)_{i=0}^{n}$ with $0 = t_0 < t_1 < \cdots < t_{n-1} < t_n = T$;
- the resulting tracking error process $C(Z, \tau) = (C_t(Z, \tau))_{t \in [0,T]}$ is given by

$$C_t(Z, \tau) := \mathbb{E}_{\mathbb{Q}}(Z | \mathcal{F}_t) - \mathbb{E}_{\mathbb{Q}} Z - \sum_{i=0}^{n-1} \nabla_x H(t_i, S_{t_i}) \cdot (S_{t_{i+1} \wedge t} - S_{t_i \wedge t}).$$

12.5.1 Weak Limits of Error Processes

Weak limits of stochastic processes have been intensively studied in the literature; see, for instance, [22, 23, 26]. For the particular problem of the weak convergence of the tracking error, the reader is referred to [13, 17, 18, 30]. To formulate our results, we let $\widetilde{W} = (\widetilde{W}_t)_{t\ge 0}$ be a standard Brownian motion starting at zero defined on some auxiliary probability space, where we may and do assume that all paths are continuous. In the following, $\Longrightarrow_{C[0,s]}$ stands for the weak convergence in $C[0,s]$ for some $s > 0$.

In this paragraph we assume that $T = 1$[2] and that S is the standard geometric Brownian motion, i.e., the setting of (GBM) and $\mathbb{P} = \mathbb{Q}$. The following result is the starting point of this section:

Theorem 12.14 [17] *Let $\tau_n = (i/n)_{i=0}^n$ be the equidistant time nets, and let $Z := \mathbf{1}_{[K,\infty)}(S_1)$ be the payoff of a digital option with strike price $K > 0$. Then one has that*

$$\sqrt{n} C_1(Z, \tau_n) \Longrightarrow \widetilde{W}_{\frac{1}{2}\int_0^1 |S_t^2 \frac{\partial^2 H}{\partial x^2}(t,S_t)|^2 dt},$$

where $\Longrightarrow$ denotes the weak convergence as n goes to infinity.

The remarkable fact is that the weak limit is not square-integrable. In the following we describe a way to increase the integrability of the weak limit. This is of particular interest for risk management purposes, as a higher integrability gives better tail estimates. The idea is to use adapted time nets that are more concentrated close to maturity. They are defined as follows: Given a parameter $\theta \in (0,1]$, we define the nets $\tau^{n,\theta}$ by

$$t_k^{n,\theta} := 1 - \left(1 - \frac{k}{n}\right)^{\frac{1}{\theta}}.$$

For $\theta = 1$, we have the equidistant time nets, i.e., $t_k^{n,1} = \frac{k}{n}$. Now we have the following:

Theorem 12.15 [13] *Let $0 < \theta \le 1$, $Z = h(S_1) \in L_2(\mathbb{P})$, and $0 \le s < 1$. Then*

$$\left(\sqrt{n} C_t\left(Z, \tau^{n,\theta}\right)\right)_{t\in[0,s]} \Longrightarrow_{C[0,s]} \left(\widetilde{W}_{\int_0^t \frac{(1-r)^{1-\theta}}{2\theta} |S_r^2 \frac{\partial^2 H}{\partial x^2}(r,S_r)|^2 dr}\right)_{t\in[0,s]}.$$

Moreover, the following assertions are equivalent:

(i) *One has $h(S_1) \in \widetilde{\mathbb{B}}_{2,2}^{\theta}$ for $0 < \theta < 1$ or $h(S_1) \in \widetilde{\mathbb{B}}_{2,\infty}^{1}$ for $\theta = 1$.*
(ii) *On some stochastic basis there exists a continuous square-integrable martingale $M = (M_t)_{t\in[0,1]}$ such that $\sqrt{n} C(Z, \tau^{n,\theta}) \Longrightarrow_{C[0,1]} M$.*
(iii) *For*

$$A := \int_0^1 \frac{(1-t)^{1-\theta}}{2\theta} \left| S_t^2 \frac{\partial^2 H}{\partial x^2}(t, S_t)\right|^2 dt,$$

[2] With $T = 1$, we are in accordance with the quoted literature that used Hermite polynomials. Of course, we could do a rescaling to $T > 0$ afterwards.

one has that $\mathbb{E}A < \infty$ *and*

$$\sqrt{n}C\big(Z,\tau^{n,\theta}\big) \Longrightarrow_{C[0,1]} \big(\widetilde{W}_{\chi_{\{A<\infty\}}\int_0^t \frac{(1-r)^{1-\theta}}{2\theta}|S_r^2 \frac{\partial^2 H}{\partial x^2}(r,S_r)|^2 dr}\big)_{t\in[0,1]}.$$

The theorem above gives us one way to consider the L_p-setting for $2 \le p < \infty$. Given a differentiable function $\psi : (0,\infty) \to \mathbb{R}$, we let

$$(A\psi)(x) := x\psi'(x) - \psi(x).$$

In the following, $AH(t,x)$ means that A acts on the x-variable of the function $H(t,x)$.

Definition 12.16 For $h(S_1) \in L_2(\mathbb{P})$, $0 < \theta < 1$, and $0 \le t < 1$, we let

$$D_t^{S,\theta} h(S_1) := \frac{1-\theta}{2} \int_0^1 (1-u)^{-\frac{1+\theta}{2}} \big[AH(u \wedge t, S_{u\wedge t}) - AH(0,S_0)\big]\, du.$$

For $\theta = 1$ and $t \in [0,1)$, we let $D_t^{S,1} h(S_1) := AH(t,S_t) - AH(0,S_0)$.

The process $D^{S,\theta}h(S_1) = (D_t^{\theta} h(S_1))_{t\in[0,1)}$ is a quadratic integrable martingale on the half open time interval $[0,1)$. Using the Riemann–Liouville operator of partial integration, the process $D^{S,\theta}h(S_1)$ can be interpreted as a fractional differentiation of order θ in x (see [13]). The point of the construction of $D^{S,\theta}h(S_1)$ is that we may have L_p-singularities of $S_t \frac{\partial H}{\partial x}(t,S_t)$ as $t \uparrow 1$, whereas $D^{S,\theta}h(S_1)$ remains L_p-bounded.

Theorem 12.17 [13] *For* $2 \le p < \infty$, $0 < \theta \le 1$, *and* $Z = h(S_1) \in L_2(\mathbb{P})$, *the following assertions are equivalent*:

(i) *On some stochastic basis there exists a continuous* $L_p(\mathbb{P})$*-integrable martingale* M *such that* $\sqrt{n}C(Z,\tau^{n,\theta}) \Longrightarrow_{C[0,1]} M$.
(ii) *The martingale* $D^{S,\theta}h(S_1)$ *is bounded in* $L_p(\mathbb{P})$.

12.5.2 L_2-estimates of the Tracking Error

In this section we work in the one-dimensional martingale case assuming (SDE) with $\sigma(t,x) = \sigma(x)$ and $b(t,x) = -\frac{1}{2}\sigma^2(x)$ (meaning $\mathbb{P} = \mathbb{Q}$). The payoff function h is polynomially bounded, and the option maturity is $T > 0$. We remind the reader about the time nets $\tau^{n,\theta}$ given by

$$t_k^{n,\theta} := T\left(1 - \left(1 - \frac{k}{n}\right)^{\frac{1}{\theta}}\right)$$

and that for $\theta = 1$, we obtain the equidistant nets. Let us first check what quadratic hedging error one can expect at all if the portfolio is rebalanced n-times. The answer is the rate $1/\sqrt{n}$ as shown by the following:

Theorem 12.18 [7, Theorem 2.5] *Assume that there are no constants $c_0, c_1 \in \mathbb{R}$ such that $h(S_T) = c_0 + c_1 S_T$ a.s. Then*

$$\inf_{\substack{n=1,2,\ldots \\ 0=t_0<\cdots<t_n=T}} n^{\frac{1}{2}} \left\| C\left(h(S_T), (t_k)_{k=0}^n\right) \right\|_{L_2(\mathbb{P})} > 0,$$

where the infimum is taken over deterministic time nets.

This was extended to the case of random time nets in [8] in the case of the geometric Brownian motion.

Now we continue with the case of equidistant time nets that are often used in discretizations.

Equidistant Time Nets Here a starting point is the following result of Zhang:

Theorem 12.19 [36, Theorem 2.4.1] *Assume that $h : \mathbb{R} \to \mathbb{R}$ is a Lipschitz function. Then we have that*

$$\lim_n n^{\frac{1}{2}} \left\| C\left(h(S_T), \tau^{n,1}\right) \right\|_{L_2(\mathbb{P})} \in [0, \infty).$$

This is the result one would expect: Given a Lipschitz payoff, the L_2-rate of the error is $1/2$ for equidistant nets. But this is not the case in general as shown in the following:

Theorem 12.20 [17, Theorem 1] *For $h(x) = \mathbf{1}_{[K,\infty)}(x)$ with some $K > 0$, we have that*

$$\lim_n n^{\frac{1}{4}} \left\| C\left(h(S_T), \tau^{n,1}\right) \right\|_{L_2(\mathbb{P})} \in (0, \infty).$$

This means that the L_2-approximation rate for the binary option is $n^{1/4}$ if one uses equidistant nets. The two above results also hold for appropriate $\mathbb{Q} \neq \mathbb{P}$ (i.e., S is not a martingale) where the outer L_2-norm is computed w.r.t. the historical probability $\mathbb{P}$ (cf. the remarks after Theorem 12.4).

Theorems 12.19 and 12.20 lead naturally to two questions: What is the reason for the rate $1/4$, and, secondly, can one improve the rate $1/4$? Both questions can be answered by the usage of the concept of fractional smoothness.

Theorem 12.21 [7, Theorems 2.3 and 2.8] *For $0 < \theta \leq 1$ and a polynomially bounded $h : (0, \infty) \to \mathbb{R}$, the following assertions are equivalent:*

(i) $h(S_T) \in \widetilde{\mathbb{B}}_{2,\infty}^{\theta}$;

(ii) $\sup_n n^{\frac{\theta}{2}} \| C(h(S_T), \tau^{n,1}) \|_{L_2(\mathbb{P})} < \infty$.

In particular, it turns out that $h(S_T) \in \mathbb{D}_{1,2}$ if and only if

$$\sup_n n^{\frac{1}{2}} \left\| C\left(h(S_T), \tau^{n,1}\right) \right\|_{L_2(\mathbb{P})} < \infty,$$

see [7, Theorem 2.6], where $\mathbb{D}_{1,2}$ is the Malliavin Sobolev space obtained from the construction in Sect. 12.3 with $H = L_2[0, T]$ and $g_h := \int_0^T h(t)\, dW_t$.

For the binary option, one has in Theorem 12.21 that $\theta = 1/2$ (cf. Example 3 in Sect. 12.4). This recovers the rate $1/4$ obtained in Theorem 12.20.

Nonequidistant Time Nets Next we show how to obtain the optimal rate $n^{1/2}$ by a suitable choice of the trading dates (see question (Q4) in Sect. 12.1). We can combine [7, Lemmas 3.2 and 5.3] and [12, Lemma 3.8] to get the following:

Theorem 12.22 *For $0 < \theta \leq 1$ and a polynomially bounded $h : (0, \infty) \to \mathbb{R}$, the following assertions are equivalent*:

(i) $\int_0^T (T-t)^{1-\theta} \mathbb{E}|S_t^2 \frac{\partial^2 H}{\partial x^2}(t, S_t)|^2\, dt < \infty$;
(ii) $\sup_n n^{\frac{1}{2}} \|C(h(S_T), \tau^{n,\theta})\|_{L_2(\mathbb{P})} < \infty$.

For $0 < \theta < 1$ (and at least a bounded h), condition (i) of Theorem 12.22 is equivalent to

(i′) $h(S_T) \in \widetilde{\mathbb{B}}^\theta_{2,2}$,

which can be checked by using Theorem 12.3. For the binary option, this gives that

$$\sup_n n^{\frac{1}{2}} \left\| C\left(\chi_{[K,\infty)}(S_T), \tau^{n,\eta}\right) \right\|_{L_2(\mathbb{P})} < \infty$$

for any strike $K > 0$ and $0 < \eta < 1/2$.

For the next two theorems, we assume that $T = 1$, that $S_t = e^{W_t - \frac{t}{2}}$, and that h may be general, i.e., not polynomially bounded. The formulation of Theorem 12.22 in the language of the interpolation spaces introduced in Sect. 12.3 gives the following:

Theorem 12.23 [12, Theorem 3.2] *For $0 < \theta \leq 1$ and $h(S_1) \in L_2(\mathbb{P})$, the following assertions are equivalent*:

(i) $h(e^{\cdot-(1/2)}) \in \mathbb{B}^\theta_{2,2}(\gamma_1)$ *if* $0 < \theta < 1$ *and* $h(e^{\cdot-(1/2)}) \in \mathbb{D}_{1,2}(\gamma_1)$ *if* $\theta = 1$;
(ii) $\sup_n n^{\frac{1}{2}} \|C(h(S_1), \tau^{n,\theta})\|_{L_2(\mathbb{P})} < \infty$.

Theorem 12.21 can be extended in this context to the full scale of real interpolation spaces as follows.

Theorem 12.24 [12, Theorem 3.5] *For $1 \leq q \leq \infty$, $0 < \theta < 1$, and $h(S_1) \in L_2(\mathbb{P})$, the following assertions are equivalent*:

(i) $h(e^{\cdot-(1/2)}) \in \mathbb{B}^\theta_{2,q}(\gamma_1)$;
(ii) $\|(n^{\frac{\theta}{2}-\frac{1}{q}} a_n)_{n=1}^\infty\|_{\ell_q} < \infty$ *for* $a_n := \|C(h(S_1), \tau^{n,1})\|_{L_2(\mathbb{P})}$.

Concluding Remarks

(i) The higher-dimensional case for X was considered in the literature as well. Roughly speaking, one can analogously obtain upper bounds; however precise lower bounds as in the one-dimensional case are still missing. This is due to the fact that a characterization of the L_2-error proved in [11, Theorem 4.4] and [7, Lemma 3.2] is missing for higher dimensions. However, after Zhang [36] started with the regular case, Temam [34] extended results from [17] to higher dimensions, and Hujo [20] used nonuniform time nets to improve the approximation rates for certain irregular payoffs to the optimal rate $1/\sqrt{n}$ in this setting.

(ii) Seppälä [31] found a criterion to characterize, under certain conditions, that there is a constant $c > 0$ such that

$$\inf_{\substack{\tau=(t_i)_{i=0}^n \\ 0=t_0<\cdots<t_n=1}} \left\| C\big(h(S_1), \tau\big) \right\|_{L_2(\mathbb{P})} \leq \frac{c}{\sqrt{n}},$$

where deterministic time nets are taken. It should be noted that one has a *non-linear* approximation problem as the time nets may change for fixed n from payoff to payoff h.

(iii) In the above discussion, the time nets τ are deterministic. Alternatively, one can allow the time nets to be stochastic and adapted. This issue has been handled by [28] using optimal stopping tools. The estimation of convergence rates is an open question. However, it was shown in [8] that the random time nets do not improve the best possible approximation rate $1/\sqrt{n}$ in the case (GBM) when in the nth approximation a sequence of n stopping times is used.

(iv) Similar studies can be performed when studying the Delta–Gamma hedging strategies. Instead of hedging the payoff using only the asset, we use other traded options written on the same asset. For a one-dimensional asset, if the price of the additional option is $(P(t, S_t))_{0\leq t\leq T}$, the numbers of options P and assets to hold at time t_i are respectively equal to

$$\delta_{t_i}^P := \frac{\partial_S^2 H(t_i, S_{t_i})}{\partial_S^2 P(t_i, S_{t_i})} \quad \text{and} \quad \delta_{t_i}^S := \partial_S H(t_i, S_{t_i}) - \frac{\partial_S^2 H(t_i, S_{t_i})}{\partial_S^2 P(t_i, S_{t_i})} \partial_S P(t_i, S_{t_i}).$$

In [14, Theorem 6], considering a multidimensional Black–Scholes model, it is established that for an exponentially bounded payoff such that $g(X_T) \in \widetilde{\mathbb{B}}_{2,\infty}^{\theta}$ for some $0 < \theta < 1$, the use of equidistant time nets leads to the same convergence rate $1/n^{\theta/2}$ as for the delta hedging strategy. On the contrary, the use of nonequidistant time nets $\tau^{n,\eta}$ with $0 < \eta < \theta/2$ enables us to obtain the improved convergence rate $1/n$.

12.6 Further Developments

12.6.1 Backward Stochastic Differential Equations

Makhlouf and the second author applied in [15] the concept of fractional smoothness to backward stochastic differential equations of the type

$$Y_t = g(X_T) + \int_t^T f(s, X_s, Y_s, Z_s)\, ds - \int_t^T Z_s\, dW_s,$$

where $X = (X_t)_{t \in [0,T]}$ is our forward diffusion, and the generator f is continuous in its four arguments, continuously differentiable in (x, y, z) with uniformly bounded derivatives. These equations are particularly useful in stochastic finance, since they allow one to take into account market frictions and constraints (we refer to [25] for a more complete account on this subject).

Solving numerically this type of equation is a challenging issue since it concerns a nonlinear problem (due to the generator f), generally defined in a multidimensional setting. One possible approach consists in approximating the BSDE using a discrete-time dynamic programming equation (see [5, 27, 37] among others). One of the main error contributions is related to the L_2-regularity on Z, defined by

$$\mathcal{E}(Z, \tau) = \sum_{i=1}^{n} \int_{t_{i-1}}^{t_i} \|Z_t - Z_{t_{i-1}}\|_{L_2(\mathbb{P})}^2\, dt.$$

If f is equal to 0, then the Z-component is given by $z_t = \nabla_x u(t, X_t)\sigma(t, X_t)$, where $u(t, x) = \mathbb{E}(g(X_T)|X_t = x)$. Studying the L_2-regularity of z is thus very similar to the analysis of the tracking error presented in Sect. 12.5. Additionally, using BSDE techniques, one can prove explicit upper bounds for the difference $Z - z$.

Theorem 12.25 [15, Corollary 2.1] *Assume* (SDE) *and* $g(X_T) \in \widetilde{\mathbb{B}}_{2,\infty}^{\theta}$ *for* $0 < \theta \le 1$. *Then, for some* $c > 0$, *one has that*

$$|Z_t - z_t| \le c \int_t^T \frac{\sqrt{\mathbb{E}[|g(X_T) - \mathbb{E}(g(X_T)|\mathcal{F}_s)|^2|\mathcal{F}_t]}}{T - s}\, ds + c(T - t),$$

$$\mathbb{E}|Z_t - z_t|^2 \le c(T - t)^{\theta}.$$

Taking the advantage of this approximation result close to the time singularity, we can prove that the estimate of $\mathcal{E}(z, \tau)$ (linear case) transfers to $\mathcal{E}(Z, \tau)$ (nonlinear case) and get the following:

Theorem 12.26 [15, Theorem 3.2] *Assume* (SDE), $g(X_T) \in \widetilde{\mathbb{B}}_{2,\infty}^{\theta}$, *and that* $0 < \eta < \theta < 1$ *or* $\eta = \theta = 1$. *Then one has that*

$$\mathcal{E}\big(Z, \tau^{n,\eta}\big) \le \frac{c}{n}.$$

In [9], extensions of the above in different directions are discussed.

12.6.2 Lévy Processes

An extension of the results of [12] to Lévy processes is done by C. Geiss, S. Geiss and Laukkarinen in [10]. Moreover, Tankov and Brodén [33] proved results along the line of [17].

12.6.3 Multigrid Monte Carlo Methods

In the context of Multigrid Monte Carlo Methods, it turned out that the concept of fractional smoothness is useful as well. The reader is referred to the papers of Avikainen [1, 2].

References

1. R. Avikainen, On irregular functionals of SDEs and the Euler scheme. Finance Stoch. **13**, 381–401 (2009)
2. R. Avikainen, On generalized bounded variation and approximation of SDEs. Preprint, Department of Mathematics and Statistics, University of Jyväskylä (2009). Available at www.math.jyu.fi/research/pspdf/383.pdf
3. C. Bennet, R. Sharpley, *Interpolation of Operators* (Academic Press, San Diego, 1988)
4. J. Bergh, J. Lőfstrőm, *Interpolation Spaces. An Introduction* (Springer, Berlin, 1976)
5. B. Bouchard, N. Touzi, Discrete time approximation and Monte Carlo simulation of backward stochastic differential equations. Stoch. Process. Appl. **111**, 175–206 (2004)
6. A. Friedman, *Partial Differential Equations of Parabolic Type* (Prentice-Hall, New York, 1964)
7. C. Geiss, S. Geiss, On approximation of a class of stochastic integrals and interpolation. Stoch. Stoch. Rep. **76**(4), 339–362 (2004)
8. C. Geiss, S. Geiss, On an approximation problem for stochastic integrals where random time nets do not help. Stoch. Process. Appl. **116**, 407–422 (2006)
9. C. Geiss, S. Geiss, E. Gobet, Generalized fractional smoothness and L_p-variation of BSDEs with non-Lipschitz terminal condition. Available at ArXiv
10. C. Geiss, S. Geiss, E. Laukkarinen, A note on Malliavin fractional smoothness and approximation for exponential Levy processes (2011, in preparation)
11. S. Geiss, Quantitative approximation of certain stochastic integrals. Stoch. Stoch. Rep. **73**(3–4), 241–270 (2002)
12. S. Geiss, M. Hujo, Interpolation and approximation in $L_2(\gamma)$. J. Approx. Theory **144**, 213–232 (2007)
13. S. Geiss, A. Toivola, Weak convergence of error processes in discretizations of stochastic integrals and Besov spaces. Bernoulli **15**(4), 925–954 (2009)
14. E. Gobet, A. Makhlouf, The tracking error rate of the Delta–Gamma hedging strategy. Mathematical Finance. Available at hal.archives-ouvertes.fr/hal-00401182/fr/
15. E. Gobet, A. Makhlouf, L_2-time regularity of BSDEs with irregular terminal functions. Stoch. Process. Appl. **120**, 1105–1132 (2010)
16. E. Gobet, R. Munos, Sensitivity analysis using Itô–Malliavin calculus and martingales. Application to stochastic control problem. SIAM J. Control Optim. **43**(5), 1676–1713 (2005)
17. E. Gobet, E. Temam, Discrete time hedging errors for options with irregular pay-offs. Finance Stoch. **5**(3), 357–367 (2001)

18. T. Hayashi, P. Mykland, Evaluating hedging errors: an asymptotic approach. Math. Finance **15**(2), 309–343 (2005)
19. F. Hirsch, Lipschitz functions and fractional Sobolev spaces. Potential Anal. **11**(4), 415–429 (1999)
20. M. Hujo, On discrete time hedging in d-dimensional option pricing models. Preprint 317, Department of Mathematics and Statistics, University of Jyväskylä (2005). Available at www.math.jyu.fi/research/pspdf/317.pdf
21. M. Hujo, Is the approximation rate for European pay-offs in the Black–Scholes model always $1/\sqrt{n}$? J. Theor. Probab. **19**, 190–203 (2006)
22. J. Jacod, On continuous conditional Gaussian martingales and stable convergence in law, in *Séminaire de Probabilités XXXI*. Lecture Notes in Mathematics, vol. 1655 (Springer, Berlin, 1997)
23. J. Jacod, A.N. Shiryaev, *Limit Theorems for Stochastic Processes* (Springer, Berlin, 2003)
24. I. Karatzas, S.E. Shreve, *Methods of Mathematical Finance* (Springer, New York, 1998)
25. N. El Karoui, S.G. Peng, M.C. Quenez, Backward stochastic differential equations in finance. Math. Finance **7**(1), 1–71 (1997)
26. T.G. Kurtz, P. Protter, Weak limit theorems for stochastic integrals and stochastic differential equations. Ann. Probab. **19**(3), 1035–1070 (1991)
27. J.P. Lemor, E. Gobet, X. Warin, Rate of convergence of an empirical regression method for solving generalized backward stochastic differential equations. Bernoulli **12**(5), 889–916 (2006)
28. C. Martini, C. Patry, Variance optimal hedging in the Black–Scholes model for a given number of transactions (1999). Available at hal.inria.fr/docs/00/07/28/95/PDF/RR-3767.pdf
29. D. Nualart, *Malliavin Calculus and Related Topics* (Springer, Berlin, 2006)
30. H. Rootzen, Limit distributions for the error in approximations of stochastic integrals. Ann. Probab. **8**(2), 241–251 (1980)
31. H. Seppälä, Optimal approximation rate of certain stochastic integrals. J. Approx. Theory. Preprint 375, Department of Mathematics and Statistics, University of Jyväskylä (2008). Available at www.math.jyu.fi/research/pspdf/375.pdf
32. N. Taleb, *Dynamic Hedging: Managing Vanilla and Exotic Options*. Wiley Finance Editions (Wiley, New York, 1997)
33. P. Tankov, M. Brodén, Errors from discrete hedging in exponential Lévy models: the L_2 approach. Preprint (2009)
34. E. Temam, Analysis of error with Malliavin calculus: application to hedging. Math. Finance **13**(1), 201–214 (2003)
35. S. Watanabe, Fractional order Sobolev spaces on Wiener space. Probab. Theory Relat. Fields **95**(2), 175–198 (1993)
36. R. Zhang, Couverture approchée des options Européennes. Ph.D. Thesis, Ecole Nationale des Ponts et Chaussées (1999). Available at cermics.enpc.fr/theses/99/zhang-ruotao.ps.gz
37. J. Zhang, A numerical scheme for BSDEs. Ann. Appl. Probab. **14**(1), 459–488 (2004)

Chapter 13
Liquidity Models in Continuous and Discrete Time

Selim Gökay, Alexandre F. Roch, and H. Mete Soner

Abstract We survey several models of liquidity and liquidity-related problems such as optimal execution of a large order, hedging and super-hedging options for a large trader, utility maximization in illiquid markets, and price impact models with price manipulation strategies.

Keywords Liquidity risk · Optimal execution · Hedging · Price impact · Large trader models · Utility maximization

Mathematics Subject Classification (2010) 60H30 · 91G10 · 91G20 · 91G80

13.1 What Is Illiquidity?

The study of liquidity in financial markets either invokes the ease with which financial securities can be bought and sold, or addresses the ability to trade without triggering important changes in asset prices. More specifically, one can think of liquidity as an exogenous measure of the added costs per transaction associated to trading large quantities of the asset. This is the approach advocated by Çetin et al. [9], in which an exogenously defined supply curve gives the price per share as a function of transaction size. On the other hand, one can take this idea a step further and recognize that these added costs are the product of imbalances in the supply

Research partly supported by the European Research Council under the grant 228053-FiRM. Financial support from the ETH Foundation, Swiss Finance Institute and by the National Centre of Competence in Research "Financial Valuation and Risk Management" (NCCR FINRISK) are also gratefully acknowledged.

S. Gökay (✉) · A.F. Roch · H.M. Soner
Departement Mathematik, ETH Zürich, 8092, Zürich, Switzerland
e-mail: selim.gokay@math.ethz.ch

A.F. Roch
e-mail: alexandre.f.roch@gmail.com

H.M. Soner
e-mail: hmsoner@ethz.ch

G. Di Nunno, B. Øksendal (eds.), *Advanced Mathematical Methods for Finance*,
DOI 10.1007/978-3-642-18412-3_13,

and demand of the asset due to the trading of large quantities. If the imbalance is temporary and only affects the current price paid, we are effectively in the previous setting, and the transaction costs depend mainly on the size of the trade. On the other hand, these imbalances can have a lasting effect in such a way that future prices will be affected by previous trades. For instance, Jarrow [21, 22] considers the price per share as a function of the holdings of the large trader. As we can see, these two notions are closely related, and one approach can be more convenient or realistic than the other depending on the setting.

There are four main themes present in the current mathematical literature on liquidity. The first one pertains to the problem of optimal execution of large orders. Consider the situation in which a trader plans to sell a large number of units of a risky asset before a predetermined time horizon. Since the size of the order is large, this trader may find it more optimal to work the order in several smaller slices to minimize her impact on prices by trading during times of higher liquidity and taking advantage of the resilience of the supply and demand of the asset. On the other hand, delaying the orders for too long increases the exposure to other risks. The goal is to find the right balance between liquidity risk and other market risks. Many papers have been written on this question, and we survey some of the main results in Sect. 13.2.

The second theme we discuss in this survey relates to the familiar problem of option pricing. On one hand, the existence of a supply curve that governs the liquidity cost of a transaction clearly suggests that the hedging of derivatives will be more costly that in the classical frictionless setting. On the other hand, the hedger's capacity to have an impact on prices may influence her into manipulating prices in her favor. The classical hedging problem gains a new level of complexity as the hedger's strategy, which is chosen in terms of the option payoff, has a repercussion on the future evolution of prices on which the option payoff is calculated. The different approaches commonly used in this setting are reviewed in Sects. 13.3 and 13.4. In Sect. 13.3 we review the results on hedging for a large trader, including the papers of Cvitanic and Ma [14], Platen and Schweizer [27], Bank and Baum [7], and Roch [28]. In Sect. 13.4 we introduce the supply curve model introduced in [9], discuss the superreplication problem in this context, and focus on the works of Çetin, Soner, and Touzi [11] and Gökay and Soner [18].

The third theme is related to the expected utility maximization problem with permanent or temporary price impacts. We briefly summarize some of the main results in this line of research in Sect. 13.5.

The introduction of price impacts on the evolution of the price processes evokes the possibility of *price manipulations*, defined as trading strategies with negative expected execution costs. For instance, by making the price go up after a purchase, a large trader has the possibility of making higher profits than average by reselling the shares purchased if the average impact on prices is smaller for sell orders than buy orders. This is only one example of a price manipulation, and it has led some authors to investigate these types of irregularities in terms of the price impact functions. It is the focus of Sect. 13.6.

13.2 Optimal Execution Problem

The optimal execution problem consists in allocating a large buy or sell order of a risky asset over a fixed time horizon with the aim of minimizing the expected cost of the order due to the relative illiquidity of the asset. The main challenge in this kind of allocation is to choose a trading program which is executed on a period of time short enough to reduce the risk of the uncertainty of future prices while dividing the large order in smaller ones distributed over time to reduce the liquidity costs associated to this trading program.

There are mainly two approaches in the literature which we summarize in this section. The first approach, proposed in the papers of Bertsimas and Lo [8], Almgren [4], Almgren and Chriss [5, 6], and Schied and Schöneborn [30], measures the associated cost of a sequence of transactions in terms of a permanent price impact and/or a temporary price impact which are exogenously determined and depend on the size of the transaction and the speed of change of the position in the asset. On the other hand, the second approach presumes the existence of a limit order book through which the orders of the large trader are executed. In this setting, the cost of an execution strategy depends on endogenous variables such as the density of the number of shares being offered at each price and the resilience of the order book. The main references that we will summarize for this approach are the papers of Obizhaeva and Wang [26], Alfonsi, Fruth, and Schied [1, 2], and Alfonsi and Schied [3].

13.2.1 The First Approach

In the optimal execution problem, the investor wants to liquidate a certain number $X_0 > 0$ of units of an asset before a fixed finite time horizon T. Dividing the trading period $[0, T]$ into N equal intervals of length $\tau = T/N$, the investor chooses quantities $\xi_k \geq 0$ to sell at discrete times $t_k = k\tau$ for $k = 1, \dots, N$ such that $\sum_{k=1}^{N} \xi_k = X_0$. The number of units still held by the investor at time t_k is given by $X_k = X_0 - \sum_{j=1}^{k} \xi_j$. Note that the case $X_0 < 0$ can be treated in a similar way.

Bertsimas and Lo [8] approach this problem by minimizing expected execution costs, whereas Almgren [4] and Almgren and Chriss [5, 6] extend this idea by also incorporating the risk into the execution problem using the variance of the associated costs.

Bertsimas and Lo [8] propose a general formulation for the price process of the asset, of which two special cases stand out. One special case proposed in [8] gives a stock price of the form

$$\tilde{S}_k = \tilde{S}_{k-1} - \gamma \xi_k + \epsilon_k, \quad \gamma > 0, \tag{13.1}$$

in which $\{\epsilon_k\}_{k=1}^{N}$ is a sequence of independent and identically distributed random variables with mean zero and variance σ_ϵ^2, whereas ξ_k is the size of the transaction at time t_k. The profit obtained from a strategy, also commonly called the *capture*, is given by $\sum_{k=1}^{n} \xi_k \tilde{S}_k$. The *total cost of trading* associated to a strategy X is defined as the difference between the book value $X_0 S_0$ and the capture, and is computed as

$$C(X) = X_0 S_0 - \sum_{k=1}^{N} \xi_k \tilde{S}_k.$$

In this setup, the goal is to minimize the expected execution cost

$$\min_{\{\xi_k\}_{k=1}^{N}} E\big[C(X)\big]$$

subject to the constraint

$$\sum_{k=1}^{N} \xi_k = X_0.$$

The price impact due to the trade ξ_k is said to be permanent in (13.1) since the price at time t_k is defined in terms of the price at time t_{k-1}, which is also affected by the trade ξ_{k-1} at time t_{k-1}. For this special case, there exists an explicit optimal strategy. It is called the naive strategy and is obtained by dividing the total order X_0 into N equal slices, i.e., $\xi_k = \frac{X_0}{N}$.

Bertsimas and Lo [8] also consider a linear temporary price impact model. In this setup the execution price $\tilde{S}_k$ at time k, i.e., the price paid for the transaction at time k, is decomposed into an exogenous unaffected price S_k and a price impact as a function of the trade size. The unaffected price, also called publicly available price, can be interpreted as the price that would be obtained in absence of price impacts. The execution price at time t_k is a function of ξ_k and assumed to be given by

$$\tilde{S}_k(\xi_k) = S_k - (\eta \xi_k + \gamma Y_k) S_k, \quad \eta > 0,$$

in which Y is an adapted process. In the special case that the unaffected price process $\{S_k\}_{k=1}^{N}$ follows

$$S_k = S_{k-1} \exp(\alpha_k),$$

and the state vector $\{Y_k\}_{k=1}^{N}$ satisfies

$$Y_k = \rho Y_{k-1} + \zeta_k, \quad \rho > 0,$$

in which $\{\zeta_k\}_{k=1}^{N}$ and $\{\alpha_k\}_{k=1}^{N}$ are i.i.d. normal random variables with mean 0, the authors show that the best execution strategy consists in trade sizes which are linear functions of the remaining number of shares X_k and the state variable Y_k.

The implicit assumption in the paper of Bertsimas and Lo [8] is that the investor is not risk averse as she only aims to minimize the expected cost of the execution.

In the optimal execution model of Almgren [4] and Almgren and Chriss [5, 6], the investor's tolerance for risk influences her trading decisions. To illustrate this point, consider the two following execution strategies. On one hand, a risk averse agent may choose to trade everything now. The advantage of this strategy is that the cost is known and all risks regarding the future prices of the asset are eliminated. On the other hand, the cost is high, and the investor may be willing to take some risk by dividing her orders and executing them through time in order to have a lower expected cost. Almgren and Chriss characterize this trade-off between the cost and variance of optimal execution strategies by an efficient frontier. They show that the points on the frontier are determined by the level of risk aversion of the agent. They argue that the optimal strategies for the execution problem are static, i.e., these decisions can be fully determined at the beginning of the trading period and give explicit solutions for some specific cases.

In addition to the above mathematical setup, we denote by $v_k = \frac{\xi_k}{\tau}$ the speed of trades on the kth interval. In [5], the publicly available price per share S_k is modeled as follows. Let $\{\zeta_k\}_{k=1}^N$ be i.i.d. random variables with zero mean and unit variance. We assume that

$$S_k = S_{k-1} + \sigma\sqrt{\tau}\zeta_k - \tau g(v_k), \quad k = 1, \ldots, N,$$

where $\sigma > 0$ is a volatility parameter, and $g : \mathbf{R} \to \mathbf{R}$ is a permanent impact function. The price per share paid by the investor at time k is

$$\tilde{S}_k(\xi_k) = S_{k-1} - h\left(\frac{\xi_k}{\tau}\right), \quad k = 1, \ldots, N,$$

where h is a given function, called the temporary impact function. The capture is computed as

$$\begin{aligned} C(X) &= X_0 S_0 - \sum_{k=1}^{N} \xi_k \tilde{S}_k(\xi_k) \\ &= \sum_{k=1}^{N} \tau X_k g(v_k) + \sum_{k=1}^{N} \tau v_k h(v_k) - \sigma\sqrt{\tau} \sum_{k=1}^{N-1} X_k \zeta_k, \end{aligned} \tag{13.2}$$

with expected value and variance at time 0 given by

$$E\big(C(X)\big) = \sum_{k=1}^{N} \tau X_k g(v_k) + \sum_{k=1}^{N} \tau v_k h(v_k), \qquad \mathrm{Var}\big(C(X)\big) = \sum_{k=1}^{N} \tau\sigma^2 X_k^2$$

when the strategy X is deterministic.

A strategy is called *efficient* if there is no strategy that has a lower expected value for a level of variance which is equal or lower. The family of efficient strategies is given by the solutions $X^*(\lambda)$ of the optimization problem

$$\min_X \big\{ E\big(C(X)\big) + \lambda\, \mathrm{Var}\big(C(X)\big) \big\}$$

for different values of $\lambda \geq 0$. The family of solutions $(X^*(\lambda))_{\lambda \geq 0}$ is called the efficient frontier. The parameter λ measures the risk aversion of the investor. Every point on the frontier corresponds to a pair

$$\big(\mathrm{Var}\big(C\big(X^*(\lambda)\big)\big), E\big(C\big(X^*(\lambda)\big)\big)\big)$$

for some λ. The efficient frontier gives rise to a smooth and convex function, which we denote by $\mathcal{E}(V)$, assigning the optimal expected cost $\mathcal{E}(V)$ to each possible value of the variance V, i.e., there exists $\lambda \geq 0$ such that $(V, \mathcal{E}(V)) = (\mathrm{Var}(C(X^*(\lambda))), E(C(X^*(\lambda))))$.

In [6], the permanent impact function is taken to be linear, i.e., $g(v) = \gamma v$ (with $\gamma > 0$), and the temporary price impact function consists of the sum of a fixed cost function and a linear impact function so that

$$h(v) = \theta \,\mathrm{sign}(v) + \eta v \quad (v \in \mathbf{R}) \tag{13.3}$$

for some positive constants $\theta, \eta > 0$. In this case, it is easy to see that the expectation of the cost becomes

$$E\big(C(X)\big) = \frac{1}{2}\gamma X_0^2 + \theta \sum_{k=1}^{N} |\xi_k| + \frac{\eta - \frac{1}{2}\gamma\tau}{\tau} \sum_{k=1}^{N} \xi_k^2.$$

Almgren and Chriss [6] show that the optimal solution for the case of g linear and h given by (13.3) can be written in terms of $\lambda > 0$ as

$$X_j^* = \frac{\sinh(\kappa(T - t_j))}{\sinh(\kappa T)} X_0, \quad j = 0, \dots, N,$$

where

$$\kappa \sim \sqrt{\frac{\lambda\sigma^2}{\eta}} + O(\tau), \quad \tau \to 0.$$

If the agent is risk-neutral ($\lambda = 0$), she only wants to minimize the expected cost. Then her optimal strategy is the naive strategy $\xi_k = \frac{X_0}{N}$ as we have seen earlier. In this case, the expected cost and variance of this strategy are given by

$$E_0 := \frac{1}{2}\gamma X_0^2 + \theta X_0 + \left(\eta - \frac{1}{2}\gamma\tau\right)\frac{X_0^2}{T},$$

$$V_0 := \frac{1}{3}\sigma^2 X_0^2 T\left(1 - \frac{1}{N}\right)\left(1 - \frac{1}{2N}\right).$$

The naive strategy corresponds to the minimal point of the efficient frontier, in the sense that $\frac{d\mathcal{E}}{dV}$ evaluated at (V_0, E_0) is equal to zero. Thus for (V, E) in the vicinity of (V_0, E_0),

$$E - E_0 \approx \frac{1}{2}(V - V_0)^2 \frac{d^2\mathcal{E}}{dV^2}\bigg|_{V=V_0},$$

where $d^2\mathcal{E}/dV^2|_{V=V_0}$ is positive by the convexity of the efficient frontier.

13.2.2 Continuous-Time Models

Let us now consider nonlinear impact functions and analyze the continuous-time limit of the previous model as $\tau \to 0$. Let $(\Omega, \mathcal{F}, (\mathcal{F}_t)_{t\geq 0}, \mathbf{P})$ be a given filtered probability space on which a Brownian motion W is defined. In the continuous setup, the publicly available price will be assumed to be given by

$$S_t = \sigma W_t - \int_0^t g\big(\dot{X}(t)\big)\,dt. \tag{13.4}$$

Here, $\dot{X}_t$ is the derivative of X_t with respect to t and corresponds to v_k in the previous discrete setup. The proceeds associated to a trading strategy X and an initial position of X_0 in the risky asset and y in the riskless asset are given by

$$\begin{aligned}\mathcal{R}_T(X) = X_0 S_0 + y - \int_0^T X_t g(\dot{X}_t)\,dt \\ - \int_0^T \dot{X}_t h(\dot{X}_t)\,dt + \sigma \int_0^T X_t\,dW_t.\end{aligned} \tag{13.5}$$

The cost of the strategy X is defined as $C(X) := X_0 S_0 + y - \mathcal{R}_T(X)$. This can be formally obtained as a limit of (13.2). The expectation and variance of the cost are given by

$$E\big(C(X)\big) = \int_0^T X(t)g(\dot{X}_t) + \dot{X}_t h(\dot{X}_t)\,dt, \qquad \mathrm{Var}\big(C(X)\big) = \int_0^T \sigma^2 X_t^2\,dt$$

when X is deterministic. The problem then consists in finding a deterministic absolutely continuous strategy $(X_t)_{t\in[0,T]}$ that minimizes $E(C(X)) + \lambda\,\mathrm{Var}(C(X))$ for a given risk-aversion level λ.

To obtain explicit solutions to the above minimization problem, Almgren [4] considers a linear permanent impact $g(v) = \gamma v$ and a temporary impact in the form of a power law $h(v) = \eta v^k$ with $k > 0$. For each trading horizon T, there is an optimal strategy. Almgren finds that the optimal strategy which takes the longest to execute can be expressed as

$$\frac{X_t}{X_0} = \begin{cases} (1 - \frac{k-1}{k+1}\frac{t}{T_*})^{\frac{k+1}{k-1}} & \text{if } k \neq 1, \\ \exp(-\frac{t}{T_*}) & \text{if } k = 1, \end{cases}$$

where T_*, called the *characteristic time*, is given by

$$T_* = \left(\frac{k\eta X_0^{k-1}}{\lambda\sigma^2}\right)^{1/(k+1)}.$$

For the linear case, $k = 1$, the characteristic time is independent of the initial portfolio size X_0 and corresponds to the amount of time needed for the portfolio position to decrease by a factor of e^{-1}. If $k < 1$, volatility risk dominates the expected cost as the portfolio size increases and the speed of trading decreases with time. When $k > 1$, the trading cost dominates volatility risk.

When $k \le 1$, the execution time is infinite, i.e., $X_t > 0$ for all $t < \infty$. On the other hand, when $k > 1$, the trading stops after a finite time given by

$$T = \frac{k+1}{k-1} T_*.$$

Next consider the same wealth equation as (13.5) with $T = \infty$, $h(x) = \lambda x$, and $g(x) = \gamma x$. This is the setup considered by Schied and Schöneborn [30]. The admissible portfolios $(X_t)_{t \ge 0}$ considered are more general than in the previous setups as they are assumed to satisfy the following conditions:

- X_t is absolutely continuous, and $\xi(t) := -\dot{X}(t)$,
- $X_T = 0$,
- ξ is progressively measurable with respect to the filtration $(\mathcal{F}_t)_{t \ge 0}$ with $\int_0^t \xi_s^2 \, ds < \infty$ for all $t > 0$,
- $X_t(\omega)$ is uniformly bounded in t and ω.

The class of admissible strategies starting with X_0 units of the risky asset and y shares in the riskless asset is denoted by $\mathcal{X}(X_0, r)$ in which $r = X_0 S_0 + y - \frac{\gamma}{2} X_0^2$. The goal is to maximize the expected utility of the capture $\mathcal{R}_t(X)$ over the class of admissible strategies. Assume that the utility function u is smooth with risk aversion factor

$$A(r) = -\frac{u_{rr}(r)}{u_r(r)},$$

satisfying

$$0 < A_{\min} := \inf_{r \in \mathbf{R}} A(r) \le \sup_{r \in \mathbf{R}} A(r) := A_{\max} < \infty.$$

We consider two different maximization problems. The first problem is given by the following value function:

$$v_1(x, r) = \sup_{X \in \mathcal{X}(x,r)} E\big[u\big(R_\infty(X)\big)\big],$$

where

$$R_\infty(X) = r + \sigma \int_0^\infty X_s \, dB_s - \lambda \int_0^\infty \dot{X}_s^2 \, ds.$$

In the above equation we avoid the technical limiting argument and the associated admissibility class. The second problem involves the value function

$$v_2(x, r) = \sup_{X \in \mathcal{X}(x,r)} \lim_{t \to \infty} E\big[u\big(\mathcal{R}_t(X)\big)\big],$$

where

$$R_t(X) = r + \sigma \int_0^t X_s \, dB_s - \lambda \int_0^t \dot{X}_s^2 \, ds.$$

It can be shown that v_1 and v_2 are equal and solve the Hamilton–Jacobi–Bellman equation

$$-\frac{1}{2}\sigma^2 x^2 v_{rr} + \inf_c \{\lambda v_r c^2 + v_x c\} = 0 \quad \text{for } x > 0,\ r \in \mathbf{R}, \tag{13.6}$$

together with the boundary condition

$$v(0, r) = u(r), \quad r \in \mathbf{R}.$$

The unique optimal control $\dot{X}_t^*$ is Markovian and is given in feedback form by

$$\dot{X}_t^* = c\big(X_t^*, \mathcal{R}_t(X^*)\big), \tag{13.7}$$

where $c(x, r) = -\frac{v_x(x,r)}{2\lambda v_r(x,r)}$.

To prove the above statements, the authors show that there exists a sufficiently smooth solution $\tilde{c} : (y, r) \in \mathbf{R}_0^+ \times \mathbf{R} \to \tilde{c}(y, r) \in \mathbf{R}$ of the partial differential equation

$$\tilde{c}_y = -\frac{3}{2}\lambda \tilde{c}\tilde{c}_r + \frac{\sigma^2}{4\tilde{c}}\tilde{c}_{rr}$$

with initial value

$$\tilde{c}(0, r) = \sqrt{\frac{\sigma^2 A(r)}{2\lambda}}.$$

Moreover, the solution satisfies

$$\sqrt{\frac{\sigma^2 A_{\min}}{2\lambda}} \le \tilde{c}(y, r) \le \sqrt{\frac{\sigma^2 A_{\max}}{2\lambda}}. \tag{13.8}$$

Also, there exists a sufficiently smooth solution $\tilde{w} : \mathbf{R}_0^+ \times \mathbf{R} \to \mathbf{R}$ of the transport equation

$$\tilde{w}_y = -\lambda \tilde{c} \tilde{w}_r$$

with initial value

$$\tilde{w}(0, r) = u(r).$$

Then the function $w(x, r) := \tilde{w}(x^2, r)$ solves the HJB equation (13.6), and the unique minimum is attained at

$$c(x, r) := \tilde{c}\big(x^2, r\big)x.$$

A verification argument concludes that the solution of the HJB equation (13.6) must be equal to the value functions v_1 and v_2, and the unique optimal control satisfies (13.7) where $c(x,r) = -\frac{v_x(x,r)}{2\lambda v_r(x,r)}$. Then in view of (13.7), the asset position $X_t^{\hat{\xi}}$ at time t under the optimal control $\hat{\xi}_t$ is given as

$$X_t^{\hat{\xi}} = X_0 \exp\left(-\int_0^t \tilde{c}\big((X_s^{\hat{\xi}})^2, R_s^{\hat{\xi}}\big)\,ds\right),$$

and because of (13.8), it is bounded as follows:

$$X_0 \exp\left(-t\sqrt{\frac{\sigma^2 A_{\max}}{2\lambda}}\right) \le X_t^{\hat{\xi}} \le X_0 \exp\left(-t\sqrt{\frac{\sigma^2 A_{\min}}{2\lambda}}\right).$$

In the case with constant absolute risk aversion $A = A_{\min} = A_{\max}$, the optimal adaptive liquidation strategy is static and is given by

$$X_t^{\hat{\xi}} = X_0 \exp\left(-t\sqrt{\frac{\sigma^2 A}{2\lambda}}\right).$$

Since the absolute risk aversion of the utility function determines the initial condition of the partial differential equation for $\tilde{c}$, it is a key factor for the optimal trading strategy. In particular, the optimal strategy inherits monotonicity properties of the absolute risk aversion. Let u^0 and u^1 be two utility functions with corresponding absolute risk aversions $A^0(r)$ and $A^1(r)$. If $A^1(r) \ge A^0(r)$ for all r, then an investor with utility function u^1 liquidates the same portfolio X_0 faster than an investor with utility function u^0. More precisely, we get

$$c^1 \ge c^0 \quad \text{and} \quad \hat{\xi}_t^1 \ge \hat{\xi}_t^0 \quad \mathbf{P}\text{-a.s.},$$

where c^i and $\hat{\xi}_t^i$ are obtained from the utility function u^i with $i \in \{0, 1\}$. As a corollary, it follows that $c(x,r)$ is increasing (decreasing) in r for all values of x if and only if the absolute risk aversion parameter $A(r)$ is increasing (decreasing) in r. Therefore, an investor with increasing absolute risk aversion $A(r)$ would sell faster when prices rise, since an increase in prices lead to an increase in r. In this case, the investor is called aggressive in-the-money. On the other hand, an investor having a decreasing absolute risk aversion $A(r)$ is passive in-the-money, i.e., she would sell slower when prices increase.

13.2.3 Models of Limit Order Books

We now analyze the limit order book (LOB) models and focus on the papers by Obizhaeva and Wang [26] and Alfonsi, Fruth, and Schied [1, 2]. As before, we take the point of view of a large trader who needs to liquidate a certain fixed number of units of a risky asset. In limit order books, as opposed to modeling the price process directly, one models the dynamics of supply and demand for the asset in the market and its impact on the execution cost. Then the supply and demand levels determine the magnitude of price impacts.

A limit order is an order to sell or buy a certain number of shares of an asset at a specified price. The limit order book consists of the collection of all sell and buy limit orders. A market order is an order to buy or sell a certain number of shares at the most favorable price available in the limit order book. The lowest specified price in the LOB for a sell order is called the best ask price, whereas the highest price of a buy order in the LOB is the best bid price. A market order to buy (resp. sell) is executed against the limit orders to sell (resp. buy). In LOB models, the dynamics of the LOB is assumed to only be affected by noise traders when the large trader is inactive, and their actions determine the unaffected best ask price A_t^0 and the unaffected bid price B_t^0. The processes $A^0 = (A_t^0)_{t \geq 0}$ and $B^0 = (B_t^0)_{t \geq 0}$ are adapted, exogenously defined stochastic processes on the filtered probability space $(\Omega, \mathcal{F}, (\mathcal{F}_t)_{t \geq 0}, \mathbf{P})$. Clearly, a natural condition to impose on these two processes is $A_t^0 \geq B_t^0$ for all $t \geq 0$. We denote the density of the LOB at the price $A_t^0 + x$ (resp. $B_t^0 + x$) by $f(x)$ for $x > 0$ (resp. $x < 0$), i.e., the number of shares offered at the price $A_t^0 + x$ (resp. $B_t^0 + x$) is given by $f(x)\,dx$. It is assumed that $f : \mathbf{R} \to (0, \infty)$ is a bounded continuous function, called the shape function of the LOB. The large trader makes buy and sell orders, thereby temporarily depleting parts of the LOB. We denote by F the antiderivative of f, i.e.,

$$F(y) = \int_0^y f(x)\,dx.$$

The actual best ask price at time t, denoted by A_t, takes into account the price impacts of the previous market orders of the large trader. The positive difference between the actual and the unaffected best ask prices $D_t^A = A_t - A_t^0$ is called the extra spread. A buy order of size $\xi > 0$ at time t consumes all shares in the LOB from the actual best ask price A_t to

$$A_{t+} = A_t + D_t^A(\xi) - D_t^A,$$

where $D_t^A(\xi)$ is determined by the relation

$$\int_{D_t^A}^{D_t^A(\xi)} f(x)\,dx = \xi.$$

The process D^A specifies the market impact of orders on the best ask price. For a general shape function f, the market impact $D_t^A(\xi) - D_t^A$ is nonlinear. However, if we assume a block shaped LOB, i.e., an LOB in which the shape function is equal to a constant q above the actual best ask price, then the market impact $D_t^A(\xi) - D_t^A$ is linear and equal to ξ/q.

We now describe the admissible trading strategies for the large trader. Assume that the trader wants to buy $x > 0$ shares in $N + 1$ trades within the time interval $[0, T]$. The trading strategies considered by Alfonsi and Schied [3] are simple strategies of the form

$$X_t = \xi_0 + \sum_{n=1}^{N} \xi_n \mathbf{1}_{\{t \geq \tau_n\}} \quad (0 \leq t < T),$$

where $\tau_0, \ldots, \tau_N$ are stopping times satisfying $0 = \tau_0 \leq \tau_1 \leq \cdots \leq \tau_N$, and every ξ_n is bounded below and measurable with respect to $\mathcal{F}_{\tau_n}$. The quantity ξ_n represents the size of the market order placed at time τ_n. We denote this set of admissible strategies by $\mathcal{X}_N$. In [1, 2], the admissible strategies considered are special cases of the above setup, i.e., the trading times are not stopping times, but they are predetermined. For convenience, we denote by $X_t^+ = \xi_0 + \sum_{i=1}^N \xi_n \mathbf{1}_{\{t \geq \tau_n, \xi_n > 0\}}$ the cumulative buy orders and by $X_t^- = X_t - X_t^+$ the cumulative sell orders.

It is assumed that the market impact decays with time as the result of new sell orders coming in the order book. This phenomenon is known as the resilience of the LOB. In [2, 3] there are two different approaches to model resilience. Either the volume of the order book consumed by the large trader, denoted at time t by E_t^A, is assumed to recover exponentially, or the extra spread D_t^A decays exponentially. The assumption regarding resilience is stated as follows: there is a deterministic rate process $(\rho_t)_{t \geq 0}$ such that either

$$dE_t^A = -\rho_t E_t^A \, dt + dX_t^+$$

or

$$dD_t^A = -\rho_t D_t^A \, dt + D_t^A \left(\Delta X_t^+ \right)$$

In the specific case of a block-shaped LOB, it can be shown that

$$D_t^A = \frac{1}{q} \sum_n e^{-\int_{\tau_n}^t \rho_s \, ds} \xi_n \mathbf{1}_{\{\tau_n \leq t, \xi_n > 0\}}. \tag{13.9}$$

It is easy to see that these two approaches of resilience coincide for block-shaped LOBs. The dynamics of the bid side of the LOB are modeled identically. As before, the density of the number of shares offered at the price $B_t^0 + x$ for $x < 0$ are given by the shape function f. The extra spread D_t^B is the difference between the actual best bid price and the best unaffected bid price $D_t^B = B_t - B_t^0$, which is nonpositive. A sell order of size $\xi < 0$ will move the actual best bid price to

$$B_{t^+} = B_t + D_t^B(\xi) - D_t^B,$$

where $D_t^B(\xi)$ is defined, as before, by

$$\xi = \int_{D_t^B}^{D_t^B(\xi)} f(x) \, dx.$$

As before, the resilience is either modeled in terms of the volume consumed by the large trader or the extra spread as follows:

$$dE_t^B = -\rho_t E_t^B \, dt + dX_t^-, \quad \text{or}$$

$$dD_t^B = -\rho_t D_t^B\,dt + D_t^B\big(\Delta X_t^-\big).$$

The difference $A_t - B_t$ between the best ask and the best bid price is called the bid-ask spread.

A buy order of size $\xi > 0$ at time t consumes the $f(x)\,dx$ available shares at price $A_t^0 + x$, where x ranges from D_t^A to $D_t^A(\xi)$. The cost associated to this transaction is given by

$$\pi_t(\xi) = \int_{D_t^A}^{D_t^A(\xi)} \big(A_t^0 + x\big) f(x)\,dx = A_t^0\xi + \int_{D_t^A}^{D_t^A(\xi)} x f(x)\,dx.$$

Similarly for a sell order $\xi < 0$, we have

$$\pi_t(\xi) = \int_{D_t^B}^{D_t^B(\xi)} \big(B_t^0 + x\big) f(x)\,dx = B_t^0\xi + \int_{D_t^B}^{D_t^B(\xi)} x f(x)\,dx.$$

The expected cost $\mathcal{C}(X)$ of an admissible strategy X can then be obtained by

$$\mathcal{C}(X) = E\left[\sum_{n=0}^{N} \pi_{\tau_n}(\xi_n)\right].$$

The goal is then to minimize $\mathcal{C}(X)$ among all admissible strategies X. Note that, in contrast with the works of Almgren [4] and Almgren and Chriss [5, 6], intermediate sell orders (resp. buy orders) are allowed for execution orders of $x > 0$ (resp. $x < 0$) shares.

In [3], it is established that minimizing $\mathcal{C}(X)$ over the set of admissible strategies $\mathcal{X}_N$ is equivalent, under some assumptions on the density function f, to minimizing $\mathcal{C}(X)$ under the constraint that the trading times sequence $(\tau_0, \tau_1, \dots, \tau_n)$ is given by the time spacing $\mathcal{T}^* = (t_0^*, t_1^*, \dots, t_N^*)$ defined by

$$\int_{t_{i-1}^*}^{t_i^*} \rho_s\,ds = \frac{1}{N}\int_0^T \rho_s\,ds, \quad i = 1, \dots, N.$$

The unique optimal strategy for the first model of resilience is given by

$$\xi_1^* = \cdots = \xi_{N-1}^* = \xi_0^*\big(1 - a^*\big),$$

where

$$a^* = \exp\left(-\frac{1}{N}\int_0^T \rho_s\,ds\right),$$

and ξ_0^* solves

$$F^{-1}\big(x - N\xi_0^*\big(1 - a^*\big)\big) = \frac{F^{-1}(\xi_0^*) - a^* F^{-1}(a^*\xi_0^*)}{1 - a^*}.$$

The last order ξ_N^* is determined so that

$$\xi_N^* = X_0 - \xi_0^* - (N-1)\big(1-a^*\big)\xi_0^*.$$

When f is constant,

$$\xi_0^* = \frac{x}{(N-1)(1+a^*)+2}.$$

In the asymptotic limit, i.e., as $N \to \infty$, of the block-shaped LOB, the optimal execution strategy is a combination of discrete and continuous trades when the resilience factor ρ is constant. The initial and final trades are discrete, whereas the intermediate ones are continuous. The optimal strategy is given by

$$\xi_0^* = \xi_T^* = \frac{X_0}{\rho T + 2}, \qquad \frac{d}{dt}\xi_t^* = \frac{\rho X_0}{\rho T + 2}.$$

Note that in the LOB price impact model described above, the impact of a trade is not permanent: the extra spread decays with time. Alfonsi et al. [1] and Obizhaeva and Wang [26] include an additional permanent impact factor in the block-shaped LOB model. More specifically, they let the density function $f = q \in \mathbf{R}$ and assume that the extra spread D^A caused by a strategy X satisfies

$$D_t^A = \gamma \sum_n \xi_n \mathbf{1}_{\{\tau_n \le t, \xi_n > 0\}} + \kappa \sum_n \exp\left(-\int_{\tau_n}^t \rho_s \, ds\right) \xi_n \mathbf{1}_{\{\tau_n \le t, \xi_n > 0\}},$$

where $0 \le \gamma \le 1/q$ is the permanent effect factor, and $\kappa = 1/q - \gamma$ is the proportion of the market impact that decays with time. Similar dynamics holds for sell orders. Comparing this to (13.9), we see that a proportion $\frac{\gamma}{1/q}$ of the consumed volume by the large trader does not recover in the long run. It turns out that the minimization problem with permanent impact has the same optimal trading strategy as the minimization problem with $\gamma = 0$.

In [1], Alfonsi et al. consider this problem under convex constraints and obtain closed-form solutions. The set of strategies considered is however smaller than $\mathcal{X}_N$ as trading is only permitted on a predetermined time grid $t_0, t_1, \ldots, t_n$. The aim is to reduce the constrained optimization problem to the minimization of a positive definite quadratic form on a convex subset of Euclidean space. As a special case, they obtain closed-form solutions for the unconstrained problem.

13.3 Option Hedging for Large Traders

In this section we survey the large trader models for hedging options. The trades of the large trader are assumed to have an impact on the prices so that she has to take this effect into account when considering hedging options. There are various approaches to incorporate the trading decisions of the large trader into the price process of the underlying. Jarrow [21, 22] considers the price process expressed in terms of reaction functions of the holdings of the large trader. This turns out to be a generalization of Huberman and Stanzl's model for price manipulation. In [14] and [13], the coefficients of the price process depend exogenously on the large trader's portfolio. Platen and Schweizer [27], Frey and Stremme [16], and Sircar and Papanicolau [31] use an equilibrium approach to derive the reaction function for the price process. Frey [15] assumes that this reaction function describing the price process as a function of the holdings of the large trader is exogenously given. Bank and Baum [7] model the price process of the risky asset in terms of a smooth family of semimartingales $(S^z)_{z\in\mathbf{R}}$, where S^z describes the evolution of the stock price process for constant z, which represents the size of the large trader's holdings. Roch [28] considers a setup similar to the limit order book models described above in which the parameter of the linear permanent impact function is given by a stochastic process.

Throughout the remaining sections, we work with a filtered probability space $(\Omega, \mathcal{F}, \mathbf{F}, \mathbf{P})$, which supports a standard Brownian motion $(W_t)_{0\leq t\leq T}$. We also fix a finite time horizon $T > 0$. Unless otherwise specified, there is one risky asset and one riskless asset in the market. We normally think of the risky asset as a stock and the riskless asset as a money market account. The money market account is taken to be a numéraire so that its price is normalized to unity. The discounted price of the stock process at time t is denoted by S_t. There are two types of traders in the economy, one large trader and reference traders. The large trader can be a speculator, a program trader, or a portfolio insurer. The reference traders are typically noise traders or arbitrage-based speculators. Let X_t be the number of money market units, Y_t the book value of the stock position, and Z_t be the number of stocks the large trader holds at time t. The processes X, Y, and Z are assumed to be adapted to the filtration $\mathbf{F}$.

In classical settings based on the Black–Scholes model, the stock price process S_t is modeled as a solution of a linear stochastic differential equation (SDE). The drift and volatility coefficients of the SDE are not influenced by the agents portfolio and wealth processes. This is based on the assumption that agents are price takers in this framework. Cvitanic and Ma [14] model the price process of the underlying asset by an SDE taking into account that large trader's decisions have a price impact. In particular, they assume that the drift and volatility coefficients depend on the large trader's portfolio and wealth process. The authors consider a market with d risky assets (stocks) and one riskless asset (money market account). Let S_t^0 be the price process of the money market account, and S_t^i be the price process of the ith stock. Then the dynamics of these processes are given as

$$dS_t^0 = S_t^0 r(t, Y_t, Z_t)\,dt, \quad 0 \leq t \leq T,\ S_0^0 = 1,$$

$$dS_t^i = b_i(t, S_t, Y_t, Z_t)\,dt + \sum_{j=1}^{d} \sigma_{ij}(t, S_t, Y_t, Z_t)\,dW_t^j, \quad 0 \leq t \leq T,\ S_t^i = s_i > 0,$$

$$dY_t = \hat{b}(t, S_t, Y_t, Z_t)\,dt + \hat{\sigma}(t, S_t, Y_t, Z_t)\,dW_t, \quad 0 \leq t \leq T,\ Y_0 = y > 0,$$

where

$$\hat{b}(t,s,y,z) = \left(y - \sum_{i=1}^{d} s_i z_i\right) r(t,y,z) + \sum_{i=1}^{d} z_i b_i(t,s,y,z),$$

$$\hat{\sigma}_j(t,s,y,z) = \sum_{i=1}^{d} z_i \sigma_{ij}(t,s,y,z), \quad j = 1,\dots,d.$$

Under additional assumptions on the coefficients of the above SDEs, the authors show that the replication of European options with payoff in the form $g(S_T)$ has a solution. The method is based on forward–backward stochastic differential equations and the well-known four-step scheme of Ma et al. [25].

Platen and Schweizer [27], Frey and Stremme [16], Frey [15], and Sircar and Papanicolau [31] do not model the price process explicitly as in [13] and [14]. However, they follow a microeconomic equilibrium approach to understand the feedback effects from hedging strategies. As before, there are two types of investors in the market, a large trader and reference traders. The aggregate demand of the reference trader at time t is given by $D(t, F_t, S_t)$, where $F = (F_t)_{0 \le t \le T}$ is the fundamental state process, and S_t is the price for stock. The fundamental state process can represent various things, for instance, noise or misspecifications in the model, demand for liquidity, or aggregated income of the reference trader. Supposing that at time t the large trader possesses a fraction α_t of the total supply of the stock, then the market clearing condition states that

$$D(t, F_t, S_t) + \alpha_t = 1.$$

Under some assumptions, it can be shown that there is a unique solution for S_t in terms of t, α_t, and F_t, i.e., we can express $S_t = \psi(t, F_t, \alpha_t)$. The function ψ is called the *reaction* function.

Frey and Stremme [16] investigate the impact of dynamic hedging on the price process in a general discrete-time economy with the equilibrium model. They pass to the diffusion limit and investigate the continuous-time equilibrium price process and its volatility. The price process is still represented by an Itô process, but the volatility increases and becomes time- and price-dependent.

Sircar and Papanicolau [31] analyze the increases in market volatility of asset prices. Many investors use Black–Scholes trading strategies to hedge derivatives. The use of these hedging strategies is so extensive that they have an impact on the price of the asset, which in turn influences the price of the derivative. In their framework, there is an interaction between reference traders and large traders who follow a dynamic Black–Scholes hedging strategy. Following an equilibrium analysis, they derive a stochastic process for the price of the asset that depends on the hedging strategy of the large trader. Then they derive a nonlinear partial differential equation for the derivative price and the hedging strategy. They observe that the increase in volatility can be attributed to the feedback effect of Black–Scholes hedging strategies.

Platen and Schweizer [27] aim to study the implied volatility structure in the above reaction setup. In other words, instead of taking an exogenously given price process, they develop a diffusion model for stock prices that incorporates the technical demand induced by the hedgers. The diffusion model is endogenously determined by the trading decisions in the economy. With their modeling, they can explain volatility smiles and skews as a result of feedback effects from hedging derivatives. They consider the following specification of the demand function:

$$D(t, F_t, S_t) = F_t + \gamma\big(\log(S_t) - \log(S_0)\big),$$

where $F_t = vW_t + mt$ is a random error term, and $\gamma > 0$ represents how reference traders react to changes in logarithmic stock prices. The last term can be interpreted as the demand created by trading decisions of hedging options. The option hedgers work under the assumption that the stock price $S^{(0)}$ is given by a geometric Brownian motion with constant drift μ_0 and volatility σ_0 to hedge a given number of call options with different maturities and strikes. This determines the term α_t in the market clearing equation. Then the market equilibrium condition determines the resulting price process $S_t^{(1)}$ by

$$dS_t^{(1)} = S_t^{(1)}\big(\sigma\big(S_t^{(1)}\big)\,dW_t + \mu\big(S_t^{(1)}\big)\,dt\big),$$

where

$$\sigma(s) = -\frac{v}{\gamma + \xi'(\log(s))},$$

$$\mu(s) = \frac{m}{v}\sigma(s) + \frac{1}{2}\sigma^2(s) + \frac{\xi''(\log(s))}{2v}\sigma^3(s),$$

and the term $\xi(\log(s))$ represents the hedging demand created in the market. Observe that we started with a model $S_t^{(0)}$ for stock price process and derived another model $S_t^{(1)}$ by equilibrium approach that incorporates the hedging decisions of the large trader. However, the sophisticated large traders could also use the model $S_t^{(1)}$ to hedge derivatives so that we would obtain another model $S_t^{(2)}$ in equilibrium. In general, one can start from $S^{(k)}$ and use this to compute option values and hedging strategies. The equilibrium argument will yield a new model $S^{(k+1)}$. In the end, one wonders if there exists a fixed point $S^{(\infty)}$ of this transformation. Such a model $S^{(\infty)}$ would be used by the hedgers to compute their hedging strategy and would also be the one obtained in equilibrium.

Frey [15] takes the reaction function $S_t = \psi(t, F_t, \alpha_t)$ as given. He considers replicating the payoff of certain non-path-dependent derivatives. In this continuous-time setup, there is a nonlinear partial differential equation for the hedge of the option replication problem. In particular, these hedging strategies take the feedback effect of their implementation on the price process into account. Therefore, Frey argues that the existence of these hedging strategies for certain payoffs corresponds to the fixed point of the volatility transformation introduced in [27].

Bank and Baum [7] assume that there exists a smooth family of semimartingales S^z for $z \in \mathbf{R}$ that specify the price process of the risky asset when the large trader's holdings are kept at a constant size z. For fixed z, the semimartingale S^z can be interpreted as the fluctuations of the asset prices when the large trader is not active in the market. If the large trader follows a semimartingale strategy $(Z_t)_{0\le t\le T}$, then the asset price obtained is given by

$$S_t = S_t^{Z_t} =: S(Z_t, t).$$

The self-financing portfolio strategies are characterized by

$$X_t = X_{0-} - \int_0^t S(Z_{s-}, s)\, dZ_s - \big[S(Z, \cdot), Z\big]_t.$$

Bank and Baum assume that asset prices are nondecreasing with respect to the position of the large trader, i.e., for $z \le z'$, we have $S^z \le S^{z'}$. In an illiquid market, there are many possible ways to value the large trader's portfolio. One can consider the book value Y_t of the portfolio evaluated at current prices,

$$Y_t = X_t + S(Z_t, t) Z_t,$$

or the real wealth achieved by the trading strategy Z until time t given by

$$V_t = X_t + L(Z_t, t),$$

where

$$L(z, t) = \int_0^z S(x, t)\, dx.$$

The term $L(z, t)$ represents the liquidation value of z shares by splitting the order into infinitesimally small packages and selling them over an infinitesimally small time period. By the Itô–Wentzell formula for smooth family of semimartingales, the real wealth process has the dynamics

$$V_t = V_{0-} + \int_0^t L(Z_{s-}, ds) - \frac{1}{2}\int_0^t S_z(Z_{s-}, s)\, d[Z]_s^c$$
$$- \sum_{0\le s\le t} \int_{Z_{s-}}^{Z_s} \big\{S(Z_s, s) - S(x, s)\big\}\, dx.$$

The term $\int_0^t L(Z_{s-}, ds)$ represents the profit or loss coming from price fluctuations caused by exogenous random shocks. The term $\frac{1}{2}\int_0^t S'(Z_{s-}, s)\, d[Z]_s^c$ gives the transaction costs due to continuous trading, and

$$\sum_{0\le s\le t} \int_{Z_{s-}}^{Z_s} \big\{S(Z_s, s) - S(x, s)\big\}\, dx$$

sums up the transaction costs due to discrete block orders. These two transaction terms disappear if one follows trading strategies that are continuous and of bounded variation. As in [21, 22], Bank and Baum investigate the possibility of arbitrage opportunities for the large trader. On one hand, the large trader has the power to influence the market prices, and, on the other hand, her trading incurs transaction costs, i.e., her orders affect the stock price before they are exercised. If there exists a measure $\mathbf{P}^* \approx \mathbf{P}$ which is a local martingale measure for all the processes P^θ simultaneously, then there are no arbitrage opportunities for the investor.

A natural problem in this setting is to describe the set of payoffs the large trader can attain with continuous strategies of bounded variation. To answer this question, Bank and Baum introduce two definitions. A contingent claim $H \in L^0(\mathcal{F}_T)$ is *attainable modulo transaction costs* for initial capital v if

$$H = v + \int_0^T L(Z_s, ds)$$

almost surely for some L-integrable predictable process Z such that $\int_0^\cdot L(Z_s, ds)$ is uniformly bounded from below. The claim H is *approximately attainable* for initial capital v if for any $\epsilon > 0$, there exists a self-financing strategy Z^ϵ such that $\int_0^\cdot L(Z_s^\epsilon, ds)$ is uniformly bounded from below, and

$$\left|H - V_T^\epsilon\right| \leq \epsilon$$

holds $\mathbf{P}$ almost surely, in which V_T^ϵ is the real wealth process associated to strategy Z^ϵ. To this end, the authors establish an approximation scheme for stochastic integrals. Let $\epsilon > 0$. If Z is an L-integrable, predictable process with $Z_0 \in L^0(\mathcal{F}_0)$ and $Z_T \in L^0(\mathcal{F}_{T-})$, then there exists a predictable process Z^ϵ with continuous paths of bounded variation such that $Z_0^\epsilon = Z_0$, $Z_T^\epsilon = Z_T$ and

$$\sup_{0 \leq t \leq T} \left| \int_0^t L(Z_s, ds) - \int_0^t L\big(Z_s^\epsilon, ds\big) \right| \leq \epsilon \quad \mathbf{P}\text{-a.s.}$$

From this it is easy to see that any contingent claim $H \in L^0(\mathcal{F}_T)$ which is attainable modulo transaction costs is approximately attainable with the same initial capital. Furthermore, under some further assumptions, the attainable claims in a suitable small investor model become approximately attainable for the large trader. Moreover, the authors show that to compute the superreplication cost of a claim $H(\omega, Z_T(\omega)) \in \mathcal{F}_{T-} \otimes \mathcal{B}(\mathbf{R})$, one first determines the terminal position Z_T^* which minimizes the payoff, i.e., $Z_T^*(\omega) = \arg\min_{z \in \mathbf{R}} H(\omega, z)$, and then compute the small investor superreplication price of the claim $H(\omega, Z_T^*(\omega))$.

Roch [28] extends the linear version of the liquidity risk model of Çetin et al. [9] to allow for price impacts. The author considers the hedging problem faced by a large trader who makes market order through a limit order book with stochastic density. More specifically, it is assumed that the limit order book has a constant density at time t given by $\frac{1}{2M_t}$, in which M is an adapted stochastic process. Liquidity becomes a risk factor when the magnitude of the impact of these phenomena changes randomly over time. We denote by S the observed marginal price process, i.e., S_t is the price per share for an infinitesimal order size at time t. By the constant density property of the LOB, it is clear that a transaction of size ΔZ_t at time t has a cost of $\Delta Z_t(S_t + M_t \Delta Z_t)$. The model proposed in [28] is based on a well-documented feature of asset prices that volatility is high when liquidity is low, and low when liquidity is high. Since M is a measure of illiquidity, we can expect the instantaneous variance of the log-returns of the stock price to be in part correlated with M. Consequently, we let F denote the unaffected marginal price process. It is the equilibrium (or fundamental) price process observed in absence of large traders. It is defined by the following stochastic volatility model:

$$dF_t = \Sigma_t F_t \, dW_{1,t},$$

where W_1 is the first component of the three-dimensional Brownian motion W, and Σ_t is the stochastic volatility. We are working directly under a risk-neutral measure $\mathbf{Q}$ for unaffected prices, and hence F has no drift term. We model M and Σ as follows. Define V and U as the solutions of

$$\begin{aligned} dU_t &= \gamma(U_t + \eta)\,dt + \Phi(U_t)\,dW_{2,t}, \\ dV_t &= \alpha(V_t + a)\,dt + \Theta(V_t)\,dW_{3,t}, \end{aligned}$$

where $W = (W_{j,t})_{j \le 3, t \le T}$ is a three-dimensional Brownian motion, and $\alpha, \gamma, \eta, a \in \mathbf{R}$. We define $\Sigma_t^2 = U_t + V_t$ and let $M = \varepsilon \Gamma(U)$, where Γ is strictly increasing and twice continuously differentiable and $\epsilon > 0$. Φ and Θ are given real-valued functions. We are using a three-dimensional Brownian motion since there are three different sources of risk in this model, namely the stock price, the liquidity level, and the volatility, which is, in practice, only partially dependent on the level of liquidity.

The specification of the process S is similar to the one of the LOB models described above. Indeed, it is assumed that the observed marginal price process S is obtained from the unaffected process F by directly adding the impact of the large trader as follows:

$$S_{t+} = F_t + 2\lambda \int_0^t M_{u-}\,dZ_u + 2\lambda \int_0^t d[M, Z]_u \quad (t \le T)$$

for a semimartingale trading strategy Z. S_{t+} denotes the observed price after the trade at time t. λ is a resilience parameter and should be taken between 0 and 1. It measures the proportion of buy (resp. sell) limit orders versus sell (resp. buy) limit orders that come in to fill up the LOB after a market order to buy (resp. sell).

It can be shown that the money market account position X and the position Z in the stock satisfy the following identity:

$$\begin{aligned} &X_t + Z_t\big(S_{t+} - \lambda M_t Z_t\big) \\ &\quad = X_{t_0-} + Z_{t_0-}\big(S_{t_0} - \lambda M_{t_0} Z_{t_0-}\big) + \int_{t_0}^T Z_{u-}\,dF_u \end{aligned}$$

$$-\lambda \int_{t_0}^{T} Z_{u-}^2 \, dM_u - \int_{t_0}^{T} (1-\lambda) M_u \, d[Z,Z]_u. \tag{13.10}$$

One can think of $Y_t + x(S_t - \lambda M_t x)$ as the liquidation value of a portfolio with x shares at time t. Similar to the infinitely liquid case ($M = 0$), (13.10) states that the difference in the liquidation values between time t_0 and t is equal to the cumulative gains in the risky asset $\int_{t_0}^{t} Z_{u-} dF_u$, except that in this case there are added costs coming from the finite liquidity of the asset. First note that if $\lambda = 0$, we get a linear version of the CJP model. The integral with respect to M is related to the impact of trading. If $\lambda = 0$, the limit order book is automatically refilled after a market order, as in the CJP model. At the other extreme, when $\lambda = 1$, the impact of trading is at its fullest. It is interesting to notice that whatever the trading strategy is used, an investor always has a partial benefit from the asset becoming more liquid. Indeed, as M_t decreases, the associated integral is positive no matter what the sign of Z_t is. To understand this, it is important to remember that the hedger's trades have a permanent impact on the quoted price which is proportional to the level of liquidity. If the liquidity is low when he purchases a share and high when she sells it, the price goes up higher after her purchase, and then it comes down after the sale. As a result, the hedger has a partial gain from this trade. This is a typical characteristic of large trader models. Note that, unless the hedger uses a trading strategy with zero quadratic variation, this is only a partial benefit because there is always a liquidity cost associated to her trades.

Equation (13.10) allows us to obtain a sufficient condition to rule out arbitrage opportunities in this setting. Indeed, Roch [28] shows that the existence of an equivalent measure $\mathbf{Q}$ under which the unaffected price process F is a local martingale and M is a local submartingale suffices to exclude the existence of arbitrage opportunities. For a precise statement, we refer the reader to Definition 2.5 and Theorem 2.6 of [28]. The advantage of this result is that it is stated in terms of the exogenously defined processes F and M. Note that in the terminology of Sect. 13.2 the impact of the hedger's trade in the above model is linear, i.e., a trade of size ΔZ_t at time t is of the form $g_t(\Delta Z_t) = 2\lambda M_t \Delta Z_t$. The case of M_t constant corresponds to the linear permanent impact models of Huberman and Stanzl [20], Almgren and Chriss [5], and others. In this case, M clearly is a local submartingale under any risk-neutral measure for S. In this sense, the no-arbitrage condition in [28] extends the results of Huberman and Stanzl [20] in the case of a stochastic linear permanent impact function.

We now turn to the replication problem. The relation between liquidity and volatility risk is a key observation which enables us to hedge derivatives. Indeed, we will see that one can hedge against the liquidity risk by trading variance swaps. Since volatility is one of the most correlated quantities to liquidity risk, this is a very natural choice. We thus consider contingent claims denoted by G_i ($i = 1, 2$) for which the payoff at time $T_i > T$ ($T_1 \neq T_2$) equals the difference between the realized variance over the time interval $[0, T_i]$ and a strike K_i, i.e.,

$$G_{i,T_i} = \int_0^{T_i} \Sigma_s^2 \, ds - K_i$$
$$= \int_0^{T_i} (U_s + V_s) \, ds - K_i.$$

To rule out arbitrage opportunities, we assume the unaffected price processes G^i are $\mathbf{Q}$-martingales ($i = 1, 2$).

Let h be the payoff function of a European option with maturity T. Suppose that h is a Lipschitz function. For $x \in \mathbf{R}$, define $\widetilde{S}_T^x := F_T - 2\lambda \int_0^T x \hat{Z}_{u-} \, dM_u$, where $\hat{Z}$ is the solution of the replication problem in the case $\lambda = 0$, $\varepsilon = 0$, and $x = 1$. It can be shown that $\widetilde{S}_T^x$ is an approximation of the observed price process S obtained when the large trader hedges the option with payoff h. Jarrow [22] used a similar approach and interpreted $\hat{Z}_t$ as the market's perception of the option's "delta" Z_t. The main result of the paper states that $xh(\widetilde{S}_T^x)$ can be approximately replicated in L^2 for all $x \in \mathbf{R}$ in the sense that there exists a sequence of trading strategies Z^n for which the terminal wealth X_T after liquidation converges to $xh(\widetilde{S}_T^x)$ in L^2.

Due to the nonadditivity of liquidity costs, it is clear that the replicating cost of x units of the option h is not x times the replicating price of one unit. Let $H_t^n(x)$ denote the approximate-replication cost per unit of x units of h; then $H_t^n(0) = \mathbf{E}(h(S_T)|\mathcal{F}_t)$, and $H_t^n(x)$ is a.s. differentiable at $x = 0$. Furthermore, it can be shown that

$$\lim_{n\to\infty} \frac{d}{dx} H_t^n(0) = \lambda \mathbf{E}\left(\int_t^T \mu(M_s) \hat{Z}_s^2 \, ds \middle| \mathcal{F}_t \right)$$
$$- 2\lambda \mathbf{E}\left(h'(S_T) \left(\int_t^T \hat{Z}_s \, dM_s \right) \middle| \mathcal{F}_t \right)$$

when h is differentiable everywhere except at a finite number of points.

Jarrow et al. [23] have used ideas from the above setup to construct a liquidity-based model for financial bubbles which explains both bubble formation and bubble bursting. In contrast with the classical approach to bubbles based on local martingales, the authors define the asset's fundamental price process exogenously, and asset price bubbles are endogenously determined by market trading activity. More specifically, they assume that the stock price is governed by the following dynamic:

$$S_t = F_t + 2 \int_0^t \Lambda_u M_{u-} \, dZ_u \quad (t \leq T),$$

where F is the fundamental price process, Λ is a process version of the resilience parameter λ in [28], and Z represents the signed volume of aggregate market orders (volume of market buy orders minus volume of market sell orders). The bubble at time t is then defined by $S_t - F_t$, the difference between the market price of the stock and its fundamental value. In their model, the impact of trading activity on the fundamental price process—derived in terms of a liquidity risk process M, the resilience process Λ, and the market orders—is what generates price bubbles. They study conditions under which asset price bubbles are consistent with no arbitrage opportunities.

13.4 Supply Curve Models

Çetin, Jarrow, and Protter [9] model illiquidity with a supply curve model. This supply curve incorporates the temporary impact of the trade size into the price of the security. Assume that the marginal price process S is given. Then the price deviation at time t from S_t is determined by the supply curve in terms of the size of the trade. We denote the price per share for a trade of ν shares at time t by $\mathbf{S}(t, S_t, \nu)$. For instance, for a supply curve of the form

$$\mathbf{S}(t, S_t, \nu) = S_t \exp(\Lambda \nu), \tag{13.11}$$

a trade of size ν would deviate from the marginal price process by a factor of $\exp(\Lambda \nu)$. Since Λ measures the price impact, it is called the liquidity parameter of the market. $\Lambda = 0$ corresponds to a infinitely liquid market. Investors are price-takers with respect to the curve, and their trading decisions affect the price only instantaneously, and hence they have no lasting impact. Therefore, the Çetin–Jarrow–Protter model (henceforth called CJP model) belongs to a temporary price impact setting. An order of size $\nu > 0$ is a buy and of size $\nu < 0$ is a sell. $\mathbf{S}(t, S_t, 0)$ is equal to the marginal price S_t. Apart from measurability and smoothness assumptions, we assume that $\mathbf{S}(t, S_t, \nu)$ is monotone in ν.

Consider a finite horizon economy with $T > 0$. Take a filtered probability space $(\Omega, \mathcal{F}, \mathbf{F}, \mathbf{P})$ satisfying the usual conditions. We let $(W_t)_{0 \le t \le T}$ be a standard Brownian motion with respect to the filtration $\mathbf{F} = (\mathcal{F}_t)_{0 \le t \le T}$. Assume that there are two assets in the economy, one risk-free asset, and one risky asset. We consider a money market account as the risk-free asset and normalize its price to unity. The risky asset is by convention a stock, and the price per share of stock is $\mathbf{S}(t, S_t, \nu)$ with the marginal price process S_t. Let X_t and Z_t represent the holdings of the trader at time t in the money market account and in the stock, respectively. There are various ways to value the wealth process of the investor. One way is to look at the block liquidation value

$$X_t + Z_t \mathbf{S}(t, S_t, -Z_t).$$

Another way is to consider the book or paper value of the portfolio

$$Y_t := X_t + Z_t S_t$$

evaluated at the marginal process S. It is shown in [33] that this value Y_t also corresponds to infinitesimal liquidation value. In the remainder of the section we focus on the book value Y_t and specify its dynamics. It is natural to define the self-financing condition for simple strategies of the form $Z_t = \sum_{i=1}^{N} \Delta Z_{\tau_i} \mathbf{1}_{\{t \geq \tau_i\}}$ with a sequence of stopping times $0 = \tau_0 < \tau_1 < \cdots < \tau_N = T$ by

$$X_{\tau_{k+1}} = X_{\tau_k} - \Delta Z_{\tau_{k+1}} \mathbf{S}(\tau_{k+1}, S_{\tau_{k+1}}, \Delta Z_{\tau_{k+1}}), \tag{13.12}$$

where $\Delta Z_{\tau_{k+1}} = (Z_{\tau_{k+1}} - Z_{\tau_k})$. Then the dynamics of the book value Y for simple strategies is described as

$$\begin{aligned} Y_{\tau_{k+1}} &= Y_{\tau_k} + Z_{\tau_k}(S_{\tau_{k+1}} - S_{\tau_k}) \\ &\quad - \Delta Z_{\tau_{k+1}} \big[\mathbf{S}(\tau_{k+1}, S_{\tau_{k+1}}, \Delta Z_{\tau_{k+1}}) - S_{\tau_{k+1}} \big]. \end{aligned} \tag{13.13}$$

Formally, for general semimartingale strategies Z, one can pass to the limit as $N \to \infty$ to obtain the dynamics

$$Y_t = y + \int_0^t Z_{u-} \, dS_u - \sum_{0 \leq u \leq t} \Delta Z_u \big[\mathbf{S}(u, S_u, \Delta Z_u) - S_u \big] \tag{13.14}$$

$$- \int_0^t \frac{\partial \mathbf{S}}{\partial \nu}(u, S_u, 0) \, d[Z, Z]_u^c \tag{13.15}$$

for $0 \leq t \leq T$. The term $\int_0^t Z_{u-} \, dS_u$ represents the capital gains and losses. The other terms in the above equation appear because of liquidity effects, the first one is a result of block orders, and the second one of continuous trading. These liquidity costs can be eliminated by using continuous strategies of finite variation. Furthermore, Çetin et al. [9] prove that for any appropriately integrable predictable process Z, there exists a sequence $\{Z^n\}_{n \geq 0}$ of predictable continuous strategies of finite variation such that

$$\int_0^T Z_u^n \, dS_u \to \int_0^T Z_u \, dS_u \quad \text{in } L^2.$$

This approximation also follows from the Bank and Baum [7] result as well.

Çetin et al. find sufficient conditions to rule out arbitrage in the CJP model. They generalize the first fundamental theorem of asset pricing to their setting. They show that there is no free lunch with vanishing risk in their framework if and only if there exists an equivalent local martingale measure for the marginal price process S_t. They also establish that if there exists an equivalent local martingale measure $\mathbf{Q}$ for the marginal price process S, then any appropriately integrable claim C can be attained in the L^2 sense. Then the above approximation result shows that all liquidity costs can be avoided in this setting and the value of the claim is the classical one given by $E^{\mathbf{Q}}[C]$.

The previous result is sometimes seen as a shortcoming of the CJP model. As a response, Rogers and Singh consider a temporary price impact model in [29] in which the liquidity cost cannot be avoided by the use of continuous strategies of finite variation. In their setup, the admissible portfolio processes $Z = (Z_t)_{0 \leq t \leq T}$ are taken to be absolutely continuous with density $\dot{Z} = (\dot{Z}_t)_{0 \leq t \leq T}$. The cost of liquidity enters into their wealth dynamics $Y = (Y_t)_{0 \le t \le T}$ as a penalization of the speed of trading like in the framework of Almgren and Chriss [6]:

$$dY_t = Z_t \, dS_t - S_t l(\dot{Z}_t) \, dt.$$

They take S_t as a geometric Brownian motion with zero drift and l a convex, non-negative function with $l(0) = 0$. In [7] and [9], all transaction costs due to illiquidity can be eliminated by using continuous strategies of finite variation. However, in the setup of Rogers and Singh [29], the use of these strategies induces a liquidity cost. Assume that an investor holds Z_0 number of shares, x units of money market account, and she wants to replicate a European contingent claim with payoff $g(S_T)$. Since the Black–Scholes hedge $\theta(t, S_t)$ of a European contingent claim is of infinite variation, it will incur infinite liquidity costs. As a result, the authors propose to minimize the mean squared hedging error and the associated liquidity costs incurred over portfolio processes $Z = (Z_t)_{0 \le t \le T}$,

$$\frac{1}{2} E\left[\left(x + Z_0 S_0 + \int_0^T Z_t \, dS_t - g(S_T)\right)^2\right] + E\left[\int_0^T S_t l(\dot{Z}_t) \, dt\right].$$

They solve the Hamilton–Jacobi–Bellman equation for the associated optimal control problem in almost closed form and study it numerically.

Çetin, Soner, and Touzi [11] study the superreplication problem using the CJP model under the additional constraint on the boundedness of the quadratic variation and the absolute continuous parts of the portfolio processes. Their driving motivation is the lack of liquidity premium, i.e., the extra amount one has to pay due to illiquidity, in the papers by Bank and Baum [7] and Çetin et al. [9] as a result of using continuous strategies of bounded variation. They link the absence of the liquidity premium to the choice of admissible strategies and show that one can find a nonzero liquidity premium in continuous time for a set of admissible strategies appropriately defined. Their results and the justification for the set of admissible strategies they consider are well supported by a convergence result of the discrete-time setting in [18]. In fact, there are no restrictions on the portfolio strategies in [18]. As the dynamics of the paper value of the portfolio Y in (13.14) is obtained as a limit of the natural discrete-time self-financing conditions, this is a justification of the validity of the constraints placed on the portfolio processes in [11]. In particular, Gökay and Soner [18] analyze the asymptotic limit of the Binomial version of the CJP model both numerically and theoretically. Although there are no constraints placed on the portfolio processes in their model, Gökay and Soner recover the same superreplicating cost as in [11] in the limit and hence show that the liquidity premium persists in the continuous time.

Çetin et al. [11] consider a marginal process S satisfying the stochastic differential equation

$$S_r = s + \int_t^r S_u \sigma(u, S_u) \, dW_u^0,$$

which has a strong solution denoted by $S_\cdot^{t,s}$ with the initial condition $S_t = s$. Moreover, they take the portfolio process Z to be of the form

$$Z_r = \sum_{n=0}^{N-1} z_n \mathbf{1}_{\{r \geq \tau_{n+1}\}} + \int_t^r \alpha_u \, du + \int_t^r \Gamma_u \, dS_u^{t,s},$$

where $t = \tau_0 < \tau_1 < \cdots$ is an increasing sequence of $[t, T]$-valued $\mathbf{F}$-stopping times, the random variable

$$N := \inf\{n \in \mathbf{N} : \tau_n = T\}$$

indicates the number of jumps, and z_n is $\mathcal{F}(\tau_n)$-measurable. The infinite variation part of this trading strategy consists of an integral with respect to the marginal price process S, where the integrand is the gamma $\Gamma = (\Gamma_t)_{0 \leq t \leq T}$ of the portfolio. The integrands α and Γ are $\mathbf{F}$-progressively measurable processes. Moreover, there are additional constraints imposed on the processes Z, α, and Γ, similar to those in [12] and [32]. Then the authors consider superreplicating a European contingent claim with payoff g. The payoff g is continuous, nonnegative, and satisfies $g(s) \leq C(1 + s)$ for some constant C. If the supply curve is of the form (13.11), then the superreplicating cost $\phi(t, s)$ is the unique viscosity solution of the dynamic programming equation

$$-\phi_t(t,s) + \sup_{\beta \geq 0} \left(-\frac{1}{2} s^2 \sigma^2 \big(\phi_{ss}(t,s) + \beta\big) - \Lambda s^2 \sigma^2(t,s) \big(\phi_{ss}(t,s) + \beta\big)^2 \right) = 0$$

and satisfies the terminal condition $\phi(T, \cdot) = g(\cdot)$ along with the growth condition $0 \leq \phi(t, s) \leq C(1 + s)$ for some constant C. With constant volatility σ, one can rewrite it as

$$-\phi_t(t,s) - s^2 \sigma^2 H\big(\phi_{ss}(t,s) + \beta\big) = 0, \tag{13.16}$$

where

$$H(\gamma) = \begin{cases} \frac{1}{2}\gamma + \Lambda \gamma^2, & \gamma \geq -\frac{1}{4\Lambda}, \\ -\frac{1}{16\Lambda}, & \gamma < -\frac{1}{4\Lambda}, \end{cases}$$

and the liquidity parameter Λ is given as $\frac{\partial \mathbf{S}}{\partial v}(t, 0)$. For $\Lambda = 0$, one recovers the Black–Scholes setting. In fact, if ϕ_{BS} is the Black–Scholes value of the claim g, then by a maximum principle argument one concludes that $\phi(t, s) \geq \phi_{BS}(t, s)$. Moreover, ϕ and ϕ_{BS} coincide if and only if the payoff is an affine function. This implies that there exists a liquidity premium, a difference between the superreplicating cost ϕ and the Black–Scholes value ϕ_{BS}, for nontrivial claims g. This result conflicts with the statement that in an illiquid market all liquidity costs can be avoided by approximating with continuous strategies with finite variation. The intuitive reasoning is that such an approximation neutralizes the gamma of the portfolio process; however it makes α infinitely large in the limit so that it no longer satisfies the imposed constraints.

Çetin et al. [11] also study the associated superhedging strategy under liquidity costs. They characterize a set $\mathcal{C}$ such that outside $\mathcal{C}$, the hedging strategy is given by $\phi_s(t,s)$ and in $\mathcal{C}$ the strategy is a mixture of dynamically replicating an auxiliary function ψ and applying a buy and hold strategy to $\phi - \psi$. The set $\mathcal{C}$ is determined by a level of concavity on the value function ϕ.

Gökay and Soner [18] study a discrete version of the supply curve model. For a fixed step size $h > 0$, they divide the trading period $[0, T]$ into equal intervals of length h. The evolution of the marginal price process is given by a Binomial tree, i.e., at any node (t, S_t) it either goes up by a factor of $1 + \sigma\sqrt{h}$ or down by a factor of $1 - \sigma\sqrt{h}$. We use the notation

$$S_{t+h} = S_t(1 \pm \sigma\sqrt{h}).$$

The filtration $\mathbf{F}$ is generated by the marginal price process S, and the portfolio process Z is taken to be adapted with respect to $\mathbf{F}$. They consider a supply curve of the form

$$\mathbf{S}(t, s, \nu) = S_t + \Lambda\nu$$

with liquidity parameter Λ. Observe that this supply curve may take negative values, so one may consider $\mathbf{S}(t, S_t, \nu) = (S_t + \Lambda\nu)^+$; however, the analysis in [18] shows that both supply curves yield the same partial differential equation in the limit. The self-financing condition is given as in (13.12), and the book value Y has the dynamics of (13.13). We introduce the notation $Z_\cdot^{t,z}$ to denote the portfolio process with initial condition $Z_t = z$ and $Y_t^{t,y,Z}$ the book value that starts $Y_t = y$ and uses the control $Z_\cdot$. The authors study the superreplication problem of a European contingent claim with payoff g. As in [11], the payoff g is continuous, nonnegative, and satisfies the linear growth condition $g(s) \le C(1+s)$ for some $C > 0$. The minimal superreplicating cost $\phi^h(t,s)$ at time t and $S_t = s$ is given by

$$\phi^h(t,s) = \inf\{y \mid \exists\mathbf{F}\text{-adapted } \{Z_\cdot\} \text{ such that } Y_T^{t,y,Z_\cdot} \ge g(S_T^{t,s}) \text{ a.s.}\}.$$

The main observation is that dynamic programming approach fails for the value function $\phi^h(t,s)$, therefore to restore dynamic programming one needs introduce the dependence of the value function on the portfolio position z in addition to the current stock price and time. So we define

$$\begin{aligned} v^h(t,s,z) := \inf\big\{y \mid\ & \exists\mathbf{F}\text{-adapted}\{Z_\cdot\} \\ & \text{such that } Z_t = z \text{ and } Y_T^{t,y,Z_\cdot} \ge g(S_T^{t,s}) \ \text{ a.s.}\big\}. \end{aligned}$$

Clearly,

$$\phi^h(t,s) = \inf_z v^h(t,s,z).$$

The following dynamic programming is the key element of the analysis of Gökay and Soner [18]:

$$\begin{aligned} v^h(t,s,z) = \inf\big\{y \mid\ & \exists\mathbf{F}\text{-adapted } \{Z_\cdot\} \\ & \text{s.t. } Z_t = z \text{ and } Y_\tau^{t,y,Z_\cdot} \ge v^h(\tau, S_\tau^{t,s}, Z_\tau) \text{ a.s.}\big\}, \end{aligned}$$

where $t = nh < \tau = mh \le T$ for some $n, m \in \mathbf{N}$. In particular, for $\tau = t + h$, we have the following form:

$$v^h(t,s,z) = \max\left(\min_a \{ v^h(t+h, su, z+a) - zs\sigma\sqrt{h} + \Lambda a^2 \}, \right.$$
$$\left. \min_b \{ v^h(t+h, sd, z+b) + zs\sigma\sqrt{h} + \Lambda b^2 \} \right).$$

This equation is complemented by the terminal data

$$v^h(T,s,z) = g(s).$$

Using the theory of viscosity solutions, the authors pass to the limit by letting the time step $h \downarrow 0$. In particular, they show that $v^h(t,s,z)$ converges to the solution $\phi(t,s)$ of the partial differential equation (13.16) locally uniformly as $h \downarrow 0$. To this aim, they consider the standard upper and lower relaxed limits in the theory of viscosity solutions

$$\phi^*(t,s,z) = \limsup_{\substack{h\to 0\\ (t',s',z')\to(t,s,z)}} v^h(t',s',z'),$$
$$\phi_*(t,s,z) = \liminf_{\substack{h\to 0\\ (t',s',z')\to(t,s,z)}} v^h(t',s',z').$$

The authors prove that $\phi^*(t,s,z)$ is independent of z and set

$$\phi^*(t,s) := \phi^*(t,s,z).$$

However, it is difficult to derive directly a similar claim for $\phi_*(t,s,z)$. In fact, the challenge in proving this convergence result is that in discrete-time the value function $v^h(t,s,z)$ depends on the initial portfolio value z, whereas this dependence becomes irrelevant in the limit $\phi(t,s)$. Therefore, the authors overcome this difficulty by defining

$$\phi_*(t,s) = \inf_z \Big\{ \liminf_{\substack{h\to 0\\ (t',s',z')\to(t,s,z)}} v^h(t',s',z') \Big\}$$

and developing further the idea of corrector functions as in the applications of viscosity solutions to homogenization. The authors proceed by showing that the upper semi-continuous relaxed limit $\phi^*(t,s)$ is a viscosity subsolution and the lower semi-continuous relaxed limit $\phi_*(t,s)$ is a viscosity supersolution of the partial differential equation (13.16). Moreover, both ϕ_* and ϕ^* are growing almost linearly and attain $\phi_*(T,s) = \phi^*(T,s) = g(s)$. So by the comparison argument established in [11], they conclude that $\phi_* = \phi^*$ and it is equal to the unique viscosity solution of (13.16). Now the local uniform convergence of $v^h(t,s,z)$ to $\phi(t,s)$ will follow from the definitions of ϕ_* and ϕ^*.

13.5 Expected Utility Maximization in Illiquid Markets

In this section, we briefly review some results regarding the problem of expected utility maximization in illiquid markets. We first consider the permanent price impact setting of Ly Vath et al. [24] and then the setup of temporary price impacts in discrete time, as done in Çetin and Rogers [10].

Ly Vath et al. [24] solve the expected utility maximization problem with permanent price impacts in continuous time with admissible strategies of the form

$$Z_t = \xi_0 + \sum_{i=1}^{N} \xi_n \mathbf{1}_{\{t \geq \tau_n\}} \quad (0 \leq t < T), \tag{13.17}$$

where $\{\tau_n\}_{n\geq 1}$ is a sequence of stopping times, and $\xi_n \in \mathcal{F}_{\tau_n}$ for all $n \geq 1$. A trade of size ξ at time t is assumed to have a permanent impact of the exponential form. Furthermore, they assume that the stock price evolves as a geometric Brownian motion between trades, i.e.,

$$dS_t = \mu S_{t-}\,dt + \sigma S_{t-}\,dW_t + \lambda S_{t-}\,dZ_t$$

for some positive constants λ, $\sigma > 0$, and $\mu \in \mathbf{R}$. Each time a transaction is made, the investor pays a fixed transaction cost k so that the money market account obeys the following equation:

$$X_t = \int_0^t rX_{s-}\,ds - \int_0^t S_{s-}e^{\lambda \Delta Z_s}\,dZ_s - \sum_{i\geq 1} k\mathbf{1}_{\{\tau_n \leq t\}}.$$

A strategy Z belongs to the set of admissible strategies $\mathcal{A}(t,x,z,s)$ started at time t with $X_t = x$, $Z_t = z$, and $S_t = s$ if it satisfies the solvency constraint

$$X_s + S_{s-}e^{-\lambda Z_s} Z_s - k \geq 0$$

for all $t \leq s \leq T$. The second term in the above inequality is the liquidation value of a position of size Z_t in the risky asset S. The solvency constraint states that the liquidation value of an admissible portfolio is always positive. Due to this fixed cost at each transaction, the authors show that the optimal trading strategy which maximizes the expected utility is indeed of the form of (13.17), and they describe the optimal trading times τ_n in terms of the value of the money market account, the position in the risky asset, and the current price. Their main result is to show that the value function

$$v(t,x,z,s) = \sup_{Z\in\mathcal{A}(t,x,z,s)} \mathbf{E}U\big(X_T + S_{T-}e^{-\lambda Z_T} Z_T - k\big)$$

is a viscosity solution of the following quasi-variational Hamilton–Jacobi–Bellman inequality:

$$\min\left\{-\frac{\partial v}{\partial t} - rx\frac{\partial v}{\partial x} - \mathcal{L}v,\, v - \mathcal{H}v\right\} = 0,$$

where $\mathcal{L}$ is the infinitesimal generator of a geometric Brownian motion, and $\mathcal{H}$ is an impulse generator of the form

$$\mathcal{H}v(t,x,z,s) = \sup_{\xi} v\big(t, x - se^{\lambda\xi}\xi, z - \xi, se^{\lambda\xi}\big)$$

with the supremum taken over the set of transactions that satisfy the solvency condition.

Çetin and Rogers [10] study the discrete-time utility maximization problem using a supply curve of the form

$$\mathbf{S}(t, S_t, \nu) = \varphi(\nu)S_t,$$

where φ is a strictly increasing and strictly convex function. Their objective is to maximize utility from terminal liquidation value $Y_N = X_N + Z_N S_N$, where $Z_N = 0$, and U is a strictly concave and strictly increasing utility function. They show that this problem has a solution. Moreover, the marginal utility of optimal terminal wealth $U'(Y_N)$ is an equivalent martingale measure, and the process $M_n = \varphi'(\Delta Z_n)S_n$ becomes a martingale under this measure.

13.6 Price Manipulation strategies in Price Impact Models

So far, there is one fundamental notion of finance we have not addressed: arbitrage from price manipulations. The assumption that the large trader has a temporary and permanent impact on the prices clearly suggests the possibility that she can manipulate the prices in her favor. In Sect. 13.2, this issue has been partly avoided by either assuming a priori that the execution of the large sell (resp. buy) order is restricted to smaller sell (buy) orders or that this condition is satisfied a posteriori as a consequence of the assumptions made. Indeed, in the former case, arbitrage is not possible since a sell order makes the price lower so that the next sell order will come at a less favorable price. In more general models, however, there sometimes exists weaker version of the arbitrage condition. For instance, the widespread concept of quasi-arbitrage and price manipulations which correspond to strategies with a negative expected cost is often considered in the literature. This particular approach can be found in the papers of Huberman and Stanzl [19], Gatheral [17], and Jarrow [21, 22].

To make the notion of quasi-arbitrage more precise, Huberman and Stanzl [19] define the notion of a *round trip*, a trading strategy that starts with zero shares and terminates with zero shares of the risky asset. They consider a model in discrete time, with n time steps. There are noise traders, and we denote by η_k the number of shares of the risky asset they purchase at time k. As before, ξ_k denotes the trade size of the large trader at time k. Let $\{\zeta_k\}_{k=1,\dots,N}$ be i.i.d. random variables with zero expectation. We also assume that $\{\eta_k\}_{k=1,\dots,N}$ are i.i.d. random variables with zero expectation. The authors consider the following dynamic for the marginal price of the risky asset:

$$S_k = S_{k-1} + g(\xi_k + \eta_k) + \zeta_k.$$

They also hypothesize the existence of a temporary price impact function h, so that the large trader pays a total of $\xi_k(S_k + h(\xi_k + \eta_k))$ at time k. The temporary impact includes the noise traders' trading volume η_k, and the η_k's are assumed to be unknown by the large trader at the moment of the transaction at time k. The profit of a round trip is given by $\pi(\xi) = -\sum_{k=1}^{n} \xi_k(S_k + h_k(\xi_k + \eta_k))$. Huberman and Stanzl [19] define a *price manipulation* as a round trip with positive expected value. They also define a *quasi-arbitrage* as a sequence of round trips $\xi^m = \{\xi_k^m\}_{k=1,\dots,n}$ for $m = 1, 2, \dots$ such that $\lim_{m\to\infty} \mathbf{E}\pi(\xi^m) = \infty$ and

$$\lim_{m\to\infty} \frac{\mathbf{E}\pi(\xi^m)}{\sqrt{\mathrm{Var}(\pi(\xi^m))}} = \infty.$$

Their main result states that if $\mathbf{P}(\eta_k = 0) = 1$ $(k = 1, \dots, n)$ or the η_k's are normally distributed, then the absence of price manipulation implies that the permanent impact function g is linear. On the other hand, no restriction is required on the temporary impact function h.

Gatheral [17] considers models for stock prices with price impacts that decay with time. More specifically, he focuses on models on the following form:

$$S_t = S_0 + \int_0^t g(\dot{X}_s) G(t-s)\, ds + \sigma W_t,$$

where g is the permanent impact function, and G is the decay factor. In words, the permanent impact of a trade at time t decays with time due to the function G. The setting is the same as in (13.4) when $G = 1$. The author finds a relationship between the shape of the market impact function g and the resilience function G under the no-dynamic-arbitrage assumption. In particular, he obtains similar results to Huberman and Stanzl [19] regarding the linearity of the price impact function.

In [21], Jarrow considers a discrete-time economy. In his model, the stock price can be expressed in terms of a sequence $\{g_{t_k}\}_{0\le k\le N}$ with $g_{t_k} : \Omega \times \mathbf{R}^{t+1} \to \mathbf{R}$ such that

$$S_{t_k}(\omega) = g_{t_k}\big(\omega, Z_{t_k}(\omega), \dots, Z_0(\omega)\big) \quad \forall \omega \in \Omega,\ 0 \le k \le N.$$

The functions $\{g_{t_k}\}_{0\le k\le N}$ are the *reaction* functions that reflect how the participants of the market react to large trader's portfolio decisions. Particular cases of these functions are the permanent and temporary impact function described in Sect. 13.2. These reaction functions provide the reduced form equilibrium relationship between relative prices and the large trader's trades. In [21], Jarrow concentrates on market manipulation strategies for the large trader. In Jarrow's terminology, a *market manipulation strategy* is a strategy that can generate positive real wealth for the large trader without taking any risk. The real wealth for the large trader is characterized as the value of her portfolio after liquidation. Market manipulation strategies are shown to sometimes exist in this economy. Sufficient conditions are provided that restrict the market manipulation strategies. These conditions include the requirement that the stock price process is independent of the past holdings of the large trader and depends only on her instantaneous holdings, i.e.,

$$S_{t_k}(\omega) = g_{t_k}\big(\omega, Z_{t_k}(\omega)\big)$$

and that if the large trader is not active in the time interval $[t_k, t_{k+1}]$, then there are no arbitrage opportunities available for the reference traders in this time period. In [22], Jarrow extends this framework for markets that include a derivative security. He finds sufficient conditions to exclude market manipulation strategies, after showing that market manipulation strategies can exist after the introduction of the derivative security. To avoid market manipulation strategies, the market must be in synchrony. This means that the number of shares, whether bought in the stock market or acquired jointly in the stock and derivative market, should yield the same stock price. Moreover, Jarrow shows that one can hedge options using the standard method based on the binomial model with random volatilities.

References

1. A. Alfonsi, A. Fruth, A. Schied, Constrained portfolio liquidation in a limit order book model. Banach Cent. Publ. **83**, 9–25 (2008)
2. A. Alfonsi, A. Fruth, A. Schied, Optimal execution strategies in limit order books with general shape functions. Quant. Finance **10**(2), 143–157 (2010)
3. A. Alfonsi, A. Schied, Optimal execution and absence of price manipulations in limit order book models. doi:10.1137/090762786
4. R. Almgren, Optimal execution with nonlinear impact functions and trading-enhanced risk. Appl. Math. Finance **10**, 1–18 (2003)
5. R. Almgren, N. Chriss, Value under liquidation. Risk **12**(12), 61–63 (1999)
6. R. Almgren, N. Chriss, Optimal execution of portfolio transactions. J. Risk **3**, 5–39 (2000)
7. P. Bank, D. Baum, Hedging and portfolio optimization in financial markets with a large trader. Math. Finance **14**, 1–18 (2004)
8. D. Bertsimas, A. Lo, Optimal control of execution costs. J. Financ. Mark. **1**, 1–50 (1998)
9. U. Çetin, R. Jarrow, P. Protter, Liquidity risk and arbitrage pricing theory. Finance Stoch. **8**, 311–341 (2004)
10. U. Çetin, L.C.G. Rogers, Modelling liquidity effects in discrete time. Math. Finance **17**, 15–29 (2007)
11. U. Çetin, H.M. Soner, N. Touzi, Option hedging for small investors under liquidity costs. Finance Stoch. doi:10.1007/s00780-009-0116-x
12. P. Cheridito, H.M. Soner, N. Touzi, N. Victoir, Second order backward stochastic differential equations and fully non-linear parabolic PDE's. Commun. Pure Appl. Math. **60**(7), 1081–1110 (2007)
13. D. Cuoco, J. Cvitanic, Optimal consumption choices for a large investor. J. Econ. Dyn. Control **22**, 401–436 (1998)
14. J. Cvitanic, J. Ma, Hedging options for a large investor and forward backward sdes. Ann. Appl. Probab. **6**, 370–398 (1996)
15. R. Frey, Perfect option hedging for a large trader. Finance Stoch. **2**, 115–141 (1998)

16. R. Frey, A. Stremme, Market volatility and feedback effects from dynamic hedging. Math. Finance **7**, 351–374 (1997)
17. J. Gatheral, No-dynamic-arbitrage and market impact. Quant. Finance. doi:10.1080/14697680903373692
18. S. Gökay, H.M. Soner, Liquidity in a binomial market. doi:10.1111/j.1467-9965.2010.00462.x
19. G. Huberman, W. Stanzl, Price manipulation and quasi-arbitrage. Econometrica **72**(4), 1247–1275 (2004)
20. G. Huberman, W. Stanzl, Optimal liquidity trading. Rev. Finance **9**, 165–200 (2005)
21. R. Jarrow, Market manipulation, bubbles, corners and short squeezes. J. Financ. Quant. Anal. **27**(3), 311–336 (1992)
22. R. Jarrow, Derivative securities markets, market manipulation and option pricing theory. J. Financ. Quant. Anal. **29**(2), 241–261 (1994)
23. R. Jarrow, P. Protter, A. Roch, A liquidity based model for asset price bubbles. Preprint (2010)
24. V. Ly Vath, M. Minf, H. Pham, A model of optimal portfolio selection under liquidity risk and price impact. Finance Stoch. **11**, 51–90 (2007)
25. J. Ma, P. Protter, J. Yong, Solving forward-backward stochastic differential equations explicitly—a four step scheme. Probab. Theory Relat. Fields **98**, 339–359 (1994)
26. A. Obizhaeva, J. Wang, Optimal trading strategy and supply/demand dynamics. Preprint (2005)
27. E. Platen, M. Schweizer, On feedback effects from hedging derivatives. Math. Finance **8**, 67–84 (1998)
28. A. Roch, Liquidity risk, price impacts and the replication problem. Finance Stoch. (2011, to appear)
29. L.C.G. Rogers, S. Singh, The cost of illiquidity and its effects on hedging. doi:10.1111/j.1467-9965.2010.00413.x
30. A. Schied, T. Schöneborn, Risk aversion and the dynamics of optimal liquidation strategies in illiquid markets. Finance Stoch. **13**(2), 181–204 (2009)
31. R. Sircar, G. Papanicolaou, General Black–Scholes models accounting for increased market volatility from hedging strategies. Appl. Math. Finance **5**, 45–82 (1998)
32. H.M. Soner, N. Touzi, Superreplication under gamma constraints. SIAM J. Control Optim. **39**(1), 73–96 (2000)
33. H.M. Soner, N. Touzi, The dynamic programming equation for second order stochastic target problems. SIAM J. Control Optim. **48**(4), 2344–2365 (2009)

Chapter 14
Some New BSDE Results for an Infinite-Horizon Stochastic Control Problem

Ying Hu and Martin Schweizer

Abstract We study in a continuous filtration a quadratic BSDE with an unbounded generator and an infinite time horizon. This equation comes from a stochastic control problem in the context of robust utility maximisation. We prove the existence and uniqueness, in a suitable class, of a solution to the BSDE, and we show that the BSDE characterises the dynamic value process of the stochastic control problem.

Keywords Backward stochastic differential equations · Infinite horizon · Quadratic BSDE · Unbounded solution · Stochastic control · Robust utility maximisation

Mathematics Subject Classification (2010) 60H10 · 60H20 · 60G44 · 91B16

14.1 Introduction

This paper studies a stochastic control problem arising in the context of robust utility maximisation and proves new results via BSDE techniques. A particular feature is that the problem is formulated and solved for an infinite horizon and that we also obtain new results on a certain infinite-horizon BSDE with quadratic generator.

In loose terms, the basic problem we should like to tackle has the form

$$\text{find } \sup_{\pi} \inf_{Q} \mathbf{U}(\pi, Q),$$

Y. Hu
IRMAR, Université Rennes 1, Campus de Beaulieu, 35042 Rennes Cedex, France
e-mail: ying.hu@univ-rennes1.fr

M. Schweizer (✉)
Departement Mathematik, ETH Zürich, 8092 Zürich, Switzerland
e-mail: martin.schweizer@math.ethz.ch

M. Schweizer
Swiss Finance Institute, Walchestrasse 9, 8006 Zürich, Switzerland

G. Di Nunno, B. Øksendal (eds.), *Advanced Mathematical Methods for Finance*,
DOI 10.1007/978-3-642-18412-3_14,

where $\mathbf{U}$ is some utility functional, π runs through a set of investment and consumption strategies, and Q through a set of models given by probability measures. In a first step, we focus only on the inner minimisation problem; thus we think of π as being fixed and look for a worst-case model Q. The functional $\mathbf{U}(\pi, Q)$ we consider has the form

$$\mathbf{U}(\pi, Q) = E_Q\big[\mathcal{U}^\delta_{0,\infty} + \beta\mathcal{R}^\delta_{0,\infty}(Q)\big],$$

where $\mathcal{U}^\delta_{0,\infty} = \alpha\int_0^\infty S^\delta_s U_s\,ds$ stands for a discounted utility term (whose dependence on the fixed π is suppressed), and $\mathcal{R}^\delta_{0,\infty}(Q) = \int_0^\infty S^\delta_s \log Z^Q_s\,ds$ is an entropic penalty term. A precise formulation is given later.

The finite-horizon version of this problem has been studied in Bordigoni/Matoussi/Schweizer [2], who have characterised the dynamic value process $V = V^{(T)}$ of the resulting stochastic control problem as the unique solution of the BSDE

$$dY_t = (\delta_t Y_t - \alpha U_t)\,dt + \frac{1}{2\beta}\,d\langle M\rangle_t + dM_t \tag{14.1a}$$

with a final condition at time T. We generalise these results here to an infinite-horizon setting with the terminal condition

$$\lim_{t\to\infty} Y_t = 0. \tag{14.1b}$$

In an unpublished Ph.D. thesis, G. Bordigoni has already shown that V for the infinite-horizon problem is a solution of the BSDE (14.1a), (14.1b); but uniqueness and the required integrability (and hence the characterisation of V by the BSDE) remained open. We close this gap here.

In contrast to Bordigoni [1], our approach and main results here are on the side of BSDE theory. Equations (14.1a), (14.1b) are a quadratic BSDE in a continuous filtration and have an unbounded generator (due to the presence of U) and an infinite horizon. For the finite-horizon case, the classical results of Kobylanski [10] on the existence and uniqueness of a bounded solution for quadratic BSDEs in a Brownian filtration have been extended to unbounded solutions in the Brownian setting by Briand/Hu [5, 6], and to bounded solutions in a continuous filtration by Morlais [12]. In infinite-horizon settings, Briand/Hu [4] and later Royer [13] have studied bounded solutions for BSDEs with a Lipschitz generator, and Briand/Confortola [3] have extended these results to bounded solutions for a quadratic generator. The methods in all these papers rely on Girsanov techniques.

Our approach here is quite different. To prove the existence of a solution to (14.1a), (14.1b), we have adapted the localisation method from Briand/Hu [5], while for uniqueness, we have applied the θ-difference method from Briand/Hu [6]. Both these techniques have so far only been used in finite-horizon settings.

Finally, let us also mention two closely related papers from the finance and economics literature. Schroder/Skiadas [14] study the same BSDE as we do and obtain the existence and uniqueness of unbounded solutions, but in a Brownian filtration and with a finite time horizon. Hansen/Sargent/Turmuhambetova/Williams [9] study robustness aspects for infinite-horizon utility maximisation problems; their main ideas and problems are similar to ours, but the approach is rather heuristic, using Hamilton–Jacobi–Bellman equations and formal manipulations in a Markovian setting. For a more detailed discussion of additional references to the literature, we refer to Sect. 6 of Bordigoni/Matoussi/Schweizer [2].

The paper is structured as follows. After some preliminaries and notation in Sect. 14.2, we study in Sect. 14.3 the BSDE on a finite horizon. This serves as preparation for the infinite-horizon BSDE studied in Sect. 14.4 and gives to that end fairly precise estimates for the solution Y. Section 14.4 establishes the existence and uniqueness of a solution (Y, M) for the infinite-horizon BSDE (14.1a), (14.1b) and gives a sufficient condition for $\mathcal{E}(-\frac{1}{\beta}M)$ to be a martingale. In Sect. 14.5, we prove by general arguments as in Bordigoni/Matoussi/Schweizer [2] and Bordigoni [1] the existence of a solution to our stochastic control problem and show that its value process V satisfies the boundary condition $\lim_{t\to\infty} V_t = 0$. Finally, Sect. 14.6 uses the BSDE results to characterise V as the unique solution, in a suitable space, for the BSDE (14.1a), (14.1b), and in particular establishes that V has the required good integrability properties.

14.2 Preliminaries and Overview

In this section, we introduce all required notation, the basic BSDEs and the basic optimisation problems. We start with a probability space $(\Omega, \mathcal{F}, P)$ and a time horizon $T \in (0, \infty]$. The filtration $\mathbb{F} = (\mathcal{F}_t)_{t\geq 0}$ satisfies the usual conditions of right-continuity and P-completeness, $\mathcal{F}_0$ is P-trivial, and we set $\mathcal{F}_\infty := \bigvee_{t\geq 0} \mathcal{F}_t$. The basic ingredients for our optimisation problems are

- parameters $\alpha, \alpha' \in [0, \infty)$ and $\beta \in (0, \infty)$;
- progressively measurable processes $\delta = (\delta_t)_{t\geq 0}$ and $U = (U_t)_{t\geq 0}$;
- an $\mathcal{F}_T$-measurable random variable U'_T, with $U'_\infty := 0$ for $T = \infty$.

With these, we can formulate the BSDEs studied here. On the one hand, for a finite horizon $T < \infty$, we introduce the BSDE

$$dY_t = (\delta_t Y_t - \alpha U_t)\,dt + \frac{1}{2\beta}\,d\langle M\rangle_t + dM_t, \qquad Y_T = \alpha' U'_T. \tag{14.2}$$

On the other hand, for an infinite horizon $T = \infty$, the BSDE is

$$dY_t = (\delta_t Y_t - \alpha U_t)\,dt + \frac{1}{2\beta}\,d\langle M\rangle_t + dM_t, \qquad \lim_{t\to\infty} Y_t = 0. \tag{14.3}$$

A solution of (14.2) or (14.3) is a pair (Y, M) satisfying (14.2) or (14.3), respectively, where Y is a P-semimartingale and M is a locally P-square-integrable local P-martingale null at 0.

For the optimisation problems, we first define the discounting process

$$S_t^\delta := \exp\biggl(-\int_0^t \delta_s\, ds\biggr), \quad t \ge 0,$$

and for $T < \infty$ the auxiliary quantities, for $0 \le t \le T$,

$$\begin{aligned} \mathcal{U}_{t,T}^\delta &:= \alpha \int_t^T \frac{S_s^\delta}{S_t^\delta} U_s\, ds + \alpha' \frac{S_T^\delta}{S_t^\delta} U_T' \\ &= \int_t^T \alpha e^{-\int_t^s \delta_r\, dr} U_s\, ds + \alpha' e^{-\int_t^T \delta_r\, dr} U_T', \end{aligned} \tag{14.4}$$

$$\mathcal{R}_{t,T}^\delta(Q) := \int_t^T \delta_s \frac{S_s^\delta}{S_t^\delta} \log \frac{Z_s^Q}{Z_t^Q}\, ds + \frac{S_T^\delta}{S_t^\delta} \log \frac{Z_T^Q}{Z_t^Q}, \tag{14.5}$$

for $Q \ll P$ on $\mathcal{F}_T$ with density process Z^Q on $[0,T]$. We consider the cost functional

$$c_T(Q) := \mathcal{U}_{0,T}^\delta + \beta \mathcal{R}_{0,T}^\delta(Q),$$

and the basic stochastic control problem on a finite horizon is to minimise the functional

$$Q \mapsto \Gamma_T(Q) := E_Q\bigl[c_T(Q)\bigr]$$

over a suitable class of probability measures $Q \ll P$ on $\mathcal{F}_T$. In a classical way, we can choose an adapted RCLL process $V = (V_t)_{0 \le t \le T}$ such that

$$V_t = \operatorname*{ess\,inf}_Q E_Q\bigl[\mathcal{U}_{t,T}^\delta + \beta \mathcal{R}_{t,T}^\delta(Q) \big| \mathcal{F}_t\bigr], \quad 0 \le t \le T.$$

For $T = \infty$, we define similarly, for $t \ge 0$,

$$\mathcal{U}_{t,\infty}^\delta := \alpha \int_t^\infty \frac{S_s^\delta}{S_t^\delta} U_s\, ds = \int_t^\infty \alpha e^{-\int_t^s \delta_r\, dr} U_s\, ds,$$

$$\mathcal{R}_{t,\infty}^\delta(Q) := \int_t^\infty \delta_s \frac{S_s^\delta}{S_t^\delta} \log \frac{Z_s^Q}{Z_t^Q}\, ds$$

for $Q \overset{\text{loc}}{\ll} P$ with density process Z^Q. We consider the cost functional

$$c_\infty(Q) := \mathcal{U}_{0,\infty}^\delta + \beta \mathcal{R}_{0,\infty}^\delta(Q)$$

and in principle want to minimise the functional

$$Q \mapsto \Gamma_\infty(Q) := E_Q\bigl[c_\infty(Q)\bigr]$$

over a suitable class of probability measures $Q \overset{\text{loc}}{\ll} P$. (For the precise formulation, we refer to Sect. 14.5.) In a similar manner as for $T < \infty$, we can choose an adapted RCLL process $V = (V_t)_{t \ge 0}$, which is again called the dynamic value process of our stochastic control problem. Of course, to be accurate, we should distinguish in notation between $V^{(T)}$ and $V^{(\infty)}$.

Our main results in this paper are:

(a) existence and uniqueness results for the above BSDEs (both with finite and infinite horizon), and
(b) a characterisation of the value process $V = V^{(\infty)}$ for the infinite-horizon setting as the solution of the BSDE (14.3).

14.3 The BSDE on a Finite Horizon

The main goal of this section is to prove the existence of a solution to the finite-horizon BSDE under weak conditions. This slightly extends previous work and, above all, serves as preparation for the infinite-horizon case. So we fix $T \in (0, \infty)$ and view U and δ as processes on $[0, T]$.

Hypothesis 14.1 *Throughout this section, we impose the standing assumptions*

$\mathbb{F}$ is a *continuous filtration*, i.e. all local $(P, \mathbb{F})$-martingales are continuous. (14.6a)

$\delta \geq 0$ is uniformly bounded (by $\bar{\delta}$, say). (14.6b)

Precise assumptions on U, U'_T will be specified below. Now we introduce the quantities

$$B := \alpha \int_0^T S_s^\delta U_s \, ds + \alpha' S_T^\delta U'_T = \mathcal{U}_{0,T}^\delta,$$

$$B_- := \alpha \int_0^T S_s^\delta U_s^- \, ds + \alpha' S_T^\delta (U'_T)^-,$$

$$B_+ := \alpha \int_0^T S_s^\delta U_s^+ \, ds + \alpha' S_T^\delta (U'_T)^+.$$

The BSDE (14.2) under study is

$$dY_t = (\delta_t Y_t - \alpha U_t)\, dt + \frac{1}{2\beta} d\langle M \rangle_t + dM_t, \qquad Y_T = \alpha' U'_T. \tag{14.7}$$

Definition 14.2 A *solution* of (14.7) is a pair of processes (Y, M) satisfying (14.7), where Y is a P-semimartingale and M is a locally P-square-integrable local P-martingale null at 0.

Due to the standing assumption (14.6a), M and then Y are continuous for any solution (Y, M) of (14.7). Our proof of existence applies the localisation method originally developed in a Brownian setting by Briand/Hu [5]. To that end, we need to establish precise a priori estimates in the bounded case. Note that $1/S^\delta = S^{-\delta}$.

Proposition 14.3 (A priori estimates) *Suppose that $\int_0^T |U_s|\,ds$ and U_T' are bounded random variables. Then there exists a unique solution (Y, M) to (14.7) such that Y is a bounded process. Moreover, we have for $0 \le t \le T$ the estimates*

$$S_t^{-\delta}\underline{Y}_t - \int_0^t \alpha e^{\int_s^t \delta_r\,dr} U_s\,ds \le Y_t \le S_t^{-\delta}\overline{Y}_t - \int_0^t \alpha e^{\int_s^t \delta_r\,dr} U_s\,ds, \qquad (14.8)$$

where

$$\underline{Y}_t := -\beta e^{-\bar{\delta}T} \log E\big[e^{-\frac{1}{\beta}e^{\bar{\delta}T} B}\big|\mathcal{F}_t\big] \quad \text{and} \quad \overline{Y}_t := E[B|\mathcal{F}_t]. \qquad (14.9)$$

Proof The existence and uniqueness of a solution with Y bounded are immediate from Theorems 2.5 and 2.6 of Morlais [12]. From the definitions of $\underline{Y}$ and $\overline{Y}$ and from Itô's formula, it is clear that there exist $\underline{M}$ and $\overline{M}$ such that

$$d\underline{Y}_t = \frac{1}{2\beta} e^{\bar{\delta}T}\, d\langle \underline{M}\rangle_t + d\underline{M}_t, \qquad \underline{Y}_T = B, \qquad (14.10)$$

$$d\overline{Y}_t = d\overline{M}_t, \qquad \overline{Y}_T = B. \qquad (14.11)$$

If we first set $Y_t^1 := S_t^\delta Y_t$ and $M_t^1 := \int_0^t S_s^\delta\,dM_s$, then the BSDE (14.7) is transformed to

$$dY_t^1 = -\alpha S_t^\delta U_t\,dt + \frac{1}{2\beta} S_t^{-\delta}\, d\langle M^1\rangle_t + dM_t^1, \qquad Y_T^1 = \alpha' S_T^\delta U_T'. \qquad (14.12)$$

If we next put $Y_t^2 := Y_t^1 + \int_0^t \alpha S_s^\delta U_s\,ds$ and $M^2 := M^1$, the BSDE (14.12) becomes

$$dY_t^2 = \frac{1}{2\beta} S_t^{-\delta}\, d\langle M^2\rangle_t + dM_t^2, \qquad Y_T^2 = \alpha' S_T^\delta U_T' + \alpha \int_0^T S_s^\delta U_s\,ds = B. \quad (14.13)$$

Because $0 \le S_t^{-\delta} \le e^{\bar{\delta}T}$, we deduce by comparison of (14.10), (14.13) and (14.11) that

$$\underline{Y}_t \le Y_t^2 \le \overline{Y}_t, \quad 0 \le t \le T.$$

Returning to Y by the formula $Y_t = S_t^{-\delta} Y_t^2 - \int_0^t \alpha e^{\int_s^t \delta_r\,dr} U_s\,ds$, we conclude the proof. □

We now apply the localisation method to get the following existence result.

Theorem 14.4 (Existence of solution) *Let us suppose that*

$$E\big[e^{\frac{1}{\beta}e^{\bar{\delta}T} B_-}\big] + E[B_+] < \infty. \qquad (14.14)$$

Then the BSDE (14.7) *admits a solution* (Y, M) *which satisfies*

$$S_t^{-\delta}\underline{Y}_t - \int_0^t \alpha e^{\int_s^t \delta_r\,dr} U_s\,ds \le Y_t \le S_t^{-\delta}\overline{Y}_t - \int_0^t \alpha e^{\int_s^t \delta_r\,dr} U_s\,ds \tag{14.15}$$

for $0 \le t \le T$, *with* $\underline{Y}, \overline{Y}$ *given in* (14.9).

Proof (1) We first assume that U'_T and U are nonnegative; then $\underline{Y}$ is also nonnegative. For each $n \in \mathbb{N}$, we consider $U_t^n := U_t \mathbf{1}_{\{\int_0^t U_s\,ds \le n\}}$, $0 \le t \le T$, and $U_T^{\prime,n} := U'_T \wedge n$. According to Proposition 14.3, the BSDE

$$dY_t = \left(\delta_t Y_t - \alpha U_t^n\right)dt + \frac{1}{2\beta}\,d\langle M\rangle_t + dM_t, \qquad Y_T = \alpha' U_T^{\prime,n},$$

admits a unique solution (Y^n, M^n) such that Y^n is a bounded process and, by (14.8), extended from $U_T^{\prime,n}$ and U^n to U'_T and U thanks to nonnegativity,

$$-\int_0^t \alpha e^{\int_s^t \delta_r\,dr} U_s\,ds \le Y_t^n \le S_t^{-\delta} E[B|\mathcal{F}_t].$$

Since $U_T^{\prime,n} \le U_T^{\prime,n+1}$ and $U^n \le U^{n+1}$, the sequence (Y^n) is nondecreasing by a comparison result; this can be obtained similarly as in the proof of Theorem 8 in Mania/Schweizer [11]. For $k \in \mathbb{N}$, define the stopping times

$$\tau_k := \inf\left\{t \in [0, T] : \int_0^t \alpha e^{\int_s^t \delta_r\,dr} U_s\,ds + S_t^{-\delta} E[B|\mathcal{F}_t] \ge k\right\} \wedge T$$

and note that $(\tau_k)_{k\in\mathbb{N}}$ increases to T stationarily. By construction, the stopped processes $Y^{n;k} := (Y^n)^{\tau_k}$, $n \in \mathbb{N}$, are uniformly bounded by k. Setting $M^{n;k} := (M^n)^{\tau_k}$, we have

$$Y_t^{n;k} = Y_{\tau_k}^n - \int_t^T \mathbf{1}_{\{s\le\tau_k\}}\left(\delta_s Y_s^{n;k} - \alpha U_s^n\right)ds - \int_t^T \frac{1}{2\beta}\,d\langle M^{n;k}\rangle_s - \int_t^T dM_s^{n;k}.$$

We now take the supremum over n and apply the monotonic stability theorem (see e.g. Lemma 3.3 in Morlais [12]) to obtain, for each k, a solution (Y^k, M^k) to the BSDE

$$Y_t^k = \xi_k - \int_t^{\tau_k}\left(\delta_s Y_s^k - \alpha U_s\right)ds - \int_t^{\tau_k} \frac{1}{2\beta}\,d\langle M^k\rangle_s - \int_t^{\tau_k} dM_s^k \quad \text{with } \xi_k := \sup_{n\in\mathbb{N}} Y_{\tau_k}^n.$$

More precisely, that result shows that as $n \to \infty$, $M_T^{n;k}$ converges to M_T^k in L^2, so that both $M_t^{n;k}$ and $\langle M^{n;k}\rangle_t$ converge uniformly over $t \in [0, T]$ in probability to M_t^k and $\langle M^k\rangle_t$, respectively. Moreover, we have $\tau_k \le \tau_{k+1}$ by construction; hence $Y_{t\wedge\tau_k}^{n;k+1} = Y_t^{n;k}$, and so we have the localisation property

$$Y_{t\wedge\tau_k}^{k+1} = Y_t^k \quad \text{and} \quad M_{t\wedge\tau_k}^{k+1} = M_t^k.$$

So if we set $\tau_0 := 0$ and define the processes Y and M on $[0, T]$ by

$$Y_t := Y_0^1 + \sum_{k=1}^{\infty} Y_t^k \mathbf{1}_{(\tau_{k-1},\tau_k]}(t) \quad \text{and} \quad M_t := \sum_{k=1}^{\infty} M_t^k \mathbf{1}_{(\tau_{k-1},\tau_k]}(t),$$

the last BSDE can be rewritten as

$$Y_t = \xi_k - \int_t^{\tau_k} (\delta_s Y_s - \alpha U_s)\, ds - \int_t^{\tau_k} \frac{1}{2\beta}\, d\langle M\rangle_s - \int_t^{\tau_k} dM_s.$$

Finally we observe that P-a.s., $\tau_k = T$ for k large enough. This allows us to send $k \to \infty$ in the previous equation and hence to prove that (Y, M) is a solution to (14.7). Inequality (14.15) is satisfied by the process Y since it holds for each Y^n in view of Proposition 14.3.

(2) If U'_T and U are not necessarily nonnegative, we use a double approximation by introducing the quantities $U_t^{n,p} := U_t^+ \mathbf{1}_{\{\int_0^t |U_s|\, ds \le n\}} - U_t^- \mathbf{1}_{\{\int_0^t |U_s|\, ds \le p\}}$ and $U_T^{\prime,n,p} := (U'_T)^+ \wedge n - (U'_T)^- \wedge p$. Condition (14.14) is used here to extend (14.8) from the truncated to the general case and to ensure that $\overline{Y}, \underline{Y}$ remain well defined. In some more detail, we define τ_k (with $|U|$ and $|B|$) and $Y^{n,p;k}$ and $M^{n,p;k}$ analogously as before. Then $Y^{n,p;k}$ is increasing in n and decreasing in p, and it remains bounded by k. Arguing as before, we set $Y^k := \inf_p \sup_n Y^{n,p;k}$ to get the existence of M^k such that $\lim_{p\to\infty} \lim_{n\to\infty} M^{n,p;k}_{s\wedge\tau_k} = M_s^k$ and (Y^k, M^k) still solves the BSDE. The rest of the proof is unchanged. □

In connection with the stochastic control problem, it will be important to know when the stochastic exponential $\mathcal{E}(-\frac{1}{\beta}M)$ is a true martingale, where M comes from the solution of the BSDE (14.7).

Theorem 14.5 *Suppose that there exists a constant $\lambda > 1 + \frac{e^{\bar{\delta}T}-1}{e^{\bar{\delta}T}}$ such that*

$$E\big[e^{\lambda \frac{1}{\beta} e^{\bar{\delta}T} B_+}\big] + E\big[e^{\lambda \frac{1}{\beta} e^{\bar{\delta}T} B_-}\big] < \infty. \tag{14.16}$$

Then the stochastic exponential $\mathcal{E}(-\frac{1}{\beta}M)$ is bounded in $L\log L(P)$ and hence a (uniformly integrable) martingale on $[0, T]$.

Proof Since (14.16) clearly implies (14.14), the existence of a solution is ensured by Theorem 14.4. For any stopping time τ with values in $[0, T]$, the BSDE (14.7) gives

$$\begin{aligned}\mathcal{E}\Big(-\frac{1}{\beta}M\Big)_\tau &= \exp\Big(-\frac{1}{\beta}\Big(M_\tau + \frac{1}{2\beta}\langle M\rangle_\tau\Big)\Big)\\ &= \exp\Big(-\frac{1}{\beta}\Big(Y_\tau - Y_0 - \int_0^\tau \delta_s Y_s\, ds + \alpha \int_0^\tau U_s\, ds\Big)\Big).\end{aligned}$$

Hence it suffices to prove that there exists a constant $\lambda_1 > 1$ such that

$$E\bigg[\exp\bigg(-\lambda_1 \frac{1}{\beta}\bigg(Y_\tau - \int_0^\tau \delta_s Y_s\,ds + \alpha \int_0^\tau U_s\,ds\bigg)\bigg)\bigg] \le C,$$

where $C \in (0, \infty)$ is a constant independent of τ.

From the estimate (14.15) we have

$$\begin{aligned} Y_\tau + \alpha \int_0^\tau U_s\,ds - \int_0^\tau \delta_t Y_t\,dt &\ge S_\tau^{-\delta} \underline{Y}_\tau + \int_0^\tau \alpha\big(1 - e^{\int_s^\tau \delta_r\,dr}\big) U_s\,ds \\ &\quad - \int_0^\tau \delta_t \bigg(S_t^{-\delta} \overline{Y}_t - \int_0^t \alpha e^{\int_s^t \delta_r\,dr} U_s\,ds\bigg)\,dt. \end{aligned}$$

But Fubini's theorem gives $\int_0^\tau \delta_t \int_0^t \alpha e^{\int_s^t \delta_r\,dr} U_s\,ds\,dt = \int_0^\tau \alpha(e^{\int_s^\tau \delta_r\,dr} - 1) U_s\,ds$, and so we deduce that

$$Y_\tau - \int_0^\tau \delta_s Y_s\,ds + \alpha \int_0^\tau U_s\,ds \ge S_\tau^{-\delta} \underline{Y}_\tau - \int_0^\tau \delta_t S_t^{-\delta} \overline{Y}_t\,dt. \tag{14.17}$$

Now pick $p > 1$ with $\lambda > p > 1 + \frac{e^{\bar{\delta}T} - 1}{e^{\bar{\delta}T}}$ and set $\lambda_1 = \frac{\lambda}{p} > 1$. Using (14.17) and (14.9) and setting $L_t^* := \sup_{0 \le s \le t} E[B_+ | \mathcal{F}_s]$, we obtain with (14.6b) that

$$\begin{aligned} &E\bigg[\exp\bigg(-\lambda_1 \frac{1}{\beta}\bigg(Y_\tau - \int_0^\tau \delta_s Y_s\,ds + \alpha \int_0^\tau U_s\,ds\bigg)\bigg)\bigg] \\ &\le E\bigg[\exp\bigg(-\lambda_1 \frac{1}{\beta}\bigg(S_\tau^{-\delta} \underline{Y}_\tau - \int_0^\tau \delta_t S_t^{-\delta} \overline{Y}_t\,dt\bigg)\bigg)\bigg] \\ &\le E\bigg[\exp\big(\lambda_1 \log E[e^{\frac{1}{\beta} e^{\bar{\delta}T} B_-} | \mathcal{F}_\tau]\big) \exp\bigg(\lambda_1 \frac{1}{\beta} L_T^* \int_0^T \delta_t S_t^{-\delta}\,dt\bigg)\bigg] \\ &\le E\big[\big(E[e^{\frac{1}{\beta} e^{\bar{\delta}T} B_-} | \mathcal{F}_\tau]\big)^{\lambda_1} e^{\frac{\lambda_1}{\beta}(e^{\bar{\delta}T} - 1) L_T^*}\big]. \end{aligned}$$

Finally, we set $q := \frac{p}{p-1}$ and $r := (p-1)\frac{e^{\bar{\delta}T}}{e^{\bar{\delta}T} - 1} > 1$ and use Hölder's inequality, $\lambda_1 p = \lambda$, Jensen's inequality and Doob's inequality in L^r to get

$$\begin{aligned} &E\bigg[\exp\bigg(-\lambda_1 \frac{1}{\beta}\bigg(Y_\tau - \int_0^\tau \delta_s Y_s\,ds + \alpha \int_0^\tau U_s\,ds\bigg)\bigg)\bigg] \\ &\le E\big[\big(E[e^{\frac{1}{\beta} e^{\bar{\delta}T} B_-} | \mathcal{F}_\tau]\big)^{\lambda_1 p}\big]^{1/p} E\big[e^{\frac{\lambda_1 q}{\beta}(e^{\bar{\delta}T} - 1) L_T^*}\big]^{1/q} \\ &\le C_r' E\big[e^{\lambda \frac{1}{\beta} e^{\bar{\delta}T} B_-}\big]^{1/p} E\big[e^{\frac{\lambda_1 qr}{\beta}(e^{\bar{\delta}T} - 1) B_+}\big]^{\frac{1}{qr}} = C < \infty, \end{aligned}$$

because $\lambda_1 qr(e^{\bar{\delta}T} - 1) = \lambda e^{\bar{\delta}T}$. □

14.4 The BSDE on an Infinite Horizon

In this section, we use BSDE techniques to prove the existence and uniqueness of a solution to the infinite-horizon BSDE under suitable conditions.

Hypothesis 14.6 *Throughout this section, we impose the standing assumptions*

$$\mathbb{F} \text{ is a } \textit{continuous filtration}, \text{ i.e. all local } (P, \mathbb{F})\text{-martingales are continuous.} \tag{14.18a}$$

$$\delta \geq 0 \text{ is uniformly bounded (by } \bar{\delta}\text{, say).} \tag{14.18b}$$

Again, the assumptions on $U = (U_t)_{t \geq 0}$ will be specified later; U'_T does not appear here. The BSDE (14.3) under study is now

$$dY_t = (\delta_t Y_t - \alpha U_t)\,dt + \frac{1}{2\beta}\,d\langle M\rangle_t + dM_t, \qquad \lim_{t\to\infty} Y_t = 0 \quad P\text{-a.s.}, \tag{14.19}$$

and as before, a *solution* of (14.19) is a pair (Y, M) satisfying (14.19), where Y is a P-semimartingale and M is a locally P-square-integrable local P-martingale null at 0.

The first step in tackling (14.19) is to obtain a priori estimates for the finite-horizon version with terminal condition $Y_T = 0$. But in contrast to Sect. 14.3, we now need the bounds to be uniform in T, and so we need stronger assumptions on U.

Definition 14.7 We say that a random variable X is in $D^{\exp}$ if $E[e^{\lambda|X|}] < \infty$ for all $\lambda > 0$. A progressively measurable process $U = (U_t)_{t\geq 0}$ is in $D^{\exp}_{1,T}$ for $T \in (0, \infty]$ if $\int_0^T |U_s|\,ds$ is in $D^{\exp}$, and an RCLL process $Y = (Y_t)_{t\geq 0}$ is in $D^{\exp}_{0,T}$ for $T \in (0, \infty]$ if $Y^*_T := \sup_{0\leq t\leq T} |Y_t|$ is in $D^{\exp}$. (By convention, $Y^*_\infty := \sup_{t\geq 0} |Y_t|$.)

Let us now consider the BSDE

$$dY_t = (\delta_t Y_t - \alpha U_t)\,dt + \frac{1}{2\beta}\,d\langle M\rangle_t + dM_t, \qquad Y_T = 0. \tag{14.20}$$

Proposition 14.8 (A priori estimates) *Suppose that $\int_0^T |U_s|\,ds$ is a bounded random variable. Then there exists a unique solution (Y, M) to (14.20) such that Y is a bounded process. Moreover, we have the estimate*

$$|Y_t| \leq \beta \log E\left[\exp\left(\frac{1}{\beta}\int_t^T \alpha|U_s|\,ds\right)\middle|\mathcal{F}_t\right], \quad 0 \leq t \leq T. \tag{14.21}$$

Proof The existence and uniqueness of a solution with Y bounded follow as for Proposition 14.3 from Theorems 2.5 and 2.6 of Morlais [12]. Applying Tanaka's formula then first yields

$$d|Y_t| = \operatorname{sign}(Y_t)\,dY_t + dL_t = \operatorname{sign}(Y_t)\bigg((\delta_t Y_t - \alpha U_t)\,dt + \frac{1}{2\beta}\,d\langle M\rangle_t + dM_t\bigg) + dL_t,$$

where L is the local time at 0 of the continuous semimartingale Y. Next, applying Itô's formula to the bounded process $Z_t := \exp(\frac{1}{\beta}(|Y_t| + \int_0^t \alpha|U_s|\,ds))$, we obtain

$$\begin{aligned} dZ_t &= \frac{1}{\beta} Z_t \operatorname{sign}(Y_t)\bigg((\delta_t Y_t - \alpha U_t)\,dt + \frac{1}{2\beta}\,d\langle M\rangle_t + dM_t\bigg) \\ &\quad + \frac{1}{\beta} Z_t\,dL_t + \frac{1}{\beta} Z_t \alpha |U_t|\,dt + \frac{1}{2\beta^2} Z_t\,d\langle M\rangle_t \\ &\geq \frac{1}{\beta} Z_t \operatorname{sign}(Y_t)\,dM_t, \end{aligned}$$

in the sense that the difference of the terms on the two sides of the inequality is an increasing process. So Z is a submartingale, which gives

$$\exp\bigg(\frac{1}{\beta}\bigg(|Y_t| + \int_0^t \alpha|U_s|\,ds\bigg)\bigg) \leq E\bigg[\exp\bigg(\frac{1}{\beta}\int_0^T \alpha|U_s|\,ds\bigg)\bigg|\mathcal{F}_t\bigg]$$

since $Y_T = 0$, and (14.21) follows. □

From this a priori estimate and the localisation method we obtain the existence of a solution to the infinite-horizon BSDE (14.19).

Theorem 14.9 (Existence of solution) *Let us suppose that*

$$E\bigg[\exp\bigg(\frac{1}{\beta}\int_0^\infty \alpha|U_s|\,ds\bigg)\bigg] < \infty, \quad \textit{i.e.} \quad \exp\bigg(\frac{1}{\beta}\int_0^\infty \alpha|U_s|\,ds\bigg) \in L^1. \tag{14.22}$$

Then the BSDE (14.19) *admits a solution* (Y, M) *which satisfies*

$$|Y_t| \leq \beta \log E\bigg[\exp\bigg(\frac{1}{\beta}\int_t^\infty \alpha|U_s|\,ds\bigg)\bigg|\mathcal{F}_t\bigg], \quad t \geq 0. \tag{14.23}$$

If $\exp(\frac{1}{\beta}\int_0^\infty \alpha|U_s|\,ds)$ *is in* L^r *for some* $r > 1$, *then so is* $\exp(\frac{1}{\beta}Y^*_\infty)$. *If* $U \in D^{\exp}_{1,\infty}$, *then* $Y \in D^{\exp}_{0,\infty}$.

Proof (1) We first assume that U is nonnegative and set $U^n_t := U_t \mathbf{1}_{\{\int_0^t U_s\,ds \leq n\}}$ for each $n \in \mathbb{N}$. According to Proposition 14.8, the BSDE

$$dY_t = \big(\delta_t Y_t - \alpha U^n_t\big)\,dt + \frac{1}{2\beta}\,d\langle M\rangle_t + dM_t, \qquad Y_n = 0,$$

on $[0, n]$ admits a unique solution (Y^n, M^n) with Y^n bounded, and by (14.21),

$$\big|Y^n_t\big| \leq \beta \log E\bigg[\exp\bigg(\frac{1}{\beta}\int_t^\infty \alpha|U_s|\,ds\bigg)\bigg|\mathcal{F}_t\bigg], \quad t \in [0, n].$$

If we set $Y_t^n = 0$ and $M_t^n = M_n^n$ for $t > n$, then (Y^n, M^n) also satisfies on $[0, n+1]$ the BSDE

$$dY_t = \left(\delta_t Y_t - \alpha \mathbf{1}_{\{t \le n\}} U_t^n\right) dt + \frac{1}{2\beta} d\langle M\rangle_t + dM_t, \qquad Y_{n+1} = 0.$$

Because U is nonnegative, $\mathbf{1}_{\{t\le n\}} U_t^n \le U_t^{n+1}$ for $t \in [0, n+1]$, and so the sequence (Y^n) is nondecreasing by the comparison theorem. For each $k \in \mathbb{N}$, we define the stopping time

$$\tau_k := \inf\left\{t \ge 0 : \beta \log E\left[\exp\left(\frac{1}{\beta}\int_t^\infty \alpha |U_s|\, ds\right)\middle|\mathcal{F}_t\right] \ge k\right\} \wedge k.$$

Introducing the stopped processes $Y^{n;k} := (Y^n)^{\tau_k}$ and $M^{n;k} := (M^n)^{\tau_k}$, we can argue exactly as in the proof of Theorem 14.4 to construct processes Y and M, now on $[0, \infty)$, satisfying for each T the BSDE

$$Y_t = Y_T - \int_t^T (\delta_s Y_s - \alpha U_s)\, ds - \int_t^T \frac{1}{2\beta} d\langle M\rangle_s - \int_t^T dM_s.$$

Since each Y^n satisfies estimate (14.23) by Proposition 14.8, it follows from the construction that so does Y, and this implies, due to (14.22), that

$$\lim_{t\to\infty} Y_t = 0 \quad P\text{-a.s.}$$

(2) In the general case where U need not be nonnegative, we use the double approximation $U_t^{n,p} := U_t^+ \mathbf{1}_{\{\int_0^t |U_s|\, ds \le n\}} \mathbf{1}_{\{t\le n\}} - U_t^- \mathbf{1}_{\{\int_0^t |U_s|\, ds \le p\}} \mathbf{1}_{\{t \le p\}}$, $t \ge 0$, and denote by $(Y^{n,p}, M^{n,p})$ the solution to

$$dY_t = \left(\delta_t Y_t - \alpha U_t^{n,p}\right) dt + \frac{1}{2\beta} d\langle M\rangle_t + dM_t, \qquad Y_{n\vee p} = 0.$$

Then the proof goes like for Theorem 14.4, using that $Y^{n,p}$ increases in n and decreases in p.

(3) The integrability assertions about Y follow from (14.23) and Doob's inequality. □

To get a uniqueness result for the infinite-horizon BSDE (14.19), we need a stronger assumption.

Theorem 14.10 (Uniqueness of solution) *Suppose that U is in $D_{1,\infty}^{\exp}$. Then the BSDE* (14.19) *admits a unique solution (Y, M) with $Y \in D_{0,\infty}^{\exp}$.*

Proof The existence is clear from Theorem 14.9. For uniqueness, let (Y, M) and (Y', M') be two such solutions and note that the martingale part is always unique by the uniqueness of the canonical decomposition of a special semimartingale. Fix $\theta \in (0,1)$ and set $\hat{Y} := Y - \theta Y'$ and $\hat{M} := M - \theta M'$. Then

$$d\hat{Y}_t = \big(\delta_t \hat{Y}_t - \alpha(1-\theta)U_t\big)\,dt + \frac{1}{2\beta}\,d\big(\langle M\rangle_t - \theta\langle M'\rangle_t\big) + d\hat{M}_t. \tag{14.24}$$

Noting that convexity gives

$$d\langle M\rangle_t = d\Big\langle \theta M' + (1-\theta)\frac{\hat{M}}{1-\theta}\Big\rangle_t \le \theta\, d\langle M'\rangle_t + \frac{1}{1-\theta}\,d\langle \hat{M}\rangle_t, \tag{14.25}$$

we rewrite (14.24) as

$$\begin{aligned} d\hat{Y}_t &= \big(\delta_t \hat{Y}_t - \alpha(1-\theta)U_t\big)\,dt + d\hat{M}_t \\ &\quad + \frac{1}{2\beta}\,d\Big(\langle M\rangle_t - \theta\langle M'\rangle_t - \frac{1}{1-\theta}\langle\hat{M}\rangle_t\Big) + \frac{1}{2\beta(1-\theta)}\,d\langle\hat{M}\rangle_t. \end{aligned} \tag{14.26}$$

Now Tanaka's formula yields $d\hat{Y}_t^- = -\mathbf{1}_{\{\hat{Y}_t \le 0\}}\,d\hat{Y}_t + \frac{1}{2}\,d\hat{L}_t$, where $\hat{L}$ is the local time at 0 of the process $\hat{Y}$. Applying next Itô's formula to the process

$$Z_t := \exp\bigg(\frac{1}{\beta(1-\theta)}\Big(\hat{Y}_t^- + \int_0^t \alpha(1-\theta)|U_s|\,ds\Big)\bigg), \quad t \ge 0,$$

we get

$$\begin{aligned} dZ_t &= \frac{1}{\beta(1-\theta)} Z_t \Big(-\mathbf{1}_{\{\hat{Y}_t \le 0\}}\,d\hat{Y}_t + \frac{1}{2}\,d\hat{L}_t + \alpha(1-\theta)|U_t|\,dt\Big) \\ &\quad + \frac{1}{2\beta^2(1-\theta)^2} Z_t \mathbf{1}_{\{\hat{Y}_t \le 0\}}\,d\langle\hat{M}\rangle_t \\ &\ge -\frac{1}{\beta(1-\theta)} Z_t \mathbf{1}_{\{\hat{Y}_t \le 0\}}\,d\hat{M}_t, \end{aligned}$$

where the last inequality uses (14.26) and (14.25). Thus Z is a local submartingale, and so there exists an increasing sequence of stopping times $\tau_n \nearrow \infty$ such that

$$\exp\Big(\frac{1}{\beta(1-\theta)}\hat{Y}^-_{t\wedge\tau_n}\Big) \le E\bigg[\exp\Big(\frac{1}{\beta(1-\theta)}\hat{Y}^-_{T\wedge\tau_n} + \frac{1}{\beta}\int_{t\wedge\tau_n}^{T\wedge\tau_n} \alpha|U_s|\,ds\Big)\bigg|\mathcal{F}_t\bigg].$$

Because $\hat{Y} \in D^{\exp}_{0,\infty}$, $U \in D^{\exp}_{1,\infty}$ and $\lim_{t\to\infty}\hat{Y}_t = 0$, we obtain for $n \to \infty$ and $T \to \infty$ that

$$\exp\Big(\frac{1}{\beta(1-\theta)}\hat{Y}_t^-\Big) \le E\bigg[\exp\Big(\frac{1}{\beta}\int_t^\infty \alpha|U_s|\,ds\Big)\bigg|\mathcal{F}_t\bigg],$$

which is equivalent to

$$(Y_t - \theta Y_t')^- = \hat{Y}_t^- \le \beta(1-\theta)\log E\bigg[\exp\Big(\frac{1}{\beta}\int_t^\infty \alpha|U_s|\,ds\Big)\bigg|\mathcal{F}_t\bigg].$$

Letting $\theta \to 1$, we deduce that $Y_t \geq Y'_t$, and since a symmetrical argument gives the reverse inequality, the proof is complete. □

As in Sect. 14.3, we again want to know when the stochastic exponential $\mathcal{E}(-\frac{1}{\beta}M)$ is a true martingale, where M now comes from the solution of the BSDE (14.19). However, we only expect to obtain this here on the open interval $[0,\infty)$, and the proof below shows why $T=\infty$ causes a difficulty.

Theorem 14.11 *Suppose that U is in $D^{\exp}_{1,\infty}$ and denote by (Y,M) the solution to* (14.19) *from Theorem* 14.9. *Then for every finite T and every $r<\infty$, the stochastic exponential $(\mathcal{E}(-\frac{1}{\beta}M)_t)_{0\leq t\leq T}$ is bounded in L^r, and so $\mathcal{E}(-\frac{1}{\beta}M)$ is a martingale on $[0,\infty)$.*

Proof Fix $T\in(0,\infty)$ and let τ be a stopping time with values in $[0,T]$. As in the proof of Theorem 14.5, the BSDE (14.19) gives

$$\begin{aligned}\mathcal{E}\Big(-\frac{1}{\beta}M\Big)_\tau &= \exp\Big(-\frac{1}{\beta}\Big(Y_\tau - Y_0 - \int_0^\tau \delta_s Y_s\,ds + \alpha\int_0^\tau U_s\,ds\Big)\Big)\\ &\leq \exp\Big(\frac{1}{\beta}Y_0\Big)\exp\Big(\frac{1}{\beta}Y^*_T(1+\bar\delta T)+\frac{1}{\beta}\int_0^\infty \alpha|U_s|\,ds\Big),\end{aligned}$$

and the conclusion follows from Theorem 14.9 because $Y\in D^{\exp}_{0,\infty}$. □

14.5 The Stochastic Control Problem on an Infinite Horizon

In this section, we prove the existence and uniqueness of a solution to the infinite-horizon stochastic control problem.

Let us first give a precise formulation. We recall from Sect. 14.2 the underlying filtered probability space $(\Omega,\mathcal{F},\mathbb{F},P)$ with $\mathbb{F}=(\mathcal{F}_t)_{0\leq t\leq\infty}$, the parameters $\alpha\geq 0$, $\beta>0$ and the processes $\delta=(\delta_t)_{t\geq 0}$ and $U=(U_t)_{t\geq 0}$. We denote by $\mathcal{Q}$ the set of all probability measures $Q\overset{\text{loc}}{\ll}P$ and by $Z^Q=(Z^Q_t)_{t\geq 0}$ an RCLL version of the density process of Q with respect to P. Since $\mathcal{F}_0$ is trivial, we have

$$\begin{aligned}&\{Z^Q\mid Q\in\mathcal{Q}\}\\ &\quad\subseteq\{\text{all RCLL } P\text{-martingales } Z=(Z_t)_{t\geq 0} \text{ with } Z\geq 0 \text{ and } Z_0=1\}=:\mathcal{Z}.\end{aligned}$$

Now define for any $t\geq 0$ and $Z\in\mathcal{Z}$, in analogy to (14.4) and (14.5),

$$\tilde{\mathcal{U}}^\delta_{t,\infty}(Z):=\alpha\int_t^\infty \frac{S^\delta_s}{S^\delta_t}\frac{Z_s}{Z_t}U_s\,ds,$$

$$\tilde{\mathcal{R}}^\delta_{t,\infty}(Z):=\int_t^\infty \delta_s\frac{S^\delta_s}{S^\delta_t}\frac{Z_s}{Z_t}\log\frac{Z_s}{Z_t}\,ds$$

and the cost functional

$$\tilde{c}_\infty(Z) := \tilde{\mathcal{U}}^\delta_{0,\infty}(Z) + \beta\tilde{\mathcal{R}}^\delta_{0,\infty}(Z).$$

The stochastic control problem studied here is to minimise the functional

$$Z \mapsto \Gamma_\infty(Z) := E_P\big[\tilde{c}_\infty(Z)\big]$$

over a subset $\mathcal{Z}_f$ of $\mathcal{Z}$, defined below. Note that by the minimum principle for supermartingales, $Z \in \mathcal{Z}$ remains 0 if it ever hits 0; so both summands of $\tilde{c}_\infty(Z)$ are well defined.

Hypothesis 14.12 *Throughout this section, we impose the standing assumptions*

There exists some $T_0 \in (0,\infty)$ such that for all $\gamma > 0$,

$$E_P\left[\exp\left(\gamma\int_0^{T_0}|U_s|\,ds\right)\right] + E_P\left[\int_{T_0}^\infty \exp\big(\gamma|U_s|\big)\mathbf{1}_{\{U_s\neq 0\}}\,ds\right] < \infty. \quad (14.27a)$$

$$0 < \underline{\delta} \leq \delta_t \leq \bar{\delta} < \infty, \text{ uniformly in } (t,\omega), \text{ for constants } \underline{\delta}, \bar{\delta}. \quad (14.27b)$$

The first condition in (14.27a) says that U is in $D^{\exp}_{1,T_0}$; we remark that the indicator function in the second term fixes an obvious oversight in (4.4) of Bordigoni [1]. Condition (14.27b) is natural for an infinite-horizon problem. Note that we do not assume here that $\mathbb{F}$ is a continuous filtration.

Definition 14.13 $\mathcal{Z}_f$ denotes the set of all martingales $Z \in \mathcal{Z}$ satisfying $E_P[\tilde{\mathcal{R}}^\delta_{0,\infty}(Z)] < \infty$.

Our stochastic control problem is slightly more general than the one studied in Chap. 4 of Bordigoni [1] since we do not insist on working on the canonical (Skorokhod) path space. The presentation here is linked to Bordigoni [1] and to the slightly different formulation in Sect. 14.2 as follows. For any Q with $Z^Q \in \mathcal{Z}_f$, we have under (14.27a), for any $t \geq 0$,

$$E_P\big[\tilde{\mathcal{R}}^\delta_{t,\infty}(Z^Q)\big|\mathcal{F}_t\big] = E_P\left[\int_t^\infty \delta_s \frac{S^\delta_s}{S^\delta_t}\frac{Z^Q_s}{Z^Q_t}\log\frac{Z^Q_s}{Z^Q_t}\,ds\,\middle|\,\mathcal{F}_t\right] = E_Q\big[\mathcal{R}^\delta_{t,\infty}(Q)\big|\mathcal{F}_t\big],$$

$$E_P\big[\tilde{\mathcal{U}}^\delta_{t,\infty}(Z^Q)\big|\mathcal{F}_t\big] = E_P\left[\int_t^\infty \frac{S^\delta_s}{S^\delta_t}\frac{Z^Q_s}{Z^Q_t}U_s\,ds\,\middle|\,\mathcal{F}_t\right] = E_Q\big[\mathcal{U}^\delta_{t,\infty}\big|\mathcal{F}_t\big];$$

this is proved in Lemma 4.6 and Remark 4.10 of Bordigoni [1], essentially by using Bayes' rule. In particular, this shows that

$$\Gamma_\infty(Z^Q) := E_P\big[\tilde{c}_\infty(Z^Q)\big] = E_Q\big[c_\infty(Q)\big] =: \Gamma_\infty(Q).$$

Expressing everything under P and working with martingales $Z \in \mathcal{Z}_f$ turns out to be a bit more flexible than working with probability measures $Q \overset{\text{loc}}{\ll} P$. From now on, all expectations without subscript are under P.

Remark 14.14 Under (14.27b), $\int_t^\infty \delta_s S_s^\delta \, ds = S_t^\delta$ for every $t \geq 0$, and hence also

$$E\left[\int_t^\infty \delta_s \frac{S_s^\delta}{S_t^\delta} \frac{Z_s}{Z_t} \, ds \middle| \mathcal{F}_t\right] = 1 \quad \text{for every } t \geq 0 \text{ and any } Z \in \mathcal{Z}.$$

We start with some auxiliary estimates. These are true for any $Z \in \mathcal{Z}$, with the understanding that we set $E[\tilde{\mathcal{R}}_{0,\infty}^\delta(Z)] = +\infty$ for $Z \in \mathcal{Z} \setminus \mathcal{Z}_f$.

Lemma 14.15 *For every $Z \in \mathcal{Z}$ and every $T \in (0, \infty)$,*

$$E[Z_T \log Z_T] \leq \frac{1}{\underline{\delta}} e^{\bar{\delta}(T+1)} \left(E\left[\tilde{\mathcal{R}}_{0,\infty}^\delta(Z)\right] + e^{-1}\right).$$

Proof Since $t \mapsto E[Z_t \log Z_t]$ is increasing and $z \log z \geq -e^{-1}$, Fubini and (14.27b) give

$$\begin{aligned} E[Z_T \log Z_T] &\leq E\left[\int_T^{T+1} \frac{\delta_s}{\underline{\delta}} S_s^\delta e^{\bar{\delta} s} \left(Z_s \log Z_s + e^{-1}\right) ds\right] \\ &\leq \frac{1}{\underline{\delta}} e^{\bar{\delta}(T+1)} E\left[\int_0^\infty \delta_s S_s^\delta \left(Z_s \log Z_s + e^{-1}\right) ds\right] \\ &= \frac{1}{\underline{\delta}} e^{\bar{\delta}(T+1)} \left(E\left[\tilde{\mathcal{R}}_{0,\infty}^\delta(Z)\right] + e^{-1}\right) \end{aligned}$$

by Remark 14.14. □

Proposition 14.16 *There is a constant $C < \infty$ such that for every $Z \in \mathcal{Z}$,*

$$E\left[\int_0^\infty S_s^\delta Z_s |U_s| \, ds\right] \leq C\left(1 + E\left[\tilde{\mathcal{R}}_{0,\infty}^\delta(Z)\right]\right) \leq C\left(1 + \Gamma_\infty(Z)\right). \tag{14.28}$$

Proof This is a simplified version of the proofs for Lemma 4.9 and Proposition 4.11 in Bordigoni [1]. In the sequel, we use several times the elementary inequality $xy \leq e^x + y \log y$ for $x \in \mathbb{R}$, $y \geq 0$, typically applied to $xy = \gamma x \frac{1}{\gamma} y$. Starting with $\tilde{c}_\infty(Z)$, we split $\tilde{\mathcal{U}}_{0,\infty}^\delta(Z)$ into an integral from 0 to T_0 and another from T_0 to ∞ to write first

$$\begin{aligned} E\left[\int_0^{T_0} S_s^\delta Z_s |U_s| \, ds\right] &\leq E\left[Z_{T_0} \int_0^{T_0} |U_s| \, ds\right] \\ &\leq E\left[\exp\left(\gamma \int_0^{T_0} |U_s| \, ds\right)\right] + \frac{1}{\gamma} E\left[Z_{T_0}\left(\log Z_{T_0} + |\log \gamma|\right)\right]. \end{aligned} \tag{14.29}$$

Next, we have due to $S^\delta \le 1$ and Remark 14.14 that

$$
\begin{aligned}
E\bigg[\int_{T_0}^{\infty} S_s^\delta Z_s |U_s|\, ds\bigg]
&\le E\bigg[\int_{T_0}^{\infty} e^{\gamma |U_s|} \mathbf{1}_{\{U_s \neq 0\}}\, ds\bigg] + E\bigg[\int_{T_0}^{\infty} \frac{1}{\gamma} S_s^\delta Z_s \log\Big(\frac{1}{\gamma} S_s^\delta Z_s\Big) \mathbf{1}_{\{U_s \neq 0\}}\, ds\bigg] \\
&\le E\bigg[\int_{T_0}^{\infty} e^{\gamma |U_s|} \mathbf{1}_{\{U_s \neq 0\}}\, ds\bigg] + \frac{1}{\gamma} E\bigg[\int_0^{\infty} \frac{\delta_s}{\underline{\delta}} S_s^\delta \big(Z_s \log Z_s + e^{-1}\big)\, ds\bigg] \\
&\quad + \frac{1}{\gamma} |\log \gamma| E\bigg[\int_0^{\infty} \frac{\delta_s}{\underline{\delta}} S_s^\delta Z_s\, ds\bigg] \\
&= E\bigg[\int_{T_0}^{\infty} e^{\gamma |U_s|} \mathbf{1}_{\{U_s \neq 0\}}\, ds\bigg] + \frac{1}{\gamma \underline{\delta}} \big(E\big[\tilde{\mathcal{R}}_{0,\infty}^\delta(Z)\big] + e^{-1} + |\log \gamma|\big).
\end{aligned} \tag{14.30}
$$

Combining (14.29) with Lemma 14.15 and (14.30) gives the left inequality in (14.28) in the form

$$
E\bigg[\int_0^{\infty} S_s^\delta Z_s |U_s|\, ds\bigg] \le C + E\big[\tilde{\mathcal{R}}_{0,\infty}^\delta(Z)\big] \frac{1}{\gamma \underline{\delta}} \big(e^{\bar{\delta}(T_0+1)} + 1\big), \tag{14.31}
$$

where the constant C depends on γ and also on U via (14.27a). By definition, then,

$$
\begin{aligned}
\Gamma_\infty(Z) &= E\big[\tilde{\mathcal{U}}_{0,\infty}^\delta(Z)\big] + \beta E\big[\tilde{\mathcal{R}}_{0,\infty}^\delta(Z)\big] \\
&\ge -\alpha E\bigg[\int_0^{\infty} S_s^\delta Z_s |U_s|\, ds\bigg] + \beta E\big[\tilde{\mathcal{R}}_{0,\infty}^\delta(Z)\big] \\
&\ge -\alpha C + E\big[\tilde{\mathcal{R}}_{0,\infty}^\delta(Z)\big] \bigg(\beta - \frac{\alpha}{\gamma \underline{\delta}} \big(1 + e^{\bar{\delta}(T_0+1)}\big)\bigg),
\end{aligned}
$$

and so the right inequality in (14.28) follows by taking γ large enough and choosing a new constant appropriately. □

The above argument also shows that $\Gamma_\infty(Z) \le C(1 + E[\tilde{\mathcal{R}}_{0,\infty}^\delta(Z)])$ for all $Z \in \mathcal{Z}$, with a suitable constant. Another direct consequence is the following result that will be used later.

Corollary 14.17 *For every $T \ge T_0$, every $Z \in \mathcal{Z}$ and every $\gamma > 0$,*

$$
\begin{aligned}
E\bigg[\int_T^{\infty} S_s^\delta Z_s U_s\, ds\bigg] &\le E\bigg[\int_T^{\infty} e^{\gamma |U_s|} \mathbf{1}_{\{U_s \neq 0\}}\, ds\bigg] \\
&\quad + \frac{1}{\gamma \underline{\delta}} \big(E\big[\tilde{\mathcal{R}}_{0,\infty}^\delta(Z)\big] + e^{-1} + |\log \gamma|\big).
\end{aligned}
$$

With these preparations, we are ready to prove the existence and uniqueness of a solution to our stochastic control problem.

Theorem 14.18 *Under Hypothesis* 14.12, *there exists a unique* $Z^* \in \mathcal{Z}_f$ *that minimises the cost functional* $Z \mapsto \Gamma_\infty(Z)$ *over all* $Z \in \mathcal{Z}_f$.

Proof This again follows closely the arguments in Bordigoni [1]; see there the proof of Theorem 4.15. Since we optimise over Z instead of Q, we need not work on path space and can simplify some arguments.

First of all, the uniqueness is clear because $Z \mapsto \Gamma_\infty(Z)$ is strictly convex like $z \mapsto z \log z$. The existence is proved in several steps.

(1) Since $Z \equiv 1$ is in $\mathcal{Z}_f$ and $E[\tilde{\mathcal{R}}^\delta_{0,\infty}(1)] = 0$, (14.27a) and Proposition 14.16 imply that $-\infty < \inf_{Z \in \mathcal{Z}_f} \Gamma_\infty(Z) < \infty$. So we can take a sequence $(Z^n)_{n\in\mathbb{N}}$ in $\mathcal{Z}_f$ such that $\Gamma_\infty(Z^n)$ decreases to $\inf_{Z\in\mathcal{Z}_f} \Gamma_\infty(Z)$ as $n \to \infty$. Combining the well-known Komlós-type result in Lemma A1.1 in Delbaen/Schachermayer [7] with a diagonalisation argument produces a sequence $(\bar{Z}^n)_{n\in\mathbb{N}}$ with $\bar{Z}^n \in \operatorname{conv}(Z^n, Z^{n+1}, \dots)$ for all n and such that with probability 1,

$$\lim_{n\to\infty} \bar{Z}^n_r =: \bar{Z}^\infty_r \quad \text{exists in } [0,\infty] \text{ for all } r \in \mathbb{Q}^+.$$

Since each $\bar{Z}^n$ is like the Z^n a martingale ≥ 0 with expectation 1, Fatou's lemma yields that each $\bar{Z}^\infty_r$ is integrable and $(\bar{Z}^\infty_r)_{r\in\mathbb{Q}^+}$ is a supermartingale. By a standard argument (see Dellacherie/Meyer [8], Theorem VI.2), we can therefore extend $(\bar{Z}^\infty_r)_{r\in\mathbb{Q}^+}$ to a process $Z^* = (Z^*_t)_{t\geq 0}$ with RCLL trajectories and such that Z^* is a supermartingale ≥ 0 (now over $[0,\infty)$ instead of $\mathbb{Q}^+$). In fact, we can take $Z^*_t := \lim_{r\searrow t,\, r\in\mathbb{Q}^+} \bar{Z}^\infty_r$.

(2) In order to show that Z^* is even a martingale and in $\mathcal{Z}_f$, we first use Lemma 14.15 to obtain for each $r \in \mathbb{Q}^+$ that

$$\sup_{n\in\mathbb{N}} E\big[\bar{Z}^n_r \log \bar{Z}^n_r\big] \leq \frac{1}{\underline{\delta}} e^{\bar{\delta}(r+1)} \Big(\sup_{n\in\mathbb{N}} E\big[\tilde{\mathcal{R}}^\delta_{0,\infty}(\bar{Z}^n)\big] + e^{-1} \Big) < \infty,$$

because by Proposition 14.16 and the convexity of $Z \mapsto \Gamma_\infty(Z)$,

$$\begin{aligned} E\big[\tilde{\mathcal{R}}^\delta_{0,\infty}(\bar{Z}^n)\big] &\leq C\big(1 + \Gamma_\infty(\bar{Z}^n)\big) \\ &\leq C\Big(1 + \sup_{m\geq n} \Gamma_\infty(Z^m)\Big) \leq C\big(1 + \Gamma_\infty(Z^1)\big) < \infty. \end{aligned}$$

So $(\bar{Z}^n_r)_{n\in\mathbb{N}}$ is uniformly integrable for each $r \in \mathbb{Q}^+$, and this implies

$$E\big[\bar{Z}^\infty_r\big] = \lim_{n\to\infty} E\big[\bar{Z}^n_r\big] = 1 \quad \text{for all } r \in \mathbb{Q}^+,$$

which means that the supermartingale $\bar{Z}^\infty$ is a martingale (over $\mathbb{Q}^+$). Using Doob's maximal inequality and the fact that $(\bar{Z}^n_m)_{n\in\mathbb{N}}$ converges to Z^*_m in L^1 for every $m \in \mathbb{N}$ next shows that with probability 1, $(\bar{Z}^n)_{n\in\mathbb{N}}$ converges to Z^* uniformly on compact subsets of $[0,\infty)$), and the same uniform integrability argument as above then yields that also Z^* is a martingale (over $[0,\infty)$), hence in $\mathcal{Z}$. Finally, Fatou's lemma, Remark 14.14 and Proposition 14.16 give

$$\begin{aligned} E\big[\tilde{\mathcal{R}}^{\delta}_{0,\infty}(Z^*)\big] &= E\bigg[\int_0^\infty \delta_s S^\delta_s Z^*_s \log Z^*_s\, ds\bigg] \\ &\le \liminf_{n\to\infty} E\big[\tilde{\mathcal{R}}^{\delta}_{0,\infty}(\bar{Z}^n)\big] \le \sup_{n\in\mathbb{N}} C\big(1+\Gamma_\infty(\bar{Z}^n)\big) < \infty, \end{aligned}$$

so that Z^* is in $\mathcal{Z}_f$.

(3) To show that Z^* is optimal, we want to prove that $Z \mapsto \Gamma_\infty(Z)$ is lower semicontinuous along the sequence $(\bar{Z}^n)_{n\in\mathbb{N}}$, because we then get by convexity

$$\Gamma_\infty(Z^*) \le \liminf_{n\to\infty} \Gamma_\infty(\bar{Z}^n) \le \liminf_{n\to\infty} \Gamma_\infty(Z^n) = \inf_{Z\in\mathcal{Z}_f} \Gamma_\infty(Z).$$

We have just seen in step (2) that $E[\tilde{\mathcal{R}}^{\delta}_{0,\infty}(Z^*)] \le \liminf_{n\to\infty} E[\tilde{\mathcal{R}}^{\delta}_{0,\infty}(\bar{Z}^n)]$, so that it only remains to prove the analogous inequality for the part with $\tilde{\mathcal{U}}^{\delta}_{0,\infty}(Z)$. Now $U \in D^{\exp}_{1,T_0}$ by (14.27a), and so we can use the finite-horizon results in Bordigoni/Matoussi/Schweizer [2] to obtain

$$\begin{aligned} E\bigg[\int_0^{T_0} S^\delta_s Z^*_s U_s\, ds\bigg] &= E\bigg[Z^*_{T_0}\int_0^{T_0} S^\delta_s U_s\, ds\bigg] \\ &\le \liminf_{n\to\infty} E\bigg[\bar{Z}^n_{T_0}\int_0^{T_0} S^\delta_s U_s\, ds\bigg] = \liminf_{n\to\infty} E\bigg[\int_0^{T_0} S^\delta_s \bar{Z}^n_s U_s\, ds\bigg]; \end{aligned}$$

see step (4) in the proof of Theorem 9 in Bordigoni/Matoussi/Schweizer [2]. We then split the remaining integral from T_0 to ∞ into one integral from T_0 to $T \ge T_0$ and another from T to ∞. The integral over the finite interval $(T_0, T]$ is again treated as above, using that (14.27a) also gives $U \in D^{\exp}_{1,T}$ by Jensen's inequality. Finally, Corollary 14.17 gives

$$\begin{aligned} E\bigg[\int_T^\infty S^\delta_s \bar{Z}^n_s U_s\, ds\bigg] &\le E\bigg[\int_T^\infty e^{\gamma|U_s|}\mathbf{1}_{\{U_s\neq 0\}}\, ds\bigg] \\ &\quad + \frac{1}{\gamma\underline{\delta}}\Big(\sup_{n\in\mathbb{N}} E\big[\tilde{\mathcal{R}}^{\delta}_{0,\infty}(\bar{Z}^n)\big] + e^{-1} + |\log\gamma|\Big) \end{aligned}$$

for all $\gamma > 0$ and all $n \in \mathbb{N}$, and the same estimate holds for Z^* instead of $\bar{Z}^n$ as well. Choosing first γ large to make the second summand above small and then, using (14.27a), T large to get the first summand small as well, we deduce that

$$\lim_{T\to\infty} \sup\bigg\{E\bigg[\int_T^\infty S^\delta_s Z_s U_s\, ds\bigg] : Z = Z^* \text{ or } Z = \bar{Z}^n \text{ for some } n\in\mathbb{N}\bigg\} = 0,$$

and so we obtain after putting everything together that

$$E\left[\int_0^\infty S_s^\delta Z_s^* U_s\,ds\right] \le \liminf_{n\to\infty} E\left[\int_0^\infty S_s^\delta \bar{Z}_s^n U_s\,ds\right].$$

This completes the proof. □

As in the finite-horizon case treated in Bordigoni/Matoussi/Schweizer [2], one can show that the optimal Z^* from Theorem 14.18 is strictly positive. If there exists Q^* with density process Z^* (e.g. as in Bordigoni [1] if one works on path space), this translates into saying that $Q^* \overset{\text{loc}}{\approx} P$. The proof of positivity can be found in Bordigoni [1], Theorem 4.18, and largely parallels that of Theorem 12 in Bordigoni/Matoussi/Schweizer [2].

Also as in Bordigoni/Matoussi/Schweizer [2], one can show that the *martingale optimality principle* holds in our setting; see Proposition 4.19 and Corollary 4.20 in Bordigoni [1] for details of this standard argument. As a consequence, the optimal Z^* from Theorem 14.18 is also conditionally optimal at any time t or even stopping time τ. To properly formulate this, we denote by $V = (V_t)_{t\ge 0}$ an RCLL version of the process

$$V_t := \operatorname*{ess\,inf}_{Z\in\mathcal{Z}_f} E\big[\tilde{\mathcal{U}}_{t,\infty}^\delta(Z) + \beta\tilde{\mathcal{R}}_{t,\infty}^\delta(Z)\big|\mathcal{F}_t\big] =: \operatorname*{ess\,inf}_{Z\in\mathcal{Z}_f} J_t(Z), \quad t\ge 0. \tag{14.32}$$

Then conditional optimality says that

$$V_t = J_t\big(Z^*\big) \quad P\text{-a.s. for all } t\ge 0,$$

and we now use this to describe the behaviour of V_t as $t\to\infty$.

Proposition 14.19 *Under Hypothesis* 14.12,

$$\lim_{t\to\infty} V_t = 0 \quad P\text{-a.s.}$$

Proof This is analogous to the proof of Lemma 4.22 in Bordigoni [1]. Since $Z\equiv 1$ is in $\mathcal{Z}_f$, S^δ is decreasing and $E[\tilde{\mathcal{R}}_{t,\infty}^\delta(Z)|\mathcal{F}_t]\ge 0$, we have

$$V_t = J_t\big(Z^*\big) \le J_t(1) \le \alpha E\left[\int_t^\infty |U_s|\,ds\,\middle|\,\mathcal{F}_t\right],$$

which yields $\limsup_{t\to\infty} V_t \le 0$ due to (14.27a). To get a lower bound for V_t, we use analogous arguments as in the proof of Proposition 14.16 to first obtain

$$\left|E\left[\int_t^\infty \frac{S_s^\delta}{S_t^\delta}\frac{Z_s^*}{Z_t^*}U_s\,ds\,\middle|\,\mathcal{F}_t\right]\right| \le E\left[\int_t^\infty e^{\gamma|U_s|}\mathbf{1}_{\{U_s\neq 0\}}\,ds\,\middle|\,\mathcal{F}_t\right] + \frac{1}{\gamma\underline{\delta}}\big(E\big[\tilde{\mathcal{R}}_{t,\infty}^\delta\big(Z^*\big)\big|\mathcal{F}_t\big] + e^{-1} + |\log\gamma|\big).$$

Hence,

$$V_t = J_t(Z^*) \geq -\alpha E\left[\int_t^\infty e^{\gamma|U_s|}\mathbf{1}_{\{U_s\neq 0\}}\,ds\middle|\mathcal{F}_t\right]$$
$$-\frac{\alpha}{\gamma\underline{\delta}}(e^{-1}+|\log\gamma|)+\left(\beta-\frac{\alpha}{\gamma\underline{\delta}}\right)E[\tilde{\mathcal{R}}^\delta_{t,\infty}(Z^*)|\mathcal{F}_t],$$

and taking γ so large that $\beta - \frac{\alpha}{\gamma\underline{\delta}} \geq 0$, we get from $E[\tilde{\mathcal{R}}^\delta_{t,\infty}(Z^*)|\mathcal{F}_t] \geq 0$ and (14.27a) that

$$\liminf_{t\to\infty} V_t \geq -\frac{\alpha}{\gamma\underline{\delta}}(e^{-1}+|\log\gamma|) \quad P\text{-a.s.}$$

Since γ is arbitrary, we conclude that $\liminf_{t\to\infty} V_t \geq 0$ P-a.s., which completes the proof. □

All results in this section so far hold for a general filtration. If $\mathbb{F}$ is continuous, one can in addition show as in Bordigoni/Matoussi/Schweizer [2] that V obeys the dynamics

$$dV_t = (\delta_t V_t - \alpha U_t)\,dt + \frac{1}{2\beta}\,d\langle M\rangle_t + dM_t$$

for some (continuous) local martingale M; see Theorem 4.27 in Bordigoni [1] for a detailed proof. Together with Proposition 14.19, this explains where the infinite-horizon BSDE (14.3) comes from. Since the above derivation uses no essential new ideas in comparison with Bordigoni/Matoussi/Schweizer [2], we refrain from giving more details.

14.6 Solving the Stochastic Control Problem via the BSDE

Our goal in this section is to use the results on the infinite-horizon BSDE (14.19) for a characterisation of the dynamic value process V for the stochastic control problem from Sect. 14.5. As just mentioned, we could have shown that V solves (14.19), but this is not enough: The uniqueness result in Theorem 14.10 only holds for solutions (Y, M) with $Y \in D^{\exp}_{0,\infty}$, and we do not know at this point how to argue directly that V from the control problem is in $D^{\exp}_{0,\infty}$. The BSDE techniques developed so far will enable us to prove this. To that end, we now show how one can construct from a (particular) solution to the BSDE (14.19) a (and actually the, by uniqueness) solution for the infinite-horizon stochastic control problem.

Hypothesis 14.20 *Throughout this section, we impose the standing assumptions*

$\mathbb{F}$ is a *continuous filtration*, i.e. all local $(P,\mathbb{F})$-martingales are continuous. (14.33a)

$0 < \underline{\delta} \leq \delta_t \leq \bar{\delta} < \infty$, uniformly in (t,ω), for constants $\underline{\delta}, \bar{\delta}$. (14.33b)

Conditions on U will be specified below, when we successively treat three cases.

Our arguments rely substantially on the finite-horizon results proved in Bordigoni/Matoussi/Schweizer [2], so that we very briefly recall these here. Fix $T < \infty$ and consider on $[0, T]$ the BSDE

$$dY_t = (\delta_t Y_t - \alpha U_t)\,dt + \frac{1}{2\beta}\,d\langle M\rangle_t + dM_t, \qquad Y_T = \alpha' U'_T. \tag{14.34}$$

Recall from (14.4) and (14.5) the definitions of $\mathcal{U}^{\delta}_{t,T}$ and $\mathcal{R}^{\delta}_{t,T}$, and assume that U (as a process on $[0, T]$) is in $D^{\exp}_{1,T}$ and U'_T is in $D^{\exp}$. Then Theorem 17 of Bordigoni/Matoussi/Schweizer [2] states that (14.34) has a unique solution (Y, M) in $D^{\exp}_{0,T} \times \mathcal{M}_{0,\mathrm{loc}}(P)$, that $\bar{Z} := \mathcal{E}(-\frac{1}{\beta}M)$ is a martingale on $[0, T]$ with $E[\bar{Z}_T \log \bar{Z}_T] < \infty$, and that for any martingale $Z \ge 0$ on $[0, T]$ with $Z_0 = 1$ and $E[Z_T \log Z_T] < \infty$, we have for any stopping time $\tau \le T$ that

$$Y_\tau = E\bigg[\frac{\bar{Z}_T}{\bar{Z}_\tau}\mathcal{U}^{\delta}_{\tau,T} + \beta\tilde{\mathcal{R}}^{\delta}_{\tau,T}(\bar{Z})\bigg|\mathcal{F}_\tau\bigg] \le E\bigg[\frac{Z_T}{Z_\tau}\mathcal{U}^{\delta}_{\tau,T} + \beta\tilde{\mathcal{R}}^{\delta}_{\tau,T}(Z)\bigg|\mathcal{F}_\tau\bigg].$$

(This reformulates the statement that the dynamic value process of the finite-horizon stochastic control problem is the unique solution of (14.34).) For $\tau \equiv 0$, this reduces to

$$\begin{aligned} Y_0 \le {} & E\bigg[Z_T\alpha\int_0^T S^{\delta}_s U_s\,ds + Z_T\,\alpha' S^{\delta}_T U'_T\bigg] \\ & + \beta E\bigg[\int_0^T \delta_s S^{\delta}_s Z_s \log Z_s\,ds + S^{\delta}_T Z_T \log Z_T\bigg], \end{aligned} \tag{14.35}$$

with equality for $Z = \bar{Z}$.

14.6.1 The Bounded Case

Let us now first study the case where $\int_0^\infty |U_s|\,ds$ is bounded. This is of course a restrictive assumption, but it allows fairly simple arguments and provides a basic building block. We shall see that the proofs in more general cases follow the same scheme.

Proposition 14.21 *Suppose that $\int_0^\infty |U_s|\,ds$ is a bounded random variable. For any solution (Y, M) to the infinite-horizon BSDE* (14.19) *with Y bounded, we then have, for any $t \ge 0$,*

$$Y_t = \operatorname*{ess\,inf}_{Z\in\mathcal{Z}_f} E\big[\tilde{\mathcal{U}}^{\delta}_{t,\infty}(Z) + \beta\tilde{\mathcal{R}}^{\delta}_{t,\infty}(Z)\big|\mathcal{F}_t\big].$$

Proof Without loss of generality, we prove the result for $t = 0$.

(1) Let us start by arguing that for any $Z \in \mathcal{Z}_f$, we have

$$Y_0 \le E\big[\tilde{\mathcal{U}}^\delta_{0,\infty}(Z) + \beta\tilde{\mathcal{R}}^\delta_{0,\infty}(Z)\big]. \tag{14.36}$$

We first note that the nonnegative function $g(s) := E[\delta_s S^\delta_s(Z_s \log Z_s + e^{-1})]$ satisfies, by Remark 14.14 and Fubini,

$$\int_0^\infty g(s)\,ds = E\bigg[\int_0^\infty \delta_s S^\delta_s\big(Z_s \log Z_s + e^{-1}\big)\,ds\bigg] = E\big[\tilde{\mathcal{R}}^\delta_{0,\infty}(Z)\big] + e^{-1} < \infty.$$

This implies that there exists a sequence of deterministic times $T_n \nearrow \infty$ such that

$$\lim_{n\to\infty} E\big[S^\delta_{T_n}\big(Z_{T_n}\log Z_{T_n} + e^{-1}\big)\big] \le \lim_{n\to\infty} \frac{1}{\underline{\delta}}\, g(T_n) = 0$$

and therefore also

$$\lim_{n\to\infty} E\big[S^\delta_{T_n} Z_{T_n} \log Z_{T_n}\big] = 0, \tag{14.37}$$

since $S^\delta_{T_n} \le e^{-\underline{\delta} T_n} \to 0$. Moreover, because Y is bounded and $Z \ge 0$ is a martingale, we have

$$\lim_{n\to\infty} \big|E\big[S^\delta_{T_n} Z_{T_n} Y_{T_n}\big]\big| \le \lim_{n\to\infty} e^{-\underline{\delta} T_n}\|Y\|_\infty E[Z_{T_n}] = 0. \tag{14.38}$$

Now Y is bounded, hence in $D^{\exp}_{0,T_n}$, and satisfies the finite-horizon BSDE (14.34) with final value $\alpha' U'_{T_n} := Y_{T_n}$. Moreover, $Z \in \mathcal{Z}_f$ verifies $E[Z_{T_n} \log Z_{T_n}] < \infty$ due to Lemma 14.15, and so the finite-horizon results tell us that

$$\begin{aligned} Y_0 \le{}& E\bigg[\alpha \int_0^{T_n} S^\delta_s Z_s U_s\,ds + S^\delta_{T_n} Z_{T_n} Y_{T_n}\bigg] \\ &+ \beta E\bigg[\int_0^{T_n} \delta_s S^\delta_s Z_s \log Z_s\,ds + S^\delta_{T_n} Z_{T_n} \log Z_{T_n}\bigg]. \end{aligned} \tag{14.39}$$

On the right-hand side, the second and the fourth summands tend to 0 as $n \to \infty$ by (14.38) and (14.37), respectively. Next, Fatou's lemma yields

$$\begin{aligned} E\bigg[\int_0^\infty S^\delta_s Z_s |U_s|\,ds\bigg] &\le \liminf_{n\to\infty} E\bigg[Z_{T_n} \int_0^{T_n} S^\delta_s |U_s|\,ds\bigg] \\ &\le \liminf_{n\to\infty} E\bigg[Z_{T_n} \int_0^\infty |U_s|\,ds\bigg] < \infty \end{aligned}$$

since $\int_0^\infty |U_s|\,ds$ is bounded and $Z \ge 0$ is a martingale. By the dominated convergence theorem, we thus deduce that the first summand in (14.39) converges to $E[\alpha \int_0^\infty S^\delta_s Z_s U_s\,ds] = E[\tilde{\mathcal{U}}^\delta_{0,\infty}(Z)]$ as $n \to \infty$. Finally,

$$\int_0^{T_n} \delta_s S^\delta_s Z_s \log Z_s\,ds = \int_0^{T_n} \delta_s S^\delta_s\big(Z_s \log Z_s + e^{-1}\big)\,ds - e^{-1}\big(1 - S^\delta_{T_n}\big)$$

implies via monotone integration and by using $\int_0^\infty \delta_s S_s^\delta \, ds = 1$ that the third summand in (14.39) converges to $\beta E[\int_0^\infty \delta_s S_s^\delta Z_s \log Z_s \, ds] = \beta E[\tilde{\mathcal{R}}_{0,\infty}^\delta(Z)]$ as $n \to \infty$. Putting everything together gives (14.36).

(2) Now define $\bar{Z} := \mathcal{E}(-\frac{1}{\beta}M)$. Since $\int_0^\infty |U_s| \, ds$ is bounded, U is in $D_{1,\infty}^{\exp}$; and since (Y, M) solves (14.19), Theorem 14.11 tells us that $\bar{Z}$ is a martingale on $[0, \infty)$, so that $\bar{Z} \in \mathcal{Z}$. We want to prove that $\bar{Z}$ is even in $\mathcal{Z}_f$. To that end, we apply the finite-horizon results to write

$$
\begin{aligned}
Y_0 = {} & E\bigg[\bar{Z}_{T_n} \alpha \int_0^{T_n} S_s^\delta U_s \, ds + S_{T_n}^\delta \bar{Z}_{T_n} Y_{T_n}\bigg] \\
& + \beta E\bigg[\int_0^{T_n} \delta_s S_s^\delta \bar{Z}_s \log \bar{Z}_s \, ds + S_{T_n}^\delta \bar{Z}_{T_n} \log \bar{Z}_{T_n}\bigg].
\end{aligned}
$$

Using Fatou's lemma and $z \log z \geq -e^{-1}$ therefore gives

$$
\begin{aligned}
& \beta E\big[\tilde{\mathcal{R}}_{0,\infty}^\delta(\bar{Z})\big] \\
& = \beta E\bigg[\int_0^\infty \delta_s S_s^\delta \bar{Z}_s \log \bar{Z}_s \, ds\bigg] \\
& \leq \liminf_{n\to\infty} \beta E\bigg[\int_0^{T_n} \delta_s S_s^\delta \bar{Z}_s \log \bar{Z}_s \, ds\bigg] \\
& = \liminf_{n\to\infty}\bigg(Y_0 - \beta E\big[S_{T_n}^\delta \bar{Z}_{T_n} \log \bar{Z}_{T_n}\big] - E\bigg[\bar{Z}_{T_n} \alpha \int_0^{T_n} S_s^\delta U_s \, ds + S_{T_n}^\delta \bar{Z}_{T_n} Y_{T_n}\bigg]\bigg) \\
& \leq \liminf_{n\to\infty}\bigg(Y_0 + \beta e^{-1} + \alpha \bigg\| \int_0^\infty |U_s| \, ds \bigg\|_{L^\infty} E[\bar{Z}_{T_n}] + \|Y_\infty^*\|_{L^\infty} E[\bar{Z}_{T_n}]\bigg) < \infty
\end{aligned}
$$

because Y and $\int_0^\infty |U_s| \, ds$ are bounded and $\bar{Z} \geq 0$ is a martingale. Hence $\bar{Z}$ is indeed in $\mathcal{Z}_f$.

(3) Since $\bar{Z}$ is in $\mathcal{Z}_f$ by step (2) and satisfies (14.35) with equality, the same argument as in step (1) shows that the inequality in (14.36) becomes an equality for $Z = \bar{Z}$. Hence $\bar{Z}$ attains the infimum, and the proof is complete. □

14.6.2 The Positive Case

We now turn to the case where U is nonnegative and satisfies some integrability condition. Note that $U \geq 0$ is a fairly natural assumption. Indeed, if we think of a full-fledged robust control problem for utility maximisation, then U_t typically represents the utility $\mathbf{U}(c_t)$ from consumption at time t, where we still optimise over c in a second step. As a consumption rate, $c_t \geq 0$; so $U_t = \mathbf{U}(c_t) \geq 0$ for any nonnegative utility function $\mathbf{U}$ on $[0, \infty)$, like e.g. the power utility $\mathbf{U}(x) = \frac{1}{\gamma}x^\gamma$ for $\gamma \in (0, 1)$.

Theorem 14.22 *Suppose that $U \geq 0$ and U is in $D^{\exp}_{1,\infty}$. For the solution (Y, M) to* (14.19) *from Theorem* 14.9, *we then have, for any $t \geq 0$,*

$$Y_t = \operatorname*{ess\,inf}_{Z \in \mathcal{Z}_f} E\big[\tilde{\mathcal{U}}^{\delta}_{t,\infty}(Z) + \beta \tilde{\mathcal{R}}^{\delta}_{t,\infty}(Z) \,\big|\, \mathcal{F}_t\big].$$

Proof Without loss of generality, we again argue for $t = 0$. The overall structure of the proof is like for Proposition 14.21, but we first need to recall the construction of (Y, M). For each $n \in \mathbb{N}$, set $U^n_t := U_t \mathbf{1}_{\{\int_0^t U_s\,ds \leq n\}}$ and denote by (Y^n, M^n) with Y^n bounded the solution to the BSDE

$$dY_t = \big(\delta_t Y_t - \alpha U^n_t\big)\,dt + \frac{1}{2\beta}\,d\langle M\rangle_t + dM_t, \qquad Y_n = 0,$$

on $[0, n]$. Extending (Y^n, M^n) to $[0, \infty)$ by setting $Y^n_t = 0$, $M^n_t = M^n_n$ for $t > n$, we then get on $[0, \infty)$ a solution (Y^n, M^n) to the BSDE

$$dY_t = \big(\delta_t Y_t - \alpha \mathbf{1}_{\{t \leq n\}} U^n_t\big)\,dt + \frac{1}{2\beta}\,d\langle M\rangle_t + dM_t, \qquad \lim_{t\to\infty} Y_t = 0, \tag{14.40}$$

and $Y_t = \nearrow\text{-}\lim_{n\to\infty} Y^n_t$ for all $t \geq 0$. We first prove that, for any $Z \in \mathcal{Z}_f$,

$$Y_0 \leq E\big[\tilde{\mathcal{U}}^{\delta}_{0,\infty}(Z) + \beta \tilde{\mathcal{R}}^{\delta}_{0,\infty}(Z)\big]. \tag{14.41}$$

Indeed, applying Proposition 14.21 to the process $(\mathbf{1}_{\{t\leq n\}} U^n_t)_{t\geq 0}$ and the solution to (14.40) gives

$$Y^n_0 \leq E\bigg[\alpha \int_0^\infty S^\delta_s Z_s \mathbf{1}_{\{s\leq n\}} U^n_s\,ds + \beta \int_0^\infty \delta_s S^\delta_s Z_s \log Z_s\,ds\bigg],$$

and (14.41) follows by monotone integration since $U \geq 0$.

Now set $\bar{Z} := \mathcal{E}(-\frac{1}{\beta}M)$, so that $\bar{Z} \in \mathcal{Z}$ by Theorem 14.11; this uses the integrability assumption on U. To prove that $\bar{Z}$ is even in $\mathcal{Z}_f$, we first note that by Proposition 14.21 and its proof, we have equality in (14.41) for the choice $Z = \bar{Z}^n := \mathcal{E}(-\frac{1}{\beta}M^n)$, so that

$$Y^n_0 = E\bigg[\alpha \int_0^\infty S^\delta_s \bar{Z}^n_s \mathbf{1}_{\{s\leq n\}} U^n_s\,ds + \beta \int_0^\infty \delta_s S^\delta_s \bar{Z}^n_s \log \bar{Z}^n_s\,ds\bigg].$$

But by construction, $(Y^n_0)_{n\in\mathbb{N}}$ increases to Y_0, and $(M^n)_{n\in\mathbb{N}}$ and $(\langle M^n\rangle)_{n\in\mathbb{N}}$ converge to M and $\langle M\rangle$ locally uniformly in probability, so that also $\bar{Z}^n \to \bar{Z}$ locally uniformly in probability as $n \to \infty$. Hence Fatou's lemma yields

$$Y_0 \geq E\bigg[\alpha \int_0^\infty S^\delta_s \bar{Z}_s U_s\,ds + \beta \int_0^\infty \delta_s S^\delta_s \bar{Z}_s \log \bar{Z}_s\,ds\bigg] = E\big[\tilde{\mathcal{U}}^{\delta}_{0,\infty}(\bar{Z}) + \beta \tilde{\mathcal{R}}^{\delta}_{0,\infty}(\bar{Z})\big],$$

and so $\bar{Z} \in \mathcal{Z}_f$ because $U \geq 0$. Since (14.41) gives the converse inequality, we actually have equality in (14.41) for $Z = \bar{Z}$, and this completes the proof. □

14.6.3 The General Case

Finally, we study a situation where U can be real-valued. Then we need slightly stronger integrability assumptions.

Theorem 14.23 *Suppose that U is in $D^{\mathrm{exp}}_{1,\infty}$ and that U also satisfies* (14.27a), *i.e. there exists some $T_0 \in (0,\infty)$ such that, for all $\gamma > 0$,*

$$E_P\left[\exp\left(\gamma \int_0^{T_0} |U_s|\,ds\right)\right] + E_P\left[\int_{T_0}^{\infty} \exp\big(\gamma |U_s|\big)\mathbf{1}_{\{U_s \neq 0\}}\,ds\right] < \infty.$$

For the solution (Y,M) to (14.19) *from Theorem* 14.9, *we then have, for any $t \geq 0$,*

$$Y_t = \operatorname*{ess\,inf}_{Z \in \mathcal{Z}_f} E\big[\tilde{\mathcal{U}}^{\delta}_{t,\infty}(Z) + \beta \tilde{\mathcal{R}}^{\delta}_{t,\infty}(Z)\big|\mathcal{F}_t\big].$$

Proof As already in the last proof, we argue for $t=0$ without loss of generality and again first recall from the proof of Theorem 14.9 the construction of (Y,M). For $n, p \in \mathbb{N}$, set

$$U^{n,p}_t := U^+_t \mathbf{1}_{\{\int_0^t |U_s|\,ds \leq n\}} \mathbf{1}_{\{t \leq n\}} - U^-_t \mathbf{1}_{\{\int_0^t |U_s|\,ds \leq p\}} \mathbf{1}_{\{t \leq p\}}$$

and denote by $(Y^{n,p}, M^{n,p})$ with $Y^{n,p}$ bounded the solution to the BSDE

$$dY_t = (\delta_t Y_t - \alpha U_t)\,dt + \frac{1}{2\beta}\,d\langle M\rangle_t + dM_t, \qquad Y_{n\vee p} = 0,$$

on $[0, n \vee p]$. We extend $(Y^{n,p}, M^{n,p})$ to $[0,\infty)$ by setting $Y^{n,p}_t = 0$ and $M^{n,p}_t = M^{n,p}_{n\vee p}$ for $t > n \vee p$ to get on $[0,\infty)$ a solution to the BSDE

$$dY_t = \big(\delta_t Y_t - \alpha \mathbf{1}_{\{t \leq n\vee p\}} U^{n,p}_t\big)\,dt + \frac{1}{2\beta}\,d\langle M\rangle_t + dM_t, \qquad \lim_{t\to\infty} Y_t = 0. \quad (14.42)$$

Then $Y_t = \nearrow\text{-}\lim_{n\to\infty} \searrow\text{-}\lim_{p\to\infty} Y^{n,p}_t$ and $M_t = \lim_{n\to\infty}\lim_{p\to\infty} M^{n,p}_t$. We proceed on a familiar path.

(1) First we prove that, for any $Z \in \mathcal{Z}_f$,

$$Y_0 \leq E\big[\tilde{\mathcal{U}}^{\delta}_{0,\infty}(Z) + \beta \tilde{\mathcal{R}}^{\delta}_{0,\infty}(Z)\big]. \quad (14.43)$$

Like in Sect. 14.6.2, using Proposition 14.21 gives

$$Y^{n,p}_0 \leq E\left[\alpha \int_0^{\infty} S^{\delta}_s Z_s \mathbf{1}_{\{s \leq n\vee p\}} U^{n,p}_s\,ds + \beta \int_0^{\infty} \delta_s S^{\delta}_s Z_s \log Z_s\,ds\right],$$

and (14.43) follows by letting $p \to \infty$ and then $n \to \infty$, provided that we can use dominated convergence. But this is ensured by the first estimate in Proposition 14.16; indeed, (14.27a) yields $E[\int_0^{\infty} S^{\delta}_s Z_s |U_s|\,ds] < \infty$ for $Z \in \mathcal{Z}_f$.

(2) Thanks to the first integrability assumption on U, Theorem 14.11 implies that the process $\bar{Z} := \mathcal{E}(-\frac{1}{\beta}M)$ is in $\mathcal{Z}$. To show that it is even in $\mathcal{Z}_f$, we use Proposition 14.21 for $\bar{Z}^{n,p} := \mathcal{E}(-\frac{1}{\beta}M^{n,p})$ to get

$$Y_0^{n,p} = E\left[\alpha \int_0^\infty S_s^\delta \bar{Z}_s^{n,p} \mathbf{1}_{\{s\leq n\vee p\}} U_s^{n,p}\, ds + \beta \int_0^\infty \delta_s S_s^\delta \bar{Z}_s^{n,p} \log \bar{Z}_s^{n,p}\, ds\right]. \tag{14.44}$$

But (14.31) in the proof of Proposition 14.16 gives

$$\begin{aligned}\left| E\left[\alpha \int_0^\infty S_s^\delta \bar{Z}_s^{n,p} \mathbf{1}_{\{s\leq n\vee p\}} U_s^{n,p}\, ds\right]\right| &\leq E\left[\alpha \int_0^\infty S_s^\delta \bar{Z}_s^{n,p} |U_s|\, ds\right] \\ &\leq C_{\gamma,U} + E\big[\tilde{\mathcal{R}}_{0,\infty}^\delta\big(\bar{Z}^{n,p}\big)\big]\frac{1}{\gamma\underline{\delta}}\big(e^{\bar{\delta}(T_0+1)}+1\big),\end{aligned}$$

where the constant $C_{\gamma,U}$ depends on γ and U via (14.27a), but not on n and p. Plugging this estimate with a minus sign into (14.44) and taking γ big enough yields

$$\sup_{n,p\in\mathbb{N}} E\big[\tilde{\mathcal{R}}_{0,\infty}^\delta\big(\bar{Z}^{n,p}\big)\big] \leq C\Big(1 + \sup_{n,p\in\mathbb{N}} Y_0^{n,p}\Big) < \infty, \tag{14.45}$$

because applying the a priori estimate (14.23) from Theorem 14.9 to (14.42) tells us that

$$\big|Y_0^{n,p}\big| \leq \beta \log E\left[\exp\left(\frac{1}{\beta}\int_0^\infty \alpha |U_s|\, ds\right)\right] < \infty$$

for all n and p, by using the definition of $U^{n,p}$. As $n \to \infty$ and $p \to \infty$, we have locally uniformly in probability $M^{n,p} \to M$ and $\langle M^{n,p}\rangle \to \langle M\rangle$, hence also $\bar{Z}^{n,p} \to \bar{Z}$, and so $\bar{Z} \in \mathcal{Z}_f$ because $E[\tilde{\mathcal{R}}_{0,\infty}^\delta(\bar{Z})] < \infty$ by Fatou's lemma and (14.45).

(3) To prove that we have equality in (14.43) for $Z = \bar{Z}$, we start from the equality in (14.44). As $n \to \infty$ and $p \to \infty$, $Y_0^{n,p}$ tends to Y_0, and we have $\liminf_{n\to\infty} \liminf_{p\to\infty} E[\tilde{\mathcal{R}}_{0,\infty}^\delta(\bar{Z}^{n,p})] \geq E[\tilde{\mathcal{R}}_{0,\infty}^\delta(\bar{Z})]$ by Fatou's lemma. Because $Z \mapsto E[\tilde{\mathcal{R}}_{0,\infty}^\delta(Z)]$ is bounded along the sequence $(\bar{Z}^{n,p})_{n,p\in\mathbb{N}}$ by (14.45), almost the same argument as in step 3) of the proof of Theorem 14.18 gives

$$\begin{aligned}\liminf_{n\to\infty}\liminf_{p\to\infty} E\left[\alpha \int_0^\infty S_s^\delta \bar{Z}_s^{n,p} \mathbf{1}_{\{s\leq n\vee p\}} U_s^{n,p}\, ds\right] &\geq E\left[\alpha \int_0^\infty S_s^\delta \bar{Z}_s U_s\, ds\right] \\ &= E\big[\tilde{\mathcal{U}}_{0,\infty}^\delta(\bar{Z})\big].\end{aligned}$$

Note that this exploits the integrability assumption (14.27a). Therefore (14.44) implies that $Y_0 \geq E[\tilde{\mathcal{U}}_{0,\infty}^\delta(\bar{Z}) + \beta\tilde{\mathcal{R}}_{0,\infty}^\delta(\bar{Z})]$, and so we must have equality due to (14.43) since $\bar{Z} \in \mathcal{Z}_f$. This completes the proof. □

14.6.4 Consequences for the Stochastic Control Problem

As in (14.32), denote by $V = (V_t)_{t \geq 0}$ the dynamic value process of the infinite-horizon stochastic control problem. We already mentioned at the end of Sect. 14.5 that if $\mathbb{F}$ is continuous, then V satisfies the infinite-horizon BSDE (14.19). For the proof, we referred to Bordigoni [1]; let us just note here that the required assumptions are Hypothesis 14.12 plus continuity of $\mathbb{F}$, i.e. Hypothesis 14.20 plus (14.27a). Under a slightly stronger condition, we can now even prove a BSDE characterisation for V.

Theorem 14.24 *Assume Hypothesis* 14.20 *and that U is in $D^{\exp}_{1,\infty}$. If in addition either $U \geq 0$ or U satisfies* (14.27a), *then V is the first component of the unique solution in $D^{\exp}_{0,\infty} \times \mathcal{M}_{0,\mathrm{loc}}(P)$ to the infinite-horizon BSDE* (14.19). *In particular, $V \in D^{\exp}_{0,\infty}$.*

Proof By Theorem 14.10 (14.19) has a unique solution (Y, M) with $Y \in D^{\exp}_{0,\infty}$; and by the definition of V in (14.32) and either Theorem 14.22 or Theorem 14.23, Y coincides with V. □

Remark 14.25

(1) The second case of Theorem 14.24 is the infinite-horizon analogue to the finite-horizon Theorem 17 in Bordigoni/Matoussi/Schweizer [2], with assumptions and conclusions almost exactly parallel. The only difference lies in the conditions on U: In (14.27a), we need $U \in D^{\exp}_{1,T_0}$, but also an exponential moment control over U on the infinite time interval $[T_0, \infty)$. See Remark 4.28 in Bordigoni [1] for a more detailed comment on this point. The result for $U \geq 0$ has no precedent.
(2) Our approach for $T = \infty$ here is different from Bordigoni [1] in that we show for the solution of the BSDE (14.19) that it satisfies the defining property (14.32) of the value process V. As a bonus, we are able to deduce that V is indeed in $D^{\exp}_{0,\infty}$; this was conjectured, but not proved, in Bordigoni [1].
(3) Again in remarkable analogy to the finite-horizon results in Bordigoni/Matoussi/Schweizer [2], we obtain the existence of a solution to the stochastic control problem for a general filtration $\mathbb{F}$. But the integrability property $V \in D^{\exp}_{0,\infty}$ is only known for continuous $\mathbb{F}$, since its proof exploits the BSDE results. Like in Bordigoni/Matoussi/Schweizer [2], we do not know if $V \in D^{\exp}_{0,\infty}$ also holds for general $\mathbb{F}$.

Acknowledgements The work of Ying Hu was partly supported by funds from the Marie Curie ITN grant, Controlled Systems, GA 213841/2008. Much of this work was done during a visit of Martin Schweizer to the Department of Mathematics at the University of Rennes, whose hospitality was greatly appreciated. Martin Schweizer gratefully acknowledges financial support by the National Centre of Competence in Research "Financial Valuation and Risk Management" (NCCR FINRISK), Project D1 (Mathematical Methods in Financial Risk Management). The NCCR FINRISK is a research instrument of the Swiss National Science Foundation.

References

1. G. Bordigoni, Stochastic control and BSDEs in a robust utility maximization problem with an entropic penalty term. Ph.D. Thesis D01509, Politecnico di Milano (2007)
2. G. Bordigoni, A. Matoussi, M. Schweizer, A stochastic control approach to a robust utility maximization problem, in *Stochastic Analysis and Applications. Proceedings of the Second Abel Symposium*, ed. by F.E. Benth et al., Oslo, 2005 (Springer, Berlin, 2007), pp. 125–151
3. P. Briand, F. Confortola, Quadratic BSDEs with random terminal time and elliptic PDEs in infinite dimension. Electron. J. Probab. **13**, 1529–1561 (2008)
4. P. Briand, Y. Hu, Stability of BSDEs with random terminal time and homogenization of semilinear elliptic PDEs. J. Funct. Anal. **155**, 455–494 (1998)
5. P. Briand, Y. Hu, BSDE with quadratic growth and unbounded terminal value. Probab. Theory Relat. Fields **136**, 604–618 (2006)
6. P. Briand, Y. Hu, Quadratic BSDEs with convex generators and unbounded terminal conditions. Probab. Theory Relat. Fields **141**, 543–567 (2008)
7. F. Delbaen, W. Schachermayer, A general version of the fundamental theorem of asset pricing. Math. Ann. **300**, 463–520 (1994)
8. C. Dellacherie, P.-A. Meyer, *Probabilities and Potential B. Theory of Martingales* (North-Holland, Amsterdam, 1982)
9. L.P. Hansen, T.J. Sargent, G. Turmuhambetova, N. Williams, Robust control and model misspecification. J. Econ. Theory **128**, 45–90 (2006)
10. M. Kobylanski, Backward stochastic differential equations and partial differential equations with quadratic growth. Ann. Probab. **28**, 558–602 (2000)
11. M. Mania, M. Schweizer, Dynamic exponential utility indifference valuation. Ann. Appl. Probab. **15**, 2113–2143 (2005)
12. M.-A. Morlais, Quadratic BSDEs driven by a continuous martingale and applications to the utility maximization problem. Finance Stoch. **13**, 121–150 (2009)
13. M. Royer, BSDEs with a random terminal time driven by a monotone generator and their links with PDEs. Stoch. Stoch. Rep. **76**, 281–307 (2004)
14. M. Schroder, C. Skiadas, Optimal consumption and portfolio selection with stochastic differential utility. J. Econ. Theory **89**, 68–126 (1999)

Chapter 15
Functionals Associated with Gradient Stochastic Flows and Nonlinear SPDEs

B. Iftimie, M. Marinescu, and C. Vârsan

Abstract In this paper we construct and provide a representation for a classical solution of some nonlinear SPDE driven by Fisk–Stratonovich stochastic integral. Our main assumption is the commuting property of the drift and diffusion vector fields with respect to the usual Lie bracket. This result is next applied for a system of Burgers equations with stochastic perturbations and also to the computations of some expectations of functionals depending on the final value of some non-Markovian process.

Keywords Stochastic partial differential equations · Fisk–Stratonovich stochastic integral · Stochastic flow · Gradient representation · Hamilton–Jacobi equations

Mathematics Subject Classification (2010) 60H15 · 60H30 · 35F20 · 35Q35

15.1 Introduction

The investigation of evolution equations with stochastic perturbations serves a large variety of areas of applicability, among which mathematical finance as well. Nonlinear SPDEs have for instance applications in modelling of interest rates, in stochastic control with partial information (as it is specified in Lions and Souganidis [15]). Other applications of SPDEs (including finance) may be found in Da Prato and Tubaro [6].

The research of the authors was supported by *CNCSIS* under the research grant 3-RNP/AMaMeF, under the authority of European Science Foundation.

B. Iftimie (✉) · M. Marinescu
Department of Mathematics, Academy of Economic Sciences, 6 Piaţa Romană, 010374, Bucharest, Romania
e-mail: Bogdan.Iftimie@csie.ase.ro

C. Vârsan
Institute of Mathematics of the Romanian Academy "Simion Stoilow", P.O. Box 1–764, Calea Griviţei, 70700, Bucharest, Romania
e-mail: constantin.varsan@imar.ro

DOI 10.1007/978-3-642-18412-3_15, © Springer-Verlag Berlin Heidelberg 2011

During the last three decades, SPDEs of the form

$$\begin{cases} du(t,x) = L(t,x,u(t,x),\nabla u(t,x),D^2u(t,x))\,dt \\ \qquad\qquad + \sum_{i=1}^n P_i(t,x,u(t,x),\nabla u(t,x))\,dW_i(t), \\ u(0,x) = \varphi(x), \quad x \in \mathbb{R}^n, \end{cases}$$

have been intensively studied, under suitable assumptions imposed on the coefficients. Here $W(t)$ stands for a standard n-dimensional Wiener process, and the first-order operators $P_i(t,x,u,p)$ are linear with respect to u, p, i.e. $P_i(t,x,u,p) = \langle b_i(t,x), p\rangle + c_i(t,x)u$.

The case when L is a linear differential operator

$$L(t,x,u,p,q) = \sum_{i,j=1}^n a_{ij}(t,x)q_{ij} + \sum_{i=1}^n a_i(t,x)p_i + a_0(t,x)u$$

and $c_i(t,x) = 0$ was studied in [19]. The idea is to transform the SPDE into a linear parabolic equation with random parameter ω (for which the author uses a semi-group approach based on the Kato–Tanabe theory), via the *stochastic characteristics* method (which was introduced by Kunita for first order SPDEs, see [13]). More precisely, the solution $\xi(t,x)$ of the SDE

$$\begin{aligned} \xi(t) &= x + \frac{1}{2}\sum_{i=1}^n \int_0^t \nabla b_i\big(s,\xi(s)\big)\cdot b_i\big(s,\xi(s)\big)\,ds - \sum_{i=1}^n \int_0^t b_i\big(s,\xi(s)\big)\,dW_i(s) \\ &= x + \sum_{i=1}^n \int_0^t \big(-b_i\big(s,\xi(s)\big)\big)\circ dW_i(s), \end{aligned} \tag{15.1}$$

called the *stochastic characteristics*, is a diffeomorphism with respect to x and if $\eta(t,x)$ denotes its inverse, then the random mapping $v(t,x,\omega) := u(t,\xi(t,x,\omega),\omega)$ solves a linear parabolic PDE with random parameter. Next, the unique solution u is obtained as $u(t,x) = v(t,\eta(t,x))$. Notice that the stochastic integral appearing in the last line of the formula (15.1) has to be understood in the Fisk–Stratonovich sense.

The case when the operator L is semilinear, i.e.,

$$L(t,x,u,p,q) = \sum_{i,j=1}^n a_{ij}(t,x)q_{ij} + a_0(t,x,u,p),$$

was treated by several authors (see, for instance, [4]), while the quasilinear case

$$L(t,x,u,p,q) = \sum_{i,j=1}^n a_{ij}(t,x,p)q_{ij} + a_0(t,x,u,p)$$

was studied only by a few authors (see, e.g., [1], where a *splitting up* method was used for operators L in divergence form, or [7], where a semigroup approach was preferred). In all the situations from above, the mappings $a_0(t,x,u,p), a_{ij}(t,x,u,p)$ were assumed (locally) Lipschitz continuous with respect to u, p.

In [5] the authors treat the case of a fully nonlinear operator $L(t,x,u,p,q)$ under suitable assumptions, and in particular L is locally Lipschitz continuous with respect to u, p and q, uniformly with respect to t,x. The presence of the additional term $c(t,x)u$ in the diffusion part of the SPDE does not modify too much the approach used in [19]. Indeed, with the notation from above, by the transformation $w(t,x) := \rho(t,x)u(t,\xi(t,x))$, where

$$\rho(t,x) = \exp\left[-\sum_{i=1}^{n}\int_0^t c_i\big(s,\xi(s,x)\big)\,dW_i(s) + \frac{1}{2}\sum_{i=1}^{n}\int_0^t c_i^2\big(s,\xi(s,x)\big)\,ds\right],$$

the SPDE is transformed in a deterministic nonlinear equation with random parameter, for the unknown $w(t,x,\omega)$. Finally, u is easily obtained from w and ρ via the diffeomorphism η.

Buckdahn and Ma are treating in [2] nonlinear SPDEs driven by Fisk–Stratonovich integrals of the form

$$\begin{cases} du(t,x) = \tilde{L}(t,x,u(t,x),\nabla u(t,x),D^2u(t,x))\,dt \\ \qquad\qquad + \sum_{i=1}^{m} g_i(t,x,u(t,x)) \circ dW_i(t), \\ u(0,x) = u_0(x), \quad t\in[0,T],\ x\in\mathbb{R}^n, \end{cases}$$

where

$$\tilde{L}(t,x,u,p,q) = \sum_{i,j=1}^{n} a_{ij}(x)q_{ij} + \sum_{i=1}^{n} b_i(x)p_i + f\big(t,x,u,\sigma^*(x)p\big),$$

where $a=\sigma\sigma^*$, under the assumption that f is Lipschitz with respect to u, p. Under the notation from above,

$$L(t,x,u,p,q) = \tilde{L}(t,x,u,p,q) + \frac{1}{2}\sum_{i=1}^{m} \nabla_u g_i(t,x,u)\cdot g_i(t,x,u).$$

They prove, under weak conditions on the coefficients, the existence (and the uniqueness in a latter paper) of the so-called *stochastic viscosity solution*, introduced by Lions and Souganidis for a general class of SPDEs in [15], via the corresponding *stochastic flow* $\xi(t,x,y)$, solution of the SDE determined by the stochastic perturbation of the SPDE, i.e.,

$$\xi(t,x,y) = y + \frac{1}{2}\sum_{i=1}^{m}\int_0^t \nabla_u g_i\big(s,x,\xi(s,x,y)\big)\cdot g_i\big(s,x,\xi(s,x,y)\big)\,ds$$

$$+\sum_{i=1}^{m}\int_0^t g_i\big(s,x,\xi(s,x,y)\big)\,dW_i(s)$$

$$=y+\sum_{i=1}^{m}\int_0^t g_i\big(s,x,\xi(s,x,y)\big)\circ dW_i(s).$$

This is not the usual stochastic characteristic, which cannot be associated here since the diffusion part of the SPDE depends only on u (nonlinearly), being independent of its gradient.

In this paper we are dealing with the Cauchy problem associated to the first-order nonlinear SPDE, considered in the strong sense,

$$\begin{cases} du(t,x)=\langle\nabla u(t,x),g_0(x)\rangle\,u(t,x)\,dt+\sum_{i=1}^{m}\langle\nabla u(t,x),g_i(x)\rangle\circ dW_i(t),\\ u(0,x)=\varphi(x),\quad t\in[0,T],\ x\in\mathbb{R}^n, \end{cases}\tag{15.2}$$

or, equivalently,

$$u(t,x)=\varphi(x)+\int_0^t\big\langle\nabla u(s,x),g_0(x)\,u(s,x)\big\rangle\,ds$$
$$+\sum_{i=1}^{m}\int_0^t\big\langle\nabla u(s,x),g_i(x)\big\rangle\circ dW_i(s),\tag{15.3}$$

where the stochastic integral is understood in the Fisk–Stratonovich sense. In our case $P_i(t,x,u,p):=\langle g_i(x),p\rangle$, and the drift $L(t,x,u,p,q)$ contains the term $\langle g_0(x),p\rangle u$ which *is not Lipschitz* with respect to u, p, and this is the main difference between the results we mentioned above and our case. Thus, if we try to reduce the SPDE to a random PDE, using the stochastic characteristics, then the existence of a solution is not easy to obtain. In order to overcome this issue, we adopt another approach by considering the system of characteristics defined by (15.2) (which is defined in analogy to the characteristics associated to deterministic PDEs). This leads us to a system of SDEs and ODEs for which existence of solutions is not hard to prove. The technique of considering the system of characteristics associated to a parabolic SPDE was already used by Iftimie and Vârsan in [10].

A main assumption is the commuting property of the vector fields g_i, $i=0,\ldots,m$, with respect to the usual Lie bracket (see Assumption (A.4)). Kunita also made this hypothesis (see [13], pp. 236 and 238, and also [12]), and some authors are referring to it as a compatibility condition concerning the mentioned vector fields (see [2], Remark 3.3). Under this hypothesis, we obtain a gradient representation for the stochastic flow associated with the stochastic differential equation obtained by means of the system of characteristics defined by the SPDE (15.2) and the corresponding fundamental solution $\psi(t,x)$ of the same SPDE. $\psi(t,x)$ will be described as the composition between the fundamental solution of some deterministic nonlinear Hamilton–Jacobi equations (see Lemma 15.13 below) and the fundamental solution of a reduced SPDE (see (15.17)).

The turbulent fluid motion can be described by the Burgers equation

$$\frac{\partial u}{\partial t}(t,x) = \nu \frac{\partial^2 u}{\partial x^2}(t,x) + u(t,x)\frac{\partial u}{\partial x}(t,x),$$

where $u(t,x)$ stands for the velocity field, and the positive constant ν is the viscosity. Burgers equations with a forcing term given by a random perturbation are more realistic. In [3], the authors establish an existence result for mild solutions of the Cauchy problem with additive space-time white noise, given by

$$\frac{\partial u}{\partial t}(t,x) = \nu \frac{\partial^2 u}{\partial x^2}(t,x) - u(t,x)\frac{\partial u}{\partial x}(t,x) + \varepsilon \frac{\partial^2 W}{\partial t \partial x}(t,x),$$

where $W(t,x)$ is a space-time white noise, and the partial derivative $\frac{\partial^2 W}{\partial t \partial x}(t,x)$ has to be understood in the generalized sense. Using the so called Cole–Hopf transformation, the initial integral equation obtained by convolution with the heat kernel is transformed into a linear SPDE driven by a Fisk–Stratonovich stochastic integral.

Another class of stochastic Burgers equations can be obtained from the SPDEs

$$\begin{aligned}\frac{\partial u}{\partial t}(t,x) = {} & \frac{\partial^2 u}{\partial x^2}(t,x) + f\big(t,x,u(t,x)\big) + \frac{\partial g}{\partial x}\big(t,x,u(t,x)\big)\\ & + \sigma\big(t,x,u(t,x)\big)\frac{\partial^2 W}{\partial t \partial x}(t,x),\end{aligned}$$

where again $W(t,x)$ is a space-time white noise, and the mapping $g(t,x,u)$ has quadratic growth with respect to u. The existence of a unique generalized solution (which is proved to be a mild solution also) of the associated Cauchy problem is shown, based on several estimates of the heat kernel partial derivatives.

Being motivated by these results, we study a system of Burgers equations with stochastic perturbations defined by a Wiener process, for which constant vector fields, both in the drift and diffusion part, are used in order to derive the existence of a global classical solution. We hope that this example would be of some interest for specialists in the field.

Another application consists in computing expectations of functionals depending on the terminal value of some non-Markovian process, obtained via the solution of the SDE obtained by writing down the system of characteristics associated to system (15.2). Usually, these types of expectations are related to Kolmogorov backward parabolic equations, but this procedure is not applicable in our case due to the non-Markovian nature of the process involved. Under appropriate conditions and avoiding SPDE techniques, the parameterized conditional expectation is the solution of a backward parabolic equation of Kolmogorov type with parameter.

15.2 Preliminaries

Let $\{W(t), t \geq 0\}$ be an m-dimensional Wiener process on a complete filtered probability space $\{\Omega, \mathcal{F}, \{\mathcal{F}_t\}, P\}$, where the filtration $\{\mathcal{F}_t\}$ stands for the augmentation under P of the natural filtration $\{\mathcal{F}_t^W\}$ generated by the Brownian motion W. T is a fixed time horizon. Let the following assumptions be in force:

(A.1) The vector fields $g_1, \dots, g_m$ belong to $\mathcal{C}^2(\mathbb{R}^n; \mathbb{R}^n)$ and have bounded derivatives of first two orders; $g_0 \in \mathcal{C}_b^1(\mathbb{R}^n; \mathbb{R}^n)$, and its partial derivatives are bounded.

(A.2) The initial condition $\varphi \in \mathcal{C}^2(\mathbb{R}^n)$ and admits bounded first-order partial derivatives.

(A.3) $\rho := TMK < 1$, where $M := \sup\{|\nabla\varphi(x)|, x \in \mathbb{R}^n\}$ and $K := \sup\{|g_0(x)|, x \in \mathbb{R}^n\}$.

Throughout this paper we shall use the notations $\langle , \rangle$ for the inner product and ∇h for the gradient with respect to x of some (vector) function $h(t, x)$.

If $Y(t)$ and $X(t)$ are continuous one-dimensional semimartingales, then the Fisk–Stratonovich integral of $Y(t)$ with respect to $X(t)$ is defined as

$$\int_0^t Y(s) \circ dX(s) := \int_0^t Y(s)\, dX(s) + \frac{1}{2}\langle Y, X\rangle_t, \tag{15.4}$$

where the stochastic integral on the right-hand side is the usual Itô integral, and $\langle Y, X\rangle_t$ stands for the quadratic variation of the processes $(Y(t))$ and $(X(t))$. If $Y(t)$ is d-dimensional, we can still define the integral $\int_0^t Y(s) \circ dX(s) := (\int_0^t Y_i(s) \circ dX(s))_{1\le i\le d}$. We state the Itô's formula involving the Fisk–Stratonovich integral (see, e.g., [11], Problem 3.14, p. 156, or [17], Theorem 34, p. 82).

Proposition 15.1 *Let $Y(t)$ be a d-dimensional continuous semimartingale, and $f : \mathbb{R}^d \to \mathbb{R}^k$ a vector function with the components belonging to $\mathcal{C}^3(\mathbb{R}^d)$. Then*

$$f\big(Y(t)\big) = f\big(Y(0)\big) + \sum_{i=1}^d \int_0^t \frac{\partial f}{\partial x_i}\big(Y(s)\big) \circ dY_i(s). \tag{15.5}$$

We shall also need the following result.

Lemma 15.2 *Let $X(t)$ and $Y(t)$ be continuous semimartingales with decompositions $X(t) = X(0) + A(t) + \int_0^t M(s)\,dW(s)$ and $Y(t) = Y(0) + B(t) + \int_0^t N(s)\,dW(s)$, where $A(t)$, $B(t)$ are adapted, continuous processes with bounded variation, and the processes defined by the stochastic integrals are (local) martingales (this decomposition holds for any continuous semimartingale, since the filtration $(\mathcal{F}_t)$ stands for the completion of the natural filtration generated by W, see* [17], *Theorem* 43, *Chap.* IV). *Then*

$$\int_0^t X(s) \circ d\bigg(\int_0^s Y(r) \circ dW(r)\bigg) = \int_0^t X(s)Y(s) \circ dW(s). \tag{15.6}$$

Proof The first term of the left-hand side of the formula can be written as

$$\int_0^t X(s) \circ d\bigg(\int_0^s Y(r) \circ dW(r)\bigg)$$

$$
\begin{aligned}
&= \int_0^t X(s) \circ d\Big(\int_0^s Y(r)\,dW(r) + \frac{1}{2}\int_0^s N(r)\,dr\Big) \\
&= \int_0^t X(s)\,d\Big(\int_0^s Y(r)\,dW(r) + \frac{1}{2}\int_0^s N(r)\,dr\Big) + \frac{1}{2}\int_0^t M(s)Y(s)\,ds \\
&= \int_0^t X(s)Y(s)\,dW(s) + \frac{1}{2}\int_0^t X(s)N(s)\,ds + \frac{1}{2}\int_0^t M(s)Y(s)\,ds \\
&= \int_0^t X(s)Y(s)\,dW(s) + \frac{1}{2}\langle XY, W\rangle_t,
\end{aligned}
$$

where the formula of integration by parts (for semimartingales) was also used, in order to derive the martingale part of $X(t)Y(t)$. □

The corresponding system of characteristics (see, e.g., [14], Chap. 6) is given by

$$
\begin{cases}
d\hat{x}(t;\lambda) = -\hat{u}(t;\lambda)g_0(\hat{x}(t;\lambda))\,dt + \sum_{i=1}^m (-g_i)(\hat{x}(t;\lambda)) \circ dW_i(t); \\
\hat{x}(0,\lambda) = \lambda; \\
d\hat{u}(t,\lambda) = 0, \quad \hat{u}(0,\lambda) = \varphi(\lambda),\ \lambda \in \mathbb{R}^n.
\end{cases} \tag{15.7}
$$

Remark 15.3 Notice that the integrals $\int_0^t (-g_i)(\hat{x}(s;\lambda)) \circ dW_i(s)$ and $-\int_0^t g_i(\hat{x}(s;\lambda)) \circ dW_i(s)$ are not equal.

We deduce that $\hat{u}(t,\lambda) = \varphi(\lambda)$ and $\hat{x}$ is the solution of the SDEs

$$
\begin{aligned}
\hat{x}(t;\lambda) &= \lambda - \varphi(\lambda)\int_0^t g_0\big(\hat{x}(s;\lambda)\big)\,ds + \sum_{i=1}^m \int_0^t (-g_i)\big(\hat{x}(s;\lambda)\big) \circ dW_i(s) \\
&= \lambda - \int_0^t \Big[\varphi(\lambda)g_0\big(\hat{x}(s;\lambda)\big) - \frac{1}{2}\nabla g_i\big(\hat{x}(s;\lambda)\big)\cdot g_i\big(\hat{x}(s;\lambda)\big)\Big]\,ds \\
&\quad - \sum_{i=1}^m \int_0^t g_i\big(\hat{x}(s;\lambda)\big)\,dW_i(s).
\end{aligned} \tag{15.8}
$$

According to formula (15.4), the (local) martingale part of $\int_0^t (-g_i)(\hat{x}(s;\lambda)) \circ dW_i(s)$ is given by $-\int_0^t g_i(\hat{x}(s;\lambda))\,dW_i(s)$, which is also the (local) martingale part of the process $\hat{x}(t;\lambda)$ (see (15.7)). Hence, by virtue of Itô's lemma, the martingale part of $(-g_i)(\hat{x}(t;\lambda))$ is $\int_0^t \nabla g_i(\hat{x}(s;\lambda)) \cdot g_i(\hat{x}(s;\lambda))\,dW_i(s)$, and this implies that

$$
\big\langle(-g_i^j)\big(\hat{x}(\cdot;\lambda)\big), W_i(\cdot)\big\rangle_t = \int_0^t \big(\nabla g_i\big(\hat{x}(s;\lambda)\big)\cdot g_i\big(\hat{x}(s;\lambda)\big)\big)^j\,ds
$$

for $j = 1,\dots,n$.

The assumptions imposed on the coefficients g_i, $i = 0,\dots,m$, ensure the existence of a unique solution $\hat{x}_\varphi(t;\lambda)$ of system (15.8). Under the same assumptions, the vector fields g_i $i = 0, 1, \dots, m$, are complete, i.e., they generate globally defined flows $G_i(t, x) = G_i(t)(x)$ satisfying

$$\frac{\partial G_i}{\partial t}(t, x) = g_i\big(G_i(t, x)\big) \quad \text{for all } t \in \mathbb{R},\ x \in \mathbb{R}^n;\ G_i(0, x) = x.$$

It is well known that for each t, $G_i(t)(\cdot)$ is a diffeomorphism, the map $(t, x) \in \mathbb{R} \times \mathbb{R}^n \mapsto G_i(t, x)$ is smooth, and $G_i(t_1 + t_2, x) = G_i(t_1)(G_i(t_2, x))$. The last property implies that $(G_i(t))^{-1}(\cdot) = G_i(-t)(\cdot) := H_i(t)(\cdot)$. We define $G(p)(x)$, $p = (t_1, \dots, t_m) \in \mathbb{R}^m$, $x \in \mathbb{R}^n$, as the composition of the flows associated to $g_1, \dots, g_m$, i.e.,

$$G(p)(x) = G(p, x) := G_1(t_1) \circ \cdots \circ G_m(t_m)(x). \tag{15.9}$$

We assume from now on that the vector fields $g_0, \dots, g_m$ commute with respect to the usual Lie bracket, i.e.,

(A.4) $[g_i, g_j](x) := \nabla g_i(x) g_j(x) - \nabla g_j(x) g_i(x) = 0$,

and this means that $G_i(t_i) \circ G_j(t_j) = G_j(t_j) \circ G_i(t_i)$ for $0 \le i, j \le m$. Under this assumption, the composition of flows $G(p, x)$ is the solution of the gradient system defined by the original vector fields, i.e.,

$$\frac{\partial G}{\partial t_i}(p, x) = g_i\big(G(p, x)\big).$$

Set also $H(p, x) := G(-p, x)$ for $p = (t_1, \dots, t_m)$.

15.3 Gradient Representation of Stochastic Flow and Construction of a Solution of Nonlinear SPDE

The next lemma provides a gradient representation for the stochastic flow $\hat{x}_\varphi(t; \lambda)$.

Lemma 15.4 *The stochastic flow generated by the solution of the SDEs* (15.8) *can be represented as*

$$\hat{x}_\varphi(t; \lambda) = G\big(-W(t)\big) \circ G_0\big(-t\varphi(\lambda)\big)(\lambda) = H\big(W(t)\big) \circ H_0\big(t\varphi(\lambda)\big)(\lambda). \tag{15.10}$$

Proof Set $v(t, y) := G(-y) \circ G_0(-t\varphi(\lambda))(\lambda)$. It is obvious that $v \in \mathcal{C}^{1,3}(\mathbb{R} \times \mathbb{R}^m; \mathbb{R}^n)$, and a slightly modified version of Proposition 15.1 leads us to

$$\begin{aligned} v\big(t, W(t)\big) &= \lambda + \int_0^t \frac{\partial v}{\partial t}\big(s, W(s)\big)\, ds + \sum_{i=1}^m \int_0^t \frac{\partial v}{\partial y_i}\big(s, W(s)\big) \circ dW_i(s) \\ &= \lambda - \varphi(\lambda) \int_0^t g_0\big(v\big(s, W(s)\big)\big)\, ds \end{aligned}$$

$$+\sum_{i=1}^{m}\int_0^t (-g_i)\big(v\big(s, W(s)\big)\big)\circ dW_i(s).$$

The result follows by the uniqueness of solutions of SDEs. □

The next step consists in finding the inverse mapping of the diffeomorphism $\lambda \to \hat{x}_\varphi(t;\lambda)$, i.e., we solve the equation

$$\hat{x}_\varphi(t;\lambda) = x \tag{15.11}$$

with respect to the unknown λ. Taking into account the formula (15.10) and the properties of the flows G_i (which are preserved by G), this is equivalent with

$$G_0\big(-t\varphi(\lambda)\big)(\lambda) = G\big(W(t)\big)(x) := z(t,x).$$

We first consider the equation $G_0(-t\varphi(\lambda))(\lambda) = z$ for arbitrary $t \in [0,T]$ and $z \in \mathbb{R}^n$, which can be rewritten as

$$G_0\big(t\varphi(\lambda)\big)(z) = \lambda. \tag{15.12}$$

Set $V(t,z,\lambda) := G_0(t\varphi(\lambda))(z)$.

Lemma 15.5 *Equation* (15.12) *admits a unique solution given by a* (*deterministic*) *smooth mapping* $\hat{\psi}(t,z) \in \mathcal{C}^{1,1}([0,T]\times\mathbb{R}^n;\mathbb{R}^n)$, *which satisfies the estimate*

$$\big|\hat{\psi}(t,z) - z\big| \le \frac{TK}{1-\rho}\big|\varphi(z)\big|.$$

In addition, $\hat{\psi}(t,z)$ *is the unique solution of the Hamilton–Jacobi equations*

$$\begin{cases} \frac{\partial\hat{\psi}}{\partial t}(t,z) = \nabla\hat{\psi}(t,z)\, g_0(z)\, \varphi(\hat{\psi}(t,z)), \\ \hat{\psi}(0,z) = z. \end{cases} \tag{15.13}$$

Proof Notice that the mapping $\lambda \in \mathbb{R}^n \mapsto V(t,z,\lambda)$ is a contractive mapping, uniformly with respect to $(t,z) \in [0,T]\times\mathbb{R}^n$, since

$$\big|\nabla_\lambda V(t,z,\lambda)\big| = \big|g_0\big(V(t,z,\lambda)\big)\big|\big|t\nabla\varphi(\lambda)\big| \le \rho < 1, \tag{15.14}$$

according to assumption (A3). The sequence $(\lambda_k)(t,z)$ defined by

$$\lambda_0(t,z) = z, \qquad \lambda_{k+1}(t,z) = V\big(t,z,\lambda_k(t,z)\big)$$

satisfies

$$\big|\lambda_{k+1}(t,z) - \lambda_k(t,z)\big| \le \rho^k\big|\lambda_1(t,z) - \lambda_0(t,z)\big|$$

and

$$\big|\lambda_1(t,z) - \lambda_0(t,z)\big| = \big|V(t,z,z) - z\big| \le TK\big|\varphi(z)\big|.$$

A standard procedure leads us to the first part of the lemma. Furthermore, using the properties of flows, we get

$$\hat{\psi}(0,z)=V\big(0,z,\hat{\psi}(0,z)\big)=G_0(0,z)=z$$

and

$$V\big(t,G_0\big(-t\varphi(\lambda),\lambda\big),\lambda\big)=G_0\big(t\varphi(\lambda),G_0\big(-t\varphi(\lambda),\lambda\big)\big)=\lambda.$$

A straight differentiation with respect to t leads us to

$$\frac{\partial V}{\partial t}\big(t,G_0\big(-t\varphi(\lambda),\lambda\big),\lambda\big)-\nabla_z V\big(t,G_0\big(-t\varphi(\lambda),\lambda\big),\lambda\big)g_0\big(G_0\big(-t\varphi(\lambda),\lambda\big)\big)\,\varphi(\lambda)$$
$$=0,$$

and in particular, for $\lambda=\hat{\psi}(t,z)$, it yields

$$\frac{\partial V}{\partial t}\big(t,z,\hat{\psi}(t,z)\big)-\nabla_z V\big(t,z,\hat{\psi}(t,z)\big)g_0(z)\,\varphi\big(\hat{\psi}(t,z)\big)=0. \tag{15.15}$$

On the other hand, differentiation with respect to t, λ in the equality $V(t,z,\hat{\psi}(t,z)) = \hat{\psi}(t,z)$ yields

$$\frac{\partial\hat{\psi}}{\partial t}(t,z)=\frac{\partial V}{\partial t}\big(t,z,\hat{\psi}(t,z)\big)+\nabla_\lambda V\big(t,z,\hat{\psi}(t,z)\big)\frac{\partial\hat{\psi}}{\partial t}(t,z)$$

and

$$\nabla\hat{\psi}(t,z)=\nabla_z V\big(t,z,\hat{\psi}(t,z)\big)+\nabla_\lambda V\big(t,z,\hat{\psi}(t,z)\big)\nabla\hat{\psi}(t,z).$$

Taking into account estimate (15.14), it is easy to see that the matrix $I_n - \nabla_\lambda V(t,z,\hat{\psi}(t,z))$ is invertible (here I_n stands for the $(n\times n)$-identity matrix), and it holds

$$\frac{\partial\hat{\psi}}{\partial t}(t,z)=\big[I_n-\nabla_\lambda V\big(t,z,\hat{\psi}(t,z)\big)\big]^{-1}\frac{\partial V}{\partial t}\big(t,z,\hat{\psi}(t,z)\big)$$

and

$$\nabla\hat{\psi}(t,z)=\big[I_n-\nabla_\lambda V\big(t,z,\hat{\psi}(t,z)\big)\big]^{-1}\nabla_z V\big(t,z,\hat{\psi}(t,z)\big).$$

Finally, combining the last two formulas with (15.15), we get the last conclusion of the lemma. □

The next result is straightforward.

Corollary 15.6 *The flow equation* (15.11) *allows a unique solution* $\lambda=\psi(t,x)$, *which can be represented as* $\psi(t,x):=\hat{\psi}(t,z(t,x))$, *where recall that* $z(t,x)=G(W(t))(x)$. *Moreover, the mapping* $\psi(t,x)$ *is smooth with respect to* (t,x) *and is* $(\mathcal{F}_t)$*-adapted for every fixed* x.

Notice now that the composition of flows $G(p,x)$ is the solution of the following Hamilton–Jacobi equations

$$\frac{\partial G}{\partial t_i}(p,x)=\nabla G(p,x)g_i(x), \qquad G(0,x)=x. \tag{15.16}$$

For notational convenience, let us prove this formula only for $m=1$. Obviously, $G_1(t,G_1(-t,x))=x$, and differentiation with respect to t yields

$$\frac{\partial G_1}{\partial t}\big(t,G_1(-t,x)\big)-\nabla G_1\big(t,G_1(-t,x)\big)g_1\big(G_1(t,x)\big)=0.$$

Replacing now x with $G_1(t,x)$, we get the desired result. Since $z(t,x)=G(W(t),x)$, by virtue of Proposition 15.1 and formula (15.16), we obtain

$$\begin{aligned} z(t,x)&=x+\sum_{i=1}^{m}\int_0^t\frac{\partial G}{\partial t_i}\big(W(s),x\big)\circ dW_i(s)\\ &=x+\sum_{i=1}^{m}\int_0^t\nabla G\big(W(s),x\big)g_i(x)\circ dW_i(s)\\ &=x+\sum_{i=1}^{m}\int_0^t\nabla z(s,x)g_i(x)\circ dW_i(s). \end{aligned}\tag{15.17}$$

Recall that the vector fields g_i, $i=0,\dots,m$, are commuting. Hence,

$$G_0\big(t_0,z(t,x)\big)=z\big(t,G_0(t_0,x)\big),$$

and differentiation with respect to t_0 yields

$$g_0\big(G_0\big(t_0,z(t,x)\big)\big)=\nabla z\big(t,G_0(t_0,x)\big)g_0\big(G_0(t_0,x)\big).$$

By replacing x with $G_0(-t_0,x)$ we get

$$g_0\big(z(t,x)\big)=\nabla z(t,x)g_0(x). \tag{15.18}$$

We are now in position to state the main result of this section.

Theorem 15.7 *Set $u(t,x):=\varphi(\psi(t,x))$. Then, under assumptions* (A.1)–(A.4), *$u(t,x)$ is a classical solution of the nonlinear SPDE* (15.2).

Proof The stochastic rule of derivation stated in Proposition 15.1 applied to $u(t,x)=\varphi(\hat{\psi}(t,z(t,x)))$ reads

$$\begin{aligned} du(t,x)=&\left\langle\nabla\varphi\big(\hat{\psi}\big(t,z(t,x)\big)\big),\frac{\partial\hat{\psi}}{\partial t}\big(t,z(t,x)\big)\right\rangle dt\\ &+\big(\nabla\varphi\big(\hat{\psi}\big(t,z(t,x)\big)\big)\big)^{*}\nabla\hat{\psi}\big(t,z(t,x)\big)\circ dz(t,x). \end{aligned}$$

Taking into account the system of PDEs (15.13) satisfied by $\hat{\psi}(t,z)$ and formula (15.18), notice that the first term from the right-hand side is equal to

$$\begin{aligned}&\big\langle\nabla\varphi\big(\hat{\psi}\big(t,z(t,x)\big)\big),\nabla\hat{\psi}\big(t,z(t,x)\big)g_0\big(z(t,x)\big)\big\rangle\,\varphi\big(\hat{\psi}\big(t,z(t,x)\big)\big)\\&\quad=\big\langle\nabla u(t,x),g_0(x)\big\rangle\,u(t,x),\end{aligned}$$

while by virtue of Lemma 15.2 and formula (15.17), the second term from the r.h.s. is rewritten as

$$\begin{aligned}&\sum_{i=1}^{m}\big\langle\big(\nabla\varphi\big(\hat{\psi}\big(t,z(t,x)\big)\big)\big)^*\nabla\hat{\psi}\big(t,z(t,x)\big)\nabla z(t,x),g_i(x)\big\rangle\circ dW_i(t)\\&\quad=\sum_{i=1}^{m}\big\langle\nabla u(t,x),g_i(x)\big\rangle\circ dW_i(t).\end{aligned}$$

The proof is complete. □

Remark 15.8 The random smooth vector function $\psi(t,x)$ is a fundamental solution of the SPDE (15.2) constructed via n linearly independent solutions. It is obtained as the composition between the deterministic smooth mapping $\hat{\psi}(t,x)$ (which satisfies the Hamilton–Jacobi equations (15.13)) and $z(t,x)$, the fundamental solution of the reduced SPDE (15.17). It fulfills the nonlinear SPDE

$$\begin{cases}d\psi(t,x)=\langle\nabla\psi(t,x),g_0(x)\rangle\,\varphi(\psi(t,x))\,dt\\\qquad\qquad+\sum_{i=1}^{m}\langle\nabla\psi(t,x),g_i(x)\rangle\circ dW_i(t),\\\psi(0,x)=x,\quad t\in[0,T],\ x\in\mathbb{R}^n.\end{cases}\tag{15.19}$$

Remark 15.9 If we drop the commuting property of the vector fields $g_0,g_1,\dots,g_m$, one step forward would consist in assuming that the vector fields $g_1,\dots,g_m$ are in involution over $\mathbb{R}$, i.e.,

$$[g_i,g_j](x)=\sum_{k=1}^{m}\alpha_k g_k(x)\quad\forall x\in\mathbb{R}^n,$$

with the scalars α_k depending on g_i,g_j. In this case, we have a global gradient representation of the form

$$\nabla_p G(p,x)=\big(g_1\big(G(p,x)\big),\dots,g_m\big(G(p,x)\big)\big)A(p),$$

where $A(p)$ is a nonsingular $(m\times m)$-matrix for every $p\in\mathbb{R}^m$, not depending on the origin x.

According to [20], there exist smooth vector fields $q_j(p)$, $j=1,\dots,m$, such that $\nabla_p G(p,x)q_j(p)=g_j(G(p,x))$, which implies

$$\nabla_p H(p,x)q_j(p)=-\nabla_x H(p,x)g_j(x).$$

Consider the stochastic differential system

$$y(t) = \lambda + \sum_{j=1}^{m} \int_0^t (-g_j)\big(y(s)\big) \circ dW_j(s),$$

which is obtained by taking only the "diffusion part" of (15.7). When solving the auxiliary SDE

$$p(t) = -\sum_{j=1}^{m} \int_0^t q_j\big(p(s)\big) \circ dW_j(s),$$

notice that the diffusion fields are not Lipschitz and do not have linear growth. Define a $\mathcal{C}_0^\infty$ function $\rho(p)$ which is equal to 1 in the closed ball $\{p \in \mathbb{R}^m \mid |p| \le M\}$, where M is an arbitrary positive number. Set $\tilde{q}_j(p) := \rho(p)q_j(p)$. The SDE $p(t) = -\sum_{j=1}^m \int_0^t \tilde{q}_j(p(s)) \circ dW_j(s)$ satisfies the conditions of existence and uniqueness of the solution, and let $\tilde{p}(t)$ be its solution. Define now the stopping time $\tau := \inf\{t \in [0, T] \mid |\tilde{p}(t)| \ge M\}$. It follows that the stopped process $\hat{p}(t) := \tilde{p}(t \wedge \tau)$ takes values in B_M and satisfies

$$\hat{p}(t) = -\sum_{j=1}^{m} \int_0^{t\wedge\tau} q_j\big(\hat{p}(s)\big) \circ dW_j(s).$$

If we assume that g_0 commutes with each g_j, $j = 1, \ldots, m$, it is easy to check that the gradient representation for the stochastic flow $\hat{x}_\varphi(t; \lambda)$ is given by $\hat{x}_\varphi(t; \lambda) = G(-\hat{p}(t)) \circ G_0(-t\varphi(\lambda))(\lambda)$ for $t \in [0, \tau]$. The results stated in Lemma 15.5 and Theorem 15.7 remain valid, but the differential equations appearing there are satisfied only for $t \in [0, \tau]$.

Using the same type of arguments, we can extend our analysis to SDEs of the form

$$y(t) = \lambda + \sum_{i=1}^{d} \int_0^t \varphi_i(\lambda) f_i\big(y(s)\big)\, ds + \sum_{j=1}^{m} \int_0^t g_j\big(y(s)\big) \circ dW_j(s),$$
$$t \in [0, T]. \tag{15.20}$$

Here we assume that

- $f_1, \ldots, f_d, g_1, \ldots, g_m$ commute with respect to the usual Lie bracket;
- the vector fields $g_1, \ldots, g_m$ belong to $\mathcal{C}^2(\mathbb{R}^n; \mathbb{R}^n)$ with bounded partial derivatives of first and second orders; $f_1, \ldots, f_d \in \mathcal{C}_b^1(\mathbb{R}^n; \mathbb{R}^n)$, and their partial derivatives are bounded;
- $\varphi_1, \ldots, \varphi_d \in \mathcal{C}^2(\mathbb{R}^n)$ and admit bounded first-order partial derivatives.
- $\rho := TMK < 1$, where $M := \sup\{|\nabla\varphi_i(x)|, x \in \mathbb{R}^n, i = 1, \ldots d\}$ and $K := \sup\{|f_i(x)|, x \in \mathbb{R}^n,\ i = 1, \ldots, d\}$.

Denote by $G_j(t,x) = G_j(t)(x)$ the global flow associated to each complete vector field $g_j(x)$ and by $F_i(t,x) = F_i(t)(x)$ the global flow generated by the complete vector field $f_i(x)$. Set also the compositions of flows $G(p)(x) = G(p,x) := G_1(t_1) \circ \cdots \circ G_m(t_m)(x)$ for $p = (t_1, \ldots, t_m) \in \mathbb{R}^m$ and $F(q)(x) = F(q,x) := F_1(t_1) \circ \cdots \circ F_p(t_d)(x)$ for $q = (t_1, \ldots, t_d) \in \mathbb{R}^d$.

The following lemma can be proved by combining the same type of arguments with those leading to Lemma 15.5.

Lemma 15.10 *There exists a (unique) smooth mapping $\hat{\psi}(t,z)$ such that*

$$G\big(p\big(t, \hat{\psi}(t,z)\big)\big)\big(\hat{\psi}(t,z)\big) = z, \qquad \hat{\psi}(0,z) = z.$$

Moreover,

$$\big|\hat{\psi}(t,z) - z\big| \le \frac{TK}{1-\rho}\big|\varphi(z)\big|,$$

and $\hat{\psi}(t,z)$ satisfies the Hamilton–Jacobi equation

$$\frac{\partial \hat{\psi}}{\partial t}(t,z) + \sum_{i=1}^{d} \nabla \hat{\psi}(t,z) \cdot f_i(z)\, \varphi_i\big(\hat{\psi}(t,z)\big) = 0. \tag{15.21}$$

We are in position to state the following:

Theorem 15.11 *Under the assumptions from above, the stochastic flow associated to SDE* (15.20) *has the form*

$$\hat{y}(t;\lambda) = G\big(W(t)\big) \circ F\big(p(t,\lambda)\big)(\lambda), \quad t \in [0,T],\ \lambda \in \mathbb{R}^n, \tag{15.22}$$

where $p(t,\lambda) := (t\varphi_1(\lambda), \ldots, t\varphi_d(\lambda))$. In addition, the flow equation $\hat{y}(t;\lambda) = x$ admits the solution $\lambda = \psi(t,x) := \hat{\psi}(t, z(t,x))$, where $z(t,x) := G(-W(t))(x)$.

The proof of this theorem is quite similar to the proof of Theorem 15.7.

15.4 Applications

15.4.1 Pathwise Solutions of Burgers Equations with Stochastic Perturbations

In this section we provide a construction of a solution of a system of Burgers equations with stochastic perturbations by using the results obtained in the previous section. The SPDEs under consideration are given by

$$\begin{cases} du_i(t,x) = [\frac{1}{2}\Delta u_i(t,x) + \langle \nabla u_i(t,x), u(t,x)\rangle]\,dt \\ \qquad\qquad + \sum_{k=1}^{n} \frac{\partial u_i}{\partial x_k}(t,x)\,dW_k(t), \quad t\in[0,T], \\ u_i(0,x) = \varphi_i(x), \quad x\in\mathbb{R}^n,\ i=1,\dots,n. \end{cases} \tag{15.23}$$

Here $(W(t))$ is a standard n-dimensional Brownian motion defined on a complete filtered probability space $\{\Omega, \mathcal{F}, \{\mathcal{F}_t\}, P\}$, $\varphi_i \in \mathcal{C}^2(\mathbb{R}^n)$ with bounded first-order derivatives, and the stochastic integral is the usual Itô integral.

We are looking for smooth solutions with respect to the space variable and which are $(\mathcal{F}_t)$-adapted for fixed x. Differentiation with respect to x_l yields

$$\frac{\partial u_i}{\partial x_l}(t,x) = \frac{\partial \varphi_i}{\partial x_l}(x) + \sum_{k=1}^{n} \int_0^t \left[\frac{1}{2}\frac{\partial^3 u_i}{\partial x_l \partial x_k^2}(s,x) + \frac{\partial^2 u_i}{\partial x_l \partial x_k}(s,x) u_k(s,x) \right.$$
$$\left. + \frac{\partial u_i}{\partial x_k}(s,x)\frac{\partial u_k}{\partial x_l}(s,x)\right] ds + \sum_{k=1}^{n} \int_0^t \frac{\partial^2 u_i}{\partial x_l \partial x_k}(s,x)\,dW_k(s),$$

where the derivatives with respect to x_l have to be understood in the L^2 sense, and since the mapping $u(t,\cdot)$ is smooth, they coincide with the classical ones. We deduce

$$\left\langle \frac{\partial u_i}{\partial x_l}(\cdot,x), W_l(\cdot)\right\rangle_t = \int_0^t \frac{\partial^2 u_i}{\partial x_l^2}(s,x)\,ds.$$

Therefore, using formula (15.4), it is easy to see that system (15.23) can be rewritten as

$$\begin{cases} du_i(t,x) = \langle \nabla u_i(t,x), \sum_{k=1}^{n} u_k(t,x) e_k\rangle\,dt \\ \qquad\qquad + \sum_{k=1}^{n} \langle \nabla u_i(t,x), e_k\rangle \circ dW_k(t), \\ u_i(0,x) = \varphi_i(x), \end{cases} \tag{15.24}$$

where the system $\{e_1,\dots,e_n\}$ stands for the canonical basis of $\mathbb{R}^n$. Proceeding in a similar way we did for (15.2), we associate the following system of characteristics

$$\begin{cases} d\hat{x}(t;\lambda) = -\sum_{k=1}^{n} \hat{u}_k(t;\lambda) e_k\,dt - \sum_{k=1}^{n} e_k \circ dW_k(t), \\ \hat{x}(0,\lambda) = \lambda \in \mathbb{R}^n; \\ d\hat{u}_i(t,\lambda) = 0, \quad t\in[0,T],\ \hat{u}_i(0,\lambda) = \varphi_i(\lambda). \end{cases} \tag{15.25}$$

It yields $\hat{u}_i(t,\lambda) = \varphi_i(\lambda)$, and $\hat{x}(t;\lambda)$ satisfies the system of SDEs

$$\hat{x}(t;\lambda) = \lambda - \sum_{k=1}^{n} \int_0^t \varphi_i(\lambda) e_k\,ds - \sum_{k=1}^{n} \int_0^t e_k \circ dW_k(s)$$

$$= \lambda + \sum_{k=1}^{n} \int_0^t \varphi_i(\lambda)(-e_k)\,ds$$

$$+ \sum_{k=1}^{n} \int_0^t (-e_k) \circ dW_k(s), \quad 0 \le t \le T. \tag{15.26}$$

Assume that $TK = \rho < 1$, where $K := \sup\{|\nabla\varphi_i(\lambda|; \lambda \in \mathbb{R}^n, i = 1, \dots, n\}$. In the setting of Theorem 15.11, $d = m = n$ and $f_i(y) = g_i(y) = -e_i$ for $1 \le i \le n$. Hence, $F_i(t, x) = G_i(t, x) = -te_i + x$, and for $t = (t_1, \dots, t_n)$,

$$F(t, x) = G(t, x) = -\sum_{i=1}^{n} t_i e_i + x = -t + x.$$

Set $p(t, \lambda) := t\varphi(\lambda)$, where $\varphi(\lambda) = (\varphi_1(\lambda), \dots, \varphi_n(\lambda))$. Then

$$G\big(W(t)\big) \circ F\big(p(t, \lambda)\big)(\lambda) = -W(t) + F\big(t\varphi(\lambda)\big)(\lambda) = -W(t) - t\varphi(\lambda) + \lambda.$$

We apply now Theorem 15.11 and obtain the following:

Theorem 15.12 *The stochastic flow $\hat{x}(t; \lambda)$ can be represented as $\hat{x}(t; \lambda) = \lambda - t\varphi(\lambda) - W(t)$, and the flow equation $\hat{x}(t; \lambda) = x$ has a unique solution given by $\lambda = \psi(t, x) = \hat{\psi}(t, x + W(t))$, where $\hat{\psi}(t, z)$ is the solution of the Hamilton–Jacobi equations*

$$\begin{cases} \frac{\partial\hat{\psi}}{\partial t}(t, z) = \sum_{i=1}^{n} \frac{\partial\hat{\psi}}{\partial z_i}(t, z)\varphi_i(\hat{\psi}(t, z)), \\ \hat{\psi}(0, z) = z. \end{cases}$$

Set $u_i(t, x) := \varphi_i(\psi(t, x))$ for $i = 1, \dots, n$. Then $u(t, x) = (u_1(t,), \dots, u_n(t, x)) = \varphi(\psi(t, x))$ is a solution of the system of stochastic Burgers equations (15.24).

15.4.2 A Filtering Problem for SDEs Associated with Parameterized Backward Parabolic Equations

In the setting of Sect. 15.3, we consider the (slightly modified) SDE

$$\begin{cases} dx(t) = \varphi(\lambda)g_0(x(t))\,dt + \sum_{i=1}^{m} g_i(x(t)) \circ dW_i(t), \\ x(0) = \lambda \end{cases} \tag{15.27}$$

(which is obtained from the SDE (15.8) by simply replacing g_i with $-g_i$) and admitting as a solution the stochastic flow $\hat{x}_\varphi(t, \lambda)$. The flow equation $\hat{x}_\varphi(t, \lambda) = x$, with respect to the unknown λ, has a unique solution $\lambda = \psi(t, x)$.

Set $\hat{x}_{\varphi}(s;t,x),\ t \le s \le T$, the stochastic flow associated to the SDE

$$dx(s) = \varphi\big(\psi(t,x)\big) g_0\big(x(s)\big)\, ds + \sum_{i=1}^{m} g_i\big(x(s)\big) \circ dW_i(s), \tag{15.28}$$

obtained by means of the SDE (15.27) with parameter $\lambda = \psi(t,x)$.

Our goal is to compute expectations of the form $E(h(\hat{x}_{\varphi}(T;t,x)))$, involving the non-Markovian process $\hat{x}_{\varphi}(s;t,x)$.

Usually, when one wants to compute $u(t,x)$ defined as the expectation of some functional depending on the final value $\xi(T;t,x)$ of some diffusion process $\xi(s;t,x),\ t \le s \le T$, starting at time t from the point x, it is well known that the function $u(t,x)$ is the solution of some backward parabolic equation, called the Kolmogorov equation (see, for instance, [8], Theorem 6.1). This procedure cannot be applied in our case if we take into account the non-Markovian nature of the process involved.

Let $h \in \mathcal{C}^2(\mathbb{R}^n)$ with bounded first-order partial derivatives. In order to compute the expectation $E(h(\hat{x}_{\varphi}(T;t,x)))$, we consider the conditional expectation $v(t,x) := E[h(\hat{x}_{\varphi}(T;t,x)) | \psi(t,x)]$. A direct computation of $v(t,x)$ involves the knowledge of the process $\psi(t,x)$, for which a favorable situation indicates a nonlinear SPDE. A more suitable description of $v(t,x) = u(t,x,\psi(t,x))$ is obtained by using the parameterized version $u(t,x,\lambda)$, which is the solution of a backward Kolmogorov equation with parameter.

Using the results obtained in Sect. 15.3, notice that the gradient representation of the stochastic flow $\hat{x}_{\varphi}(T;t,x)$ is given by

$$\hat{x}_{\varphi}(T;t,x) = G\big(W(T) - W(t)\big) \circ G_0\big((T-t)\varphi\big(\psi(t,x)\big)\big)(x). \tag{15.29}$$

Set $v(t,x) := E[h(\hat{x}_{\varphi}(T;t,x)) | \psi(t,x)]$ and $y_{\varphi}(s;t,x,\lambda) := G(W(s) - W(t)) \circ G_0((s-t)\varphi(\lambda))(x)$ for $t \le s \le T$.

Since $\psi(t,x) = \hat{\psi}(t, G(-W(t),x))$ (see Corollary 15.6) and $\hat{\psi}(t,z)$ is deterministic (recall Lemma 15.5), it follows that the random variables $\psi(t,x)$ and $y_{\varphi}(T;t,x,\lambda)$ are independent. Notice that $\hat{x}_{\varphi}(T;t,x) = y_{\varphi}(T;t,x,\psi(t,x))$. Therefore, the Independence Lemma (see [18], Lemma 2.3.4) leads us to the representation

$$v(t,x) = E\big[h\big(y_{\varphi}(T;t,x,\lambda)\big)\big]\big|_{\lambda=\psi(t,x)}.$$

Define $u(t,x;\lambda) := E[h(y_{\varphi}(T;t,x,\lambda))]$. Obviously, $y_{\varphi}(s;t,x,\lambda)$ is the solution of the SDE

$$y(s) = x + \varphi(\lambda) \int_t^s g_0\big(y(r)\big)\, dr + \sum_{i=1}^{m} \int_t^s g_i\big(y(r)\big) \circ dW_i(r), \quad s \in [t,T].$$

Clearly, $y_{\varphi}(s;t,x,\lambda)$ is a Markovian process. Applying now Theorem 6.1 of [8], it is straightforward that $u(t,x;\lambda)$ satisfies the parameterized Kolmogorov backward parabolic equation

$$\begin{cases} \frac{\partial u}{\partial t}(t,x;\lambda) + \langle \nabla u(t,x;\lambda), g(x,\lambda)\rangle \\ \quad + \frac{1}{2}\sum_{i=1}^{m}\langle D^2 u(t,x;\lambda) g_i(x), g_i(x)\rangle = 0, \\ u(T,x;\lambda) = h(x), \quad t \in [0,T], \end{cases} \tag{15.30}$$

where $g(x,\lambda) := g_0(x)\varphi(\lambda) + \frac{1}{2}\sum_{i=1}^{m} \nabla g_i(x) g_i(x)$, and $D^2 u$ stands for the Jacobian matrix of u. The analysis from above can be summarized in the next statement

Theorem 15.13 *Under assumptions* (A.1)–(A.4), *the conditional expectation* $v(t,x) = E[h(\hat{x}_\varphi(T;t,x)) | \psi(t,x)]$ *can be represented as*

$$v(t,x) = u(t,x;\lambda)|_{\lambda=\psi(t,x)},$$

where $u(t,x;\lambda)$ *is the solution of the backward parabolic equation* (15.30). *In addition, the expectation* $E(h(\hat{x}_\varphi(T;t,x)))$ *can be computed as*

$$E\big(h\big(\hat{x}_\varphi(T;t,x)\big)\big) = E\big(v(t,x)\big).$$

References

1. A. Bensoussan, Some existence results for stochastic partial differential equations, in *Stochastic Partial Differential Equations and Applications*, ed. by G. Da Prato, L. Tubaro. Pitman Res. Notes Math. Ser., vol. 268 (1992), pp. 37–53
2. R. Buckdahn, J. Ma, Stochastic viscosity solutions for nonlinear stochastic partial differential equations. Part I. Stoch. Process. Appl. **93**, 181–204 (2001)
3. L. Bertini, N. Cancrini, G. Jona-Lasinio, The stochastic Burgers equation. Commun. Math. Phys. **165**, 211–232 (1994)
4. G. Da Prato, J. Zabczyk, *Stochastic Differential Equations in Infinite Dimensions*. Encyclopedia of Mathematics and Its Applications (Cambridge University Press, Cambridge, 1992)
5. G. Da Prato, L. Tubaro, Fully nonlinear stochastic partial differential equations. SIAM J. Math. Anal. **27**(1), 40–55 (1996)
6. G. Da Prato, L. Tubaro, *Stochastic Partial Differential Equations and Applications*. Lecture Notes in Pure and Applied Mathematics, vol. 227 (2002)
7. Y.L. Dalecky, N.Y. Goncharuk, On a quasilinear stochastic equation of parabolic type. Stoch. Anal. Appl. **12**, 103–129 (1994)
8. A. Friedman, *Stochastic Differential Equations and Applications*, vol. I (Academic Press, San Diego, 1975)
9. I. Gyöngy, D. Nualart, On the stochastic Burgers equation in the real line. Ann. Probab. **27**(2), 782–202 (1999)
10. B. Iftimie, C. Vârsan, A pathwise solution for nonlinear parabolic equations with stochastic perturbations. Cent. Eur. J. Math. **3**, 367–381 (2003)
11. I. Karatzas, S. Shreve, *Brownian Motion and Stochastic Calculus*, 2nd edn. (Springer, Berlin, 1991)
12. H. Kunita, On the decomposition of solutions of stochastic differential equations, in *Proc. Durham Conf. Stochastic Integrals*. Lecture Notes, vol. 851 (1980), pp. 213–255
13. H. Kunita, First order stochastic partial differential equations, in *Proc. Taniguchi Int. Sym on Stochastic Analysis*, Japan, 1982. North Holland Math Libr, vol. 32 (1986), pp. 249–269

14. H. Kunita, *Stochastic Flows and Stochastic Differential Equations*, vol. 24 (Cambridge University Press, Cambridge, 1990)
15. P.-L. Lions, P.E. Souganidis, Fully nonlinear stochastic partial differential equations. C. R. Acad. Sci. Paris **1**(326), 1085–1092 (1998)
16. E. Pardoux, Stochastic partial differential equations and filtering of diffusion processes. Stochastics **3**(2), 127–167 (1979)
17. P.E. Protter, *Stochastic Integration and Differential Equations*, 2nd edn. (Springer, Berlin, 2005)
18. S. Shreve, *Stochastic Calculus for Finance II. Continuous-Time Models*. Springer Finance (Springer, Berlin, 2004)
19. L. Tubaro, Some results on stochastic partial differential equations by the stochastic characteristic method. Stoch. Anal. Appl. **62**, 217–230 (1988)
20. C. Vârsan, *Applications of Lie Algebras to Hyperbolic and Stochastic Differential Equations* (Kluwer Academic, Dordrecht, 1999)

Chapter 16
Pricing and Hedging of Rating-Sensitive Claims Modeled by $\mathbb{F}$-doubly Stochastic Markov Chains

Jacek Jakubowski and Mariusz Niewęgłowski

Abstract In this paper, we achieve two goals. First we give a formula describing prices of defaultable rating-sensitive claims of general type. Secondly, we solve the problem of replication of an arbitrary rating-sensitive claim on a market on which we can trade in default free assets and a fixed number of defaultable general rating-sensitive claims. The credit rating migration process is modeled by $\mathbb{F}$-doubly stochastic Markov chains, a broad class of processes which contains Markov chains and is fully characterized by some martingale property.

Keywords Credit derivatives · Ex-dividend price · Cumulative price · Cash flow · Rating migration · Hedging · $\mathbb{F}$-doubly stochastic Markov chain

Mathematics Subject Classification (2010) Primary 91G40 · Secondary 91G20 · 60H30

16.1 Introduction

In this paper, we are interested in pricing and hedging of rating-sensitive claims of general form.

The problem of modeling credit risk taking into consideration rating migration was proposed by Jarrow, Lando, and Turnbull [17]. They took Markov chains to model time evolution of credit ratings. Jarrow et al. [17] considered both the discrete- and continuous-time cases, and within this framework, they derived a val-

---

Research partially supported by MNiSW grant N N201 547838.

J. Jakubowski (✉)
Institute of Mathematics, University of Warsaw, Banacha 2, 02-097 Warszawa, Poland
e-mail: jakub@mimuw.edu.pl

J. Jakubowski · M. Niewęgłowski
Faculty of Mathematics and Information Science, Warsaw University of Technology, Plac Politechniki 1, 00-661 Warszawa, Poland

M. Niewęgłowski
e-mail: M.Nieweglowski@mini.pw.edu.pl

G. Di Nunno, B. Øksendal (eds.), *Advanced Mathematical Methods for Finance*, DOI 10.1007/978-3-642-18412-3_16,

uation formula for defaultable bonds expressed through risk-neutral transition probabilities. Subsequently, Lando [19] extended the framework of Jarrow et al. [17] by constructing a rating migration process which follows a conditional Markov chain (see also Bielecki and Rutkowski [5] for a precise definition of conditional Markov property). In Lando [19] and Bielecki and Rutkowski [5], the generator of the credit rating process follows a matrix-valued stochastic process. We also stress that Lando [19] has considered the problem of providing explicit formulas for some credit derivatives connected with ratings, which is also of interest to us. Lando has shown that, under the assumption that the generator matrix process has eigenvectors constant in time, it is possible to solve the conditional Kolmogorov equation and obtain explicit formulas for bond prices and rating-dependent payoffs. However, the structure of payoffs considered in [19] was very simple compared to ours: only a terminal payoff contingent on rating at a terminal date was considered. We also mention the recent work of Bielecki et al. [7], which deals with the problem of pricing basket derivatives with rating migrations in a very efficient Markovian setting. Recently Hurd and Kuznetsov [12, 13] introduced so-called affine Markov chains models for valuation of basket credit derivatives with rating migrations. They constructed rating processes as continuous Markov chains with time change via an independent affine process. They showed how to price efficiently simple instruments such as defaultable bonds and more complicated ones like CDO's tranches.

The topic of hedging of credit risk was started by Blanchet and Jeanblanc [9] and also Belanger, Shreve, and Wong [3]. In Blanchet and Jeanblanc [9], the martingale representation for some class of payoffs is derived, and then a self-financing trading strategy under the H hypothesis is calculated. Bielecki, Jeanblanc, and Rutkowski [6] examine the problem of pricing and hedging of defaultable claims within Markovian setting. This allows them to use PDE techniques for describing prices and hedging strategies for defaultable claims which are attainable. Bielecki, Jeanblanc, and Rutkowski [8] derive the dynamic of a general defaultable claim under a martingale measure without assuming the H hypothesis. Then they show that in the market with a family of single names CDS, the value of a replication portfolio of an attainable claim coincides with the cumulated price of this claim. Therefore the risk-neutral pricing of claims is supported by replication of these claims by dynamic trading strategies. This holds under the H hypothesis. Subsequently, they generalized these results to a market with several correlated credit names and to pricing and hedging of the so-called first-to-default claims. We emphasize that all these results are proved in the case without rating migrations. In Sect. 16.2, we consider the general notion of pricing and hedging of payment stream. We introduce the notions of ex-dividend price, cumulative price, D-financing portfolio, admissibility for a process D describing total cash flows. To model rating migration, we use $\mathbb{F}$-DS Markov chains. This class of processes contains Markov chains and other processes usually used to model rating migrations. Also the constructions given by Lando [19] and Bielecki and Rutkowski [5] are $\mathbb{F}$-DS Markov chains. We describe $\mathbb{F}$-DS Markov chains in Sect. 16.3. In Sect. 16.4, we consider the problem of pricing defaultable rating-sensitive claims. By a defaultable rating-sensitive claim we mean a classical one broadened by a migration process and payoffs connected with rating changes.

We give a general formula for the ex-dividend price process of defaultable rating-sensitive claim in terms of processes defining this claim and characteristics of the rating migration process (see Theorem 16.38). This generalizes the known results obtained for the case without rating migration (see, e.g., Bielecki, Jeanblanc, and Rutkowski [8]). As an example, we give formulas for some known claims such as a defaultable bond with fractional recovery of par value, Credit Sensitive Note and Credit Default Swap. Section 16.5 is devoted to problems of hedging general rating-sensitive claims. We prove that by trading in default free assets and fixed number of general defaultable rating-sensitive claims, we can replicate an arbitrary rating-sensitive claim. We find appropriate martingale representations, which allows us to construct replication strategies.

This paper significantly extends the results of [14], where only the problem of pricing rating-sensitive claims was considered.

16.2 General Notion of Hedging a Payment Stream

We consider processes on a complete probability space $(\Omega, \mathcal{F}, \mathbb{P})$ with filtration $\mathbb{G}$ satisfying the "usual conditions."

Let D be a given stochastic process which represents total cash flows (dividends) generated by some unspecified claim up to time t. We make the following assumptions concerning D:

1. D is a càdlàg process of finite variation.
2. $D_0 = 0$, that is, there are no payments at $t = 0$.
3. $D_t = D_{t \wedge T}$, which means that D matures at T.

Now, for a given cash flow process D, we introduce a counterpart of the notion of self-financing strategy. We assume that an investor can invest in a money account with price process denoted by B and satisfying

$$dB_t = r_t B_t \, dt, \qquad B_0 = 1,$$

and some m dividend paying assets with price process (given by the market) denoted by $Y = (Y^1, \ldots, Y^m)^\top$, which is assumed to be a semimartingale. While holding assets, we get some cash flows (dividends). By $F = (F^1, \ldots, F^m)^\top$ we denote the process of cumulated cash flows. The process F is supposed to satisfy the same conditions as D. By $V_t(\phi)$ we denote the wealth of the portfolio held by the investor at time t, i.e., $V_t(\phi) := \psi_t B_t + \langle \varphi_t, Y_t \rangle$, where $\varphi = (\varphi^1, \ldots, \varphi^m)^\top$ is a $\mathbb{G}$-predictable stochastic process, and ψ is a $\mathbb{G}$-adapted stochastic process. Considering dividend paying assets, it is convenient to introduce the following process:

$$Y_t^c := Y_t + B_t \int_{]\!]0,t]\!]} \frac{1}{B_u} \, dF_u,$$

we refer to Y^c as the cumulative price process (see also Bielecki et al. [8]). This cumulative price process represents the (ex-dividend) price plus the dividends invested in the money account. Let $Y_t^{c*} := B_t^{-1} Y_t^c$ denote the discounted cumulative prices process. The following definition of martingale measure generalizes the notion of martingale measure for the market with dividend paying assets (see also Duffie [11, Sect. 6.L]).

Definition 16.1 A probability measure $\mathbb{Q}$ equivalent to $\mathbb{P}$ is called a *martingale measure (spot)* for the market $(B, (Y, F))$ if the discounted cumulative price process Y^{c*} is a $\mathbb{G}$-martingale under $\mathbb{Q}$.

Assumption EMM In this paper, we assume that the set of equivalent martingale measures for the market $(B, (Y, F))$ is nonempty.

Remark 16.2 If primary assets do not pay dividends, i.e., $F \equiv 0$, then the above notion of martingale measure is well known from classical arbitrage pricing theory. If primary assets pay dividends, then the definition of martingale measure implies that for each $t \le T$, we have

$$Y_t^* + \int_{]\!]0,t]\!]} \frac{1}{B_u} dF_u = \mathbb{E}_{\mathbb{Q}}\left(Y_T^* + \int_{]\!]0,T]\!]} \frac{1}{B_u} dF_u \middle| \mathcal{G}_t \right),$$

and this gives

$$Y_t = B_t \mathbb{E}_{\mathbb{Q}}\left(\frac{Y_T}{B_T} + \int_{]\!]t,T]\!]} \frac{1}{B_u} dF_u \middle| \mathcal{G}_t \right). \tag{16.1}$$

Throughout the rest of the paper, we make the following assumption:

Assumption INT-1 For a given payment process D maturing at $T < \infty$, we assume that for some martingale measure $\mathbb{Q}$,

$$\mathbb{E}_{\mathbb{Q}}\left(\int_{]\!]0,T]\!]} \frac{1}{B_u} d|D|_u \right) < \infty,$$

where $|D|$ denotes the total-variation process of D.

In the following, we are going to consider dividend paying securities; for such securities, it is common to define their price at time $t \in [\![0, T]\!]$ as the conditional expectation of the integral over the time interval $]\!]t, T]\!]$ of the discount factor process with respect to the dividend flow process D (see, e.g., Duffie [11] or Bielecki et al. [8]). The natural idea that calculating the value at time t we take only discounted future cash flows (from the time interval $]\!]t, T]\!]$) goes back to Lucas [21] (see also a recent paper of Aase [1]).

Definition 16.3 The *ex-dividend price* S associated with the payment process D is defined, for every $t \in [0, T]$, by setting

$$S_t = B_t \mathbb{E}_{\mathbb{Q}}\left(\int_{]\!]t,T]\!]} \frac{1}{B_u} dD_u \middle| \mathcal{G}_t \right).$$

Assumption INT-1 immediately implies that the ex-dividend price is well defined for each $t \in [\![0,T]\!]$, and moreover $S_T = 0$. As was observed in Bielecki et al. [4], if assets mature at T and their value comes only from their cash flows, this means that if $Y_t = 0$ for $t \geq T$, we see immediately by (16.1) that the price of the asset given by the market is equal to the ex-dividend price of the asset. In the following, it is also convenient to introduce the notion of cumulative price of the payment process D.

Definition 16.4 The *cumulative price* S^c for the payment process D equals, for every $t \in [0,T]$,

$$S_t^c = S_t + B_t \int_{]\!]0,t]\!]} \frac{1}{B_u}\, dD_u = B_t \mathbb{E}_{\mathbb{Q}}\left(\int_{]\!]0,T]\!]} \frac{1}{B_u}\, dD_u \Big| \mathcal{G}_t \right). \tag{16.2}$$

Remark 16.5

(a) From Assumption INT-1 it follows that

$$S_t^{c,*} := S_t^c B_t^{-1} = \mathbb{E}_{\mathbb{Q}}\left(\int_{]\!]0,T]\!]} \frac{1}{B_u}\, dD_u \Big| \mathcal{G}_t \right) \tag{16.3}$$

is a $\mathbb{G}$-martingale under $\mathbb{Q}$.

(b) Consider a single payment X at T. Then the corresponding dividend process D equals $D_t = X \mathbb{1}_{[\![T,\infty[\![}(t)$, and the ex-dividend price process S is different from the cumulative price process S^c only at maturity time, i.e., we have

$$S_t = S_t^c \mathbb{1}_{[\![0,T[\![}(t).$$

We will show that the existence of a replication portfolio for D implies that at given t, the value of replication portfolio equals the ex-dividend price. Before we do this, we need to introduce appropriate notions of self-financing strategy and replication of the payment process D. The following definition generalizes the standard idea of self-financing portfolio to the case of assets with cash flows (see also Karatzas and Shreve [18, Sect. 2]).

Definition 16.6 We call $\phi = (\psi, \varphi)$ a *D-financing portfolio* for the payment process D if the following condition holds:

$$V_t(\phi) = V_0(\phi) + \int_0^t \psi_u\, dB_u + \int_{]\!]0,t]\!]} \langle \varphi_u, dY_u + dF_u \rangle - D_t. \tag{16.4}$$

The class of all D-financing trading strategies (portfolios) for the payment process D is denoted by $\Phi(D)$.

Notice that for $\phi \in \Phi(D)$, we have

$$V_T(\phi) - V_t(\phi) = \int_{]\!]t,T]\!]} \psi_u\, dB_u + \int_{]\!]t,T]\!]} \langle \varphi_u, dY_u + dF_u \rangle - (D_T - D_t). \tag{16.5}$$

Intuitively, this D-financing condition means that the change in the portfolio value is due to gains from investing in primary assets and that only inflows and outflows of funds are made through the dividend process D. Note also that $V_t(\phi)$ should be viewed as the wealth of the portfolio after paying liabilities (or receiving payments). If we take the point of view of the seller of D who is obliged to deliver payments according to D, then the process of cumulated cash flows D plays a role of consumption.

We have the following condition equivalent to the D-financing condition.

Lemma 16.7 *A portfolio ϕ is D-financing iff*

$$V_t^*(\phi) := \frac{V_t(\phi)}{B_t} = V_0^*(\phi) + \int_{]\!]0,t]\!]} \langle \varphi_u, dY_u^{c*} \rangle - \int_{]\!]0,t]\!]} \frac{1}{B_u} dD_u.$$

Proof Assume that ϕ is a D-financing portfolio. Then Itô's lemma yields

$$\begin{aligned} d\left(\frac{V_t(\phi)}{B_t}\right) &= \frac{1}{B_t} dV_t(\phi) + V_t(\phi) d\left(\frac{1}{B_t}\right) \\ &= \frac{1}{B_t}\left(\psi_t dB_t + \langle \varphi_t, dY_t + dF_t \rangle - dD_t\right) - \frac{V_t(\phi)}{B_t} r_t\, dt, \end{aligned}$$

and hence we obtain

$$\begin{aligned} d\left(\frac{V_t(\phi)}{B_t}\right) =& \left(\psi_t r_t dt + \left\langle \varphi_t, dY_t^* + \frac{1}{B_t} dF_t \right\rangle + \frac{\langle \varphi_t, Y_t \rangle}{B_t} r_t\, dt - \frac{1}{B_t} dD_t\right) \\ & - \frac{V_t(\phi)}{B_t} r_t\, dt = \left\langle \varphi_t, dY_t^* + \frac{1}{B_t} dF_t \right\rangle - \frac{1}{B_t} dD_t, \end{aligned}$$

which completes the proof of the first part. To prove the converse, we proceed in a analogous way. □

Remark 16.8 Lemma 16.7 implies that if we know an investment strategy φ for risky assets and the initial value x of the D-financing portfolio $\phi = (\psi, \varphi)$, then an investment ψ in a money account is uniquely determined by the formula

$$\psi_t = V_t^*(\phi) - \langle \varphi_t, Y_t^* \rangle = V_0^*(\phi) + \int_{]\!]0,t]\!]} \langle \varphi_u, dY_u^{c*} \rangle - \int_{]\!]0,t]\!]} \frac{1}{B_u} dD_u - \langle \varphi_t, Y_t^* \rangle$$

and the condition that $V_0^*(\phi) = x$.

Following an idea of Karatzas and Shreve [18] and Møller [22], we can now define what we mean by replication of a given stream D.

Definition 16.9 We say that a portfolio $\phi = (\psi, \varphi)$ *replicates* a payments stream D if

1. the portfolio ϕ is D-financing.
2. $V_T(\phi) = 0$ $\mathbb{P}$ a.s.

If there exists a portfolio that replicates a given payment stream D, then we say that the stream D (or just claim D) is *attainable*.

Remark 16.10 We stress that the first notion of hedging in this spirit can be found in Sect. 2 of Karatzas and Shreve [18]. They introduced the general European contingent claim which corresponds to our dividend process D. A portfolio *hedges* D if there exists a $(-D)$-financed portfolio with terminal wealth $V_T(\phi) \geq 0$ such that an agent who has sold the claim D is still solvent at the maturity of the claim. However, Karatzas and Shreve worked within the setting of a multidimensional complete Black–Scholes market, and they considered only nondividend paying primary assets. Møller [22] defined attainability of the *payment process* in the form given by the above Definition 16.9 (see also Schweizer [23]). Møller also treated nondividend paying primary assets, and he was mainly concerned with risk minimization of the payment stream and not with the notion of attainability in the context of replication. Definition 16.9 appears also in Dana and Jeanblanc [10, Definition 7.3.1] for the market described by Itô processes.

Remark 16.11 Definitions 16.6 and 16.9 imply that if we start trading from an initial investment $V_0(\phi)$ and trade according to ϕ, then at maturity T, we will cover all liabilities, i.e.,

$$D_T = V_0(\phi) + \int_{]\!]0,T]\!]} \psi_u \, dB_u + \int_{]\!]0,T]\!]} \langle \varphi_u, dY_u + dF_u \rangle,$$

and hence $V_0(\phi)$ can be viewed as the replication price for the payment stream D viewed from time $t = 0$.

Now, we give a condition which is equivalent to condition (2) of Definition 16.9, provided that $\phi \in \Phi(D)$. This equivalent condition should be viewed as a dynamic version of the property described in Remark 16.11.

Lemma 16.12 *Assume that ϕ is a D-financing portfolio. Then the following conditions are equivalent*:

1. *ϕ replicates D.*
2. *For every $t \in [0, T]$, we have*

$$D_T - D_t = V_t(\phi) + \int_{]\!]t,T]\!]} \psi_u \, dB_u + \int_{]\!]t,T]\!]} \langle \varphi_u, dY_u + dF_u \rangle.$$

Proof The equivalence is obvious from (16.5) and Definition 16.9. □

The preceding lemma says that for a D-financing portfolio ϕ that replicates D, the value $V_t(\phi)$ can be viewed, at each time t, as the replication price for the process of remaining payments, i.e., for $(D_s - D_t)_{s \geq t}$.

Definition 16.13 Let $\mathbb{Q}$ be a martingale measure. We say that a D-financing portfolio $\phi = (\psi, \varphi)$ is $\mathbb{Q}$-*admissible* if the process of discounted gains

$$G_t^*(\phi) := \int_{]\!]0,t]\!]} \langle \varphi_s, dY_s^{c*} \rangle$$

follows a martingale under $\mathbb{Q}$. The class of all $\mathbb{Q}$-admissible trading strategies which are D-financing is denoted by $\Phi(D, \mathbb{Q})$.

If there exists a replication portfolio for D which is $\mathbb{Q}$-admissible, then its value process is given by the risk-neutral valuation formula. To prove this fact, we need Assumption INT-1 on integrability of the discounted cash flow process D.

Proposition 16.14 *Assume that $\phi = (\psi, \varphi)$ is D-financing. These statements are equivalent*:

(i) *ϕ is $\mathbb{Q}$-admissible and replicates D*,
(ii) *for every $t \in [0, T]$*,

$$V_t(\phi) = B_t \mathbb{E}_{\mathbb{Q}}\Big(\int_{]\!]t,T]\!]} \frac{1}{B_u} dD_u \Big| \mathcal{G}_t \Big). \tag{16.6}$$

Proof Let us observe that by Assumption INT-1 the RHS of (16.6) is well defined for every $t \in [0, T]$.

(i) $\Rightarrow$ (ii) Using Lemma 16.7, we have

$$V_T^*(\phi) - V_t^*(\phi) = \int_{]\!]t,T]\!]} \langle \varphi_u, dY_u^{c*} \rangle - \int_{]\!]t,T]\!]} \frac{1}{B_u} dD_u,$$

and taking expectation yields

$$\mathbb{E}_{\mathbb{Q}}\big(V_T^*(\phi) - V_t^*(\phi) \big| \mathcal{G}_t\big) = \mathbb{E}_{\mathbb{Q}}\Big(\int_{]\!]t,T]\!]} \langle \varphi_u, dY_u^{c*} \rangle \Big| \mathcal{G}_t \Big) - \mathbb{E}_{\mathbb{Q}}\Big(\int_{]\!]t,T]\!]} \frac{1}{B_u} dD_u \Big| \mathcal{G}_t \Big).$$

Since ϕ replicates D, we have $V_T(\phi) = 0$. Admissibility of φ implies that the first term on the RHS vanishes, which yields (16.6).

(ii) $\Rightarrow$ (i) By (16.6), $V_T(\phi) = 0$, so ϕ replicates D, and

$$V_t^*(\phi) + \int_{]\!]0,t]\!]} \frac{1}{B_u} dD_u = \mathbb{E}_{\mathbb{Q}}\Big(\int_{]\!]0,T]\!]} \frac{1}{B_u} dD_u \Big| \mathcal{G}_t \Big).$$

Since the RHS is a martingale, application of Lemma 16.7 completes the proof. □

This proposition gives a justification of the risk-neutral valuation formula based on the replication arguments. Usually using the term *price* for the RHS of (16.6) is justified by assuming that the asset with a given dividend process D is tradable and by considering a self-financing buy-and-hold trading strategy (see, e.g., Bielecki and Rutkowski [4, Sect. 2.1.3]).

We have proved that under a fixed martingale measure $\mathbb{Q}$, the value process of a D-financing $\mathbb{Q}$-admissible portfolio that replicates the payment stream D is uniquely determined. We also expect that in the case of several martingale measures and a claim which is attainable under each of them, the price of the claim is independent of the choice of an equivalent martingale measure. To prove this conjecture, we need, as in the classical case, additional assumptions and a different class of admissible trading strategies.

Definition 16.15 We call a D-financing strategy $\phi \in \Phi(D)$ *admissible* if

(i) $\phi \in \Phi(D, \mathbb{Q})$ for some equivalent martingale measure $\mathbb{Q}$;
(ii) ϕ is tame for D, i.e., for some $K \in \mathbb{R}$ and for each $t \in [0, T]$, we have $V_t^*(\phi) + \int_{]\!]0,t]\!]} \frac{1}{B_u} dD_u \geq K$.

Proposition 16.16 *Let $\mathbb{Q}_1$ and $\mathbb{Q}_2$ be two equivalent martingale measures, and D a payment process. Assume that there exist tame strategies $\phi^1 \in \Phi(D, \mathbb{Q}_1)$ and $\phi^2 \in \Phi(D, \mathbb{Q}_2)$ which replicate D. Then*

$$\mathbb{E}_{\mathbb{Q}_1}\left(\int_{]\!]t,T]\!]} \frac{1}{B_u} dD_u \middle| \mathcal{G}_t\right) = \mathbb{E}_{\mathbb{Q}_2}\left(\int_{]\!]t,T]\!]} \frac{1}{B_u} dD_u \middle| \mathcal{G}_t\right).$$

Proof Follows by standard arguments. □

Thus, we have proved that the price of a claim which is replicated by an admissible strategy is independent of the choice of an equivalent martingale measure.

Similarly to the classical case, a representation of the martingale S^{c*}, i.e., the discounted cumulated price of D, as a stochastic integral with respect to $Y^{c,*}$, gives a replication strategy for D.

Lemma 16.17 *Suppose that the martingale S^{c*} given by* (16.3) *has a representation*

$$S_t^{c*} = S_0^{c*} + \int_{]\!]0,t]\!]} \langle \varphi_u, dY_u^{c*} \rangle, \tag{16.7}$$

where φ is a $\mathbb{G}$-predictable process. Then the portfolio $\phi = (\psi, \varphi)$, where

$$\psi_t = S_t^{c*} - \langle \varphi_t, Y_t^* \rangle - \int_{]\!]0,t]\!]} \frac{1}{B_u} dD_u,$$

replicates the payment stream D.

Proof Notice that the portfolio ϕ is D-financing, since

$$\begin{aligned} V_t^*(\phi) &= \left(\left(S_t^{c*} - \int_{]\!]0,t]\!]} \frac{1}{B_u} dD_u\right) - \langle \varphi_t, Y_t^* \rangle\right) + \langle \varphi_t, Y_t^* \rangle \\ &= S_0^{c*} + \int_{]\!]0,t]\!]} \langle \varphi_u, dY_u^{c*} \rangle - \int_{]\!]0,t]\!]} \frac{1}{B_u} dD_u. \end{aligned}$$

Moreover, $V_T(\phi)=0$. Indeed,

$$\begin{aligned}V_T(\phi)&=\langle\varphi_T,Y_T\rangle+\psi_T B_T\\&=\langle\varphi_T,Y_T\rangle+\left(S_T^{C*}-\int_{]\!]0,T]\!]}\frac{1}{B_u}\,dD_u\right)B_T-\langle\varphi_T,Y_T^*\rangle B_T=0,\end{aligned}$$

by (16.3). □

We emphasize that we consider hedging of a general dividend process in a market with dividend paying primary assets, which is not a standard case. We also stress that a replication strategy for a single payment in the sense of Definition 16.9 coincides with a replication strategy in the classical sense for $t<T$. These two strategies differ at $t=T$ in the money account.

16.3 Doubly Stochastic Markov Chains

In this section we describe doubly stochastic Markov chains and present several results that are used in the subsequent sections. For the convenience of the reader, we repeat the relevant material from Jakubowski and Niewęgłowski [16] without proofs, thus making our exposition self-contained. By $\mathbb{F}^X$ we denote the filtration generated by a process X. We always assume that $\mathbb{F}^X$ satisfies the "usual conditions."

Definition 16.18 A càdlàg process C is called an $\mathbb{F}$-doubly stochastic Markov chain with state space $\mathcal{K}\subset\mathbb{Z}=\{\dots,-1,0,1,2,\dots\}$ if there exists a family of stochastic matrices $P(s,t)=(p_{i,j}(s,t))_{i,j\in\mathcal{K}}$ for $0\le s\le t$ such that

1. the matrix $P(s,t)$ is $\mathcal{F}_t$-measurable, and $P(s,\cdot)$ is $\mathbb{F}$-progressively measurable,
2. for any $t\ge s\ge 0$ and every $i,j\in\mathcal{K}$, we have

$$\mathbb{P}\big(C_t=j\mid\mathcal{F}_\infty\vee\mathcal{F}_s^C\big)\mathbb{1}_{\{C_s=i\}}=\mathbb{1}_{\{C_s=i\}}p_{i,j}(s,t).\qquad(16.8)$$

The process P is called the conditional transition probability process of C.

The class of $\mathbb{F}$-DS Markov chains contains many well-known processes.

Example 16.19 A Markov chain is an $\mathbb{F}$-DS Markov chain with respect to the trivial filtration $\mathbb{F}$.

Example 16.20 Let X be a compound Poisson process with jumps in $\mathbb{Z}$. By standard calculations we see that X is an $\mathbb{F}$-DS Markov chain with respect to the trivial filtration $\mathbb{F}$.

Example 16.21 (Cox process) A Cox process C with intensity λ is an $\mathbb{F}^{\lambda}$-DS Markov chain with $\mathcal{K} = \mathbb{N}$. Using the definition of a Cox process, we can calculate explicitly transition probabilities

$$p_{i,j}(s,t) = \begin{cases} \frac{(\int_s^t \lambda_u \, du)^{j-i}}{(j-i)!} e^{-\int_s^t \lambda_u \, du} & \text{for } j \geq i, \\ 0 & \text{for } j < i. \end{cases}$$

Example 16.22 (Time-changed discrete Markov chain) Assume that $\bar{C}$ is a discrete-time Markov chain with values in $\mathcal{K} = \{1, \ldots, K\}$, N is a Cox process, and the processes $(\bar{C}_k)_{k \geq 0}$ and $(N_t)_{t \geq 0}$ are independent and conditionally independent given $\mathcal{F}_\infty$. Then the process $C_t := \bar{C}_{N_t}$ is an $\mathbb{F}$-DS Markov chain (see Jakubowski and Niewęgłowski [15, Theorems 7 and 9]).

Example 16.23 The process C obtained by the canonical construction in Bielecki and Rutkowski [4] is an $\mathbb{F}$-DS Markov chain.

For $\mathbb{F}$-DS Markov chains, we introduce the concept of intensity, analogous to that for continuous-time Markov chains.

Definition 16.24 We say that an $\mathbb{F}$-DS Markov chain C has an intensity if there exists an $\mathbb{F}$-adapted matrix-valued process $\Lambda = (\Lambda(s))_{s \geq 0} = (\lambda_{i,j}(s))_{s \geq 0}$ such that:

(1) Λ is locally integrable, i.e., for any $T > 0$,

$$\int_{]\!]0,T]\!]} \sum_{i \in \mathcal{K}} \left|\lambda_{i,i}(s)\right| ds < +\infty; \tag{16.9}$$

(2) Λ satisfies the conditions:

$$\lambda_{i,j}(s) \geq 0 \quad \forall i, j \in \mathcal{K}, i \neq j, \qquad \lambda_{i,i}(s) = -\sum_{j \neq i} \lambda_{i,j}(s) \quad \forall i \in \mathcal{K}, \tag{16.10}$$

the Kolmogorov backward equation: for all $v \leq t$,

$$P(v,t) - \mathbb{I} = \int_v^t \Lambda(u) P(u,t) \, du, \tag{16.11}$$

the Kolmogorov forward equation: for all $v \leq t$,

$$P(v,t) - \mathbb{I} = \int_v^t P(v,u) \Lambda(u) \, du. \tag{16.12}$$

A process Λ satisfying the above conditions is called an intensity of the $\mathbb{F}$-DS Markov chain C.

Remark 16.25

(a) The solution to ODEs (16.11) and (16.12) is an invertible matrix, and by $Q(s,t)$ we denote the inverse of $P(s,t)$. The inverse matrix $Q(s,t)$ is a solution of the following (random) ODE:

$$\frac{\partial Q(s,t)}{\partial t} = -\Lambda_t Q(s,t)\,dt, \qquad Q(s,s) = \mathbb{I}.$$

(b) The conditional transition matrices $(P(t,u))_{0\le t\le u}$ satisfy the Chapman–Kolmogorov equations, i.e., for $0 \le u \le t$, we have

$$P(0,u) = P(0,t)P(t,u).$$

If C has an intensity matrix Λ, then $P(0,t)$ is invertible, which implies that we can write

$$P(t,u) = Q(0,t)P(0,u). \tag{16.13}$$

(c) If C is an $\mathbb{F}$-DS Markov chain, then $\mathbb{F}$ is immersed in $\mathbb{F} \vee \mathbb{F}^C$, i.e., every $\mathbb{F}$-martingale is an $\mathbb{F} \vee \mathbb{F}^C$-martingale (for proof, see [16, Proposition 3.4]).

Theorem 16.26 *Let $(\Lambda(t))_{t\ge 0}$ be an arbitrary $\mathbb{F}$-adapted matrix-valued stochastic process which satisfies conditions* (16.9) *and* (16.10). *Then there exists an $\mathbb{F}$-DS Markov chain with intensity $(\Lambda(t))_{t\ge 0}$.*

Further, the following theorem will be crucial:

Theorem 16.27 *Let $\widetilde{\mathbb{G}} = (\widetilde{\mathcal{G}}_t)_{t\ge 0}$, where $\widetilde{\mathcal{G}}_t := \mathcal{F}_\infty \vee \mathcal{F}_t^C$. Suppose that $(C_t)_{t\ge 0}$ is a $\mathcal{K}$-valued stochastic process and $(\Lambda(t))_{t\ge 0}$ is a matrix-valued process satisfying* (16.9) *and* (16.10). *The following conditions are equivalent*:

1. *The process C is an $\mathbb{F}$-DS Markov chain with an intensity Λ.*
2. *The process M defined by*

$$M_t := H_t - \int_0^t \Lambda_u^\top H_u\,du, \tag{16.14}$$

where $H_t^i = \mathbb{1}_{\{C_t=i\}}$, and $H_t := (H_t^1, \ldots, H_t^K)^\top$ is a $\widetilde{\mathbb{G}}$-local martingale.

3. *The process L defined by*

$$L_t := Q^\top(0,t)H_t = H_0 + \int_{]\!]0,t]\!]} Q^\top(0,u)\,dM_u \tag{16.15}$$

is a $\widetilde{\mathbb{G}}$-local martingale.

Remark 16.28 In a view of the above characterization, the class of $\mathbb{F}$-DS Markov chains can be described as the class of processes that *behave* like time-inhomogeneous Markov chains when conditioned on $\mathcal{F}_\infty$. The word *behave* should be understood by means of a behavior of $\mathcal{F}_\infty$-conditional transition probabilities.

We say that a given random time ρ avoids $\mathbb{F}$-stopping times if $\mathbb{P}(\rho = \sigma) = 0$ for every $\mathbb{F}$-stopping time σ.

Proposition 16.29 *Let C be an $\mathbb{F}$-DS Markov chain with an intensity, $\tau_0 = 0$, and let*

$$\tau_k := \inf\{t > \tau_{k-1} : C_t \neq C_{\tau_{k-1}}\}. \tag{16.16}$$

For $k \geq 1$, the stopping time τ_k avoids $\mathbb{F}$-stopping times, provided that $\tau_k < \infty$ a.s.

16.4 Valuation of Defaultable Rating-Sensitive Claims with Ratings Given by a Doubly Stochastic Markov Chain

16.4.1 Description of Claims

We consider an arbitrage-free market with finite horizon on which defaultable instruments are also traded. We denote by $\mathbb{F}$ the reference filtration corresponding to observation of the market without credit rating, i.e., a filtration corresponding to the interest rate risk and other market factors that drive the credit risk. C is a credit rating process which takes values in the set of rating classes $\mathcal{K} = \{1, \ldots, K\}$ with the unique absorbing state K. If $K = 2$, then it is understood that there are only two states, nondefault and default. We assume that the process C is càdlàg. Let $\mathcal{G}_t = \mathcal{F}_t \vee \mathcal{F}_t^C$. By a defaultable rating-sensitive claim we mean a classical one broadened by a migration process.

Definition 16.30 By a *defaultable rating-sensitive claim* maturing at T we mean a quintuple (X, A, Z, C, τ), where $X = (X^1, \ldots, X^{K-1})^\top$ is a $(K-1)$-dimensional vector of $\mathcal{F}_T$-measurable random variables, $A = (A^1, \ldots, A^{K-1})^\top$ is a $(K-1)$-dimensional vector-valued $\mathbb{F}$-progressively measurable stochastic process of finite variation, Z is an $\mathbb{F}$-predictable $K \times K$-dimensional matrix-valued process with zero on the diagonal, C is a càdlàg process with values in $\mathcal{K}$, and τ is a positive random variable defined by

$$\tau := \inf\{t \geq 0 : C_t = K\}.$$

In this definition, X describes the promised payoff which is contingent on rating at maturity T, i.e., the payoff is equal to X^i, provided that $\{C_T = i\}$; the process A models the process of promised dividends which can depend on current credit rating; the processes $Z^{i,j}$ describe the payments at times when the rating changes from i to j; in particular, $Z^{j,K}$ specifies the recovery payment at the default time τ, provided that before the default time, we are in the state j; and C is the credit rating process. This definition of claim is very general and covers many different types of claims.

Remark 16.31 If we put $X^i = Y$ for each i, where Y is $\mathcal{F}_T$-measurable, bounded random variable, then the promised payment depends only on the default time:

$$\sum_{i=1}^{K-1} X^i \mathbb{1}_{\{C_T = i\}} = Y \sum_{i=1}^{K-1} \mathbb{1}_{\{C_T = i\}} = Y \mathbb{1}_{\{C_T \neq K\}} = Y \mathbb{1}_{\{\tau > T\}}.$$

Remark 16.32 Since

$$\sum_{i=1}^{K-1}\int_{]\!]0,t\wedge T]\!]} Z_u^{i,K}\,dH_u^{i,K} = \sum_{i=1}^{K-1} Z_\tau^{i,K}\mathbb{1}_{\{0<\tau\le t\wedge T, C_{\tau-}=i\}} = Z_\tau^{C_{\tau-},K}\mathbb{1}_{\{0<\tau\le t\wedge T\}},$$

the recovery process allows recovery depending on the rating of the bond before the default time τ.

Now, we define the dividend process which describes the cash flows from the claim in the interval $[0, T]$.

Definition 16.33 The dividend process $D = (D_t)_{t\ge 0}$ of the claim (X, A, Z, C, τ) maturing at T equals, for $t \ge 0$,

$$D_t = \sum_{i=1}^{K-1}\left(X^i H_T^i \mathbb{1}_{[\![T,+\infty[\![}(t) + \int_{]\!]0,t\wedge T]\!]} H_u^i\,dA_u^i + \sum_{j\ne i\in\mathcal{K}}\int_{]\!]0,t\wedge T]\!]} Z_u^{i,j}\,dH_u^{i,j}\right). \tag{16.17}$$

Remark 16.34 For fixed i, if at time t the rating process changes from state i to state j, then the promised dividend $A_t^i - A_{t-}^i$ is not passed over to the holder of the claim, and if the rating process changes from some j to i, then the promised dividend $A_t^i - A_{t-}^i$ is passed over to the holder of the claim.

Example 16.35 Consider a defaultable bond with fractional recovery of par value. In this case, the bond's holder receives at maturity time T its face value (say 1 unit of cash), provided that default did not occur before or at T. If the default occurred before or at time T, the recovery $\delta_{C_{\tau-}}$ is paid at the default time τ to the bond holder. So, the recovery payment depends on the predefault rating $C_{\tau-}$, and it is assumed that the recovery $\delta_i \in [0, 1)$ is a fixed number for each $i \in \mathcal{K}\setminus K$. By taking

$$X^i = 1,\quad A^i = 0,\quad Z^{i,K} = \delta_i \quad \text{for } i = 1,\dots,K-1,\qquad Z^{i,j} = 0 \quad \text{for } j \ne K,$$

we see that a defaultable bond is a claim in the sense of our definition. The dividend process for such a claim equals

$$D_t = \mathbb{1}_{\{\tau>T\}}\mathbb{1}_{[\![T,+\infty[\![}(t) + \delta_{C_{\tau-}}\mathbb{1}_{\{0<\tau\le t\wedge T\}}.$$

Example 16.36 Another example is a defaultable credit-sensitive note. It is a corporate bond with coupons linked to the rating of corporation. The coupons of this note are paid at prespecified coupon dates $0 < T_1 < T_2 < \cdots < T_n$ if default does not arise and the value of the coupon is contingent on rating corporate at the coupon date. If a default occurred before or at time T, the recovery $\delta_{C_{\tau-}}$ is paid at the default time τ to the bond holders. It is assumed that $\delta_i \in [0, 1)$ is fixed for each $i \in \mathcal{K}\setminus K$. So

$$X^i = 1,\quad Z^{i,K} = \delta_i \quad \text{for } i \le K-1,\qquad Z^{i,j} = 0 \quad \text{for } j \ne K,$$

$$A_t^i = \sum_{j=1}^{n} \mathbb{1}_{\{t \geq T_j\}} d_{i,j},$$

where $d_{i,j}$ are fixed constants chosen in advance, and the dividend process of this note is given by

$$D_t = \mathbb{1}_{\{\tau > T\}} \mathbb{1}_{[\![T,+\infty[\![}(t) + \sum_{i=1}^{K-1} \int_{]\!]0,t\wedge T]\!]} \delta_i \, dH_u^{i,K} + \sum_{i=1}^{K-1} \int_{]\!]0,t\wedge T]\!]} H_u^i \, dA_u^i.$$

Example 16.37 We can consider modification of the example above with $A_t^i = d_i t$, which corresponds to continuous payments at the rate d_i, provided that at time t the rating is equal to i. The dividend process of this note is given by

$$D_t = \mathbb{1}_{\{\tau > T\}} \mathbb{1}_{[\![T,+\infty[\![}(t) + \sum_{i=1}^{K-1} \int_{]\!]0,t\wedge T]\!]} \delta_i \, dH_u^{i,K} + \sum_{i=1}^{K-1} \int_{]\!]0,t\wedge T]\!]} H_u^i \, d_i \, du.$$

16.4.2 Pricing of Rating-Sensitive Claims

Now, we consider the problem of pricing of rating-sensitive claims. We put ourselves in an arbitrage-free framework, which means that we assume existence of a spot martingale measure $\mathbb{Q}$ for the underlying market. We fix $\mathbb{Q}$ in what follows. As usual, the spot martingale measure $\mathbb{Q}$ is a measure related to the choice of the saving account B as a numéraire. Then the price process discounted by B of any tradable security (nondividend paying) is a martingale under $\mathbb{Q}$. We assume that the dynamics of saving account B is given by

$$dB_t = r_t B_t \, dt, \qquad B_0 = 1,$$

where r is a nonnegative $\mathbb{F}$-progressively measurable stochastic process. So B^{-1} is a discount factor. In what follows, we assume that the process C of rating migration is an $\mathbb{F}$-DS Markov chain under $\mathbb{Q}$ with intensity process Λ which satisfies the following integrability condition:

$$\mathbb{E}_{\mathbb{Q}}\left(\int_{]\!]0,T]\!]} \sum_{i \in \mathcal{K}} \left| \lambda_{i,i}(s) \right| ds \right) < \infty. \tag{16.18}$$

In this subsection, conditional expectations are calculated under spot martingale measure $\mathbb{Q}$; hence, for short, we shall write $\mathbb{E}$ instead of $\mathbb{E}_{\mathbb{Q}}$.

The main theorem of this subsection gives a convenient form of the ex-dividend price process S of a defaultable rating-sensitive claim. It generalizes the results of Bielecki et al. [8] obtained for $K = 2$.

Theorem 16.38 *Let (X, A, Z, τ, C) be a defaultable rating-sensitive claim with X, Z, A bounded. For $t < T$, the value of ex-dividend price is given by the formula*

$$S_t \mathbb{1}_{\{C_t=i\}} = \mathbb{1}_{\{C_t=i\}} \sum_{j=1}^{K-1} B_t \mathbb{E}\Big(\frac{X^j p_{i,j}(t,T)}{B_T} + \int_{]\!]t,T]\!]} B_u^{-1} p_{i,j}(t,u)\, dA_u^j$$

$$+ \int_{]\!]t,T]\!]} \sum_{k=1}^{K} \frac{Z_u^{j,k}}{B_u} p_{i,j}(t,u) \lambda_{j,k}(u)\, du \Big| \mathcal{F}_t \Big). \tag{16.19}$$

In the proof of theorem, we need the following lemmas which generalize results of Bielecki et al. [4, Corollary 5.1.1, Propositions 5.1.1, and 5.1.2] to the case of multiple ratings. In the proofs, we will use the immersion property of filtrations connected with $\mathbb{F}$-DS Markov chains.

Lemma 16.39 *Let X be a bounded $\mathcal{F}_T$-measurable random variable, and $j \in \mathcal{K} \setminus K$. Then*

$$\mathbb{E}(X \mathbb{1}_{\{C_T=j\}} | \mathcal{G}_t) = \sum_{i=1}^{K-1} \mathbb{1}_{\{C_t=i\}} \mathbb{E}\big(X p_{i,j}(t,T) \big| \mathcal{F}_t\big).$$

Proof This is an easy consequence of the definition of an $\mathbb{F}$-DS Markov chain and the immersion property (see [16, Proposition 3.4]). Indeed,

$$\begin{aligned}
\mathbb{E}(X \mathbb{1}_{\{C_T=j\}} | \mathcal{G}_t) &= \mathbb{E}\big(X \mathbb{E}\big(\mathbb{1}_{\{C_T=j\}} \big| \mathcal{F}_\infty \vee \mathcal{F}_t^C\big) \big| \mathcal{G}_t\big) \\
&= \sum_{i=1}^{K-1} \mathbb{1}_{\{C_t=i\}} \mathbb{E}\big(X p_{i,j}(t,T) \big| \mathcal{G}_t\big) \\
&= \sum_{i=1}^{K-1} \mathbb{1}_{\{C_t=i\}} \mathbb{E}\big(X p_{i,j}(t,T) \big| \mathcal{F}_t\big).
\end{aligned}$$

□

Lemma 16.40 *Let Z be a bounded $\mathbb{F}$-predictable stochastic process, and $j \in \mathcal{K} \setminus K$. Under condition* (16.18), *for $k \neq j$, we have*

$$\mathbb{E}\Big(\int_{]\!]t,T]\!]} Z_u\, dH_u^{j,k} \Big| \mathcal{G}_t \Big) = \sum_{i=1}^{K-1} \mathbb{1}_{\{C_t=1\}} \mathbb{E}\Big(\int_{]\!]t,T]\!]} Z_u p_{i,j}(t,u) \lambda_{j,k}(u)\, du \Big| \mathcal{F}_t \Big). \tag{16.20}$$

Proof Fix $k \neq j \in \mathcal{K}$. Because

$$\int_{]\!]t,T]\!]} Z_u\, dH_u^{j,k} = \int_{]\!]t,T]\!]} Z_u\, dM_u^{j,k} + \int_{]\!]t,T]\!]} Z_u H_u^j \lambda_{j,k}(u)\, du,$$

and $M^{j,k}$ is a martingale and Z a bounded process, we have

$$\mathbb{E}\biggl(\int_{]\!]t,T]\!]} Z_u \, dH_u^{j,k} \bigg| \mathcal{G}_t\biggr) = \mathbb{E}\biggl(\int_{]\!]t,T]\!]} Z_u H_u^j \lambda_{j,k}(u) \, du \bigg| \mathcal{G}_t\biggr) = I.$$

Using the conditional version of Fubini's theorem (see, e.g., Applebaum [2, p. 12]), the definition of an $\mathbb{F}$-DS Markov chain, hypothesis H, and taking $\widetilde{\mathcal{G}}_t := \mathcal{F}_\infty \vee \mathcal{F}_t^C$, we have

$$\begin{aligned}
I &= \int_{]\!]t,T]\!]} \mathbb{E}\bigl(Z_u H_u^j \lambda_{j,k}(u) \big| \mathcal{G}_t\bigr) \, du = \int_{]\!]t,T]\!]} \mathbb{E}\bigl(\mathbb{E}\bigl(Z_u H_u^j \lambda_{j,k}(u) \big| \widetilde{\mathcal{G}}_t\bigr) \big| \mathcal{G}_t\bigr) \, du \\
&= \int_{]\!]t,T]\!]} \mathbb{E}\Biggl(Z_u \lambda_{j,k}(u) \Biggl(\sum_{i=1}^{K-1} \mathbb{1}_{\{C_t=i\}} p_{i,j}(t,u)\Biggr) \Bigg| \mathcal{G}_t\Biggr) \, du \\
&= \sum_{i=1}^{K-1} \mathbb{1}_{\{C_t=i\}} \int_{]\!]t,T]\!]} \mathbb{E}\bigl(Z_u p_{i,j}(t,u) \lambda_{j,k}(u) \big| \mathcal{F}_t\bigr) \, du \\
&= \sum_{i=1}^{K-1} \mathbb{1}_{\{C_t=i\}} \mathbb{E}\biggl(\int_{]\!]t,T]\!]} Z_u p_{i,j}(t,u) \lambda_{j,k}(u) \, du \bigg| \mathcal{F}_t\biggr),
\end{aligned}$$

and this completes the proof. □

Lemma 16.41 *Let A be an $\mathbb{F}$-predictable bounded stochastic process of finite variation. Under condition* (16.18), *for any $j \in \mathcal{K} \setminus K$, we have*

$$\mathbb{E}\biggl(\int_{]\!]t,v]\!]} H_u^j \, dA_u \bigg| \mathcal{G}_t\biggr) = \sum_{i=1}^{K-1} \mathbb{1}_{\{C_t=i\}} \mathbb{E}\biggl(\int_{]\!]t,v]\!]} p_{i,j}(t,u) \, dA_u \bigg| \mathcal{F}_t\biggr).$$

Proof We follow the idea from Bielecki and Rutkowski's book [4], in which the case with two states (default and no default) is considered. Fix t. Define $\widetilde{A}_u := A_u - A_t$ for $u \in [t, v]$. Obviously, this is an $\mathbb{F}$-predictable bounded process of finite variation, and $\widetilde{A}_t = 0$. The integrals with respect to A and $\widetilde{A}$ are equal, and therefore,

$$\begin{aligned}
\mathbb{E}\biggl(\int_{]\!]t,v]\!]} H_u^j \, dA_u \bigg| \mathcal{G}_t\biggr) &= \mathbb{E}\biggl(\int_{]\!]t,v]\!]} H_u^j \, d\widetilde{A}_u \bigg| \mathcal{G}_t\biggr) \\
&= \mathbb{E}\biggl(\widetilde{A}_v H_v^j - \widetilde{A}_t H_t^j - \int_{]\!]t,v]\!]} \widetilde{A}_{u-} \, dH_u^j \bigg| \mathcal{G}_t\biggr) \\
&= \mathbb{E}\biggl(\widetilde{A}_v H_v^j - \int_{]\!]t,v]\!]} \widetilde{A}_{u-} \, dH_u^j \bigg| \mathcal{G}_t\biggr) = I_1 + I_2,
\end{aligned}$$

where

$$I_1 := \mathbb{E}\bigl(\widetilde{A}_v H_v^j \big| \mathcal{G}_t\bigr), \qquad I_2 := \mathbb{E}\biggl(\int_{]\!]t,v]\!]} \widetilde{A}_{u-} \, dH_u^j \bigg| \mathcal{G}_t\biggr).$$

Since $\widetilde{A}_v$ is $\mathcal{F}_\infty$-measurable, it follows that

$$I_1 = \mathbb{E}\big(\widetilde{A}_v \mathbb{E}\big(H_v^j | \mathcal{F}_\infty \vee \mathcal{F}_t^C\big) | \mathcal{G}_t\big) = \sum_{i=1}^{K-1} \mathbb{1}_{\{C_t=i\}} \mathbb{E}\big(\widetilde{A}_v p_{i,j}(t,v) | \mathcal{F}_t\big).$$

Now we calculate I_2. By boundedness of $\widetilde{A}$, using martingale property of M^j, the conditional Fubini theorem, hypothesis H, and the Kolmogorov forward equation (16.12), we have

$$\begin{aligned}
I_2 &= \mathbb{E}\left(\int_{]\!]t,v]\!]} \widetilde{A}_{u-}\, dM_u^j + \int_{]\!]t,v]\!]} \widetilde{A}_{u-} \lambda_{C_u,j}(u)\, du \,\middle|\, \mathcal{G}_t\right) \\
&= \mathbb{E}\left(\int_{]\!]t,v]\!]} \widetilde{A}_{u-} \sum_{k=1}^{K-1} H_u^k \lambda_{k,j}(u)\, du \,\middle|\, \mathcal{G}_t\right) = \int_{]\!]t,v]\!]} \mathbb{E}\left(\widetilde{A}_{u-} \sum_{k=1}^{K-1} H_u^k \lambda_{k,j}(u) \,\middle|\, \mathcal{G}_t\right) du \\
&= \int_{]\!]t,v]\!]} \mathbb{E}\left(\widetilde{A}_{u-} \sum_{k=1}^{K-1} \mathbb{E}\big(H_u^k | \mathcal{F}_\infty \vee \mathcal{F}_t^C\big) \lambda_{k,j}(u) \,\middle|\, \mathcal{G}_t\right) du \\
&= \sum_{i=1}^{K-1} \mathbb{1}_{\{C_t=i\}} \int_{]\!]t,v]\!]} \mathbb{E}\left(\widetilde{A}_{u-} \left(\sum_{k=1}^{K-1} p_{i,k}(t,u) \lambda_{k,j}(u)\right) \,\middle|\, \mathcal{F}_t\right) du \\
&= \sum_{i=1}^{K-1} \mathbb{1}_{\{C_t=i\}} \mathbb{E}\left(\int_{]\!]t,v]\!]} \widetilde{A}_{u-} \left(\sum_{k=1}^{K-1} p_{i,k}(t,u) \lambda_{k,j}(u)\right) du \,\middle|\, \mathcal{F}_t\right) \\
&= \sum_{i=1}^{K-1} \mathbb{1}_{\{C_t=i\}} \mathbb{E}\left(\int_{]\!]t,v]\!]} \widetilde{A}_{u-}\, dp_{i,j}(t,u) \,\middle|\, \mathcal{F}_t\right).
\end{aligned}$$

Hence,

$$I_1 + I_2 = \sum_{i=1}^{K-1} \mathbb{1}_{\{C_t=i\}} \mathbb{E}\left(\widetilde{A}_v p_{i,j}(t,v) - \int_{]\!]t,v]\!]} \widetilde{A}_{u-} dp_{i,j}(t,u) \,\middle|\, \mathcal{F}_t\right),$$

and by integration by parts,

$$\begin{aligned}
I_1 + I_2 &= \sum_{i=1}^{K-1} \mathbb{1}_{\{C_t=i\}} \mathbb{E}\left(\widetilde{A}_t p_{i,j}(t,t) + \int_{]\!]t,v]\!]} p_{i,j}(t,u)\, d\widetilde{A}_u \,\middle|\, \mathcal{F}_t\right) \\
&= \sum_{i=1}^{K-1} \mathbb{1}_{\{C_t=i\}} \mathbb{E}\left(\int_{]\!]t,v]\!]} p_{i,j}(t,u)\, dA_u \,\middle|\, \mathcal{F}_t\right).
\end{aligned}$$

□

Proof of Theorem 16.38 The theorem follows immediately from (16.17) and the above lemmas applied to $\widetilde{X} := \frac{X}{B_T}$, $d\widetilde{A}_u := \frac{1}{B_u} dA_u$, $\widetilde{Z}_u := \frac{Z_u}{B_u}$. Of course, Assumption INT-1 is satisfied. □

Remark 16.42 In Theorem 16.38, we have assumed that X, A, Z are bounded. This assumption can be relaxed, e.g., one can assume weaker conditions:

$$\begin{aligned}
&\mathbb{E}\big(B_T^{-1}|X^i|\big) < \infty, \quad \mathbb{E}\Big(\int_{]\!]0,T]\!]} |A_u^i|^2 |\lambda_{C_u,i}(u)|\,du\Big) < \infty \quad \forall i \in \mathcal{K} \setminus K, \\
&\mathbb{E}\Big(\int_{]\!]0,T]\!]} |Z_u^{i,j}|^2 H_{u-}^i \lambda_{i,j}(u)\,du\Big) < \infty \quad \forall i \neq j,\ i,j \in \mathcal{K} \setminus K.
\end{aligned} \tag{16.21}$$

Under these assumptions, the results of Lemmas 16.39–16.41 remain valid. Moreover, these conditions imply that INT-1 is satisfied, so the ex-dividend price is well defined. Indeed, it is easily seen that Assumption INT-1 for a dividend process D of a defaultable rating-sensitive claim (X, A, Z, C, τ) is equivalent to

$$\begin{aligned}
&\mathbb{E}\big(B_T^{-1}|X^i|\big) < \infty, \quad \mathbb{E}\Big(\int_{]\!]0,T]\!]} H_u^i B_u^{-1} d|A^i|_u\Big) < \infty \quad \forall i \in \mathcal{K} \setminus K, \\
&\mathbb{E}\Big(\int_{]\!]0,T]\!]} |Z_u^{i,j}| B_u^{-1} H_{u-}^i\, dH_u^j\Big) < \infty \quad \forall i \neq j,\ i,j \in \mathcal{K} \setminus K,
\end{aligned}$$

and (16.21) imply these conditions.

16.4.3 Examples of Pricing of Selected Instruments

In a series of propositions, we now give examples of application of general Theorem 16.38. All these results are stated under the assumption that the migration process C is an $\mathbb{F}$-DS Markov chain with intensity process $(\Lambda_t)_{t\geq 0}$. Whenever we apply results based on Lemma 16.40, we assume that $(\Lambda_t)_{t\geq 0}$ satisfies condition (16.18).

Defaultable Bond with Fractional Recovery of Par Value

This simple example of a rating-sensitive claim is described in Example 16.35. We only stress that the recovery payment is contingent on the predefault rating, i.e., on $C_{\tau-}$.

Proposition 16.43 *The ex-dividend price D^δ of a defaultable bond with fractional recovery of par value has for $t < T$, on the set $\{C_t = i\}$, the form*

$$D^\delta(t,T) = \sum_{j=1}^{K-1} B_t \mathbb{E}\bigg(\frac{p_{i,j}(t,T)}{B_T} + \int_{]\!]t,T]\!]} \frac{\delta_j}{B_u} p_{i,j}(t,u)\lambda_{j,K}(u)\,du \,\bigg|\, \mathcal{F}_t\bigg).$$

Credit Sensitive Note (CNS)-Resetting at Coupon Payment Date

Recall that CSN are, generally speaking, corporate coupon bonds that pay coupons which are sensitive to credit rating of a firm assigned by some rating agency (see Example 16.36). Coupons are usually greater if the rating is worse.

Proposition 16.44 *The ex-dividend price of a Credit Sensitive Note with coupons resetting at coupon payment date is, for $t < T$, equal to*

$$S_t \mathbb{1}_{\{C_t=i\}} = \mathbb{1}_{\{C_t=i\}} \sum_{j=1}^{K-1} \mathbb{E}\bigg(e^{-\int_t^T r_u\,du} p_{i,j}(t,T) + \sum_{k:t<T_k} e^{-\int_t^{T_k} r_u\,du} d_{j,k} p_{i,j}(t,T_k)$$
$$+ \delta_j \int_{]\!]t,T]\!]} e^{-\int_t^u r_v\,dv} p_{i,j}(t,u) \lambda_{j,K}(u)\,du \bigg| \mathcal{F}_t \bigg).$$

Remark 16.45 Specifying d by $d_{j,k} = s_U(j - i_U)^+$ where s_U is constant, one can include a rating-triggering step-up feature to coupon payments. If the rating crosses level i_U (step-up), then the coupon will increase proportionally. The so-called Rating-Triggered Step-Up Bonds were issued by some European telecom companies, e.g., Deutsche Telecom, France Telecom; for details, see Lando and Mortensen [20].

Credit Sensitive Note—Continuous Coupon Payments

One can consider CSN with coupons that are paid continuously in time at rate d_{C_t} depending on rating state at t (see Example 16.37). This is a mathematical idealization of the previous case, rather than a real-life example, but it might be seen as approximation of discrete payments considered in the previous subsection.

Proposition 16.46 *The ex-dividend price of the Credit Sensitive Note with continuous coupon payments is, for $t \le T$, equal to*

$$S_t \mathbb{1}_{\{C_t=i\}} = \mathbb{1}_{\{C_t=i\}} \sum_{j=1}^{K-1} \mathbb{E}\bigg(e^{-\int_t^T r_u\,du} p_{i,j}(t,T) + \int_{]\!]t,T]\!]} d_j e^{-\int_t^u r_v\,dv} p_{i,j}(t,u)\,du$$
$$+ \delta_j \int_{]\!]t,T]\!]} e^{-\int_t^u r_v\,dv} p_{i,j}(t,u) \lambda_{j,K}(u)\,du \bigg| \mathcal{F}_t \bigg).$$

Credit Default Swap

Credit Default Swap is an agreement between two parties, protection buyer and protection seller. This agreement has two legs:

Premium Leg: The protection buyer agrees to pay a fixed amount κ (CDS spread) at fixed times $\mathcal{T} = \{T_1 < T_2 < \cdots < T_n\}$. He pays $\kappa \Delta_k$ at time T_k (where $\Delta_k := T_k - T_{k-1}$), provided that no default has occurred before or at T_k. Then for $t < T_n$, the value of the premium leg is equal to

$$V_P(t) = B_t \mathbb{E}\left(\frac{\kappa}{B_\tau}(\tau - T_{\beta(\tau)-1}) \mathbb{1}_{\{t<\tau\le T_n\}} + \sum_{k=\beta(t)}^{n} \frac{\kappa \Delta_k}{B_{T_k}} \mathbb{1}_{\{\tau > T_k\}} \middle| \mathcal{G}_t \right),$$

where $\beta(t) := \inf\{j : T_j \ge t\}$.

Default leg: The protection seller agrees to cover all losses on the reference bond, provided that the loss occurs before the protection horizon T_n. For $t < T_n$, the value of this default leg is equal to

$$V_D(t) = B_t \mathbb{E}\left(\frac{1 - \delta_{C_{\tau-}}}{B_\tau} \mathbb{1}_{\{t<\tau\le T_n\}} \middle| \mathcal{G}_t \right).$$

If we know the value of the spread, i.e., κ, then the CDS value at time t is the difference between the premium leg and the default leg,

$$\mathrm{CDS}(t, \mathcal{T}, \kappa) = V_P(t) - V_D(t).$$

A market CDS spread (fair spread) $\kappa = \kappa(t, \mathcal{T})$ is agreed at contract's inception (at some time $t < T_1$) in such a way that the value of the contract is 0, i.e., $\mathrm{CDS}(t, \mathcal{T}, \kappa) = 0$. The next theorem, which is an easy consequence of Theorem 16.38, provides formulas for the value of both legs expressed through the conditional transition probability process P and intensity process Λ, so we can calculate $\mathrm{CDS}(t, \mathcal{T}, \kappa)$.

Theorem 16.47 *Assume that C is an $\mathbb{F}$-DS Markov chain with intensity matrix process $(\Lambda_u)_{u\ge 0}$ and conditional transition probability process $P(s,t)$. The value of the default leg of CDS, for $t < T_n$, is equal to*

$$V_D(t) = \sum_{i=1}^{K-1} \mathbb{1}_{\{C_t = i\}} \left(\sum_{j=1}^{K-1} (1 - \delta_j) \int_t^{T_n} \mathbb{E}\left(e^{-\int_t^u r_v \, dv} p_{i,j}(t,u) \lambda_{j,K}(u) \middle| \mathcal{F}_t \right) du \right).$$

The value of the premium leg of CDS is given by

$$\begin{aligned} V_P(t) = \sum_{i=1}^{K-1} \mathbb{1}_{\{C_t = i\}} \Bigg(& \sum_{k=\beta(t)}^{n} \mathbb{E}\left(e^{-\int_t^{T_k} r_u \, du} (1 - p_{i,K}(t, T_k)) \Delta_k \middle| \mathcal{F}_t \right) \\ & + \sum_{j=1}^{K-1} \int_t^{T_n} (u - T_{\beta(u)-1}) \mathbb{E}\left(e^{-\int_t^u r_u \, du} p_{i,j}(t,u) \lambda_{j,K}(u) \middle| \mathcal{F}_t \right) du \Bigg). \end{aligned}$$

In the case of the model given in Example 16.22, one can obtain a more explicit formula for values of the default leg and the premium leg of the CDS (see Sect. 4.2 in [15]).

16.5 Hedging of Rating-Sensitive Claims

Considering the problem of hedging requires a detailed description of the market. Especially, we have to specify instruments which we use in hedging. We assume that on the market, we can put money into a saving account and trade in n nondividend paying default free assets with price Y. Let $\mathbb{F} = \mathbb{F}^{B,Y}$, C be a rating migration process, and R be an ex-dividend price process of a $(K-1)$-dimensional vector of rating-sensitive claims. To construct a hedging strategy for a given payment stream D, according to Lemma 16.17, it is convenient to have some form of martingale representation of discounted cash flows accumulated till maturity T, i.e., a martingale representation of

$$\int_{]\!]0,T]\!]} \frac{1}{B_u} \, dD_u.$$

In what follows, we find such representation. We assume throughout the rest of the paper that the following hypotheses hold:

Assumption 16.48 There exists a martingale measure $\mathbb{Q}$ for the market consisting with the saving account and the default free assets with price process Y, i.e., a measure such that $\mathbb{Q} \sim \mathbb{P}$ and $Y^* = Y/B$ is an $\mathbb{F}$-martingale. Moreover, we assume that the market is $\mathbb{F}$-complete in the sense that every square-integrable $\mathcal{F}_T$-measurable random variable is attainable.

Assumption 16.49 There exists an equivalent martingale measure for the extended market, i.e., a probability measure $\mathbb{Q}^* \sim \mathbb{P}$ such that (Y^*, R^{c*}) is a $\mathbb{G} = \mathbb{F}^{B,Y,C}$-martingale.

Assumption 16.50 The rating migration process C is an $\mathbb{F}$-doubly stochastic Markov chain under $\mathbb{Q}^*$ with intensity matrix process Λ and absorbing state K.

Remark 16.51 For $\mathbb{Q}$ and any $\mathbb{Q}^*$ satisfying the above assumptions, there holds

$$\mathbb{Q}^*|_{\mathcal{F}_T} = \mathbb{Q}|_{\mathcal{F}_T}.$$

Remark 16.52 Let C be an $\mathbb{F}$-DS Markov chain under a probability measure $\mathbb{P}$. The process C does not have to be an $\mathbb{F}$-DS Markov chain under an equivalent probability measure $\mathbb{Q}$. However, there is a large class of equivalent probability measures under which C is an $\mathbb{F}$-DS Markov chain (see [16]).

Remark 16.53 Obviously, processes obtained from M and L by removing the last Kth coordinate are also $\mathbb{G}$-local martingales. The last row of Λ contains only zeros since the state K is an absorbing state of C, and so

$$\widetilde{M}_t := \widetilde{H}_t - \int_0^t \widetilde{\Lambda}_u^\top \widetilde{H}_u \, du, \tag{16.22}$$

where $\widetilde{}$ denotes the matrix (vector) with the last row and last column (last coordinate) deleted. The same remark concerns L:

$$\widetilde{L}_t := \widetilde{Q}^\top(0,t)\widetilde{H}_t = \widetilde{H}_0 + \int_{]\!]0,t]\!]} \widetilde{Q}^\top(0,u)\,d\widetilde{M}_u, \tag{16.23}$$

where $\widetilde{Q}(0,u)$ denotes the matrix $Q(0,u)$ with the last row and last column deleted. However, for simplicity, we abuse notation and in what follows write M, L, Q, Λ for $\widetilde{M}$, $\widetilde{L}$, $\widetilde{Q}$, $\widetilde{\Lambda}$. We insist here on working with these "reduced" versions of matrices/vectors since they are in some sense "minimal." The reduced form models of credit risk can be considered as two-state rating models, and martingale representation theorems are also given with "reduced" versions of vectors/matrices (see Blanchet-Scaillet [9]). Martingales which correspond to M and L are in this case given by

$$M'_t = \mathbb{1}_{\{\tau\le t\}} - \int_0^t \mathbb{1}_{\{\tau>u\}}\lambda_u\,du, \qquad L'_t = e^{\int_0^t \lambda_u\,du}\mathbb{1}_{\{\tau>t\}}.$$

In the following, all conditional expectations are calculated under the martingale measure $\mathbb{Q}^*$; hence, for short, we shall write $\mathbb{E}$ instead of $\mathbb{E}_{\mathbb{Q}^*}$. We divide our results into two subsections.

In the first part, for simplicity of exposition, we assume that the rating-sensitive claims are of a very special form, namely, they are T maturing defaultable claims with only terminal payoffs at time T of the form $\mathbb{1}_{\{C_T=i\}}$. The vector of prices is denoted by

$$R(t,T) = \big(R_1(t,T),\ldots,R_{K-1}(t,T)\big)^\top, \tag{16.24}$$

and according to our convention, R^c denotes its cumulative price, so we have $R_i^c(T,T) = \mathbb{1}_{\{C_T=i\}}$, whereas $R_i(T,T) = 0$ (see Remark 16.5). We show how to deal with hedging of rating-sensitive claims by using rating digital options. In the subsequent subsection, we will show that similar results can be derived for more complex primary instruments with a little effort.

16.5.1 Hedging with Rating Digital Options

Our main aim in this subsection is to show that general rating-sensitive claims can be replicated by trading in some default free assets and a system of $K-1$ special rating-sensitive claims, namely the rating digital options. We will proceed in two steps. First, we provide a replication strategy for very simple rating-sensitive claims, i.e., $(X,0,0,C,\tau)$, and in the second step, we show how to deal with the case of a general rating-sensitive claim (X,A,Z,C,τ). Although from the result of the second step we can easily obtain a replication strategy for simple rating-sensitive claims, we believe that this makes the presentation of the results more transparent. We want to apply Lemma 16.17, so we need to know the dynamics of the discounted cumulative prices R^{c*} of the primary assets (which are rating-sensitive claims) under the martingale measure $\mathbb{Q}^*$. Secondly, we find a representation of the martingale S^{c*} given by formula (16.3) as a stochastic integral with respect to some martingale and then translate it into a stochastic integral with respect to discounted primary assets as in (16.7) of Lemma 16.17. The proposition below gives the dynamics of the vector-valued process R^{c*} which represents the discounted cumulative price of rating digital options (see Remark 16.2):

$$R_j^{c*}(t,T) := \mathbb{E}\left(\frac{\mathbb{1}_{\{C_T=j\}}}{B_T}\middle|\mathcal{G}_t\right),$$

where R is given by (16.24).

Proposition 16.54 *Assume that rating digital options with the price R are tradable assets. Then we have the following representation for the vector R^c of cumulative prices of T-maturing rating digital options:*

$$R^c(t,T) = \widehat{R}(t,T)^\top H_t,$$

where

$$\widehat{R}(t,T) := B_t \mathbb{E}\left(\frac{P(t,T)}{B_T}\middle|\mathcal{F}_t\right). \tag{16.25}$$

Moreover, $R^{c*}(t,T) := R^c(t,T)B_t^{-1}$ has the following dynamics:

$$dR^{c*}(t,T) = \left(B_t^{-1}\widehat{R}(t-,T)\right)^\top dM_t + \left(H_t^\top Q(0,t)\,dm_t\right)^\top,$$

where M_t is defined by (16.22), *and m_t is a matrix-valued martingale defined by the formula*

$$[m_t]_{i,j} := \mathbb{E}\left(\frac{p_{i,j}(0,T)}{B_T}\middle|\mathcal{F}_t\right). \tag{16.26}$$

Proof For the jth coordinate of $R^c(t,T)$, we have

$$\frac{R_j^c(t,T)}{B_t} = \mathbb{E}\left(\frac{\mathbb{1}_{\{C_T=j\}}}{B_T}\middle|\mathcal{G}_t\right) = \sum_{i=1}^{K-1} H_t^i \mathbb{E}\left(\frac{p_{i,j}(t,T)}{B_T}\middle|\mathcal{F}_t\right),$$

where the second equality follows from the fact that C is assumed to be an $\mathbb{F}$-DS stochastic Markov chain. So, in the matrix-vector notation, we have

$$\begin{aligned}\frac{R^c(t,T)}{B_t} &= \left(\mathbb{E}\left(\frac{P(t,T)}{B_T}\middle|\mathcal{F}_t\right)\right)^\top H_t = \left(Q(0,t)\mathbb{E}\left(\frac{P(0,T)}{B_T}\middle|\mathcal{F}_t\right)\right)^\top H_t \\ &= \left(\mathbb{E}\left(\frac{P(0,T)}{B_T}\middle|\mathcal{F}_t\right)\right)^\top L_t = m_t^\top L_t.\end{aligned}$$

By integration by parts it follows that

$$\begin{aligned} dR^{c*}(t,T) &= m_{t-}^{\top} Q^{\top}(0,t)\,dM_t + \left(dm_t^{\top}\right) Q^{\top}(0,t) H_{t-} + \Delta m_t^{\top} Q^{\top}(0,t)\Delta H_t \\ &= \left(Q(0,t) m_{t-}\right)^{\top} dM_t + \left(H_{t-}^{\top} Q(0,t)\,dm_t\right)^{\top}, \end{aligned}$$

where the last equality follows from Proposition 16.29. □

Remark 16.55

(a) We have derived formulas for cumulative prices of rating digital options. Since these instruments have only terminal payment at the maturity time T, the ex-dividend price process R is different from the cumulative price process R^c only at maturity time (see Remark 16.5).
(b) Elements of the matrix $\widehat{R}$ are conditional cumulative prices of rating digital options R, i.e., $[\widehat{R}(t,T)]_{i,j}$ is the price of the digital option on the jth rating on the set $\{C_t = i\}$.

Lemma 16.56 *For $p_{i,j}(0,T)/B_T$ which is an $\mathcal{F}_T$ measurable random variable, $i, j \in \mathcal{K} \setminus K$, there exist $\mathbb{F}$-predictable stochastic processes $\mu^{l,i,j}$, $l = 1, \ldots, n$, such that the following representation holds*:

$$\frac{p_{i,j}(0,T)}{B_T} = \mathbb{E}\left(\frac{p_{i,j}(0,T)}{B_T}\right) + \int_{]\!]0,T]\!]} \sum_{l=1}^{n} \mu_t^{l,i,j}\, dY_t^{l*}.$$

If we denote by $\mu_t^{(l)}$ the matrix $[\mu_t^{(l)}]_{i,j} := \mu_t^{l,i,j}$, then the above representation can be written in the following matrix form:

$$\frac{P(0,T)}{B_T} = \mathbb{E}\left(\frac{P(0,T)}{B_T}\right) + \int_{]\!]0,T]\!]} \sum_{l=1}^{n} \mu_t^{(l)}\, dY_t^{l*}. \tag{16.27}$$

Proof The random variable $p_{i,j}(0,T)/B_T$ is bounded, hence it is square integrable, and therefore the assertion follows by Assumption 16.48. □

Lemma 16.56 gives the representation of matrix-valued martingale $(m_t)_{t\in[\![0,T]\!]}$ defined by formula (16.26). The following proposition gives a representation of the martingale $\mathbb{E}(X^{\top} H_T/B_T|\mathcal{G}_t)$ in terms of stochastic integrals with respect to the martingales $(m_t^X)_{t\in[\![0,T]\!]}$ and $(M_t)_{t\geq 0}$.

Proposition 16.57 *Let $T > 0$, and $X = (X^1, \ldots, X^{K-1})^{\top}$ be a vector of $\mathcal{F}_T$-measurable random variables such that $\frac{X^{\top} H_T}{B_T}$ is square integrable. Then the $\mathbb{G}$-martingale defined by*

$$X^{c*}(t,T) := \mathbb{E}\left(\frac{X^{\top} H_T}{B_T}\,\middle|\,\mathcal{G}_t\right), \quad t \in [0,T], \tag{16.28}$$

has the representation

$$X^{c*}(t,T) = B_t^{-1} \widehat{X}(t,T)^\top H_t,$$

where $\widehat{X}(t,T)$ *is defined by*

$$\widehat{X}(t,T) := B_t \mathbb{E}\left(\frac{P(t,T)X}{B_T} \middle| \mathcal{F}_t \right). \tag{16.29}$$

Moreover, the martingale X^{c*} *has the following representation via stochastic integrals w.r.t.* M *and* m^X:

$$X^{c*}(t,T) = \mathbb{E}\left(\frac{P(0,T)X}{B_T} \right)^\top H_0 + \int_{]\!]0,t]\!]} \left(B_t^{-1} \widehat{X}(u-,T) \right)^\top dM_u + \int_{]\!]0,t]\!]} H_{u-}^\top Q(0,u)\, dm_u^X, \tag{16.30}$$

where m^X *is the vector-valued* $\mathbb{F}$*-martingale given by*

$$m_t^X := \mathbb{E}\left(\frac{P(0,T)X}{B_T} \middle| \mathcal{F}_t \right).$$

Proof The proof is analogous to the proof of Proposition 16.54. □

Remark 16.58

(a) The coordinates of the vector $\widehat{X}$ are conditional cumulative prices of the claim $(X, 0, 0, C, \tau)$, i.e., $[\widehat{X}(t,T)]_i$ is the price of the digital option on the set $\{C_t = i\}$, provided that $t < T$.
(b) The stochastic integrals in formula (16.30) are stopped at τ. The fact that the first integral is stopped at τ is obvious and follows from the fact that M is stopped at τ. The second integral is also stopped at τ, since H_{t-} is constant and equal to zero after τ.

Corollary 16.59 *Assume that* X/B_T *is an* $\mathbb{R}^{K-1}$*-valued vector of square-integrable random variables. Then there exist* $\mathbb{F}$*-predictable stochastic processes* $\mu^{X,l}$ *(with values in* $\mathbb{R}^{K-1}$*) such that*

$$m_T^X = m_0^X + \int_{]\!]0,T]\!]} \sum_{l=1}^{n} \mu_t^{X,l}\, dY_t^{l,*}. \tag{16.31}$$

Proof Note that $P(0,T)^\top X/B_T$ is a square-integrable random variable, and therefore Assumption 16.48 implies the desired representation. □

Representation (16.31) is a generalization of Lemma 16.56 and gives a representation of the martingale $(m_t^X)_{t\in[0,T]}$.

The theorem below yields a replication strategy for the claim $(X, 0, 0, C, \tau)$ maturing at $T > 0$.

Theorem 16.60 *Let $(X, 0, 0, C, \tau)$ be a rating-sensitive claim such that XB_T^{-1} is square integrable. Assume that for each t, $\widehat{X}(t,T) \in \mathrm{Im}(\widehat{R}(t,T))$, where $\widehat{R}$ is given by* (16.25), *and $\widehat{X}$ by* (16.29). *Then the portfolio (ψ, φ, γ) with the initial capital $V_0(\phi) = \mathbb{E}(\frac{P(0,T)X}{B_T})^\top H_0$ is a D-financing portfolio that replicates the claim $(X, 0, 0, C, \tau)$, provided that γ is given as a minimum norm solution to the following system of linear equations*:

$$\widehat{R}(t-,T)\gamma_t = \widehat{X}(t-,T), \tag{16.32}$$

and ψ, φ are defined by

$$\begin{aligned}
\varphi_t^l &= H_{t-}^\top Q(0,t)\big(\mu_t^{X,l} - \mu_t^{(l)}\gamma_t\big), \\
\psi_t &= V_0(\phi) + \int_{]\!]0,t]\!]} \sum_{l=1}^n \varphi_u^l \, dY_u^{l*} + \int_{]\!]0,t]\!]} \gamma_t^\top \, dR^{c*}(u,T) \\
&\quad - \sum_{l=1}^n \varphi_t^l Y_t^{l,*} - \gamma_t^\top R^*(t,T) - \frac{X^\top H_T}{B_T} \mathbb{1}_{[\![T,\infty[\![}(t)
\end{aligned} \tag{16.33}$$

with $\mu^{X,l}$ given by (16.31) *and $\mu_t^{(l)}$ by* (16.27).

Proof We have to show that there exists a D-financing portfolio for the dividend process $D_t := X^\top H_T \mathbb{1}_{[T,+\infty[}(t)$. From Proposition 16.57 and (16.31) it follows that

$$\begin{aligned}
X^{c*}(t,T) &= \mathbb{E}\bigg(\frac{P(0,T)X}{B_T}\bigg)^\top H_0 + \int_{]\!]0,t]\!]} \big(B_u^{-1}\widehat{X}(u-,T)\big)^\top dM_u \\
&\quad + \int_{]\!]0,t]\!]} H_{u-}^\top Q(0,u-) \sum_{l=1}^n \mu_u^{X,l} \, dY_u^{l,*}.
\end{aligned}$$

Now, if we find $\mathbb{G}$-predictable processes φ and γ such that

$$X^{c*}(t,T) = V_0^* + \int_{]\!]0,t]\!]} \sum_{l=1}^n \varphi_u^l \, dY_u^{l,*} + \int_{]\!]0,t]\!]} \gamma_u^\top \, dR^{c*}(u,T) =: I, \tag{16.34}$$

then, by Lemma 16.17, we can choose ψ in such a way that (ψ, φ, γ) is a D-financing portfolio that replicates D. Proposition 16.54 implies that the RHS of (16.34) has the form

$$\begin{aligned}
I &= V_0^* + \int_{]\!]0,t]\!]} \sum_{l=1}^n \varphi_u^l \, dY_u^{l,*} + \int_{]\!]0,t]\!]} \gamma_u^\top \big(B_u^{-1}\widehat{R}(u-,T)\big)^\top dM_u \\
&\quad + \int_{]\!]0,t]\!]} \big(H_u^\top Q(0,u) \, dm_u \gamma_u\big)^\top,
\end{aligned}$$

and Lemma 16.56 gives

$$I = V_0^* + \int_{]\!]0,t]\!]} \gamma_u^\top \big(B_u^{-1}\widehat{R}(u-,T)\big)^\top dM_u$$
$$+ \int_{]\!]0,t]\!]} \sum_{l=1}^{n} (\varphi_u^l + \widetilde{H}_u^\top Q(0,u)\mu_u^{(l)}\gamma_u)\, dY_u^{l*}.$$

If $V_0 = \mathbb{E}(\frac{P(0,T)X}{B_T})^\top H_0$ and φ, γ are $\mathbb{G}$-predictable processes satisfying

$$\big(\widehat{X}(u-,T)\big)^\top = \gamma_u^\top \big(\widehat{R}(u-,T)\big)^\top, \tag{16.35}$$
$$H_{u-}^\top Q(0,u-)\mu_u^{X,l} = \big(\varphi_u^l + H_{u-}^\top Q(0,u-)\mu_u^{(l)}\gamma_u\big), \tag{16.36}$$

then (16.34) holds. Note that the system of linear equations (16.35) is equivalent to (16.32); moreover, the solution to the second system (16.36) is obviously given by (16.33). The component φ is clearly $\mathbb{G}$-predictable, and taking γ to be a minimum norm solution to (16.32) gives a $\mathbb{G}$-predictable solution to (16.32). This finishes the proof. □

Now we pass to general rating-sensitive claims. We present the way of hedging general rating-sensitive claims using digital options and default-free assets. We start by introducing matrix notation which will be very convenient in the subsequent calculations.

Notation 16.61 We will use the following notation:

$$\check{Z}_t = \begin{bmatrix} -Z_t^{1,K} & Z_t^{1,2} - Z_t^{1,K} & \dots & Z_t^{1,K-1} - Z_t^{1,K} \\ Z_t^{2,1} - Z_t^{2,K} & -Z_t^{2,K} & \dots & Z_t^{2,K-1} - Z_t^{2,K} \\ \vdots & \vdots & \ddots & \vdots \\ Z_t^{K-1,1} - Z_t^{K-1,K} & Z_t^{K-1,2} - Z_t^{K-1,K} & \dots & -Z_t^{K-1,K} \end{bmatrix},$$

$$(Z\Lambda)_t^j := \big[Z_t \Lambda_t^\top\big]_{j,j} = \sum_{k=1,k\neq j}^{K} Z_t^{j,k}\lambda_{j,k}(t) = \sum_{k=1}^{K} \check{Z}_t^{j,k}\lambda_{j,k}(t) =: (\check{Z}\Lambda)_t^j.$$

Using our notation, we can represent the vector of ex-dividend price of the rating-sensitive claim (X, A, Z, C, τ) described in Theorem 16.38 in the compact form

$$\frac{S_t}{B_t} = \mathbb{E}\bigg(\frac{P(t,T)X}{B_T}\mathbb{1}_{\{t<T\}} + \int_{]\!]t,T]\!]} \frac{P(t,u)}{B_u}\, dA_u$$
$$+ \int_{]\!]t,T]\!]} \frac{P(t,u)}{B_u}(\check{Z}\Lambda)_u\, du \bigg| \mathcal{F}_t\bigg)^\top H_t. \tag{16.37}$$

As we have seen in the case of claims of the form $(X, 0, 0, C, \tau)$, the crucial point in finding a replication strategy is a suitable martingale representation theorem for X^{c*}, i.e., for the cumulative price process of the claim $(X, 0, 0, C, \tau)$. For a general rating-sensitive claim (X, A, Z, C, τ), we will present a similar version of a martingale representation for S^{c*}.

The following martingale representation theorem allows us to separate hedging of *default risk*, associated with jumps due to rating changes, and *spread risk*, associated with uncertainty as to the predefault value of the claim.

Theorem 16.62 *Let* (X, A, Z, C, τ) *be an arbitrary rating-sensitive claim satisfying Assumption* INT-1. *Then*

$$\frac{S_t^c}{B_t} = S_0^c + \int_{]\!]0,t]\!]} \alpha_u^\top \, dm_u^S + \int_{]\!]0,t]\!]} \beta_u^\top \, dM_u, \tag{16.38}$$

where M *is given by* (16.22), m^S *is an* $\mathbb{F}$*-martingale given by*

$$m_t^S := \mathbb{E}\left(\frac{P(0,T)X}{B_T} + \int_{]\!]0,T]\!]} \frac{P(0,u)}{B_u} \, dA_u + \int_{]\!]0,T]\!]} \frac{P(0,u)}{B_u} (\check{Z}\Lambda)_u \, du \,\middle|\, \mathcal{F}_t \right), \tag{16.39}$$

and α, β *are* $\mathbb{G}$*-predictable stochastic processes defined by*

$$\alpha_t = Q^\top(0,t) H_{t-}, \qquad \beta_t = \frac{\widehat{S}(t-,T) + \check{Z}_t^\top H_{t-}}{B_t}, \tag{16.40}$$

with

$$\widehat{S}(t,T) := B_t \mathbb{E}\left(\frac{P(t,T)X}{B_T} + \int_{[\![t,T]\!]} \frac{P(t,u)}{B_u} \, dA_u + \int_{]\!]t,T]\!]} \frac{P(t,u)}{B_u} (\check{Z}\Lambda)_u \, du \,\middle|\, \mathcal{F}_t \right) \tag{16.41}$$

and $S_0^c = (m_0^S)^\top H_0$.

Proof The processes m^S and $\widehat{S}$ are well defined by Assumption INT-1. It is enough to prove the theorem for the claim of the form $(0, A, Z, C, \tau)$. Indeed, having a general rating-sensitive claim $(X, \widehat{A}, Z, C, \tau)$, observe that by letting $A_t = \widehat{A}_t + X \mathbb{1}_{[\![T,\infty[\![}(t)$ we can restrict to the case of $(0, A, Z, C, \tau)$. For such a claim, we have

$$\frac{S_t^c}{B_t} = \mathbb{E}\left(\int_{]\!]0,T]\!]} \frac{1}{B_u} \, dD_u \,\middle|\, \mathcal{G}_t \right) = I_t^{(1)} + I_t^{(2)},$$

where

$$I_t^{(1)} := \int_{]\!]0,t]\!]} \frac{1}{B_u} \, dD_u \quad \text{and} \quad I_t^{(2)} := \mathbb{E}\left(\int_{]\!]t,T]\!]} \frac{1}{B_u} \, dD_u \,\middle|\, \mathcal{G}_t \right).$$

Using (16.22) and the fact that

$$\int_{]\!]0,t]\!]} \frac{1}{B_u} (Z_u^\top \widehat{H}_{u-})^\top \, d\widehat{H}_u = \int_{]\!]0,t]\!]} \frac{1}{B_u} (\check{Z}_u^\top H_{u-})^\top \, dH_u,$$

where $\widehat{H}_t := (H_t^1, \dots, H_t^K)^\top$, we have

$$\begin{aligned} I_t^{(1)} &= \int_{]\!]0,t]\!]} \frac{1}{B_u} H_u^\top \, dA_u + \int_{]\!]0,t]\!]} \frac{1}{B_u} \big(Z_u^\top \widehat{H}_{u-}\big)^\top \, d\widehat{H}_u \\ &= \int_{]\!]0,t]\!]} \frac{1}{B_u} H_u^\top \, dA_u + \int_{]\!]0,t]\!]} \frac{1}{B_u} \big(\check{Z}_u^\top H_{u-}\big)^\top \, dM_u \\ &\quad + \int_{]\!]0,t]\!]} \frac{1}{B_u} \big(\check{Z}_u^\top H_{u-}\big)^\top \Lambda_u^\top H_{u-} \, du \\ &= \int_{]\!]0,t]\!]} \frac{1}{B_u} H_u^\top \, dA_u + \int_{]\!]0,t]\!]} \frac{1}{B_u} \big(\check{Z}_u^\top H_{u-}\big)^\top \, dM_u + \int_{]\!]0,t]\!]} \frac{1}{B_u} (\check{Z}\Lambda)_u^\top H_{u-} \, du, \end{aligned}$$

where in the last equality, we have used

$$\big(\check{Z}_u^\top H_{u-}\big)^\top \Lambda_u^\top H_{u-} = H_{u-}^\top \check{Z}_u \Lambda_u^\top H_{u-} = (\check{Z}\Lambda)_u^\top H_{u-}.$$

Equation (16.19) implies that

$$\begin{aligned} I_t^{(2)} = \sum_{i=1}^{K-1} H_t^i \sum_{j=1}^{K-1} \mathbb{E}\bigg(&\int_{]\!]t,T]\!]} B_u^{-1} p_{i,j}(t,u) \, dA_u^j \\ &+ \int_{]\!]t,T]\!]} \sum_{k=1}^{K} \frac{Z_u^{j,k}}{B_u} p_{i,j}(t,u) \lambda_{j,k}(u) \, du \,\bigg|\, \mathcal{F}_t \bigg). \end{aligned}$$

Using Notation 16.61, we can write $I^{(2)}$ in the form

$$\begin{aligned} I_t^{(2)} &= \bigg(\mathbb{E}\bigg(\int_{]\!]t,T]\!]} \frac{P(t,u)}{B_u} \, dA_u + \int_{]\!]t,T]\!]} \frac{P(t,u)}{B_u} (\check{Z}\Lambda)_u \, du \,\bigg|\, \mathcal{F}_t \bigg) \bigg)^\top H_t \\ &= \bigg(\mathbb{E}\bigg(\int_{]\!]t,T]\!]} \frac{P(0,u)}{B_u} \, dA_u + \int_{]\!]t,T]\!]} \frac{P(0,u)}{B_u} (\check{Z}\Lambda)_u \, du \,\bigg|\, \mathcal{F}_t \bigg) \bigg)^\top Q^\top(0,t) H_t \\ &= \bigg(m_t - \int_{]\!]0,t]\!]} \frac{P(0,u)}{B_u} \, dA_u - \int_{]\!]0,t]\!]} \frac{P(0,u)}{B_u} (\check{Z}\Lambda)_u \, du \bigg)^\top L_t, \end{aligned}$$

where $L_t = Q^\top(0,t) H_t$ (so L is given by (16.23)), and m is defined by

$$m_t := \mathbb{E}\bigg(\int_{]\!]0,T]\!]} \frac{P(0,u)}{B_u} \, dA_u + \int_{]\!]0,T]\!]} \frac{P(0,u)}{B_u} (\check{Z}\Lambda)_u \, du \,\bigg|\, \mathcal{F}_t \bigg). \tag{16.42}$$

Hence,

$$dI_t^{(2)} = \bigg(m_{t-} - \int_{]\!]0,t[\![} \frac{P(0,u)}{B_u} \, dA_u - \int_{]\!]0,t[\![} \frac{P(0,u)}{B_u} (\check{Z}\Lambda)_u \, du \bigg)^\top dL_t$$

$$+\left(dm_t - \frac{P(0,t)}{B_t}\,dA_t - \frac{P(0,t)}{B_t}(\check{Z}\Lambda)_t\,dt\right)^\top L_{t-}$$

since the continuous martingale part of L is zero. Thus, by the continuity of $Q(0,\cdot)$ and (16.23) we have

$$\begin{aligned}
dI_t^{(2)} &= \left(Q(0,t)\left(m_{t-} - \int_{]\!]0,t[\![} \frac{P(0,u)}{B_u}\,dA_u - \int_{]\!]0,t[\![} \frac{P(0,u)}{B_u}(\check{Z}\Lambda)_u\,du\right)\right)^\top dM_t \\
&\quad + \left(Q(0,t)\,dm_t - Q(0,t)\frac{P(0,t)}{B_t}\,dA_t - Q(0,t)\frac{P(0,t)}{B_t}(\check{Z}\Lambda)_t\,dt\right)^\top H_{t-} \\
&= \left(Q(0,t)\left(m_{t-} - \int_{]\!]0,t[\![} \frac{P(0,u)}{B_u}\,dA_u - \int_{]\!]0,t[\![} \frac{P(0,u)}{B_u}(\check{Z}\Lambda)_u\,du\right)\right)^\top dM_t \\
&\quad + \left(Q(0,t)\,dm_t - \frac{1}{B_t}\,dA_t - \frac{1}{B_t}(\check{Z}\Lambda)_t\,dt\right)^\top H_{t-}.
\end{aligned}$$

Finally, since $I_0^{(2)} = S_0^c$, we have

$$\begin{aligned}
\frac{S_t^c}{B_t} &= S_0^c + \int_{]\!]0,t]\!]} H_{u-}^\top Q(0,u)\,dm_u + \int_{]\!]0,t]\!]} \frac{1}{B_u}\left(\check{Z}_u^\top H_{u-}\right)^\top dM_u \\
&\quad + \int_{]\!]0,t]\!]} \left(Q(0,u)\left(m_{u-} - \int_{]\!]0,u[\![} \frac{P(0,v)}{B_v}\,dA_v \right.\right. \\
&\quad \left.\left. - \int_{]\!]0,u]\!]} \frac{P(0,v)}{B_v}(\check{Z}\Lambda)_v\,dv\right)\right)^\top dM_u
\end{aligned}$$

with $S_0^c = m_0^\top H_0$. Hence, to obtain representation (16.38), it is enough to observe that (16.42) implies

$$Q(0,t)\left(m_t - \int_{]\!]0,t[\![} \frac{P(0,v)}{B_v}\,dA_v - \int_{]\!]0,t]\!]} \frac{P(0,v)}{B_v}(\check{Z}\Lambda)_v\,dv\right) = \frac{\widehat{S}(t,T)}{B_t}.$$

The process β is $\mathbb{G}$-predictable since $\frac{\widehat{S}(t-,T)}{B_t}$ and Z are $\mathbb{G}$-predictable processes by definition. □

Remark 16.63 We note that, in general, the components of the vector $\widehat{S}(t,T)$ defined in Theorem 16.62 do not represent the ex-dividend price of claim (X, A, Z, C, τ) given by (16.37) on the sets $\{C_{t-} = i\}$. There are two differences. The first is that $\widehat{S}(t,T)$ includes possible payment of promised dividend at time t. The second is at time T, when the ex-dividend price is by definition equal to zero.

Now we introduce another assumption on rating sensitive claims.

Assumption INT-2 We assume that a defaultable rating-sensitive claim satisfies

$$\mathbb{E}\bigg(\bigg|\frac{P(0,T)X}{B_T}\bigg|^2 + \bigg|\int_{]\!]0,T]\!]} \frac{P(0,u)}{B_u}\,dA_u\bigg|^2 + \bigg|\int_{]\!]0,T]\!]} \frac{P(0,u)}{B_u}(\check{Z}\Lambda)_u\,du\bigg|^2\bigg) < \infty.$$

As an immediate consequence of Assumption 16.48, we obtain:

Remark 16.64 If a claim (X, A, Z, C, τ) satisfies INT-2, then there exist $(K-1)$-dimensional $\mathbb{F}$-predictable stochastic processes $\mu^{S,l}$, $l = 1, \ldots, n$, such that m^S given by (16.39) has the representation

$$m_T^S = m_0^S + \int_{]\!]0,T]\!]} \sum_{l=1}^{n} \mu_t^{S,l}\,dY_t^{l*}, \tag{16.43}$$

where

$$m_0^S := \mathbb{E}\bigg(\frac{P(0,T)X}{B_T} + \int_{]\!]0,T]\!]} \frac{P(0,u)}{B_u}\,dA_u + \int_{]\!]0,T]\!]} \frac{P(0,u)}{B_u}(\check{Z}\Lambda)_u\,du\bigg).$$

The following theorem yields replication strategy for the general rating-sensitive claims.

Theorem 16.65 *Let (X, A, Z, C, τ) be an arbitrary rating-sensitive claim satisfying Assumption* INT-2. *Assume that, for each t,*

$$\big(\widehat{S}(t-,T) + \check{Z}_t^\top H_{t-}\big) \in \mathrm{Im}\big(\widehat{R}(t-,T)\big),$$

where $\widehat{R}$ is given by (16.25), *and $\widehat{S}$ by* (16.41). *Then the portfolio (ψ, φ, γ) with the initial capital*

$$V_0(\phi) = \mathbb{E}\bigg(\frac{P(0,T)X}{B_T} + \int_{]\!]0,T]\!]} \frac{P(0,u)}{B_u}\,dA_u + \int_{]\!]0,T]\!]} \frac{P(0,u)}{B_u}(\check{Z}\Lambda)_u\,du\bigg)^\top H_0$$

is a D-financing portfolio which replicates claim (X, A, Z, C, τ), provided that γ is given as a minimum norm solution to the following system of linear equations:

$$\widehat{R}(t-,T)\gamma_t = \widehat{S}(t-,T) + \check{Z}_t^\top H_{t-},$$

and φ, ψ are defined by

$$\varphi_t^l = H_{t-}^\top Q(0,t)\big(\mu_t^{S,l} - \mu_t^{(l)}\gamma_t\big),$$

$$\psi_t = V_0(\phi) + \int_{]\!]0,t]\!]} \sum_{l=1}^{n} \varphi_u^l\,dY_u^{l*} + \int_{]\!]0,t]\!]} \gamma_t^\top\,dR^{c*}(u,T) - \sum_{l=1}^{n} \varphi_t^l Y_t^l - \gamma_t^\top R(t,T)$$
$$- \int_{]\!]0,t]\!]} \frac{1}{B_u}\,dD_u$$

with $\mu^{S,l}$ given by (16.43) *and $\mu^{(l)}$ by* (16.27).

Proof For the proof, it is sufficient to repeat the arguments from Theorem 16.60 and use Theorem 16.62. □

16.5.2 Hedging with General Rating-Sensitive Claims

In the previous subsection, we have solved problem of replication of general rating-sensitive claims by trading in default-free assets and in the system of digital rating options. This result is unsatisfactory for practitioners since we have used in a replication the digital rating options which are not traded on the real market. In this subsection, we give a generalization of the previous results by allowing to trade in an arbitrary (to some extent) system of tradable general rating-sensitive claims. We point out that this system includes defaultable bonds, Credit Sensitive Notes, CDS, etc.

Now we will consider the case where an investor can trade in m general rating-sensitive claims, where $m \geq K-1$. In the other words, we consider the case where, on the market, we can trade in general rating-sensitive claims $(X^{(i)}, A^{(i)}, Z^{(i)}, C, \tau)$, $i = 1, \dots, m$, and we are interested in providing sufficient conditions for the existence of replication strategy for an arbitrary claim (X, A, Z, C, τ). Before we proceed further, we introduce another convenient matrix notation.

Notation 16.66

$$
\begin{aligned}
&\mathbf{X} := \big[X^{(1)}, \dots, X^{(m)}\big], \qquad \mathbf{A}_t := \big[A_t^{(1)}, \dots, A_t^{(m)}\big],\\
&[\check{\mathbf{Z}}\mathbf{H}]_t := \big[\big(\check{Z}_t^{(1)}\big)^\top H_{t-}, \dots, \big(\check{Z}_t^{(m)}\big)^\top H_{t-}\big],\\
&[\check{\mathbf{Z}}\boldsymbol{\Lambda}]_t := \big[\big(\check{Z}^{(1)}\Lambda\big)_t, \dots, \big(\check{Z}^{(m)}\Lambda\big)_t\big].
\end{aligned}
$$

In the next proposition, we give the dynamics of the discounted cumulative prices of a system of $m \geq K-1$ rating-sensitive claims. We will use the same idea and proceed as before.

Proposition 16.67 *Assume that a system of m traded general rating-sensitive claims* $((X^{(i)}, A^{(i)}, Z^{(i)}, C, \tau))_{i=1}^m$ *satisfies Assumption* INT-1. *Then the discounted cumulative price process* R^{c*} *of these claims has the dynamics*

$$
\frac{R^c(t,T)}{B_t} = R^c(0,T) + \int_{]\!]0,t]\!]} \big(\alpha_u^\top d\mathbf{m}_u\big)^\top + \int_{]\!]0,t]\!]} \boldsymbol{\Theta}_u^\top dM_u, \tag{16.44}
$$

where $\mathbf{m}$ *is an* $\mathbb{F}$*-martingale with values in the space of matrices of dimension* $(K-1) \times m$, *given by*

$$
\mathbf{m}_t := \mathbb{E}\bigg(\frac{P(0,T)\mathbf{X}}{B_T} + \int_{]\!]0,T]\!]} \frac{P(0,u)}{B_u}\, d\mathbf{A}_u + \int_{]\!]0,T]\!]} \frac{P(0,u)}{B_u} [\check{\mathbf{Z}}\boldsymbol{\Lambda}]_u\, du \,\bigg|\, \mathcal{F}_t\bigg), \tag{16.45}
$$

and α, β are $\mathbb{G}$-predictable stochastic processes defined by

$$\alpha_t = Q^\top(0,t)H_{t-}, \qquad \mathbf{\Theta}_t = \frac{\widehat{\mathbf{R}}(t-,T) + [\check{\mathbf{Z}}\mathbf{H}]_t}{B_t}, \tag{16.46}$$

with $R^c(0,T) = \mathbf{m}_0^\top H_0$ and

$$\widehat{\mathbf{R}}(t,T) := B_t \mathbb{E}\bigg(\frac{P(t,T)\mathbf{X}}{B_T} + \int_{[\![t,T]\!]} \frac{P(t,u)}{B_u}\, d\mathbf{A}_u + \int_{]\!]t,T]\!]} \frac{P(t,u)}{B_u}[\check{\mathbf{Z}}\mathbf{\Lambda}]_u\, du \bigg| \mathcal{F}_t\bigg). \tag{16.47}$$

Proof The thesis follows immediately from the martingale representation, Theorem 16.62. □

Remark 16.68 Assume that the claims $(X^{(i)}, A^{(i)}, Z^{(i)}, C, \tau)$, $i = 1, \ldots, m$, satisfy Assumption INT-2 for every i. Then Assumption 16.48 implies that $\mathbf{m}_T$ given by (16.45) has the following representation

$$\mathbf{m}_T = \mathbf{m}_0 + \int_{]\!]0,T]\!]} \sum_{l=1}^{n} \mu_t^{(l)}\, dY_t^{l*}, \tag{16.48}$$

where $\mu^{(l)}$, $l = 1, \ldots, n$, are $\mathbb{F}$-predictable stochastic processes with values in matrices of dimension $(K-1) \times m$ and

$$\mathbf{m}_0 := \mathbb{E}\bigg(\frac{P(0,T)\mathbf{X}}{B_T} + \int_{]\!]0,T]\!]} \frac{P(0,u)}{B_u}\, d\mathbf{A}_u + \int_{]\!]0,T]\!]} \frac{P(0,u)}{B_u}[\check{\mathbf{Z}}\mathbf{\Lambda}]_u\, du\bigg).$$

Now we present the theorem which gives sufficient conditions for a replication of a general rating-sensitive claim.

Proposition 16.69 *Let (X, A, Z, C, τ) be an arbitrary rating-sensitive claim satisfying Assumption* INT-2. *Assume that, for each t,*

$$\widehat{S}(t-,T) + \check{Z}_t^\top H_{t-} \in \operatorname{Im}\big(\widehat{\mathbf{R}}(t-,T) + [\check{\mathbf{Z}}\mathbf{H}]_t\big),$$

where $\widehat{\mathbf{R}}$ is given by (16.47) *and $\widehat{S}$ by* (16.41). *Then the portfolio (ψ, φ, γ) with initial capital*

$$V_0(\phi) = \mathbb{E}\bigg(\frac{P(0,T)X}{B_T} + \int_{]\!]0,T]\!]} \frac{P(0,u)}{B_u}\, dA_u + \int_{]\!]0,T]\!]} \frac{P(0,u)}{B_u}(\check{Z}\Lambda)_u\, du\bigg)^\top H_0$$

is a D-financing portfolio that replicates claim (X, A, Z, C, τ), provided that γ is given as a minimum norm solution to the following system of linear equations:

$$\big(\widehat{\mathbf{R}}(t-,T) + [\check{\mathbf{Z}}\mathbf{H}]_t\big)\gamma_t = \widehat{S}(t-,T) + \check{Z}_t^\top H_{t-},$$

and φ, ψ are defined by

$$\begin{aligned}\varphi_t^l &= H_{t-}^\top Q(0,t)\big(\mu_t^{S,l} - \mu_t^{(l)}\gamma_t\big),\\ \psi_t &= V_0(\phi) + \int_{]\!]0,t]\!]} \sum_{l=1}^n \varphi_u^l \, dY_u^{l*} + \int_{]\!]0,t]\!]} \gamma_u^\top \, dR^{c*}(u,T)\\ &\quad - \sum_{l=1}^n \varphi_t^l Y_t^l - \gamma_t^\top R(t,T) - \int_{]\!]0,t]\!]} \frac{1}{B_u}\, dD_u,\end{aligned} \tag{16.49}$$

where $\mu^{(l)}$ are $\mathbb{F}$-predictable stochastic processes that appear in representation (16.48), *and $\mu^{S,l}$ are given in representation* (16.43).

Proof According to Lemma 16.17, we are looking for process (φ, γ) such that

$$S_t^{c*} = V_0(\phi) + \int_{]\!]0,t]\!]} \sum_{l=1}^n \varphi_u^l \, dY_u^{l*} + \int_{]\!]0,t]\!]} \gamma_u^\top \, dR^{c*}(u,T). \tag{16.50}$$

Lemma 16.17 also implies that by taking ψ given by (16.49) we obtain a D-financing portfolio which replicates (X, A, Z, C, τ). Equation (16.44) and representation (16.48) imply that

$$\begin{aligned}S_t^{c*} &= V_0(\phi) + \int_{]\!]0,t]\!]} \sum_{l=1}^n \varphi_u^l \, dY_u^{l*} + \int_{]\!]0,t]\!]} \gamma_u^\top \big(\alpha_u^\top d\mathbf{m}_u\big)^\top + \int_{]\!]0,t]\!]} \gamma_u^\top \mathbf{\Theta}_u^\top \, dM_u\\ &= V_0(\phi) + \int_{]\!]0,t]\!]} \sum_{l=1}^n \big(\varphi_u^l + \big(\alpha_u^\top \mu_u^{(l)} \gamma_u\big)^\top\big)\, dY_u^{l*} + \int_{]\!]0,t]\!]} (\mathbf{\Theta}_u \gamma_u)^\top \, dM_u.\end{aligned} \tag{16.51}$$

Moreover, by (16.38) and representation (16.43),

$$S_t^{c*} = S_0^c + \int_{]\!]0,t]\!]} \sum_{l=1}^n \alpha_u^\top \mu_u^{S,l} \, dY_u^{*,l} + \int_{]\!]0,t]\!]} \beta_u^\top \, dM_u. \tag{16.52}$$

Comparing coefficients in (16.51) and (16.52), we see that (16.50) holds, provided that γ, φ satisfies the system of linear equations

$$\begin{aligned}\mathbf{\Theta}_t \gamma_t &= \beta_t,\\ \varphi_t^l + \alpha_t^\top \mu_t^{(l)} \gamma_t &= \alpha_t^\top \mu_t^{S,l} \quad \text{for } l = 1, \ldots, n,\end{aligned}$$

where β and α are given by (16.40). The proof is complete. □

Remark 16.70 We assume that primary rating-sensitive instruments have the same maturity $T > 0$ as the rating-sensitive claim we want to replicate. This assumption can be easily relaxed. We could allow the primary rating-sensitive instruments to have different maturities T_i but greater or equal to T, the maturity of the rating-sensitive claim we want to replicate.

Remark 16.71 If the number of primary rating-sensitive assets is large, then system of equations

$$\boldsymbol{\Theta}_t \gamma_t = \beta_t$$

can have infinitely many solutions. One can ask the question whether we can choose γ in such a way that additional n constrains will be satisfied? More precisely, can we choose γ such that

$$\mu_t^{(l)} \gamma_t = \mu_t^{S,l}$$

for $l = 1, \ldots, n$? The positive answer implies that we do not need nondefaultable primary assets in the portfolio, i.e., $\varphi^l = 0$ for each $l = 1, \ldots, n$. A sufficient condition for this to hold is that the matrix

$$G_t := \begin{bmatrix} \boldsymbol{\Theta}_t \\ \mu_t^{(1)} \\ \vdots \\ \mu_t^{(n)} \end{bmatrix}$$

of dimensions $(K-1)(n+1) \times m$ has the rank equal to $(K-1)(n+1)$. In this case, for an arbitrary rating-sensitive claim (X, A, Z, C, τ), we can solve the system of equations

$$G_t \gamma_t = \left[\beta_t, \mu_t^{S,1}, \ldots, \mu_t^{S,n}\right]^\top$$

and obtain portfolio with $\varphi = 0$. A necessary condition for this gives trivially that the number of primary rating-sensitive assets m should satisfy $m \geq (K-1)(n+1)$. This condition can be viewed as a nonsingularity condition for the volatility matrix.

References

1. K.K. Aase, On the consistency of the Lucas pricing formula. Math. Finance **18**(2), 293–303 (2008)
2. D. Applebaum, *Lévy Processes and Stochastic Calculus*, 2nd edn. Cambridge Studies in Advanced Mathematics, vol. 116 (Cambridge University Press, Cambridge, 2009), xxx, 460 p.
3. A. Bélanger, S.E. Shreve, D. Wong, A general framework for pricing credit risk. Math. Finance **14**(3), 317–350 (2004)
4. T.R. Bielecki, M. Rutkowski, *Credit Risk: Modelling, Valuation and Hedging*. Springer Finance (Springer, Berlin, 2002), xviii, 500 p.
5. T.R. Bielecki, M. Rutkowski, Modeling of the defaultable term structure: Conditionally Markov approach. IEEE Trans. Autom. Control **49**, 361–373 (2004)
6. T.R. Bielecki, M. Jeanblanc, M. Rutkowski, PDE approach to valuation and hedging of credit derivatives. Quant. Finance **5**(3), 257–270 (2005)

7. T.R. Bielecki, S. Crépey, M. Jeanblanc, M. Rutkowski, Valuation of basket credit derivatives in the credit migrations environment, in *Handbooks in Operations Research and Management Science*, ed. by J. Birge, V. Linetsky. Financial Engineering, vol. 15 (2008), pp. 471–510
8. T.R. Bielecki, M. Jeanblanc, M. Rutkowski, Pricing and trading credit default swaps in a hazard process model. Ann. Appl. Probab. **18**(6), 2495–2529 (2008)
9. C. Blanchet-Scalliet, M. Jeanblanc, Hazard rate for credit risk and hedging defaultable contingent claims. Finance Stoch. **8**(1), 145–159 (2004)
10. R.-A. Dana, M. Jeanblanc, *Financial Markets in Continuous Time. Translated from French by Anna Kennedy*. Springer Finance (Springer, Berlin, 2003), xi, 324 p.
11. D. Duffie, *Dynamic Asset Pricing Theory*, 3rd edn. (Princeton University Press, Princeton, 2001), xix, 465 p.
12. T.R. Hurd, A. Kuznetsov, Fast CDO computations in the affine Markov chain model. Working Paper (2006)
13. T.R. Hurd, A. Kuznetsov, Affine Markov chain model of multifirm credit migration. J. Credit Risk **3**, 3–29 (2007)
14. J. Jakubowski, M. Niewęgłowski, Doubly stochastic Markov chain—a new class of processes for modeling credit rating migration processes and valuation of claims. IMPAN Preprint 698 (2008)
15. J. Jakubowski, M. Niewęgłowski, Pricing bonds and CDS in the model with rating migration induced by a Cox process, in *Advances in Mathematics of Finance*, ed. by Stettner, Łukasz. Banach Center Publications, vol. 83 (2008), pp. 159–182
16. J. Jakubowski, M. Niewęgłowski, A class of $\mathbb{F}$-doubly stochastic Markov chains. Electron. J. Probab. **15**, 1743–1771 (2010)
17. R.A. Jakubowski, D. Lando, S.T. Turnbull, A Markov model for the term structure of credit risk spreads. Rev. Financ. Stud. **10**, 481–523 (1997)
18. I. Karatzas, S.E. Shreve, *Methods of Mathematical Finance*. Applications of Mathematics (Springer, Berlin, 1998), xv, 407 p.
19. D. Lando, On Cox processes and credit risky securities. Rev. Deriv. Res. **2**, 99–120 (1998)
20. D. Lando, A. Mortensen, On the pricing of step-up bonds in the European telecom sector. J. Credit Risk **1**, 71–110 (2004)
21. R.E. Lucas Jr., Asset prices in an exchange economy. Econometrica **46**, 1429–1445 (1978)
22. T. Møller, Risk-minimizing hedging strategies for insurance payment processes. Finance Stoch. **5**(4), 419–446 (2001)
23. M. Schweizer, Local risk-minimization for multidimensional assets and payment streams, in *Advances in Mathematics of Finance*, ed. by Stettner, Łukasz. Banach Center Publications, vol. 83 (2008), pp. 213–229

Chapter 17
Exotic Derivatives under Stochastic Volatility Models with Jumps

Aleksandar Mijatović and Martijn Pistorius

Abstract In equity and foreign exchange markets the risk-neutral dynamics of the underlying asset are commonly represented by stochastic volatility models with jumps. In this paper we consider a dense subclass of such models and develop analytically tractable formulae for the prices of a range of first-generation exotic derivatives. We provide closed-form formulae for the Fourier transforms of vanilla and forward starting option prices as well as a formula for the slope of the implied volatility smile for large strikes. A simple explicit approximation formula for the variance swap price is given. The prices of volatility swaps and other volatility derivatives are given as a one-dimensional integral of an explicit function. Analytically tractable formulae for the Laplace transform (in maturity) of the double-no-touch options and the Fourier–Laplace transform (in strike and maturity) of the double knock-out call and put options are obtained. The proof of the latter formulae is based on extended matrix Wiener–Hopf factorisation results. We also provide convergence results.

Keywords Double-barrier options · Volatility surface · Volatility derivatives · Forward starting options · Stochastic volatility models with jumps · Fluid embedding · Complex matrix Wiener–Hopf factorisation

Mathematics Subject Classification (2010) 60K15 · 91G20

17.1 Introduction

A key step in the valuation and hedging of exotic derivatives in financial markets is to decompose these in terms of simpler securities, e.g. vanilla options, which trade in larger volumes, are generally very liquid and therefore have a well-defined price. Such a decomposition is often achieved in two steps. First a model for the

A. Mijatović (✉)
Department of Statistics, University of Warwick, Coventry, UK
e-mail: a.mijatovic@warwick.ac.uk

M. Pistorius
Department of Mathematics, Imperial College London, London, UK
e-mail: m.pistorius@imperial.ac.uk

G. Di Nunno, B. Øksendal (eds.), *Advanced Mathematical Methods for Finance*,
DOI 10.1007/978-3-642-18412-3_17,

underlying asset under a risk-neutral measure is calibrated to the implied volatility surface. In this step the current state of the market, as described by the prices of vanilla derivatives, is expressed in terms of the parameter values of the model. In other words the chosen model is used to impose a structure on the option prices. The second step consists of pricing the exotic derivative of interest in the calibrated model.

It is well known that in equity and foreign exchange markets stochastic volatility models with jumps can be used to accomplish the first step described above (see e.g. [17, 24]). In the present paper we consider forward starting vanilla options, volatility derivatives and barrier options, which are among the most widely traded exotic derivatives in the equity and foreign exchange markets.

The desired properties of the model in each of the two steps described above place diametrically opposite restrictions on the choice of modelling framework. This is because in the calibration step one requires a flexible stochastic process that can describe well the current state of the vanilla market (i.e. can calibrate accurately to the observed implied volatility surface), while such flexibility can be a source of problems in the second step, where one needs to compute expectations of path-dependent functionals of the process. A more rigid modelling framework with, say, continuous trajectories and some distributional properties (e.g. independence of increments) could yield the structure of the process needed to establish efficient pricing algorithms for exotic derivatives.

We investigate two families of stochastic volatility models with jumps: the time-changed exponential Lévy models and the stochastic volatility models driven by Lévy processes, where the volatility process is independent of the Lévy driver. In these two families of models, the pricing of European derivatives is well understood, and efficient calibration methods have been developed (see for example [12] and [24]), i.e. the first step of the two-stage procedure outlined above. However once the model is calibrated, the problem of pricing the first-generation exotic derivatives (e.g. barrier options) is quite involved. The law of the first-exit time from a bounded interval in stochastic volatility models with jumps, for instance, is not usually available in analytically tractable form. Because of the lack of structural properties that can be exploited to find the laws of the path-dependent functionals of interest, one would typically need to resort to Monte Carlo methods for the pricing of such derivatives in this setting. It is well known that these methods are time-consuming and yield unstable results, especially when used to calculate the sensitivities of derivative securities. The method proposed in this paper to calculate the prices of such contracts consists of two steps: (i) a Markov chain approximation of the volatility process and (ii) an analytically tractable solution of the value function of the contract of interest in the approximating model. We provide proofs for the convergence of option prices under this approximation and derive explicit expressions of Laplace and/or Fourier transforms of the value functions under the approximating model.

The approximating class of stochastic volatility processes with jumps considered in this paper retains the structural properties required for the semi-analytic pricing (i.e. up to an integral transform) of forward starting options, volatility derivatives and barrier options. In the case of double-barrier option prices we will show that the process considered here admits explicit formulae for the Laplace/Fourier transforms in terms of the solutions of certain quadratic matrix equations. The main mathematical contribution of the present paper, which underpins the derivation of these closed form formulae, is the proof of the existence and uniqueness of the matrix Wiener–Hopf factorisation of a class of complex-valued matrices related to the approximating model (see Theorem 17.21). These results extend those of [21] where the corresponding results for the real-valued case are established. In the context of noisy fluid flow models, the matrix Wiener–Hopf factorisation for the case of a regime-switching Brownian motion is studied by [4, 28].

It should be noted that the matrix Wiener–Hopf factorisation results developed in this paper can also be applied to the pricing of American call and put options in model (17.19) as follows. First we apply the main result in [21] to obtain the price of the perpetual American call or put and then, via the randomization algorithm introduced in [11], find the actual price of the option.

Related Markov chain mixture models, which are special cases of the model considered in this paper, have been studied before in the mathematical finance literature. In [19] and [22] explicit formulae were derived for the price of a perpetual American put option under a regime-switching Brownian motion model. The same process was used in [18] to model stochastic dividend rates where the problem of the pricing of barrier options on equity was considered. Finite maturity American put options were considered in [10] under a regime-switching Brownian motion model. More generally in [7, 8] numerical algorithms were developed in the case of regime-switching Lévy processes. Furthermore extensive work has been done on derivative pricing under stochastic volatility models with and without jumps (see the standard references [14, 17, 24] and [9]).

The remainder of the paper is organized as follows. In Sect. 17.2 we state and prove the properties of continuous-time Markov chains and phase-type distributions that are needed to define the class of stochastic volatility models with jumps studied in this paper. In Sect. 17.3 we give a precise definition of this class of models and describe an explicit construction of an approximating sequence of models, based on Markov chains, which converges to the given stochastic volatility model with jumps. The models are set against the backdrop of a foreign exchange market which allows us to include naturally the stochastic foreign and domestic discount factors. In Sect. 17.4 we provide explicit formulae for the Fourier transforms in model (17.19) for vanilla and forward starting options. This section also gives an approximate explicit formula for the pricing of variance swaps, a one-dimensional integral representation of the price of a volatility swap and formulae for the asymptotic behaviour of the implied volatility smile for large strikes. Section 17.5 is devoted to the first-passage times of regime-switching processes. Section 17.6 discusses the pricing of double-no-touch and double-barrier knock-out options. It provides a formula for the single Laplace transform (in maturity) and the Laplace–Fourier transform (in strike and maturity) of the double-no-touch and the double-barrier knock-out options respectively in terms of the quantity that can be obtained from the complex matrix Wiener–Hopf factorisation (see Theorem 17.25). Section 17.7 describes the fluid embedding of the model in (17.19), which plays a central role in the Wiener–Hopf factorisation. The key mathematical results of the paper, which allow us to price barrier options in the setting of stochastic volatility, are contained in Sect. 17.8, where matrix Wiener–Hopf factorisation is defined, and the theorems asserting its uniqueness and existence are stated.

17.2 Markov Chains and Phase-Type Distributions

17.2.1 Finite-State Markov Chains

We start by collecting some useful and well-known properties of finite-state Markov chains that will be important in the sequel (see e.g. [16]). For completeness, we will also present the proofs. Throughout the paper we will denote by

$$M(i,j) = M_{ij} = e_i' M e_j, \quad m(j) = m_j = m' e_j, \quad i,j = 1,\dots,n,$$

the ijth element of an $n \times n$ matrix M and the jth element of an n-dimensional vector m, where the vectors e_i, $i = 1,\dots,n$, denote the standard basis of $\mathbb{C}^n$, and where $'$ means transposition. Throughout the paper I will denote an identity matrix of appropriate size, and $\mathbb{R}_+ = [0,\infty)$ the nonnegative real line.

Lemma 17.1 *Let Z be a Markov chain on a state space $E^0 := \{1,\dots,N_0\}$, where $N_0 \in \mathbb{N}$, and let $B : E^0 \to \mathbb{C}$ be any function. If Q denotes the generator of Z and Λ_B is a diagonal matrix of size N_0 with diagonal elements equal to $B(i)$, $i = 1,\dots,N_0$, then it holds that*

$$\mathbb{E}_i\left[\exp\left(\int_0^t B(Z_s)\,ds\right) I_{\{Z_t=j\}}\right] = \exp\big(t(Q+\Lambda_B)\big)(i,j) \quad \text{for any } i,j \in E^0,\ t \geq 0, \tag{17.1}$$

where $\mathbb{E}_i[\cdot] = \mathbb{E}[\cdot|Z_0 = i]$, $\mathbb{P}_i[\cdot] = \mathbb{P}[\cdot|Z_0 = i]$, and $I_{\{\cdot\}}$ is the indicator of a set $\{\cdot\}$.

Proof Let $(P_t)_{t\geq 0}$ be a family of N_0-dimensional square matrices with entries $P_t(i,j)$, $i,j = 1,\dots,N_0$, given by the left-hand side of (17.1). It is clear that $P_0 = I$, where I is the N_0-dimensional identity matrix. The Markov property of the chain Z yields the Chapman–Kolmogorov equation $P_{t+s} = P_s P_t = P_t P_s$ for all $s,t \geq 0$. If we show that the family of matrices $(P_t)_{t\geq 0}$ satisfies the system of ODEs with constant coefficients

$$\frac{dP_t}{dt} = (Q + \Lambda_B)P_t, \qquad P_0 = I, \tag{17.2}$$

then the lemma will follow, since (17.2) is well known to have a unique solution given by the right-hand side of (17.1). The Chapman–Kolmogorov equation implies that $P_{t+h} - P_t = (P_h - I)P_t$, and it is therefore enough to show $\lim_{h\to 0}(P_h - I)/h = Q + \Lambda_B$. In other words, we need to prove

$$\lim_{h\to 0}\big(P_h(i,j) - I(i,j)\big)/h = \begin{cases} Q(i,j) & \text{if } i \neq j, \\ Q(i,j) + B(j) & \text{if } i = j. \end{cases} \tag{17.3}$$

The random variables $B(Z_s)$ are bounded uniformly in s, and hence the Taylor expansion of the exponential yields

$$P_h(i,j) = \mathbb{E}_i\left[I_{\{Z_h=j\}}\left(1 + \int_0^h B(Z_s)\,ds\right)\right] + o(h) \quad \text{for all } i, j \in \{1,\dots,N_0\}.$$

It is clear that

$$\lim_{h\to 0} I_{\{Z_h=j\}}\frac{1}{h}\int_0^h B(Z_s)\,ds = I_{\{Z_0=j\}}B(Z_0) \quad \mathbb{P}_i\text{-a.s.} \quad \text{for all } i, j \in \{1,\dots,N_0\},$$

since the paths of Z are $\mathbb{P}_i$-a.s. constant for exponentially distributed amount of time. The dominated convergence theorem and the well-known fact $\mathbb{E}_i[I_{\{Z_h=j\}}] = h(I(i,j) + Q(i,j)) + o(h)$ therefore imply (17.3). This concludes the proof. □

We now apply Lemma 17.1 to establish a simple but important property of the spectrum of a discounted generator.

Lemma 17.2 *Let Q be a generator of a Markov chain with $N_0 \in \mathbb{N}$ states, and let D be a complex diagonal matrix of dimension N_0. Then every eigenvalue $\lambda \in \mathbb{C}$ of the matrix $Q - D$ (i.e. a solution of the equation $(Q - D)x = \lambda x$ for some nonzero element x in $\mathbb{C}^{N_0}$) satisfies the inequality*

$$\Re(\lambda) \le -\min\big\{\Re(d_i) : i = 1,\dots,N_0\big\},$$

where $d_i = D(i,i)$, $i = 1,\dots,N_0$, are diagonal elements of D. In particular, if $\min\{\Re(d_i) : i = 1,\dots,N_0\} > 0$, then the matrix $Q - D$ is invertible. Furthermore, the real part of every eigenvalue of Q is nonpositive.

Proof Let λ be an eigenvalue of the matrix $Q - D$ that corresponds to the eigenvector $x \in \mathbb{C}^{N_0}$. Then x is also an eigenvector with eigenvalue $\exp(\lambda)$ of the matrix $\exp(Q - D)$. Lemma 17.1 implies that if Z is the chain generated by Q, then the following identity holds:

$$e_i' \exp(Q - D)x = \sum_{j=1}^{N_0} x_j \mathbb{E}_i\left[\exp\left(-\int_0^1 d_{Z_t}\,dt\right) I_{\{Z_1=j\}}\right], \quad i = 1,\dots,N_0, \tag{17.4}$$

where e_i (resp. d_i) denotes the ith basis vector in $\mathbb{C}^{N_0}$ (resp. diagonal element of the matrix D).

Assume now without loss of generality that the norm $\|x\|_\infty := \max\{|x_i| : i = 1, \ldots, N_0\}$ of the vector x is one. Then identity (17.4) implies the estimate

$$\exp(\Re(\lambda)) = |\exp(\lambda)| = \|\exp(Q - D)x\|_\infty \le \exp(-\min\{\Re(d_i) : i = 1, \ldots, N_0\}),$$

which proves the lemma. □

Lemma 17.3 *Let $q \in \mathbb{C}$ be such that $\Re(q) > 0$, and M a matrix whose eigenvalues all have nonpositive real part. Then the matrix $qI - M$ is invertible, and the following formula holds*:

$$\int_0^\infty e^{-qt} \exp(tM)\, dt = (qI - M)^{-1}. \tag{17.5}$$

Proof The following identity holds for any $T \in (0, \infty)$ by the fundamental theorem of calculus:

$$\int_0^T \exp((M - qI)t)\, dt = (M - qI)^{-1}(\exp((M - qI)T) - I),$$

and, since the real part of the spectrum of the matrix $M - qI$ is strictly negative by Lemma (17.2), in the limit as $T \to \infty$, we obtain

$$\int_0^\infty \exp((M - qI)t)\, dt = (qI - M)^{-1}.$$ □

17.2.2 (Double) Phase-Type Distributions

In this section we review basic properties of phase-type distributions, as these will play an important role in the sequel. We refer to Neuts [27] and Asmussen [2] for further background on phase-type distributions.

A distribution function $F : \mathbb{R}_+ \to [0, 1]$ is called *phase-type* if it is the distribution of the absorption time of a continuous-time Markov chain on $(m + 1)$ states, for some $m \in \mathbb{N}$, with one state absorbing and the remaining states transient. The distribution F is uniquely determined by the matrix $A \in \mathbb{R}^{m\times m}$, which is the generator of the chain restricted to the transient states, and the initial distribution of the chain on the transient states $\alpha \in \mathbb{R}^m$ (i.e. the coordinates of α are nonnegative, and the inequalities $0 \le \alpha'\mathbf{1} \le 1$ hold, where $\mathbf{1}$ is the m-dimensional vector with each coordinate equal to one and $'$ denotes transposition). The notation $X \sim \mathrm{PH}(\alpha, A)$ is commonly used for a random variable X with cumulative distribution function F. Note also that the law of the original chain on the entire state space is given by

$$\text{the initial distribution } \begin{pmatrix} \alpha \\ 1 - \alpha'\mathbf{1} \end{pmatrix} \text{ and the generator matrix } \begin{pmatrix} A & (-A)\mathbf{1} \\ \mathbf{0} & 0 \end{pmatrix},$$

where $\mathbf{0}$ denotes a row of m zeros. It is clear from this representation that the cumulative distribution function F and its density f are of the form

$$F(t) = 1 - \alpha' e^{tA}\mathbf{1} \quad \text{and} \quad f(t) = -\alpha' e^{tA} A\mathbf{1} \quad \text{for any } t \in \mathbb{R}_+. \tag{17.6}$$

Note also that 0 is an atom of the distribution if and only if $\alpha'\mathbf{1} < 1$, in which case the function f is a density of a subprobability measure on $(0, \infty)$. The nth moment of the random variable $X \sim \mathrm{PH}(\alpha, A)$ is given by

$$\mathbb{E}\big[X^n\big] = n!\alpha'(-A)^{-n}\mathbf{1}.$$

It follows from the definition the phase-type distribution that the matrix A can be viewed as a generator of a killed continuous-time Markov chain on m states. Therefore we can express the matrix A as $A = Q - D$, where Q is the generator of a chain on m states, and D is a diagonal matrix with nonnegative diagonal elements that are equal to the coordinates of the vector $-A\mathbf{1}$. Lemma 17.2 therefore implies that the real part of each eigenvalue of A is nonpositive. The next proposition gives a characterisation of the existence of exponential moments of a phase-type distribution in terms of the eigenvalues of the matrix A.

Proposition 17.4 *Let* $X \sim \mathrm{PH}(\alpha, A)$ *be a phase-type random variable as defined above, and let* λ_0 *be the eigenvalue of the matrix* A *with the largest real part, i.e.* $\Re(\lambda_0) = \max\{\Re(\lambda) : \lambda \text{ eigenvalue of } A\}$. *Then, for any* $u \in \mathbb{C}$, *the exponential moment* $\mathbb{E}[\exp(uX)]$ *exists and is finite if and only if* $\Re(u) < -\Re(\lambda_0)$, *in which case the following formula holds*:

$$\mathbb{E}\big[\exp(uX)\big] = \alpha'(A + uI)^{-1}A\mathbf{1} + (1 - \alpha'\mathbf{1}),$$

where I *denotes an* m*-dimensional identity matrix.*

Proof It is clear that the identity

$$\mathbb{E}\big[\exp(uX)\big] = \mathbb{P}(X = 0) + \int_0^\infty \exp(tu) f(t)\, dt \tag{17.7}$$

must hold for all $u \in \mathbb{R}$, where f is the density of X on the interval $(0, \infty)$. Hence the question of existence of $\mathbb{E}[\exp(uX)]$ is equivalent to the question of convergence of the integral. Using formula (17.6) for the density f, the fact $\exp(t(A + uI)) = \exp(tA)\exp(tu)$ for all $u \in \mathbb{C}$ and the Jordan canonical decomposition of the matrix A, we can conclude that

$$\begin{aligned}\mathbb{E}\big[\big|\exp(uX)\big|\big] < \infty \quad &\Longleftrightarrow \quad -\alpha'\bigg(\int_0^\infty \exp\big(\big(A + \Re(u)I\big)t\big)\, dt\bigg)A\mathbf{1} < \infty \\ &\Longleftrightarrow \quad \Re(\lambda_0 + u) < 0,\end{aligned}$$

where λ_0 is as defined above. This proves the equivalence in the proposition.

Note that the condition $\Re(u) < -\Re(\lambda_0)$ implies, by Lemma (17.2), that the matrix $A + uI$ is invertible. For any $T \in \mathbb{R}_+$ the fundamental theorem of calculus therefore yields the matrix identity

$$\int_0^T \exp\bigl((A+uI)t\bigr)\,dt = (A+uI)^{-1}\bigl[\exp\bigl((A+uI)T\bigr) - I\bigr]. \tag{17.8}$$

Since all the eigenvalues of $A + uI$ have a strictly negative real part, it follows from Jordan canonical decomposition of $A+uI$ that $\lim_{T\to\infty} \exp((A+uI)T) = 0$. Therefore identities (17.7) and (17.8) conclude the proof of the proposition. □

More generally, a *double phase-type* jump distribution $\mathrm{DPH}(p, \beta^+, B^+, \beta^-, B^-)$ is defined to have density

$$\begin{aligned} &f(x) := pf^+(x)I_{(0,\infty)}(x) + (1-p)f^-(-x)I_{(-\infty,0)}(x) \quad \text{such that} \\ &p \in [0,1], \quad f^\pm \sim \mathrm{PH}\bigl(\beta^\pm, B^\pm\bigr), \quad f^\pm(x) = -\bigl(\beta^\pm\bigr)' e^{xB^\pm} B^\pm \mathbf{1} \quad \text{and} \\ &\mathbf{1}'\beta^\pm = 1, \end{aligned} \tag{17.9}$$

where the phase-type distributions $\mathrm{PH}(\beta^\pm, B^\pm)$ are as described above, $\mathbf{1}$ is a vector of the appropriate size with all coordinates equal to 1, and as usual I_A denotes the indicator of a set A. The condition $\mathbf{1}'\beta^\pm = 1$ ensures that the distribution of jump sizes has no atom at zero.

The class of double phase-type distributions is vast. Not only does it contain double exponential distributions

$$\begin{aligned} &f(x) := p\alpha^+ e^{-x\alpha^+} I_{(0,\infty)}(x) + (1-p)\alpha^- e^{x\alpha^-} I_{(-\infty,0)}(x) \\ &\quad \text{where } \alpha^\pm > 0 \text{ and } p \in [0,1], \end{aligned} \tag{17.10}$$

mixtures of double exponential distributions and Erlang distributions, but this class is in fact dense in the sense of weak convergence in the space of all probability distributions on $\mathbb{R}$.

Proposition 17.5 *Let F be a probability distribution function on $\mathbb{R}$. Then there exists a sequence $(F_n)_{n\in\mathbb{N}}$ of double-phase-type distributions F_n such that $F_n \Rightarrow F$ as $n \to \infty$.*[1]

This result directly follows from the three observations that (a) any probability distribution on the real line can be approximated in distribution arbitrarily closely by a random variable taking only finitely many values and (b) any constant random variable is the limit in distribution of Erlang or the negative of Erlang random variables, and (c) a mixture of Erlang distributions is a phase-type distribution.

[1] We write $F_n \Rightarrow F$ for a sequence of distribution functions F_n and a distribution function F if F_n converges in distribution to F, that is, $\lim_{n\to\infty} F_n(x) = F(x)$ for all x where F is continuous.

An important property of exponential distributions is the lack-of-memory property, which can be generalised to stopping times as follows:

Lemma 17.6 *Let* $(\mathcal{F}_t)_{t\geq 0}$ *a filtration, and let* ρ *be any stopping time*[2] *with respect to this filtration. Let* $\mathbf{e}_q$ *be an exponentially distributed random variable with parameter* $q > 0$ *which is independent of the* σ*-algebra generated by* $\bigcup_{t\geq 0}\mathcal{F}_t$. *Then the equality*

$$\mathbb{E}\big[I_{\{\rho<\mathbf{e}_q\}}\exp\big(-\lambda(\mathbf{e}_q-\rho)\big)\big|\mathcal{F}_\rho\big] = \frac{q}{\lambda+q}e^{-q\rho} \quad \textit{holds for all } \lambda \geq 0,$$

and hence the positive random variable $\mathbf{e}_q - \rho$ *defined on the event* $\{\rho < \mathbf{e}_q\}$ *is, conditional on* $\mathcal{F}_\rho$, *exponentially distributed with parameter* q.

Remarks

(i) This lemma can be viewed as a generalisation of the lack of memory property,

$$\mathbb{P}(\mathbf{e}_q > t+s|\mathbf{e}_q > s) = \mathbb{P}(\mathbf{e}_q > t),$$

of the exponential random variable $\mathbf{e}_q$ when the constant time s is substituted by a stopping time ρ. Note also that it follows from the lemma that the conditional probability of the event $\{\rho < \mathbf{e}_q\}$ equals

$$\mathbb{P}(\mathbf{e}_q > \rho|\mathcal{F}_\rho) = \exp(-q\rho).$$

(ii) Phase-type distributions enjoy a similar property that can be seen as a generalisation of the lack-of-memory of the exponential distribution. More specifically, let T follow a PH(α, B) distribution independent of the σ-algebra generated by $\bigcup_{t\geq 0}\mathcal{F}_t$. Then for any stopping time ρ with respect to $(\mathcal{F}_t)_{t\geq 0}$, the random variable $T-\rho$ defined on the event $\{\rho < T\}$, conditional on $\mathcal{F}_\rho$, is PH(α_ρ, B) distributed where

$$\alpha_\rho = \big(\alpha' e^{\rho B}\mathbf{1}\big)^{-1}\alpha'\exp\{\rho B\},$$

since the identity

$$\mathbb{E}\big[I_{\{\rho<T\}}\exp\big(-\lambda(T-\rho)\big)\big|\mathcal{F}_\rho\big] = \alpha'\exp(\rho B)(B-\lambda I)^{-1}B\mathbf{1}$$

holds for all $\lambda \geq 0$. This follows by the same argument as in the proof of Lemma 17.6. Furthermore we have the following expression for the conditional probability of the event $\{\rho < T\}$:

$$\mathbb{P}(T > \rho|\mathcal{F}_\rho) = \alpha'\exp(\rho B)\mathbf{1}.$$

---

[2]By definition the stopping time ρ takes values in $[0,\infty]$ and satisfies the condition $\{\rho \leq t\} \in \mathcal{F}_t$ for all $t \in [0,\infty)$. The σ-algebra $\mathcal{F}_\rho$ consists of all events A such that $A \cap \{\rho \leq t\} \in \mathcal{F}_t$ for all $t \in [0,\infty)$.

Proof The following direct calculation based on Fubini's theorem, which is applicable since all the functions are nonnegative,

$$\mathbb{E}\big[I_A I_{\{\rho<\mathbf{e}_q\}} \exp\big(-\lambda(\mathbf{e}_q-\rho)\big)\big] = \mathbb{E}\bigg[I_A I_{\{\rho<\infty\}} e^{\lambda\rho} q \int_\rho^\infty e^{-(\lambda+q)t}\,dt\bigg]$$
$$= \mathbb{E}\bigg[I_A \frac{q}{\lambda+q} e^{-q\rho}\bigg], \quad \text{where } A \in \mathcal{F}_\rho,$$

proves the identity in the lemma for all nonnegative λ. Since the Laplace transform uniquely determines the distribution of a random variable, the lemma follows. □

17.3 Stochastic Volatility Models with Jumps

We next describe in detail the two classes of stochastic volatility models with jumps that we will consider.

Let $v = \{v_t\}_{t\geq 0}$ be a Markov process that takes positive values, modelling the underlying stochastic variance, and let X be a Lévy process[3] which drives the noise in the log-price process. The processes are taken to be mutually independent and are both defined on some probability space $(\Omega, \mathcal{F}, \mathbb{P})$.

The law of X is determined by its characteristic exponent $\psi : \mathbb{R} \to \mathbb{R}$ which is according to the Lévy–Khintchine formula given by

$$\mathbb{E}\big[e^{\mathrm{i}uX_t}\big] = e^{t\psi(u)} \tag{17.11}$$

with

$$\psi(u) = \mathrm{i}cu - \frac{\sigma^2}{2}u^2 + \int_{-\infty}^{\infty} \big[e^{\mathrm{i}ux} - 1 - \mathrm{i}uxI_{\{|x|\leq 1\}}\big]\nu(dx), \tag{17.12}$$

where $\sigma^2 \geq 0$ and c are constants, and ν is the Lévy measure that satisfies the integrability condition $\int_{\mathbb{R}}(1 \wedge x^2)\nu(dx) < \infty$. The triplet (c, σ^2, ν) is also called the characteristic triplet of X.

To guarantee that the option prices be finite, we impose the usual restriction that X admits (positive) exponential moments; more precisely, we assume that for some $p > 1$,

$$\int_1^\infty e^{px}\nu(dx) < \infty, \tag{17.13}$$

which implies that $\mathbb{E}[e^{pX_t}] < \infty$ for all $t \geq 0$. In this case identity (17.11) remains valid for all u in the strip $\{u \in \mathbb{C} : \Im(u) \in (-p, 0]\}$ in the complex plan, where the function ψ is analytically extended to this strip.

[3]A Lévy process $X = \{X_t\}_{t\geq 0}$ is a stochastic process that has independent and stationary increments, and has right-continuous paths with left limits with $X_0 = 0$.

In this setting a candidate stochastic volatility process with jumps $S = \{S_t\}_{t\in[0,T]}$, where $T > 0$ denotes a maturity or time-horizon, is given by

$$S_t := S_0 \exp\left((r-d)t + \int_0^t \sqrt{v_u}\, dX_u - \int_0^t \psi(-\mathrm{i}\sqrt{v_s})\, ds\right), \qquad S_0 = s > 0, \tag{17.14}$$

where r and d are the instantaneous interest rate and dividend yield respectively. Here we assume that the variance process v satisfies the following integrability condition:

$$\int_0^T \left|\psi(-\mathrm{i}\sqrt{v_s})\right| ds < \infty \quad \text{a.s.} \tag{17.15}$$

It is easy to see by conditioning on the filtration generated by the variance process v that the integrability condition in (17.15) implies the martingale property of the discounted process $\{e^{-(r-d)t} S_t\}_{t\in[0,T]}$.

Note that if we take for example X to be a Brownian motion with drift and v an independent square-root process, the process S reduces to a Heston model with zero correlation between the driving Brownian factors (see e.g. [17]). The class of models described by (17.14) is quite flexible and contains for example the stochastic volatility models with jumps described in Lipton [24], as long as there is no correlation between the driving Brownian motions.

A related class of models that has been proposed in the literature is the one where the effect of stochasticity of volatility is achieved by randomly changing the time-scale (see e.g. Carr et al. [12]); in the setting above the price process $\{S_t\}_{t\in[0,T]}$ is defined by

$$S_t := S_0 \exp\big((r-d)t + X_{V_t} - \psi(-\mathrm{i})V_t\big), \qquad S_0 = s > 0, \text{ where } V_t := \int_0^t v_u\, du, \tag{17.16}$$

and we assume that v satisfies the integrability condition

$$V_T < \infty \quad \text{a.s.} \tag{17.17}$$

Also in this case the discounted process $\{e^{-(r-d)t} S_t\}_{t\in[0,T]}$ is a martingale.

It is clear from the definitions that in the case where X is a Brownian motion with drift, the classes of models in (17.14) and (17.16) coincide, due to the scaling property of Brownian motion. Whereas the effect of the variance process v on the Brownian motion with drift is the same in both classes of models, the effect of the process v on the behaviour of jumps is different. In (17.16) the Markov process v modulates only the intensity of the jumps of X, while in model (17.14) the volatility scales the distribution of size of the jumps but does not affect the intensity.

In the next section we will describe a modelling framework in which any model in the classes given by (17.14) and (17.16) can be approximated. The approximation in Sect. 17.3.2 retains the structural properties required for the semi-analytic pricing (i.e. up to an integral transform) of barrier options, forward starting options and volatility derivatives.

17.3.1 A Class of Regime-Switching Models

Let the set $E^0 := \{1, \ldots, N_0\}$ be the state space of a continuous-time Markov chain $Z = (Z_t)_{t\geq 0}$, and let the process $W = (W_t)_{t\geq 0}$ denote a standard Brownian motion which is independent of the chain Z. For each $i \in E^0$, let the process $J^i := (J^i_t)_{t\geq 0}$ be a compound Poisson process with intensity $\lambda_i \geq 0$ and jump-sizes distributed according to a double-phase-type distribution $\mathrm{DPH}(p_i, \beta_i^+, B_i^+, \beta_i^-, B_i^-)$. In particular the jump-size distributions have no atom at zero, i.e. $(\beta_i^+)'\mathbf{1} = 1$ for all $i \in E^0$ such that $\lambda_i p_i > 0$ and analogously for β_i^-. Assume further that the processes J^i are mutually independent as well as independent from the Brownian motion W and the chain Z.

In this setting consider the following model for the underlying price process $S = (S_t)_{t\geq 0}$, the (domestic) money market account $B^D = (B_t^D)_{t\geq 0}$ and the cumulative dividend yield $B^F = (B_t^F)_{t\geq 0}$:

$$B_t^D := \exp\left(\int_0^t R_D(Z_s)\,ds\right), \qquad B_t^F := \exp\left(\int_0^t R_F(Z_s)\,ds\right), \tag{17.18}$$
$$S_t := \exp(X_t),$$

where

$$X_t := x + \int_0^t \mu(Z_s)\,ds + \int_0^t \sigma(Z_s)\,dW_s + \sum_{i\in E^0} \int_0^t I_{\{Z_s=i\}}\,dJ_s^i. \tag{17.19}$$

In the case of the Foreign Exchange market the process B^F can be interpreted as a foreign money market account. The point $x \in \mathbb{R}$ is the starting value of the process X, and $R_D, R_F, \mu, \sigma : E^0 \to \mathbb{R}$ are given real-valued functions on E^0 such that R_D, R_F are nonnegative and σ is strictly positive. To price derivatives in our model, we need to understand the law of the Markov process (X, Z), which is determined by the characteristic matrix exponent K, defined as follows.

Definition The *characteristic matrix exponent* $K : \mathbb{R} \to \mathbb{C}^{N_0\times N_0}$ of (X, Z) is given by

$$K(u) = Q + \Lambda(u),$$

where Q denotes the generator of the chain Z, and, for $u \in \mathbb{R}$, $\Lambda(u)$ is a diagonal matrix of size $N_0 \times N_0$, where the ith diagonal element equals the characteristic exponent of the process X in regime i, given by

$$\begin{aligned}\psi_i(u) := \mathtt{i}u\mu_i - \sigma_i^2 u^2/2 + \lambda_i\big[p_i(\beta_i^+)'\big(B_i^+ + \mathtt{i}uI\big)^{-1}B_i^+\mathbf{1} \\ + (1-p_i)(\beta_i^-)'\big(B_i^- - \mathtt{i}uI\big)^{-1}B_i^-\mathbf{1} - 1\big],\end{aligned} \tag{17.20}$$

where I and $\mathbf{1}$ are an identity matrix and a vector with all coordinates equal to one of the appropriate dimensions.

Remarks

(i) Note that the functions ψ_i defined in (17.20) can be analytically extended to the strip in the complex plane $\Im(u) \in (-\alpha_i^+, \alpha_i^-)$, where

$$\alpha_i^{\pm} := \min\{-\Re(\lambda) : \lambda \text{ eigenvalue of } B_i^{\pm}\} \quad \text{for any state } i \in E^0. \qquad (17.21)$$

(ii) In the special case that the jumps follow a double exponential distribution, the diagonal elements of the matrix $\Lambda(u)$ take the simpler form

$$\psi_i(u) = \mathrm{i}u\mu_i - \sigma_i^2 u^2/2 + \lambda_i p_i\left(\frac{\beta_i^+}{\beta_i^+ - \mathrm{i}u} - 1\right) + \lambda_i(1-p_i)\left(\frac{\beta_i^-}{\beta_i^- + \mathrm{i}u} - 1\right),$$

where $\beta_i^{\pm}$ and p_i are the parameters of the double exponential distribution.

(iii) Throughout the paper we will use $\mathbb{E}_{x,i}[\cdot]$ to denote the conditional expectation $\mathbb{E}[\cdot|X_0 = x, Z_0 = i]$ and on occasion $\mathbb{E}_i[\cdot]$ to represent $\mathbb{E}_{0,i}[\cdot]$.

We now define two matrices the will play an important role in the sequel.

Definition The *discount rate matrix* Λ_D is the diagonal matrix with elements $\Lambda_D(i,i) := R_D(i)$ where $i \in E^0$. The *dividend yield matrix* Λ_F is the diagonal matrix given by $\Lambda_F(i,i) := R_F(i)$ for $i \in E^0$.

Theorem 17.7 *The discounted characteristic function of the Markov process* (X, Z) *is given by the formula*

$$\mathbb{E}_{x,i}\left[\frac{\exp(\mathrm{i}uX_t)}{B_t^D} I_{\{Z_t=j\}}\right] = \exp(\mathrm{i}ux) \cdot \exp\big(t\big(K(u) - \Lambda_D\big)\big)(i,j) \qquad (17.22)$$

for all $u \in \mathbb{R}$.

Remarks

(i) The left-hand side is finite for all $u \in \mathbb{C}$ in the strip $\Im(u) \in (-\alpha_*^+, \alpha_*^-)$ where

$$\begin{aligned} \alpha_*^+ &= \min\{\alpha_k^+ : \lambda_k p_k > 0, k \in E^0\} \quad \text{and} \\ \alpha_*^- &= \min\{\alpha_k^+ : \lambda_k(1-p_k) > 0, k \in E^0\}, \end{aligned} \qquad (17.23)$$

the quantities $\alpha_i^{\pm}$ are defined in (17.21), and the minimum over the empty set is taken to be $+\infty$. It follows, by analytical continuation, that identity (17.22) remains valid for all u in this strip. Furthermore, the slope of the implied volatility smile in model (17.19) is determined by α_*^+ and α_*^- (see Sect. 17.4.2).

(ii) The Markov property and Theorem 17.7 imply that the process $\{S_t B_t^F / B_t^D\}_{t \geq 0}$ is a martingale if the following two conditions hold:

$$1 < \alpha_k^+ \quad \text{for all } k \in E^0 \text{ such that } \lambda p_k > 0, \qquad (17.24)$$

$$\Lambda(-\mathrm{i}) = \Lambda_D - \Lambda_F. \qquad (17.25)$$

Condition (17.24) ensures that $\mathbb{E}_{i,x}[S_T]$ is finite for all $T \geq 0$ and hence by Theorem 17.7 takes the form $\mathbb{E}_{i,x}[S_T] = e^x[\exp(TK(-\mathtt{i}))\mathbf{1}](i)$. The equality in (17.25) guarantees that S has instantaneous drift given by the rates $\Lambda_D - \Lambda_F$. Any model from the class (17.18)–(17.19) that satisfies conditions (17.24) and (17.25) can be taken as a specification of the price process of the risky asset under a pricing measure. From now on we assume that model (17.19) is specified under the pricing measure given by condition (17.24)–(17.25).

(iii) For later reference we record that, under a pricing measure, the price at time s of a zero-coupon bond maturing at time $t \geq s$ is given by

$$\mathbb{E}_i\left[\frac{1}{B_t^D}\middle|\mathcal{F}_s^{(X,Z)}\right] = \frac{1}{B_s^D} \cdot \big(\exp\big((t-s)(Q - \Lambda_D)\big)\mathbf{1}\big)(Z_s), \tag{17.26}$$

where $\mathcal{F}_s^{(X,Z)} = \sigma\{(X_u, Z_u)\}_{u \leq s}$ denotes the standard filtration generated by (X, Z). In particular, at time 0 the price is given by

$$\mathbb{E}_i\big[\big(B_t^D\big)^{-1}\big] = \big(\exp\big(t(Q - \Lambda_D)\big)\mathbf{1}\big)(i).$$

(iv) The infinitesimal generator $\mathcal{L}$ of the Markov process (X, Z) acts on sufficiently smooth functions[4] $f : \mathbb{R} \times E^0 \to \mathbb{R}$ as

$$\begin{aligned}\mathcal{L}f(x,i) &= \frac{\sigma^2(i)}{2} f''(x,i) + \mu(i) f'(x,i) \\ &\quad + \lambda(i)\left[\int_{\mathbb{R}} f(x+z,i) g_i(z)\,dz - f(x,i)\right] \\ &\quad + \sum_{j \in E^0} q_{ij}\big[f(x,j) - f(x,i)\big],\end{aligned} \tag{17.27}$$

where g_i is the density of the double phase-type distribution DPH($p_i, \beta_i^+, B_i^+, \beta_i^-, B_i^-$), q_{ij} is the ijth element of Q, and $'$ denotes differentiation with respect to x.

(v) For a specific regime-switching model (namely the case where the Markov chain Z has two states only), the calibration is studied in [26].

Proof It is clear from the definition of (X, Z) that it is a Markov process. Let $\mathcal{F}_t^Z := \sigma(Z_s : s \in [0, t])$ be the σ-algebra generated by the chain Z up to time t. Since the compound Poisson processes and Brownian motion in model (17.19) are mutually independent as well as independent of the σ-algebra $\mathcal{F}_t^Z$, it is easy to see that by

[4]For example, functions f with $f(\cdot, i) \in C_c^2(\mathbb{R})$ for $i \in E^0$, where $C_c^2(\mathbb{R})$ are the twice continuously differentiable functions with compact support.

conditioning on $\mathcal{F}_t^Z$ for any $i \in E^0$ we obtain

$$\begin{aligned}
&\mathbb{E}_{x,i}\big[\exp(uX_t)\big|\mathcal{F}_t^Z\big] \\
&\quad = \exp\bigg(ux + u\int_0^t \mu(Z_s)\,ds + \frac{u^2}{2}\int_0^t \sigma(Z_s)^2\,ds + \int_0^t \nu(Z_s,u)\,ds\bigg), \qquad (17.28)\\
&\nu(i,u) := \lambda_i\big[\mathbb{E}\big[\exp(uJ_i)\big] - 1\big] \\
&\qquad = \lambda_i\big[p_i\big(\beta_i^+\big)'\big(B_i^+ + uI\big)^{-1}B_i^+\mathbf{1} + (1-p_i)\big(\beta_i^-\big)'\big(B_i^- - uI\big)^{-1}B_i^-\mathbf{1} - 1\big],
\end{aligned}$$

where the random variable J_i denotes the size of jumps of the compound Poisson process J^i. The last equality in this calculation is a consequence of the choice (17.9) of the distribution of jump sizes and Proposition 17.4. Therefore the complex number u must be contained in all intervals $(-\alpha_k^-, \alpha_k^+)$, $k \in E^0$, where $\alpha_k^\pm$ are defined in Theorem 17.7. The identity in (17.28) holds more generally for any jump-distribution that admits a moment-generating function. The well-known identity from the theory of Markov chains given in Lemma 17.1 can now be applied to obtain the expectations of the expressions on both sides of (17.28). This concludes the proof of Theorem 17.7. □

From Theorem 17.7 one may obtain an explicit expression for the marginal distributions of (X, Z) by inverting the Fourier transform (17.22):

Proposition 17.8 *For any $T > 0$, the joint distribution $q_T^{x,i}(y,j) = \frac{d}{dy}\mathbb{P}_{x,i}[X_T \le y, Z_T = j]$ of (X_T, Z_T) is given by*

$$q_T^{x,i}(y,j) = \frac{1}{2\pi}\int_{\mathbb{R}} e^{\mathrm{i}\xi(x-y)}\exp\big(K(\xi)T\big)(i,j)\,d\xi, \quad y \in \mathbb{R},\ i,j \in E^0. \qquad (17.29)$$

In particular, X_T is a continuous random variable with probability density function $q_T^{x,i}(y) = \frac{\mathbb{P}_{x,i}[X_T \in dy]}{dy}$ given by

$$q_T^{x,i}(y) = \frac{1}{2\pi}\int_{\mathbb{R}} e^{\mathrm{i}\xi(x-y)}\big[\exp(K(\xi)T)\mathbf{1}\big](i)\,d\xi, \quad y \in \mathbb{R},\ i \in E^0. \qquad (17.30)$$

Proof It is well known that a probability law on the real line $\mathbb{R}$ has a density with respect to the Lebesgue measure if its characteristic function is in $L^1(\mathbb{R})$. The characteristic function of X_T by Theorem 17.7 equals

$$\mathbb{E}_{x,j}\big[\exp(\mathrm{i}\xi X_T)\big] = e^{\mathrm{i}\xi x}\big[\exp\big(K(\xi)T\big)\mathbf{1}\big](j). \qquad (17.31)$$

We now show that this characteristic function is asymptotically equal to $\exp(-c\xi^2)$, as $|\xi| \to \infty$, for some positive constant c. Note first that the volatility vector σ in model (17.19) has nonzero coordinates by assumption and the spectra of matrices $B_i^{\pm}$, $i = 1, \ldots, N_0$, do not contain any points of the form $\mathtt{i}\xi$, for $\xi \in \mathbb{R}$, by Lemma 17.2. Therefore the functions $\xi \mapsto \psi_i(\xi)$, $i = 1, \ldots, N_0$, defined in (17.20) are asymptotically equal to downward facing parabolas. A further application of Lemma 17.2 implies that the characteristic function has the desired asymptotic behaviour. This further implies that the density $q_T^{x,i}$ exists and is given by the inversion formula (17.30).

$$\xi \mapsto e^{\mathtt{i}\xi x} \exp\bigl(TK(\xi)\bigr)(i,j) = \mathbb{E}_{x,i}\bigl[\exp(\mathtt{i}\xi X_T) I_{\{Z_T = j\}}\bigr]$$

is in $L^1(\mathbb{R})$ and that the two equalities hold. Therefore the Fourier inversion formula is valid, and the identity in (17.29) follows. □

17.3.2 Two-Step Approximation Procedure

The construction of a regime-switching Lévy process with jump sizes distributed according to a double phase-type distribution that approximates a given stochastic volatility process with jumps from either of the two-classes (17.14) and (17.16) takes place in two steps:

(i) Approximation of the variance process v by a finite-state continuous-time Markov chain and
(ii) Approximation of the Lévy process X by a Lévy process with double-phase-type jumps.

By approximating the variance process by a finite-state Markov chain the resulting approximating process is a regime-switching Lévy process. The approximation of the jump part of X by a compound Poisson process with double phase-type jumps will enable us to employ matrix Wiener–Hopf factorisation results, needed to obtain tractable formulae for the prices of barrier-type options. The two steps will be described in detail in the present section.

Markov Chain Approximation of the Variance Process

The first step of the approximation procedure that was outlined above is to approximate the variance process v by a finite-state continuous-time Markov chain on some grid contained in the positive real line. We will restrict ourselves to the case that the variance process v is a Feller process on the state space $\mathbb{R}_+ = [0, \infty)$. This assumption implies that v is a Markov process satisfying some regularity properties.

The Feller property is phrased in terms of the semi-group $(P_t)_{t \geq 0}$ of v acting on $C_0(\mathbb{R}_+)$, the set of continuous functions on $\mathbb{R}_+$ that tend to zero at infinity. Recall that, for any Borel function f on $\mathbb{R}_+$ and $t \geq 0$, the map $P_t f : \mathbb{R}_+ \to \mathbb{R}$ is given by

$$P_t f(x) = \mathbb{E}_x\bigl[f(v_t)\bigr].$$

Assumption 17.9 *The Markov process* $v = \{v_t\}_{t\ge 0}$ *is a Feller process; that is, for any* $f \in C_0(\mathbb{R}_+)$, *the family* $(P_t f)_{t\ge 0}$ *satisfies the following two properties*:

(i) $P_t f \in C_0(\mathbb{R}_+)$ *for any* $t > 0$;
(ii) $\lim_{t\downarrow 0} P_t f(v) = f(v)$ *for any* $v \in \mathbb{R}_+$.

An approximating Markov chain Z with generator Q on a state-space $E^0 = \{x_1, \dots, x_{N_0}\}$ can be constructed by choosing E^0 to be some appropriate (nonuniform) grid in $\mathbb{R}_+$, and specifying the generator Q such that an appropriate set of instantaneous (local) moments of the chain Z and the target process v are matched. See [25] for details on this procedure.

Denote by $\mathcal{G} : \mathcal{D} \to C_0(\mathbb{R}_+)$ the infinitesimal generator of v defined on its domain $\mathcal{D}$, and let $Z^{(n)}$ be a sequence of Markov chains with generators $Q^{(n)}$ and state spaces $E^{0(n)} = \{x_1^{(n)}, \dots, x_{N^{(n)}}^{(n)}\}$, and denote by $Q^{(n)} f_n$ the vector with coordinates

$$Q^{(n)} f_n(x_i) = \sum_{x_j \in E^{0(n)}} Q^{(n)}(x_i, x_j) f(x_j), \qquad x_i \in E^{0(n)}.$$

The sequence $Z^{(n)}$ weakly approximates the variance process v if the range of the state spaces $E^{0(n)}$ grows sufficiently fast as n tends to infinity and if, for all regular functions f, $Q^{(n)} f_n$ converges uniformly to $\mathcal{G} f$, that is, $\epsilon_n(f) \to 0$, where

$$\epsilon_n(f) := \max_{x \in (E^{0(n)})^o} \left| Q^{(n)} f_n(x) - \mathcal{G} f(x) \right|,$$

and $(E^{0(n)})^o$ is equal to $E^{0(n)}$ without the smallest and the largest elements. The precise statement reads as follows:

Theorem 17.10 *Assume that the following two conditions are satisfied for any function in a core*[5] *of* $\mathcal{L}$:

$$\epsilon_n(f) \to 0 \quad \text{as } n \to \infty, \tag{17.32}$$

$$\text{either (i)} \lim_{y \searrow 0} \mathcal{G} f(y) = 0 \text{ or (ii)} \lim_{n\to\infty} \mathbb{P}_x\big[\tau^{(n)}_{(E^{0(n)})^o} > T\big] = 1, \tag{17.33}$$

where for any set $G \subset \mathbb{R}_+$, *we define* $\tau_G^{(n)} = \inf\{t \ge 0 : X_t^{(n)} \notin G\}$. *Then it holds that, as* $n \to \infty$, $v^{(n)} \overset{\mathcal{L}}{\Rightarrow} v$.[6]

[5] A core $\mathcal{C}$ of the operator $\mathcal{L}$ is a subspace of the domain of $\mathcal{L}$ that is (i) dense in $C_0(\mathbb{R}_+)$ and (ii) there exists $\lambda > 0$ such that the set $\{(\lambda - \mathcal{L}) f : f \in \mathcal{C}\}$ is dense in $C_0(\mathbb{R}_+)$.

[6] By $v^{(n)} \overset{\mathcal{L}}{\Rightarrow} v$ we denote the convergence in law of $v^{(n)}$ to v in the Skorokhod topology, i.e. convergence of the distributions of $v^{(n)}$ to those of v in the set of probability measures on the Skorokhod space endowed with the Skorokhod topology.

Remark

(i) Note that the convergence in law implies in particular that

$$\mathbb{E}_x\big[g\big(Z_T^{(n)}\big)\big] \longrightarrow \mathbb{E}_x\big[g(\upsilon_T)\big]$$

for any bounded continuous function g.

(ii) A proof of this statement can be found in [25].

Approximation of a Lévy Process

The second stage of the aforementioned approximation procedure amounts to an approximation of a Lévy process by a compound Poisson process with double phase-type jumps.

Proposition 17.11 *For any Lévy process X, there exists a sequence $(X^{(n)})_{n\in\mathbb{N}}$ of Lévy processes with double phase-type jumps such that $X^{(n)} \overset{\mathcal{L}}{\Rightarrow} X$ as $n \to \infty$.*

Remarks

(i) A proof of this result can be found in e.g. Jacod and Shiryaev [20, Sect. VII.3]. It is based on the fact that a sequence $(X^{(n)})_{n\in\mathbb{N}}$ of Lévy processes weakly converges to a given Lévy process X if and only if $X_1^{(n)}$ converges in distribution to X_1 (see e.g. [20, Corollary VII.3.6] for a proof).

(ii) Furthermore [20, Corollary VII.3.6] implies that a sufficient condition to guarantee that $X_1^{(n)}$ converges in distribution to X_1 is that the characteristic triplets (c_n, σ_n^2, ν_n) of $X^{(n)}$ converges to the triplet (c, σ^2, ν) of X as follows as $n \to \infty$: for some $a > 0$ that is a continuity point of $\nu(dx)$ and $\nu(-dx)$, it holds that

$$c_n \to c, \qquad \sigma_n^2 + \int_{(-a,a)} x^2 \nu_n(dx) \to \sigma^2 + \int_{(-a,a)} x^2 \nu(dx), \tag{17.34}$$

$$\int_{(0,\infty)} \big(x^2 \wedge a\big)\big|\overline{\nu}_n(x) - \overline{\nu}(x)\big|\, dx + \int_{(-\infty,0)} \big(|x|^2 \wedge a\big)\big|\underline{\nu}_n(x) - \underline{\nu}(x)\big|\, dx \to 0 \tag{17.35}$$

where, for any measure m on $\mathbb{R}$, $\overline{m}$ and $\underline{m}$ are the left and right tails, $\overline{m}(x) = m([x, \infty))$, $\underline{m}(x) = m((-\infty, x])$.

Suppose now that the Lévy process X is a model for the log of a stock price. Then X has a triplet (c, σ^2, ν) satisfying the exponential moment condition in (17.13). An example of a sequence of Lévy processes with DPH distributed jumps that weakly converges to X is then given as follows. Let $\lambda_n = \nu((-1/n, 1/n)^c)$, and F_n be a double phase-type distribution that approximates in distribution the probability measure $\widetilde{F}_n(dx) = I_{\{|x|\geq 1/n\}}\nu(dx)/\lambda_n$, and define the measure ν_n by $\nu_n(dx) = \lambda_n F_n(dx)$. Here the F_n and σ_n^2 are to be chosen such that (17.35) and the second requirement in (17.34) hold true. See e.g. [3] for a fitting procedure based on the EM algorithm. Then the sequence of Lévy processes $X^{(n)}$ with triplets $(c, \sigma_n^2, \lambda_n F_n)$ satisfies the conditions (17.35) and thus approximates X in law as n tends to infinity.

17.3.3 Convergence of the Approximation Procedure

Combining the two steps in the approximation, we can now identify candidate sequences of regime-switching processes that converge to either of the stochastic volatility processes with jumps (17.14) and (17.16) and establish the convergence.

Let $(X^{(n)})_{n\in\mathbb{N}}$ be a sequence of Lévy processes with DPH jumps, and let $(Z^{(n)})_{n\in\mathbb{N}}$ be a sequence of Markov chains that is independent of X and $(X^{(n)})_{n\in\mathbb{N}}$. Let ψ_n denote the characteristic exponent of $X^{(n)}$, and (c_n, σ_n^2, ν_n) the characteristic triplet. Consider then the sequences of stochastic processes $(S^{(int\text{-}n)})_{n\in\mathbb{N}}$ and $(S^{(tc\text{-}n)})_{n\in\mathbb{N}}$ with $S^{(int\text{-}n)} = \{S_t^{(int\text{-}n)}\}_{t\in[0,T]}$ and $S^{(tc\text{-}n)} = \{S_t^{(tc\text{-}n)}\}_{t\in[0,T]}$ given by

$$S_t^{(int\text{-}n)} := S_0 \exp\left((r-d)t + \int_0^t \sqrt{Z_s^{(n)}}\, dX_s^{(n)} - \int_0^t \psi_n\left(-\mathrm{i}\sqrt{Z_s^{(n)}}\right) ds\right),$$

$$S_t^{(tc\text{-}n)} := S_0 \exp\left((r-d)t + X^{(n)}\left(V_t^{(n)}\right) - \psi_n(-\mathrm{i}) V_t^{(n)}\right),$$

$$\text{where } V_t^{(n)} = \int_0^t Z_s^{(n)}\, ds.$$

The processes $S^{(int\text{-}n)}$ and $S^{(tc\text{-}n)}$ are in law equal to exponential Lévy processes:

Proposition 17.12 *For $n \in \mathbb{N}$, $\log S^{(int\text{-}n)}$ and $\log S^{(tc\text{-}n)}$ are in law equal to regime-switching Lévy processes of the form* (17.19).

Proof Note first that since $X^{(n)} = (X_t^{(n)})_{t\geq 0}$ is a Lévy process with DPH jumps, it is of the form

$$X_t^{(n)} = \mu^{(n)} t + \sigma^{(n)} W_t + J_t^{(n)} \quad \text{with } J_t^{(n)} = \sum_{i=1}^{M_t^{(n)}} U_i, \tag{17.36}$$

where $\mu^{(n)}$, $\sigma^{(n)}$ are constants, $M^{(n)} = (M_t^{(n)})_{t\geq 0}$ Poisson processes with jump-rates $\lambda^{(n)}$, and U_i are i.i.d. random variables following a DPH distribution. Then it is clear that $X^{(int\text{-}n)} = \log(S^{(int\text{-}n)}/S_0)$ is in law equal to the process $\widetilde{X}^{(int\text{-}n)} = (\widetilde{X}_t^{(int\text{-}n)})_{t\geq 0}$ given by

$$\widetilde{X}_t^{(int\text{-}n)} = \int_0^t \left(r - d + \mu^{(n)}\sqrt{Z_s^{(n)}} - \psi^{(n)}\left(-\mathrm{i}\sqrt{Z_s^{(n)}}\right)\right) ds + \int_0^t \sigma^{(n)} \sqrt{Z_s^{(n)}}\, dW_s$$
$$+ \sum_{j=1}^{N^{(n)}} \int_0^t I_{\{Z_s^{(n)} = x_j^{(n)}\}}\, d\widetilde{J}_s^{(n,j)},$$

where $\widetilde{J}^{(n,j)}$, $j = 1, \ldots, N^{(n)}$, are independent compound Poisson processes that are in law equal to the processes $J^{(n,j)} = (J_t^{(n,j)})_{t \geq 0}$ with $J_t^{(n,j)} = \sum_{i=1}^{N_t^{(n)}} \sqrt{x_j^{(n)}} U_i$, respectively. Since, for any constant $c \neq 0$, cU_i follows a DPH distribution, $\widetilde{X}^{(int\text{-}n)}$ is a regime-switching Lévy process of the form (17.19).

As a consequence of the scaling property of Brownian motion, it follows that $X^{(tc\text{-}n)} = \log(S^{(tc\text{-}n)}/S_0)$ is in law equal to the process $\widetilde{X}^{(tc\text{-}n)} = (\widetilde{X}_t^{(tc\text{-}n)})_{t \geq 0}$ given by

$$\widetilde{X}_t^{(tc\text{-}n)} = \int_0^t \left(r - d + [\mu^{(n)} - \psi^{(n)}(-\mathrm{i})] Z_s^{(n)}\right) ds + \int_0^t \sigma^{(n)} \sqrt{Z_s^{(n)}}\, dW_s$$
$$+ \sum_{j=1}^{N^{(n)}} \int_0^t I_{\{Z_s^{(n)} = x_j^{(n)}\}}\, d\widehat{J}_s^{(n,j)},$$

where $\widehat{J}^{(n,j)}$, $j = 1, \ldots, N^{(n)}$, are independent compound Poisson processes that are in law equal to the processes $J^{(n,j)} = (J_t^{(n,j)})_{t \geq 0}$ with $J_t^{(n,j)} = \sum_{i=1}^{M_t^{(n,j)}} U_i$, respectively, where $M^{(n,j)}$ is a Poisson process with jump-rate $\lambda \cdot x_j^{(n)}$. Here we used that, conditional on $Z^{(n)}$, the process $X^{(tc\text{-}n)}$ has independent increments, so that the law of $X^{(tc\text{-}n)}$ conditional on $Z^{(n)}$ is determined by the conditional characteristic functions of $X_t^{(tc\text{-}n)}$, $t \geq 0$. A straightforward calculation verifies that the conditional characteristic functions of $X_t^{(tc\text{-}n)}$ and $\widetilde{X}_t^{(tc\text{-}n)}$ are equal for $t \geq 0$. □

To establish the convergence in law of $(S^{(int\text{-}n)})_{n \in \mathbb{N}}$ and $(S^{(tc\text{-}n)})_{n \in \mathbb{N}}$, we will first show the convergence of the finite-dimensional distributions.[7]

Proposition 17.13 *Assume that* $X^{(n)} \stackrel{\mathcal{L}}{\Rightarrow} X$ *and* $Z^{(n)} \stackrel{\mathcal{L}}{\Rightarrow} v$ *as n tends to infinity. Then the following holds true*:

(a) $(Z^{(n)}, S^{(int\text{-}n)})_n \stackrel{fidis}{\Rightarrow} (v, S)$, *where S is the model given in* (17.14).
(b) $(V^{(n)}, S^{(tc\text{-}n)})_n \stackrel{fidis}{\Rightarrow} (V, S)$, *where S is the time-change model given in* (17.16).

Remark The convergence of European put option prices (and hence, by put–call parity, also of European call option prices) under the approximating models to those under the limiting models is a direct consequence of the convergence in finite-dimensional distributions. To establish the convergence of path-dependent option prices such as barrier option prices, it is required to prove that the approximating models converge in law.

[7] A sequence of processes $(Y_n)_{n \in \mathbb{N}}$ converges in finite-dimensional distributions to the process $Y = \{Y_t\}_{t \in [0,T]}$ if, for any partition $t_1 < \cdots < t_m$ of $[0, T]$, $\mathbb{P}(Y_{t_1}^{(n)} \leq x_1, \ldots, Y_{t_m}^{(n)} \leq x_m) \to \mathbb{P}(Y_{t_1} \leq x_1, \ldots, Y_{t_m} \leq x_m)$. We will denote this convergence by $Y_n \stackrel{fidis}{\Rightarrow} Y$.

Proof (b) To prove the convergence of the finite-dimensional distributions, it suffices, in view of the Markov property, to show that, for each fixed $t \in [0, T]$, the characteristic functions χ_n of $(V_t^{(n)}, X^{(n)}(V_t^{(n)}))$ converge pointwise to the characteristic function χ of $(V_t, X(V_t))$ as n tends to infinity. By conditioning and using the independence of $V^{(n)}$ from $X^{(n)}$ and of V from X we find that

$$\chi_n(u, v) = \mathbb{E}_{x,i}\big[\exp\big\{\big(\mathrm{i}u + \psi_n(v)\big)V_t^{(n)}\big\}\big],$$
$$\chi(u, v) = \mathbb{E}_{x,i}\big[\exp\big(\big(\mathrm{i}u + \psi(v)\big)V_t\big)\big].$$

Since $v^{(n)}$ converges in law to v in the Skorokhod topology, the Skorokhod representation theorem implies that on some probability space $Z^{(n)} \to v$, almost surely, with the convergence with respect to the Skorokhod metric. Since, for any $t \in [0, T]$, the map $i_t : D_{\mathbb{R}}[0, T] \to \mathbb{R}$ given by $i_t : x \mapsto \int_0^t x(s)\,ds$ is continuous in the Skorokhod topology, we deduce that $V_t^{(n)} \to V_t$ almost surely. In particular, χ_n converges pointwise to χ.

The proof of (a) is similar and omitted. □

The next result concerns the convergence in law of the sequences $(S^{(int\text{-}n)})_{n\in\mathbb{N}}$ and $(S^{(tc\text{-}n)})_{n\in\mathbb{N}}$:

Theorem 17.14 *The following statements hold true*:

(a) *Assume that* (v, S) *is a Feller process, where* S *is the model given in* (17.14), *that* $Z^{(n)}$ *satisfies the conditions in Theorem* 17.10, *and that the characteristics of* $X^{(n)}$ *satisfy conditions* (17.34) *and* (17.35). *Then, as* $n \to \infty$,

$$S^{(int\text{-}n)} \overset{\mathcal{L}}{\Rightarrow} S.$$

(b) *Assume that* $X^{(n)} \overset{\mathcal{L}}{\Rightarrow} X$ *and* $Z^{(n)} \overset{\mathcal{L}}{\Rightarrow} v$ *as* $n \to \infty$. *Then it holds that*

$$S^{(tc\text{-}n)} \overset{\mathcal{L}}{\Rightarrow} S,$$

where S *is the time-change model given in* (17.16).

Remark The convergence in law stated above carries over to the convergence of barrier option prices under the respective models if the boundaries are continuity points of the limiting model. For instance, if we denote by $\tau_A = \inf\{t \geq 0 : S_t \notin A\}$ the first time that S leaves the set $A := [\ell, u]$, and $\mathbb{P}(S_T \in \{\ell, u\}) = 0$, then, for any bounded continuous pay-off functions $g, h : \mathbb{R}_+ \to \mathbb{R}$, we have that, as $n \to \infty$, the double knock-out option and rebate option prices under the approximating models converge to those under the limiting model:

$$\mathbb{E}_x\big[g\big(S_T^{(n)}\big)I_{\{\tau_A^{(n)} > T\}}\big] \longrightarrow \mathbb{E}_x\big[g(S_T)I_{\{\tau_A > T\}}\big],$$
$$\mathbb{E}_x\big[e^{-r\tau_A^{(n)}}h\big(S_{\tau_A^{(n)}}^{(n)}\big)I_{\{\tau_A^{(n)} \leq T\}}\big] \longrightarrow \mathbb{E}_x\big[e^{-r\tau_A}h(S_{\tau_A})I_{\{\tau_A \leq T\}}\big],$$

where $S^{(n)}$ denotes $S^{(int\text{-}n)}$ or $S^{(tc\text{-}n)}$. A proof of this result was given in [25].

Proof In view of Proposition 17.13, it suffices[8] to verify that the sequences $(S^{(int\text{-}n)})_{n\in\mathbb{N}}$ and $(S^{(tc\text{-}n)})_{n\in\mathbb{N}}$ are relatively compact in $D_{\mathbb{R}}[0,T]$.

(a) We will establish the relative compactness of the sequence $(X^{(int\text{-}n)})_{n\in\mathbb{N}} = (\log S^{(int\text{-}n)})_{n\in\mathbb{N}}$. Let $X' = \log S$. It is straightforward to check that the set of functions $f : \mathbb{R}_+ \times \mathbb{R}_+ \to \mathbb{R}$ of the form $f(x,v) = g(x)h(v)$ with h in the core of $\mathcal{G}$, the infinitesimal generator of v, and with $g \in C_c^\infty(\mathbb{R})$[9] is dense in $C_0(\mathbb{R}_+ \times \mathbb{R})$ and is contained in the domain $\mathcal{D}(\mathcal{L}')$ of the infinitesimal generator $\mathcal{L}'$ of (v, X'). Furthermore, $\mathcal{L}'$ acts on such f as

$$\begin{aligned}\mathcal{L}' f(x,v) &= \frac{1}{2} v h(v)\sigma^2 g''(x) + \big[(r-d) + c\sqrt{v} - \psi(-\mathrm{i}\sqrt{v})\big]h(v)g'(x)\\ &\quad + h(v)\int_{\mathbb{R}} \big[g(x+z\sqrt{v}) - g(x) - z\sqrt{v}g'(x)I_{\{|z|\le 1\}}\big]\nu(dz)\\ &\quad + g(x)\mathcal{G}h(v), \qquad x\in\mathbb{R}, v>0,\end{aligned}$$

since by construction the stochastic integral $\{\int_0^t \sqrt{v_s}\,dX_s\}_{t\ge 0}$ jumps if and only if the Lévy process X jumps and, if the jump occurs at time t, the quotient of the jump sizes equals $\sqrt{v_t}$. On the other hand, the infinitesimal generator $\mathcal{L}^{(n)}$ of the regime-switching processes $(Z^{(n)}, X^{(int\text{-}n)})$ acts on $f(x,v) = g(x)h(v)$ as

$$\begin{aligned}\mathcal{L}^{(n)} f(x,v) &= \frac{1}{2} v \sigma_n^2 h(v) g''(x) + \big[(r-d) + c_n\sqrt{v} - \psi_n(-\mathrm{i}\sqrt{v})\big]h(v)g'(x)\\ &\quad + h(v)\int_{\mathbb{R}} \big[g(x+z\sqrt{v}) - g(x) - z\sqrt{v}g'(x)I_{\{|z|\le 1\}}\big]\nu_n(dz)\\ &\quad + g(x)Q^{(n)}h(v), \qquad x\in\mathbb{R}, v\in E^{0(n)}.\end{aligned}$$

$\mathcal{L}^{(n)} f$ converges to $\mathcal{L}' f$ uniformly as $n\to\infty$:

$$\epsilon_n'(f) := \sup_{x\in\mathbb{R}, v\in E^{0(n)}} \big|\mathcal{L}' f(x,v) - \mathcal{L}^{(n)} f(x,v)\big| \to 0. \tag{17.37}$$

To see why this is true, note that the triangle inequality and integration by parts imply that

$$\begin{aligned}\epsilon_n'(f) &\le \big\|\mathcal{G}h - Q^{(n)}h\big\|_n \|g\|_\infty + C_1\big|\overline{\nu}(a) - \overline{\nu}_n(a)\big| + C_2 \max\big\{\big|\sigma_n^2 - \sigma^2\big|, |c_n - c|\big\}\\ &\quad + C_3\bigg\{\int_a^\infty \big|\overline{\nu}(z) - \overline{\nu}_n(z)\big|\,dz + \int_{-\infty}^{-a} \big|\underline{\nu}(z) - \underline{\nu}_n(z)\big|\,dz\bigg\}\\ &\quad + C_4\bigg\{\int_{(0,a)} z^2\big|\overline{\nu}(z) - \overline{\nu}_n(z)\big|\,dz + \int_{(-a,0)} z^2\big|\underline{\nu}(z) - \underline{\nu}_n(z)\big|\,dz\bigg\},\end{aligned} \tag{17.38}$$

[8]See e.g. Theorem 3.7.8 in Ethier and Kurtz [15] for a proof of this well-known fact.

[9]$C_c^\infty(\mathbb{R}_+)$ denotes the set of infinitely differentiable functions with compact support contained in $\mathbb{R}_+$.

where a is a continuity point of the measures $\nu(dx)$ and $\nu(-dx)$, and C_1, $\ldots, C_4$, are certain finite constants independent of n, and we denoted $\|f\|_n = \sup_{x\in E^{0(n)}} |f(x)|$ and $\|f\|_\infty = \sup_{x\in\mathbb{R}} |f(x)|$. In view of conditions (17.34) and (17.35), $\epsilon'_n(f)$ tends to zero as n tends to infinity. Corollary 4.8.6 in Ethier and Kurtz [15] implies then that $(Z^{(n)}, X^{(int\text{-}n)})_{n\in\mathbb{N}}$ and hence $(Z^{(n)}, S^{(int\text{-}n)})_{n\in\mathbb{N}}$ is relatively compact in $D_{\mathbb{R}^2}[0, T]$.

(b) Denote by $\widetilde{X}^{(n)} = \{\widetilde{X}_t^{(n)}\}_{t\geq 0}$ and $\widetilde{X} = \{\widetilde{X}_t\}_{t\geq 0}$ the Lévy processes given by

$$\widetilde{X}_t^{(n)} = X_t^{(n)} - \psi^{(n)}(-\mathrm{i})t \quad \text{and} \quad \widetilde{X}_t = X_t - \psi(-\mathrm{i})t.$$

We will verify[10] the relative compactness of the sequence $(Y^{(n)})_{n\in\mathbb{N}}$ with $Y_t^{(n)} = \widetilde{X}^{(n)}(V_t^{(n)})$. In view of the Skorokhod embedding theorem and the convergence in law of $(Z^{(n)}, \widetilde{X}^{(n)})$ to $(v, \widetilde{X})$, we may and shall assume that $(Z^{(n)}, \widetilde{X}^{(n)})_{n\in\mathbb{N}}$ and $(v, \widetilde{X})$ are defined on the same probability space $(\Omega', \mathcal{F}', \mathbb{P}')$, and that, $\mathbb{P}'$-almost surely, $(Z^{(n)}, X^{(n)}) \to (v, X)$ with respect to the Skorokhod metric. Fix an $\omega \in \Omega'$ for which this convergence holds true.

Observe that, for any $U > 0$, $\overline{z} = \sup_n Z_U^{(n)}(\omega)$ is finite, as $v_U(\omega)$ is finite and $|Z_U^{(n)}(\omega) - v_U(\omega| \to 0$ as n tends to infinity. For $x \in D_{\mathbb{R}}[0, U]$, $\delta > 0$, $U > 0$, denote by $w'(x, \delta, U)$ the modulus of continuity

$$w'(x, \delta, U) = \inf_{\{t_i\}} \max_i \sup_{s,t\in[t_{i-1},t_i)} \left|x(s) - x(t)\right|,$$

where $\{t_i\}$ ranges over all partitions $0 = t_0 < t_1 < \cdots < t_{n-1} < U \leq t_n$ with $\min_{1\leq i\leq n} |t_i - t_{i-1}| > \delta$. Note that $w'(x, \delta, U)$ is nondecreasing in δ and U. Therefore it is straightforward to check that

$$w'_n(U) := w'\big(Y^{(n)}(\omega), \delta, U\big) \leq w'\big(\widetilde{X}^{(n)}(\omega), \delta\overline{z}, U\overline{z}\big). \tag{17.39}$$

Furthermore, observe that

$$\mathcal{Y}_n(\omega) := \big\{Y_s^{(n)}(\omega) : s \leq U\big\} \subset \big\{\widetilde{X}_s^{(n)}(\omega) : s \leq U\overline{z}\big\}. \tag{17.40}$$

Since $\{\widetilde{X}^{(n)}(\omega)\}_{n\in\mathbb{N}}$ is convergent in $D_{\mathbb{R}}[0, \infty)$, it follows that (I) for every rational $t \in [0, U]$, there exists a compact set $C_t \subset \mathbb{R}$ such that $Y_t^{(n)}(\omega) \in C_t$ for all n and (II) for every $U > 0$, $\lim_{\delta\to 0} \sup_n w'_n(U) = 0$ and, as a consequence,[11] $\mathcal{Y}_n$ is relatively compact in $D_{\mathbb{R}}[0, \infty)$. □

---

[10]The proof draws on and combines a number of results from the theory of weak convergence of probability measures that can be found in Ethier and Kurtz [15, Chaps. 3, 6]

[11]Both applications follow from the fact that conditions (I) and (II) are necessary and sufficient for the relative compactness of $(Y^{(n)})_{n\in\mathbb{N}}$ (Ethier and Kurtz [15, Theorem 3.6.3]).

17.4 European and Volatility Derivatives

17.4.1 Call and Put Options

We first turn to the valuation of a call option price. In the model under consideration a closed-form expression is available, in terms of the original parameters, for the Fourier transform c_T^* in log-strike $k = \log K$ of the call prices $C_T(K)$ with maturity T,

$$c_T^*(\xi) = \int_{\mathbb{R}} e^{\mathrm{i}\xi k} C_T(e^k)\, dk \quad \text{where } \Im(\xi) < 0.$$

Proposition 17.15 *Define, for any* $\xi \in \mathbb{C}\backslash\{0, \mathrm{i}\}$, $x \in \mathbb{R}$ *and* $j \in E^0$, *the value* $D(\xi, x, j)$ *by the formula*

$$D(\xi, x, j) := \frac{e^{(1+\mathrm{i}\xi)x}}{\mathrm{i}\xi - \xi^2} \cdot \big[\exp\big\{T\big(K(1+\mathrm{i}\xi) - \Lambda_D\big)\big\}\mathbf{1}\big](j). \tag{17.41}$$

Then if $\Im(\xi) < 0$, *it holds that*

$$c_T^*(\xi) = D(\xi, x, j),$$

where $x = \log S_0$ *is the log-price at the current time, and* $Z_0 = j$ *the initial level of the volatility.*

Remarks

(i) The call option price can now be calculated using the method described in Carr–Madan [13] by evaluating the integral

$$\begin{aligned} C_T(K) &= \frac{\exp(-\alpha k)}{2\pi} \int_{-\infty}^{\infty} e^{-\mathrm{i}sk} c_T^*(s - \mathrm{i}\alpha)\, ds \\ &= \frac{\exp(-\alpha k)}{\pi} \int_0^{\infty} \Re\big[e^{-\mathrm{i}sk} D(s - \mathrm{i}\alpha, \log S_0, Z_0)\big]\, ds \end{aligned} \tag{17.42}$$

for $k = \log(K)$ and any strictly positive α. The integral in (17.42) can be approximated efficiently by a finite sum using the FFT algorithm (see [13]). Since in our model we have an explicit formula for the transform $c_T^*(s)$ given by (17.41), the pricing of European call options is immediate.

(ii) A simple calculation shows that the put option price $P_T(K) = \mathbb{E}_{x,j}[(B_T^D)^{-1} \times (K - S_T)^+]$ can be expressed in terms of the formula for $D(\xi, x, j)$ in (17.41) and any strictly negative constant α in the following way:

$$P_T(K) = \frac{\exp(-\alpha k)}{\pi} \int_0^{\infty} \Re\big[e^{-\mathrm{i}sk} D(s - \mathrm{i}\alpha, \log S_0, Z_0)\big]\, ds,$$

where $k = \log(K)$.

Proof To find the European call option price $C_T(K) = \mathbb{E}_{x,j}[(B_T^D)^{-1}(S_T - K)^+]$ in model (17.19), we first need to find the Fourier transform in the log-strike $k = \log(K)$ of the function

$$c_T(k) = \exp(\alpha k)\mathbb{E}_{x,j}\big[\big(B_T^D\big)^{-1}\big(S_T - \exp(k)\big)^+\big],$$

where α is some strictly positive constant. Fubini's theorem and the form of the characteristic function (17.31) imply the following for $\xi = v - \mathrm{i}\alpha$:

$$\begin{aligned} c_T^*(\xi) &= \int_{\mathbb{R}} \exp\big((\mathrm{i}v + \alpha)k\big)\mathbb{E}_{x,j}\big[\big(B_T^D\big)^{-1}\big(S_T - \exp(k)\big)^+\big]dk \\ &= \mathbb{E}_{x,j}\Big[\big(B_T^D\big)^{-1}\int_{\mathbb{R}} \exp\big((\mathrm{i}v + \alpha)k\big)\big(S_T - \exp(k)\big)^+ dk\Big] \\ &= \mathbb{E}_{x,j}\big[\big(B_T^D\big)^{-1}\exp\big((1 + \alpha + \mathrm{i}v)X_T\big)\big]/\big(\alpha^2 + \alpha - v^2 + \mathrm{i}(2\alpha + 1)v\big) \\ &= \frac{e^{x(1+\alpha+\mathrm{i}v)}}{\alpha^2 + \alpha - v^2 + \mathrm{i}(2\alpha + 1)v}\big[\exp\big(T\big(K(1 + \alpha + \mathrm{i}v) - \Lambda_D\big)\big)\mathbf{1}\big](j). \end{aligned}$$

This concludes the proof. □

17.4.2 Implied Volatility at Extreme Strikes

The *implied volatility* $\sigma_{x,i}(K, T)$ for a given strike K and maturity T is uniquely defined by the identity

$$C^{\mathrm{BS}}\big(S_0, K, T, \sigma_{x,i}(K, T)\big) = \mathbb{E}_{x,i}\big[\big(B_T^D\big)^{-1}(S_T - K)^+\big], \qquad (17.43)$$

where $C^{\mathrm{BS}}(S_0, K, T, \sigma)$ is the Black–Scholes formula, and $S_0 = \exp(x)$. The results in Lee [23] and refinements in Benaim and Friz [6] imply that in model (17.19) the slope of the volatility smile is uniquely determined by the quantities $\alpha_i^{\pm}$, $i = 1, \ldots, n$, defined in (17.21). In the particular case where the distribution of jumps is double exponential, $\alpha_i^{\pm}$ are in fact the reciprocals of mean-jump sizes in model (17.19).

In order to state the precise result, define

$$\begin{aligned} q_+ &:= \sup\big\{u : \mathbb{E}_{x,i}\big[S_T^{1+u}\big] < \infty \text{ for all } i \in E^0\big\}, \\ q_- &:= \sup\big\{u : \mathbb{E}_{x,i}\big[S_T^{-u}\big] < \infty \text{ for all } i \in E^0\big\}. \end{aligned}$$

If the chain Z is irreducible, the quantities $q_\pm$ can be identified explicitly to be equal to

$$q_+ = \min\big\{\alpha_i^+ - 1 : i \in \{1, \ldots, N_0\} \,\&\, p_i\lambda_i > 0\big\}, \qquad (17.44)$$

$$q_- = \min\big\{\alpha_i^- : i \in \{1, \ldots, N_0\} \,\&\, (1 - p_i)\lambda_i > 0\big\}. \qquad (17.45)$$

As noted above, the quantities q_+ and q_- depend only on the mean-jump sizes of the compound Poisson processes in model (17.19).

Denote the forward price by $F_T := \mathbb{E}_{x,i}[S_T]$. Then the asymptotic behaviour for the implied volatility is described as follows:

Proposition 17.16 *Suppose that Z is irreducible. For $T > 0$ and $K > 0$ and with $q_\pm$ given in* (17.44)–(17.45), *it holds that*

$$\lim_{K\to\infty} \frac{T\sigma_{x,i}(K,T)^2}{\log(K/F_T)} = 2 - 4\Big(\sqrt{q_+^2 + q_+} - q_+\Big),$$

$$\lim_{K\to 0} \frac{T\sigma_{x,i}(K,T)^2}{|\log(K/F_T)|} = 2 - 4\Big(\sqrt{q_-^2 + q_-} - q_-\Big).$$

Remark Note that if the Markov chain Z is irreducible, the asymptotic slope of the implied volatility smile for large and small strikes depends neither on the spot $S_0 = e^x$ nor on the starting volatility regime i.

17.4.3 Forward Starting Options and the Forward Smile

A forward starting call option is a call option whose strike is fixed at a later date as a proportion of the value of the underlying at that moment. More precisely, the pay-off of a T_1-forward starting call option at maturity $T_2 > T_1$ is given by

$$(S_{T_2} - \kappa S_{T_1})^+, \quad \kappa \in \mathbb{R}_+.$$

Denote the current value of this forward starting option by $F_{T_1,T_2}(\kappa)$ and let

$$F^*_{T_1,T_2}(\xi) = \int_{\mathbb{R}} e^{\mathrm{i}\xi k} F_{T_1,T_2}\big(e^k\big)\, dk, \quad \text{where } \Im(\xi) < 0,$$

be its Fourier transform in the forward log-strike $k = \log \kappa$.

Proposition 17.17 *For ξ with $\Im(\xi) < 0$, it holds that*

$$F^*_{T_1,T_2}(\xi) = \frac{e^{(1+\mathrm{i}\xi)x}}{\mathrm{i}\xi - \xi^2} \cdot \big[\exp\big(T_1(Q - \Lambda_F)\big)\exp\big\{(T_2 - T_1)\big(K(1+\mathrm{i}\xi) - \Lambda_D\big)\big\}\mathbf{1}\big](j), \tag{17.46}$$

where $x = \log S_0$ is the log-spot price, and $Z_0 = j$ the initial level of the volatility.

Remark An inversion formula, analogous to the one in (17.42), can be used to obtain the value $F_{T_1,T_2}(\kappa)$ from Proposition 17.17.

Proof The price of a T_1-forward starting option is given by the expression

$$F_{T_1,T_2}(\kappa) = \mathbb{E}_{x,i}\big[\big(B^D_{T_2}\big)^{-1}(S_{T_2} - \kappa S_{T_1})^+\big], \quad \kappa \in \mathbb{R}_+.$$

The process (X, Z) is Markov, and therefore, by conditioning on the σ-algebra generated by the process up to time T_1, employing the form (17.31) of the characteristic function of X_T and using the spatial homogeneity of the log-price $X_t = \log S_t$ in our model, we obtain the following expression for the price of the forward starting option

$$
\begin{aligned}
F_{T_1,T_2}(\kappa) &= \mathbb{E}_{x,i}\big[\big(B^D_{T_2}\big)^{-1}(S_{T_2} - \kappa S_{T_1})^+\big] && (17.47)\\
&= \mathbb{E}_{x,i}\left[\frac{S_{T_1}}{B^D_{T_1}}\mathbb{E}_{0,Z_{T_1}}\big[\big(B^D_{T_2-T_1}\big)^{-1}(S_{T_2-T_1} - \kappa)^+\big]\right]\\
&= \sum_{j\in E^0}\mathbb{E}_{x,i}\left[\frac{S_{T_1}}{B^D_{T_1}}I_{\{Z_{T_1}=j\}}\right]\mathbb{E}_{0,j}\big[\big(B^D_{T_2-T_1}\big)^{-1}(S_{T_2-T_1} - \kappa)^+\big]\\
&= S_0\sum_{j\in E^0} e_i'\exp\big(T\big(K(-\mathtt{i}) - \Lambda_D\big)\big)e_j\mathbb{E}_{0,j}\big[\big(B^D_{T_2-T_1}\big)^{-1}(S_{T_2-T_1} - \kappa)^+\big]\\
&= S_0 e_i'\exp\big(T\big(K(-\mathtt{i}) - \Lambda_D\big)\big)C_{T_2-T_1}(\kappa; 1), && (17.48)
\end{aligned}
$$

where $C_{T_2-T_1}(\kappa; 1)$ is a vector (of call option prices) with the jth component equal to $\mathbb{E}_{0,j}[(B^D_{T_2-T_1})^{-1}(S_{T_2-T_1} - \kappa)^+]$. The martingale condition in (17.25) and Proposition 17.15 conclude the proof. □

Remarks

(i) A quantity of great interest in the derivatives markets is the *forward implied volatility* $\sigma^{fw}_{x,i}(S_T, \kappa, T)$ at a future time T implied by the model. It is defined as the unique solution to the equation

$$
C^{\mathrm{BS}}\big(S_{T_1}, \kappa S_{T_1}, T_2 - T_1, \sigma^{fw}_{x,i}(S_{T_1}, \kappa, T_1)\big) = \mathbb{E}_{x,i}\left[\frac{B^D_{T_1}}{B^D_{T_2}}(S_{T_2} - \kappa S_{T_1})^+\middle| S_{T_1}\right], \quad (17.49)
$$

where the left-hand side denotes the Black–Scholes formula with strike κS_{T_1} and spot S_{T_1}. The reason for the importance of the forward implied volatility $\sigma^{fw}_{x,i}(S_T, \kappa, T)$ lies in the problem of hedging of exotic derivatives using vanilla options. If at a future time T the spot trades at the level S_T, then the trader needs to know where, according to the model, would the vanilla surface be trading at. This is of particular importance when hedging a barrier contract that knocks out at the level S_T, because conditional on this event the trader is left with a portfolio of vanilla options that was created as a semi-static hedge for the exotic derivative.

(ii) In the model given by (17.18) we can compute the right-hand side of (17.49). In view of the Markov property of the process (X, Z), this equation is equivalent to

$$
C^{\mathrm{BS}}\big(S_{T_1}, \kappa S_{T_1}, T_2 - T_1, \sigma^{fw}_{x,i}(S_{T_1}, \kappa, T_1)\big)
$$

$$= S_{T_1} \mathbb{E}_{x,i}\big[\mathbb{E}_{0,Z_{T_1}}\big[\big(B^D_{T_2-T_1}\big)^{-1}(S_{T_2-T_1}-\kappa)^+\big]\big|S_{T_1}\big]$$
$$= S_{T_1} \sum_{j\in E^0} \mathbb{P}_{x,i}[Z_{T_1}=j|S_{T_1}]\mathbb{E}_{0,j}\big[\big(B^D_{T_2-T_1}\big)^{-1}(S_{T_2-T_1}-\kappa)^+\big]$$
$$= S_{T_1} f^{x,i}(X_{T_1},T_1)' C_{T_2-T_1}(\kappa,1),$$

where the coordinates of the vector $f^{x,i}(y,T)$ are defined by

$$f^{x,i}_j(y,T) := \mathbb{P}_{x,i}[Z_T=j|X_T=y] \tag{17.50}$$

and $C_{T_2-T_1}(\kappa;1)$ is as defined in the line following (17.48).

(iii) The vector $C_{T_2-T_1}(\kappa;1)$ can be computed by formula (17.42), and, in the light of definition (17.50), Proposition 17.8 and formulae (17.29) and (17.30) for $q^{x,i}_T(y,j)$ and $q^{x,i}_T(y)$, it follows that

$$f^{x,i}_j(y,T)q^{x,i}_T(y) = q^{x,i}_T(y,j).$$

This yields the quantity in (17.50) and hence a formula for the forward implied volatility in our model.

17.4.4 Volatility Derivatives

An option on the realized variance is a derivative security that delivers $\phi(\Sigma_T)$ at expiry T, where $\phi:\mathbb{R}_+\to\mathbb{R}$ is some measurable payoff function, and Σ_T is the quadratic variation up to time T of the process $\log S = X$. More formally, for a refining sequence of partitions[12] $(\Pi_n)_{n\in\mathbb{N}}$ of the interval $[0,T]$, Σ_T is given by

$$\Sigma_T := \lim_{n\to\infty} \sum_{t^n_i\in\Pi_n, i\geq 1} \log\left(\frac{S_{t^n_i}}{S_{t^n_{i-1}}}\right)^2.$$

It is well known that the sequence on the right-hand side converges in probability, uniformly on compact time intervals (see Jacod and Shiryaev [20], Theorem 4.47) and the limit is given by

$$\Sigma_T = \int_0^T \sigma(Z_t)^2\,dt + \sum_{i\in E^0}\sum_{t\leq T} I_{\{Z_t=i\}}\big(\Delta J^i_t\big)^2, \tag{17.51}$$

where $\Delta J^i_t := J^i_t - J^i_{t-}$. The process $\{\Sigma_t\}_{t\geq 0}$ is called the quadratic variation or realized variance process of X, and its law is explicitly characterised as follows:

[12] The sets $\Pi_n = \{t^n_0, t^n_1, \ldots, t^n_n\}$, $n\in\mathbb{N}$, consist of increasing sequences of times such that $t^n_0=0$, $t^n_n=T$, $\Pi_n\subset\Pi_{n+1}$ for all $n\in\mathbb{N}$ and $\lim_{n\to\infty}\max\{|t^n_i-t^n_{i-1}|: i=1,\ldots,n\}=0$.

Proposition 17.18

(i) *The process* $\{(\Sigma_t, Z_t)\}_{t\geq 0}$ *is a Markov process with*

$$\Sigma_t = \int_0^t \sigma(Z_s)^2\, ds + \sum_{i\in E^0} \int_0^t I_{\{Z_s=i\}}\, d\widetilde{J}_s^i,$$

where $\widetilde{J}^i$, $i \in E^0$, *is a compound Poisson process with intensity* λ_i *and positive jump sizes* K_i *with probability density*

$$g_i(x) = \frac{1}{2\sqrt{x}}\big[p_i\beta_i^+ e^{\sqrt{x}B_i^+}(-B_i^+)\mathbf{1} + (1-p_i)\beta_i^- e^{\sqrt{x}B_i^-}(-B_i^-)\mathbf{1}\big]I_{(0,\infty)}(x).$$

(ii) *The discounted Laplace transform of* Σ_t *is given by*

$$\mathbb{E}_i\left[\frac{\exp(-u\Sigma_t)}{B_t^D}\right] = \big[\exp\big(t\big(K_\Sigma(u) - \Lambda_D\big)\big)\mathbf{1}\big](i), \quad u > 0, \tag{17.52}$$

where $K_\Sigma(u) = Q + \Lambda_\Sigma(u)$ *with* $\Lambda_\Sigma(u)$ *an* $N_0 \times N_0$ *diagonal matrix with* i*th element given by*

$$\psi_i^\Sigma(u) := -u\sigma_i^2 + \lambda_i\big(\mathbb{E}\big[\exp(-uK_i)\big] - 1\big) \tag{17.53}$$

with

$$\mathbb{E}\big[\exp(-uK_i)\big] = \sqrt{\frac{\pi}{u}}\bigg(p_i\beta_i^+\Phi\bigg(\frac{1}{\sqrt{2u}}B_i^+\bigg)(-B_i^+)$$
$$+ (1-p_i)\beta_i^-\Phi\bigg(\frac{1}{\sqrt{2u}}B_i^-\bigg)(-B_i^-)\bigg)\mathbf{1},$$

where $\Phi(x) := \exp(x^2/2)\mathcal{N}(x)$ *with the cumulative normal distribution function* $\mathcal{N}$.

Remarks

(i) As a given matrix M in practice typically[13] admits a spectral decomposition $M = UDU^{-1}$ where D is a diagonal matrix, $\Phi(M)$ can be evaluated by $\Phi(M) = U\Phi(D)U^{-1}$, where $\Phi(D)$ is the diagonal matrix with ith element $\Phi(D_{ii})$.

(ii) It is important to note that the realized variance process Σ does not possess exponential moments of any order. This follows directly from the fact that the distribution of jumps g_i given in (17.60) decays at the rate $e^{-c\sqrt{x}}$, for some positive constant c, and implies that the left-hand side in formula (17.52) will be infinite for complex numbers u with negative imaginary part.

[13] This is the case since the set of all square matrices that do not possess a diagonal decomposition is of codimension one in the space of all square matrices and therefore has Lebesgue measure zero.

(iii) The expression (17.52) for the discounted Laplace transform can be employed to obtain explicit results for the values of volatility derivatives. The buyer of a swap on the realized variance pays premiums at a certain rate (the swap rate) to receive at maturity a pay-off $\phi(\Sigma_T)$ that is a function ϕ of the realized variance Σ_T, with as most common examples the volatility and the variance swap. In the case of a variance swap this function is linear ($\phi(x) = x/T$), whereas for a volatility swap, it is a square root ($\phi(x) = \sqrt{x/T}$). The swap rates are determined such that at initiation the value of the swap is nil.

Corollary 17.19 *Suppose that $Z_0 = j$. Then the variance and volatility swap rates $\varsigma_{\text{var}}(T, j)$ and $\varsigma_{\text{vol}}(T, j)$ are given as follows:*

$$\varsigma_{\text{vol}}(T, j) = \frac{1}{2\sqrt{\pi T}} \int_0^\infty \big\{\big[\exp\big(T(Q - \Lambda_D)\big) - \exp\big(T\big(K_\Sigma(u) - \Lambda_D\big)\big)\big]\mathbf{1}\big\}(j)\frac{du}{u^{3/2}}, \tag{17.54}$$

$$\varsigma_{\text{var}}(T, j) = \frac{1}{T}\int_0^T \big[\exp\{t(Q - \Lambda_D)\}\Lambda_V \exp\{(T-t)(Q - \Lambda_D)\}\mathbf{1}\big](j)\,dt \tag{17.55}$$

$$= \frac{1}{Th}\big[\big\{\exp\big(T(Q - \Lambda_D)\big) - \exp\big(T\big(K_\Sigma(h) - \Lambda_D\big)\big)\big\}\mathbf{1}\big](j) + o(h), \quad h \downarrow 0, \tag{17.56}$$

where Λ_V is an $N_0 \times N_0$ diagonal matrix with ith element given by

$$V(i) = \sigma_i^2 + 2\lambda_i\big(p_i\big(\beta_i^+\big)'\big(B_i^+\big)^{-2} + (1 - p_i)\big(\beta_i^-\big)'\big(B_i^-\big)^{-2}\big)\mathbf{1}. \tag{17.57}$$

Remarks

(i) The Laplace transform $\widehat{\sigma}_{\text{var}}(q, j)$ of $\sigma_{\text{var}}(\cdot, j) : T \mapsto T\varsigma_{\text{var}}(T, j)$ is explicitly given by

$$\widehat{\sigma}_{\text{var}}(q, j) = \big[(qI - \Lambda_D - Q)^{-1}\Lambda_V(qI - \Lambda_D - Q)^{-1}\mathbf{1}\big](j). \tag{17.58}$$

(ii) It is clear from the definition of $K_\Sigma(u)$ that the integral in (17.54) converges at the rate proportional to $1/\sqrt{U}$, where U is an arbitrary upper bound used in the numerical integration in (17.54).

Proof of Proposition 17.18 It is clear from the representation (17.51) that the increment $\Sigma_t - \Sigma_s$, for any $t > s \geq 0$, satisfies the equation

$$\Sigma_t - \Sigma_s = \int_s^t \sigma^2(Z_u)\,du + \sum_{i \in E^0}\int_s^t I_{\{Z_u = i\}}\,d\widetilde{J}_u^i, \tag{17.59}$$

where $\widetilde{J}^i$, $i \in E^0$, is a compound Poisson process with intensity λ_i and positive jump sizes K_i distributed as $(U^i)^2$ where U_i follows a DPH$(p_i, \beta_i^+, B_i^+, \beta_i^-, B_i^-)$

distribution. In particular, K_i is distributed according to the density

$$g_i(x) = \frac{1}{2\sqrt{x}}\big(f_i(\sqrt{x}) + f_i(-\sqrt{x})\big), \quad x > 0, \tag{17.60}$$

where the probability density function f_i is given by

$$f_i(x) = p_i\big(\beta_i^+\big)' e^{xB_i^+}\big(-B_i^+\big)\mathbf{1}I_{(0,\infty)}(x) + (1-p_i)\big(\beta_i^-\big)' e^{xB_i^-}\big(-B_i^-\big)\mathbf{1}I_{(-\infty,0)}(x).$$

As the $\widetilde{J}^i$ have independent increments, and Z is a Markov chain, it directly follows from (17.59) that (Σ_t, Z_t) is a Markov process and, moreover, a regime-switching subordinator. The form of the discounted Laplace transform can be derived as in Theorem 17.7. □

Proof of Corollary 17.19 Employing the following representation for the square root,

$$\sqrt{x} = \frac{1}{2\sqrt{\pi}}\int_0^\infty \big[1 - \exp(-ux)\big]\frac{du}{u^{3/2}} \quad \text{for any } x \geq 0,$$

as well as the form of the discounted characteristic function given in (17.52) and Fubini's theorem yields that the volatility swap rate can be calculated via a single one-dimensional integral

$$\mathbb{E}_i\left[\frac{\sqrt{\frac{1}{T}\Sigma_T}}{B_T^D}\right] = \frac{1}{2\sqrt{\pi T}}\int_0^\infty e_i'\big[\exp\big(T(Q - \Lambda_D)\big) - \exp\big(T\big(K_\Sigma(u) - \Lambda_D\big)\big)\big]\mathbf{1}\frac{du}{u^{3/2}}.$$

The derivation of the variance swap rate formula (17.55) rests on a conditioning argument. Indeed, by conditioning on the sigma algebra $\mathcal{F}_T^Z = \sigma(\{Z_t\}_{t\leq T})$ generated by Z up to time T, it follows that

$$\begin{aligned}
&\mathbb{E}_{x,i}\left[\frac{1}{B_T^D}\left\{\int_0^t \sigma(Z_s)^2\,ds + \sum_{i\in E^0}\int_0^t I_{\{Z_s=i\}}\,d\widetilde{J}_s^i\right\}\right] \\
&\quad = \sum_{j\in E^0}\mathbb{E}_{x,i}\left[\frac{1}{B_T^D}\int_0^T I_{\{Z_s=j\}}\,ds\right]w(j),
\end{aligned}$$

where $w(j) := \sigma^2(j) + \mathbb{E}[\widetilde{J}_1^j]$. From the definition of $\widetilde{J}^j$ it is easily checked that $\mathbb{E}[\widetilde{J}_1^j]$ is equal to λ_j times the second moment of the density f_i. One verifies by a straightforward calculation that $w(j)$ is equal to $V(j)$ given in (17.57). Furthermore, the Markov property of Z applied at time t yields that

$$\mathbb{E}_{x,i}\left[\frac{1}{B_T^D}\int_0^T I_{\{Z_s=j\}}\,ds\right] = e_i'\exp\big(t(Q - \Lambda_D)\big)e_j' e_j\exp\big((T-t)(Q - \Lambda_D)\big)\mathbf{1}.$$

Equation (17.55) follows then by an application of Fubini's theorem. Furthermore, (17.56) follows by noting that the expectation $\mathbb{E}_{x,i}[\Sigma_T / B_T^D]$ can also be obtained by calculating the negative of the derivative of the Laplace transform at zero:

$$\mathbb{E}_{x,i}\left[\frac{\Sigma_T}{B_T^D}\right] = -\frac{d}{du}\{[\exp(T(K_\Sigma(u) - \Lambda_D))\mathbf{1}](i)\}\bigg|_{u=0}. \qquad \square$$

17.5 First Passage Times for Regime-Switching Processes

17.5.1 Three Key Matrices

The characteristics of the process (X, Z) can be summarised in terms of three matrices Q_0, Σ and V that will shortly be specified. Given those three matrices, we will show how to reconstruct (X, Z) in Sect. 17.7.

The matrices Q_0, Σ and V will be specified in nine-block matrices using block notation; the middle block of Q_0, Σ and V describes the rates of regime-switches, and the volatility and drift of the process in the different regimes, while the upper left and lower right blocks of Q_0 specify the distribution of up-ward and down-ward jumps in the different regimes, in terms of the (phase-type) generators. More precisely, we define the three key matrices Q_0, Σ and V, in block notation, by

$$Q_0 := \begin{pmatrix} B^+ & b^+ & O \\ A^+ & Q - \Lambda_\lambda & A^- \\ O & b^- & B^- \end{pmatrix}, \tag{17.61}$$

$$\Sigma := \begin{pmatrix} O & O & O \\ O & \Lambda_S & O \\ O & O & O \end{pmatrix}, \qquad V := \begin{pmatrix} I & O & O \\ O & \Lambda_M & O \\ O & O & -I \end{pmatrix}. \tag{17.62}$$

Here Q is the generator matrix of the chain Z, and Λ_λ, Λ_S, Λ_M denote the $N_0 \times N_0$ diagonal matrices with elements

$$\Lambda_\lambda(i,i) := \lambda_i, \qquad \Lambda_S(i,i) := \sigma(i) \quad \text{and} \quad \Lambda_M(i,i) := \mu(i). \tag{17.63}$$

Further, O and I are zero and identity matrices of appropriate sizes such that Q_0, Σ and V are square matrices of the same dimension. In block notation $A^\pm$, $B^\pm$ and $b^\pm$ are given by

$$A^\pm := \begin{pmatrix} \lambda_1^\pm \beta_1^{\pm\prime} & & \\ & \ddots & \\ & & \lambda_N^\pm \beta_N^{\pm\prime} \end{pmatrix}, \qquad B^\pm := \begin{pmatrix} B_1^\pm & & \\ & \ddots & \\ & & B_N^\pm \end{pmatrix},$$
$$b^\pm := \begin{pmatrix} -B_1^\pm \mathbf{1} & & \\ & \ddots & \\ & & -B_N^\pm \mathbf{1} \end{pmatrix}, \tag{17.64}$$

where $\lambda_i^+ := \lambda_i p_i$ and $\lambda_i^- := \lambda_i(1 - p_i)$.

Remark The matrix Q_0 is in fact the generator matrix of a Markov chain, as it has nonnegative off-diagonal elements and zero row sums. We denote the state space of this Markov chain by E. In the sequel we will frequently use the following partition of the set E:

$$\begin{aligned} E^{\pm} &= \{i \in E : \Sigma_{ii} \neq 0\}, \qquad E^{+} = \{i \in E : \Sigma_{ii} = 0, V_{ii} > 0\}, \\ E^{-} &= \{i \in E : \Sigma_{ii} = 0, V_{ii} < 0\}. \end{aligned} \tag{17.65}$$

Note further that $E^{\pm}$ can and will be identified with the state space E^0 of the chain Z. See Sect. 17.7 for further properties of the Markov chain defined by the generator Q_0.

17.5.2 Matrix Wiener–Hopf Factorisation

For a given vector of discount rates $h : E \to \mathbb{C}$, the matrix Wiener–Hopf factorisation associates to the matrix

$$Q_h := Q_0 - \Lambda_h,$$

where Λ_h is a diagonal matrix with ith diagonal element $\Lambda_h(i,i) = h(i)$, a quadruple of matrices which, as we will show below, characterises the distributions of the running maximum and minimum of X.

Let us briefly describe the sets of matrices of which this quadruple are elements. Denote by $\mathbb{D}(n)$ the set of $n \times n$ square matrices whose eigenvalues all have nonpositive real part. Note that by Lemma 17.2, $\mathbb{D}(n)$ includes the set $\mathbb{G}(n)$ of $n \times n$ subgenerator matrices (i.e. matrices with nonnegative off-diagonal elements and nonpositive rows). Recall that $\mathbb{C}^{n \times m}$ denotes the set of $n \times m$ matrices with complex entries. Denote by $\mathbb{H}(n, m)$ the set of $n \times m$ subprobability matrices (i.e. matrices with nonnegative elements and row sums smaller or equal to one).

Denote by N, N^+, $\underline{N}^+$, N^- and $\underline{N}^-$ the number of elements of the sets E, $E^0 \cup E^+$, E^+, $E^0 \cup E^-$ and E^-, respectively. Also, let $\mathcal{H}$ denote the set

$$\mathcal{H} = \left\{ h : E \to \mathbb{C} : \min_{i \in E} \Re\big(h(i)\big) \geq 0,\ \min_{i \in E^0} \Re\big(h(i)\big) > 0 \right\}.$$

Definition 17.20 Let $h \in \mathcal{H}$, and let W^+, G^+, W^- and G^- be elements of the sets $C^{N \times N^+}$, $\mathbb{D}(N^+)$, $\mathbb{C}^{N \times N^-}$ and $\mathbb{D}(N^-)$, respectively. A quadruple (W^+, G^+, W^-, G^-) is called a *matrix Wiener–Hopf factorisation* of Q_h if the following matrix equations are satisfied:

$$\frac{1}{2} \Sigma^2 W^+ (G^+)^2 - V W^+ G^+ + Q_h W^+ = O^+, \tag{17.66}$$

$$\frac{1}{2} \Sigma^2 W^- (G^-)^2 + V W^- G^- + Q_h W^- = O^-, \tag{17.67}$$

where O^+ and O^- are zero matrices of sizes $N \times N^+$ and $N \times N^-$.

Theorem 17.21

(i) *For any $h \in \mathcal{H}$, there exists a unique matrix Wiener–Hopf factorisation of Q_h, denoted by $(\eta_h^+, Q_h^+, \eta_h^-, Q_h^-)$.*
(ii) *If $h = \Re(h)$, then $Q_h^+ \in \mathbb{G}(N^+)$ and $Q_h^- \in \mathbb{G}(N^-)$ are subgenerator matrices, and $\eta_h^+ \in \mathbb{H}(N, N^+)$ and $\eta_h^- \in \mathbb{H}(N, N^-)$ are in block notation given by*

$$\eta_h^+ = \begin{pmatrix} I_+ \\ \underline{\eta}_h^+ \end{pmatrix} \quad \text{and} \quad \eta_h^- = \begin{pmatrix} \underline{\eta}_h^- \\ I_- \end{pmatrix} \tag{17.68}$$

for some matrices $\underline{\eta}_h^+ \in \mathbb{H}(\underline{N}^-, N^+)$ and $\underline{\eta}_h^- \in \mathbb{H}(\underline{N}^+, N^-)$, and identity matrices I_+ and I_- of sizes $N^+ \times N^+$ and $N^- \times N^-$.

Remarks

(i) For h given by $h(i) = q I_{E^0}(i)$ with $q > 0$, we will also write $(\eta_q^+, Q_q^+, \eta_q^-, Q_q^-)$. The proof of Theorem 17.21 will be given in Sect. 17.8.
(ii) We allow the vector h to take complex values to be able to deal with a Laplace inversion using a Browmich integral, which involves the integration of the resulting first-passage quantities over a curve in the complex plane. Note that, for any real-valued $h \in \mathcal{H}$, i.e. $h = \Re(h)$, the matrix Q_h is a transient generator matrix, since the off-diagonal elements of Q_h are nonnegative and $Q_h \mathbf{1} \neq 0$.

17.5.3 First-Passage into a Half-Line

The marginal distributions of the maximum and minimum as well as the distributions of the first-passage times into a half-line can be described explicitly in terms of the matrix Wiener–Hopf factorisation. Denote by $\overline{X}_t = \sup_{0\leq s\leq t} X_s$ the running maximum of X at time t and by $\underline{X}_t = \inf_{0\leq s\leq t} X_s$ the corresponding running minimum, and let T_a^+ and T_a^- be the first passage times of X into a half-line,

$$T_a^+ = \inf\{t \geq 0 : X_t \in (a, \infty)\}, \qquad T_a^- = \inf\{t \geq 0 : X_t \in (-\infty, a)\}.$$

The distributions of those random variables are related via

$$\begin{aligned} \Gamma^+(t) &= \mathbb{P}_{x,i}\big(T_a^+ < t\big) = \mathbb{P}_{x,i}(\overline{X}_t > a), \\ \Gamma^-(t) &= \mathbb{P}_{x,i}\big(T_a^- < t\big) = \mathbb{P}_{x,i}(-\underline{X}_t > a). \end{aligned}$$

If $\mathbf{e}_q$ denotes a random time that is exponentially distributed with parameter $q > 0$ and that is independent of (X, Z), then the running maximum and minimum at time $\mathbf{e}_q$ follow a phase-type distribution, with parameters explicitly given in terms of the matrix Wiener–Hopf factorisation $(\eta_q^+, Q_q^+, \eta_q^-, Q_q^-)$.

For any $h \in \mathcal{H}$, define the $N \times N^+$ and $N \times N^-$ matrices $\Phi^+_{h,a}(x)$ and $\Phi^-_{h,a}(x)$ by

$$\Phi^+_{h,a}(x,i,j) = \big[\eta^+_h \exp\big((a-x)Q^+_h\big)\big](i,j) I_{(-\infty,a]}(x) + \delta_{ij} I_{(a,\infty)}(x),$$
$$i \in E,\ j \in E^+ \cup E^0, \tag{17.69}$$
$$\Phi^-_{h,a}(x,i,j) = \big[\eta^-_h \exp\big((x-a)Q^-_h\big)\big](i,j) I_{[a,\infty)}(x) + \delta_{ij} I_{(-\infty,a)}(x),$$
$$i \in E,\ j \in E^- \cup E^0, \tag{17.70}$$

where δ_{ij} is the Kronecker delta (i.e. $\delta_{ij} = I_{\{i\}}(j)$). For h given by $h(i) = qI_{E^0}(i)$ with $q > 0$, we will also denote $\Phi^\pm_{h,a}(x)$ by $\Phi^\pm_{q,a}(x)$.

Proposition 17.22 *For $q > 0$, it holds that under $\mathbb{P}_{x,i}$*

$$\overline{X}_{\mathbf{e}_q} - x \sim \mathrm{PH}\big(\eta^+_q(i), Q^+_q\big), \quad x \in \mathbb{R},\ i \in E^0,$$
$$-\underline{X}_{\mathbf{e}_q} + x \sim \mathrm{PH}\big(\eta^-_q(i), Q^-_q\big), \quad x \in \mathbb{R},\ i \in E^0,$$

where $\eta^+_q(i)$ and $\eta^-_q(i)$ are the ith rows of η^+_q and η^-_q. In particular the Laplace transforms $\widehat{\Gamma}^\pm(q) := \int_0^\infty e^{-qt}\Gamma^\pm(t)\,dt$, for $q > 0$, are given by the formulae

$$\widehat{\Gamma}^+(q) = \frac{1}{q}\sum_{j\in E^0} \mathbb{E}_{x,i}\big[e^{-qT^+_a} I_{\{Z_{T^+_a}=j\}}\big] = \frac{1}{q}\sum_{j\in E^0} \Phi^+_{q,a}(x,i,j), \quad x, a \in \mathbb{R},\ i \in E^0,$$

$$\widehat{\Gamma}^-(q) = \frac{1}{q}\sum_{j\in E^0} \mathbb{E}_{x,i}\big[e^{-qT^-_a} I_{\{Z_{T^-_a}=j\}}\big] = \frac{1}{q}\sum_{j\in E^0} \Phi^-_{q,a}(x,i,j), \quad x, a \in \mathbb{R},\ i \in E^0.$$

Remarks

(i) The proof of Proposition 17.22 will be given in Sect. 17.8.
(ii) Denote by $\Phi^+_{q,\upsilon}(x,i)$ and $\Phi^-_{q,\ell}(x,i)$ the N^+-dimensional and N^--dimensional row vectors with jth elements $\Phi^+_{q,\upsilon}(x,i,j)$ and $\Phi^-_{q,\ell}(x,i,j)$, respectively. Then Proposition 17.22 implies that, under the probability measure $\mathbb{P}_{x,i}$, $x \in \mathbb{R}$, $i \in E^0$, the processes $M^+ = \{M^+_t\}_{t\geq 0}$ and $M^- = \{M^-_t\}_{t\geq 0}$, defined by

$$M^+_t = e^{-q(t\wedge T^+_\upsilon)} \Phi^+_{q,\upsilon}(X_{t\wedge T^+_\upsilon}, Z_{t\wedge T^+_\upsilon}),$$
$$M^-_t = e^{-q(t\wedge T^-_\ell)} \Phi^-_{q,\ell}(X_{t\wedge T^-_\ell}, Z_{t\wedge T^-_\ell}),$$

are row-vectors of bounded martingales. Indeed, the Markov property of (X, Z) implies that

$$\mathbb{E}_{x,i}\big[e^{-qT^+_\upsilon} I_{\{Z_{T^+_\upsilon}=j\}} \big| \mathcal{F}^{(X,Z)}_t\big]$$
$$= I_{\{t<T^+_\upsilon\}} e^{-qt} \mathbb{E}_{X_t,Z_t}\big[e^{-qT^+_\upsilon} I_{\{Z_{T^+_\upsilon}=j\}}\big] + I_{\{t\geq T^+_\upsilon\}} e^{-qT^+_\upsilon} I_{\{Z_{T^+_\upsilon}=j\}}$$

$$= I_{\{t<T_\upsilon^+\}} e^{-qt} \mathbb{E}_{X_t, Z_t}\big[e^{-qT_\upsilon^+} I_{\{Z_{T_\upsilon^+}=j\}}\big]$$
$$+ I_{\{t\ge T_\upsilon^+\}} e^{-qT_\upsilon^+} \mathbb{E}_{X_{T_\upsilon^+}, Z_{T_\upsilon^+}}\big[e^{-qT_\upsilon^+} I_{\{Z_{T_\upsilon^+}=j\}}\big]$$
$$= e^{-q(t\wedge T_\upsilon^+)} \mathbb{E}_{X_{t\wedge T_\upsilon^+}, Z_{t\wedge T_\upsilon^+}}\big[e^{-qT_\upsilon^+} I_{\{Z_{T_\upsilon^+}=j\}}\big]$$
$$= e^{-q(t\wedge T_\upsilon^+)} \Phi_{q,\upsilon}^+(X_{t\wedge T_\upsilon^+}, Z_{t\wedge T_\upsilon^+}) e_j^+ = M_t^+ e_j^+,$$

where e_j^+ denotes the jth standard basis vector in $\mathbb{R}^{N^+}$, and we used that $\mathbb{E}_{x,i}[e^{-qT_\upsilon^+} I_{\{Z_{T_\upsilon^+}=j\}}] = e_j^+(i)$ if $x \ge \upsilon$ and $i \in E^0$, which directly follows from the definition (17.69).

(iii) Suppose that $Q_0 = O$. This corresponds to a model in which there are no jumps and no switches between the regimes (i.e. with probability one the process stays in the starting regime and evolves as a Brownian motion with drift). In this case we can identify the matrix Wiener–Hopf factorisation in closed form. Note that we have $N^+ = N^- = N$, and hence the matrices $G^\pm$ and $W^\pm$ are of dimension $N \times N$. If we take $W^\pm$ to be equal to the identity matrix and $h(i) = R_D(i) + q$, the matrix equations (17.66)–(17.67) reduce to

$$\frac{1}{2}\Sigma^2 (G^+)^2 - VG^+ = (\Lambda_D + qI) = \frac{1}{2}\Sigma^2 (G^-)^2 + VG^-,$$

where I is the $N \times N$ identity matrix (recall that the discount rate matrix Λ_D is diagonal and satisfies $\Lambda_D(i,i) = R_D(i)$ for all $i \in E^0$). These equations are satisfied by the diagonal matrices

$$G^+ = \operatorname{diag}(-\omega_i^+, i = 1, \dots, N), \qquad G^- = \operatorname{diag}(-\omega_i^-, i = 1, \dots, N), \tag{17.71}$$

where

$$\omega_i^\pm = \mp \frac{\mu_i}{\sigma_i^2} + \sqrt{\left(\frac{\mu_i}{\sigma_i^2}\right)^2 + \frac{2(q + r_i)}{\sigma_i^2}} \quad \text{and} \quad r_i := R_D(i). \tag{17.72}$$

In particular, we obtain the well-known fact that the maximum of a Brownian motion with drift at an independent exponential time is exponentially distributed:

$$\mathbb{P}_{x,i}(\overline{X}_{\mathbf{e}_{q+r_i}} > a) = e^{-\omega_i^+(a-x)} \quad \text{for } x < a.$$

17.5.4 Joint Distribution of the Maximum and Minimum

In the previous section we have shown how the marginal distribution of the maximum and of the minimum can be explicitly expressed in terms of the matrix Wiener–

Hopf factorisation. Also the *joint* distribution of the running maximum and minimum,

$$\psi_{x,i}(t) = \mathbb{P}_x(\underline{X}_t > \ell, \overline{X}_t < \upsilon),$$

can be explicitly identified in terms of the matrix Wiener–Hopf factorisation, by considering appropriate linear combinations of the functions Φ^+ and Φ^- defined in (17.69).

To formulate the result, introduce the matrices $Z^+ \in \mathbb{C}^{N^- \times N^+}$ and $Z^- \in \mathbb{C}^{N^+ \times N^-}$ by

$$Z^+(i,j) = \left[\eta_h^+ e^{Q_h^+(\upsilon-\ell)}\right](i,j), \quad i \in E^0 \cup E^-,\ j \in E^0 \cup E^+,$$
$$Z^-(i,j) = \left[\eta_h^- e^{Q_h^-(\upsilon-\ell)}\right](i,j), \quad i \in E^0 \cup E^+,\ j \in E^0 \cup E^-,$$

and define, for any $h \in \mathcal{H}$ and any $x \in \mathbb{R}$, the $N \times N^+$ and $N \times N^-$ matrices $\Psi^+_{h,\ell,\upsilon}(x)$ and $\Psi^-_{h,\ell,\upsilon}(x)$:

$$\Psi^+_{h,\ell,\upsilon}(x) = \left(\eta_h^+ e^{Q_h^+(\upsilon-x)} - \eta_h^- e^{Q_h^-(x-\ell)} Z^+\right)\left(I - Z^- Z^+\right)^{-1} I_{[\ell,\upsilon]}(x)$$
$$+ \Delta^+ I_{(\upsilon,\infty)}(x), \tag{17.73}$$
$$\Psi^-_{h,\ell,\upsilon}(x) = \left(\eta_h^- e^{Q_h^-(x-\ell)} - \eta_h^+ e^{Q_h^+(\upsilon-x)} Z^-\right)\left(I - Z^+ Z^-\right)^{-1} I_{[\ell,\upsilon]}(x)$$
$$+ \Delta^- I_{(-\infty,\ell)}(x), \tag{17.74}$$

where Δ^+ and Δ^- are the $N \times N^+$ and $N \times N^-$ matrices with elements

$$\Delta^+(i,j) = \delta_{ij}, \quad \Delta^-(i,k) = \delta_{ik}, \quad i \in E, j \in E^+ \cup E^0, k \in E^0 \cup E^-.$$

For h given by $h(i) = q I_{E^0}(i)$ with $q > 0$, we will also write $\Psi^{\pm}_{h,\ell,\upsilon}(x) = \Psi^{\pm}_{q,\ell,\upsilon}(x)$.

Proposition 17.23 *Let $i \in E^0$ and*

$$\psi^+_{x,i}(t) = \mathbb{P}_{x,i}(T_{\ell,\upsilon} < t, X_{T_{\ell,\upsilon}} \geq \upsilon), \qquad \psi^-_{x,i}(t) = \mathbb{P}_{x,i}(T_{\ell,\upsilon} < t, X_{T_{\ell,\upsilon}} \leq \ell),$$

where

$$T_{\ell,\upsilon} := \inf\{t \geq 0 : X_t \notin [\ell,\upsilon]\} = T_\ell^- \wedge T_\upsilon^+.$$

For any $q > 0$, the Laplace transforms in t of $\psi^+_{x,i}$ and $\psi^-_{x,i}$ are given by

$$\widehat{\psi}^+_{x,i}(q) = \frac{1}{q}\left(\Psi^+_{q,x}\mathbf{1}\right)(i), \qquad \widehat{\psi}^-_{x,i}(q) = \frac{1}{q}\left(\Psi^-_{q,x}\mathbf{1}\right)(i).$$

The Laplace transform of $\psi_{x,i}(t) = \mathbb{P}_{x,i}(\underline{X}_t > \ell, \overline{X}_t < \upsilon)$ is hence of the form

$$\widehat{\psi}_{x,i}(q) = \widehat{\psi}^+_{x,i}(q) + \widehat{\psi}^-_{x,i}(q). \tag{17.75}$$

Remarks

(i) If $Q_0 = O$, then $\eta_q^{\pm}$ are identity matrices, and the following identities hold:

$$\Psi_{q,x}^{+} = \big(e^{Q_q^{+}(\upsilon - x)} - e^{Q_q^{-}(x-\ell)} e^{Q_q^{+}(\upsilon-\ell)}\big)\big(I - e^{Q_q^{-}(\upsilon-\ell)} e^{Q_q^{+}(\upsilon-\ell)}\big)^{-1},$$
$$\Psi_{q,x}^{-} = \big(e^{Q_q^{+}(\upsilon - x)} - e^{Q_q^{+}(x-\ell)} e^{Q_q^{-}(\upsilon-\ell)}\big)\big(I - e^{Q_q^{+}(\upsilon-\ell)} e^{Q_q^{-}(\upsilon-\ell)}\big)^{-1},$$

where $Q_q^{\pm}$ are diagonal matrices given in (17.71). In particular, we find the well-known two-sided exit identity for Brownian motion with drift:

$$\widehat{\psi}_{x,i}(q) = \mathbb{E}_{x,i}\big[e^{-q\tau_{\ell,\upsilon}} I_{\{\tau_\upsilon < \tau_\ell\}}\big] = \frac{e^{\omega_i^{+}(x-\ell)} - e^{\omega_i^{-}(x-\ell)}}{e^{\omega_i^{+}(\upsilon-\ell)} - e^{\omega_i^{-}(\upsilon-\ell)}},$$

where $\omega_i^{\pm}$ is given in (17.72) with $r_i = 0$.

(ii) In the case that no volatility is present ($\Sigma \equiv 0$) the identities simplify, and we find the expressions in [5].

(iii) Under $\mathbb{P}_{x,i}$, $x \in \mathbb{R}$, $i \in E^0$, the process $\widetilde{M}^{+} = \{\widetilde{M}_t^{+}\}_{t\geq 0}$ defined by

$$\widetilde{M}_t^{+} = e^{-q(t\wedge T_{\ell,\upsilon})} \Psi_{q,\ell,\upsilon}^{+}(X_{t\wedge T_{\ell,\upsilon}}, Z_{t\wedge T_{\ell,\upsilon}})$$

is a row-vector of bounded martingales, where we denoted by $\Psi_{q,\ell,\upsilon}^{+}(x,i)$ the N^{+}-dimensional row vector with jth element $\Psi_{q,\ell,\upsilon}^{+}(x,i,j)$.

Proof Let $q > 0$ and define

$$g^{+}(x,i) = \Psi_{q,\ell,\upsilon}^{+}(x,i).$$

In view of the fact that $\widetilde{M}^{+}$ is a bounded martingale, it holds that

$$\begin{aligned} g^{+}(x,i) &= \sum_{j\in E^0\cup E^{+}} \mathbb{E}_{x,i}\big[e^{-q\tau}\Psi_{q,\ell,\upsilon}^{+}(X_\tau, Z_\tau, j) I_{\{\tau<\infty\}}\big] \\ &= \sum_{j\in E^0\cup E^{+}} \mathbb{E}_{x,i}\big[e^{-q\tau} I_{\{X_\tau \geq \upsilon, Z_\tau = j\}}\big] \\ &= q\int_0^\infty e^{-qt}\mathbb{P}_{x,i}(\tau < t, X_\tau \geq \upsilon)\, dt = q\widehat{\psi}_{x,i}^{+}(q). \end{aligned}$$

Here we used that, in view of the definitions (17.73) and (17.68) of $\Psi_{q,\ell,\upsilon}^{+}$ and η_q^{+}, the function g^{+} satisfies:

$$\begin{aligned} g^{+}(x,i) &= 1 \quad \text{if } x \geq \upsilon,\ i \in E^0, \\ g^{+}(x,i) &= 0 \quad \text{if } x \leq \ell,\ i \in E^0. \end{aligned}$$

The expression for $\widehat{\psi}_{x,i}^{-}$ can be derived by a similar reasoning. □

17.5.5 Valuing a Double-Barrier Rebate Option

A double-barrier rebate option pays a constant rebate L at the moment τ one of the barrier levels is crossed, if this happens before maturity T. By standard arbitrage pricing arguments the Laplace transform $\widehat{v}_{\text{reb}}$ in maturity of the price of v_{reb} of such an option is given by

$$\widehat{v}_{\text{reb}}(q) = \frac{1}{q}\mathbb{E}_{x,i}\big[\big(B_\tau^D\big)^{-1}\exp(-q\tau)\big]. \tag{17.76}$$

We will find below the following more general quantity, which will also be employed in the sequel:

$$H^{x,i}(q,u,j) := \mathbb{E}_{x,i}\big[\big(B_\tau^D\big)^{-1}\exp(\mathrm{i}uX_\tau - q\tau)I_{\{Z_\tau=j\}}\big] \quad \text{for } j=1,\dots,N. \tag{17.77}$$

In the following result an explicit expression is given for the quantity $H_j^{x,i}$ in terms of the matrix Wiener–Hopf factorisation, which is an extension of identity (17.75).

We denote by E_i^+ and E_i^-, for $i \in E^0 = \{1,\dots,N\}$, the parts of the state space E corresponding to the blocks B_i^+ and B_i^- in the matrix Q_0 in (17.61). Recall that the definition of $\alpha_i^\pm$ was given in (17.21).

Theorem 17.24 *For any $h \in \mathcal{H}$, $i,j \in E^0$ and $u \in \mathbb{C}$ that satisfies $\Im(u) \in (-\alpha_j^+, \alpha_j^-)$, it holds that*

$$H_j^{x,i}(h,u) = \big(\Psi_{h,x}^+ k_{j,u}^+\big)(i) + \big(\Psi_{h,x}^- k_{j,u}^-\big)(i), \tag{17.78}$$

where the column vectors $k_{j,u}^+ = (k_{j,u}^+(i), i \in E^+ \cup E^0)$ and $k_{j,u}^- = (k_{j,u}^-(i), i \in E^0 \cup E^-)$ are given by

$$k_{j,u}^+(i) = e^{uv}\cdot\begin{cases}1 & \text{if } i=j\in E^0,\\ ((-uI_j^+ - B_j^+)^{-1}(-B_j^+)\mathbf{1})(i) & \text{if } i \in E_j^+,\\ 0 & \text{otherwise},\end{cases}$$

$$k_{j,u}^-(i) = e^{u\ell}\cdot\begin{cases}1 & \text{if } i=j\in E^0,\\ ((uI_j^- - B_j^-)^{-1}(-B_j^-)\mathbf{1})(i) & \text{if } i \in E_j^-,\\ 0 & \text{otherwise},\end{cases}$$

where I_j^+ and I_j^- are $|E_j^+|\times|E_j^+|$ and $|E_j^-|\times|E_j^-|$ identity matrices.

The proof is given in Sect. 17.8.

17.6 Double-No-Touch and Other Barrier Options

In this section we show how the prices of double-no-touch and double knock-out call options can be expressed in terms characteristic function H of the process at the corresponding first passage time, which was identified in Theorem 17.24.

A *double-no-touch* is a derivative security that pays one unit of the underlying asset at expiry T if the underlying asset price does not leave the interval $[\exp(\ell), \exp(\upsilon)]$ during the time period $[0, T]$, where $\ell < \upsilon$. Similarly a *double knock-out call option* struck at K delivers the payoff $(S_T - K)^+ := \max\{S_T - K, 0\}$ if throughout the life of the option the asset price stays within the interval $[\exp(\ell), \exp(\upsilon)]$. The arbitrage-free prices for a double-no-touch and a double knock-out call options are respectively given by

$$D_{x,i}(T) = \mathbb{E}_{x,i}\left[\frac{I_{\{\tau>T\}}}{B_T^D}\right], \tag{17.79}$$

$$C_{x,i}(k,T) = \mathbb{E}_{x,i}\left[\frac{I_{\{\tau>T\}}}{B_T^D}(S_T - K)^+\right], \tag{17.80}$$

where $k = \log K$ is the log-strike, and τ is the first time the process S leaves the interval $[\exp(\ell), \exp(\upsilon)]$ or equivalently

$$\tau := \inf\{t \geq 0 : X_t \notin [\ell, \upsilon]\}. \tag{17.81}$$

We will find it more convenient to consider the double-touch-in and the knock-in call options whose values υ_{dti} and υ_{kic}, as functions of maturity T and log strike $k = \log K$, are given by

$$\upsilon_{\mathrm{dti}}(T) := \mathbb{E}_{x,i}\left[\frac{I_{\{\tau\leq T\}}}{B_T^D}\right] = \mathbb{E}_i\left[\left(B_T^D\right)^{-1}\right] - D_{x,i}(T), \tag{17.82}$$

$$\begin{aligned} \upsilon_{\mathrm{kic}}(T,k) &:= \mathbb{E}_{x,i}\left[\frac{I_{\{\tau\leq T\}}}{B_T^D}(S_T - K)^+\right] \\ &= \mathbb{E}_{x,i}\left[\left(B_T^D\right)^{-1}\left(S_T - e^k\right)^+\right] - C_{x,i}(k,T). \end{aligned} \tag{17.83}$$

Since the zero-coupon bond price and the call option price have already been identified, the problem of calculating the double-no-touch and knock-out call prices thus reduces to identifying υ_{dti} and υ_{kic}. In general, no closed-form expressions are known for these functions in terms of elementary functions. Below we show that the Laplace transform in T of $\upsilon_{\mathrm{dti}}(T)$ as well as the joint Fourier–Laplace transform in (k, T) of υ_{kic} can be identified explicitly in terms of the parameters that define the log-price process X. Both transforms will be identified in terms of the Laplace transform $\widehat{F}_{x,i}(u, q)$ in T of the function $T \mapsto F_{x,i}(u, T)$ given by

$$F_{x,i}(u,T) := \mathbb{E}_{x,i}\left[\frac{I_{\{\tau\leq T\}}}{B_T^D}\exp(\mathrm{i}uX_T)\right]. \tag{17.84}$$

The Laplace transform $\widehat{F}_{x,i}(u,q)$ in turn will be given in terms of the N_0-vector $H^{x,i}(q,u)$ whose coordinates are given by

$$H_j^{x,i}(q,u) := \mathbb{E}_{x,i}\left[\exp\left(\mathrm{i}uX_\tau - \int_0^\tau \big(R_D(Z_s)+q\big)\,ds\right) I_{\{Z_\tau=j\}}\right]$$

for $j=1,\dots,N_0$.

Theorem 17.25 *For any $q>0$ and ξ with $\Im(\xi)<0$, it holds that*

$$\widehat{v}_{\mathrm{dti}}(q) = \widehat{F}_{x,i}(0,q), \tag{17.85}$$

$$\widehat{v}^*_{\mathrm{kic}}(q,\xi) = \frac{1}{\mathrm{i}\xi - \xi^2}\widehat{F}_{x,i}(\xi - \mathrm{i}, q). \tag{17.86}$$

Here, $\widehat{F}_{x,i}(u,q)$ is given by

$$\widehat{F}_{x,i}(u,q) = \big(H^{x,i}(q,u)\big)'\big(qI + \Lambda_D - K(u)\big)^{-1}\mathbf{1} \tag{17.87}$$

for all $q\in\mathbb{C}$ such that $\Re(q) > q^* = \max\{\Re(\psi_i(u)) - R_D(i) : i=1,\dots,N_0\}$.

Remarks

(i) Note that if $u=0$, then $q^*\le 0$, as the interest rates $R_D(i)$ are assumed to be nonnegative.
(ii) The Laplace transform in (17.87) can be inverted by evaluating the Bromwich integral

$$F_{x,i}(u,t) = \frac{1}{2\pi}\int_{c-\mathrm{i}\infty}^{c+\mathrm{i}\infty} e^{tq}\big(H^{x,i}(q,u)\big)'\big(qI+\Lambda_D - K(u)\big)^{-1}\mathbf{1}\,dq \tag{17.88}$$

for any $c>q^*$. An efficient algorithm to approximate this integral can e.g. be found in Abate and Whitt [1].

Proof First note that by a calculation similar to the one in the proof of Proposition 17.15 we find that the Fourier transform in log-strike k of $e^{\alpha k}v_{\mathrm{kic}}(T,k)$ can be expressed in terms of F by

$$\begin{aligned}&\int_{\mathbb{R}} e^{(\mathrm{i}v+\alpha)k} v_{\mathrm{kic}}(T,k)\,dk\\ &\quad= \frac{1}{(\alpha+\mathrm{i}v)(1+\alpha+\mathrm{i}v)}\mathbb{E}_{x,i}\left[\frac{I_{\{\tau\le T\}}}{B_T^D}\exp\big\{(1+\alpha+\mathrm{i}v)X_T\big\}\right]\end{aligned}$$

for any $\alpha>0$.

Assume next that $q>0$ is real with $q>q^*$. The Laplace transform $\widehat{F}_{x,i}(u,q)$ has the following well-known equivalent probabilistic representation

$$\widehat{F}_{x,i}(u,q) = \mathbb{E}\big[F_{x,i}(u,e_q)\big]/q, \tag{17.89}$$

where e_q is an exponential random variable with parameter q which is independent of the process (X, Z).

By conditioning on $\mathcal{F}_\tau$ and applying strong Markov property at τ of the process (X, Z) together with Lemma 17.6, we find that

$$\begin{aligned}\mathbb{E}\big[F_{x,i}(u, e_q)\big] &= \mathbb{E}_{x,i}\left[\frac{\exp(\mathrm{i}uX_\tau)}{B_\tau^D}\mathbb{E}\left[I_{\{\tau\le e_q\}}\exp\left(-\int_\tau^{e_q} R_D(Z_s)\,ds\right.\right.\right.\\ &\qquad \left.\left.\left.+\,\mathrm{i}u(X_{e_q}-X_\tau)\right)\middle|\mathcal{F}_\tau\right]\right]\\ &= \mathbb{E}_{x,i}\left[\frac{\exp(\mathrm{i}uX_\tau)}{B_\tau^D}h(Z_\tau, q, u)e^{-q\tau}\right]\\ &= \sum_j H_j^{x,i}(q,u)h(j,q,u), \end{aligned} \tag{17.90}$$

where the value $h(j, q, u)$ for each state $j \in E^0$ of the Markov chain Z is given by

$$\begin{aligned} h(j,q,u) &:= \mathbb{E}_{0,j}\left[\exp\left(-\int_0^{e'_q} R_D(Z_s)\,ds + \mathrm{i}uX_{e'_q}\right)\right]\\ &= \left[\left(qI + \Lambda_D - K(u)\right)^{-1}\mathbf{1}\right](j) \end{aligned} \tag{17.91}$$

for some exponential random variable e'_q with parameter q that is independent of the Markov process (X, Z), where the second equality follows from Lemma 17.3 and Theorem 17.7.

A key observation that follows from the representation (17.90) is that the function

$$q \mapsto \mathbb{E}\big[F_{x,i}(u, e_q)\big]/q$$

has a holomorphic extension to the complex half-plane $\{q \in \mathbb{C} : \Re(q) > q^*\}$ which therefore[14] coincides on this domain with the Laplace transform. Thus, (17.90) holds for q in this domain, and the proof is complete. □

17.7 Embedding of the Process (X, Z)

In this section and the next we will provide a proof of the matrix Wiener–Hopf factorisation and its corollaries derived in previous sections. We will proceed in two steps:

[14] If two holomorphic functions defined on a connected open set Ω in $\mathbb{C}$ coincide on a subset with at least one accumulation point in Ω, then they coincide on the entire Ω. For a proof of this well-known statement see [29], p. 208, Theorem 10.18.

(i) Reduction of the first-passage problems of X over constant levels, which will involve overshoots and undershoots due to the jumps of X, to the first-hitting problem of a constant level by a regime-switching Brownian motion, employing a classical argument for the embedding of phase-type jumps (see Asmussen [2]), which we review in this section.
(ii) Solution of the first-passage problem of regime-switching Brownian motion via a characterisation of the dynamics of the ladder processes, which is carried out in Sect. 17.8.

Let Y be a continuous-time Markov chain with finite state space $E \cup \partial$ and generator restricted to E given by Q_0, where ∂ is an absorbing cemetery state, and E and Q_0 will be specified shortly, and denote by $\xi = \{\xi_t, t \geq 0\}$ the Markov-modulated Brownian motion given by

$$\xi_t = x + \int_0^t s(Y_s)\,dW_s + \int_0^t m(Y_s)\,ds, \tag{17.92}$$

where $x \in \mathbb{R}$ is the starting point of ξ, and s and m are functions from $E \cup \partial$ to $\mathbb{R}$ also to be specified shortly. The couple (ξ, Y) defined as such is a two-dimensional strong Markov process. In the sequel we will denote by $\widetilde{\mathbb{P}}_{x,i}$ and $\widetilde{\mathbb{E}}_{x,i}$ the conditional probability $\widetilde{\mathbb{P}}_{x,i} = \widetilde{\mathbb{P}}[\cdot|\xi_0 = x, Y_0 = i]$ and conditional expectation $\widetilde{\mathbb{E}}_{x,i} = \widetilde{\mathbb{E}}[\cdot|\xi_0 = x, Y_0 = i]$, respectively.

Let the state space E be as in Sect. 17.5. In other words, $E = E^- \cup E^0 \cup E^+$, where E^0 is the state space of the chain Z, and E^+ and E^- are given in (17.65). The generator Q_0 is given in (17.61), and $m(i) := \Lambda_V(i,i)$, $s(i) := \Lambda_\Sigma(i,i)$ for all $i \in E$, where the matrices Λ_V, Λ_Σ are defined in (17.62). Thus, while the chain Y is in state $j \in E^0$, ξ evolves as a Brownian motion with drift $m(j)$ and volatility $s(j)$, and while Y takes values in E^+ and E^-, the path of ξ is linear with slope $+1$ or -1. Informally, a path of X can be obtained from a path of ξ by replacing these stretches of unit slope by jumps of the same length, as is illustrated in Fig. 1. These linear increasing and decreasing stretches of path of ξ thus correspond to the positive and negative jumps of X, respectively.

More formally, by appropriately time-changing (ξ, Y) a stochastic process can be constructed that has the same law as (X, Z). Denote by

$$T_0(t) = \int_0^t I_{\{Y_s \in E^0\}}\,ds \quad \text{and} \quad T_0^{-1}(u) = \inf\{t \geq 0 : T_0(t) > u\},$$

the time before t spent by the chain Y in E^0 and its right-continuous inverse respectively. It is clear from the definition of the generator Q_0 in (17.61) that when the chain Y jumps from any of the states in $E^+ \cup E^-$ to a state $j \in E^0$, it must have been in the state j just before it had left E^0. Therefore the form of the matrix (17.61) implies that the process $Y \circ T_0^{-1}$, which simply ignores all excursions of Y into $E^+ \cup E^-$, is a Markov chain with the same generator as Z. Furthermore it is straightforward to verify that $\xi \circ T_0^{-1}$ is a regime-switching jump-diffusion and the law of the process $(\xi \circ T_0^{-1}, Y \circ T_0^{-1})$ under $\widetilde{\mathbb{P}}_{x,i}$ is equal to that of (X, Z) under $\mathbb{P}_{x,i}$ for all $x \in \mathbb{R}$ and $i \in E^0$.

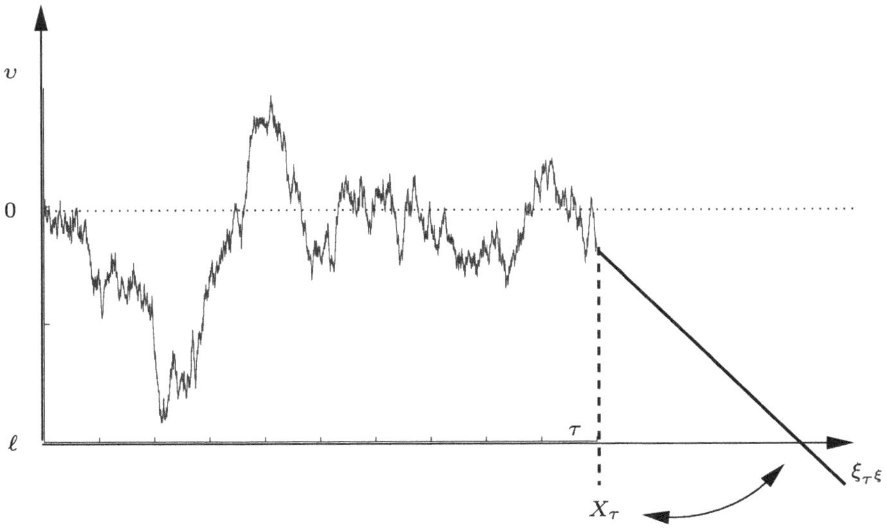

Fig. 1 Shown is a sample path of X until the first time τ that X exits the interval $[\ell, \upsilon]$. The process ξ has no positive jumps and always hits a level at first-passage

If we define the first-passage time

$$\tau^{\xi} = \inf\{t \geq 0 : \xi_t \notin [\ell, \upsilon] \text{ and } Y_t \in E^0\}, \tag{17.93}$$

it follows that random variables $T_0(\tau^{\xi})$ and the stopping time τ defined in (17.81) have the same distribution. In particular, with the extension of h to E that puts $h(i) = 0$ for $i \notin E^0$ and that we will also denote by h, it holds that

$$\left(\xi_{\tau^{\xi}}, \int_0^{\tau^{\xi}} h(Y_s)\, ds, Y_{\tau^{\xi}}\right) \text{ under } \widetilde{\mathbb{P}}_{x,i} \text{ has the same distribution as}$$
$$\left(X_{\tau}, \int_0^{\tau} h(Z_s)\, ds, Z_{\tau}\right) \text{ under } \mathbb{P}_{x,i}$$

for $x \in \mathbb{R}$ and $i \in E^0$. The function $H_j^{x,i}$ defined in (17.77) can thus be expressed in terms of the embedding (ξ, Y) as follows:

$$H_j^{x,i}(h, u) = \widetilde{\mathbb{E}}_{x,i}\left[\exp\left(u\xi_{\tau^{\xi}} - \int_0^{\tau^{\xi}} h(Y_s)\, ds\right) I_{\{Y_{\tau^{\xi}} = j\}}\right], \quad i, j \in E^0. \tag{17.94}$$

17.8 Ladder Processes

This section is devoted to the proof of existence and uniqueness of the h-matrix Wiener–Hopf factorisation for any $h \in \mathcal{H}$. For real $h \in \mathcal{H}$, the vector h has the probabilistic interpretation of a vector of state-dependent rates of discounting, and in this case the matrix Wiener–Hopf factorisation has a probabilistic interpretation in terms of ladder processes. The existence and uniqueness results are extended to the case of general complex $h \in \mathcal{H}$ by analytical continuation arguments. It is of interest to consider the case of complex entries, as the Laplace transforms of barrier option prices are expressed in terms of the matrix Wiener–Hopf factorisation as we saw above, and some widely used Laplace transform inversion algorithms are based on the Bromwich (complex) integral representation.

A classical probabilistic approach to characterisation of the joint distribution of τ_a^+ and the position of Y at τ_a^+ is to consider the up-crossing ladder process of (ξ, Y), defined as follows:

Definition 17.26 The *up-crossing ladder process* $Y^+ = \{Y_t^+\}_{t \geq 0}$ of (ξ, Y) is given by

$$Y_a^+ := \begin{cases} Y(\tau_a^+) & \text{if } \tau_a^+ < \infty, \\ \partial & \text{otherwise,} \end{cases} \tag{17.95}$$

where ∂ is a graveyard state, and

$$\tau_a^+ = \inf\{s \geq 0 : \xi_s > a\}$$

with $\inf \emptyset = \infty$.

Remark In the case that the original chain is killed at (state-dependent, real-valued) rate h, the up-crossing ladder process can be defined as follows. Recall from Markov chain theory that the chain Y^h with state space $E \cup \{\partial\}$ and generator matrix

$$\begin{pmatrix} Q_h & -\Lambda_h \mathbf{1} \\ \mathbf{0} & 0 \end{pmatrix}, \quad \text{where } Q_h := Q_0 - \Lambda_h, \text{ and } \Lambda_h \text{ is the diagonal matrix}$$

with $(\Lambda_h)_{ii} = h_i$,

has the same distribution as the chain Y killed (i.e. sent to the graveyard state ∂) independently at rate $h(i)$ when $Y_t = i$. In particular, for $i, j \in E$ and $t \geq 0$, it holds that

$$\widetilde{\mathbb{E}}_i\left[e^{-\int_0^t h(Y_s)\,ds} I_{\{Y_t = j\}}\right] = \widetilde{\mathbb{P}}\left(Y_t^h = j \middle| Y_0^h = i\right) = \exp(Q_h t)(i, j). \tag{17.96}$$

If we denote by $\xi^h = \{\xi_t^h\}_{t \geq 0}$ the process defined by (17.92) driven by the "killed" chain Y^h instead of Y, then the up-crossing ladder process Y^{h+} is given by

$$Y_t^{h+} = \begin{cases} Y_{\tau_t^{h+}}^h & \text{if } \tau_t^{h+} < \infty, \\ \partial & \text{otherwise,} \end{cases}$$

where τ_a^{h+} is the stopping time defined by

$$\tau_a^{h+} = \inf\left\{s \geq 0 : \xi_s^h > a\right\}$$

with $\inf \emptyset = \infty$.

Proposition 17.27 *The process Y^+ is a Markov chain with state space $E^0 \cup E^+ \cup \{\partial\}$.*

Remarks

(i) We denote by Q_0^+ the generator of Y^+ restricted to $E^0 \cup E^+$, and by η_0^+ the initial distribution given by

$$\eta_0^+(i,j) = \widetilde{\mathbb{P}}_{0,i}\big[Y_0^+ = j, \tau_0^+ < \infty\big] \quad \text{for } i \in E, j \in E^+ \cup E^0. \tag{17.97}$$

(ii) More generally, for any real-valued $h \in \mathcal{H}$, it holds that Y^{h+} is a Markov chain with generator restricted to $E^0 \cup E^+$, denoted by Q_h^+, and initial (sub)probability distribution by $\eta_h^+ = (\eta_h^+(i,j),\ i \in E,\ j \in E^+ \cup E^0)$ with

$$\eta_h^+(i,j) = \widetilde{\mathbb{E}}_{0,i}\big[e^{-\int_0^{\tau_0^+} h(Y_s)\,ds} I_{\{Y_0^+ = j, \tau_0^+ < \infty\}}\big] \quad \text{for } i \in E, j \in E^+ \cup E^0. \tag{17.98}$$

We have the following identities analogous to (17.96):

$$\begin{aligned}
&\widetilde{\mathbb{E}}\big[e^{-\int_0^{\tau_t^+} h(Y_s)\,ds} I_{\{Y_{\tau_t^+} = j, \tau_t^+ < \infty\}} \big| Y_0 = i\big] \\
&\quad = \widetilde{\mathbb{E}}\big[e^{-\int_{\tau_0^+}^{\tau_t^+} h(Y_s)\,ds} I_{\{Y_{\tau_t^+} = j, \tau_t^+ < \infty\}} \big| Y_{\tau_0^+} = i\big] \\
&\quad = \widetilde{\mathbb{P}}\big(Y^h_{\tau_t^{h+}} = j, \tau_t^{h+} < \infty \big| Y^h_{\tau_0^{h+}} = i\big) \\
&\quad = \widetilde{\mathbb{P}}\big(Y_t^{h+} = j \big| Y_0^{h+} = i\big) \\
&\quad = \exp\big(Q_h^+ t\big)(i,j), \quad i,j \in E^0 \cup E^+,
\end{aligned} \tag{17.99}$$

where we wrote $\widetilde{\mathbb{P}}[\cdot|A] := \widetilde{\mathbb{P}}[\cdot|A \cap \{\xi_0 = 0\}]$ and $\widetilde{\mathbb{E}}[\cdot|A] := \widetilde{\mathbb{E}}[\cdot|A \cap \{\xi_0 = 0\}]$ to simplify the notation. Note that the first equality holds since $\tau_0^+ = 0$ if the chain Y starts at $i \in E^0 \cup E^+$. In particular, from (17.99) we find that, for $i, j \in E^0 \cup E^+$,

$$Q_h^+(i,j) = \lim_{t \downarrow 0} \frac{1}{t} \widetilde{\mathbb{E}}_{0,i}\big[e^{-\int_0^{\tau_t^+} h(Y_s)\,ds} I_{\{Y_{\tau_t^+} = j, \tau_t^+ < \infty\}}\big], \quad i \neq j, \tag{17.100}$$

$$Q_h^+(i,i) = \lim_{t \downarrow 0} \frac{1}{t} \big\{\widetilde{\mathbb{E}}_{0,i}\big[e^{-\int_0^{\tau_t^+} h(Y_s)\,ds} I_{\{Y_{\tau_t^+} = i, \tau_t^+ < \infty\}}\big] - 1\big\}. \tag{17.101}$$

Proof Let f be any bounded real-valued Borel function, and let $b < a$. Denote by $\mathcal{F}_a = \sigma\{Y_u^+\}_{u \le a}$ and $\mathcal{G}_t = \sigma\{Y_s\}_{s \le t}$ the sigma algebras generated by Y_u^+ up to time a and by Y_s up to time t. Observing that $\mathcal{F}_b \subset \mathcal{G}_{\tau_b}$ and using the spatial homogeneity and continuity of ξ and the strong Markov property of (ξ, Y), we find that

$$\begin{aligned}\widetilde{\mathbb{E}}_{x,i}\big[f\big(Y_a^+\big)\big|\mathcal{F}_b\big] &= \widetilde{\mathbb{E}}_{x,i}\big[\widetilde{\mathbb{E}}_{x,i}\big[f(Y_{\tau_a^+})\big|\mathcal{G}_{\tau_b}\big]\big|\mathcal{F}_b\big] \\ &= \widetilde{\mathbb{E}}_{x,i}\big[\widetilde{\mathbb{E}}_{b,Y_{\tau_b^+}}\big[f(Y_{\tau_a^+})\big]\big|\mathcal{F}_b\big] \\ &= \widetilde{\mathbb{E}}_{x,i}\big[\widetilde{\mathbb{E}}_{Y_b^+}\big[f\big(Y_{a-b}^+\big)\big]\big|\mathcal{F}_b\big] = \widetilde{\mathbb{E}}_{Y_b^+}\big[f\big(Y_{a-b}^+\big)\big],\end{aligned}$$

where we wrote $\widetilde{\mathbb{E}}_j = \widetilde{\mathbb{E}}_{0,j}$. □

For any $h \in \mathcal{H}$, we define the square-matrix (resolvent) functions $u \mapsto R_h^+(u)$ of dimension N^+ on the complex half-plane $\mathbb{C}_{>0}$ by

$$R_h^+(u)(i,j) = \widetilde{\mathbb{E}}_{0,i}\left[\int_0^\infty e^{-uy-\int_0^{\tau_y^+} h(Y_s)\,ds} I_{\{Y_y^+=j,\tau_y^+<\infty\}}\,dy\right]$$

$$\text{for } i,j \in E^+ \cup E^0 \text{ and } u \in \mathbb{C}_{>0}. \tag{17.102}$$

The matrix Q_h^+ can then be defined for any $h \in \mathcal{H}$ as follows:

Lemma 17.28 *Let $h \in \mathcal{H}$ and $a \geq 0$ and define the matrix $Q_h^+ \in \mathbb{D}(N^+)$ by*

$$Q_h^+(i,j) = -\omega_i^+ I_{\{i=j\}} + \begin{cases} \frac{2}{\sigma_i^2}(\widetilde{Q}\eta_h^+ R_h^+(\omega_i^-))(i,j) & \text{if } i \in E^0, \\ (\widetilde{Q}\eta_h^+)(i,j) & \text{if } i \in E^+, \end{cases} \tag{17.103}$$

where $q_i = -Q_0(i,i)$, $h_i = h(i)$, $\widetilde{Q} = Q_0 + \operatorname{diag}\{q_i : i \in E\}$, and

$$\omega_i^+ = \begin{cases} F\big(\frac{\mu_i}{\sigma_i^2}, \frac{q_i+h_i}{\sigma_i^2}\big) & \text{if } i \in E^0, \\ q_i + h_i & \text{if } i \in E^\pm, \end{cases} \tag{17.104}$$

where

$$F(\nu,\theta) := -\nu + \sqrt{\nu^2 + 2\theta}, \quad \nu \in \mathbb{R},\ \theta \in \mathbb{C}_{>0}. \tag{17.105}$$

Then it holds that

$$\widetilde{\mathbb{E}}_{0,i}\big[e^{-\int_0^{\tau_a^+} h(Y_s)\,ds} I_{\{Y_a^+=j,\tau_a^+<\infty\}}\big] = \exp\big(Q_h^+ a\big)(i,j), \quad i,j \in E^0 \cup E^+. \tag{17.106}$$

Remarks

(i) Note that Proposition 17.22 follows as a direct consequence of (17.106) since

$$\mathbb{P}_{x,i}(\overline{X}_{\mathbf{e}_q} > a) = \mathbb{P}_{x,i}\big(Y_\zeta^{q+} > a\big),$$

where ζ is the life time of the killed up-crossing ladder process.

(ii) Equation (17.106) yields in particular the joint Laplace transform of the vector

$$\mathbf{Z}_a^+ = \left(\int_0^{\tau_a^+} I_{\{Y_s=i\}}\, ds,\ i \in E \right), \tag{17.107}$$

whose components record the length of time spent by Y in each of the states in E, until the moment τ_a^+ of first-passage.

(iii) The *down-crossing ladder process* Y^-, defined as the up-crossing ladder process of $(-\xi, Y)$, is a Markov chain with generator restricted to $E_0 \cup E^-$ denoted by Q_0^-. Analogously to (17.103), for any $h \in \mathcal{H}$, a matrix Q_h^- can be defined, which satisfies

$$\widetilde{\mathbb{E}}_{0,i}\left[e^{-\int_0^{\tau_a^-} h(Y_s)\,ds} I_{\{Y_a^-=j,\tau_a^-<\infty\}}\right] = \exp\left(Q_h^- a\right)(i, j), \quad i, j \in E^0 \cup E^-. \tag{17.108}$$

17.8.1 Proof of Lemma 17.28

We will show that the limits in (17.100)–(17.101) in fact exist for any $h \in \mathcal{H}$, and identify these. Let

$$\rho = \inf\{t \geq 0 : Y_t \neq Y_0\}$$

be the first time that the chain Y jumps, and $\tau_a^i = \inf\{t \geq 0 : X_t^i > a\}$ the first-passage time of $X_t^i := \mu_i t + \sigma_i W_t$ (recall that if $i \in E^\pm$, then $\mu_i = \pm 1$ and $\sigma_i = 0$) over the level a, and let $\mathbf{e}_i$ be an exponential random time with mean $1/q_i$ (with $q_i = -Q_0(i, i)$) that is independent of X^i. In view of the definition of ξ (i.e. the first jump time ρ of Y is independent of X^i), it follows that

$$\begin{aligned}
&\widetilde{\mathbb{E}}_{0,i}\left[e^{-\int_0^{\tau_t^+} h(Y_s)\,ds} I_{\{Y_t^+=j,\tau_t^+<\infty,\tau_t^+<\rho\}}\right] \\
&= I_{\{i=j\}}\widetilde{\mathbb{E}}\left[e^{-\tau_t^i h(i)} I_{\{\tau_t^i<\mathbf{e}_i\}}\right] = I_{\{i=j\}}\widetilde{\mathbb{E}}\left[e^{-\tau_t^i (h_i+q_i)}\right] \\
&= I_{\{i=j,i\in E^+\}} \exp\left(-t(h_i + q_i)\right) \\
&\quad + I_{\{i=j,i\in E^0\}} \exp\left(-t\sigma_i^{-2}\left(-\mu_i + \sqrt{\mu_i^2 + 2(q_i + h_i)\sigma_i^2}\right)\right) \\
&= I_{\{i=j\}}\left[1 - \omega_i^+ t + o(t)\right] \quad \text{as } t \downarrow 0,
\end{aligned}$$

where ω_i^+ was defined in Lemma 17.28, and we used the fact that the Laplace transform of τ_a^i is given by

$$\widetilde{\mathbb{E}}\left[e^{-q\tau_a^i}\right] = \exp\left(-a\sigma_i^{-2}\left(-\mu_i + \sqrt{\mu_i^2 + 2q\sigma_i^2}\right)\right), \quad a > 0,\ q \in \mathbb{C}_{>0}.$$

For every $h \in \mathcal{H}$, we can define the following matrices:

$$\widetilde{P}_t^{h+}(i,j) := \widetilde{\mathbb{E}}_{0,i}\big[e^{-\int_0^{\tau_t^+} h(Y_s)\,ds} I_{\{Y_t^+=j,\tau_t^+<\infty\}}\big] \quad \text{for } i,j \in E^+ \cup E^0. \tag{17.109}$$

Note that the definition in (17.102) implies that the identity $R_h^+(u)(i,j) = \int_0^\infty e^{-uy} \widetilde{P}_y^{h+}(i,j)dy$ holds for all $u \in \mathbb{C}_{>0}$. For every $h \in \mathcal{H}$, $x \in (-\infty, t]$ and $m \in E$, we have the following identity:

$$\begin{aligned}
&\widetilde{\mathbb{E}}_{x,m}\big[e^{-\int_0^{\tau_t^+} h(Y_s)\,ds} I_{\{Y_t^+=j,\tau_t^+<\infty\}}\big] \\
&= \widetilde{\mathbb{E}}_{0,m}\big[e^{-\int_0^{\tau_{t-x}^+} h(Y_s)\,ds} I_{\{Y_{t-x}^+=j,\tau_{t-x}^+<\infty\}}\big] \\
&= \big(\eta_h^+ \widetilde{P}_{t-x}^{h+}\big)(m,j) \quad \text{for } j \in E^+ \cup E^0,
\end{aligned} \tag{17.110}$$

where η_h^+ is defined in (17.98). The first equality in (17.110) is a consequence of the spacial homogeneity of the process ξ.

Let $\overline{\xi}_u = \sup_{s\le u} \xi_s$ denote the running supremum of the stochastic process ξ. The strong Markov property applied at the first jump time ρ of the chain Y and (17.110) imply that

$$\begin{aligned}
&\widetilde{\mathbb{E}}_{0,i}\big[e^{-\int_0^{\tau_t^+} h(Y_s)\,ds} I_{\{Y_t^+=j,\tau_t^+<\infty,\tau_t^+\ge\rho\}}\big] \\
&= \widetilde{\mathbb{E}}_{0,i}\big[e^{-h(i)\rho} I_{\{\overline{\xi}_\rho\le t\}} \big(\eta_h^+ \widetilde{P}_{t-\xi_\rho}^{h+}\big)(Y_\rho,j)\big] \\
&= \sum_{m\in E\setminus\{i\}} \frac{Q_0(i,m)}{q_i} \widetilde{\mathbb{E}}_{0,i}\big[e^{-h(i)\rho} I_{\{\overline{\xi}_\rho\le t\}} \big(\eta_h^+ \widetilde{P}_{t-\xi_\rho}^{h+}\big)(m,j)\big].
\end{aligned} \tag{17.111}$$

The last equality is a consequence of the fact that Y_ρ is independent of the random vector $(\rho, \xi_\rho, \overline{\xi}_\rho)$ and takes values in the set $E\setminus\{i\}$ with $\widetilde{\mathbb{P}}_{0,i}(Y_\rho = m) = Q_0(i,m)/q_i$. Since ρ is the first jump time of the chain Y, the vector $(\rho, \xi_\rho, \overline{\xi}_\rho)$ has the same distribution as the vector $(\mathbf{e}_i, X^i_{\mathbf{e}_i}, \overline{X}^i_{\mathbf{e}_i})$, where, as above, $\mathbf{e}_i$ is an exponential random variable with mean $1/q_i$, independent of the Brownian motion with drift $X^i_t = \mu_i t + \sigma_i W_t$. The symbol $\overline{X}^i_{\mathbf{e}_i}$ denotes the maximum of X^i at the independent exponential time $\mathbf{e}_i$. Note that if $i \in E^+$, then $\overline{X}^i_t = X^i_t = t$ for all $t \in \mathbb{R}_+$ and the expectation in (17.111) is very easy to compute.

Assume now that $i \in E^0$. Then the Wiener–Hopf factorisation implies that the random variables $\overline{X}^i_{\mathbf{e}_u}$ and $\overline{X}^i_{\mathbf{e}_u} - X^i_{\mathbf{e}_u}$ are exponentially distributed with parameters

$$\sigma_i^{-2}\big(-\mu_i + \sqrt{\mu_i^2 + 2u\sigma_i^2}\big) \quad \text{and} \quad \sigma_i^{-2}\big(\mu_i + \sqrt{\mu_i^2 + 2u\sigma_i^2}\big)$$

respectively and independent for any exponential random variable $\mathbf{e}_u$ with parameter $u > 0$, which is independent of X^i. Therefore the joint density $f_{\overline{X}^i_t, \overline{X}^i_t - X^i_t}$ satisfies the following identity for any $x, y \in (0, \infty)$:

$$\int_0^\infty ue^{-ut} f_{\overline{X}^i_t, \overline{X}^i_t - X^i_t}(x, y)\, dt = \int_0^\infty ue^{-ut} f_{\overline{X}_t}(x)\, dt \int_0^\infty ue^{-ut} f_{\overline{X}^i_t - X^i_t}(y)\, dt$$

$$\text{for all } u > 0, \tag{17.112}$$

where $f_{\overline{X}^i_t - X^i_t}$ and $f_{\overline{X}^i_t}$ are the densities of the corresponding random variables. It is clear that there exists a unique extension to the complex half-plane $\mathbb{C}_{>0}$ of both sides of the formula in (17.112) and that the following formulae hold:

$$\int_0^\infty e^{-ut} f_{\overline{X}_t}(x)\, dt = \frac{\sigma_i^{-2}(-\mu_i + \sqrt{\mu_i^2 + 2u\sigma_i^2})}{u} e^{-x\sigma_i^{-2}(-\mu_i + \sqrt{\mu_i^2 + 2u\sigma_i^2})},$$

$$x > 0, u \in \mathbb{C}_{>0}, \tag{17.113}$$

$$\int_0^\infty e^{-ut} f_{\overline{X}^i_t - X^i_t}(y)\, dt = \frac{\sigma_i^{-2}(\mu_i + \sqrt{\mu_i^2 + 2u\sigma_i^2})}{u} e^{-y\sigma_i^{-2}(\mu_i + \sqrt{\mu_i^2 + 2u\sigma_i^2})},$$

$$y > 0, u \in \mathbb{C}_{>0}. \tag{17.114}$$

By substituting identity (17.112) into the expectation (17.111), applying the formulae in (17.113) and (17.114) and taking the limit as t tends to zero, we see that the limits in (17.100) and (17.101) are as stated for any $h \in \mathcal{H}$. Furthermore the formula in (17.103) holds.

The strong Markov property of (ξ, Y) next implies that for any $h \in \mathcal{H}$ and any $t > 0$, the matrices $\widetilde{P}^{h+}_t = (\widetilde{P}^{h+}_t(i, j)_t, i, j \in E^0 \cup E^+)$ with $\widetilde{P}^{h+}_t(i, j)$ defined by (17.109) satisfy the system of ordinary differential equations

$$\frac{d}{dt} \widetilde{P}^{h+}_t = \widetilde{P}^{h+}_t Q^+_h, \quad t > 0, \qquad P^{h+}_0 = \mathbb{I},$$

where $\mathbb{I}$ denotes the $N^+ \times N^+$ identity matrix, the unique solution of which is given by

$$\widetilde{P}^{h+}_t = \exp(Q^+_h t).$$

This proves that the matrix Q^+_h identified above satisfies (17.106). Since $\widetilde{P}^{h+}_t(i, j) \to 0$ for all $i, j \in E^0 \cup E^+$ as $t \to \infty$, it follows that all eigenvalues of Q^+_h must have nonpositive real parts and therefore $Q^+_h \in \mathbb{D}(N^+)$. The proof of the existence of a matrix $Q^-_h \in \mathbb{D}(N^-)$ satisfying (17.108) is similar and omitted.

17.8.2 Proof of Theorem 17.21

(Existence) For any $h \in \mathcal{H}$ and $x, \ell \in \mathbb{R}$, define the matrices $\Phi^\pm_\ell(x)$ by

$$\Phi^+_\ell(x) = \eta^+_h \exp(Q^+_h(\ell - x)), \qquad \Phi^-_\ell(x) = \eta^-_h \exp(Q^-_h(x - \ell)). \tag{17.115}$$

The proof of existence rests on the martingale property of $M^+ = \{M_t^+\}_{t\geq 0}$ and $M^- = \{M_t^-\}_{t\geq 0}$ given by

$$\begin{aligned} M_t^+ &= e^{-\int_0^{t\wedge\tau_\ell^+} h(Y_s)\,ds} f_+(Y_{t\wedge\tau_\ell^+}, \xi_{t\wedge\tau_\ell^+}) \quad \text{and} \\ M_t^- &= e^{-\int_0^{t\wedge\tau_\ell^-} h(Y_s)\,ds} f_-(Y_{t\wedge\tau_\ell^-}, \xi_{t\wedge\tau_\ell^-}) \end{aligned} \tag{17.116}$$

with

$$f_+(i,x) := e_i'\Phi_\ell^+(x)k_+, \qquad f_-(i,x) := e_i'\Phi_\ell^-(x)k_-, \tag{17.117}$$

where k_+ and k_- are N^+- and N^--column vectors, respectively.

The martingale property of M^+ follows from the equality

$$M_t^+ = \widetilde{\mathbb{E}}_{x,i}\big[e^{-\int_0^{\tau_\ell^+} h(Y_s)\,ds} k_+(Y_\ell^+) I_{\{\tau_\ell^+<\infty\}} \big| \mathcal{G}_t\big],$$

where $\{\mathcal{G}_t\}_{t\geq 0}$ denotes the filtration generated by (ξ, Y). To verify this identity, observe first that the Markov property of (ξ, Y) yields that

$$\begin{aligned} &\widetilde{\mathbb{E}}_{x,i}\big[e^{-\int_0^{\tau_\ell^+} h(Y_s)\,ds} k_+(Y_\ell^+) I_{\{\tau_\ell^+<\infty\}} \big| \mathcal{G}_t\big] \\ &\quad = e^{-\int_0^{t\wedge\tau_\ell^+} h(Y_s)\,ds} \widetilde{\mathbb{E}}_{x,i}\big[e^{-\int_0^{\tau_\ell^+} h(Y_s)\,ds} k_+(Y_\ell^+) I_{\{\tau_\ell^+<\infty\}}\big]\big|_{(x,i)=(\xi_{t\wedge\tau_\ell^+}, Y_{t\wedge\tau_\ell^+})}. \end{aligned}$$

Further, in view of the strong Markov property and spatial homogeneity of ξ, the expectation on the right-hand side of the previous display is for $x \leq \ell$ given by

$$\begin{aligned} &\widetilde{\mathbb{E}}_{x,i}\big[e^{-\int_0^{\tau_\ell^+} h(Y_s)\,ds} k_+(Y_\ell^+) I_{\{\tau_\ell^+<\infty\}}\big] \\ &\quad = \widetilde{\mathbb{E}}_{0,i}\big[e^{-\int_0^{\tau_{\ell-x}^+} h(Y_s)\,ds} k_+(Y_{\ell-x}^+) I_{\{\tau_{\ell-x}^+<\infty\}}\big] \\ &\quad = \sum_{j\in E^0\cup E^+} \widetilde{\mathbb{E}}_{0,i}\big[e^{-\int_0^{\tau_0^+} h(Y_s)\,ds} I_{\{Y_0^+=j,\tau_0^+<\infty\}}\big] \\ &\qquad \times \widetilde{\mathbb{E}}\big[e^{-\int_0^{\tau_{\ell-x}^+} h(Y_s)\,ds} k_+(Y_{\ell-x}^+) I_{\{\tau_{\ell-x}^+<\infty\}} \big| Y_0 = j, \xi_0 = 0\big] \\ &\quad = e_i'\eta_h^+ \exp\big(Q_h^+(\ell - x)\big) = e_i'\Phi_\ell^+(x)k_+ = f_+(i,x), \end{aligned} \tag{17.118}$$

where the last line follows by definitions (17.106) and (17.98) of Q_h^+ and η_h^+.

As M^+ is a martingale, an application of Itô's lemma shows that $f_+ = (f_+(i,u), i \in E)$ satisfies, for all $u < \ell$,

$$\frac{1}{2}s(i)^2 f_+''(i,u) + m(i)f_+'(i,u) + \sum_j q_{ij}\big(f_+(j,u) - f_+(i,u)\big) = 0, \tag{17.119}$$

where f'_+ and f''_+ denote the first and second derivatives of f_+ with respect to u. By substituting the expressions (17.115)–(17.117) into (17.119) we find, since k^+ was arbitrary, that Q_h^+ and η_h^+ satisfy the first set of equations of system (17.66). The proof for Q_h^- and η_h^- is analogous and omitted.

(Uniqueness) Now we turn to the proof of the uniqueness of the Wiener–Hopf factorization. To this end, let (W^+, G^+, W^-, G^-) be a complex matrix Wiener–Hopf factorization and define the function $\widetilde{f}$ as f_+ in (17.117), but replacing η^+ and Q^+ by W^+ and G^+, respectively. Since the pair (W^+, G^+) satisfies (17.66), it follows by an application of Itô's lemma that $M'_t = e^{-\int_0^t h(Y_s)ds}\widetilde{f}(Y_t, \xi_t)$ is a local martingale. In view of the facts that $G^+ \in \mathbb{D}(N^+)$ and $h \in \mathcal{H}$, it follows that M' is in fact bounded on $\{t \le \tau_\ell^+\}$. An application of Doob's optional stopping theorem then yields that

$$
\begin{aligned}
\widetilde{f}(j,x) &= \widetilde{\mathbb{E}}_{x,j}\big[e^{-\int_0^{t\wedge\tau_\ell^+} h(Y_s)\,ds}\widetilde{f}(Y_{t\wedge\tau_\ell^+}, \xi_{t\wedge\tau_\ell^+})\big] \\
&= \widetilde{\mathbb{E}}_{x,j}\big[e^{-\int_0^{\tau_\ell^+} h(Y_s)\,ds}\widetilde{f}(Y_\ell^+, \xi_{\tau_\ell^+})I_{\{\tau_\ell^+<\infty\}}\big] \\
&\quad + \lim_{t\to\infty}\widetilde{\mathbb{E}}_{x,j}\big[e^{-\int_0^t h(Y_s)\,ds}\widetilde{f}(Y_t, \xi_t)I_{\{\tau_\ell^+=\infty\}}\big].
\end{aligned}
\tag{17.120}
$$

By the definition of $\widetilde{f}$, the absence of positive jumps of ξ and (17.118), the first expectation in (17.120) is equal to

$$
\begin{aligned}
&\widetilde{\mathbb{E}}_{x,j}\big[e^{-\int_0^{\tau_\ell^+} h(Y_s)\,ds}\widetilde{f}(Y_\ell^+, \ell)I_{\{\tau_\ell^+<\infty\}}\big] \\
&\quad = \widetilde{\mathbb{E}}_{x,j}\big[e^{-\int_0^{\tau_\ell^+} h(Y_s)\,ds}k_+(Y_\ell^+)I_{\{\tau_\ell^+<\infty\}}\big] = f_+(j,x)
\end{aligned}
$$

for $x \le \ell$. The second term in (17.120) is zero, since $\int_0^t I_{\{Y_s\in E_0\}}\,ds \to \infty$ $\widetilde{\mathbb{P}}_{x,j}$ almost surely on the event $\{\tau_\ell^+ = \infty\}$ for all $j \in E$ and $x \in \mathbb{R}$, and $\min_{i\in E_0}\Re(h(i)) > 0$. Thus $f_+ = \widetilde{f}$ for all N^+-column vectors k_+, and we deduce that $G^+ = Q_h^+$ and $W^+ = \eta_h^+$. Similarly, one can show that $G^- = Q_h^-$ and $W^- = \eta_h^-$, and the uniqueness is proved.

17.8.3 Proof of Theorem 17.24

Applying the strong Markov property at $\bar{\tau}$ and noting that $\bar{\tau} \le \tau^\xi$ yields, in view of the representation (17.94), that

$$
\begin{aligned}
H_j^{x,i}(h,u) &= \widetilde{\mathbb{E}}_{x,i}\big[e^{-\int_0^{\bar{\tau}} h(Y_s)\,ds}I_{\{Y_{\bar{\tau}}=j\}}F(\xi_{\bar{\tau}}, Y_{\bar{\tau}})\big] \\
&= \widetilde{\mathbb{E}}_{x,i}\big[e^{-\int_0^{\bar{\tau}} h(Y_s)\,ds}I_{\{Y_{\bar{\tau}}=j,\tau_\upsilon^+<\tau_\ell^-\}}F(\upsilon, Y_{\bar{\tau}})\big]
\end{aligned}
$$

$$+\widetilde{\mathbb{E}}_{x,i}\big[e^{-\int_0^{\bar{\tau}} h(Y_s)\,ds} I_{\{Y_{\bar{\tau}}=j,\tau_{\upsilon}^{+}>\tau_{\ell}^{-}\}} F(\ell, Y_{\bar{\tau}})\big],$$

where

$$F(x,i)=\widetilde{\mathbb{E}}_{x,i}\big[e^{-\int_0^{\tau^{\xi}} \widetilde{h}(Y_s)\,ds+u\xi_{\tau^{\xi}}} I_{\{Y_{\tau^{\xi}}=j\}}\big]=\widetilde{\mathbb{E}}_{x,i}\big[e^{u\xi_{\tau^{\xi}}} I_{\{Y_{\tau^{\xi}}=j\}}\big]$$

since $\widetilde{h}(i)=h(i)I_{\{i\in E^0\}}$, by the definition of $\widetilde{h}$. From the definition of τ^{ξ} it is straightforward to check that

$$\widetilde{\mathbb{P}}_{x,i}\big[\tau^{\xi}=0\big]=1 \quad \text{for } i\in E^0 \text{ and } x\in\{\ell,\upsilon\},$$

so that $F(\upsilon,i)=e^{u\upsilon}\delta_{ij}$ and $F(\ell,i)=e^{u\ell}\delta_{ij}$ if $i\in E^0$, where δ_{ij} denotes the Kronecker delta.

Moreover, in view of the form (17.61)–(17.62) of Q_h and the definition of phase-type distribution, it is clear that, conditionally on $Y_0=i\in E_j^+$ and $\xi_0=\upsilon$, $Y_{\tau^{\xi}}=j$ and $\tau^{\xi}\sim \mathrm{PH}(\delta_i, B_j^+)$, where δ_i is the vector with elements $\delta_i=(\delta_{ik})$. Therefore we find, using (17.4), that, for u with $\Re(u)<\alpha_j^+$ and $i\in E_j^+$,

$$\begin{aligned}F(\upsilon,i)&=\widetilde{\mathbb{E}}_{\upsilon,i}\big[e^{u\xi_{\tau^{\xi}}}\big]\\&=e^{u\upsilon}\widetilde{\mathbb{E}}_{\upsilon,i}\big[e^{u\tau^{\xi}}\big]=e^{u\upsilon}\big[\big(-uI_j^+-B_j^+\big)^{-1}\big(-B_j^+\big)\mathbf{1}\big](i).\end{aligned}$$

Similarly, it follows that, for $i\in E_j^-$ and u with $\Re(u)>-\alpha_j^-$,

$$F(\ell,i)=e^{u\ell}\big[\big(uI_j^- - B_j^-\big)^{-1}\big(-B_j^-\big)\mathbf{1}\big](i).$$

Acknowledgements Research supported by EPSRC grant EP/D039053. Most of the work was carried out while the first author was at the Department of Mathematics of Imperial College London. We thank an anonymous referee for the useful suggestions that improved the presentation of the paper.

References

1. J. Abate, W. Whitt, Numerical inversion of Laplace transforms of probability distributions. ORSA J. Comput. **7**, 36–43 (1995)
2. S. Asmussen, *Ruin Probabilities* (World Scientific, Singapore, 2000)
3. S. Asmussen, O. Nerman, M. Olsson, Fitting phase–type distributions via the EM algorithm. Scand. J. Stat. **23**, 419–441 (1996)
4. S. Asmussen, Stationary distributions for fluid flow models with or without Brownian noise. Commun. Stat. Stoch. Models **11**(1), 21–49 (1995)
5. F. Avram, M.R. Pistorius, M. Usabel, The two barriers ruin problem via a Wiener Hopf decomposition approach. An. Univ. Craiova, Ser. Mat. Inform. **30**(1), 38–44 (2003)
6. S. Benaim, P. Fritz, Smile asymptotics II: Models with known MGF. J. Appl. Probab. **45**(1), 16–32 (2008)
7. S.I. Boyarchenko, S.Z. Levendorskii, American options in Lévy models with stochastic interest rates. J. Comput. Finance **12**(4), Summer 2009. http://ssrn.com/abstract=1015409

8. S.I. Boyarchenko, S.Z. Levendorskii, American options in regime-switching models: FFT realisations. Preprint (2008). http://ssrn.com/abstract=1127562
9. S.I. Boyarchenko, S.Z. Levendorskii, *Non-Gaussian Merton–Black–Scholes Theory* (World Scientific, Singapore, 2002)
10. J. Buffington, R.J. Elliott, American options with regime switching. Int. J. Theor. Appl. Finance **5**, 497–514 (2002)
11. P. Carr, Randomization and the American put. Rev. Financ. Stud. **11**, 597–626 (1998)
12. P. Carr, H. Geman, D. Madan, M. Yor, Stochastic volatility for Lévy processes. Math. Finance **13**(3), 345–382 (2003)
13. P. Carr, D. Madan, Option pricing and the fast Fourier transform. Comput. Finance **2**(4), 61–73 (1999)
14. R. Cont, P. Tankov, *Financial Modelling With Jump Processes* (Chapman & Hall, London, 2003)
15. S.N. Ethier, T.G. Kurtz, *Markov Processes: Characterization and Convergence* (Wiley, New York, 1986)
16. D. Friedman, *Markov Chains* (Holden-Day, San Francisco, 1971)
17. J. Gatheral, *The Volatility Surface: A Practitoner's Guide* (Wiley, New York, 2006)
18. G. Di Graziano, L.C.G. Rogers, Barrier option pricing for assets with Markov-modulated dividends. Comput. Finance **9**, 75–87 (2006)
19. X. Guo, Q. Zhang, Closed-form solutions for perpetual American put options with regime switching. Siam J. Appl. Math. (2004)
20. J. Jacod, A.N. Shiryaev, *Limit Theorems for Stochastic Processes*, 2nd edn. A Series of Comprehensive Studies in Mathematics, vol. 288 (Springer, Berlin, 2003)
21. Z. Jiang, M.R. Pistorius, On perpetual American put valuation and first passage in a regime-switching model with jumps. Finance Stoch. **12**, 331–355 (2008)
22. A. Jobert, L.C.G. Rogers, Option pricing with Markov-modulated dynamics. SIAM J. Control Optim. **44**(6), 2063–2078 (2006)
23. R.W. Lee, The moment formula for implied volatility at extreme strikes. Math. Finance **14**(3), 496–480 (2004)
24. A. Lipton, *Mathematical Methods for Foreign Exchange* (World Scientific, Singapore, 2001)
25. A. Mijatović, M. Pistorius, Continuously monitored barrier options under Markov processes. Working Paper, Imperial College London (2009). http://arxiv.org/abs/0908.4028
26. A. Mijatović, M. Pistorius, Double-no-touch options and the volatility smile. Math. Finance (to appear)
27. M. Neuts, *Matrix-Geometric Solutions in Stochastic Models* (John Hopkins University Press, Baltimore, 1981)
28. L.C.G. Rogers, Fluid models in queueing theory and Wiener–Hopf factorization of Markov chains. Ann. Appl. Probab. **4**(2), 390–413 (1994)
29. W. Rudin, *Real and Complex Analysis*, 3rd edn. (McGraw-Hill, New York, 1987)

Chapter 18
Asymptotics of HARA Utility from Terminal Wealth under Proportional Transaction Costs with Decision Lag or Execution Delay and Obligatory Diversification

Lukasz Stettner

Abstract In the paper optimal asymptotics of HARA utility from terminal wealth under proportional transaction costs is considered. The asset prices are modeled as exponents of diffusion with jumps whose parameters depend on a finite-state Markov process of economic factors. An obligatory portfolio diversification is introduced, according to which it is required to invest at least a fixed small portion of our wealth in each asset. Since we are looking for optimal strategies within the class of impulse controls, two kinds of delay are introduced: decision lag, when successive portfolio changes are separated by a fixed time lag h, and execution delay, when portfolio is changed after h units of time following the decision.

Keywords HARA utility · Terminal weath asymptotics · Proportional transaction costs · Log Levy asset prices

Mathematics Subject Classification (2010) 91G10 · 91G80

18.1 Introduction

Assume that we are given a market with d assets. The prices $S_i(t)$ of the ith asset are of the form

$$S_i(t) = S_i(0)e^{X_i(t)} \tag{18.1}$$

for $i = 1, 2, \ldots, d$, with the vector $X(t)$ being solution to the equation

$$dX(t) = \alpha(z_t)\,dt + \sigma(z_t)\,dB(t) + \int_{R^d} \gamma(z_t, u)\tilde{N}(dt, du) \tag{18.2}$$

with $X(0) = 0$, and where on a given complete probability space $(\Omega, \mathcal{F}, (\mathcal{F}_t), P)$, we have $(\mathcal{F}_t)$-adapted independent processes: a Brownian motion $(B(t))$, a compensated Poisson measure $\tilde{N}$, and a Markov process (z_t) of economic factors on

Research supported by MNiSzW grant no. N N201 371836.

L. Stettner (✉)
Institute of Mathematics, Polish Acad. Sci., Sniadeckich 8, 00-956 Warsaw, Poland
e-mail: stettner@impan.gov.pl

G. Di Nunno, B. Øksendal (eds.), *Advanced Mathematical Methods for Finance*,
DOI 10.1007/978-3-642-18412-3_18,

the finite state space D. We furthermore assume that $\gamma_{ik}(z, y_k)$ depends on the kth coordinate y_k of $y \in R^d$ and

$$\int_{R^d} |\gamma_{ik}|^\iota(z, y_k)\nu_k(dy_k) < \infty \tag{18.3}$$

for $\iota = 1, 2$, $i, k = 1, 2, \ldots, d$, with ν_k being the Lévy measure corresponding to $\tilde{N}(dt, du)$.

Denote by W_t and W_t^- respectively the wealth processes after and before transaction at time t, and similarly by $N_i(t)$ and $N_i^-(t)$ the numbers of assets at time t in portfolio after and before possible transaction. Then $\pi_i^-(t) = \frac{N_i^-(t)S_i(t)}{W_t^-}$ and $\pi_i(t) = \frac{N_i(t)S_i(t)}{W_t}$ are the portions of the capital invested in the ith asset before and after transaction at time t. Clearly, $\pi(t) \in \mathcal{S} = \{\nu \in R^d : \nu_i \geq 0, \sum_{i=1}^d \nu_i = 1\}$.

In the paper we shall assume either decision lag h, which means that the next decision can be taken after h units of time, following the previous decision or execution delay under which the portfolio change decision is executed with a fixed delay h.

As a consequence of these assumptions, we shall consider impulse control strategies $V = \{(\tau_i, \pi^i)\}$, consisting of increasing stopping times (τ_i) such that $\tau_{i+1} \geq \tau_i + h$ and the portions of capital π^i we would like to have after transaction at time τ_i in the case of decision lag, or at time $\tau_i + h$ in the case of execution delay, respectively.

For a given $\delta < \frac{1}{d}$, let

$$\mathcal{S}_\delta^0 = \left\{\nu \in R^d : \nu_i > \delta, \sum_{i=1}^d \nu_i = 1\right\}.$$

In what follows we shall impose an obligatory portfolio diversification: when $\pi(t)$ leaves $\mathcal{S}_\delta^0$, we change portfolio choosing new portfolio from $\mathcal{S}_{\delta'} = \{\nu \in R^d : \nu_i \geq \delta', \sum_{i=1}^d \nu_i = 1\}$ with $\frac{1}{d} > \delta' > \delta$. Let

$$T^{\delta^0} = \inf\{s \geq 0 : \pi(s) \in \mathcal{S} \setminus \mathcal{S}_\delta^0\}. \tag{18.4}$$

Clearly, $\tau_1 \leq T^{\delta^0}$.

Remark 18.1 Notice that (for details, see Remark 1 in [4]) by the Law of Large Numbers if there is a unique invariant measure μ for (z_t), we have

$$\lim_{t\to\infty} \frac{1}{t} X_i(t) = \lim_{t\to\infty} \frac{1}{t} \int_0^t \alpha_i(z_s)\,ds = \int \alpha_i(z)\mu(dz) = r_i.$$

Therefore, if there is k such that $r_k > r_i$ for $i \neq k$, then $\pi(t) \to \pi(\infty) = \delta_k$, provided that $\pi_k(0) > 0$, and the limit portfolio is degenerate since we keep all capital in the asset k.

Given a utility function U, i.e., an increasing concave function, we would like to find the best asymptotics of utility U from terminal wealth W_T letting $T \to \infty$. Our expectation is that $E[U(W_T)] \sim_{T\to\infty} U(e^{\lambda T})$ with some constant λ. The problem is to determine this λ and to find optimal investment strategies under which such optimal growth λ is achieved. In particular cases of HARA utility functions, i.e., $U(x) = \ln x$ or $U(x) = x^\gamma$ with $\gamma \in (0, 1)$, we respectively might expect to have linear or exponential growth asymptotics:

$$E\big[\ln(W_T)\big] \sim \lambda T \quad \text{or} \quad E\big[(W_T)^\gamma\big] \sim e^{\lambda_r \gamma T}.$$

The constants λ or λ_r can be determined from maximization of the following cost functionals respectively:

$$J(V) = \liminf_{T\to\infty} \frac{1}{T} E\big[\ln(W_T)\big], \tag{18.5}$$

$$J_\gamma^r(V) = \liminf_{T\to\infty} \frac{1}{\gamma T} E\big[(W_T)^\gamma\big]. \tag{18.6}$$

The optimal portfolio strategy for the cost functional (18.5) is called growth optimal portfolio (GOP), while the optimal strategy for (18.6) is called power utility optimal portfolio (PUOP).

Remark 18.2 Very often we consider normalized utility functions $U(x) = \frac{x^\gamma}{\gamma}$. This function is a utility function also for $\gamma < 0$. Such normalization is not important when we study asymptotics. When we consider the cost functional $\frac{1}{\gamma} \ln E[(W_T)^\gamma]$ with $\gamma < 0$, we "measure" not only the average portfolio growth but also its variance with negative weight γ, plus higher moments of W_T, with weights being powers of γ. Such cost functional is therefore called risk sensitive. There is a vast literature concerning risk-sensitive control problems (see [3, 9, 11], and references therein). Notice moreover that, for $\gamma > 0$,

$$\frac{-1}{\gamma} \ln E\big[(W_T)^{-\gamma}\big] \le E\big[\ln(W_T)\big] \le \frac{1}{\gamma} \ln E\big[(W_T)^\gamma\big],$$

and letting $\gamma \to 0$, the external terms converge to the middle term.

In the case of HARA utility functions portfolio optimization with cost functionals (18.5) and (18.6) leads to stationary strategies depending on the values of economic factors only. In this paper we consider a more general case of proportional transaction costs that in particular cover the case of no transaction costs. There is an intensive literature concerning portfolio optimization with proportional transaction costs and cost functionals of the form (18.5) or (18.6), see [1, 4, 5, 8, 9, 11, 12, 14, 15] for lognormal or discrete-time asset prices. This paper generalizes former papers by considering log Lévy asset prices with existence of economic factors. In the papers [5, 14, 15], and [4] an additional fixed proportional transaction cost was introduced to have optimal impulse strategies. Here, for this purpose, we impose decision lag or execution delay. The existence of such delay creates additional difficulties in the model. We use a number of analytic properties of the log Lévy asset prices shown in [4]. The part of the paper concerning the cost functional (18.6) exploits new arguments, and such continuous-time probabilistic approach seems to be completely novel. In the paper we prove the existence of solutions to the suitable Bellman equations and show the form of optimal strategies, which is a standard step in the further construction of nearly optimal strategies. In the final section we introduce a general form of the Bellman equation, which covers both logarithmic and power utility functions. As is shown in [7], unfortunately, the class of utility functions to which this Bellman equation can be studied is practically limited to logarithmic and power utility functions.

Under proportional transaction costs, the change of portfolio from $\pi_i^-(t) = \frac{N_i^-(t)S_i(t)}{W_t^-}$ to $\pi_i(t) = \frac{N_i(t)S_i(t)}{W_t}$, or from $N^-(t)$ to $N(t)$ assets in portfolio results in the change of the wealth process W_t^- to W_t, where

$$W_t^- = W_t + W_t^- c\big(\pi' - \pi^-(t)\big) \tag{18.7}$$

with $\pi' \in \mathcal{S}^0 = \{\nu \in R^d : \nu_i \geq 0, \sum_{i=1}^d \nu_i \leq 1\}$ such that $\pi(t) = g(\pi')$, where $g_i(\pi') = \frac{\pi_i'}{\sum_{j=1}^d \pi_j'}$ and

$$c(\nu) = \sum_{i=1}^d c_i^+(\nu_i)^+ + \sum_{i=1}^d c_i^-(\nu_i)^-. \tag{18.8}$$

We have the following (see Lemma 1 of [11]):

Lemma 18.3 *Given $\pi, \bar{\pi} \in \mathcal{S}$, there is a continuous $e(\pi, \bar{\pi}) \in (0, 1]$ such that $F^{\pi,\bar{\pi}}(e(\pi, \bar{\pi})) = 1$ with $F^{\pi,\bar{\pi}}(\delta) = \delta + c(\delta\bar{\pi} - \pi)$. Consequently, the change of portfolio at time t from $\pi^-(t)$ to $\pi(t)$ results in the change of the wealth W_t^- to W_t, where $W_t^- = e(\pi^-(t), \pi(t))W_t$. Furthermore, if there are no transactions in the time interval $(s, s + t)$, then*

$$W_{t+s}^- = W_t\pi(t) \cdot e^{X(t+s)-X(s)}$$

and

$$\pi^-(t + s) = g\big(\pi(t) \diamond e^{X(t+s)-X(t)}\big),$$

where $\pi(t) \diamond e^{X(t+s)-X(s)} = (\pi_i(t)e^{X_i(t+s)-X_i(t)})$.

By the form of (18.2), it is clear that the pair $(\pi(t), z_t)$ and the triple $(\pi(t), z_t, W_t)$ are Markov processes with transition operators Π_t and Π_t^e, respectively.

Since $(z_t, t \geq 0)$ is a finite-state continuous-time, time-homogeneous Markov process, its evolution can be described in the following form:

$$\begin{aligned} \varsigma_1 &= \inf\{s \geq 0: \; z_s \neq z_0\}, \\ \varsigma_{n+1} &= \inf\{s \geq 0: \; z_{s+\varsigma_n} \neq z_{\varsigma_n}\}, \end{aligned} \tag{18.9}$$

and for $z_0 = z$,

$$\begin{aligned} P_z[\varsigma_1 \leq t] &= \int_0^t n(z,s)\,ds, \\ P_z\big[P_{z_{\varsigma_n}}[\varsigma_{n+1} \leq t]\big] &= E_z\bigg[\int_0^t n(z_{\varsigma_n}, s)\,ds\bigg], \\ P_z[z_{\varsigma_1} = z'] &= P(z,z'). \end{aligned}$$

We shall assume that $n(z,s) > 0$ for $z \in D$ and $s > 0$ and the matrix $P(z,z')$ is ergodic, i.e., from one state we can enter the other state with probability one (the states are communicative). It is assumed further that

(A) the matrix $\sigma(z)\sigma(z)^T$ is uniformly elliptic, i.e., there is $\epsilon > 0$ such that for all $b \in R^d$ and $z \in D$,

$$b^T \sigma(z)\sigma(z)^T b \geq \epsilon b^T b.$$

We have (see Lemma 1 of [4]) the following:

Lemma 18.4 *Under* (A), *the solution to* (18.2) *with the initial condition* $X(0) = x$ *has a continuous density* p_t *for each fixed* $z_t = z$ *with respect to the Lebesgue measure* l^d *at time* $t > 0$, *i.e., for a Borel set* $A \subset R^d$,

$$P_{xz}\big\{X^z(t) \in A\big\} = \int_A p_t^z(x,x')\,l^d(dx'), \tag{18.10}$$

where $X^z(t)$ *is a solution to* (18.2) *with* $z_t \equiv z$, *and* $p_t^z(x,x')$ *is a continuous function of* x *and* x'. *Furthermore, given* (A1), *for Borel sets* $B \subset R^d$ *and* $z' \in D$,

$$P_{x,z}\big\{X(t) \in B, z_t = z'\big\} = \int_B p_t(x,x',z,z')\,l^d(dx'), \tag{18.11}$$

where

$$\begin{aligned} & p_t\big(x,x',z,z'\big) \\ &\quad = 1_{z=z'}\bigg(1 - \int_0^t n(z,s)\,ds\bigg) p_t^z\big(x,x'\big) + \sum_{k=1}^{\infty} \sum_{z^1 \in D} \sum_{z^2 \in D} \cdots \sum_{z^{k-1} \in D} \int_0^t n(z,s_1) \\ &\qquad \times \int_{R^d} p_{s_1}^z(x,x_1) P\big(z,z^1\big) \int_0^{t-s_1} n\big(z^1,s_2\big) \int_{R^d} p_{s_2}^{z^1}(x_1,x_2) P\big(z^1,z^2\big) \ldots \end{aligned}$$

$$\times \int_0^{t-s_1-\cdots-s_{k-1}} n\big(z^{k-1}, s_k\big)$$

$$\times \int_{R^d} p_{s_k}^{z^{k-1}}(x_{k-1}, x_k) P\big(z^{k-1}, z'\big) p_{t-s_1-\cdots-s_k}^{z'}(x_k, y)$$

$$\times \int_{t-s_1-\cdots-s_k}^{t} n(z', u)\, du\, l(dx_k)\, ds_k\, l(dx_{k-1})\, ds_{k-1} \dots l(dx_1)\, ds_1. \quad (18.12)$$

The following continuity properties are crucial in further investigations (see Proposition 1, Lemma 7, Lemma 4 of [4] and also [2]):

Proposition 18.5 *Under* (A), *the operator* Π_t^e *is continuous in variation norm for* $(\pi, z, W) \in \mathcal{S}_\delta \times D \times (0, \infty)$, *i.e., for* $(\pi_{(n)}, z, W_{(n)}) \to (\pi, z, W) \in \mathcal{S}_\delta \times D \times (0, \infty)$, *where* $(\pi_{(n)}, n \geq 1)$ *is a sequence in* $\mathcal{S}_\delta$, *and* $(W_{(n)}, n \geq 1)$ *is a sequence in* $(0, \infty)$ *with* $W > 0$, *it follows that*

$$\sup_{A \in B(\mathcal{S} \times D \times (0,\infty))} \big| \Pi_t^e(\pi_{(n)}, z, W_{(n)}, A) - \Pi_t^e(\pi, z, W, A) \big| \to 0 \quad (18.13)$$

as $n \to \infty$, *with* $B(\mathcal{S} \times D \times (0, \infty))$ *denoting the family of Borel subsets of* $\mathcal{S} \times D \times (0, \infty)$. *In particular,*

$$\sup_{A \in B(\mathcal{S} \times D)} \big| \Pi_t(\pi_{(n)}, z, A) - \Pi_t(\pi, z, A) \big| \to 0, \quad (18.14)$$

and there is a positive continuous density of operator Π_t. *Moreover, for any continuous bounded function* $F : \mathcal{S} \times D \times (0, \infty) \to R$ *and* $t > 0$, *it follows that the mappings*

$$\mathcal{S} \times D \times (0, \infty) \ni (\pi, z, W) \mapsto E_{\pi z W}\big[\chi_{t \leq T^{\delta 0}} F\big(\pi(t), z_t, W_t\big)\big] \quad (18.15)$$

and

$$\mathcal{S} \times D \times (0, \infty) \ni (\pi, z, W) \mapsto E_{\pi z W}\big[F\big(\pi\big(t \wedge T^{\delta^0}\big), z_{t \wedge T^{\delta 0}}, W_{t \wedge T^{\delta 0}}\big)\big] \quad (18.16)$$

are continuous, and for sufficiently small $\gamma > 0$,

$$\sup_{z \in D} \sup_{\pi \in \mathcal{S}_\delta^0} E_{\pi z}\big[e^{\gamma T^{\delta^0}}\big] < \infty. \quad (18.17)$$

18.2 Discounted GOP

In this section we shall consider the model with decision lag pointing out at the end of the section differences in the case of execution delay. To obtain the existence of the solutions to the Bellman equation corresponding to the cost functional (18.5), we introduce the first discounted cost functional of the form

$$J^{\alpha}_{\pi,z}(V) = E_{\pi,z}\left[\sum_{i=1}^{\infty} e^{-\alpha\tau_i}\left[\ln\left(\pi(\tau_{i-1})\cdot e^{X(\tau_i)-X(\tau_{i-1})}\right) + \ln e\left(\pi^-(\tau_i),\pi^i\right)\right]\right] \tag{18.18}$$

and the value function corresponding to this functional,

$$w^{\alpha}(\pi,z) = \sup_V J^{\alpha}_{\pi,z}(V). \tag{18.19}$$

We have the following Bellman equation corresponding to the cost functional (18.18)

$$\begin{aligned} w^{\alpha}(\pi,z) = \sup_{\tau} E_{\pi,z}\Big[e^{-\alpha\tau\wedge T^{\delta 0}}\Big[\ln\big(\pi\cdot e^{X(\tau\wedge T^{\delta 0})}\big) \\ + M^{\alpha}_h w^{\alpha}\big(\pi\big(\tau\wedge T^{\delta 0}\big), z_{\tau\wedge T^{\delta 0}}\big)\Big]\Big] \end{aligned} \tag{18.20}$$

with

$$M^{\alpha}_h w(\pi,z) = \sup_{\pi'\in\mathcal{S}_{\delta'}} \Big[\ln e(\pi,\pi') + E_{\pi',z}\Big[e^{-\alpha h}\Big(\ln\big(\pi'\cdot e^{X(h)}\big) + w\big(\pi(h),z_h\big)\Big)\Big]\Big]. \tag{18.21}$$

From the very definition we immediately have that

$$\sup_{\pi,z}\left|M^{\alpha}_h w_1(\pi,z) - M^{\alpha}_h w_2(\pi,z)\right| \le e^{-\alpha h}\|w_1 - w_2\| \tag{18.22}$$

with $\|.\|$ standing for the supremum norm. Therefore, one could expect to find a solution to (18.20) using Banach contraction principle in the space of continuous functions of the first coordinate. For this purpose, however, we need the continuity of the right-hand side of (18.20), which follows from the following lemma

Lemma 18.6 *If the function $g : \mathcal{S}\times D \mapsto R$ is continuous and bounded, then*

$$v^{\alpha}(\pi,z) := \sup_{\tau} E_{\pi,z}\Big[e^{-\alpha\tau\wedge T^{\delta 0}}\Big[\ln\big(\pi\cdot e^{X(\tau\wedge T^{\delta 0})}\big) + g\big(\pi\big(\tau\wedge T^{\delta 0}\big), z_{\tau\wedge T^{\delta 0}}\big)\Big]\Big] \tag{18.23}$$

is also continuous and bounded.

Proof By the uniform integrability of the term $\ln(\pi\cdot e^{X(\tau\wedge T^{\delta 0})})$ (which follows from (18.17)) it suffices to show the continuity of

$$v^{\alpha}(\pi,z,W) := \sup_{\tau} E_{\pi,z}\Big[e^{-\alpha\tau\wedge T^{\delta 0}}\Big[f\big(W_{\tau\wedge T^{\delta 0}} + g\big(\pi\big(\tau\wedge T^{\delta 0}\big), z_{\tau\wedge T^{\delta 0}}\big)\big)\Big]\Big]$$

with continuous bounded function f. The last problem can be solved using the so-called penalty method, i.e., solving for $\beta > 0$ the equation

$$v^{\alpha,\beta}(\pi, z, W) = E_{\pi,z}\Big[\beta \int_0^{T^{\delta 0}} e^{-\alpha s}\big(f(W_s) + g\big(\pi(s), z_s\big) - v^{\alpha,\beta}\big(\pi(s), z_s, W_s\big)\big)^+ ds + e^{\alpha T^{\delta 0}}\big(f(W_{T^{\delta 0}}) + g\big(\pi\big(T^{\delta 0}\big), z_{T^{\delta 0}}\big)\big)\Big]. \tag{18.24}$$

By Theorem 1 of [4] (see also [13]) there is a unique continuous bounded function $v^{\alpha,\beta} : \mathcal{S}^0 \times D \times (0, \infty)$ which is a solution to the penalty equation (18.24), and letting $\beta \to \infty$, we have that $v^{\alpha,\beta}$ converges to v^{α} uniformly on compact subsets of $\mathcal{S}^0 \times D \times (0, \infty)$. □

We now have the following:

Theorem 18.7 *There is a unique continuous bounded solution w^{α} to* (18.20), *and w^{α} coincides with the value function* (18.19) *of the cost functional* (18.18).

Proof By Lemma 18.6 the operator defined by the right-hand side of (18.20) transforms the class of continuous bounded functions into itself. By (18.22) this operator is a contraction. Consequently, by the Banach contraction principle there is a unique continuous solution to (18.20). By the proof of Theorem 2 of [4] the continuous solution to (18.20) is the value function (18.19). □

Remark 18.8 In the case of execution delay h, the form of the Bellman equation is the same as (18.20) with the change in the operator M only, which is of the form

$$M_h^{\alpha} w(\pi, z) = \sup_{\pi' \in \mathcal{S}_{\delta'}} E_{\pi,z}\big[e^{-\alpha h}\big(\ln\big(\pi \cdot e^{X(h)}\big) + \ln e\big(\pi(h), \pi'\big) + w(\pi', z_h)\big)\big]. \tag{18.25}$$

By similar consideration we have the existence of a unique continuous solution to the Bellman equation, which is also the value function for the cost functional (18.18).

18.3 Long-Run GOP

In this section, using vanishing discount approach, we prove the existence of solutions to the long-run Bellman equation. We first need some auxiliary results.

Lemma 18.9 *We have*

$$\sup_{\pi,\pi' \in \mathcal{S}_{\delta}} \sup_{z,z'} \sup_{A} \big|\Pi_h(\pi, z, A) - \Pi_h(\pi', z', A)\big| := L < 1. \tag{18.26}$$

Proof Suppose that (18.26) is not satisfied. Then there are sequences $(\pi_{(n)})$, $(\pi'_{(n)})$, $(z_{(n)})$, $(z'_{(n)})$, and A_n with $\pi_{(n)}, \pi'_{(n)} \in \mathcal{S}_\delta$ such that $\Pi_h(\pi_{(n)}, z_{(n)}, A_n) \to 1$ and $\Pi_h(\pi'_{(n)}, z'_{(n)}, A_n) \to 0$. Choosing subsequences, there are π and $\pi' \in \mathcal{S}_\delta$ and $z, z' \in D$ such that (by Proposition 18.5) $\Pi_h(\pi, z, A_n) \to 1$ and $\Pi_h(\pi', z', A_n) \to 0$. Let $f^{\pi,z}(\bar{\pi}, \bar{z})$ be the density of the operator $\Pi_h(\pi, z, \cdot)$, and

$$\Gamma_\alpha^{\bar{z}} = \big\{(\bar{\pi}) : f^{\pi,z}(\bar{\pi}, \bar{z}) \geq \alpha, \text{ for } (\pi, z) \in \mathcal{S}_\delta \times D\big\}.$$

By continuity of $f^{\pi,z}$, the set $\Gamma_\alpha^{\bar{z}}$ is closed, and $l^d(\bigcap_{\bar{z}\in D} \Gamma_\alpha^{\bar{z}}) > 0$ for a sufficiently small α. If $\Pi_h(\pi, z, A_n) \to 1$, then $\Pi_h(\pi, z, \mathcal{S} \times D \setminus A_n) \to 0$, and consequently $l^d(\bigcap_{\bar{z}\in D} \Gamma_\alpha^{\bar{z}} \setminus A_n) \to 0$. Therefore, also $\Pi_h(\pi', z', \bigcap_{\bar{z}\in D} \Gamma_\alpha^{\bar{z}} \cup A_n) \to 0$. But $\Pi_h(\pi', z', \bigcap_{\bar{z}\in D} \Gamma_\alpha^{\bar{z}} \cup A_n) \geq \alpha l^d(\bigcap_{\bar{z}\in D} \Gamma_\alpha^{\bar{z}}) > 0$, a contradiction. Thus we have (18.26). □

Lemma 18.9 is crucial in the proof of the following property.

Proposition 18.10 *We have*

$$\sup_\alpha \|w^\alpha\|_{sp} := \sup_\alpha\Big(\sup_{\pi,z} w^\alpha(\pi, z) - \inf_{\pi',z'} w^\alpha(\pi', z')\Big) < \infty. \tag{18.27}$$

Proof By (18.21), using (18.26), we have

$$\begin{aligned}
&\big|M_h^\alpha w^\alpha(\pi, z) - M_h^\alpha w^\alpha(\pi', z')\big| \\
&\quad \leq \sup_{\pi''\in\mathcal{S}_{\delta'}} \Bigg[\bigg|\ln\frac{e(\pi, \pi'')}{e(\pi', \pi'')}\bigg| \\
&\qquad + \big|E_{\pi'',z}\big[e^{-\alpha h}\ln\big(\pi'' \cdot e^{X(h)}\big)\big] - E_{\pi'',z'}\big[e^{-\alpha h}\ln\big(\pi'' \cdot e^{X(h)}\big)\big]\big| \\
&\qquad + \big|E_{\pi'',z}\big[w^\alpha\big(\pi(h), z_h\big)\big] - E_{\pi'',z'}\big[w^\alpha\big(\pi(h), z_h\big)\big]\big|\Bigg] \\
&\quad \leq K + L\|w^\alpha\|_{sp}
\end{aligned} \tag{18.28}$$

with constant K independent of α. Now

$$\begin{aligned}
&w^\alpha(\pi, z) - w^\alpha(\pi', z') \\
&\quad \leq w^\alpha(\pi, z) - M_h^\alpha w^\alpha(\pi', z') \\
&\quad = \sup_\tau E_{\pi,z}\big[e^{-\alpha\tau\wedge T^{\delta 0}}\big[\ln\pi \cdot e^{X(\tau\wedge T^{\delta 0})} + M_h^\alpha w^\alpha\big(\pi\big(\tau\wedge T^{\delta 0}\big), z_{\tau\wedge T^{\delta 0}}\big) \\
&\qquad - M_h^\alpha w^\alpha(\pi', z')\big] + \big(e^{-\alpha\tau\wedge T^{\delta 0}} - 1\big)M_h^\alpha w^\alpha(\pi', z')\big].
\end{aligned} \tag{18.29}$$

Notice that by Doob's maximal inequalities (see, e.g., Theorem 3.8(iv) in [6]) we have

$$\sum_{i=1}^d \sup_{z\in D} E_z\Big[\sup_s\big|X_i\big(s\wedge T^{\delta 0}\big)\big|\Big] = K' < \infty, \tag{18.30}$$

and therefore,

$$\begin{aligned}\left|w^\alpha(\pi,z)\right| &\le \sum_{i=1}^\infty E_{\pi,z}\left[e^{-\alpha\tau_i}\left|\ln\left(\pi(\tau_i)\cdot e^{X(\tau_i)-X(\tau_{i-1})}\right)\right|\right]\\ &\le \sum_{i=1}^\infty e^{-\alpha h i}K' = \frac{K'}{1-e^{-\alpha h}}.\end{aligned} \tag{18.31}$$

Furthermore, since $\alpha x - \frac{\alpha^2 x^2}{2} \le 1 - e^{-\alpha x} \le \alpha x$ for $x > 0$, we have from (18.30) and (18.31) that

$$\begin{aligned}&\left|E_{\pi,z}\left[\left(e^{-\alpha\tau\wedge T^{\delta_0}}-1\right)M_h^\alpha w^\alpha(\pi',z')\right]\right|\\ &\quad\le E_{\pi,z}\left[\left(1-e^{-\alpha T^{\delta_0}}\right)\left(\bar{K}''+\sup_{(\pi'',z'')\in\mathcal{S}\times D}\left|w^\alpha(\pi'',z'')\right|\right)\right]\\ &\quad\le K''+E_{\pi,z}\left[T^{\delta_0}\right]\alpha\sup_{(\pi'',z'')\in\mathcal{S}\times D}\left|w^\alpha(\pi'',z'')\right|\\ &\quad\le K''+K'E_{\pi,z}\left[T^{\delta_0}\right]\frac{\alpha}{1-e^{-\alpha h}}\le K''+K'E_{\pi,z}\left[T^{\delta_0}\right]\frac{1}{h-\alpha h^2}.\end{aligned} \tag{18.32}$$

Summarizing, from (18.29) and (18.28), using (18.32), we obtain that

$$\sup_\alpha\left\|w^\alpha\right\|_{sp}\le K+K''+K'\sup_{(\pi,z)\in\mathcal{S}_{\delta'}\times D}E_{\pi,z}\left[T^{\delta_0}\right]\frac{1}{h-\alpha h^2}+L\left\|w^\alpha\right\|_{sp},$$

from which the assertion of proposition follows. □

Let

$$\bar{w}^\alpha(\pi,z)=w^\alpha(\pi,z)-\inf_{\pi'\in\mathcal{S}_\delta,z'\in D}w^\alpha(\pi',z'). \tag{18.33}$$

From (18.20) we have

$$\begin{aligned}\bar{w}^\alpha(\pi,z)=\sup_\tau E_{\pi,z}\Big[&e^{-\alpha\tau\wedge T^{\delta_0}}\left[\ln\left(\pi\cdot e^{X(\tau\wedge T^{\delta_0})}\right)\right.\\ &\left.+M_h^\alpha\bar{w}^\alpha\left(\pi\left(\tau\wedge T^{\delta_0}\right),z_{\tau\wedge T^{\delta_0}}\right)\right]\\ &+\inf_{\pi'\in\mathcal{S}_\delta,z'\in D}w^\alpha(\pi',z')\left(e^{-\alpha(\tau\wedge T^{\delta_0}+h)}-1\right)\Big].\end{aligned} \tag{18.34}$$

We let $\alpha\to 0$ in (18.34). Our main result can be formulated as follows.

Theorem 18.11 *There are a constant λ and a bounded continuous solution w which form a solution to the following Bellman equation*:

$$\begin{aligned}w(\pi,z)=\sup_\tau E_{\pi,z}\Big[&\ln\left(\pi\cdot e^{X(\tau\wedge T^{\delta_0})}\right)-\lambda\left(\tau\wedge T^{\delta_0}+h\right)\\ &+M_h w\left(\pi\left(\tau\wedge T^{\delta_0}\right),z_{\tau\wedge T^{\delta_0}}\right)\Big]\end{aligned} \tag{18.35}$$

with

$$M_h w(\pi, z) = \sup_{\pi' \in \mathcal{S}_{\delta'}} \left[\ln e(\pi, \pi') + E_{\pi', z}\left[\left(\ln\left(\pi' \cdot e^{X(h)}\right) + w\left(\pi(h), z_h\right) \right) \right] \right]. \quad (18.36)$$

Moreover,

$$\lambda = \sup_V \liminf_{T \to \infty} \frac{1}{T} E_{\pi, z} \left[\sum_{i=1}^{\infty} 1_{\tau_i \le T} \left[\ln\left(\pi(\tau_{i-1}) \cdot e^{X(\tau_i) - X(\tau_{i-1})}\right) + \ln e\left(\pi^-(\tau_i), \pi^i\right) \right] \right], \quad (18.37)$$

and the strategy $\hat{V} = (\hat{\tau}_n, \hat{\pi}^n)$ *such that*

$$\hat{\tau}(\pi) = \inf\left\{ s \ge 0 \colon w\left(\pi(s), z_s\right) = M_h w\left(\pi(s), z_s\right) \right\}, \quad (18.38)$$

$$\begin{aligned} \hat{\tau}_1 &= \hat{\tau}\left(\pi(0)\right), \\ \hat{\tau}_{n+1} &= \hat{\tau}_n + \hat{\tau}\left(\pi(\hat{\tau}_n)\right) \circ \theta_{\hat{\tau}_n}, \end{aligned} \quad (18.39)$$

and

$$\hat{\pi}^n = \hat{\pi}\left(\pi^-(\hat{\tau}_n), z_{\hat{\tau}_n}\right),$$

where $\hat{\pi} : S \times D \to S_{\delta'}$ *is a Borel function such that*

$$M_h w(\pi, z) = \ln e\left(\pi, \hat{\pi}(\pi, z)\right) + E_{\hat{\pi}(\pi, z), z}\left[\left(\ln\left(\hat{\pi}(\pi, z) \cdot e^{X(h)}\right) + w\left(\pi(h), z_h\right) \right) \right],$$

is optimal.

Proof By Propositions 18.10 and 18.5, $M_h^\alpha \bar{w}^\alpha(\pi, z)$ is equicontinuous and bounded, hence, for a suitably chosen subsequence $\alpha_n \to 0$, it converges to $g(\pi, z)$ uniformly on compact sets of $\mathcal{S}^0 \times D$. Furthermore, by (18.17)

$$\sup_\tau E_{\pi, z}\left[\frac{1}{\alpha}\left(e^{-\alpha(\tau \wedge T^{\delta 0} + h)} - 1\right) + \tau \wedge T^{\delta 0} + h \right] \to 0.$$

Choosing a further subsequence, which we denote for simplicity by (α_n), by similar arguments as in the proof of Proposition 18.10, there is a constant λ such that $\frac{1}{\alpha_n} \inf_{\pi' \in \mathcal{S}^\delta, z' \in D} w^{\alpha_n}(\pi', z') \to \lambda$ as $n \to \infty$. Then there is a function w such that

$$\sup_\tau E_{\pi, z}\left[e^{-\alpha_n \tau \wedge T^{\delta 0}} \left[\ln\left(\pi \cdot e^{X(\tau \wedge T^{\delta 0})}\right) + M_h^{\alpha_n} \bar{w}^{\alpha_n}\left(\pi\left(\tau \wedge T^{\delta 0}\right), z_{\tau \wedge T^{\delta 0}}\right) \right] + \inf_{\pi' \in \mathcal{S}^\delta, z' \in D} w^{\alpha_n}(\pi', z')\left(e^{-\alpha_n(\tau \wedge T^{\delta 0} + h)} - 1\right) \right]$$
$$\to w(\pi, z)$$

$$= \sup_{\tau} E_{\pi,z}\big[\ln\big(\pi \cdot e^{X(\tau\wedge T^{\delta^0})}\big) - \lambda\big(\tau \wedge T^{\delta^0} + h\big) + g\big(\pi\big(\tau \wedge T^{\delta^0}\big), z_{\tau\wedge T^{\delta^0}}\big)\big] \quad (18.40)$$

as $n \to \infty$. Moreover, $M_h^{\alpha_n} \bar{w}^{\alpha_n}(\pi, z) \to M_h w(\pi, z)$, and therefore from (18.40) we have that $\bar{w}^{\alpha_n} \to w$ uniformly on compact subsets of $\mathcal{S}^0 \times D$, and finally we have (18.35). Let

$$\bar{w}(\pi, z, W) := \sup_{\tau} E_{\pi,z}\big[\ln\big(W\pi \cdot e^{X(\tau\wedge T^{\delta^0})}\big) - \lambda\big(\tau \wedge T^{\delta^0} + h\big) + M_h w\big(\pi\big(\tau \wedge T^{\delta^0}\big), z_{\tau\wedge T^{\delta^0}}\big)\big]. \quad (18.41)$$

Clearly, $\bar{w}(\pi, z, W) = \ln W + w(\pi, z)$, and therefore the optimal stopping time $\hat{\tau}$ is of the form (18.71) (see also the proofs of Theorems 1 and 3 in [4]). Equality (18.37) can be justified in a standard way for impulsive control of Markov processes as Theorem V.2.1 of [10] (see also Theorem 3 in [4]), using a suitable version of Lemma II.2.2 of [10]. For completeness, we sketch below the main steps. For a given impulsive strategy $V = (\tau_n, \pi^n)$, consider the following notation: for $n = 1, 2, \ldots$, $\pi_n(\tau_n) = \pi^n$, $\pi_n(\tau_n + s) = \pi^-(\tau_n + s)$ for $s > 0$, $W_{\tau_n}^n = 1$, and $W_{\tau_n+s}^n = W_{\tau_n+s}^- = \pi^n \cdot e^{X(\tau_n+s)-X(\tau_n)}$ for $s > 0$. It is clear that, for $s \geq 0$,

$$\bar{w}(\pi, z, W) \geq E_{\pi,z}\big[\bar{w}\big(\pi^-\big(s \wedge T^{\delta^0}\big), z_{s\wedge T^{\delta^0}}, W^-_{s\wedge T^{\delta^0}}\big) - \lambda\big(s \wedge T^{\delta^0}\big)\big], \quad (18.42)$$

and therefore, for $T^{\delta^0}_{\tau_n+h} := \tau_n + h + T^{\delta^0} \circ \theta_{\tau_n+h}$, where θ is a Markov shift operator, we have that

$$Z_n(s) = \bar{w}\big(\pi_n\big((\tau_n + h + s) \wedge T^{\delta^0}_{\tau_n+h}\big), z_{(\tau_n+h+s)\wedge T^{\delta^0}_{\tau_n+h}}, W^n_{(\tau_n+h+s)\wedge T^{\delta^0}_{\tau_n+h}}\big) - \lambda\big(s \wedge \big(T^{\delta^0} \circ \theta_{\tau_n+h}\big)\big) \quad (18.43)$$

is a $\mathcal{G}_s^n = \mathcal{F}_{\tau_n+h+s}$-supermartingale. For any stopping time $\tau \geq \tau_n + h$, since

$$\{\tau - \tau_n - h \leq s\} = \{\tau \leq \tau_n + h + s\} \in \mathcal{F}_{\tau_n+h+s} = \mathcal{G}_s^n,$$

we have that $\tau - \tau_n - h$ is a $(\mathcal{G}_s^n)$-stopping time. Therefore, if additionally $\tau \leq T^{\delta^0}_{\tau_n+h}$, we have

$$E\big[Z_n(\tau - \tau_n - h)\big|\mathcal{F}_{\tau_n+h}\big] \leq Z_n(0) = \bar{w}\big(\pi_n(\tau_n + h), z_{\tau_n+h}, W^n_{\tau_n+h}\big). \quad (18.44)$$

Since by the definition of the operator M_h, using $\bar{w}(\pi, z, W) = \ln W + w(\pi, z)$, we have

$$\bar{w}\big(\pi_{n-1}(\tau_n), z_{\tau_n}, W_{\tau_n}^{n-1}\big) \geq -\lambda h + E\big[\ln W_{\tau_n}^{n-1} e\big(\pi_{n-1}(\tau_n), \pi^n\big) + \bar{w}\big(\pi_n(\tau_n + h), z_{\tau_n+h}\big), W^n_{\tau_n+h}\big)\big|\mathcal{F}_{\tau_n}\big] \quad (18.45)$$

for fixed $T > 0$, we obtain

$$\begin{aligned}
&E\big[Z_n\big(\tau_{n+1} \wedge (T+h) - \tau_n - h\big)\chi_{\tau_n \le T}\big|\mathcal{F}_{\tau_n}\big] \\
&\quad \le \chi_{\tau_n \le T} E\big[Z_n(0)\big|\mathcal{F}_{\tau_n}\big] \\
&\quad \le \chi_{\tau_n \le T}\big(\bar{w}\big(\pi_{n-1}(\tau_n), z_{\tau_n}, W_{\tau_n}^{n-1}\big) + \lambda h - E\big[\ln\big(W_{\tau_n}^{n-1} e\big(\pi_{n-1}(\tau_n), \pi^n\big)\big)\big|\mathcal{F}_{\tau_n}\big]\big).
\end{aligned}$$

Therefore,

$$\begin{aligned}
&E\big[\chi_{\tau_n \le T}\big(\bar{w}\big(\pi_n\big(\tau_{n+1} \wedge (T+h)\big), z_{\tau_{n+1}\wedge(T+h)}, W^n_{\tau_{n+1}\wedge(T+h)}\big) \\
&\quad - \bar{w}\big(\pi_{n-1}(\tau_n), z_{\tau_n}, W_{\tau_n}^{n-1}\big) + \ln\big(W_{\tau_n}^{n-1} e\big(\pi_{n-1}(\tau_n), \pi^n\big)\big)\big)\big] \\
&\le E\big[\chi_{\tau_n \le T}\lambda\big(\tau_{n+1} \wedge (T+h) - \tau_n\big)\big]. \qquad (18.46)
\end{aligned}$$

Summing up over n inequalities (18.45) and using again the identity $\bar{w}(\pi, z, W) = \ln W + w(\pi, z)$, we obtain the formula

$$\begin{aligned}
&E\Bigg[\bar{w}\big(\pi_{\zeta(T)-1}\big(\tau_{\zeta(T)} \wedge (T+h)\big), z_{\tau_{\zeta(T)}\wedge(T+h)}, 1\big) - \bar{w}(\pi, z, W) \\
&\quad + \sum_{i=0}^{\zeta(T)-1} \ln\big(W^i_{\tau_{i+1}} e\big(\pi_i(\tau_{i+1}), \pi^{i+1}\big)\big) - \ln e\big(\pi_{\zeta(T)-1}(\tau_{\zeta(T)}), \pi^{\zeta(T)}\big)\Bigg] \\
&\le \lambda E\big[\tau_{\zeta(T)} \wedge (T+h)\big], \qquad (18.47)
\end{aligned}$$

where $\zeta(T) = \inf\{n : \tau_n \ge T\}$. Notice that for the strategy $\hat{V}$ defined in (18.38) and (18.39), we have equalities in (18.45)–(18.47). Dividing both sides of (18.47) by T and letting $T \to \infty$, we obtain (18.37) and the optimality of the strategy $\hat{V}$. □

Remark 18.12 In the case of execution delay we easily obtain versions of Proposition 18.10 and Theorem 18.11. The only difference is in the form of the operator M_h, which is now

$$M_h w(\pi, z) = \sup_{\pi' \in \mathcal{S}_{\delta'}} E_{\pi,z}\big[\big(\ln\big(\pi \cdot e^{X(h)}\big) + \ln e\big(\pi(h), \pi'\big) + w(\pi', z_h)\big)\big].$$

18.4 Discounted Discrete-Time PUOP

In this section we restrict ourselves to the model with execution delay. We consider power utility function $U(x) = x^\gamma$ with $\gamma \in (0, \Gamma]$, where for $\gamma \le \Gamma \le 1$, estimate (18.17) holds. We use again a version of vanishing discount approach. Due to multiplicative form of the functional, we however have a number of difficulties to overcome. In the case of decision lag we are not able to obtain technical estimations (see Proposition 18.15 below) convenient for the vanishing discount approach, and therefore we consider the model with execution delay. We start with a discounted discrete-time problem. Assume that decisions are made in discrete times $\{0, \Delta, 2\Delta, \ldots, i\Delta, \ldots\}$. The decision is executed with a delay h, which is a multiplicity of Δ. Since delay h is fixed, we consider only such Δ for which its multiplicity forms h. Denote by $\mathcal{T}_\Delta$ the family of stopping times with values in $\{0, \Delta, 2\Delta, \ldots, i\Delta, \ldots\}$.

Discrete-time discounted problem is of the form: find a bounded function $g_\alpha^\Delta : \mathcal{S} \times D \times [0, \Gamma] \to R$ such that $g_\alpha^\Delta(\pi, x, 0) \equiv 0$ and

$$g_\alpha^\Delta(\pi, z, \gamma) = 1_{\mathcal{S}_\delta^{0c}}(\pi) M_h^{r,\alpha} g_\alpha^\Delta(\pi, z, \gamma) + 1_{\mathcal{S}_\delta^0}(\pi) \max\bigl\{\ln E_{\pi,z}\bigl[\exp\bigl\{\gamma e^{-\alpha\Delta} \times \bigl[\ln\bigl(\pi \cdot e^{X(\Delta)}\bigr)\bigr] + g_\alpha^\Delta\bigl(\pi(\Delta), z_\Delta, \gamma e^{-\alpha\Delta}\bigr)\bigr\}\bigr], M_h^{r,\alpha} g_\alpha^\Delta(\pi, z, \gamma)\bigr\} \tag{18.48}$$

with

$$M_h^{r,\alpha} g(\pi, z, \gamma) = \sup_{\pi' \in \mathcal{S}_{\delta'}} \bigl[\ln E_{\pi,z}\bigl[\exp\bigl\{\gamma e^{-\alpha h} \ln\bigl(\pi \cdot e^{X(h)}\bigr) + \gamma e^{-\alpha h} \ln e\bigl(\pi(h), \pi'\bigr) + g\bigl(\pi', z_h, \gamma e^{-\alpha h}\bigr)\bigr\}\bigr]\bigr]. \tag{18.49}$$

An equivalent form of (18.48) is

$$g_\alpha^\Delta(\pi, z, \gamma) = \sup_{\tau \in \mathcal{T}_\Delta} \ln E_{\pi,z}\left[\exp\left\{\gamma \sum_{i=1}^{\frac{\tau \wedge T_\Delta^{\delta^0}}{\Delta}} e^{-\alpha i\Delta}\bigl[\ln\bigl(\pi\bigl((i-1)\Delta\bigr) \cdot e^{X(i\Delta) - X((i-1)\Delta)}\bigr)\bigr] + M_h^{r,\alpha} g_\alpha^\Delta\bigl(\pi\bigl(\tau \wedge T_\Delta^{\delta^0}\bigr), z_{\tau \wedge T_\Delta^{\delta^0}}, \gamma e^{-\alpha\tau \wedge T_\Delta^{\delta^0}}\bigr)\right\}\right] \tag{18.50}$$

with

$$T_\Delta^{\delta^0} = \inf\bigl\{i\Delta : \pi(i\Delta) \in \mathcal{S} \setminus \mathcal{S}_\delta^0\bigr\}. \tag{18.51}$$

We have the following:

Proposition 18.13 *There is a unique bounded function $g_\alpha^\Delta : \mathcal{S} \times D \times [0, \Gamma] \to R$ such that $g_\alpha^\Delta(\pi, z, 0) \equiv 0$, which is continuous in $\mathcal{S}_\delta^0$, and for which* (18.48) *and, equivalently,* (18.50) *are satisfied. Moreover,*

$$g_\alpha^\Delta(\pi, z, \gamma) = \sup_V \ln E_{\pi,z}\left[\exp\left\{\gamma \sum_{i=1}^{\frac{\tau_1}{\Delta}} e^{-\alpha i\Delta}\bigl[\ln\bigl(\pi\bigl((i-1)\Delta\bigr) \cdot e^{X(i\Delta) - X((i-1)\Delta)}\bigr)\bigr]\right.\right.$$

$$+\gamma\sum_{j=2}^{\infty}\sum_{i=\frac{\tau_{j-1}+h}{\Delta}+1}^{\frac{\tau_j}{\Delta}} e^{-\alpha i\Delta}\big[\ln\big(\pi\big((i-1)\Delta\big)\cdot e^{X(i\Delta)-X((i-1)\Delta)}\big)\big]$$
$$+\gamma\sum_{i=1}^{\infty}e^{-\alpha(\tau_i+h)}\big[\ln\big(\pi(\tau_i)\cdot e^{X(\tau_i+h)-X(\tau_i)}\big)+\ln e\big(\pi^-(\tau_i+h),\pi^i\big)\big]\Bigg\}\Bigg].$$
(18.52)

Proof We first prove that the operator T defined by right-hand side of (18.48) as

$$Tg(\pi,z,\gamma)=1_{\mathcal{S}_\delta^{0c}}(\pi)M_h^{r,\alpha}g(\pi,z,\gamma)+1_{\mathcal{S}_\delta^0}(\pi)\max\big\{\ln E_{\pi,z}\big[\exp\big\{\gamma e^{-\alpha\Delta}$$
$$\times\big[\ln\big(\pi\cdot e^{X(\Delta)}\big)\big]+g\big(\pi(\Delta),z_\Delta,\gamma e^{-\alpha\Delta}\big)\big\}\big],M_h^{r,\alpha}g(\pi,z,\gamma)\big\}$$

transforms the class of bounded functions continuous in $\mathcal{S}_\delta^0$ into itself. Note first that by Proposition 18.5 the operator $M_h^{r,\alpha}$ transforms bounded functions into bounded continuous functions. By the uniform integrability of the term $(\pi\cdot e^{X(\tau\wedge T^{\delta 0})})^{\gamma e^{-\alpha(\tau\wedge T^{\delta 0}+h)}}$ (notice that $\Gamma\le 1$ and we have assumed (18.3)) it suffices to show the continuity of

$$G_\alpha(\pi,z,W,\gamma)=\sup_\tau\ln E_{\pi,z}\big[\exp\big\{\gamma e^{-\alpha\tau\wedge T^{\delta 0}}f(W_{\tau\wedge T^{\delta 0}})$$
$$+g\big(\pi\big(\tau\wedge T^{\delta 0}\big),z_{\tau\wedge T^{\delta 0}},\gamma e^{-\alpha\tau\wedge T^{\delta 0}}\big)\big\}\big]$$

for continuous bounded functions f and g. Now, similarly as in the proof of Lemma 18.6, we can use a penalty method (for time-dependent functions of π, z, and W studied in [4] and [13]) and obtain by (18.15) and (18.16) the continuity of the function G_α. Consequently, for a bounded function g, the function Tg is also a continuous bounded function in $\mathcal{S}_\delta^0$, and $Tg(\pi,z,0)\equiv 0$, whenever $g(\pi,z,0)\equiv 0$. Moreover by the compactness of $\mathcal{S}\times\mathcal{D}$ we have the uniform convergence of $g(\pi,z,\gamma)$ to 0 as $\gamma\to 0$. Therefore, iterations of the operator T converge to a function g_α, which is a solution to (18.48). Since the term g diminishes in the successive iterations of T, the limit does not depend on the function g, provided that g is continuous bounded in $\mathcal{S}_\delta^0$ and $g(\pi,z,0)\equiv 0$. The form of (18.50) follows by iteration from (18.48) noticing that

$$\ln\big(\pi\cdot e^{X(\tau\wedge T^{\delta 0})}\big)+\ln\big(\pi(\tau)\cdot e^{X(\tau\wedge T^{\delta 0}+h)-X(\tau\wedge T^{\delta 0})}\big)=\ln\big(\pi\cdot e^{X(\tau\wedge T^{\delta 0}+h)}\big).\quad\square$$

In what follows wee need the following lemma.

Lemma 18.14 *We have*

$$\sup_{\pi,\pi'\in\mathcal{S}_\delta,\pi''\in\mathcal{S}_{\delta'}}\ \sup_{z,z',z''\in D}\ \sup_{0\le\gamma\le\Gamma}\frac{E_{\pi,z}[\exp\{\gamma e^{-\alpha h}\ln(\pi\cdot e^{X(h)}e(\pi(h),\pi''))\}1_{z''}(z_h)]}{E_{\pi',z'}[\exp\{\gamma e^{-\alpha h}\ln(\pi'\cdot e^{X(h)}e(\pi(h),\pi''))\}1_{z''}(z_h)]}$$
$$:=L'<\infty. \tag{18.53}$$

Proof Suppose that $L' = \infty$. Then there are sequences $(\pi_{(n)})$, $(\pi'_{(n)})$, $(\pi''_{(n)})$, $\gamma_n \to 0$ and elements z, z', z'' of D (since D is finite) such that

$$\frac{E_{\pi_{(n)},z}[\exp\{\gamma_n e^{-\alpha h} \ln(\pi_{(n)} \cdot e^{X(h)} e(\pi(h), \pi''_{(n)})\} 1_{z''}(z_h)]}{E_{\pi'_{(n)},z'}[\exp\{\gamma_n e^{-\alpha h} \ln(\pi'_{(n)} \cdot e^{X(h)} e(\pi(h), \pi''_{(n)})\} 1_{z''}(z_h)]} \to \infty.$$

Since the numerator and denominator are bounded from above, it may happen only when

$$E_{\pi'_{(n)},z'}\big[\exp\big\{\gamma_n e^{-\alpha h} \ln(\pi'_{(n)} \cdot e^{X(h)} e\big(\pi(h), \pi''_{(n)}\big)\big\} 1_{z''}(z_h)\big] \to 0$$

as $n \to \infty$. Therefore, $E_{z'}[1_{z''}(z_h)] = 0$. By the form of the transition density (18.12), $E_{z'}[1_{z''}(z_h)] > 0$ for any $z' \in D$, and we have a contradiction. □

The following two estimations play crucial role in the vanishing discount approach in the next section.

Proposition 18.15 *There is a constant $L < \infty$ such that, for $\pi, \pi' \in \mathcal{S}$, $z, z' \in D$, and $\gamma \le \Gamma$, we have*

$$g_\alpha^\Delta(\pi, z, \gamma) - g_\alpha^\Delta(\pi', z', \gamma) \le L \tag{18.54}$$

and, for $\gamma_1 \le \gamma_2$,

$$\sup_{\pi \in \mathcal{S}_{\delta'}} g_\alpha^\Delta(\pi, z, \gamma_1) - g_\alpha^\Delta(\pi, z; \gamma_2) \le L \frac{\gamma_2 - \gamma_1}{1 - e^{-\alpha h}}. \tag{18.55}$$

Proof Notice first that

$$\begin{aligned}
&g_\alpha^\Delta(\pi, z, \gamma) - g_\alpha^\Delta(\pi', z', \gamma) \\
&\quad \le g_\alpha^\Delta(\pi, z, \gamma) - M_h^{r,\alpha} g_\alpha^\Delta(\pi', z', \gamma) \\
&\quad \le \sup_\tau \ln E_{\pi,z}\Bigg[\exp\Bigg\{\gamma \sum_{i=1}^{\frac{\tau \wedge T_\Delta^{\delta 0}}{\Delta}} e^{-\alpha i \Delta}\big[\ln\big(\pi\big((i-1)\Delta\big) \cdot e^{X(i\Delta) - X((i-1)\Delta)}\big)\big] \\
&\qquad + M_h^{r,\alpha} g_\alpha^\Delta\big(\pi\big(\tau \wedge T_\Delta^{\delta 0}\big), z_{\tau \wedge T_\Delta^{\delta 0}}, \gamma e^{-\alpha \tau \wedge T_\Delta^{\delta 0}}\big) - M_h^{r,\alpha} g_\alpha^\Delta\big(\pi', z', \gamma e^{-\alpha \tau \wedge T_\Delta^{\delta 0}}\big) \\
&\qquad + M_h^{r,\alpha} g_\alpha^\Delta\big(\pi', z', \gamma e^{-\alpha \tau \wedge T_\Delta^{\delta 0}}\big) - M_h^{r,\alpha} g_\alpha^\Delta(\pi', z', \gamma)\Bigg\}\Bigg],
\end{aligned} \tag{18.56}$$

and, using (18.53), we obtain

$$\begin{aligned}
&M_h^{r,\alpha} g_\alpha^\Delta(\pi, z, \gamma) - M_h^{r,\alpha} g_\alpha^\Delta(\pi', z', \gamma) \\
&\quad \le \sup_{\pi'' \in \mathcal{S}_{\delta'}} \ln \frac{E_{\pi,z}[\exp\{\gamma e^{-\alpha h} \ln(\pi \cdot e^{X(h)} e(\pi(h), \pi'')) + g_\alpha^\Delta(\pi'', z_h, \gamma e^{-\alpha h})\}]}{E_{\pi',z'}[\exp\{\gamma e^{-\alpha h} \ln(\pi' \cdot e^{X(h)} e(\pi(h), \pi'')) + g_\alpha^\Delta(\pi'', z_h, \gamma e^{-\alpha h})\}]} \\
&\quad \le L'.
\end{aligned} \tag{18.57}$$

To obtain (18.55), we have to estimate

$$M_h^{r,\alpha} g_\alpha^\Delta\big(\pi', z', \gamma e^{-\alpha\tau\wedge T_\Delta^{\delta 0}}\big) - M_h^{r,\alpha} g_\alpha^\Delta(\pi', z', \gamma).$$

Let $\gamma_2 = e^{-\alpha\tau\wedge T_\Delta^{\delta 0}}$, and $\gamma_1 = \gamma\gamma_2$. By the conditional Hölder inequality

$$E\big[e^{\gamma\gamma_2 Z}|\mathcal{F}_{\tau\wedge T_\Delta^{\delta 0}}\big] \le \big(E\big[e^{\gamma Z}|\mathcal{F}_{\tau\wedge T_\Delta^{\delta 0}}\big]\big)^{\gamma_2}$$

for a suitably integrable random variable Z and from (18.52) we have

$$\frac{1}{\gamma_2} g_\alpha^\Delta(\pi, z, \gamma_1) \le g_\alpha^\Delta(\pi, z, \gamma). \tag{18.58}$$

Moreover,

$$\begin{aligned} &M_h^{r,\alpha} g_\alpha^\Delta(\pi', z', \gamma_1) - M_h^{r,\alpha} g_\alpha^\Delta(\pi', z', \gamma) \\ &\le \sup_{\pi''\in\mathcal{S}_{\delta'}} \ln \frac{E_{\pi',z'}[\exp\{\gamma_1 e^{-\alpha h}\ln(\pi'\cdot e^{X(h)} e(\pi(h),\pi'')) + g_\alpha^\Delta(\pi'', z_h, \gamma_1 e^{-\alpha h})\}]}{E_{\pi',z'}[\exp\{\gamma e^{-\alpha h}\ln(\pi'\cdot e^{X(h)} e(\pi(h),\pi'')) + g_\alpha^\Delta(\pi'', z_h, \gamma e^{-\alpha h})\}]} \\ &= \sup_{\pi''\in\mathcal{S}_{\delta'}} a(\pi''), \end{aligned} \tag{18.59}$$

and by (18.58) (applied to $\gamma := \gamma e^{-\alpha h}$), since $\gamma_2 < 1$ and $\ln E\{Z\} \ge E\{\ln Z\}$ for a positive random variable Z, we have

$$\begin{aligned} a(\pi'') &\le \ln \frac{E_{\pi',z'}[\exp\{\gamma_1 e^{-\alpha h}\ln(\pi'\cdot e^{X(h)} e(\pi(h),\pi'')) + \gamma_2 g_\alpha^\Delta(\pi'', z_h, \gamma e^{-\alpha h})\}]}{E_{\pi',z'}[\exp\{\gamma e^{-\alpha h}\ln(\pi'\cdot e^{X(h)} e(\pi(h),\pi'')) + g_\alpha^\Delta(\pi'', z_h, \gamma e^{-\alpha h})\}]} \\ &\le \ln \frac{(E_{\pi',z'}[\exp\{\gamma e^{-\alpha h}\ln(\pi'\cdot e^{X(h)} e(\pi(h),\pi'')) + g_\alpha^\Delta(\pi'', z_h, \gamma e^{-\alpha h})\}])^{\gamma_2}}{E_{\pi',z'}[\exp\{\gamma e^{-\alpha h}\ln(\pi'\cdot e^{X(h)} e(\pi(h),\pi'')) + g_\alpha^\Delta(\pi'', z_h, \gamma e^{-\alpha h})\}]} \\ &= (\gamma_2 - 1)\ln\big(E_{\pi',z'}\big[\exp\big\{\gamma e^{-\alpha h}\ln\big(\pi'\cdot e^{X(h)} e\big(\pi(h),\pi''\big)\big) \\ &\quad + g_\alpha^\Delta\big(\pi'', z_h, \gamma e^{-\alpha h}\big)\big\}\big]\big) \\ &\le (\gamma_2 - 1) E_{\pi',z'}\big[\gamma e^{-\alpha h}\ln\big(\pi'\cdot e^{X(h)} e\big(\pi(h),\pi''\big)\big) \\ &\quad + g_\alpha^\Delta\big(\pi'', z_h, \gamma e^{-\alpha h}\big)\big]. \end{aligned} \tag{18.60}$$

It remains to estimate $(\gamma_2 - 1)\inf_{\pi\in\mathcal{S}_{\delta'}} g_\alpha^\Delta(\pi, z, \gamma)$. By (18.52), changing portfolio every time when it is allowed, i.e., changing portfolio after every h units of time, we have (with $\pi(0) = \pi$)

$$\begin{aligned} g_\alpha^\Delta(\pi, z, \gamma) \ge \ln E_{\pi,z}\Bigg[\exp\Bigg\{\gamma \sum_{i=1}^{\infty} e^{-\alpha(ih)}\big[\ln\big(\pi\big((i-1)h\big)\cdot e^{X(ih)-X((i-1)h)}\big) \\ + \ln e\big(\pi^-(ih), \pi^i\big)\big]\Bigg\}\Bigg] \end{aligned}$$

$$\geq E_{\pi,z}\Bigg[\Bigg\{\gamma\sum_{i=1}^{\infty}e^{-\alpha(ih)}\big[\ln\big(\pi\big((i-1)h\big)\cdot e^{X(ih)-X((i-1)h)}\big)$$

$$+\ln e\big(\pi^{-}(ih),\pi^{i}\big)\big]\Bigg\}\Bigg]$$

$$\geq \frac{K\gamma e^{-\alpha h}}{1-e^{-\alpha h}} \tag{18.61}$$

with $K=\inf_{\pi,\pi'\in\mathcal{S}_{\delta'},z\in D}E_{\pi,z}\{\ln(\pi\cdot e^{X(h)})+\ln e(\pi^{-}(h),\pi'))\}$. Therefore, by (18.60),

$$a(\pi'')\leq(\gamma_2-1)\gamma e^{-\alpha h}\left(K+\frac{K}{1-e^{-\alpha h}}\right)\leq(1-\gamma_2)\gamma\frac{2|K|}{1-e^{-\alpha h}}$$

$$\leq\big(\alpha\tau\wedge T_{\Delta}^{\delta^0}\big)\gamma\frac{2|K|}{1-e^{-\alpha h}}\leq\frac{\alpha}{1-e^{-\alpha h}}T_{\Delta}^{\delta^0}2\gamma|K|, \tag{18.62}$$

and summarizing (18.56)–(18.62), we obtain (18.54).

We are going now to prove (18.55). For $\gamma_1\leq\gamma_2$ and a positive random variable Z, by the Hölder inequality we have $E[Z^{\gamma_1}]\leq(E[Z^{\gamma_2}])^{\frac{\gamma_1}{\gamma_2}}$. Therefore,

$$e^{g_{\alpha}^{\Delta}(\pi,z,\gamma_1)-g_{\alpha}^{\Delta}(\pi,z,\gamma_2)}$$

$$\leq\sup_{V}\frac{E_{\pi,z}[\exp\{\gamma_1\sum_{i=1}^{\frac{\tau_1}{\Delta}}e^{-\alpha i\Delta}[\ln(\pi((i-1)\Delta)\cdot e^{X(i\Delta)-X((i-1)\Delta)})]}{E_{\pi,z}[\exp\{\gamma_2\sum_{i=1}^{\frac{\tau_1}{\Delta}}e^{-\alpha i\Delta}[\ln(\pi((i-1)\Delta)\cdot e^{X(i\Delta)-X((i-1)\Delta)})]}$$

$$\frac{+\gamma_1\sum_{j=2}^{\infty}\sum_{i=\frac{\tau_{j-1}+h}{\Delta}+1}^{\frac{\tau_j}{\Delta}}e^{-\alpha i\Delta}[\ln(\pi((i-1)\Delta)\cdot e^{X(i\Delta)-X((i-1)\Delta)})]}{+\gamma_2\sum_{j=2}^{\infty}\sum_{i=\frac{\tau_{j-1}+h}{\Delta}+1}^{\frac{\tau_j}{\Delta}}e^{-\alpha i\Delta}[\ln(\pi((i-1)\Delta)\cdot e^{X(i\Delta)-X((i-1)\Delta)})]}$$

$$\frac{+\gamma_1\sum_{i=1}^{\infty}e^{-\alpha(\tau_i+h)}[\ln(\pi(\tau_i)\cdot e^{X(\tau_i+h)-X(\tau_i)})+\ln e(\pi^{-}(\tau_i+h),\pi^{i}))]\}]}{+\gamma_2\sum_{i=1}^{\infty}e^{-\alpha(\tau_i+h)}[\ln(\pi(\tau_i)\cdot e^{X(\tau_i+h)-X(\tau_i)})+\ln e(\pi^{-}(\tau_i+h),\pi^{i}))]\}]}$$

$$\leq\Bigg(\ln E_{\pi,z}\Bigg[\exp\Bigg\{\gamma_2\sum_{i=1}^{\frac{\tau_1}{\Delta}}e^{-\alpha i\Delta}\big[\ln\big(\pi\big((i-1)\Delta\big)\cdot e^{X(i\Delta)-X((i-1)\Delta)}\big)\big]$$

$$+\gamma_2\sum_{j=2}^{\infty}\sum_{i=\frac{\tau_{j-1}+h}{\Delta}+1}^{\frac{\tau_j}{\Delta}}e^{-\alpha i\Delta}\big[\ln\big(\pi\big((i-1)\Delta\big)\cdot e^{X(i\Delta)-X((i-1)\Delta)}\big)\big]$$

$$+\gamma_2 \sum_{i=1}^{\infty} e^{-\alpha(\tau_i+h)}\Big[\ln\big(\pi(\tau_i)\cdot e^{X(\tau_i+h)-X(\tau_i)}\big)$$

$$+\ln e\big(\pi^-(\tau_i+h),\pi^i\big)\Big]\Big\}\Big]\Big)^{\frac{\gamma_1}{\gamma_2}-1}. \tag{18.63}$$

Since $\ln E\{Z\} \geq E\{\ln Z\}$ for a positive random variable Z, and in the impulse strategies V we can restrict ourselves to the Markov strategies (depending on the current values of the processes $(\pi(t))$ and $(z(t))$ only), we have

$$g_\alpha^\Delta(\pi,z,\gamma_1) - g_\alpha^\Delta(\pi,z,\gamma_2)$$

$$\leq -\inf_V\left(1-\frac{\gamma_1}{\gamma_2}\right)$$

$$\times \ln E_{\pi,z}\Bigg[\exp\Bigg\{\gamma_2 \sum_{i=1}^{\frac{\tau_1}{\Delta}} e^{-\alpha i\Delta}\Big[\ln\Big(\pi\big((i-1)\Delta\big)\cdot e^{X(i\Delta)-X((i-1)\Delta)}\Big)\Big]$$

$$+\gamma_2 \sum_{j=2}^{\infty} \sum_{i=\frac{\tau_{j-1}+h}{\Delta}+1}^{\frac{\tau_j}{\Delta}} e^{-\alpha i\Delta}\Big[\ln\Big(\pi\big((i-1)\Delta\big)\cdot e^{X(i\Delta)-X((i-1)\Delta)}\Big)\Big]$$

$$+\gamma_2 \sum_{i=1}^{\infty} e^{-\alpha(\tau_i+h)}\Big[\ln\big(\pi(\tau_i)\cdot e^{X(\tau_i+h)-X(\tau_i)}\big)+\ln e\big(\pi^-(\tau_i+h),\pi^i\big)\Big]\Bigg\}\Bigg]$$

$$\leq -\inf_V\left(1-\frac{\gamma_1}{\gamma_2}\right)E_{\pi,z}\Bigg[\gamma_2 \sum_{i=1}^{\frac{\tau_1}{\Delta}} e^{-\alpha i\Delta}\Big[\ln\Big(\pi\big((i-1)\Delta\big)\cdot e^{X(i\Delta)-X((i-1)\Delta)}\Big)\Big]$$

$$+\gamma_2 \sum_{j=2}^{\infty} \sum_{i=\frac{\tau_{j-1}+h}{\Delta}+1}^{\frac{\tau_j}{\Delta}} e^{-\alpha i\Delta}\Big[\ln\Big(\pi\big((i-1)\Delta\big)\cdot e^{X(i\Delta)-X((i-1)\Delta)}\Big)\Big]$$

$$+\gamma_2 \sum_{i=1}^{\infty} e^{-\alpha(\tau_i+h)}\Big[\ln\big(\pi(\tau_i)\cdot e^{X(\tau_i+h)-X(\tau_i)}\big)+\ln e\big(\pi^-(\tau_i+h),\pi^i\big)\Big]\Bigg]$$

$$\leq \sup_V(\gamma_2-\gamma_1)E_{\pi,z}\Bigg[1+\sum_{i=2}^{\infty} e^{-\alpha(\tau_{i-1}+h)}\Bigg](\bar{\psi}_\alpha)$$

$$\leq (\gamma_2-\gamma_1)\sum_{i=1}^{\infty} e^{-\alpha h(i-1)}(\bar{\psi}_\alpha) = \frac{\gamma_2-\gamma_1}{1-e^{-\alpha h}}(\bar{\psi}_\alpha) \tag{18.64}$$

with $\bar{\psi}_\alpha \geq 0$ of the form

$$\bar{\psi}_\alpha = -\inf_{\pi' \in \mathcal{S}, \pi'' \in \mathcal{S}^0_{\delta'}, z' \in D} \inf_\tau E_{\pi', z'}\Bigg[\sum_{i=1}^{\frac{\tau \wedge T^{\delta^0}_\Delta}{\Delta}} e^{-\alpha i \Delta}\big[\ln\big(\pi\left((i-1)\Delta\right)$$

$$\cdot e^{X(i\Delta) - X((i-1)\Delta)}\big)\big]$$

$$+ e^{-\alpha(\tau \wedge T^{\delta^0}_\Delta + h)} \ln\big(\pi\left(\tau \wedge T^{\delta^0}_\Delta + h\right) \cdot e^{X(\tau \wedge T^{\delta^0}_\Delta + h)} e\big(\pi^-\left(\tau \wedge T^{\delta^0}_\Delta + h\right), \pi''\big)\big)\Bigg].$$

This completes the proof of (18.55). □

18.5 Long-Run PUOP

In this section we first consider a discrete-time version of the long-run power utility optimal control and then by limit procedure, based on the bounds from Proposition 18.15, we obtain the continuous-time long-run Bellman equation. The approach to discrete-time long-run PUOP will be based on vanishing discount. Fix $\bar{\pi} \in \mathcal{S}_\delta$ and $\bar{z} \in D$.

Let $\bar{g}^\Delta_\alpha(\pi, z, \gamma) = g^\Delta_\alpha(\pi, z, \gamma) - g^\Delta_\alpha(\bar{\pi}, \bar{z}, \gamma)$.

Then from (18.48) and (18.50) we have

$$\begin{aligned} &\bar{g}^\Delta_\alpha(\pi, z, \gamma) \\ &\quad = 1_{\mathcal{S}^{0c}_\delta}(\pi)\big(M^{r,\alpha}_h \bar{g}^\Delta_\alpha(\pi, z, \gamma) + g^\Delta_\alpha\big(\bar{\pi}, \bar{z}, \gamma e^{-\alpha h}\big) - g^\Delta_\alpha(\bar{\pi}, \bar{z}, \gamma)\big) \\ &\qquad + 1_{\mathcal{S}^0_\delta}(\pi) \max\big\{\ln E_{\pi, z}\big[\exp\big\{\gamma e^{-\alpha\Delta}\big[\ln\big(\pi \cdot e^{X(\Delta)}\big)\big] \\ &\qquad + \big(\bar{g}^\Delta_\alpha\big(\pi(\Delta), z_\Delta, \gamma e^{-\alpha\Delta}\big) + g^\Delta_\alpha\big(\bar{\pi}, \bar{z}, \gamma e^{-\alpha\Delta}\big) - g^\Delta_\alpha(\bar{\pi}, \bar{z}, \gamma)\big)\big\}\big], \\ &\qquad M^{r,\alpha}_h \bar{g}^\Delta_\alpha(\pi, z, \gamma) + g^\Delta_\alpha\big(\bar{\pi}, \bar{z}, \gamma e^{-\alpha h}\big) - g^\Delta_\alpha(\bar{\pi}, \bar{z}, \gamma)\big\}, \end{aligned} \tag{18.65}$$

and

$$\begin{aligned} &\bar{g}^\Delta_\alpha(\pi, z, \gamma) \\ &\quad = \sup_{\tau \in \mathcal{T}_\Delta} \ln E_{\pi, z}\Bigg[\exp\Bigg\{\gamma \sum_{i=1}^{\frac{\tau \wedge T^{\delta^0}_\Delta}{\Delta}} e^{-\alpha i \Delta}\big[\ln\big(\pi\left((i-1)\Delta\right) \cdot e^{X(i\Delta) - X((i-1)\Delta)}\big)\big] \\ &\qquad + M^{r,\alpha}_h \bar{g}^\Delta_\alpha\big(\pi\left(\tau \wedge T^{\delta^0}_\Delta\right), z_{\tau \wedge T^{\delta^0}_\Delta}, \gamma e^{-\alpha \tau \wedge T^{\delta^0}_\Delta}\big) + g^\Delta_\alpha\big(\bar{\pi}, \bar{z}, \gamma e^{-\alpha(\tau \wedge T^{\delta^0}_\Delta + h)}\big) \\ &\qquad - g^\Delta_\alpha(\bar{\pi}, \bar{z}, \gamma)\Bigg\}\Bigg]. \end{aligned} \tag{18.66}$$

We have the following:

Theorem 18.16 *For each $\Delta \le \Gamma$, there are a bounded function $w^{r,\Delta}$, which is continuous in the set $\mathcal{S}_\delta^0$, and a constant $\lambda^\Delta(\gamma)$ such that we have*

$$\begin{aligned}
&w^{r,\Delta}(\pi, z, \gamma)\\
&\quad = 1_{\mathcal{S}_\delta^{0c}}(\pi)\big(M_h^r w^{r,\Delta}(\pi, z, \gamma) - \lambda^\Delta(\gamma)h\big)\\
&\qquad + 1_{\mathcal{S}_\delta^0}(\pi)\max\big\{\ln E_{\pi,z}\big[\exp\big\{\gamma\big[\ln\big(\pi\cdot e^{X(\Delta)}\big)\big]\\
&\qquad + \big(w^{r,\Delta}\big(\pi(\Delta), z_\Delta, \gamma\big) - \lambda^\Delta(\gamma)\Delta\big)\big\}\big], M_h^r w^{r,\Delta}(\pi, z, \gamma) - \lambda^\Delta(\gamma)h\big\} \quad (18.67)
\end{aligned}$$

with

$$M_h^r w(\pi, z, \gamma) = \sup_{\pi' \in \mathcal{S}_{\delta'}} \ln E_{\pi,z}\big[\exp\big\{\gamma \ln\big(\pi\cdot e^{X(h)} e\big(\pi(h), \pi'\big)\big) + w(\pi', z_h, \gamma)\big\}\big] \quad (18.68)$$

and, equivalently,

$$\begin{aligned}
&w^{r,\Delta}(\pi, z, \gamma)\\
&\quad := \sup_{\tau\in\mathcal{T}_\Delta} \ln E_{\pi,z}\Bigg[\exp\Bigg\{\gamma \sum_{i=1}^{\frac{\tau\wedge T_\Delta^{\delta 0}}{\Delta}} \big[\ln\big(\pi\big((i-1)\Delta\big)\cdot e^{X(i\Delta)-X((i-1)\Delta)}\big)\big]\\
&\qquad + M_h^r w^{r,\Delta}\big(\pi\big(\tau\wedge T_\Delta^{\delta 0}\big), z_{\tau\wedge T_\Delta^{\delta 0}}, \gamma\big) - \Delta\frac{\tau\wedge T_\Delta^{\delta 0} + h}{\Delta}\lambda^\Delta(\gamma)\Bigg\}\Bigg]. \quad (18.69)
\end{aligned}$$

Furthermore, $|w^{r,\Delta}(\pi, z, \gamma)| \le L$ *(L is the same as in* (18.54)*).*

Proof Notice first that by (18.54) we have that $|\bar{g}_\alpha^\Delta(\pi, z, \gamma)| \le L$ and by (18.55), that $g_\alpha^\Delta(\bar{\pi}, \bar{z}, \gamma e^{-\alpha h}) - g_\alpha^\Delta(\bar{\pi}, \bar{z}, \gamma)$ and $g_\alpha^\Delta(\bar{\pi}, \bar{z}, \gamma e^{-\alpha\Delta}) - g_\alpha^\Delta(\bar{\pi}, \bar{z}, \gamma)$ are bounded from above as functions of α. Therefore, by (18.59), $g_\alpha^\Delta(\bar{\pi}, \bar{z}, \gamma e^{-\alpha\Delta}) - g_\alpha^\Delta(\bar{\pi}, \bar{z}, \gamma)$ is bounded, and there is a subsequence (α_n), $\lim_{n\to\infty}\alpha_n = 0$, and constants $\lambda_m^\Delta(\gamma)$ such that, for $m = 1, 2, \ldots$,

$$\lim_{n\to\infty} g_{\alpha_n}^\Delta\big(\bar{\pi}, \bar{z}, \gamma e^{-\alpha_n m\Delta}\big) - g_{\alpha_n}^\Delta\big(\bar{\pi}, \bar{z}, \gamma e^{-\alpha_n(m-1)\Delta}\big) = -\Delta\lambda_m^\Delta(\gamma). \quad (18.70)$$

By Proposition 18.5 one can choose a further subsequence of (α_n), for simplicity again denoted by (α_n), such that, for $k = 0, 1, \ldots$,

$$M_h^{r,\alpha_n}\bar{g}_{\alpha_n}^\Delta\big(\pi, z, \gamma e^{-\alpha_n k\Delta}\big) \to w_h^\Delta(\pi, z, \gamma, k) \quad (18.71)$$

for a certain function w_h^Δ with convergence uniform on compact subsets of $\mathcal{S}_0^0 \times D$. Consequently,

$$\bar{g}^{\Delta}_{\alpha_n}(\pi, z, \gamma)$$

$$\to \sup_{\tau\in\mathcal{T}_\Delta} \ln E_{\pi,z}\Bigg[\exp\Bigg\{\gamma \sum_{i=1}^{\frac{\tau\wedge T^{\delta^0}_\Delta}{\Delta}} \big[\ln\big(\pi\big((i-1)\Delta\big)\cdot e^{X(i\Delta)-X((i-1)\Delta)}\big)\big]$$

$$+ w^{\Delta}_h\bigg(\pi\big(\tau\wedge T^{\delta^0}_\Delta\big), z_{\tau\wedge T^{\delta^0}_\Delta}, \gamma, \frac{\tau\wedge T^{\delta^0}_\Delta}{\Delta}\bigg) - \Delta \sum_{i=1}^{\frac{\tau\wedge T^{\delta^0}_\Delta+h}{\Delta}} \lambda^{\Delta}_i(\gamma)\Bigg\}\Bigg], \quad (18.72)$$

and, for $m = 1, 2, \ldots,$

$$\bar{g}^{\Delta}_{\alpha_n}\big(\pi, z, \gamma e^{-\alpha_n m\Delta}\big)$$

$$\to \sup_{\tau\in\mathcal{T}_\Delta} \ln E_{\pi,z}\Bigg[\exp\Bigg\{\gamma \sum_{i=1}^{\frac{\tau\wedge T^{\delta^0}_\Delta}{\Delta}} \big[\ln\big(\pi\big((i-1)\Delta\big)\cdot e^{X(i\Delta)-X((i-1)\Delta)}\big)\big]$$

$$+ w^{\Delta}_h\bigg(\pi\big(\tau\wedge T^{\delta^0}_\Delta\big), z_{\tau\wedge T^{\delta^0}_\Delta}, \gamma, m + \frac{\tau\wedge T^{\delta^0}_\Delta}{\Delta}\bigg) - \Delta \sum_{i=m+1}^{m+\frac{\tau\wedge T^{\delta^0}_\Delta+h}{\Delta}} \lambda^{\Delta}_i(\gamma)\Bigg\}\Bigg]$$

$$:= q^{\Delta}(\pi, z, \gamma, m) \quad (18.73)$$

as $n \to \infty$ uniformly in $\pi \in \mathcal{S}_{\delta'}$. Therefore,

$$M^{r,\alpha_n}_h \bar{g}^{\Delta}_{\alpha_n}\big(\pi, z, \gamma e^{-\alpha_n k\Delta}\big) \to M^r_{h,\Delta} q^{\Delta}(\pi, z, \gamma, k) \quad (18.74)$$

with

$$M^r_{h,\Delta} g(\pi, z, \gamma, k) = \sup_{\pi'\in\mathcal{S}_{\delta'}} \ln E_{\pi,z}\bigg[\exp\bigg\{\gamma \ln\big(\pi\cdot e^{X(h)} e\big(\pi(h), \pi'\big)\big)$$

$$+ q^{\Delta}\bigg(\pi', z_h, \gamma, k + \frac{h}{\Delta}\bigg)\bigg\}\bigg]. \quad (18.75)$$

Hence, from (18.66) we have

$$q^{\Delta}(\pi, z, \gamma, m) = \sup_{\tau\in\mathcal{T}_\Delta} \ln E_{\pi,z}\Bigg[\exp\Bigg\{\gamma \sum_{i=1}^{\frac{\tau\wedge T^{\delta^0}_\Delta}{\Delta}} \big[\ln\big(\pi\big((i-1)\Delta\big)\cdot e^{X(i\Delta)-X((i-1)\Delta)}\big)\big]$$

$$+ M^r_{h,\Delta} q^{\Delta}\bigg(\pi\big(\tau\wedge T^{\delta^0}_\Delta\big), z_{\tau\wedge T^{\delta^0}_\Delta}, \gamma, m + \frac{\tau\wedge T^{\delta^0}_\Delta}{\Delta}\bigg)$$

$$- \Delta \sum_{i=m+1}^{m+\frac{\tau \wedge T_\Delta^{\delta^0}+h}{\Delta}} \lambda_i^\Delta(\gamma) \Bigg\} \Bigg]. \tag{18.76}$$

The function $f(\gamma) := \ln E[e^{\gamma Y}]$ with a given random variable Y, whenever is defined, is convex. Consequently, g_α^Δ as a function of γ is also convex, and neglecting other variables, by convexity we have

$$\frac{g_\alpha^\Delta(\gamma e^{-\alpha(m-1)\Delta}) - g_\alpha^\Delta(\gamma e^{-\alpha m \Delta})}{\gamma e^{-\alpha(m-1)\Delta}(1 - e^{-\alpha\Delta})} \geq \frac{g_\alpha^\Delta(\gamma e^{-\alpha m \Delta}) - g_\alpha^\Delta(\gamma e^{-\alpha(m+1)\Delta})}{\gamma e^{-\alpha m \Delta}(1 - e^{-\alpha\Delta})}$$

and

$$e^{-\alpha\Delta}\big(g_\alpha^\Delta\big(\gamma e^{-\alpha(m-1)\Delta}\big) - g_\alpha^\Delta\big(\gamma e^{-\alpha m \Delta}\big)\big) \geq g_\alpha^\Delta\big(\gamma e^{-\alpha m \Delta}\big) - g_\alpha^\Delta\big(\gamma e^{-\alpha(m+1)\Delta}\big).$$

Therefore, by (18.70), $\lambda_m^\Delta(\gamma) \geq \lambda_{m+1}^\Delta(\gamma)$. Since $\lambda_m^\Delta(\gamma)\Delta$ is bounded, there is $\lambda^\Delta(\gamma)$ such that $\lim_{m\to\infty} \lambda_m^\Delta(\gamma) = \lambda^\Delta(\gamma)$. Now, $\bar{g}_{\alpha_n}^\Delta(\pi, z, \gamma e^{-\alpha_n m \Delta}) - \bar{g}_{\alpha_n}^\Delta(\pi, z, \gamma e^{-\alpha_n(m-1)\Delta})$ is bounded and is a difference of two sequences that in the limit as $\alpha_n \to 0$ are monotonic, one of them is convergent to $\lambda^\Delta(\gamma)$. Consequently the other sequence is also convergent, and there is a limit $\lim_{m\to\infty} q^\Delta(\pi, z, \gamma, m) - q^\Delta(\pi, z, \gamma, m-1) := d(\pi, z, \gamma)$. Since

$$\begin{aligned} q^\Delta(\pi, z, \gamma, m) &= \big(q^\Delta(\pi, z, \gamma, m) - q^\Delta(\pi, z, \gamma, m+1)\big) + \cdots \\ &\quad + \big(q^\Delta(\pi, z, \gamma, m+k-1) - q^\Delta(\pi, z, \gamma, m+k)\big) \\ &\quad + q^\Delta(\pi, z, \gamma, m+k), \end{aligned}$$

by the boundedness of q^Δ we clearly have that $d(\pi, z, \gamma) = 0$. By the uniform continuity of $q^\Delta(\cdot, \cdot, \gamma, m)$ it follows then that there is a continuous function $w^{r,\Delta}(\pi, z, \gamma)$ on $\mathcal{S}_\delta^0 \times B$ such that $\lim_{m\to\infty} q^\Delta(\pi, z, \gamma, m) = w^{r,\Delta}(\pi, z, \gamma)$, uniformly on compact subsets of $\mathcal{S}_\delta^0 \times B$. Letting now $m \to \infty$ in (18.76), we obtain

$$\begin{aligned} &w^{r,\Delta}(\pi, z, \gamma) \\ &\quad = \sup_{\tau\in\mathcal{T}_\Delta} \ln E_{\pi,z}\Bigg[\exp\Bigg\{\gamma \sum_{i=1}^{\frac{\tau\wedge T_\Delta^{\delta^0}}{\Delta}} \big[\ln\big(\pi\big((i-1)\Delta\big) \cdot e^{X(i\Delta)-X((i-1)\Delta)}\big)\big] \\ &\qquad + w^{r,\Delta}\big(\pi\big(\tau \wedge T_\Delta^{\delta^0}\big), z_{\tau\wedge T_\Delta^{\delta^0}}, \gamma\big) - \Delta\frac{\tau \wedge T_\Delta^{\delta^0} + h}{\Delta}\lambda^\Delta(\gamma)\Bigg\}\Bigg], \end{aligned} \tag{18.77}$$

which completes the proof. □

Basing on (18.69), we can now show the main result.

Theorem 18.17 *There are a constant λ and a bounded continuous function w^r such that the following Bellman equation is satisfied*:

$$w^r(\pi, z, \gamma) = \sup_\tau \ln E_{\pi,z}\big[\exp\big\{\gamma\big[\ln\big(\pi \cdot e^{X(\tau\wedge T^{\delta^0})}\big)\big] + M_h^r w^r\big(\pi\big(\tau\wedge T^{\delta^0}\big), z_{\tau\wedge T^{\delta 0}}, \gamma\big) - \big(\tau\wedge T^{\delta^0} + h\big)\lambda(\gamma)\big\}\big]. \quad (18.78)$$

Furthermore,

$$\gamma^{-1}\lambda(\gamma) = \sup_V J_\gamma^r(V). \quad (18.79)$$

Proof By the proof of the previous theorem we know that $w^{r,\Delta}$ and λ^Δ are bounded uniformly in Δ. Therefore, for a fixed γ, one can choose a subsequence $\Delta_n \to 0$ such that $\lambda^{\Delta_n}(\gamma) \to \lambda$ and $M_h^r w^{r,\Delta_n} \to v$ uniformly on compact subsets of $\mathcal{S}_0^0 \times D$. Now, letting $\Delta_n \to 0$ in (18.69), we obtain that $\lim_{n\to\infty} w^{r,\Delta_n} = w^r$ and $v = M_h^r w^r$. This completes the proof of (18.78). To prove the formula (18.79), by an analogy to the proof of Theorem 18.11 we have to introduce the function $\bar{w}^r$,

$$\bar{w}^r(\pi, z, W, \gamma) := \sup_\tau \ln E_{\pi,z}\big[\exp\big\{\gamma\big[\ln\big(W\pi \cdot e^{X(\tau\wedge T^{\delta^0})}\big)\big] + M_h^r w^r\big(\pi\big(\tau\wedge T^{\delta^0}\big), z_{\tau\wedge T^{\delta 0}}, \gamma\big) - \big(\tau\wedge T^{\delta^0} + h\big)\lambda(\gamma)\big\}\big]. \quad (18.80)$$

Clearly, $\bar{w}^r(\pi, z, W, \gamma) = \gamma \ln W + w^r(\pi, z, \gamma)$. For a given impulsive strategy $V = (\tau_n, \pi^n)$, consider the following notation: for $n = 1, 2, \dots$, $\pi_n(\tau_n + h) = \pi^n$, $\pi_n(\tau_n + h + s) = \pi^-(\tau_n + h + s)$ for $s > 0$. Recall that W_t^- and W_t are the wealth process before and after possible transaction at time t. It is clear that, for $s \geq 0$,

$$e^{\bar{w}(\pi,z,W,\gamma)} \geq E_{\pi,z}\big[\exp\big\{\bar{w}^r\big(\pi\big(s\wedge T^{\delta^0}\big), z_{s\wedge T^{\delta 0}}, W^-_{s\wedge T^{\delta 0}}, \gamma\big) - \lambda(\gamma)\big(s\wedge T^{\delta^0}\big)\big\}\big]. \quad (18.81)$$

Therefore,

$$\tilde{Z}_n(s) = \exp\big\{\bar{w}^r\big(\pi_n\big((\tau_n + h + s)\wedge T^{\delta 0}_{\tau_n+h}\big), z_{\tau+h+s\wedge T^{\delta 0}_{\tau_n+h}}, W^-_{\tau+h+s\wedge T^{\delta 0}_{\tau_n+h}}, \gamma\big) - \lambda(\gamma)\big(s\wedge\big(T^{\delta^0}\circ\theta_{\tau_n+h}\big)\big)\big\} \quad (18.82)$$

is a $\mathcal{G}_s^n = \mathcal{F}_{\tau_n+h+s}$-supermartingale. For any stopping time $\tau \geq \tau_n + h$, since

$$\{\tau - \tau_n - h \leq s\} = \{\tau \leq \tau_n + h + s\} \in \mathcal{F}_{\tau_n+h+s} = \mathcal{G}_s^n,$$

we have that $\tau - \tau_n - h$ is a $(\mathcal{G}_s^n)$-stopping time. Therefore, if additionally $\tau \leq T^{\delta 0}_{\tau_n+h}$ (where, as before, $T^{\delta 0}_{\tau_n+h} := \tau_n + h + T^{\delta^0}\circ\theta_{\tau_n+h}$), we have

$$E\big[\tilde{Z}_n(\tau - \tau_n - h)\big|\mathcal{F}_{\tau_n+h}\big] \leq \tilde{Z}_n(0) = e^{\bar{w}^r(\pi_n(\tau_n+h), z_{\tau_n+h}, W_{\tau_n+h}, \gamma)}. \quad (18.83)$$

By the form of the operator M_h^r, taking into account that $\bar{w}^r(\pi, z, W, \gamma) = \gamma \ln W + w^r(\pi, z, \gamma)$, we obtain

$$\begin{aligned} &e^{\bar{w}^r(\pi_{n-1}(\tau_n), z_{\tau_n}, W_{\tau_n}^-, \gamma)} \\ &\quad \geq e^{-\lambda(\gamma)h} E_{\pi,z}\big[\exp\{\bar{w}^r(\pi^n, z_{\tau_n+h}, W_{\tau_n+h}, \gamma)\}\big|\mathcal{F}_{\tau_n}\big], \end{aligned} \tag{18.84}$$

and therefore, for fixed $T > 0$, we have

$$\begin{aligned} &E\big[\exp\{\chi_{\tau_n \leq T}\big(\bar{w}^r\big(\pi_n(\tau_{n+1} \wedge (T+h)\big), z_{\tau_{n+1}\wedge(T+h)}, W^-_{\tau_{n+1}\wedge(T+h)}, \gamma\big) \\ &\quad - \lambda(\gamma)\big(\tau_{n+1} \wedge (T+h) - \tau_n - h\big)\big)\}\big|\mathcal{F}_{\tau_n}\big] \\ &\leq E\big[e^{\chi_{\tau_n \leq T} \bar{w}^r(\pi_n(\tau_n+h), z_{\tau_n+h}, W_{\tau_n+h})}\big|\mathcal{F}_{\tau_n}\big] \\ &\leq e^{\chi_{\tau_n \leq T}(\lambda(\gamma)h + \bar{w}^r(\pi_{n-1}(\tau_n), z_{\tau_n}, W^-_{\tau_n}, \gamma))}. \end{aligned}$$

Consequently,

$$\begin{aligned} &E\big[\exp\{\chi_{\tau_n \leq T}\big(\bar{w}^r\big(\pi_n(\tau_{n+1} \wedge (T+h)\big), z_{\tau_{n+1}\wedge(T+h)}, W^-_{\tau_{n+1}\wedge(T+h)}, \gamma\big) \\ &\quad - \bar{w}^r\big(\pi_{n-1}(\tau_n), z_{\tau_n}, W^-_{\tau_n}, \gamma\big) \\ &\quad - \lambda(\gamma)\big(\tau_{n+1} \wedge (T+h) - \tau_n\big)\big)\}\big|\mathcal{F}_{\tau_n}\big] \leq 1, \end{aligned} \tag{18.85}$$

and therefore,

$$\begin{aligned} &E\Bigg[\exp\Bigg\{\sum_{n=1}^{\infty} \chi_{\tau_n \leq T}\big(\bar{w}^r\big(\pi_n\big(\tau_{n+1} \wedge (T+h)\big), z_{\tau_{n+1}\wedge(T+h)}, \\ &\quad W^-_{\tau_{n+1}\wedge(T+h)}, \gamma\big) - \bar{w}^r\big(\pi_{n-1}(\tau_n), z_{\tau_n}, W^-_{\tau_n}, \gamma\big) \\ &\quad - \lambda(\gamma)\big(\tau_{n+1} \wedge (T+h) - \tau_n\big)\big)\Bigg\}\Bigg] \leq 1. \end{aligned}$$

Finally, using $\zeta(T) = \inf\{n : \tau_n \geq T\}$, we rewrite the last inequality in the form

$$\begin{aligned} &E\big[\exp\{\big(\bar{w}^r\big(\pi_{\zeta(T)-1}\big(\tau_{\zeta(T)} \wedge (T+h)\big), z_{\tau_{\zeta(T)}\wedge(T+h)}, \\ &\quad W^-_{\tau_{\zeta(T)}\wedge(T+h)}, \gamma\big) - \bar{w}^r(\pi, z, W, \gamma) - \lambda(\gamma)\big(\tau_{\zeta(T)} \wedge (T+h) - \tau_{\zeta(T)-1}\big)\big)\}\big] \leq 1 \end{aligned}$$

and

$$\begin{aligned} &E\big[\exp\{\big(w^r\big(\pi_{\zeta(T)-1}\big(\tau_{\zeta(T)} \wedge (T+h)\big), z_{\tau_{\zeta(T)}\wedge(T+h)}, \gamma\big) \\ &\quad + \gamma \ln W^-_{\tau_{\zeta(T)}\wedge(T+h)} + \bar{w}^r(\pi, z, W, \gamma) - \lambda(\gamma)\big(\tau_{\zeta(T)} \wedge (T+h) - \tau_{\zeta(T)-1}\big)\big)\}\big] \leq 1. \end{aligned} \tag{18.86}$$

Notice that for the strategy $\hat{V}$ defined by optimal stopping times, from (18.81) and portfolio changes accordingly to the selector of the operator M_h^r we have the equalities in (18.84)–(18.86). Taking the logarithm, then dividing both sides of (18.86) by T and letting $T \to \infty$, we obtain (18.81) and the optimality of the strategy $\hat{V}$. □

18.6 General Form of the Long-Run Bellman Equations

We consider in this section (for simplicity) the case with execution delay. The case with decision lag can be studied in a similar way. To simplify the notation, we neglect here obligatory diversification. We would like to find a general, unified form of the Bellman equation, which covers the cases of logarithmic and power utility function. Following [7], we can define the problem as follows: find a function w and a constant λ such that for any positive K, we have

$$\begin{aligned} & U\big(Ke^{w(W,\pi,z)}\big) \\ & \quad = \sup_\tau E_{\pi,z}\Big[\sup_{\pi'} U\big(\big(\pi^-(\tau+h)\cdot e^{X(\tau+h)}\big)e^{-\lambda(\tau+h)} \\ & \qquad \times Ke\big(\pi^-(\tau+h),\pi'\big)e^{w(W(\pi^-(\tau+h)\cdot e^{X(\tau+h)})e(\pi^-(\tau+h),\pi'),\pi',z_{\tau+h})}\big)\Big]. \end{aligned} \quad (18.87)$$

It can be shown (see Sect. 2 of [7]) that λ is an optimal utility growth. By (18.87) we see that the mapping $K \mapsto \sup_\tau U^{-1}E_{\pi,z}\{U(K\cdots)\}$ is positively homogeneous. This is satisfied in particular, when the mapping $K \mapsto U^{-1}E\{U(KZ)\}$ is positively homogeneous for any random variable Z. By Theorem 3.1 of [7] it holds whenever $U(x) = Ax^\gamma + B$ with $\gamma > 0$, or $U(x) = A\ln x + B$ with $A > 0$ and arbitrary B. Consequently, up to normalization, U should be a power or logarithmic utility function, and in fact this radically limits the use of the general Bellman equation of the form (18.87). Notice furthermore that we can rewrite (18.87) in the form

$$U\big(e^{w(W,\pi,z)}\big) = \sup_\tau E_{\pi,z}\big[U\big(\big(\pi^-(\tau)\cdot e^{X(\tau)}\big)e^{-\lambda(\tau+h)}Mw\big(W_\tau,\pi(\tau),z_\tau\big)\big)\big], \quad (18.88)$$

where

$$M(W,\pi,z) = \sup_{\pi'\in\mathcal{S}_{\delta'}} U^{-1}\big(E_{\pi',z}\big\{U\big(\pi\cdot e^{X(h)}e\big(\pi(h),\pi'\big)e^{w(W_h,\pi',z_h)}\big)\big\}\big). \quad (18.89)$$

As we can see in this paper, Bellman equations were independent of the initial value of the wealth process (W_t). Consequently, we may look for a solution to the equation

$$U\big(e^{w(\pi,z)}\big) = \sup_\tau E_{\pi,z}\big[U\big(\big(\pi^-(\tau)\cdot e^{X(\tau)}\big)e^{-\lambda(\tau+h)}Mw\big(\pi(\tau),z_\tau\big)\big)\big] \quad (18.90)$$

with

$$M(\pi,z) = \sup_{\pi'} U^{-1}\big(E_{\pi',z}\big[U\big(\pi'\cdot e^{X(h)}\big)e\big(\pi(h),\pi'\big)e^{w(\pi',z_h)}\big]\big).$$

In particular cases we obtain the following examples.

Example 18.18 If $U(x) = \ln x$, then

$$w(\pi, z) = \sup_{\tau} E_{\pi,z}\big[\ln\big(\pi \cdot e^{X(\tau)}\big) - \lambda(\tau + h) + M_h w\big(\pi(\tau), z_\tau\big)\big]$$

with

$$M_h w(\pi, z) = \sup_{\pi' \in \mathcal{S}_{\delta'}} E_{\pi',z}\big[\big(\ln(\pi \cdot e^{X(h)}) + \ln e\big(\pi(h), \pi'\big) + w(\pi', z_h)\big)\big].$$

Example 18.19 If $U(x) = x^\gamma$, then

$$e^{\gamma w(\pi,z)} = \sup_{\tau} E_{\pi,z}\big[\big(\pi^-(\tau) \cdot e^{X(\tau)}\big)^\gamma e^{-\lambda_r \gamma(\tau+h)} M_h^r w\big(\pi^-(\tau), z_\tau\big)\big]$$

with

$$M_h^r w(\pi, z) = \sup_{\pi' \in \mathcal{S}_{\delta'}} E_{\pi',z}\big[\big(\pi \cdot e^{X(h)}\big)^\gamma e\big(\pi(h), \pi'\big)^\gamma e^{\gamma w(\pi', z_h)}\big],$$

which modulo small transformations coincide with the Bellman equations considered in Remark 18.12 and Theorem 18.17. As we pointed out above, these two examples practically correspond to the only utility functions for which we could expect to find solutions to (18.90).

References

1. M. Akian, A. Sulem, M.I. Taksar, Dynamic optimization of long-term growth rate for a portfolio with transaction costs and logarithmic utility. Math. Finance **11**, 153–188 (2001)
2. D. Applebaum, *Lévy Processes and Stochastic Calculus* (Cambridge University Press, Cambridge, 2004)
3. T.R. Bielecki, S.P. Pliska, Risk sensitive dynamic asset management. Appl. Math. Optim. **39**, 337–360 (1999)
4. T. Duncan, B. Pasik-Duncan, L. Stettner, Growth optimal portfolio under proportional transaction costs with obligatory diversification. Appl. Math. Optim. **63**, 107–132 (2011). doi:101007/s00245-010-9113-x
5. A. Irle, J. Sass, *Good Portfolio Strategies under Transaction Costs: A Renewal Theoretic Approach*. Stochastic Finance (Springer, New York, 2006), pp. 321–341
6. I. Karatzas, S. Schreve, *Brownian Motion and Stochastic Calculus* (Springer, Berlin, 1991)
7. J. Matkowski, L. Stettner, On Bellman equation for asymptotics of utility from terminal wealth. Appl. Math. **37**, 89–97 (2010)
8. A.J. Morton, S.R. Pliska, Optimal portfolio management with fixed transaction costs. Math. Finance **5**, 337–356 (1995)
9. H. Nagai, Optimal strategies for risk-sensitive portfolio optimization problems for general factor models. SIAM J. Control Optim. **41**, 1779–1800 (2003)
10. M. Robin, Contrôle impulsionnel des processus de Markov. Thesis, University of Paris IX (1978)
11. L. Stettner, Discrete time risk sensitive portfolio optimization with consumption and proportional transaction costs. Appl. Math. **32**, 395–404 (2005)
12. L. Stettner, Long time growth optimal portfolio with transaction costs. Banach Cent. Publ. **83**, 231–241 (2008)

13. L. Stettner, Penalty method for finite horizon stopping problems. SIAM J. Control Optim. (to appear)
14. T. Tamura, Maximizing the growth rate of a portfolio with fixed and proportional transaction costs. Appl. Math. Optim. **54**, 277–294 (2006)
15. T. Tamura, Maximization of the long-term growth rate for a portfolio with fixed and proportional transaction costs. Adv. Appl. Probab. **40**, 673–695 (2008)